International Maritime Conference (Pacific 2017)

AA005498

Sydney, Australia
3-5 October 2017

ISBN: 978-1-5108-8305-5

Printed from e-media with permission by:

Curran Associates, Inc.
57 Morehouse Lane
Red Hook, NY 12571

CURRAN ASSOCIATES INC. proceedings.com

Some format issues inherent in the e-media version may also appear in this print version.

Copyright© (2017) by The Royal Institution of Naval Architects
All rights reserved.

Printed with permission by Curran Associates, Inc. (2020)

For permission requests, please contact The Royal Institution of Naval Architects
at the address below.

The Royal Institution of Naval Architects
8-9 Northumberland Street
London, WC2N 5DA
United Kingdom

Phone: 020 7235 4622
Fax: 020 7259 5912

publications@rina.org.uk

Additional copies of this publication are available from:

Curran Associates, Inc.
57 Morehouse Lane
Red Hook, NY 12571 USA
Phone: 845-758-0400
Fax: 845-758-2633
Email: curran@proceedings.com
Web: www.proceedings.com

THE ROYAL INSTITUTION OF NAVAL ARCHITECTS

IMarEST

ENGINEERS AUSTRALIA

PACIFIC 2017
International Maritime Conference
PROGRAM

Correct as 29 September 2017

TUESDAY 3 OCTOBER 2017 DAY 1

0945 - 1000	**Opening Ceremony**	**PYRMONT THEATRE**
1000 - 1100	**Keynotes** Chair: *John Jeremy, AM*	
1000 - 1030	**Opening Address** Capability Acquistion Reform *Rear Admiral Jonathan Mead, AM, RAN,* Head Navy Capability, Australia	
1030 - 1100	**Keynote Address:** Australia's Naval Shipbuilding Project - An Update *Rear Admiral Tony Dalton* General Manager Ships Capability Acquisition and Sustainment Group, Australia	
1100 - 1130	Morning Tea	

	Conference Room C2.1	Conference Room C2.2	Conference Room C2.3
1130 - 1300	**Session 1 - Ship Design** Chair: *Martin Renilson*	**Session 2 - Structural Integrity** Chair: *Tauhid Rahman*	**Session 20 - Autonomous Vessels** Chair: *Brett Morris*
1130 - 1200	A Ship Performance Modelling and Simulation Framework to Support Requirements Setting (Refereed) 1 *Dylan Dwyer* Defence Science and Technology, Australia	Maintaining Technical Integrity of the Royal Australian Navy's Survey Motor Boats 29 *Jon Emonson* and *Peter Webb* BAE Systems, Australia	Framing Considerations of Autonomous Naval Ships for the Royal Australian Navy (Refereed) 678 *Gabrielle Anastasio, Joseph Cole* and *Warren Smith* UNSW Canberra and Department of Defence, Australia
1200 - 1230	Design of Icebreaking Supply & Research Vessels 16 *Ken Goh* Knud E Hansen, Australia	Effects of Structural Ageing on Blast Analyses of a Naval Vessel: A Technical Review (Refereed) 19 *Daniel Clayton* Australian Maritime College, University of Tasmania, Australia	Integrating Unmanned Surface Vessels Into the Australian Defence and Border Protection Forces 689 *Derek Rogers* SAAB, Australia
1230 - 1300	Damen Naval Auxiliary Vessels: a Compromise or a Solution? 17 *Piet van Rooij* Damen Shipyards, The Netherlands	Lessons from Life of Type Extension (LOTE) for Naval Fleet and New Build 42 *Olaf Doerk* and *Georgios Spiliotis* DNV-GL, Germany and Australia	Application of the Lloyd's Register Unmanned Marine System Code to a Small Remotely Controlled Survey Vessel 700 *Jim Gorton* Lloyd's Register, United Kingdom

* (Refereed) = Conference paper has been peer-reviewed.

TUESDAY 3 OCTOBER 2017 DAY 1

1300 - 1400	Lunch		
1400 - 1600	**Session 3 - Ship Building** Chair: *Adrian Broadbent*	**Session 4 - Ship Safety/ Human Factors** Chair: *Rob Gehling*	**Session 21 - Combat Systems** Chair: *Andrew Tynan*
1400 - 1430	Awaiting the Next Revolution in Naval Affairs 66 *Simon Reay Atkinson* University of Sydney, Australia	Smart Escape Support System for Passenger Ship: Active Dynamic Signage & Real-time Escape Routing 97 *James Choi* Corners Inc., Republic of Korea	Knowledge-based Naval Mission & Combat Systems 715 *Alain Carof* Naval Group, France
1430 - 1500	First and Second Order Technical Implications of an Off-The-Shelf Approach to Naval Ship Acquisition 52 *Joseph Cole* Department of Defence, Australia	Commercial Ship Management - Strong Risk Management is Good Business 90 *David Parmeter* Teekay Shipping, Australia	The Genesis of Australia's Future Naval Enterprise Combat System Capability 739 *David Hanley and Rod Equid* Raytheon, Australia
1500 - 1530	Australian Innovation Transforming Modern Naval Shipbuilding 83 *Tim Speer* Austal, Australia	Human Impact Exposure on board High-Speed Boats: A Multi Agency Study 84 *Johan Ullman* High Speed Boat Operations Pro, Sweden	The Aegis Weapons System - Providing Proven and Extendable Maritime Capabilities in Meeting Defence White Paper Objectives 727 *Brad Hicks* Lockheed Martin, USA
1530 - 1600	Shipbuilding Lessons Learnt on the Air Warfare Destroyers 65 *Mark Lamarre* ASC Shipbuilding, Australia	The International Maritime Organization Guidelines on Fatigue Mitigation and Management: A Status Quo 108 *Michelle Grech* Australian Maritime Safety Authority, Australia	

1600 - 1800	**Session 5 - Ship Propulsion** Chair: *Chris Hodge*	**Session 6 - Ship Structures** Chair: *Dr Warren Smith*
1600 - 1630	Optimising High Speed Craft for Hybrid Propulsion 153 *Robert Hayes* Frazer-Nash Consultancy, Australia	Structural Integrity of Maritime Platforms 165 *Bruce Cartwright* The University of Newcastle, Australia
1630 - 1700	Power and Propulsion System Options for Future Frigate Programmes - Hybrid as an Enabler for Increased Mission Capability 138 *Edward Wright* Rolls Royce PLC, United Kingdom	Comparative Structural Requirements for High Speed Naval Craft: The 'ROURKE Class' Corvette 176 *John Lord* Independent, Member of RINA, Australia
1700 - 1730	Mission-centric Design of Hybrid Propulsion Systems for Multi-purpose Naval Vessels 122 *Pierpaolo Da Fieno* MAN Diesel & Turbo SE, Germany	Application of Structural Integrity Management Principles Demonstrated in Other Domains to Improve Seaworthiness and Cost of Ownership Outcomes 203 *Michael Houston* and *Stuart Cannon* QinetiQ and Defence Science and Technology, Australia
1730 - 1800	Offshore Supply Vessels: A Comparison of Propulsion System Architectures 109 *Cameron Sobie* Siemens Industry Software, Belgium	Trials and Tribulations: Load and Structural Response Measurements of a Naval Semi-planing Craft (Refereed) 219 *Teresa Magoga* Defence Science and Technology, Australia

WEDNESDAY 4 OCTOBER 2017 DAY 2

	Conference Room C2.5 & C2.6	Conference Room C2.1
0830 - 1100	**Session 7 - Ship Repair** Chair: *Alan Taylor*	**Session 8 - Submarine Operations** Chair: Tauhid Rahman
0830 - 0900	**Keynote Address:** Science and Technology Support for Naval Shipbuilding N/A *Dr David Kershaw* Chief of Maritime Division Defence Science and Technology, Australia	
0900 - 0930	Challenges in Naval Ship Maintenance 231 *Arshad Kunjumon* Abu Dhabi Shipbuilding, United Arab Emirates	Submarine Sovereignty – ASC's Collins Experience 282 *Martin Edwards* ASC Pty Ltd, Australia
0930 - 1000	3D Laser Scanning Applications for Shipbuilding and Throughlife Support 243 *Chris Deegan* Gibbs & Cox, USA	Integrated Performance Modelling of Submarines to Support Strategic Decision Making for the Future Submarine Program (Refereed) 283 *Hamid Diab and Karl Slater* Defence Science and Technology, Australia
1000 - 1030	Low Distortion Welding Processes for Naval Shipbuilding 257 *Stephen van Duin* University of Wollongong, Australia	The Operational Benefits of Covertly Refuelling a Conventional Submarine whilst at Sea (Refereed) 296 *Simon Harrison* Defence Science and Technology, Australia
1030 - 1100	High Entropy Brasses & Bronzes - New Marine Alloys 256 *Kevin Laws* University of New South Wales, Australia	Enhancement of Safety and Operational Reliability of Modern Submarines Through the Involvement of a Classification Society 271 *Christian von Olderhausen* DNV-GL, Germany
1100 - 1130	Morning Tea	
1130 - 1300	**Session 9 - Panel Discussion** Chair: *John Jeremy, AM*	**Session 10 - Submarine Hydrodynamics** Chair: *Helen Dorsett*
1130 - 1200	Off-the-Shelf Naval Ship Design through Requirements Engineering N/A *Mr Joe Cole* Naval Technical Bureau, Australia	The Influence of Appendages and their Stall on Submarine Hydrodynamic Loads (Refereed) 307 *Gregory Seil* Defence Science and Technology, Australia
1200 - 1230	*CMDR Lindsay Gordon, RAN* Navy Strategic Command, Australia *CMDR Dan Crocker, RAN* Navy Strategic Command, Australia	The Effects of Hull Form on the Optimum L/D Ratio for Minimum Resistance for Submerged Bodies (Refereed) 329 *Zhi Quan Leong* Australian Maritime College, University of Tasmania, Australia
1230 - 1300	*Dr Warren Smith* UNSW Canberra, Australia *Mr Graeme Dunk* Shoal, Australia	Sensitivity of a Submarine Manoeuvring Model to Parameter Uncertainty (Refereed) 317 *Siobhan Giles* BMT Design & Technology, Australia
1300 - 1400	Lunch	

WEDNESDAY 4 OCTOBER 2017 DAY 2

Time		
1400 - 1600	**Session 11 - Survivability** Chair: *Professor Jonathan Binns*	**Session 12 - Ship Hydrodynamics** Chair: *Phil Helmore*
1400 - 1430	**Integrated Survivability Analysis of Naval Platforms in High Threat Environments** 340 *Anthony Woolley* Defence Science and Technology, Australia	**Experiments Investigating the Mass Moment of Inertia of Fluids in Roll (Refereed)** 390 *Darris Fortescue, Martin Grimm* and *Warren Smith* Department of Defence and UNSW Canberra, Australia
1430 - 1500	**Medium Scale Experiments for the Validation of Naval Platform Vulnerability Simulations** 355 *George Yiannakopoulos* Defence Science and Technology, Australia	**The Influence of Roll Radius of Gyration Including the Effect of Inertia of Fluids in Motion Predictions (Refereed)** 404 *Martin Grimm, Warren Smith* and *Darris Fortescue* UNSW Canberra and Department of Defence, Australia
1500 - 1530	**Rapid System Vulnerability Assessment Using Fault Trees with System Location Attribution** 378 *Serap Aksu* Defence Science and Technology, Australia	**Estimation of Added Mass Moment of Inertia in Roll Motion Through Numerical Simulation (Refereed)** 431 *S.S. Kianejad* Australian Maritime College, University of Tasmania, Australia
1530 - 1600	**Rapid Vulnerability Assessment of Naval Structures subjected to Localised Blast** 366 *Luiz Bortolan Neto* Australian Nuclear Science and Technology Organisation (ANSTO), Australia	**Ship Roll Reduction Using Water-flow Induced Coriolis Effect** 419 *Mario Walker* Tokyo Institute of Technology, Japan
1600 - 1730	**Session 13 - Hydrodynamics** Chair: *Craig Boulton*	**Session 14 - Structural Integrity** Chair: *Trevor Dove*
1600 - 1630	**The Whipping Response of a Submerged Free-free Cylinder Due to Underwater Explosions (Refereed)** 474 *Steven De Candia* Australian Maritime College, University of Tasmania, Australia	**Prediction of Ultimate Strength of Locally Corroded Plates Using ANFIS Model (Refereed)** 506 *Nagi Abdussamie* Australian Maritime College, University of Tasmania, Australia
1630 - 1700	**Slow Speed Depth Control of a Submarine Under Waves** 459 *Paul Crossland* QinetiQ, United Kingdom	**Effect of Pitting Corrosion on the Ultimate Strength and Buckling Behaviour of Stiffened Aluminium Panels (Refereed)** 491 *Lenore Pedrina* Defence Science and Technology, Australia
1700 - 1730	**Innovative Experiments for Investigating Maritime Hydrodynamic Problems** 446 *Gregor Macfarlane* Australian Maritime College, University of Tasmania, Australia	**The Ungoverned Space: Call for Improved Test & Maintenance to Comply with Regulations & Go Beyond** 482 *Clare Hunter* Coltraco Ultrasonics, United Kingdom
1730 - 1800	**Underwater Glider Performance at Model- and Full-scale Reynolds Numbers (Refereed)** 608 *Max Haase* Australian Maritime College, University of Tasmania, Australia	
1800 - 1930	**Pacific 2017 International Maritime Conference Cocktail Reception** Terrace Room, Australian National Maritime Museum	

International Maritime Conference Cocktail Reception

An invitation is extended to all delegates to attend the Pacific 2017 International Maritime Conference Cocktail Reception which will be held in the Terrace Room, Australian National Maritime Museum.

DATE: Wednesday 4 October 2017

TIME: 1800 - 1930

VENUE: Terrace Room,
Australian National Maritime Museum

DRESS: Smart Casual

THURSDAY 5 OCTOBER 2017 DAY 3

	Conference Room C2.5 & C2.6	Conference Room C2.1
0830 - 1100	**Session 15 - Autonomous Underwater Vehicles** Chair: *Dr. Stuart Cannon*	**Session 16 - Ship Operations** Chair: *Ken Greig*
0830 - 0900	**Keynote Address:** How Innovation can Change a Submarine Business Performance Paradigm N/A *Dr Margaret Law* Strategy & Capability Manager ASC Pty Ltd, Australia	
0900 - 0930	A Paradigm Shift in AUV Capabilities - Boeing's Echo Voyager XLUUV 515 *Dan Tubbs* Boeing, USA	Lessons Learned from HMNZS CANTERBURYs Service in the Royal New Zealand Navy 556 *CDR Phil Bradshaw, RNZN* New Zealand Defence Force, New Zealand
0930 - 1000	Achieving Integrated Command and Control of Unmanned Systems - the Unmanned Warrior Experience 530 *Robin Campbell* QinetiQ, United Kingdom	Large Displacement Frigate: Total Ownership Cost Model 569 *Malcolm Waugh* WAMA Alliance Management Office, Australia
1000 - 1030	A Practical Demonstration of ASW Using Unmanned and Autonomous Vehicles 544 *Robin Campbell* QinetiQ, United Kingdom	Advanced Coating Solutions for the Marine Industry (Refereed) 593 *Scott Wade* Swinburne University of Technology, Australia
1030 - 1100	Northrop Grumman Demonstrates High Area Clearance Rate with Unmanned Mine Hunting System at Unmanned Warrior 526 *Dave Allan* Northrop Grumman Undersea Systems, USA	Waste Heat Recovery of Marine Diesel Engines (Refereed) 582 *Noam Olshina* The University of Melbourne, Australia
1100 - 1130	Morning Tea	
1130 - 1300	**Session 17 - Autonomous Surface Vehicles** Chair: *Trevor Blakeley*	**Session 18 - Submarine Batteries and Diesels** Chair: *Roger Turner*
1130 - 1200	Recent Advances in Utilizing Autonomous Underwater, Surface and Aerial Vehicles for Operational Field Logistics 622 *Neil Trenaman* Ocean Aero, USA	The Design & Safety Challenges of a Lithium-ion Main Storage Battery for Conventional Submarines 631 *Aidan Depetro* BMT Design & Technology, Australia
1200 - 1230	Persistent Autonomous Unmanned Surface Vessels Powered by the Ocean 630 *Robert Dane* OCIUS, Australia	Design and Qualification of Submarine Engines 649 *Arndt von Drathen* MTU Friedrichshafen GmbH, Germany
1230 - 1300		Submarine Main Propulsion using Lithium-ion Batteries 650 *Choong Yeon Chong* Kokam Co. Ltd, Republic of Korea
1300 - 1400	Lunch	
1400 - 1530	**Session 19 - SEA 5000 Future Frigate** Chair: *Karl Slater*	
1400 - 1430	Type 26 Global Combat Ship: The Next Generation Warship 662 *Chris Muskett* BAE Systems, United Kingdom	
1430 - 1500	Fincantieri SEA 5000 Presentation	
1500 - 1530	Navantia SEA5000 Presentation	
1530	Closing Ceremony	

A Ship Performance Modelling and Simulation Framework to Support Requirements Setting

Dylan Dwyer and Brett A. Morris
Defence Science and Technology Group
dylan.dwyer@dst.defence.gov.au

ABSTRACT

To support the Australian Defence Organisation (ADO) endeavours to become a Smart Buyer, Modelling and Simulation (M&S) can be used during the Risk Reduction and Requirements Setting phase of acquisitions to optimise project execution. Capability design activities conducted early in the Risk Reduction and Requirements Setting phase can assist the development of requirements. The design activities support an improved understanding and more thorough definition of requirements released to industry to constrain the technical solutions to those that will adequately meet the capability need. In the case of Royal Australian Navy (RAN) platform acquisitions, the technical solutions will be either submarines or surface ships, typically acquired using Off-the-Shelf (OTS) strategies. This paper builds upon a previously proposed model based systems engineering methodology constructed for supporting design activities during the early stages of OTS naval platform acquisitions [1].

This paper describes a Modelling and Simulation (M&S) framework for RAN platform acquisitions that enables exploration of the design space early in the Risk Reduction and Requirements Setting phase to support the development of requirements based on robust analysis. The proposed M&S framework utilises existing commercial software and analysis tools to analyse the performance of ships within a mission scenario allowing design options to be evaluated.

The paper includes an implementation of the M&S framework to explore a design space for a new hydrographic survey capability. This new capability's mission effectiveness will be driven by a technology shift towards the integration of unmanned systems into naval platforms. Key Performance Parameters (KPPs) for the capability will include the platform's seakeeping and resistance performance, reflecting the ability to launch and recover unmanned systems, as well as the ability to transit efficiently. Both KPPs are used to assess some exemplar designs' overall performance and support a process for setting requirements that address capability needs. The paper concludes that application of an M&S framework to support exploration of the design space early in Risk Reduction and Requirements Setting can optimise project execution, with observations from the test implementation and suggestions for future work.

INTRODUCTION

Following the 2015 First Principles Review of Defence, the approach to defence acquisition in the Australian Defence Organisation (ADO) was updated and set out in the Interim Capability Life Cycle Manual (ICLCM) [2]. The phases of the current ADO CLC are [2: p .4]:

1. Strategy and Concepts
2. Risk Mitigation and Requirements Setting
3. Acquisition
4. In Service and Disposal

The stage of the ADO CLC that presents the greatest opportunity for optimising project execution, and the subject of interest for the research covered in this paper is the Risk Mitigation and Requirements Setting stage. The research has been undertaken to support ADO naval platform requirements definition. Requirements definition is described as a Gate 1 activity within the ICLCM that [2: p. 28]:

> "focuses on converting the Joint Capability Needs Statement into a contractible requirement statement and preparing for the release of tender documentation (for competition or sole source)."

There is a growing understanding within the Systems Engineering discipline that the process of requirements definition should include preliminary design activities. This understanding is evidenced by the statement by Crowder, Carbone [3]: p. 105:

> "In the end, the activities which we would call design are nothing different from the activities required to create the 'to-be' requirements."

Modelling and Simulation (M&S) presents the opportunity to include preliminary design activities to improve the outcomes of the requirements definition process. Performing design activities during requirements definition could assist capability development stakeholders by providing information on how mission performance requirements impact naval platform design and vice versa. This information will support the development of fit-for purpose requirements that accurately specify a solution to the capability need and contribute to Defence's drive towards being a 'smart buyer'. This paper describes an M&S framework that has been developed to enable generation and exploration of the potential design space for a capability need. The M&S framework can be used to support the development of requirements based on robust analysis.

The paper focuses on an implementation of the M&S framework to enable exploration of a design space for a new hydrographic survey capability. Key Performance Parameters (KPPs) for the capability include the platform's seakeeping and resistance performance, reflecting the ability to launch and recover unmanned systems, as well as the ability to transit efficiently. The paper concludes with observations from the test implementation and suggestions for further work.

BACKGROUND

The research covered in this paper uses previous work by Morris and Thethy [1] as its foundation. Morris and Thethy [1] proposed a methodology for conducting Concept and Requirements Exploration in Off-the-Shelf naval platform acquisitions. The methodology used parametric and surrogate relationships between naval ship performance measures (e.g. sprint speed, cost etc.) for the capability, and physical ship design parameters (e.g. length, displacement etc.) to build a conceptual design space. This design space can provide valuable design information to acquisition stakeholders and can be used to engineer a set of requirements and constraints that accurately reflect the capability need, for release to industry in a Request-for-Tender (RFT). This will limit responses to the RFT to only those Off-the-Shelf designs which are able to meet the capability need without the need for significant design changes.

This research builds upon the previous work by developing a Modelling and Simulation (M&S) framework for higher fidelity modelling of ship performance measures. An overview of the M&S frameworks' place within the Capability Life Cycle is pictured in Figure 1. The M&S framework facilitates the generation of multiple ship hull forms, which can then be assessed for a range of performance measures in naval architecture simulation tools. This approach allows a range of ship designs to be assessed with higher accuracy than the previous parametric modelling approach.

The M&S framework comprises a process using three main tools; a rapid 3D ship hull form generation tool, a resistance prediction tool and a seakeeping prediction tool. The commercial ModelCenter® software package, central to the framework, integrates the different tools, conducts and post-processes experiments to build a design space. This paper details the integration of the 3D modelling software for hull form generation, seakeeping and resistance ship performance prediction tools to generate and explore a design space.

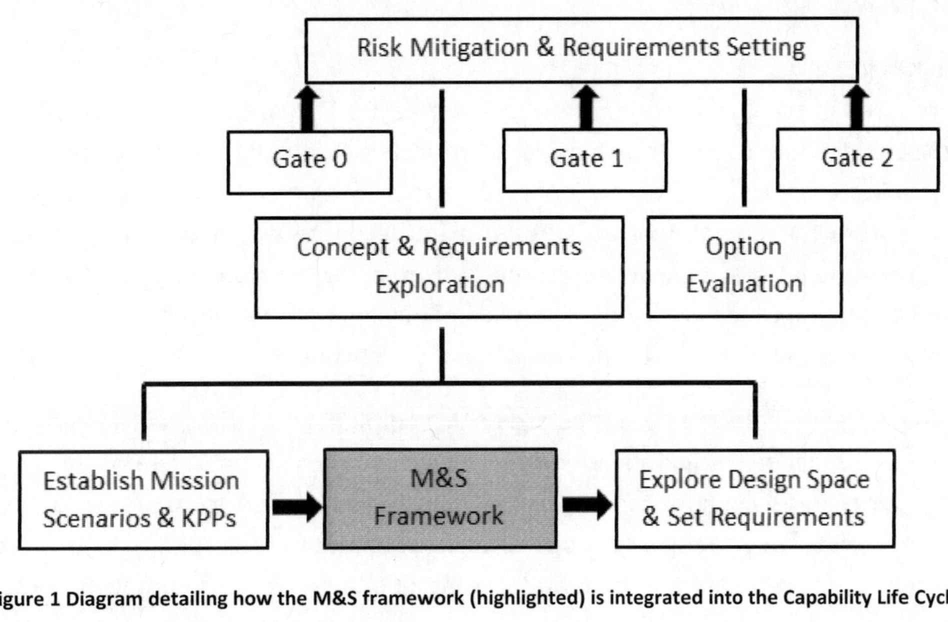

Figure 1 Diagram detailing how the M&S framework (highlighted) is integrated into the Capability Life Cycle.

TESTING THE MODELLING AND SIMULATION FRAMEWORK

The proposed framework was tested by application of a representative Hydrographic Survey (HS) capability. The test implementation is based on an exemplar capability need that employs a ship based solution in combination with an array of unmanned systems that perform the survey functions. To simplify the analysis only monohull displacement/semi-displacement hull forms were assumed. For classification of hull form characteristics, it was assumed that there was no sheer present in the deck, hence the depth at the bow and transom were equal. Also, it was assumed for seakeeping calculations that the bridge and flight deck, for HS vessel types, were located in typical locations just ahead of midships and on the aft deck respectively.

Establish the Mission Scenario

It is important that the mission scenario is able to capture all of the capability needs for the capability to be procured [1]. System development can employ a top-down approach where the system's mission is decomposed into its major functions [4], or operational activities. The indicative scenario developed for the HS capability comprises a ship transit from a base to an operational area. Once in the operational area, unmanned systems are launched to conduct hydrographic and oceanographic survey activities for a period of time. These systems are then recovered and the ship then transits back to base. In a defence acquisition project, the scenarios would be traceable to the Joint Capability Needs Statement and developed with stakeholder input, or taken from an Operational Concept Document.

Testing of the Modelling and Simulation (M&S) framework was simplified by restricting all mission activities to a single ocean area as classified by BMT [5]. Ocean area 79 encompasses the wave climate from the central eastern to the north eastern coast of Australia to approximately 1000 nm eastbound offshore. The most common wave condition from the ocean area was chosen as the environmental testing conditions with a 1.5 m waveheight and 6.5 second wavelength.

Establish KPPs

In the US Department of Defence context a Key Performance Parameter (KPP) is a "performance attribute of a system considered critical or essential to the development of an effective military capability" [6: p. D-A-1]. For the test implementation, the KPPs are established from the operational activities previously described in the mission scenario. Since the operational activities were decomposed from the mission scenario, the KPPs are traceable back to the capability needs. The seakeeping and resistance performance of the HS capability are considered critical to the operational effectiveness.

Seakeeping based KPPs were calculated by combining performance results into comprehensive seakeeping operability indices, as shown in Equation 1. Use of this equation provides a very extensive understanding of the ships seakeeping performance since it is able to capture results over ranges of operational conditions, i.e. speeds and headings. Theoretically, the practicality and benefits of using operability indices is that the measured figure represents the percentage of time a ship is able to remain fully operational in a specified wave condition/environment [7]. The Australian Defence Organisation requires use of operability indices for demonstration of a ships seakeeping performance for all Navy acquisitions. Two operability indices were used to measure the ships performance in both transiting and Launch and Recover (L&R) activities.

$$\boldsymbol{Operability} = \sum[\boldsymbol{P}(\boldsymbol{speed\ i}) \times \boldsymbol{P}(\boldsymbol{heading\ j}) \times \boldsymbol{W}(\boldsymbol{criteria\ k}) \times \boldsymbol{Operability}(\boldsymbol{criteria\ k})] \qquad 1$$

Where:

P(x) represents the probability of speed *i,* or relative wave heading *j*
W(x) represents the weighting (relative importance) assigned to each criteria *'k'*
Operability(x) represents the operability index (evaluated between 0-1) of each criteria *'k'*

The operability indices for the transit and launch and recovery operational activities differed by adopting tailored criteria and speed profiles as seen in Table 1 and Figure 2 respectively, to reflect the operational profiles of a typical HS ship in both transiting and L&R scenarios. Speed profiles were set through a brainstorming process with experienced engineers in the platform systems analysis field. Governance of individual criteria threshold values was taken from guidance set by the Royal Australian Navy's Maritime Material Requirement Set for seakeeping [7]. These threshold values assist in normalising the individual criteria's performance into an operability index from Equation 1. Additionally, certain criteria included from Table 1 were also addressed as standalone potential technical measures by JCIDS [6]. Over both operability index equations the probability of headings was evenly distributed from following seas to head seas (0 – 180 deg). Furthermore, each criteria in both the transit and L&R operability indices were given an equal weighting.

Table 1 Matrix presenting the seakeeping criteria used in Equation 1 at each location. Transit is abbreviated to 'Tr'. Launch and Recovery is abbreviated to 'L&R'

	Locations			
Criteria	Bow	Bridge	Midship Deck Edge	Aft Helo Deck
Deck Wetness	L&R, Tr		L&R	
Slamming	Tr			
MII		L&R, Tr	L&R	L&R
Vertical Velocity				L&R, Tr
Vertical Acceleration			L&R	L&R, Tr
Lat Acceleration				L&R, Tr

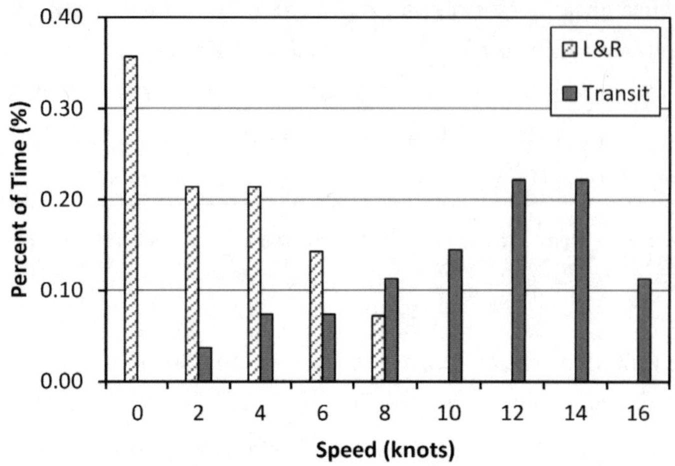

Figure 2 Speed profiles as a percentage of overall time for transit and L&R operational scenarios.

The resistance based KPP was established as a measure of the efficiency of the design transiting at a speed of 14 knots in the most prevalent sea state in the area of the mission. 14 knots was selected as the reference speed for exploration of the design space to reflect a typical service speed of current Royal Australian Navy HS capability. Due to a wide variation in ship displacement within the design space, the resistance efficiency KPP is presented in terms of total resistance per tonne of displacement R_T/Δ [8]. Hence, for a lower R_T/Δ the ship can travel further for a given speed, or, use relatively less fuel for a given range.

Ship Generation Model

The primary role of the Ship Generation Model (SGM) component of the M&S framework is facilitating the establishment of relationships between KPPs and ship design parameters. This stage of the M&S framework is critical as it allows mission performance and an early conceptual design space to be linked [9]. These relationships are elucidated within the SGM through the integration of 3D modelling software and resistance and seakeeping prediction tools in a model-based simulation environment. ModelCenter®, previously introduced; provides the model-based simulation environment where Rhino 3D modelling software has been integrated with the SHIPMO7 seakeeping code for the development of the SGM.

Rhino, along with an Orca3D plugin, was employed as the modelling software to rapidly generate each 3D hull form [10]. Orca3Ds' Hull Assistant contains a unique set of design parameters. These design parameters characterise the complex shape of each hull form and are shown in Table 2. All design parameters allow the user to have control over the hulls overall dimensions, form and bow/transom geometry.

The automated workflow of the SGM is initiated by defining and calculating the hull form design variables (DVs) used by Orca3D to generate a 3D model of the hull. After definition of sections, waterlines and buttocks for a specified load condition, Orca3D performs hydrostatic and resistance calculations on the hull form. Offsets are created by intersecting buttocks and waterlines at 21 hydrostatic stations. These offsets are used for characterisation of the hull form in an appropriate format for SHIPMO7, the program used to predict seakeeping performance. In the final step of the workflow all seakeeping results are collated to calculate the final seakeeping operability indices as well as overall resistance, introduced in the previous section.

Calm water resistance calculations are based on Holtrop and Mennen's resistance prediction method for displacement vessels [11]. Improving the practicality of the SGM's resistance calculation an added resistance due to waves prediction is added to the calm water resistance prediction. The added resistance due to waves calculation is performed within the SHIPMO7 seakeeping program, using the near field method proposed by Faltinsen and Minsaas [12].

DESIGN SPACE DEFINITION

In systems engineering language this stage of the process is similar to developing the system boundary [4]. Defining the initial design space allows for appropriate Design Variables (DVs) and constants to be established. With respect to the Ship Generation Model (SGM) the DVs, in this case, are the Orca3D design parameters. Design lanes are used to constrain the design space by applying suitable limits to the established DVs. These design lanes save time by guiding the design from the early stages of the Capability Life Cycle towards the most appropriate technical solution.

In the Hydrographic Survey (HS) capability test implementation, the global design space was constrained by ship length to ensure the Modelling and Simulation (M&S) framework results reflect design options within a realistic design space. Within the global length constraint, the design space was further decomposed by different vessel types in order to investigate the influence of vessel type on HS capability, a design activity endorsed by Defences drive towards a 'smart buyer' model. Generic hull forms representative of three vessel types were considered; a HS class, frigate and Offshore Patrol Vessel. For the purpose of this paper, generic hull forms were used to keep the test implementation of the M&S framework unclassified. However, in a real world application a database of existing vessel designs would be used in replacement to define the initial design space. For each vessel type, DVs were defined by matching Orca3D hull forms to models of the three vessel types. Constants were also defined where DVs had a value of zero and/or were not characteristic of the hull form. These constraints are given in Table 2. Limits were established to the remaining DVs by applying a ± 15% margin, forming the design lanes. The established design lanes were checked against the applicable limits for strip theory, the underlying method employed by SHIPMO7 [13]. These limits allow for the analysis of the effect individual DVs have on the design space allocated for a certain vessel type, informing the stakeholders on the influence of both vessel type and hull form.

Table 2 Design lanes detailing the limits applied to each design variable for each vessel type

Design Variable	HS Class		Generic Frigate		Generic OPV	
	Low	High	Low	High	Low	High
LOA (m)	70	95	70	95	70	95
L/B	4.05	5.15	5.98	7.08	5.48	6.58
B/T	3.27	3.97	3.88	4.59	2.96	3.66
Depth (m)	8.65	10.55	7.94	9.85	9.01	10.91
Max Area Location	0.43	0.57	0.38	0.52	0.43	0.57
Prismatic Control	0.36	0.48	0.19	0.31	0.44	0.56
Section Tightness Aft	0.50	0.71	0.40	0.61	0.90	1.00
Section Tightness Fwd	0.69	0.91	0.56	0.78	0.69	0.91
Section Tightness Mid	0.39	0.61	0.71	0.93	0.79	1.00
Deadrise Aft	0.63	0.78	0.00	0.00	0.23	0.38
Deadrise Fwd	0.29	0.51	0.79	1.00	0.79	1.00
Deadrise Mid	0.03	0.07	0.10	0.14	0.28	0.32
Side Slope Fwd	0.68	0.83	0.88	1.00	0.58	0.83
Side Slope Aft	0.08	0.12	0.18	0.22	0.08	0.12
Flare Fwd	0.69	0.91	0.79	1.00	0.39	0.61
Stem Rake (deg)	19.55	28.45	25.55	34.45	30.55	39.45
Stem Curvature	0.27	0.33	0.00	0.00	0.02	0.08
Bow Rounding	0.13	0.17	0.08	0.12	0.13	0.17
Forefoot Shape	0.26	0.34	0.00	0.00	0.46	0.54
Transom Rake (deg)	-14.95	-11.05	0.00	0.00	0.00	0.00
Transom Deck Width	0.77	1.00	0.75	1.00	0.72	0.98
Keel Rise Pt	0.62	0.78	0.3	0.46	0.45	0.61
Keel Rise Rate Aft	0.25	0.37	0.34	0.46	0.46	0.58
Keel Rise Rate Fwd	0.17	0.21	0.00	0.00	0.07	0.11

DESIGN SPACE EXPLRORATION

The Ship Generation Model (SGM) was exploited through ModelCenter® by running a Design of Experiments for all Design Variables (DVs) for each vessel type to allow the design space to be generated. To efficiently build the design space, Orthogonal Arrays (OAs) were used for the experimental designs due to their space filling properties, ensuring the designs are infiltrating the design space effectively [14]. Data from 1000 run OAs were used to develop Response Surface Models (RSMs) for analysis of the relationships between Key Performance Parameters (KPPs) and DVs. Additionally, 2D scatter plots were developed for analysis of the optimal region of the design space for each vessel type as signified by a pareto-front [15]. Exploration of the design space was conducted in two parts. First, the overall performance of each vessel type was assessed relative to each other. Secondly, the influence of DVs on KPPs was analysed for the best performing vessel type.

Vessel Type Comparison

Seakeeping and resistance performance results of the three vessel types were determined for a wave condition 1.5m in waveheight and of 6.5 second wave period, the aforementioned most common conditions in the assumed operational area of the Hydrographic Survey (HS) capability. Seakeeping based KPPs employed comprehensive operability indices calculated using Equation 1 that utilised the tailored criteria and speed profiles shown in Table 1 and Figure 2. The probability of headings was evenly distributed from head to following seas. Individual criteria threshold values were replicated from guidance set by [7]. Criteria weightings remained equally distributed. The resistance efficiency based KPP, evaluated as total resistance per tonne of displacement (R_T/Δ), concerns a typical service speed of HS capability at 14 knots, as previously described.

Scatter plots of the design space for the Launch and Recover (L&R) operability KPP have been compared against the R_T/Δ at 14 knots KPP for all three vessel types; a HS type vessel, frigate and Offshore Patrol Vessel (OPV) in Figures 3, 4 and 5 respectively. For further comparison, a colour gradient details the equivalent transit operability KPP for each design point. It can be seen by comparing Figures 3, 4 and 5 that the design space for each vessel resides in separate regions of the entire solution space. This statement is even clearer when comparing the optimal design space of each vessel type signified by the pareto-fronts.

The design space of the HS type vessel in Figure 3 is spread predominantly from a low to high L&R operability KPP though all design points are subject to a low R_T/Δ KPP. The optimal design space of the HS vessel, represented by the pareto-front (the black crosses in Figure 3), is agreeable with desired values of each assessed KPP. Hence, the four design points on the pareto-front of the HS vessel exhibit high L&R and transit operability indices as well as a low R_T/Δ. This results in a small trade-off space for the HS vessel, a desired outcome. The optimal design point on the pareto-front then becomes a function of the significance of each assessed KPP.

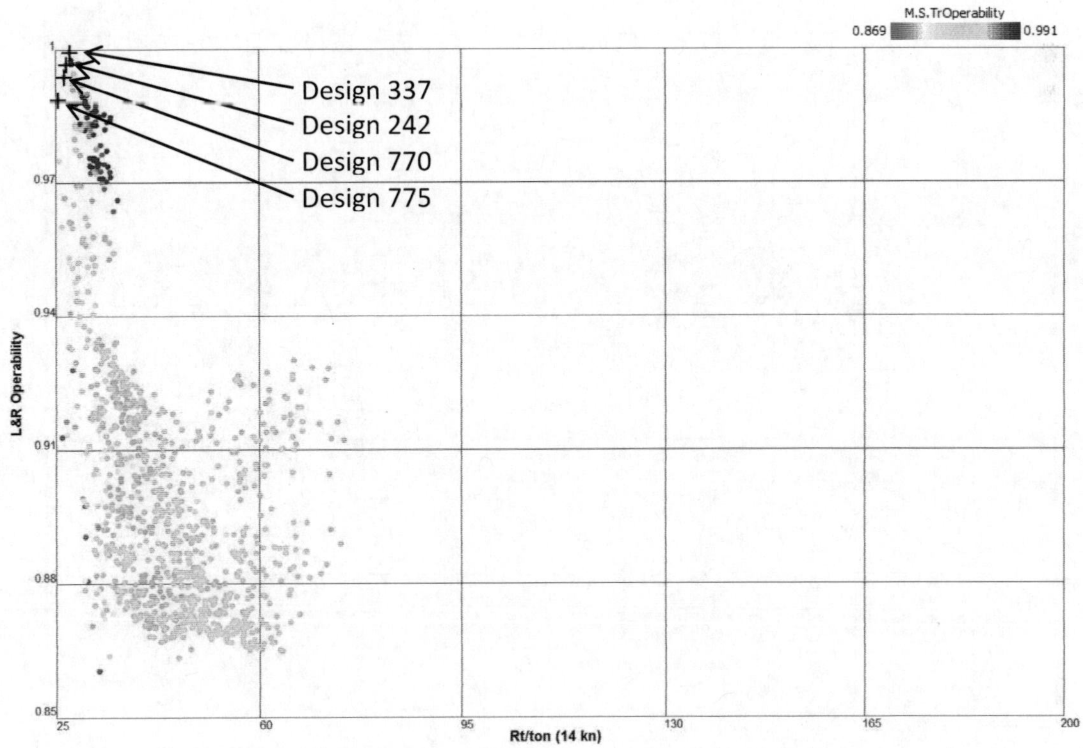

Figure 3 Design space of a hydrographic survey vessel type concerning L&R operability relative to Rt/ton (14 knots). Colour gradient signifies the equivalent transit operability value of each design point.

If both seakeeping and resistance characteristics of the capability were of equal importance then design points 770 or 242 are suitable, see Figure 3. Likewise, if the L&R seakeeping characteristics of the capability were most significant then design point 337 would become the optimal design. Otherwise, if the resistance characteristics dominated as the main requirement of the capability then design point 775 would become the optimal design. This indicates the influence of the KPPs on the trade-off space.

The design space of the frigate is spread over a large region of the entire solution space. Consequently, the pareto-front is spread predominantly from a low to high resistance KPP, though a majority of design points are subject to exhibiting a high L&R operability KPP. In comparison to Figures 3 and 5 the pareto-front is also not as defined. This indicates that small changes in a frigate's hull form design parameters lead to relatively large changes in seakeeping and resistance performance. As a result, the trade-off space for a frigate type hull form for HS capability is large, implying more compromises in performance are necessary. For example, a design with favourable L&R operability KPP will also have an unfavourable resistance KPP. A relationship between resistance and transit operability KPPs exists. For a high R_T/Δ KPP the frigate exhibits a low transit operability KPP. This relationship is favourable towards the leftmost of the design space in Figure 4 where design points exhibit low R_T/Δ KPPs and high transit operability KPPs. This relationship indicates that for a frigate vessel type the optimal solution would likely compromise L&R operability performance.

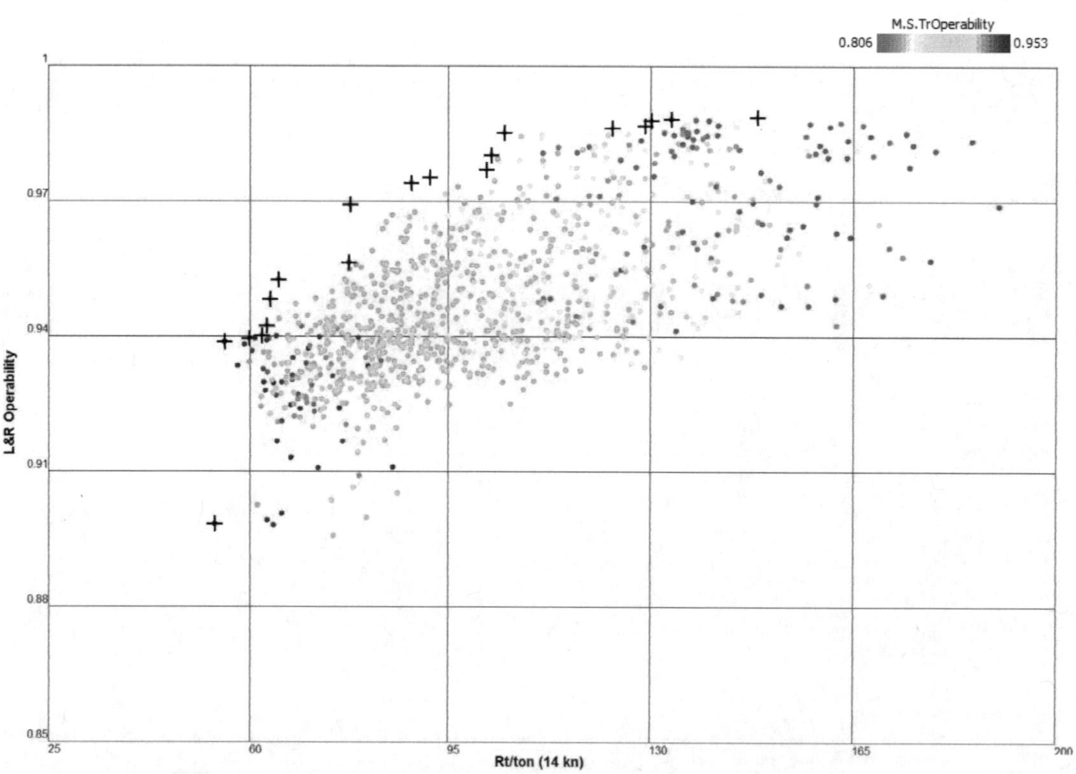

Figure 4 Design space of a frigate vessel type concerning L&R operability relative to Rt/ton (14 knots). Colour gradient signifies the equivalent transit operability value of each design point.

The design space of the OPV type hull form in Figure 5 bridges the gap between the design spaces of the HS vessel and frigate. As a result, the majority of optimal design points on the pareto-front exhibit a low to mid range R_T/Δ KPP and a mid-range L&R operability KPP.

Comparing the OPV to the frigate in Figure 4, the optimal design space for the OPV, and hence the trade-off space, is smaller. Similar to the frigate, the design space of the OPV also contains a relationship between resistance and transit operability KPPs. Concerning these KPPs, the OPV exhibits better overall resistance and transit operability performance than the frigate. However, the OPV lacks to achieve the same L&R operability performance.

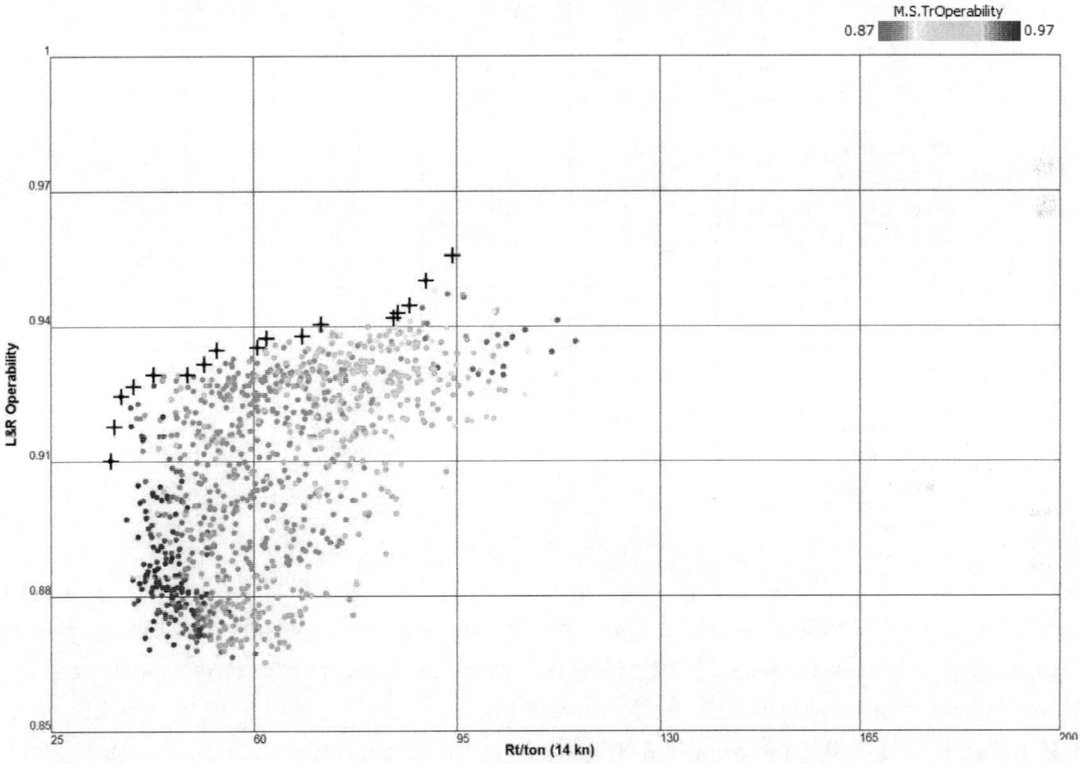

Figure 5 Design space of an OPV vessel type concerning L&R operability relative to Rt/ton (14 knots). Colour gradient signifies the equivalent transit operability value of each design point.

Comparing all three vessel types, and considering both seakeeping operability KPPs and the resistance efficiency KPP at 14 knots, the OPV and frigate are both outperformed by the HS vessel. As a result, it can be concluded that the HS vessel type is best suited to meet HS capability under the tested environmental conditions and assessed criteria.

Design Parameter Analysis

For further analysis of the design space the four designs, highlighted in Figure 3, that formed the HS vessels optimal design space, and hence performed best, were chosen for comparison of hull form design parameters. Design 337 is the non-dominated design with the maximum L&R operability KPP. Design 775 is the non-dominated design with the minimum R_T/Δ KPP. Designs 242 and 770 are non-dominated designs with good L&R operability and R_T/Δ KPPs. Table 3 lists the design parameters of each hull form while the comparison in geometry and hull form between designs 337 and 775 can be seen in Figure 6.

Table 3 Hull form design parameters of the four designs comprising of the HS vessels' pareto-front

Parameter	Design 337	Design 242	Design 770	Design 775
Length Overall (LOA)	94.20 m	90.97 m	95.00 m	95.00 m
L/B	4.26	4.08	4.05	4.12
B/T	3.47	3.27	3.72	3.34
Depth	10.00 m	10.30 m	9.63 m	10.30 m
Transom Beam	0.99	0.83	0.90	0.84
Bow Curvature	0.31	0.29	0.27	0.31
Bow Rounding	0.15	0.16	0.15	0.14
Deadrise Aft	0.63	0.70	0.70	0.68
Deadrise Fwd	0.44	0.45	0.34	0.50
Flare Fwd	0.76	0.88	0.86	0.87
Forefoot Shape	0.33	0.31	0.34	0.27
Keel Rise Point	0.64	0.70	0.67	0.75
Keel Rise Rate Aft	0.32	0.30	0.30	0.33
Keel Rise Rate Fwd	0.21	0.20	0.20	0.17
Prismatic Control	0.46	0.42	0.43	0.40
Section Tightness Aft	0.66	0.60	0.69	0.55
Section Tightness Fwd	0.87	0.86	0.88	0.75
Slope Aft	0.10	0.11	0.11	0.10
Slope Fwd	0.74	0.75	0.92	0.79
Station Maximum Area	0.57	0.57	0.48	0.46
Stem Rake	25.29 deg	27.30 deg	25.58 deg	21.27 deg
Transom Rake	-12.43 deg	-11.80 deg	-11.55 deg	-13.44 deg

From Table 3 it is evident that all designs have very similar geometric attributes, i.e. length overall, L/B ratio, B/T ratio and depth. These results indicate that in order to achieve optimal HS capability the requirements of the HS vessel must be of certain geometry. Hence, for optimal seakeeping and resistance performance the HS vessel should have a LOA between 90-95 m, a low L/B, a B/T between 3.3-3.7 and a depth of approximately 10 m. The values and small ranges of these geometric attributes suggest a larger vessel of greater displacement is better suited to meet HS capability needs. Furthermore, it is clear that the remaining hull form characteristics are of similar values. This explains further the previous discussion of the HS having a small trade-off space.

It appears that within the design space of the HS vessel a particular range of design parameters contributes towards an optimal design. When implementing the Modelling and Simulation (M&S) framework within an acquisition, it is not only able to inform the stakeholders of the most appropriate vessel type to achieve the capability. It is also able to inform the stakeholders of a particular range of hull form characteristics for the vessel type that contribute towards an optimal performance. Therefore, exploiting the M&S framework in this manner will provide stakeholders with a deeper understanding of technical solutions and help identify risks prior to requirements setting.

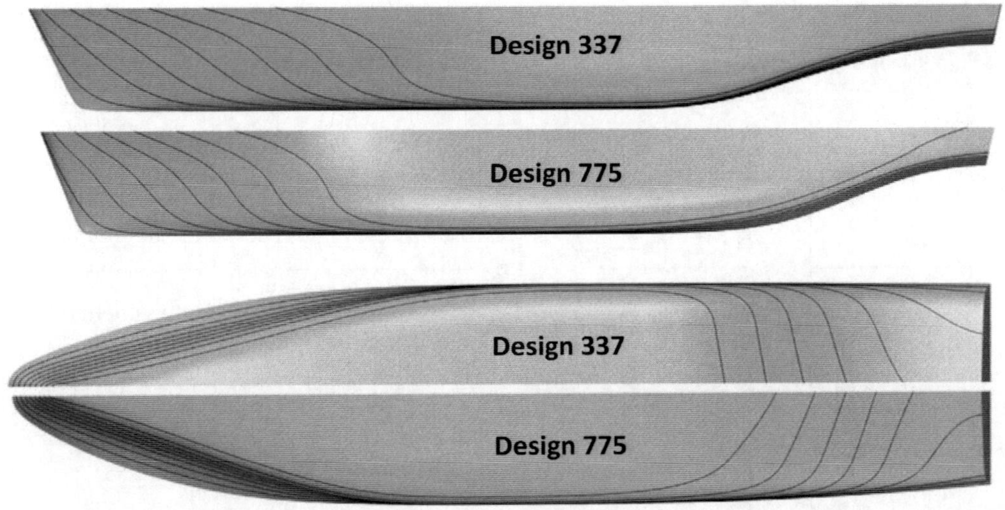

Figure 6 Comparison of design 337 and 775 geometry and hull form. (Top) profile views including buttocks lines; (Bottom) plan views of waterlines. Vessel geometry not to scale.

To enable stakeholders to develop an understanding of the influence particular hull form design parameters have on performance, RSMs can be analysed. RSMs, as seen in Figure 7, are represented by prediction profilers that graph the relationship between DVs and KPPs. The slope of the line in the graph details the change in effect the DV has on the KPP. A steep line implies a significant effect where a horizontal line implies no effect.

From Figure 7, it can be seen that L&R operability and R_T/Δ vary significantly with LOA, proving to be most sensitive. Transit operability varies between a minute range with LOA and is least sensitive. However, for all three KPPs a maximum LOA results in a maximum L&R and transit operability KPP, and a minimum R_T/Δ KPP. This supports the previous statement that a larger vessel is more capable for HS operations. It also informs the stakeholders that when considering only the KPPs highlighted throughout this paper there is little trade-off space required for length. The same cannot be stated for the remaining DVs. For a given LOA, L&R operability is maximum, and transit operability and R_T/Δ at 14 knots are at a minimum for a low L/B. With increasing L/B, L&R operability decreases significantly while transit operability improves marginally. Again, this supports the findings from Figure 3 and Table 3. L&R operability and R_T/Δ are least sensitive to B/T while transit operability is most sensitive. This indicates the importance of exploring the design space when trading off this requirement. Similarly, depth has a small influence on L&R operability and R_T/Δ though has the largest influence on transit operability. Here, the design space can be used to justify the requirement of employing a larger depth; also and outcome from Figure 3 and Table 3.

Finally, it is worth noting the discussed trends appear to have been influenced in part by the criteria and weightings used in the operability calculation. For example, the hull depth DV is at or near the upper limit for the best performing HS type hull forms in Table 3. An explanation of this could be associated with the 'deck wetness' criteria, since the higher the weatherdeck above the waterline, the less likely it is to experience wetness. The sensitivity of results to criteria is an ongoing component of the research covered in this paper.

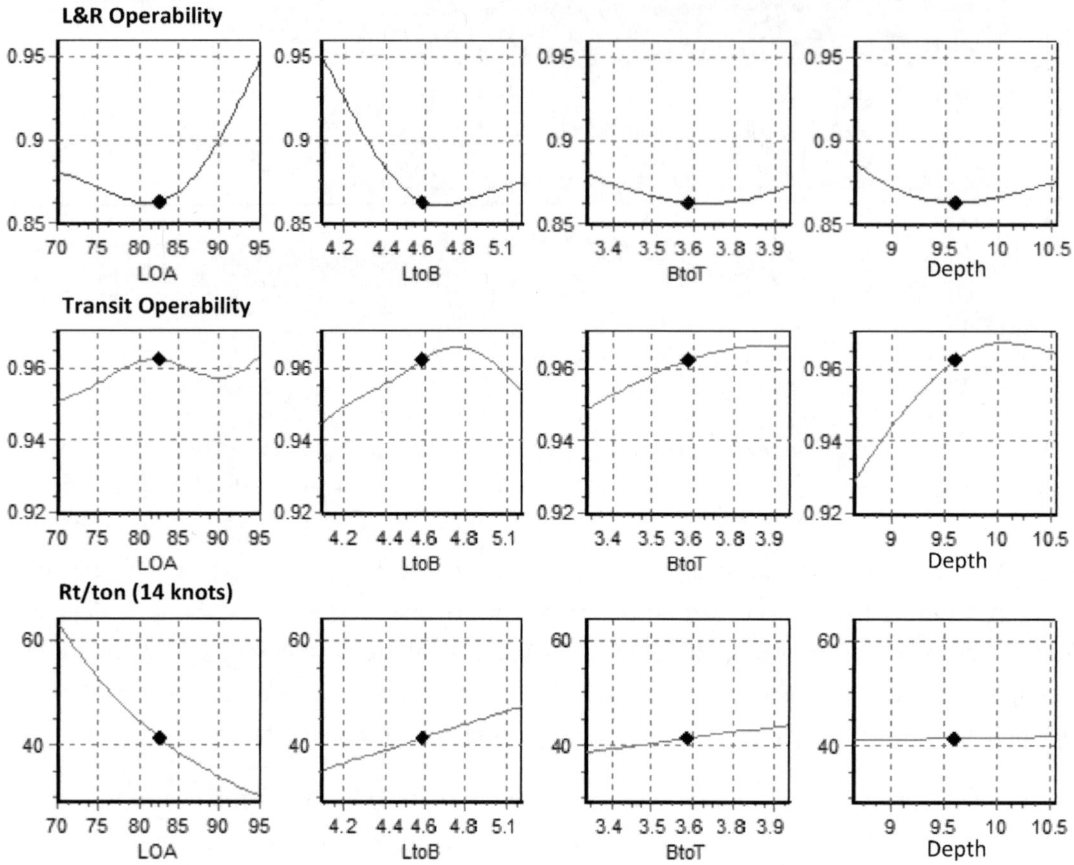

Figure 7 Design space of the HS vessel type. Prediction profiles detail the relationship between DVs and KPPs.

CONCLUSIONS AND FUTURE WORK

This paper presented a Modelling and Simulation (M&S) framework to support requirements setting for Royal Australian Navy acquisitions. The M&S framework builds upon a previously proposed methodology for Concept and Requirements Exploration by implementing higher fidelity modelling techniques. Testing of the M&S framework was performed by application of a Hydrographic Survey (HS) capability. Three representative vessel types were considered for meeting the HS capability, a HS class, frigate and Offshore Patrol Vessel. The Ship Generation Model (SGM), within the M&S framework, was exploited to generate 1000 hull forms indicative of each vessel type and assess the seakeeping and resistance performance.

Testing indicated the M&S framework is beneficial as a means of informing capability acquisition stakeholders with a possible design space. The design space can be utilised by stakeholders in an acquisition project to determine the most appropriate vessel type for meeting capability needs. Additionally, it can also be used to help inform stakeholders of hull form design parameters that contribute towards an optimal capability performance. It was shown that Response Surface Models of the design space can be employed to provide further insight into the trade-off space of a particular vessel type informing stakeholders for requirement trade-off decisions. Armed with this knowledge, acquisition stakeholders could develop a specification containing appropriate requirements and constraints on design variables, ensuring the most suitable designs were received in response to a Request-for-Tender (RFT). Setting these requirements tighter can eliminate tender responses that would otherwise be eliminated through detailed analysis during the latter stages of the Risk Reduction and Requirements Setting phase. Overall, it was shown the M&S framework can be successfully exploited in support of requirements setting early during the Capability Life Cycle.

Currently the M&S framework focuses on the SGM component to support requirements setting. Future work includes developing tools and methods that analyse the options received as an outcome of an RFT as well as assessment of different ship subsystems. The M&S framework could be improved by employing higher fidelity tools. Immediate future work includes investigating the influence of seakeeping criteria limits on operability indices.

REFERENCES

1. Morris, B.A. and B.S. Thethy, *Towards a Methodology for Naval Capability Concept and Requirements Exploration in an Off-the-Shelf Procurement Environment*, in *Pacific 2015 International Maritime Conference*. 2015: Sydney, Australia.
2. Defence, *Interim Capability Life Cycle Manual*, Defence, Editor. 2016, Department of Defence: Canberra, Australia.
3. Crowder, J.A., J.N. Carbone, and R. Demijohn, *The Overall Systems Engineering Design*, in *Multidisciplinary Systems Engineering: Architecting the Design Process*. 2016, Springer International Publishing: Cham. p. 105-127.
4. Faulconbridge, I. and M.J. Ryan, *Systems Engineering Practice*. 2014, Canberra, Australia: Argos Press.
5. BMT, *Global Wave Statistics*. 2011.
6. JCIDS, *Manual for the operation of the joint capabilities integration and development system (JCIDS).* US Department of Defense. Washington. DC, 2015.
7. Royal Australian Navy, *Part 6: Seakeeping*, in *ADF Material Requirement Set*. 2003.
8. Molland, A.F., S.R. Turnock, and D.A. Hudson, *Ship resistance and propulsion: practical estimation of propulsive power*. 2011: Cambridge university press.
9. Morris, B.A., *Blending Operations Analysis and System Development During Early Conceptual Design of Naval Ships*, in *SETE2014*. 2014: Adelaide.
10. DRS Technologies Inc, *Orca3D User Manual*. 2016.
11. Mennen, H., *An approximate power prediction method.* International Shipbuilding Progress, 1982.
12. Faltinsen, O. and M. Minsaas. *K, J. Liapis, N. Skjørdal, S, O. 1980*. in *Prediction of Resistance and Propulsion of a Ship in a Seaway. In Proceeding of 13th Symposium on Naval Hydrodynamics*. 1980.
13. Graham, R., *Applicability of Strip Theory to Vessels with Low Length to Beam Ratio: A Literature Survey*. 1984.
14. ModelCenter, *ModelCenter Help.* 2015. **11.2**.
15. Stadler, W., *Multicriteria Optimization in Engineering and in the Sciences*. Vol. 37. 2013: Springer Science & Business Media.

CONFERENCE ABSTRACT

Design of Icebreaking Supply & Research Vessels

Ken Goh, Knud E Hansen Australia, keg@knudehansen.com

KNUD E. HANSEN naval architects are at the forefront of the design of polar vessels. The company has recently been responsible for the design of the new Australian Icebreaker, which will be the world's most capable Antarctic supply and research vessel when commissioned in 2020.

The presentation will focus on the many unique requirements and challenges of polar research vessels design with reference to work on Australian and North American vessels.

The various design solutions and the design evolution from the initial proposals to the final tender design will be covered. Topics include general arrangement, hull form, propulsion systems and how virtual reality tools have been used in the design process.

DAMEN

DAMEN SHIPYARDS GROUP Pacific 2017

P.O. Box 1 4200 AA Gorinchem The Netherlands 15th August 2017

+31 183 63 99 22 info@damen.com www.damen.com

Damen announces new range of Naval Multi Role Auxiliary Vessels

Dutch shipbuilder offers flexibility and reduced total cost of ownership by means of modular mission modules with proven technologies

During the upcoming Pacific 2017 exhibition and conference in Sydney , Damen Shipyards Group will introduce a new range of Multi Role Auxiliary Vessels (MRAV) to the Pacific region. The common theme running through the series is to provide naval clients with a basic platform offering reliable and cost-effective multi role potential and hydrographic survey capabilities.

With the addition of supplementary modular mission equipment, this new family of Damen vessels can be mobilised in numerous other, mainly littoral, naval tasks such as: explosive ordnance clearance and disposal, diving operations, torpedo recovery and overhaul, ROV and UAV deployment, SAR, coastal infantry and submarine support. The largest version of the range will be able to operate worldwide, on the ocean as well as in littoral waters. This ship has additional capabilities like disaster and humanitarian relief, oceanography and naval training support.

The introduction of flexible concepts which allow as many functions as possible to be included into a range of smaller vessels without reducing the effectiveness and capacity of the fleet while maintaining the benefits of modularity: this is Damen's ambition with this new family of vessels. "To this end, plug-and-play containerised kit for many support tasks contribute considerable adaptability to a particular mission," explains Damen Shipyards Gorinchem's Principal Naval Advisor Jan van der Burg, a retired vice admiral of the Royal Netherlands Navy

One platform – multiple tasks

"The idea behind these vessels is to create a basic platform that can assist in a variety of tasks through the selection of the required mission configuration, e.g. coastal transport, submarine support or coastal infantry operations. The stimulus to switch from the traditional one-to-one replacement is to lower the total cost of ownership without losing capability and capacity."

The new range of vessels consists of three different designs: the MRAV 660, MRAV 1600, MRAV 3600. Designed for different geographic profiles, these vessels are respectively 43, 62 and 85 metres long. Hydrographic capabilities, to map the seabed for safe navigation and as a preparatory action for military operations in particular, are indispensable to navies worldwide. Depending on a naval client's specific requirements, any type of hydrographic equipment can be integrated into these three vessels.

Minimal draught

With a draught of 1.9 metres, the MRAV 660 is suited for very shallow coastal, riverine and inland water operations. In addition to shallow water hydrographic surveys, this vessel is capable of a comprehensive array of duties such as diving operations, EOCD support, ROV and UAV deployment, with a core crew of 8 and capacity for an additional 15 specialists.

"The key points are to maximise displacement, minimise resistance and optimise seakeeping characteristics for the area in which the ship will operate. The MRAV 660 has design characteristics that help achieve this: an aluminium superstructure and reduced freeboard section cut down on weight. And tunnel ducts on the underside of the hull ensure enough water reaches the propellers."

Additional roles

The MRAV 1600 is designed for littoral and regional offshore operations. Its larger size allows for greater endurance and carrying capacity of both crew, mission modules and cargo. The vessel will be manned by a core crew of 13, with capacity for an additional 30 mission specialists. The main deck can hold six standard 20-foot mission containers and the below-deck cargo hold can store two 20-foot containers and palletised cargo.

This medium-sized vessel can take on similar hydrographic and auxiliary duties to its smaller sister vessel, with the addition of torpedo recovery and overhaul tasks in support of submarines and anti-submarine warfare units. Small scale coastal transport and infantry support is also possible.

Global coverage
Intended for worldwide service, the capacity of the largest vessel in the range – the MRAV 3600 – allows for more than one specific mission during a deployment. Capable of hydrographic operations both in littoral and deeper waters, this vessel also has a helideck and substantial storage capacity for other mission configurations, equipment and cargo. Furthermore, the MRAV 3600 can serve as a base for more extensive operations such as disaster and humanitarian relief. There is accommodation for 14 core and 45 additional mission crew and enough space on board to provide emergency hospital services for 50 to 60 people.

Cost results
A modular platform is inherently flexible: this allows naval clients to better react to changes in the mission environment. Modularity also has implications on the total cost of ownership: the lifetime of an individual vessel can be efficiently extended by upgrading capabilities with new equipment modules that are not integrated into the original design.

Financial advantages are to be found in the fact that the MRAV range is commercially built and also uses commercially available components. This is made possible because of the vessels' non-combatant role.

Complementary design
The ship design itself has a proven background: "Damen has built many ships similar to the MRAV range for the offshore industry. This means that there will not be any unwanted surprises when being the first customer of this new range."

Damen Shipyards Group
Damen Shipyards Group operates 32 shipbuilding and repair yards, employing 9,000 people worldwide. Damen has delivered more than 6,000 vessels in more than 100 countries and delivers some 180 vessels annually to customers worldwide. Based on its unique, standardised ship-design concept Damen is able to guarantee consistent quality.

Damen's focus on standardisation, modular construction and keeping vessels in stock leads to short delivery times, low 'total cost of ownership', high resale values and reliable performance. Furthermore, Damen vessels are based on thorough R&D and proven technology.

Damen offers a wide range of products, including tugs, workboats, naval and patrol vessels, high speed craft, cargo vessels, dredgers, vessels for the offshore industry, ferries, pontoons and superyachts.

For nearly all vessel types Damen offers a broad range of services, including maintenance, spare parts delivery, training and the transfer of (shipbuilding) know-how. Damen also offers a variety of marine components, such as nozzles, rudders, anchors, anchor chains and steel works.

In addition to ship design and shipbuilding, Damen Shiprepair & Conversion has a worldwide network of 15 repair and conversion yards with dry docks ranging up to 420 x 80 metres. Conversion projects range from adapting vessels to today's requirements and regulations to the complete conversion of large offshore structures. DSC completes around 1,500 repair and maintenance jobs annually.

Effects of Structural Ageing on Blast Analyses of a Naval Vessel: A Technical Review

Daniel K. Clayton[1], Thomas Mitchell-Ferguson[1], Roberto Ojeda[1], Rouzbeh Abbassi[1]

[1]Australian Maritime College, University of Tasmania, Australia.

Daniel.clayton@utas.edu.au

ABSTRACT

As vessels are being asked to perform over longer service lives and under more intensive operational tempos, it is becoming more pertinent that the effects of structural aging be accounted for. Fatigue is an important aspect of structural aging of a vessel. Currently, fatigue analyses are conducted in order to determine the point at which fatigue failure might occur. However, fatigue can induce changes to the material properties long before full fatigue failure has occurred. These changes in material properties can affect the outcomes of other forms of structural failure. One such form which is of particular importance for naval vessels is failure due to blast loading.

Blast modelling utilising coupled CFD/FEA simulations are conducted when investigating the effects of an internal blast, requiring the use of a constitutive model to account for the material properties. Fatigue damage can result in a reduction of modulus of toughness of TRIP and DP steels, especially so under high strain rate loading conditions. However, these changes in material properties are not currently considered in constitutive models.

This paper reviews experimental investigations conducted into the effects of fatigue on high strain rate loading. It is shown that the effects can be significant, and that further work is required before the allowance of the effects can be introduced to blast analyses.

INTRODUCTION

Naval vessels are increasingly asked to operate over longer service lives, and for more days per year than historically required [1]. Shipbuilding steels have also progressed leading designers to adopt increasingly smaller scantling sizes that provide the same level of structural strength. These two concepts combined mean that now, more than ever, structural aging is becoming a limiting factor in design [2].

Structural aging of a vessel can be considered as the combined effects of corrosion, fatigue, and corrosion induced fatigue [3]. Fatigue of a vessel occurs due to its passage through a seaway, where cyclic alternation between hogging and sagging conditions induce cyclic bending induced stresses through the hull girder [2]. Over the lifespan of the vessel these cyclic loads induce changes to the material's properties that result in fatigue damage accumulation. This damage is accounted for at the design stage by the various rules and regulations that govern naval vessel design.

Design considerations that control the effects of fatigue damage in the structural aging of a vessel are not unique to naval platforms, however design considerations that focus on containing and limiting the damage from blast loads is predominately a military issue [4]. Modern warships are increasingly asked to engage in asymmetric warfare, whereby combatants have contrasting levels of military power, resources, and tactics [5]. Asymmetry at sea is a driving design consideration for modern and future warships, seen in the current strategies of the Islamic Revolutionary Guard Corps Navy of Iran [6, 7] and the 2000 attack on USS Cole in Aden [8]. The recent incident involving HSV-2 Swift in 2016 [9] confirms the severity of the threat that the detonation of a warhead within an internal compartment poses on the survivability of a naval platform.

The expansion of explosive gases within an internal compartment induces high strain deformation of the structure, and occurs within a very small time frame [10]. The structure's response can lead to localised failure of structure, allowing the explosive gases to cause further damage to neighbouring compartments, or even more severe global failure. Current design rules from DNV [11], DNV-GL [12], LR [13], and RINA [14] show how current blast modelling practices enable a compartment to be designed to ensure the blast does not carry through compartment bulkheads causing further damage. These modelling practices involve the use of full coupled CFD/FEA simulations of the blast event.

The constitutive models which are used in these simulations assume that the material is new, and model the material properties as such. However, the vessel is unlikely to be attacked anytime soon after initial construction. As such, it is pertinent to establish what the effect of structural aging due to fatigue will be on the structure's future ability to contain a blast.

This paper provides a review of: the current methodologies in fatigue and blast modelling separately; the effects of fatigue and high strain rate individually on the tensile properties of steel; and the effects of fatigue on the tensile properties of steel under high strain rate loading. The review aims to highlight viable simulation methodologies that incorporate the effects of structural aging due to fatigue in state-of-the-art blast containment analysis for naval surface platforms.

BLAST ANALYSIS METHODOLOGIES

There are three available approaches, experimental, analytical, or numerical modelling, adopted by researchers [10] to investigate the effects of internal blast on the vessel's structure. Experimental modelling involves the construction of a representative vessel compartment, and the detonation of an explosive charge within the compartment to study the effects. If the experiments are required to cover a wide range of blast scenarios and structural configurations, this approach can be costly in both time and resources.

The alternative approach, numerical modelling, is significantly more cost and time effective. Once the geometry, material properties, and loads are specified, simulations can be re-ran by only changing parameters as desired, rather than requiring complete recreation of the work. The limiting factor for numerical modelling is thus often the computational solution time required, rather than the construction and setup of experiments.

Two main approaches are available for numerical modelling of effects of blast depending on whether the blast is near field (such as an internal compartment blast) or far field (such as a near miss from a missile). The far field blast effect is usually modelled using empirical equations, such as the TNT equivalence method or large scale field tests [14]. The near field blast effect on structures is often investigated using a coupled CFD/FEA explicit simulation [15].

The particular case of a blast within an internal compartment is considered to be near field event [14], and as such its analysis requires computational simulations. The characteristics of an internal blast are complicated by the multiple reflections of blast wave off the multiple compartment boundaries, and the confinement of hot gases which leads to the creation of a quasi-static overpressure [16]. Hence, there is a need for full coupled simulations, rather than the application of analytical methodologies.

The numerical modelling of an internal blast requires the use of an Arbitrary Lagrangian-Eulerian (ALE) approach such as that employed in modelling of underwater explosions (UNDEX) [4]. This approach couples two numerical domains: an Eulerian domain used to perform the Computational Fluid Dynamics (CFD) calculations required to model expanding of the gas cloud an a Lagrangian domain used to perform Finite Element Analysis (FEA) calculations of the structural response. At the interface between gas and structure, the Eulerian domain maps the pressure onto the Lagrangian model, and in return the Lagrangian domain updates the shape of the boundaries of the Eulerian domain as it deforms [4]. This methodology has been adopted by previous researchers, Larcher [17] and Zakrisson et al. [18] to investigate the effects of an air blast.

ALE modelling requires the definition of the geometry of the structures, and material properties for both the gas, i.e. air, contained within the Eulerian domain, and the solid, i.e. structural material used for the Lagrangian domain. Since blast is a highly dynamic event it is critical that the material properties model used to describe the response of the Lagrangian domain correctly represents the material behaviour at the high strain rates experienced during blast loading [19]. Ngo et al. [20] review the effects of blast loading on civil structures, looking at blasts occurring due to acts of terrorism. Figure 1 illustrates the typical strain rate ranges of different sources of loading on steel reinforced concrete structures. It can be observed that blast loading subjects the structure to strain rates in the order of 10^2 to 10^4 s^{-1}. Al-Hassani et al. [21] conducted a review on high strain rate loading of materials and concluded that explosive and ballistic applications are concerned with strain rates in excess of 10^3 s^{-1}.

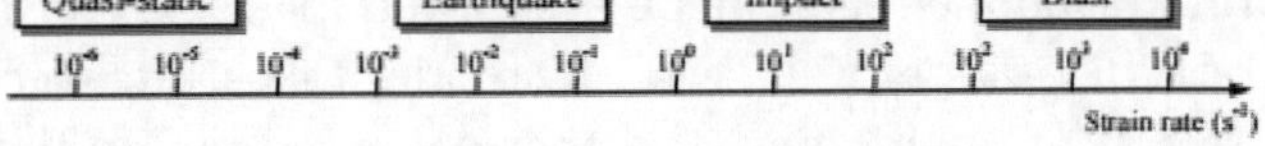

Figure 1: Strain rates associated with different types of loading [20]

From this it can be seen that blast modelling requires material constitutive models which are capable of describing high strain rate (above 10^2 strains/s) behaviour [22].

CONSTITUTIVE MODEL

A constitutive model is a mathematical representation of the stress-strain curve of the material under a given loading condition [23]. This mathematical representation is required in order to characterise the material's behaviour in a FE simulation. The complexity of the selected constitutive model is dependent upon the material properties which are most relevant to the simulation being undertaken.

Due to its common use in the naval shipbuilding industry, this review will be focussed on HSLA steel [24]. The standards for HSLA steel are outlined in ASTM [25], but general material properties of interest are a yield strength of 350 MPa, ultimate tensile strength of 490 – 620 MPa, and elongation at break of around 22%. Figure 2 depicts the tensile strength and elongation of HSLA when compared to other steel alloys.

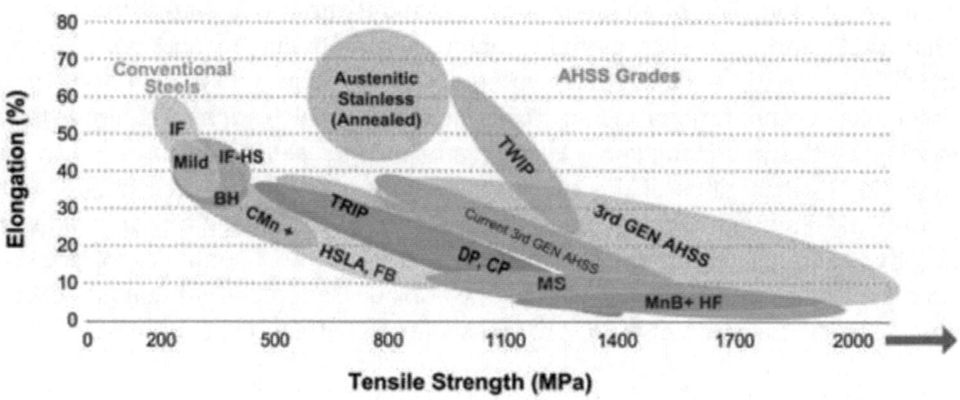

Figure 2: Comparison of tensile strength and elongation of several common steel types [26].

Investigation into the effects of high strain rate loading on materials is a well-developed and well-reviewed field. This review will thus outline the general effect which higher strain rates have on steels similar to HSLA. Health and Safety Executive [27] reviews the effect of high strain rate loading on material behaviour for steels. The most relevant finding with regards to our present problem is that the material toughness decreases with increasing strain rates. The material toughness is a measure of its energy absorption before failure, which is a direct measure of how well a given material is able to contain a blast. Similar findings have been reported by the previous researchers Al-Hassani et al. [21].

These high strain rate loading effects can be accounted for using constitutive models such as the Johnson-Cook (JC) model [28]. The JC model takes into account the effects of strain hardening, strain rate hardening, and thermal softening on the stress strain curve of a material. However, currently available JC models rely on data obtained for new material, i.e. material that has not been subjected to ageing effects such as fatigue loads.

FATIGUE

The classification of fatigue is required before its effects on material behaviour can be discussed. The distinction between low cycle and high cycle fatigue is important as it defines certain characteristic behaviours of the change in material properties during the fatigue process. The most notable difference between low cycle and high cycle fatigue is that in low cycle fatigue the plastic strain is greater, whereas the elastic strain is dominant in high cycle [29]. In general researchers consider below 10^3 cycles to failure (with high amplitude loading) as low cycle, and above 10^4 cycles to failure (with low amplitude loading) as high cycle [29, 30]. At roughly 10^8 cycles through its lifetime [1, 31] , a warship can be seen to be on the high cycle side.

Once the distinction between low and high cycle fatigue has been made, an estimate on the fatigue life of the material is required in order to quantify the fatigue history non-dimensionally. A recent report by the International Ships and Offshore Structures Congress (ISSC) [32] classifies state of the art fatigue life prediction approaches into: low-cycle, high-cycle and ultra-high cycle approaches; nominal stress approach; structural stress approach; effective notch stress approach; notch strain approach; fracture mechanics approach; and then various probabilistic approach based off either the SN or the fracture mechanics approaches.

The Palmgren-Miner [33] rule enables the non-dimensionalisation of the fatigue history of a material by defining the damage level as a linear relationship against the number of cycles applied divided by the fatigue life prediction. This approach has been often utilised for investigating the fatigue history effects on material properties. Once the fatigue history of the material is quantified, the effects which it has on the stress-strain curve of the material can be discussed. Moćko et al. [34] covered the effects of high cyclic loading on DP500 steel (material properties shown in Figure 2, DP steel type). They utilised the Palmgren-Miner rule as a damage parameter, D, where D = 0 is no damage, and D = 1 is complete failure due to fatigue crack propagation. This provides a non-dimensional representation of the fatigue history of the specimen.

The findings of the study concluded that the DP500 steel first exhibits hardening behaviour during the early stages of the cyclic testing, and then further cycles lead to a softening of the material. The testing indicated an increase in yield strength for both stress ranges (500Mpa and 520Mpa) with the higher stress range experiencing a greater increase. An increasing damage parameter saw a definite increase in the yield strength, and reduction in maximum strain (and thus reduction in modulus of toughness). Furthermore, an increase in the stress range leads to an increase in maximum yield strength, and reduction in maximum strain. It should be noted that the increase in yield strength is more sensitive to the yield range than the damage parameter. However, the decrease in maximum strain is equally sensitive to the increased damage parameter and increased stress range.

The following section discusses the effects of fatigue on material properties under high strain rate loading. It will be seen that the individual effects of high strain rate loading and fatigue on the stress strain curve are similar. The combined effects further increase the changes to the stress strain curve. These combined effects result in the reduced ability of a compartment to withstand a blast.

FATIGUE EFFECTS ON HIGH STRAIN RATE STRESS-STRAIN CURVES

A review of public domain publications on the effects of fatigue in high strain rate testing reveals a relatively new topic [35-40]. Investigation of materials such as aluminium and titanium reveals trends which can deviate to those shown for steel. Since those materials have limited application to the main structure of large surface combatant ships, the reported findings have been restricted to steel. Table 1 shows the relevant findings from the experimental work which has been done on the area. The table indicates the testing instrumentation used in each investigation which have been limited to the use of tensile Hopkinson apparatus (SHPB-T) and bend test Hopkinson apparatus (SHPB-B). The stated reductions in yield stress and elongation at break are given as percentages compared against the values for the similar strain rates, but of the as-received material. The fatigue parameters reported are: the maximum stress during fatiguing (σ_{max}), the stress ratio during fatiguing (R), the fatigue life for the given maximum stress (N_f), and the damage parameter based on Palmgren-Miners rule (D).

Table 1: Findings from literature on fatigue effects on material properties under high strain rate loading.

			Fatigue				Strain	Projectile		
Author	Test	Steel	σ_{max} (MPa)	R (-)	N_f (-)	D (-)	Rate (s^{-1})	Vel. (m/s)	% Δ Yield	% Δ Elong
Verleysen [39]	SHPB-T	TRIP700 Z	674	0.1	100,000	0.5	225		40	-42
	SHPB-T	DP600	517	0.1	100,000	0.5	750		28	-24
Sanchez [38]	SHPB-T	4140T	731	0.2	447,000	0.25		30	-1	-12
	SHPB-T					0.5		30	-2	-30
	SHPB-T					0.75		30	-2	-91
Rubio [37]	SHPB-B	4140T	650	0.2	250,000	0.25 - 0.75				
Mocko [35]	SHPB-T	DP500	400		120,000	0.5	600		7	-7
	SHPB-T		500		40,000	0.5	600		16	-25
Itabashi [36]	SHPB-T	AISI 1215	319	0		0.05, 0.2	1000			
	SHPB-T		364	0		0.05, 0.2	1000			

An interesting observation is that whilst all the above reported investigations are performed within the high cycle fatigue regime, they involve fatigue lives significantly less than the estimated 10^8 load cycles which a vessel may experience through its life [1, 31].

Figure 3, adopted from Verleysen et al. [39], illustrates the results of high strain rate testing of two samples of TRIP700 Z steel (material properties shown in Figure 2, TRIP steel type), one sample which is in the as-received condition, and one sample which has undergone fatigue loading. It can be seen that the pre-fatigued sample has had a significant reduction in modulus of toughness (approx. 30% in this case). The same results however were not seen for the investigation into DP600 steel, where it was reported that the increase in yield stress and the decrease in elongation at break resulted in a modulus of toughness of roughly the same value as the as-received condition.

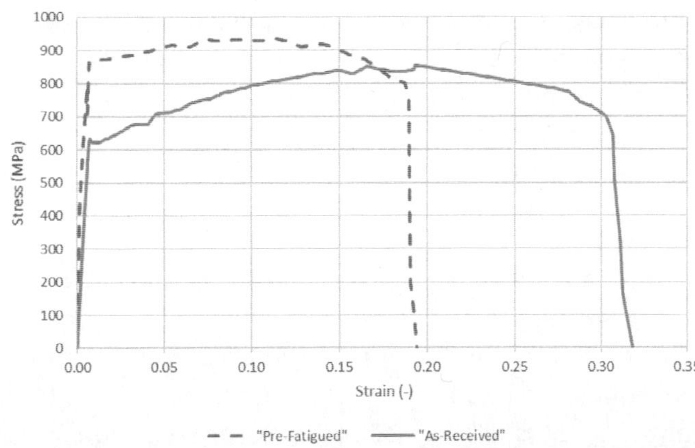

Figure 3: Stress strain curves for as-received and fatigued TRIP700 Z steel sheet, obtained at strain rates of approximately 325/s and 225/s respectively. Data reconstructed from Verleysen [39].

A reduction in modulus of toughness is seen again when considering the results of the study conducted by Sánchez-Santana et al. [38]. As can be seen from Table 1, at the highest investigated strain rate due to a 30 m/s projectile velocity, increasing the damage parameter resulted in a near constant yield stress, yet significant decrease in the elongation at failure. It must be noted that, due to the low quality of the stress strain curves presented, only a qualitative reduction in modulus of toughness can be reported.

The investigation by Rubio-González et al. [37] does not provide sufficient data on their results in order to quantify any effects of fatigue in the modulus of toughness. The study was aimed instead at any changes which might occur to the dynamic fracture toughness. A significant decrease of up to 55% was seen for the fatigued material compared to as-received, indicating that failure of the material was indicative of brittle behaviour.

The work conducted by Moćko et al. [35] focussed on the micro-scale of the fatigue effects on DP500 (material properties shown in Figure 2, DP steel type) steel. However, the macro-scale results concurred with those reported by Verleysen et al. [39]. Figure 4 adopted from previous research [35] illustrates how the effect of the pre-fatigue is synonymous to that of an increased strain rate, leading to an increase in yield stress, and decrease in ultimate plastic strain.

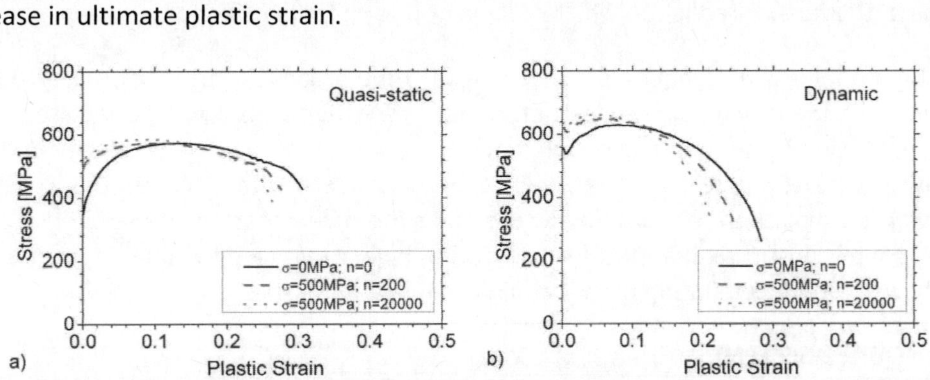

Figure 4: The results from Moćko et al. [35] show that for DP500 steel, an increase in the pre-fatigue leads to a result in line with that of an increase in the strain-rate.

The final study mentioned in Table 1, by Itabashi et al. [36], showed results inconsistent with the above reported findings. Slight increases in breaking strain were seen with an increasing damage parameter, but with significant scatter amongst the results. Further against the trend, a decrease in yield stress is reported with the presence of pre-fatigue. These results show that fatigue can have widely varying effects on the high strain rate properties depending on the material investigated, even when comparing two different types of steel. The steel investigated, AISI 1215, has an ultimate yield strength of roughly 540 MPa, and elongation at break of 10%. From this it can be seen that it sits on the opposite side of HSLA in Figure 2 than the other materials investigated in this review.

It can be seen from the above review of literature that there is cause for concern regarding a significant reduction in modulus of toughness of steel due to fatigue driven material aging. These effects result in a reduced ability of a compartment to withstand blast loading, highlighting the need for the fatigue history effects to be considered during blast loading analysis. The effects are being seen for steels which are closely related to HSLA steel. This provides an indication that it is a concern for naval applications. Of interest is that as yet, a full fitting of a fatigue inclusive, high-strain rate constitutive model has not been undertaken. This prevents the fatigue effects on blast modelling being accounted for during FEA simulations.

CONCLUSIONS

Warships are increasingly required to maintain longer service lives, and the increasingly asymmetric nature of warfare leads to an increased chance of internal detonation. These factors combined are further driving the need for accurate blast modelling to be incorporated into design. Reference to the rules, and to currently available constitutive models, shows that pre-fatigue is not currently considered with regards to its effects on a material's properties. This leads to blast modelling being conducted with material models which do not necessarily represent reality as accurately as they could.

It has been shown that the effects of pre-fatigue on high strain rate tensile properties of steel can be significant. The reduction in modulus of toughness due to pre-fatigue is synonymous with the individual effect of high strain rate loading. This effect is seen for both the quasi-static and dynamic loading conditions. Whilst a reduction in modulus of toughness can be accounted for within constitutive models due to high strain rate loading, the further reduction which can occur due to fatigue cannot.

As the modulus of toughness is a measure of the energy absorption of material prior to failure, it is the driving parameter determining whether a blast will penetrate through a bulkhead. When the modulus of toughness is overestimated during blast modelling, the ability of the compartment to contain a blast is over-estimated. For this reason, it is highlighted that there exists a need to enable the incorporation of pre-fatigue material effects into constitutive models for use during high strain rate loading applications, such as those experienced during blast loading.

ACKNOWLEDGEMENTS

This research was funded by the Defence Material Technology Centre. Special thanks are extended towards Michael Saleh of the Australian Nuclear Science and Technology Organisation, and Warren Reid of the Defence Science and Technology Group, for their assistance and advice.

REFERENCES

1. Sikora, J.P., R.W. Michaelson, and B.M. Ayyub, *Assessment of Cumulative Lifetime Seaway Loads for Ships.* Naval Engineers Journal, 2002. **114**: p. 167-180.
2. ISSC, *Committee III.2: Fatigue and Fracture*. Vol. 1. 2015, Portugal: 19th International Ship and Offshore Structures Congress.
3. Akpan, U.O., et al., *Risk assessment of aging ship hull structures in the presence of corrosion and fatigue.* Marine Structures, 2002. **15**: p. 211-231.
4. ISSC, *Committee V.5 - Naval Design*. Vol. 2. 2015, Portugal: 19th International Ship and Offshore Structures Congress.
5. Stepanova, E., *Terrorism in asymmetrical conflict: ideological and structural aspects*. 2008, New York: Oxford University Press Inc.
6. Oppenheimer, A. *Shock Tactics: Asymmetric Warfare*. 2010; Accessed on: 30/05/2017; Available from: http://www.naval-technology.com/features/feature86646/.
7. Kaplan, R.D. *Asymmetry at Sea*. 2008; Accessed on: 30/05/2017; Available from: https://www.theatlantic.com/magazine/archive/2008/10/asymmetry-at-sea/307093/.
8. Chand, N. *Asymmetric Tactics in War*. 2017; Accessed on: 30/05/2017; Available from: http://www.spsnavalforces.com/story.asp?mid=21&id=1.
9. Crawley, J., *Incat-built HSV-2 Swift battered but not sunk in rocket attack*, in *The Mercury*. 2016: Hobart.
10. Lee, S.-G., *Shock Response Analysis of Blast Hardened Bulkhead in Naval Ship under Internal Blast.* 2015.
11. Det Norske Veritas, *Design Against Accidental Loads*, in *Recommended Practice DNV-RP-C204*. 2010.
12. DNV-GL, *Rules for Classification - Naval Vessels*, in *Part 3 Chapter 1*. 2015.
13. Lloyds Register, *Rules and Regulations for the Classification of Naval Ships*. 2017.
14. RINA, *Rules for the Classification of Naval Ships*, in *Part E - Additional Class Notations*. 2011.
15. Nurick, G.N., S. Mahoi, and G.S. Langdon, *The response of plates subjected to loading arising from the detonation of different shapes of plastic explosive.* International Journal of Impact Engineering, 2016. **89**: p. 102-113.
16. ISSC, *Committee V.5 - Naval Ship Design*. Vol. 2. 2006, UK: 16th International Ship and Offshore Structures Congress.
17. Larcher, M., *Simulation of the Effects of an Air Blast Wave*. 2007, Luxembourg: Office for Official Publications of the European Communities: Joint Research Centre Technical Notes.
18. Zakrisson, B., B. Wikman, and H.-Å. Häggblad, *Numerical simulations of blast loads and structural deformation from near-field explosions in air.* International Journal of Impact Engineering, 2011. **38**: p. 597-612.
19. Longère, P., et al., *Ship structure steel plate failure under near-field air-blast loading: Numerical simulations vs experiment.* International Journal of Impact Engineering, 2013. **62**: p. 88-98.
20. Ngo, T., et al., *Blast loading and blast effects on structures - An overview.* EJSE, 2007.

21. Al-Hassani, S. and S. Reid, *The effects of high strain rates on material properties*. 1992: Health and Safety Executive.
22. Sharpe, W.N., *Springer Handbook of Experimental Solid Mechanics*. 2008: Springer.
23. Ottosen, N.S. and M. Ristinmaa, *The Mechanics of Constitutive Modeling*. 2005, Oxford: Elsevier Science Ltd.
24. Kleinosky, M., et al. *Hull Structural Material and Toughness Requirements for San Antonio-class Amphibious Transport Ships*. in *SNAME Ship Production Symposium*. 2001. Ypsilanti, MI, USA.
25. ASTM, *ASTM A131 / A131M-14*, in *Standard Specification for Structural Steel for Ships*. 2014, ASTM International: West Conshohocken, PA.
26. World Steel Auto. *Advanced High-Strength Steels Application Guidelines Version 5.0*. 2014; Accessed on: 29/05/17; Available from: http://309fbf2c62e8221fbaf0-b80c17cbaf20104b072d586b316c6210.r88.cf1.rackcdn.com/AHSS_Guidelines_V5.0_20140514.pdf.
27. Health and Safety Executive, *The behaviour of carbon steels at high strain rates and strain limits*. 2000: Health and Safety Executive.
28. Gambirasio, L. and E. Rizzi, *On the calibration strategies of the Johnson–Cook strength model: Discussion and applications to experimental data.* Materials Science and Engineering: A, 2014. **610**: p. 370-413.
29. Hosford, W.F., *Mechanical behavior of materials*. 2nd ed ed. 2010, Cambridge ; New York: Cambridge University Press. 419.
30. Collins, J.A., *Failure of materials in mechanical design: analysis, prediction, prevention*. 2nd ed ed. 1993, New York: Wiley. 654.
31. Beghin, D., *Fatigue of Ship Structural Details*. 2006, Society of Naval Architects and Marine Engineers.
32. ISSC, *Committee III.2: Fatigue and Fracture*. Vol. 1. 2009, Korea: 17th International Ship and Offshore Structures Congress.
33. Miner, M.A., *Cumulative Damage in Fatigue.* Journal of Applied Mechanics, 1945. **12**: p. 159-164.
34. Moćko, W., et al., *An influence of cyclic loading on the form of constitutive relationship for DP500 steel.* Materials and Design, 2016. **103**: p. 183-193.
35. Moćko, W., A. Brodecki, and J. Radziejewska, *Effects of pre-fatigue on the strain localization during tensile tests of DP 500 steel at low and high strain rates.* Journal of Strain Analysis for Engineering Design, 2015. **50**: p. 571-583.
36. Itabashi, M. and H. Koseki, *Mechanical characterization of pre-fatigued free-cutting steels under dynamic tension.* Engineering Transactions, 2013. **61**: p. 87-98.
37. Rubio-González, C., et al., *Dynamic fracture toughness of pre-fatigued materials.* International Journal of Fatigue, 2008. **30**: p. 1056-1064.
38. Sánchez-Santana, U., et al., *Dynamic tensile behavior of materials with previous fatigue damage.* Materials Science and Engineering A, 2008. **497**: p. 51-60.
39. Verleysen, P., et al., *High strain rate properties of fatigued advanced high strength steel sheets.* Journal De Physique, 2006. **4**: p. 1307 - 1312.
40. Xu, Z. and Y. Li. *Application of split Hopkinson tension bar technique to the study of dynamic fracture properties of materials*. in *12th International Conference on Fracture 2009, ICF-12*. 2009.

Maintaining Technical Integrity of the Royal Australian Navy's Survey Motor Boats

Jon Emonson and Peter Webb, BAE Systems Australia Hydrographic In-Service Support
jonathan.emonson@baesystems.com, peter.webb@baesystems.com

DISCLAIMER

The views and opinions expressed in this paper are the authors' own and do not necessarily reflect those of BAE Systems Australia.

ABSTRACT

The Royal Australian Navy (RAN) provides the nation's hydrographic survey capability in order to meet defence requirements and international obligations under the IMO SOLAS convention. To achieve this capability, the RAN operates two LEEUWIN and four PALUMA class vessels with a fleet of seven FANTOME class Survey Motor Boats (SMBs). The SMBs are embarked on the LEEUWIN class vessels and are typically deployed to survey areas that are difficult, dangerous or even impossible for the larger ships to navigate.

The SMBs were constructed by Pro Marine, Melbourne, in accordance with Lloyd's Register Yacht and Small Craft Rules and under survey for Hull Construction Certification. The first batch was built from 1992 to 1994 with two follow on boats delivered in 1998. They are constructed of marine grade aluminium alloy, much of which is now exceeding 25 years old.

Although a vital part of the RAN hydrographic survey capability, the SMBs represent an ongoing risk due to operational and maintenance effects, obsolescence and technological development as well as changes to legislation and their certification basis. Remediation through managed engineering change is used to control these risks and maintain the boats in a safe and operational condition.

This paper will outline the broad range of engineering activities undertaken on the SMBs including structural alterations, weight remediation, upgrades to safety systems, and updates to hydrographic survey equipment. It also describes the systems engineering approach introduced to better define requirements and to manage engineering change activities in order to maintain and assure technical integrity (i.e. seaworthiness).

BACKGROUND

Hydrography and Hydrographic Capabilities

Information to enable safe navigation has been collected and used ever since the first mariners took to the seas. This information supported knowledge of how to sail to a destination and then return safely. Safe navigation requires understanding of sea routes and locations of hazards to be avoided such as rocks, reefs and shallows as well as the impact of tides and currents. Even depth measurements and samples of seabed sediment type were used by the Phoenicians some 3000 years ago (10th century BCE) to help determine distance from land and location based on the knowledge and experience of the mariner. Initially this hydrographic information comprised personal knowledge and verbal directions, often in the form of stories to aid memory. In time, such knowledge became written down as sailing directions and charts.

Sailing directions and probably charts have existed since at least the 6th century BCE, while stereographic and orthographic projections have existed since the 2nd century BCE[1]. However the first known accurate, or realistic, nautical charts were not compiled until the medieval period, 13th to 16th centuries CE: It is noted by Nicolai [2] that "The sudden emergence of the medieval nautical charts of the Mediterranean and the Black Sea known as portolan charts is one of the most significant events in the history of cartography and without doubt the most enigmatic." Nicolai concludes that portolan charts are actually composites of earlier charts, which could have origins back to Roman times and earlier.

Hydrography capabilities and the collection of hydrographic information have generally been treated as national strategic endeavours, supporting trade advantages and naval operations. The progression from medieval portolan charts to the establishment of the British Admiralty Hydrographic Office in the early 19th century is narrated by Richie [3]. The British Admiralty's scope of responsibility included survey and production of charts for trade routes and colonies including Australia.

It took a further one hundred years before international cooperation became a reality: The International Hydrographic Bureau was formed in 1921 following agreement by twenty-four nations at a 1919 London Hydrographic Conference. This bureau became the International Hydrographic Organisation (IHO) in 1970 following agreement of an inter-governmental convention. The IHO now has eighty-five member states. Despite this long and rich history, it was not until July 2002 that the International Maritime Organisation (IMO) revised Chapter V of the international Safety of Life at Sea (SOLAS) convention to actually require Contracting Governments to provide and maintain Hydrographic services and products.

The RAN took over responsibility for this nation's hydrographic capability in 1920 and formed the Australian Hydrographic Service (AHS). The AHS provides Australia's capability to meet international hydrographic obligations under the IMO SOLAS convention and is also responsible for collection of strategic hydrographic information used for national maritime military needs. The Commonwealth of Australia's Hydroscheme 2016-2019 [4] describes the nation's hydrographic responsibilities, highlights the Australian Charting Area, and provides the current plan for surveying and charting activities.

To achieve these capabilities and deliver the Hydroscheme, the RAN operates two LEEUWIN class and four PALUMA class vessels with a fleet of seven FANTOME class Survey Motor Boats (SMBs). In addition, the RAN operates a Laser Airborne Depth Sounder (LADS system), a Deployable Geospatial Survey Team and one Antarctic Survey Vessel (ASV). One SMB is used by the hydrographic school as a training boat.

Three SMBs are embarked on each of the two LEEUWIN class vessels and are typically deployed to survey areas that are difficult, dangerous or even impossible for the larger ships to navigate. They therefore perform a critical hydrographic survey role in these high risk areas.

Survey Motor Boats

The RAN Hydrographic Service has a long history of operating small craft for inshore and estuarine survey. Initially, these duties were given to modified rescue boats fitted with modest accommodation and carrying basic mechanical survey equipment, with the first two known Survey Motor Boats (SMBs) being *Fantome* and *Endeavour* which were 28 foot (8.5m) open launches attached to HMAS *Moresby (I)* [5]. These carvel-planked open craft were constructed from Spotted Gum and fitted with a six cylinder Petrol Volvo Penta marine engine and comprised of three distinct sections – the forward section was manned by the bowman for launching, retrieval and seamanship evolutions, the midships section contained the engine and was manned by the boats mechanic and the after section was the domain of the Officer In Charge, helmsman and marine surveyor. These open boats were very basic, slow (~9 knots) and offered the operators little in way of protection from the elements other than canvas dodgers and a makeshift sun awning.

These SMBs eventually evolved into a series of multi-role wooden work boats in the early 1950s, which were 34 foot (10.3m) long and built at Garden Island out of solid Spotted Gum with Teak decking. These workboats were based upon a common hull form but built in a number of different configurations in order to meet a range of roles, such as Army training and operations, naval harbour security, Admirals' launches and as Survey Boats embarked upon the converted Frigates *Barcoo* and *Warrego*. These newer Survey boats represented a considerable step forward in terms of design and capability compared to the earlier *Fantome* and *Endeavour*, providing an enclosed cabin for operators. However, as these vessels were constructed from timber they required significant maintenance to keep in a seaworthy condition.

These boats were paid off from the RAN in the mid-1960s, where many were put into commercial service around Sydney: SMB 1393, later renamed *Geranium*, was purchased by a local Sydney family and was used as a fishing and recreational craft. She was gifted to the RAN Historical Collection in 2014 and now serves as a collection piece and as transport between the RAN Museum at Garden Island and the historical collection repository at Spectacle Island [6].

As shipbuilding and marine survey technology improved, so did the SMBs. HMAS *Moresby (II)* was fitted with 3 davits and saw the evolution of the SMBs during her career. Originally fitted with two SMBs, *Fantome* and *Herald* (assumed to be wooden boats, similar to SMB 1393 above, carried over from *Barcoo*), these boats were replaced by *Alert*, *Beagle* and *Dart* and then again by *Fly*, *Rattlesnake* and *Waterwitch* [7].

The 70's vintage SMBs were 34 ft. (10.3m) aluminium-hulled workboats constructed by deHavilland Marine to complement the new generation of naval survey vessels: *Cook, Moresby (II)* and *Flinders.* The deHavilland boats were powered by twin 100hp diesel engines with Hamilton waterjets, were capable of a maximum speed of around 18 knots and were representative of early monocline shallow V-hull powerboat design of that era. While these boats were seaworthy, they were underpowered and too slow to rapidly transit to and from the survey grounds while the waterjets made for difficult manoeuvring during survey tasks.

These boats were fitted with a "Qubit Trac V HYDLAPS" system, similar to that later fitted to the Royal Navy Survey Motor Launch *Gleaner* [8] which provided basic echo sounding data capture and analysis capabilities. Subsequently a replacement was sought in 1988 [9] that would be more manoeuvrable and capable of higher transit speeds.

In 1989, Pro Marine from Victoria won a tender to design and supply a fleet of SMBs based upon their commercially available 10.7m full cabin design, which were to be fitted with modest accommodation for four crew for 24 hours' duration, and capable of over 30 knots at full throttle care of twin Volvo Penta TAMD42 (~200hp/engine) turbo-diesel engines fitted with contra-rotating propellers on outboard legs. The first boats to be built, *Investigator, Fantome* and *Meda* were delivered in 1992 and fitted to HMAS *Moresby (II)* with *Duyfken* embarked on *Flinders*.

In order to provide training on the *Fantome* class new platforms, *Sir John Gowlland* and *Tom Thumb* were supplied to the RAN Hydrographic School at HMAS *Penguin*. At the same time, ASV *Wyatt Earp* was built to a similar but shorter hull design to provide a similar capability for the Australian Antarctic Division through the RAN Deployable Geospatial Team (DGST). In 1994, an extra two *Fantome* class SMBs were constructed, *Casuarina* and *Geographe,* with an eye to ensuring that the vessels to replace *Flinders* and *Moresby* (what became the *Leeuwin* class hydrographic ships, *Leeuwin* and *Melville*) had a full complement of SMBs.

During the development of the *Leeuwin* class Hydrographic Ships, upgrades to the survey systems fitted to the SMBs were considered and implemented in order for the SMBs to become true hydrographic survey craft in their own right. This involved fitting multi-beam and single echo sounders similar to those fitted to the mother vessels, as well as incorporating a hydrographic capture, analysis and processing computing package into the boats. During the integration of the SMBs onto the new Hydrographic Ships, it was also identified that the SMBs were too long to fit into the boat spaces as constructed, and therefore 300mm was taken off the bow at the same time as the fitting of a sonar housing, sculpted into a now distinctive shaped bow.

As a result of these changes, the *Fantome* class SMBs had the ability to conduct detailed inshore survey missions and an element of hydrographic analysis with no equal at the time.

RISK

The RAN is charged with providing the ongoing maritime defence capability of Australia including hydrographic survey and to that end safe, seaworthy, reliable and capable vessels are needed.

The SMB's represent an ongoing risk to the RAN's survey capability due to operational and maintenance effects, obsolescence and technology development, and changes to legislation and the boat's certification basis. Engineering changes can be implemented to mitigate these risks in order to maintain the boats in safe and operational condition, maintaining their seaworthiness.

Operational and Maintenance Effects

Operation of the SMBs necessarily involves wear on systems and components in addition to simple ageing. To counteract these effects, worn or aged components are replaced during maintenance. Invariably breakages and breakdowns can occur when components fail unexpectedly. Wear rates and failures are impacted by how the vessels are operated in response to capability needs, and also tend to increase with age of the components. Planned maintenance schedules include adjustment and replacement activities to reduce failures during operation.

Frequent operation at performance limits or in adverse conditions will increase wear rates and likelihood of component failure. Mission priorities may also constrain maintenance activities resulting in delays to planned maintenance that can lead to increased rates of wear. The impact of increased wear and likelihood of failure is less when components are relatively new, having been recently replaced or fitted.

An ageing platform such as the SMBs will contain a mix of ageing and newly replaced components. This complicates overall maintenance management but risk can be mitigated through use of effective maintenance planning that ensures work is carried out where and when needed, together with use of condition assessments and condition monitoring to inform materiel state.

For example, BAE Systems leverages the RAN Rolling Hull Survey program, which continuously assesses and monitors the structural and mechanical state of vessels under RAN control. This is operated through the Ship Structures cell of the Naval Technical Bureau, formerly DNPS/CME, with accredited Naval Hull Inspectors positioned strategically around the country to cover the hull survey and reporting workload.

Obsolescence and Technological Development

With ageing systems and equipment, obsolesce becomes a significant issue. This impacts maintenance as well as repair since some maintenance activities require replacement of worn or ageing components. With the SMBs being over 25 years old, nearly all systems and equipment have been affected to some degree. Obsolescence occurs for a number of reasons, e.g.: market forces and competition lead suppliers to develop new products with improved performance and reduced cost. In some cases, a manufacturer may be bought out by a competitor and rationalisation of product lines leads certain equipment be discontinued; in other cases, a manufacturer may just cease trading. Technological developments can also cause major evolutions and consequent obsolescence in product lines, particularly within electronic systems.

Legislation and Certification Basis

Over time, changes to legislation occur as well as changes to RAN's technical and operational performance requirements. These changes impact the Certification Basis of the vessels that is defined as the suite of standards (approved requirements, including legislation) against which the materiel is certified.

Some examples of significant changes, which have had an impact upon the RAN SMBs are as follows:

The OH&S Act was changed to WHS Act & Regulations in 2011 [10] as a major overhaul of work safety requirements and with an aim to drive a safety-based work culture throughout Australia. The principal impact of this change is in terminology and the introduction of a risk-based approach to the legislature. Major changes are the inclusion of Commonwealth's armed services within its scope and the emphasis on reducing risk SFAIRP (So Far As Is Reasonably Practicable). SFAIRP essentially means that an approach to "Good enough" is no longer good enough if someone else is doing more or better.

The Navigation Act of 2012 represents new legislation, albeit legislation that amalgamates and replaces a number of previous legal Acts. The Navigation Act enforces IMO conventions in Australian waters (namely MARPOL, SOLAS & IMO ColRegs). It is not applicable to Naval Vessels, and partly applicable to paramilitary (Border Force) vessels. The Naval Flag Authority choses to enforce some aspects of Navigation Act upon RAN vessels as a means of being a "good corporate citizen" in a similar way that the use of Classification Societies are used to provide confidence for mechanical and structural seaworthiness.

The Marine Safety (Domestic Commercial Vessels) National Law Act of 2012 unified the separate state marine safety organisations for Domestic Commercial Vessels (defined as vessels of under 24 metres length) and rules into one body (AMSA) and one set of rules, with the state-based USL Codes becoming the AMSA-administered NSCV. This change is not applicable to Naval/Defence Vessels, however has been taken into consideration when redefining the certification basis for the Survey Motor Boats.

RISK TREATMENT

Engineering Change Process

Systems Engineering is described by the International Council on Systems Engineering (INCOSE) as "an engineering discipline whose responsibility is creating and executing an interdisciplinary process to ensure that the customer and stakeholder's needs are satisfied in a high quality, trustworthy, cost efficient and schedule compliant manner throughout a system's entire life-cycle. This process usually comprises the following seven tasks: **S**tate the problem, **I**nvestigate alternatives, **M**odel the system, **I**ntegrate, **L**aunch the system, **A**ssess performance, and **R**e-evaluate." These functions can be summarized with the acronym SIMILAR. Re-evaluation is actually a continuous process throughout all stages and acts as a feedback mechanism to review and correct outputs [11].

BAE Systems Hydrographic In-Service Support (HISS) team and the RAN Hydrographic and Meteorological Group (HMGRP) collaborated to develop the "modify" function of the new RAN Maritime Enterprise Model. The aim was to provide an engineering change process that supports RAN business and decision making needs, and that maintains and assures technical integrity (seaworthiness). The result is a tailored systems engineering process that follows the INCOSE "SIMILAR" approach:

- Problem Report
    - A clear and concise statement of the problem including any known causes and consequences that avoids discussion about solutions. A corresponding "Problem Space" technical risk is also raised to assist prioritisation.

- Problem Assessment
    - A logical review and assessment of the problem considering affected and interfacing equipment and systems, any other changes, related risks and potential hazards.
- Operational Requirements
    - Identification of user requirements and constraints for the affected equipment or system, developed by the RAN capability manager in consultation with operators.
- Solution Types Assessment
    - Development and assessment of candidate solutions based on options to: do nothing (accepting risk); undertake minor refurbishment and repair; replace with an off-the-shelf alternative; and develop a bespoke replacement. Only high level concepts are compared at this stage with trade off conducted against the operational requirements. A corresponding "Solution Space" technical risk is then raised for the selected solution type to assist prioritisation of subsequent design activity.
- System Requirements
    - System Requirements definition is the point at which the Requirements Traceability Matrix (RTM) and Verification Cross Reference Matrix (VRCM) are generated, together with a Preliminary Hazard Log (PHL). These artefacts represent a broad first-pass identification of the system requirements that the final design has to satisfy, as well as a depiction of how their satisfaction will be judged.
- Preliminary Design
    - A first-pass refinement of the selected solution leading out of the Solution Types Assessment; development of initial system integration strategies and associated artefacts; revision of RTM, VCRM and PHL and generation of an ILS impact statement. All long-lead items to be identified at PD stage in order to pre-order for production readiness.
- Detailed Design & Draft ILS
    - A detailed second-pass detailed design solution in readiness for production and implementation, including all rip out and installation specifications, drawings and calculations. All items within VCRM and RTM should be closed and all Hazard Log items mitigated at this point. Progress past detailed design may only occur if no or minor action items (defined as an item that will be closed before production that will not impact upon the design or progress into production). First-pass drafts of the associated ILS products are presented at Detailed Design Review for end-user review at this point.
- System Verification & Validation
    - Post-implementation evaluation of integrated system against defined system and stakeholder requirements and subsequent assurance and certification.

Repair and Renewal

In 2015, the SMBs were inspected and found to be in a degraded structural state, with a considerable percentage of the bottom hull plating subject to pitting. Pitting is a problem because it not only thins the plate thickness, but also introduces sharp surface stress concentrators. Both of these act to reduce allowable stress while increasing the applied stress and lead to fatigue cracking if left unchecked.

To treat this, the SMBs have been subject to substantial structural renewal of the SMBs hull plating and structural framework. In developing the required Technical Repair Specifications (TRS), a number of important factors had to be considered: heat affected zones, engine bed plates and transducer housings all had to be carefully managed. At the same time, it was decided to improve the internal protection of the renewed shell plating in order to reduce the likelihood of further pitting through internal painting to waterline level.

Commercial / Military Off The Shelf Solutions

The use of Commercial Off The Shelf (COTS) or Military Off The Shelf (MOTS) solutions has been a growing movement worldwide since the US Department of Defence started to adopt the idea in the late 1970's, with COTS procurement of military equipment becoming mainstream in 1994 as a result of the "Perry Memo", written by the then-US Secretary of Defence, William J. Perry, which was then enforced as part of the US *Federal Acquisition Streamlining Act* of 1994. This approach has been loosely followed by the Australian Department of Defence through the DMO (now CASG).

The approach taken with the SMBs to date has been predominantly COTS with modifications to allow for successful integration, and occasionally to allow the COTS items to handle conditions or meet extra operating requirements in keeping with a defined Statement of Operating Intent. The general trend has been away from bespoke items designed to military standards, as these can result in over-runs to cost and schedule for only an equal level of quality. This has been noted as a world-wide trend, as per [12].

COTS solutions have been a mainstay of the Hydrographic Sustainment model, with the majority of engineering changes being focused upon the identification and integration of COTS solutions to meet operator requirements. This is no different for the SMBs, where the boats operate COTS systems throughout: everything from the hull, propulsion and generator systems through to the complex hydrographic survey suite, is currently COTS with some minor modifications such as custom integration items like foundations. This has provided a number of benefits, but also some technical risks:

- Advantages: Reduced development timeframe, increased product maturity, use of product families and cross-platform commonality, short term cost reduction, improved lifecycle cost, reliability (in some cases).

- Disadvantages: Design compromise (risk-based approach and gap analysis required), reliability (in some cases), and commercial aspects to supportability (i.e. industry and commercial factors impact supportability of product).

As a result of the movement towards COTS items, there has been significant benefit from the area of technology development due to the relaxation of military standards and the ability to leverage commercial advantages to reduce development cycles, which ultimately leads to a reduction in cost. However, this comes at the cost of requirements: a COTS item will generally meet the requirements of a given design, but may not meet the military requirements that allow military items to operate reliably in the field under less than ideal conditions.

Ballast Optimisation / Reduction

Ballast is a commonly used means of adjusting the centre of gravity and stability state of a vessel with marginal stability in certain load cases, however it comes at the cost of overall displacement, which means it can be a double-edged sword. The effect of added fixed ballast tends to be more marked upon smaller platforms and as such has to be engineered carefully in order to provide an optimal ballasting solution which still complies with regulations.

As built, the SMBs had a displacement of 5.9 tonnes and had no solid ballast fitted. They were individually inclined at build and were assessed and were found to be compliant to Sections 8B and 8C of the Unified Shipping Laws (USL Code). Over the life of type so far, the centre of gravity has been raised significantly and the displacement increased to the point where the vessels were no longer compliant when reassessed as a Class 2C under the NSCV in 2012.

Prior to 2010, as a result of modifications to fit the hydrographic suite and to fit aboard the *Leeuwin* class Hydrographic Ships, DNPS conducted a stability analysis of the SMBs and found that the vessels had significant stability concerns in relation to the vertical centre of gravity and subsequently and that fixed ballast should be fitted to lower the centre of gravity of the SMBs. At the same time, the vessels underwent a modest refurbishment program in order to keep the SMBs functional through their extended Life of Type. This involved changing the air conditioning system, updates to the hydrographic equipment, structural improvements and repairs and development of improved baseline ILS products to support the vessels.

After the modest refurbishment, it was noticed that the SMB's weight was now approaching the Safe Working Load of the Hydrographic Ship's SMB davits and that the boats had taken on a noticeable bow high running trim which impinged upon visibility from the helm position when operating at high transit speeds (20+knots). Additionally, the requirements baseline of the SMBs at the time was ill defined and obsolete with the introduction of NSCV to replace the state-based USL code. However, both the USL and NSCV failed to define the requirements for a specialised workboat like the SMBs where they are operated from a mother vessel and are launched and embarked via davits. Therefore, DNPS (now Hydromechanics cell of NTB) were consulted in regard to the ballasting and inclining requirements under NSCV and an experiment was conducted to determine if a "practical inclining test" [13] could be utilised in lieu of a more traditional inclining to rapidly determine the optimal ballast quantity and location in order to resolve both the weight and running trim issues.

In order to reduce the loading on the Hydrographic Ship's davits, the overall weight of the SMBs had to be reduced and a program was established to both determine the current stability state and to identify sources of weight savings. The primary source of weight savings was found to be the ballasting fitted during the 2011 modest refurbishment program; however this could only be reduced or removed if it was possible to do so without incurring loss of compliance with stability requirements. To determine if this was possible, a program of analysis and trials was conducted to find the optimal ballasting amount and configuration. Subsequently, 600kgs of ballast weight was removed from the engine space of each SMB; primarily to assist in reducing the running trim both to improve fuel efficiency and helm visibility.

Update of Lifting Points

While the davits and SMB lifting gear were assessed during the design and build of the Hydrographic Ships, the certification baseline was incomplete and certification had not been maintained leading to a lack of valid certification, which was identified and raised as an operational problem. In addition to the lack of certification, the structural arrangements of the forward and aft lifting points fitted to the SMBs was in a poor state and varied greatly from boat to boat, with anecdotal evidence that two boats were missing a critical pillar connecting the lifting hook to the keel from build. Additionally, SMB *Tom Thumb* had the forward lifting pillar relocated due to damage or misalignment.

In 2014, Lloyd's Register (LR) was tasked to assess and certify the SMB davits and SMB lifting arrangements fitted to the Hydrographic Ships in accordance with the LAME code. The output of this assessment was LR advising to cease lifting the SMBs immediately due to non-compliance, especially in regards to man riding. As this was not an acceptable outcome in terms of operational requirements, the LR LAME code was deemed unsuitable as part of the certification basis: this led to a research project to determine a more suitable rule set to be used for certification of the lifting aspects of the SMBs. To that end, NavSea NSTM Chapter 583 was identified, analysed and found to be a more suitable requirement set for the lifting structures and equipment, and subsequently the certification basis was reviewed to reflect this change.

Structural Alterations

It was clear from the 2014 engineering assessment on the SMB lifting point structures that the SMBs would require considerable structural alterations in order to rectify the known defects and also to contribute towards establishment of a common structural baseline across the SMB fleet. However, further analysis was required in order to determine the effect of the structural changes on the global structure of the SMB. To this end, a finite element model was constructed and tested which resulted in changes to the final lifting structure design, namely a change from a fully welded pillar design to a bolted connection with larger connecting flanges between the deck beams and the keel girder.

Lifting Hooks

The most important aspect of the SMB lifting structure design is the interface between the davit falls and the boat deck itself. The SMBs were originally fitted with a "Pelican Hook" that had no apparent ongoing history of maintenance and an unknown load rating. Some further Finite Element Analysis (FEA) was conducted on a model of the existing hooks, which was then verified through load testing to identify the actual failure load and subsequent factor of safety for the existing hooks as a means to justify changing to a different hook system.

The FEA indicated that the hooks would not meet the required factor of safety as per the certification basis and operator requirements, plus the apparent lack in ongoing maintenance documentation led to a statutory failure in accordance with Australian Standards. As a result, it was identified that a solution was needed that could be certified to meet the NSTM required factor of safety of 6.5, correlating to a failure load of 32.5 tonnes. A number of options were explored, and in the end a COTS item that met the requirement was found and designed into the system.

The final delivered solution removed pinch points identified in the initial design and improved the reliability of hook up. This solution was a combination of a handle to take the operator's hands away from the pinch points; swing stoppers with covers that both limit the swing angle and prevent access to the pinch area, and the addition of a nylon wedge onto the deck plate to assist in accurately aligning the hook with the connection pin.

Engine Space Fire Detection and Suppression System

When built, the RAN SMBs were only fitted with basic handheld firefighting equipment that was not likely to be of much benefit in the event of an engine space fire. Since a fire is one of the key causes of total vessel loss, this lack of fixed firefighting equipment represented a considerable oversight in the design of these boats. Subsequently, a suitable solution had to be found in order to rectify this deficiency. The main reason why this hadn't been done previously was a lack of understanding of the safety requirements for vessels of the size and capacity of the SMBs.

In 2015, the certification basis of the SMBs was updated with a view to creating a uniform configuration baseline for the SMBs and also to reflect that the USL code was obsolete. In order to bring the SMBs in line with NSCV rules and the latest version of DEF(AUST) 5000 (and also due to an alarming number of small craft fires in Queensland waters that year), the decision was made to fit the SMBs with a fixed fire detection and suppression system for the engine room. A number of options were explored during the concept and detailed design phases. These included Carbon Dioxide, Potassium Carbonate and Inergen (Argon/Nitrogen inert gas mix) suppression systems together with a range of heat and smoke detection systems.

All three suppression options were weighed up against the relatively small engine room space and the regulatory requirements for the firefighting system. This provided a limiting factor for the two gaseous agent options insofar as they would have required significant modifications to be conducted on the hulls to incorporate a separate gas cylinder stowage space. Given that SMBs are already very limited in available volume and deck space, this proved to be prohibitive. As such, both liquid and solid Potassium Carbonate options were researched. Stowage of Potassium Carbonate solution was considered, however it was found that this solution would require a greater maintenance load and potentially cause damage to electrical systems in the engine room in the event of system release.

After considerable research, a product called 'FirePro' was selected in collaboration with a remote detection system. FirePro is a solid Potassium Carbonate salt block which is activated via a small charge embedded within the block. This charge aerosolises the solid salt block and disperses the agent throughout the protected space. This system is marine type approved and compliant to NSCV fire protection requirements.

The fire detection system is as important as the selection of the fixed firefighting system: it must be easy to use, sufficiently sensitive to detect the early signs of fire ignition, and record and maintain an alarm log in order to assist investigating the cause of a fire. The system chosen for the SMB application was a VESDA – Very Early-warning Smoke Detection Aspirator, developed by Xtralis, that was integrated with a thermal detector and a bespoke control box. The main reason for the selection of this system was that it was the only system that could be mounted externally to the engine space, thus reducing the risk of damage to the detection system in the event of a fire.

Hydrographic Survey System

The SMBs were originally fitted in 1993 with a Marine Science Suite, Doppler Log and side scan sonar towfish as standard fit. With the introduction of the Leeuwin Class vessels in 2000, the SMBs were fully upgraded with a new Hydrographic Survey Suite including Atlas Fansweep 20 (200kHz) Multi Beam Echosounder (MBES), Dynabase Motion Sensor, DESO 15 Single Beam Echo Sounder (SBES) and Current Temperature Depth (CTD) system, as well as Hydromap Online and Hydromap Offline processing software and computer equipment.

By 2007/2008 the Hydromap Offline software and workstations fitted to the Leeuwin Class and the SMBs were found to be inadequate to meet latest rate of effort requirements. As a result, both classes were upgraded to CARIS HIPS/SIPS with new workstations. At the same time the SMBs were fitted with a C-Max Tow Fish side scan sonar and new davit. As part of these works, the Dynabase Motion Sensor was also upgraded to a POS-MV unit complete with its two GPS antennas fitted aft, port and starboard on the cabin roof. In addition the Terrestrial Based Positioning System was removed as it had become obsolete due to technological advancements and was no longer used or supported. Various other updates including instrumentation and communications equipment were also undertaken at this time. The DESO 15 SBES was later upgraded to a DESO 30 system in 2009, again due to obsolescence issues.

More recently, in 2015, the Hydromap Online and CTD processing hardware was replaced with a single MOXA industrial PC. Rather than adapting the existing software the new system was set up to run this original software via virtual machines with Linux and Windows 7 operating systems. Initial plans are now being made to update the Fansweep 20 MBES and other obsolescent systems.

CONCLUSION

The RAN Survey Motor Boats are the latest in a long line of inshore survey capabilities which have supported the RAN's mission to fight and win at sea, by providing detailed and accurate geospatial intelligence in restricted waters. However, the current generation of SMBs have been subject to significant risk as a result of material and obsolescence issues, predominantly due to the long life of these platforms, which has been shown through the need to implement significant maintenance and engineering changes in order to sustain their existing capability for as long as required. In order to do this effectively, BAE Systems has adapted a well proven and understood systems engineering approach, in keeping with INCOSE guidelines, that has been supported by introduction of the RAN Maritime Enterprise Model. As a result, the SMBs now provide a safer capability for personnel, the environment and remain a world leader in terms of naval small boat hydrographic capability.

REFERENCES

[1] US National Imagery and Mapping Agency, The American Practical Navigator - an Epitome of Navigation - originally by Nathaniel Bowditch, NIMA, 2002.

[2] R. Nicolai, The Enigma of the Origin of Portolan Charts: A Geodetic Analysis of the Hypothesis of a Medieval Origin, 2016.

[3] G. S. Richie, "The History of Hydrography, An Enlightened European Era 1660-1800," *International Hydrographic Review,* January 1991.

[4] Commonwealth of Australia, "Hydroscheme 2016-2019," 2016. [Online]. Available: http://www.hydro.gov.au/business-publications/HydroScheme_2016-2019_WEB_20160808.pdf. [Accessed 03 July 2017].

[5] Australian War Memorial, "SMB Endeavour, Circa 1944," 2017.

[6] L. D. Paroz, "Full circle: Heritage listed survey motor boat returns to Navy service," 2014. [Online]. Available: http://news.navy.gov.au/en/Nov2014/Events/1589/Full-circle-Heritage-listed-survey-motor-boat-returns-to-Navy-service.htm#.WV12uoSGNhE. [Accessed 05 June 2017].

[7] Navy League of Australia, "The Navy - Official Navy League Magazine," Jan - March 1998. [Online]. Available: navyleague.org.au/wp-content/uploads/2012/07/The-Navy-Vol_60_Part1-1998.pdf. [Accessed 7 July 2017].

[8] United Kingdom Hydrographic Department, "Survey Information Processing Systems in the Royal Navy," 1990. [Online]. Available: https://journals.lib.unb.ca/index.php/ihr/article/viewFile/23306/27081. [Accessed 7 July 2017].

[9] Royal Australian Navy Hydrographic Service, "Report of the Hydrographic Service, Royal Australian Navy," 30 June 1988. [Online]. Available: http://www.dtic.mil/dtic/tr/fulltext/u2/a204665.pdf. [Accessed 07 July 2017].

[10] Commonwealth of Australia, *Work Health and Safety Act and Regulations, 2011,* Canberra: Commonwealth of Australia, 2011.

[11] INCOSE, "What is Systems Engineering? A Consensus of the INCOSE Fellows," 2012. [Online]. Available: http://www.incose.org/AboutSE/WhatIsSE.html. [Accessed 03 July 2017].

[12] R. A. Lebron Jr., R. Rossi and W. Foor, "Risk-Based COTS Systems Engineering Assessment Model: A Systems Engineering Management Tool and Assessment Methodology to Cope with the Risk of Commercial Off-the-Shelf (COTS) Technology Insertion During the System Life Cycle," *RTO SCI Symposium,* 2000.

[13] Australian Maritime Safety Authority, "National Standard for Commercial Vessels," 2016. [Online]. Available: https://www.amsa.gov.au/domestic/standards/national-standards/. [Accessed 02 June 2017].

[14] Commonwealth of Australia, *ABR 6492 - Navy Technical Regulatory Manual,* Canberra: Royal Australian Navy, 2003.

[15] Commonwealth of Australia, *ABR 6303 - Navy Safety Systems Manual,* Canberra: Royal Australian Navy, 2015.

[16] Navy News, *Hydrographers' SMBs cost 1.92m,* Navy News, 1983.

Lessons from LOTE for Naval Fleet and New Build

Olaf Doerk, DNV GL /Head of Department Structures, Maritime Advisory.
Olaf.doerk@dnvgl.com
Georgios Spiliotis, Quality Production Manager Navy, DNV GL Maritime Sydney,
Georgios.spiliotis@dnvgl.com

INTRODUCTION

Today many navies are facing various challenges which are rapidly changing with the worrisome trend of increasing threats and limited budgets. In combination with very long definition, design and building phases for new projects, this often leads to the situation that the existing platforms require up-grades and in addition, need to be kept in operation much longer than originally planned and designed for.

Proving the platform's capability to fulfill the extended duties in a safe and reliable way and to assess its readiness for up-grades becomes a key task. The application of a life of type extension (LOTE) procedure provides one of the best solutions to overcome this situation. This is independent of whether an intermediate solution until new-buildings are coming into service, or long term extended operations (without a defined end of life time) is the objective.

While LOTE is a topic often thought to be directly linked and limited to the advanced ship in operation phase, it is increasingly recognized that it plays key roles too during the design and newbuilding phases, when the entire life cycle performance of the project is reviewed.

OBJECTIVES

The basic objective of any life of type extension (LOTE) activity is to ensure a safe and reliable operation of a vessel beyond the originally planned life time. Sometimes this goes along with an additional change of the operational duties or the operational area. In practical application LOTE activities are commonly limited to hull structural and machinery aspects i.e. on platform related issues, with a clear focus on the former.

The provision of sound data/information as well as a well-defined origin for the execution of a subsequent life cycle management regime, is a further closely related goal of LOTE activities.

Beside these pure technical or safety related objectives LOTE contains also a strong commercial dimension. The early application during the design and newbuilding phase target long-term cost savings by providing a robust and flexible platform with sufficient margins to cover also future requirements including system up-grades and life time extensions, the typical application during the late ship in operation phase is focused on the most efficient solution to reach the goals and targets.

DNVGL's LOTE APPROACH

The goals and boundary conditions of LOTE related navy projects are as different as the ships and their respective histories. This is reflected in DNVGL's LOTE approach by offering a range of highly flexible and levelled solution which are tailored to the particular project but covers in each case, the relevant key elements. It is the unique combination of state of the art simulations and practical experience from hundreds of naval ships in operations that enables the utilization of the whole range of LOTE capabilities.

Basically, the goals and the key elements of the selected LOTE approach must be differentiated depending on whether it is applied and aimed for a new-build or a typical ship in operation situation respectively. The respective general procedures and key elements of DNVGL's LOTE approach are illustrated in Figure 1: Key elements of DNVGL's LOTE approach for both ship in operation as well as new-building applicationfor both ship in operation as well as new-build situations. For the latter, it is not a formal LOTE assessment to be carried out but more a margin assessment. However, as the same basic procedures are applied and similar goals are aspired it is named as LOTE for new-building in this paper.

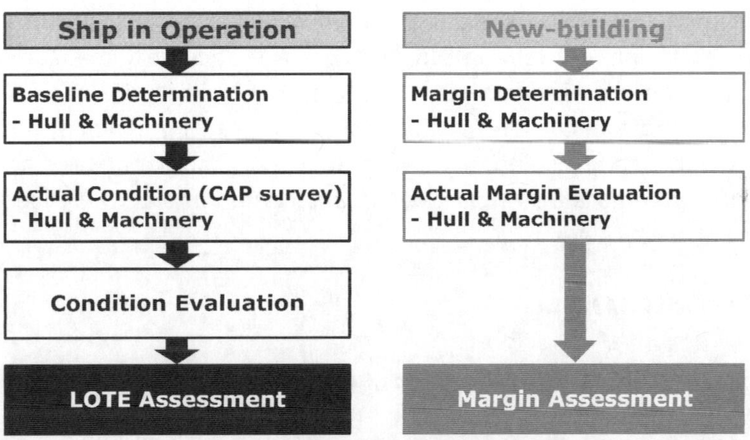

Figure 1: Key elements of DNVGL's LOTE approach for both ship in operation as well as new-building application

The level of the investigations to be carried out within a LOTE procedure depends on various factors and conditions but mainly on the level and quality of available data and information, the schedule and deadlines, the available budget and of course on the goal to be reached. In this context, the extent of the investigation is assigned to the methods applied for the different simulations e.g. if the loads are determined by first principal CFD analysis or if just standard values from the Rules e.g. /1/ are taken. For most key elements, different levelled solutions are available. However, it is not a black and white decision meaning the more sophisticated and advanced solutions are used the better the results. In practice, from a pure technical perspective this depends mainly on the quality and detail of the available information as well as a balanced cost benefit analysis.

SHIP IN OPERATION PROJECTS

The midterm or late ship in operation phase is the natural and most common application scenario for a LOTE procedure i.e. if reliable background information is required on the possibilities and efforts for keeping vessels longer in operation than designed and planned respectively. A clear definition and awareness of the LOTE objectives - e.g. determination of the required measures for reaching a new (extended) target life time or determination of the life time that can be reached without any significant up-grade or maintenance measures – is essential for the establishment of a robust plan and the subsequent successful procedure.

NEW-BUILDING PROJECTS

Although the design and new-building phase was traditionally not linked to any LOTE activities, and the application for new-building projects will not formally result in a LOTE i.e. extended life of type, the realization of LOTE procedures in these project phases, has recently gained more and more importance. Supported by the experience gained from ships in operations when reaching their design life, it becomes obvious that the early implementation of LOTE procedures provides significant benefits in terms of options, flexibility and efforts from a life cycle perspective. Besides the benefits of already setting up a life cycle management strategy, the major outcomes can be realized by already setting the right margins especially for in-service growth, future up-grades and degradation (corrosion and fatigue). In this way, the necessary measures and efforts for any later LOTE actions during mid-term and late ship in operation life can be significantly reduced and any potentially serious 'show stoppers' can be avoided. The preparation of all necessary data incl. calculation models etc. under principal consideration of future technical developments is a vital part of the application of a LOTE procedure in design or new-building phase.

DEGRADATION MECHANISMS

With respect to the aging of navy ships, corrosion and fatigue can be considered as the dominating hull structure degradation mechanisms. Although not a degradation mechanism in a classical sense, conversions and up-grades can have an important impact on the structural capabilities and utilization over time. These three mechanisms - or phenomena - are considered and assessed within DNVGL's LOTE approach.

For corrosion, it must be differentiated between *general* corrosion - resulting in a laminar thickness diminution - and *local* corrosion as pitting and grooving leading to localized thickness reductions. Both corrosion types are relevant for navy ships and should be therefore considered.

Fatigue is directly related to structural details. The actual stress concentration, the fabrication quality and metallurgical effects due to welding or thermal cutting are the determining parameters of the fatigue strength of a structural detail. In general, in terms of fatigue the applied procedures must be differentiated with respect to whether a welded joint or a structural detail consisting of a base material is to be assessed. For single hull navy ships the corners of large openings e.g. in decks, or the shell and the transitions (corners) between the superstructure and the hull are commonly subjected to a high fatigue utilization. *Figure 2: Principal stress distribution for a frigate type navy vessel resulting from a global strength analysis – fatigue areas of attention identified by red colour* - shows the distribution of principal global stresses as a typical result from a global strength analysis of a frigate type navy vessel based on a global 3D finite element (FE) model of the entire structure. Typical areas of attention regarding fatigue as described above i.e. location with high cyclic stress concentrations, are identified by red color.

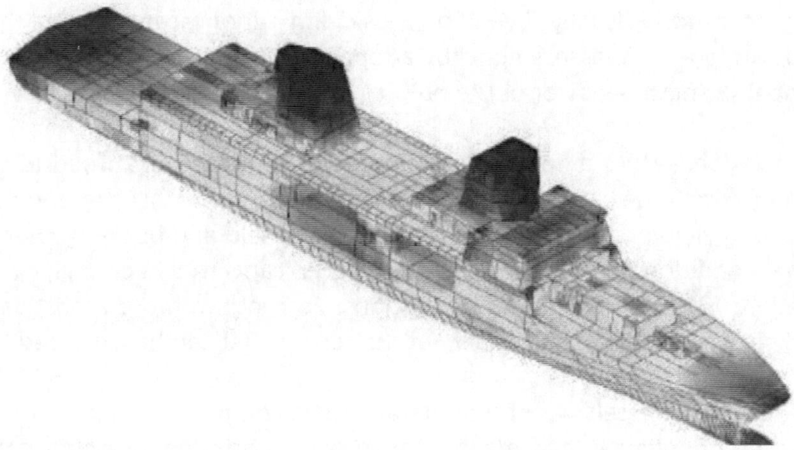

Figure 2: Principal stress distribution for a frigate type navy vessel resulting from a global strength analysis – fatlgue areas of attention identified by red colour

Baseline determination - Hull

Directly linked to the two main degradation mechanisms corrosion and fatigue the endurance limit i.e. the baseline of the design is determined by means of numerical simulations. While for corrosion the baseline is expressed by a list of the minimum thicknesses required to ensure a safe operation until the next inspection or survey the baseline with respect to fatigue, is given by the life time of the most fatigue sensitive structural details.

Loads

The calculation or derivation of appropriate design loads – ULS loads for corrosion and the required minimum thickness list and FLS for fatigue – is the initial step of any baseline determination. Depending on the available data and information, the schedule and the overall goal of the particular LOTE project different levels - ranging from first principal CFD methods to simple Rule formulas - can be applied. In practice often a combination is chosen where the global hull girder loads (ULS and FLS) are calculated by means of direct wave load analyses tailored to the vessel type to be investigated, and a Rule-based approach for local loads as per for example sea pressures, tank pressures and green water. In any case, possible impacts of conversions and/or up-grades and changes in the operational profile, have to be investigated and taken into account if revealed necessary.

Strength Limit

The main difference of strength limit calculation within a LOTE assessment or for initial design, is the hull structural condition to be considered. For the latter, as-built thicknesses and in some cases, the standard corrosion tolerances are applied while the former i.e. the LOTE assessment, is based on dedicated diminution scenarios, taking into account the impact of general corrosion beyond the standard tolerance limits as provided in the relevant rules /1/ and standards. The generation of reasonable diminution scenarios with practical relevance requires long term experience from a large base of ships and navy vessels in operation. Usually 2 – 3 of these diminution scenarios are generated and evaluated within a LOTE assessment. By adopting this approach, the impact of corrosion on the global strength behavior of the hull structure is covered.

Based on ULS loads, two different assessments have to be performed in this context. A verification needs to be carried out of the actual stresses against the permissible stresses with respect to yielding and buckling i.e. the typical yield and buckling check targeting on operational condition and an ultimate hull girder capacity check e.g. by means of the Smith method /2/ covering survival conditions. For both, a more realistic distribution of the global stresses is key for the quality of the results that can be expected.

Especially for naval vessels with their characteristic complex superstructure arrangement and large deck openings - resulting in a complex load carrying behavior of the structure – it is strongly recommended to determine the load carrying behavior and the resulting stresses by means of a 3D global strength analysis. Fully balanced wave load cases considering a realistic phasing of the different load components - e.g. according to the design wave approach /3/ - complete the advantages of a global strength analysis.

When it comes to the local loads e.g. tank pressures, green water etc. and if the efforts have to be kept within reasonable limits, it is recommended to apply simplified loads from appropriate Rules and applicable guidelines. These are based on long term practical experience and often are supported by systematic calculations and measurements. For practical applications, the high efforts of direct CFD simulation of local loads is often difficult to be justified by the to be expected benefits.

In addition to the global impact of corrosion, the local effects of corrosion must be considered within a LOTE procedure. These have to be differentiated into the local effects of general corrosion, i.e. if a plate or girder subjected to general corrosion can still withstand local loads e.g. tank pressures and local corrosion types like pitting and grooving.

While the local impact of general corrosion is assessed by means of a prescriptive net thickness based rule check, for local corrosion types a different approach should be used. DNVGL's LOTE procedure builds on a transformation of the well-established approach and acceptance limits for merchant vessels /4/ to the special conditions of navy vessels. For example, the directly attributed absolute minimum required thicknesses have to be adjusted considering the typical initial thicknesses of navy vessels. An example of this adjustment is shown for pitting type local corrosion in Table 1.

Merchant vessels /4/:	Navy (combat) vessels:
$t_{min} = 0.6 \times t_{orig}$ but not less than 6 mm t_{min} = minimum measured thickness of any individual measurement t_{orig} = as-build thickness	$t_{min} = 0.6 \times t_{orig}$ but not less than 4 mm t_{min} = minimum measured thickness of any individual measurement t_{orig} = as-build thickness

Table 1: Minimum required thicknesses for pitting corrosion (pitting intensity < 20%) for merchant and navy (combat) vessels

Areas of Attention - Fatigue

As already described above, fatigue is related to certain structural details showing high stress concentration and / or material notches. Due to the nature of the structural design of navy vessels, simplified 2D cross section based approaches are not recommended, as the accuracy and quality of the results cannot be expected to fulfill the requirements of a LOTE procedure.

Supported by experience with navy vessels, the structural details most relevant for the fatigue strength of the investigated vessel can be located and assessed within a global strength analysis. The fatigue assessment itself can be done either by application of the sub-model technique or by direct mesh refinement. In any case, so called local fatigue approaches, e.g. as described in /5/, must be applied to achieve the required quality and accuracy of the results. Figure 3 shows the global FE model and the related fine mesh sub-model of a superstructure - hull transition – a typical fatigue sensitive structural detail of navy vessels.

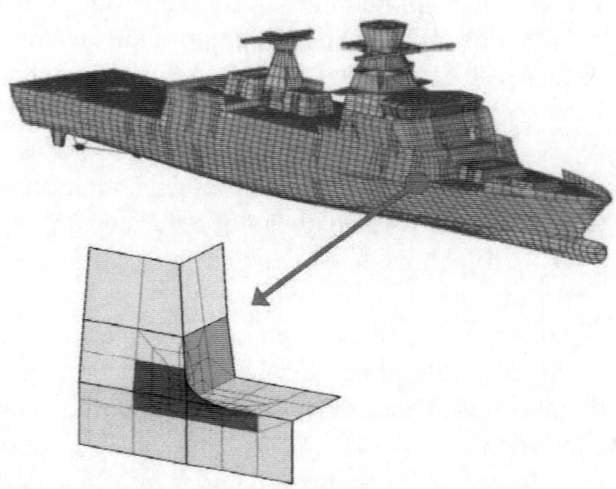

Figure 3: Global FE model and fine mesh sub-model of a superstructure – hull transition as a typical fatigue sensitive structural detail of navy vessels

Commonly for the 20 – 30 structural details most relevant with respect to fatigue, the expected life time is calculated. Influences of the operational profile as well as of upgrades and modifications are to be considered for this analysis as far as possible i.e.

depending on the available information. (Often it is difficult to obtain the required information on the individual operational profile i.e. the sailing area, days at sea, etc. for the individual vessel subjected to the LOTE procedure.)

Baseline determination - Machinery

When it comes to machinery, the situation is not as well defined as for the hull structure. A challenge in this respect, is that it is not possible to simulate or calculate the condition and its change over time for a machinery component. Currently for practical application, there is not really a viable alternative to relying on the allowable operating hours and conditions provided by the component manufacturer, without knowing the philosophy behind and the basis for assessment and safety factors. The influence of liability aspects cannot be neglected in this respect either, so that one can assume that the information provided contains a fairly good portion of implicit safety, which is unfortunately un-quantifiable.

Condition based maintenance approaches utilizing smart data evaluation currently under discussion and development, promise significant progress in this respect for the future. The basic idea is to estimate the remaining service life and necessary maintenance measures of a component by continuous reading and analysis of the relevant key parameters.

Aging of (Assessment) Procedures

Aging of platforms and its investigation, is at the core of any LOTE procedure but aging of assessment procedures and approaches, is often not in the natural focus when LOTE is discussed. Except when LOTE is applied in the new-build design phase of a project it is a topic of relevance. Continuous technical development leads to more detailed and refined assessment methods that are considered current 'state of science' or 'state of the art' respectively.

Ignoring the state of the art is of course not a reasonable solution for execution of a LOTE procedure dealing with the safety and reliability of a navy vessel. Applying state of the art procedures within a LOTE project results in two possible scenarios:

1) The methods applied during design phase were conservative or unconservative compared to the currently applied state of the art procedures. In the first case, some additional implicit margins can be explored and utilized – a quite comfortable situation.
2) The second case is a more challenging ranging from the situation that the new-build condition is revealed as not in accordance with current procedures as a worst-case scenario or that significant measures are required to fulfill the initial or extended life time.

For practical applications, simple awareness of this issue enables the establishment of possible solutions and covering procedures in an early state and therefore usually within acceptable efforts.

Actual hull condition

A dedicated navy CAP (Condition Assessment Program) survey is applied for the determination of the actual hull condition of the vessel. Based on the well-established CAP procedure for tankers the navy CAP survey is adjusted to the special conditions of military environment and navy vessels. Visual techniques including close-up visual inspection and thickness measurements (UTM) are the main elements. Key principles of the rating systematic are kept but the scope and the acceptance limits are modified for application to navy vessels.

As a result, the entire hull structure is assigned with a rating ranging from CAP1 to CAP4. A rating of CAP4 indicates that the condition is below class requirements i.e. that there is inherent need for countermeasures existing. Whilst a rating of CAP1 is linked to an excellent condition comparable to new-building/delivery.

Actual condition of the machinery

As for the hull structure, a dedicated navy CAP (Machinery) procedure is considered for the determination of the actual condition of machinery components. Current developments indicate that this will be supported by condition monitoring approaches. Beside the machinery components itself, special topics such as the aging of cables and related efforts for exchange, have to be considered when a maintenance strategy is to be developed which is a frequent consequence of such investigations.

Condition Evaluation

Within this step the results of the baseline determination are reconciled with the information on the actual condition of the vessel. This applies for both the hull structure as well as the machinery. Regarding the hull structure, the to-be-expected life-time in terms of fatigue and corrosion i.e. the time until reaching the minimum thickness limits are derived and subsequently referenced against the design and/or the actual life time respectively.

While life time is a direct result of a fatigue assessment, things are more challenging for corrosion assessment. Especially a reliable prediction of the corrosion rate i.e. the velocity of the related thickness diminution, which is of special importance to predict the time until the minimum required thickness is reached is still a subject of research and the results show very large variances /6/. So, for practical application, this is solved by applying an empirical and experienced based CAP survey systematic.

As already described the situation for machinery is not as well defined as for the hull structure meaning that a quantitative prediction of the to-be-expected life-time is extremely difficult. An intelligent combination of CAP machinery survey results and condition monitoring including a smart data analysis, offer the greatest potential to improve the situation in future. So currently the experienced based but well proven CAP machinery approach is applied to evaluate the condition of machinery components.

Often liability aspects and manufacturer's guarantees have a major impact on possible life time extensions of machinery components when a solution beyond the simple comparison of manufacturer defined and already used operating hours, is required or desired.

Life Of Type Extension - LOTE

As already indicted by the name the final step within a LOTE procedure, is benchmarking the available life time against the desired extended life time. Basically, two different situations need to be considered from the condition evaluation – the aspired life time (extension) is reached without any additional measures (Situation A) or not (Situation B). While for the former, no further action needs to be taken, for the latter reasonable countermeasures must be developed and verified to reach the (lifetime extension) goal. Figure 4 illustrates these basic situations schematically. These can range from an adjusted survey and monitoring regime, to the exchange of larger parts of the structure. Of course, the situation is often not that simple if different degradation mechanisms and disciplines show divergent results e.g. the desired lifetime extension is met for the hull structure but not for the machinery components. In any case, the decision whether a required countermeasure is meaningful under commercial aspects, will be taken by the involved naval authorities with full knowledge of the technical assessment delivered through the LOTE procedure.

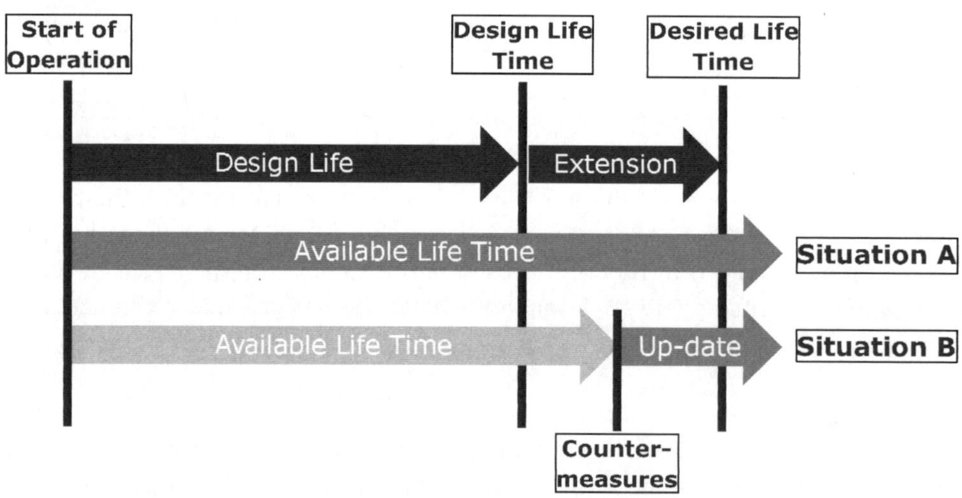

Figure 4: Basic LOTE situations – schematic illustration

In this respect, is has to be considered that if applied in new-building phase LOTE procedure will not result in a formal LOTE assessment but in a margin assessment i.e. the check of the actual available margins vs. the aspired target margins (wrt growth, operation).

IMPACT OF CLASS

Operating navy vessels under a class regime offers a perfect basis for the determination of the actual condition, as regular survey results and the results of thickness measurements - as well as many other findings – to a large extent can be re-used for LOTE purposes. The long-term experience and in-depth knowledge of the vessel obtained over many operational hours and years in service, and a sound database, cannot be underestimated in terms of value. Thus it can be said that class provides both technical and commercial benefits for a LOTE procedure execution.

CONCLUSIONS

On the one hand, very long timelines for bringing new navy projects into operation and / or budgetary restrictions on the other hand, both lead to an increasing need for navies to keep their existing vessels significantly longer in operation than initially intended. The application of LOTE procedures is increasingly a preferred solution to overcome this challenge and ensure a safe and reliable operation beyond the initial design life time.

DNV GL's systematic procedure offers a sound and flexible basis to deal with LOTE of naval vessels tailored to the individual project and its objectives. Basically, it can be applied in the design or new-building phase of a project as well as during the mid-term or late ship in operation phase. Irrespective of when it is applied, it offers clear benefits in terms of possibilities and efforts. The procedure is sound and well proven by its already successful application to several projects worldwide.

REFERENCES

/1/ DNV GL rules for classification: Naval vessels (RU-NAVAL), Part 3 Surface ships, Ch.1 Hull structures and ship equipment, Edition 2015-12

/2/ Smith, C.S., Influence of local compressive failure on ultimate longitudinal strength of a ship's hull, PRADS, Tokyo, Japan, pp 73 – 79, 1977.

/3/ Design wave approach

/4/ CG 182 Thickness measurements

/5/ Radaj D., Sonsino C.M. and Fricke W., Fatigue assessment of welded Joints by local approaches, second edition, Woodhead Publishing, 2006.

/6/ Paik J.K. and Melchers R.E. (Editors), Condition assessment of aged structures, Woodhead Publishing, 2008.

First and second order technical implications of an Off-The-Shelf approach to naval ship acquisition

Mr Joseph Cole
Department of Defence, MRINA
Joseph.cole@defence.gov.au

SUMMARY

Past reviews into defence procurement resulted in recommendations in relation to Off-The-Shelf options. These recommendations rightly set out to control risk associated with developmental solutions, however, what does not appear to be widely considered, are potential technical implications of consistently adopting an off-the-shelf strategy for new ships. This paper seeks to identify some of the immediate technical implications for acquisition programs and the ships themselves, and in a broader context for industry, academia, and engineering professionals, as well as implications for technology and innovation. The author proposes a strategy for naval ship acquisition that incorporates the advantages of off-the-shelf options, but seeks to minimise the potential adverse implications for the program, ship and broader context.

OVERVIEW OF OFF-THE-SHELF

The 2003 Defence Procurement Review [1] advised that "an off-the-shelf alternative be part of any set of options put to government to ensure that a benchmark is established against which the costs, military effects, and schedule of all proposals can be assessed" but acknowledged that Australia's operating context often requires specialist solutions. The 2008 Defence Procurement and Sustainment Review [2] took this a step further, proposing off-the-shelf as the default solution, with the recommendation that "any decision to move beyond the requirements of an off-the-shelf solution must be based on a rigorous cost-benefit analysis of the additional capability sought against the cost and risk of doing so." Subsequent defence reviews have not altered the trend in relation to off-the-shelf.

The basic premise of an off-the-shelf approach to naval ship acquisition is that risk can be controlled by the selection of a design that has been previously built and tested in service. The objective is risk mitigation in relation to aspects such as technology readiness, scope creep and design maturity of solutions, which are areas that have historically translated into schedule and cost overruns. The approach should also result in an overall cost reduction with the cost of design development largely being borne by the original customer.

While the original intent of the off-the-shelf options described in the 2003 review [1] was to only form a part of the full option set, and acknowledgment was given to the likely need for modifications to achieve unique Australian requirements, what has emerged in practice is an almost exclusive use of off-the-shelf options for naval ship acquisition. A strong emphasis on minimum change has also emerged. This is clearly illustrated by the programs for the Canberra Class, Hobart Class, and Auxiliary Oiler Replenishment Vessel.

While the general idea of off-the-self is relatively straight forward, in practice no single definition has been consistently applied for all projects. Off-the-shelf is also difficult to envision for something so specialist and complex as a naval ship, there is certainly no literal shelf with ships sitting on it. The closest may be to purchase a vessel from another nation's ongoing production program, something that Australia has done in the past. Given this is not how the term has been used in recent practice, the more straight forward term 'existing design' tends to be used in this paper, simply meaning a design developed previously for a different customer.

Minimum change is something that is also difficult to define and as a result can end up arbitrarily constrained. Typically minimum change is change constrained to a pre-determined set of systems or equipment, most commonly elements of the combat, sensor or weapon systems. For the remaining elements of a ship design a limit may be imposed on any change, regardless of the risk associated.

For the purpose of this paper the term Existing Design Minimum Change (EDMC) is introduced to refer generally to the use of a design previously produced for a different customer, with little to no change, as has been applied by the CoA for recent naval ship acquisition programs.

FIRST ORDER TECHNICAL IMPLICATIONS - ACQUISITION PROGRAMS AND THE SHIPS

The EDMC approach has direct technical implications for acquisition programs and for the ships themselves. A number of possible technical implications and risk areas can be identified on consideration of a simple systems engineering V diagram (see Figure 1). The normal project definition process creates traceability from needs, through requirements and into design and production. For a naval ship this process is highly context specific and builds in the unique needs of the customer and specific context of use, the regulatory obligations of the parent jurisdiction, and the specific trade off decisions made during design. For the new program selection of an existing solution first, without repeating the complete project definition process, inevitably creates some tension between the new customer's needs and the existing solution.

Some of the implications from a technical perspective of the EDMC approach include:

- Influencing requirements development and traceability
- Limited option set - solutions to someone else's problem
- Constraints on project engineering
- Constraints on technical risk and safety management
- Design and Equipment Obsolescence

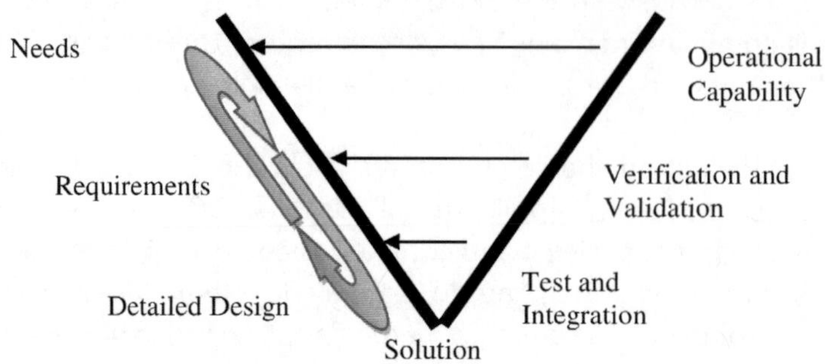

Figure 1 – Typical Systems Engineering V diagram

Requirements Development and Traceability

One of the first implications for the EDMC approach is that it may influence the requirements development process and impact on navy's ability to identify key, solution independent, capability requirements.

Within the normal project definition process, some iterations and trade off decisions will occur, however, the needs have primacy and the requirements and detailed design are derived. This ensures that the solution developed is traceable to these higher level capability requirements.

The implementation of EDMC alters the focus from the actual needs and places it on the solution. This could be seen to devalue the earlier project definition activities particularly requirements development. There is a risk that requirements end up being developed at a high level only and relate to the likely existing design candidates rather than real solution independent capability needs.

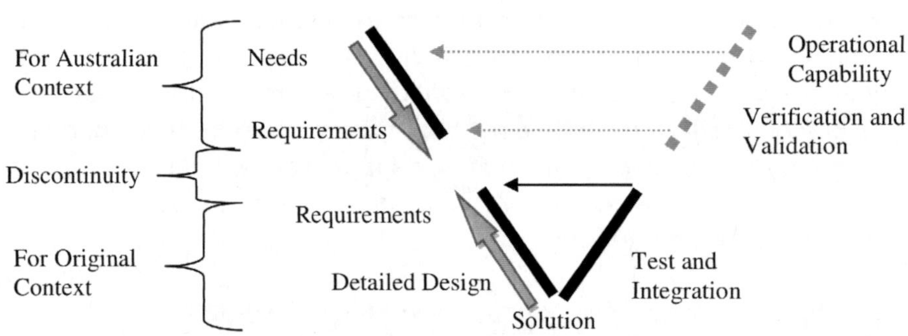

Fig. 2 – Existing Design Minimum Change Engineering V diagram

If capability requirements end up being defined at a high level only, and without the intermediate requirements layers, it is impossible to show traceability between the solution and the needs. Figure 2 shows the practical outcome of an EDMC approach and the disconnect that arises. A practical outcome of this is that while it takes in the order of thousands of requirements to describe a solution as complex as a naval vessel, recent ship programs have progressed with orders of magnitude fewer requirements.

High level requirements drive broad brush assessments and as a result less detailed information tends to be sought from suppliers. Engineers know that the devil is in the detail and that capability is built upon numerous lower level technical elements. Not considering these detailed elements upfront pushes any identification of technical risk to later stages, limiting the opportunities for resolution.

Contracts for EDMC programs typically use the original customer's ship specification as a basis, formalising the disconnect identified in Figure 2. Simply relying on the original customers' requirements and associated V&V can result in a number of technical issues downstream. This includes specific instances where the contracted V&V does not provide evidence against all requirements in the specification, making technical acceptance of the solution a challenge. A wider reaching implication would be where independent design and production assurance, typically provided by a Class Society, is excluded, as these did not form part of the original customer's requirement.

Option Set – solutions to someone else's problem

Ship design, and even more so naval ship design, is a complex multi-objective optimisation problem [3]. There are numerous individual design parameters many of which can be varied across wide ranges. This results in a virtually infinite set of possible design options for a given requirement. Figure 3 is a notional representation of this problem resolved down to two criteria, capability versus cost. It is intended to illustrate how existing designs form only a small number of discrete solutions within the larger set of all feasible solutions. These existing solutions, even if they had been optimised for the original owner, are unlikely to be optimised when re-evaluated for the RAN specific needs.

Limiting the option set to EDMC prevents the development and consideration of options specifically developed for Australian needs. This includes options that may provide greater capability at equivalent cost or equivalent capability at lower cost.

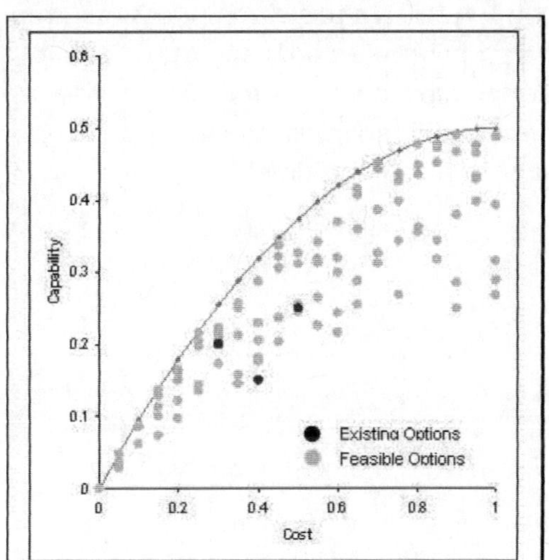

Figure 3 – Notional Design Options

Existing designs are by definition a solution developed for another customer's needs. Aside from the different capability trade off decisions, these designs have been produced in a completely different legislative and regulatory context for nations in some cases with different societal expectations in relation to safety and risk, or even in more basic elements such as habitability standards and food storage and preparation requirements. These can significantly influence the form of the solution.

Optimisation or targeting specific performance parameters can have significant implications for a ship design. Vessels with ostensibly the same role and capabilities can vary greatly dependant on some of the most fundamental design parameters. The following plots, figure 4, are illustrations of how variation of a single parameter can impact overall vessel design.

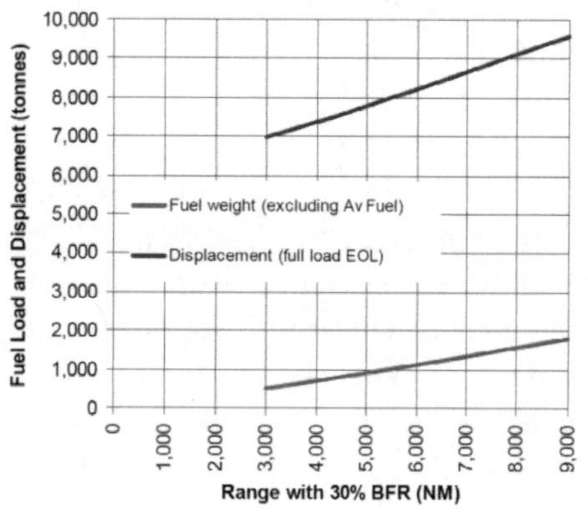

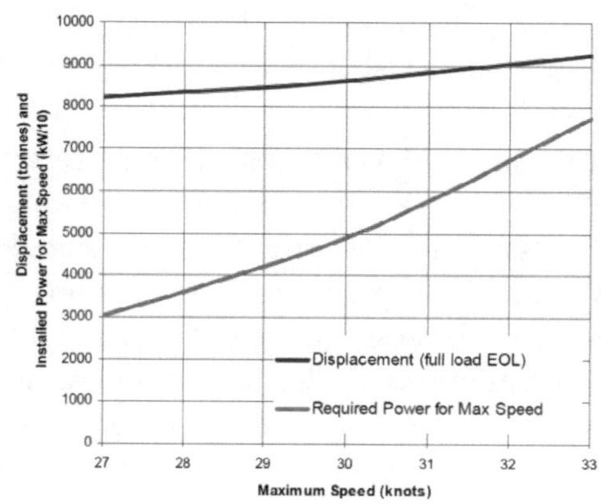

Figure 4 - Parametric variations of a notional frigate design

Constraints on project engineering

History shows that initial cost estimates, even those based on existing designs, are often inaccurate. That is, over time, complex acquisition programs will track away from the initial estimates and this can occur for a wide range of reasons [4]. A simplistic representation is given in Figure 2. In this case the cost axis represents acquisition cost and includes all cost and schedule elements and the capability axis also includes quality, reliability and cost of ownership. The figure focuses in on a single option and characterises the regions around the initial capability and cost estimate into quadrants through which a project may track with time.

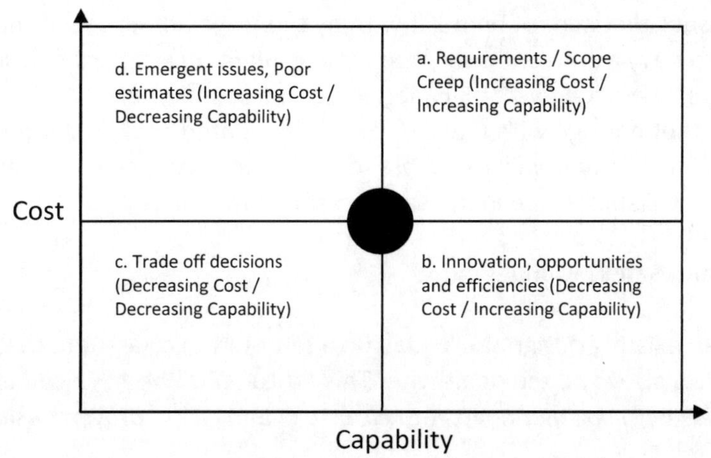

Figure 5 - Classification of changes in capability and cost

Quadrants a, b, and c are directions that may be deliberately chosen to manage cost or address capability. The fourth d., increasing cost and decreasing capability, is not a direction that would be chosen but is the one commonly reported for complex projects.

Implementing an EDMC approach severely limits the ability of a project to deliberately create progress any direction. Any change must therefore result from external or unplanned aspects. The approach is an effective control of (a) scope or requirements creep but impacts on (b) innovation and opportunities and (c) trade-off's which, as part of an active project engineering organisation, are controls that limit (d) cost escalation and capability reduction. An active project engineering organisation is able to identify and manage emergent technical issues. These are often resolved through some form of engineering change.

The constraints on change management and engineering introduced by the EDMC approach may cause a shift away from a normal program approach to the development and acquisition of naval ships to a more product centric view. This may result in an increased reliance on the designer / builder for the technical integrity of the solution, which in turn would impact on the number and role of engineering and technical personnel required by an acquisition project. A practical example might be a reduction in customers own assurance personnel for construction and a greater reliance on the builder's quality systems.
With only high level requirements defined, and a program focused on existing solutions, there is a risk that detailed technical elements can be overlooked. Many subtle capability elements exist and these require detailed and specialist analysis and cannot be assured simply in broad terms. For example a platform may contain all the obvious elements for Anti-Submarine Warfare (ASW) such as sonar systems, torpedos, organic helo capability; yet subtle elements of the capability, like its underwater acoustic signature, may render it ineffective in the ASW role because the platform is too easily detected. Detailed technical issues may be picked up and addressed by an active project engineering team but EDMC will limit this.

The technical issues that cannot be dealt with by the Project because of the EDMC approach pass to navy as the end user, and may result in operational restrictions, costly in-service changes, reduced availability and increased sustainment funding. Many technical issues that emerge with naval ships can be related back to design or production deficiencies [4]. These issues will inevitable result in some impact to capability, reduction in reliability and availability, and increases in cost of ownership.

Technical Risk and Safety Management

In selection of an existing design there may be a temptation to assume that all technical and safety risk has already been dealt with. This attitude has been voiced in sentiments like "*it's in service with another navy, it must be ok*" and "*it's a proven design*".

Just because a solution exists doesn't mean all technical or safety risk has been eliminated. There are numerous examples of existing naval ship designs that have been constructed and operated in service for a period of time after which significant issues have been identified. We must also acknowledge that there are a range of reasons why parent navies would not publicise technical issues with their existing ships, unless this was unavoidable.

It is also important to have a realistic understanding of the status of ship design and production, even for an existing design, in comparison to other industries. An off-the-shelf solution, as most probably imagine it, can only really exist in low to medium complexity systems, resulting from sophisticated high rate production, with relatively short design and production cycles. The automotive industry is an example. The high complexity, low production numbers, less sophisticated production processes, and extended design and production times associated with a naval vessel, mean that they should generally be considered more like prototypes than fully de-risked solutions. This can be seen by attempting to compare different vessels of the same class within the existing naval fleet. The process of building a ship is more akin to building a house than it is to buying a car or aircraft. This means that simply using existing design should not be assumed to be a overall control for technical and safety risk.

When considering the concept of a "*proven design*" we must remain cognisant that, from an engineering perspective, the proof of a design must largely come from the design processes and associated analysis. Design cases are often extreme cases not seen in normal service and for which normal service provides no real evidence. Hence it is not correct to assume that a design that has been constructed is necessarily more "*proven*" from a technical perspective that one that is yet to be built.

The legislative and regulative context of the parent nation has a very important role in shaping the solution and these vary significantly between nations. Different nations have different societal expectation for safety and protection of the environment. At the technical level this can manifest in issues such as commercial standards being inappropriate for military applications or the use of military exemptions in relation to environmental conventions, rather than adopting best practice solutions. One example is the requirements relating to hazardous materials, Australian legislation and defence policy prohibits the use of many substances. Other counties or defence organisations have different requirements. Identifying the elements of the design affected by these differences is a detailed technical activity that can affect virtually any element of the design and reaches down to the lowest level technical specification for materials, systems and equipment.

From a safety risk management perspective one outcome of an EDMC approach is that, in many cases, the higher order controls to eliminate or substitute a risk through engineering change are made unavailable. As a result the safety program can be limited to implementing lower order controls. This directly impacts on the effectiveness of the safety program and development of the safety case and will often result in capability limitations.

Design and Equipment Obsolescence

An existing naval ship design will typically have origins at least 15-20 years prior to the time it is being considered, hence there will usually be significant obsolescence aspects to address.

At the equipment level obsolescence is obvious. It occurs where technology has moved on and equipment is no longer available or supported. There is also equipment level obsolescence caused by changes in legislation, such as environmental legislation, where new emissions standards may drive certain equipment items to be removed from the market. These instances of obsolescence can most often be dealt with through the purchase of alternative equipment however some degree of re-design will be required to support integration.

There are also broader design related obsolesce issues. These can occur with changes in naval practice and processes, such as increasing use of deployed autonomous systems over ship based systems. There are also more subtle elements like design obsolescence issues resulting from the changes that have occurred in design standards since the original design was produced.

SECOND ORDER TECHNICAL IMPLICATIONS - INDUSTRY, RESEARCH AND INNOVATION, AND ENGINEERING PROFESSIONALS

The practice of persistent and exclusive Off-The-Shelf or EDMC procurement of Naval Ships would appear to have technical implications that reach more broadly than the specific programs and ships into industry, academia and the roles of engineering professionals.

Industry

The exclusive use of EDMC must have implications for ship designers and technology suppliers in Australian Industry. It results in all initial design work being pushed offshore to other nations and prevents Australian industry developing solutions. The longer the approach persists, the harder it must be for industry to maintain even the existing levels of naval design capability, let alone develop the capability further. Australia has professional engineers with experience in naval ship design and has the potential to grow this capability. Persistent use of EDMC will not promote growth in this part of the industry.

Existing designs will have been produced within a particular supply chain context. For a typical ship design, while many systems and pieces of equipment may be re-tendered within Australian industry, there will be numerous systems and equipment items that are integral to the design and substantial re-design may be required to open up opportunities for Australian suppliers. This may result in cases where Australian industry has superior technical solutions that cannot be integrated because of the design effort associated with the integration is not compatible with the EDMC approach.

Research and Innovation

With design work being conducted offshore, the research and development activities that feed into the designs will typically occur in these parent nations. This takes opportunities and potential funding streams away from equivalent Australian organisations.

An EDMC approach substantially limits the application path for innovative technologies. While some innovations may be at the sub-system or equipment level and more readily integrated in the design, even this will require some design change. The significant leaps in technology such as changes in construction materials, propulsion system, sensors or weapon systems can have broad reaching implications for design and should be incorporated at the early design stages, something that is excluded by EDMC. The limitations on the application path for innovative technologies in Australian naval ships must have implications for those companies and institutions involved in development of innovative technologies.

Engineering Professionals

The limitations on requirements development, design, project engineering, and research and innovation functions must result in an impact on the demand for Australian Engineering Professionals in Naval Ship Acquisition context. A direct outcome of the impact to industry and academia must be to the career paths for these professionals.

In recent times much emphasis has been placed on the blue collar workforce required for naval shipbuilding and the enabling white collar engineering design workforce has been largely overlooked. Appropriately qualified and experienced engineering professions are critical to delivering naval capability across the lifecycle. The conduct of naval ship design activities off-shore must have an impact on the employment opportunities for engineering professionals in Australia. While there are numerous Australian companies and individuals with design knowledge and skills, capability in this area has been fragmented and eroded over time.

Pushing design work off-shore must substantially limit the opportunities for Australian engineering professionals to develop the knowledge and skills associated with the design of naval ships. This knowledge is not just relevant to the initial design but to the support and upgrade of the ships across their life.

As a result of design work going off-shore, most engineering work being performed today will be in support of the existing fleet. With the reliability, cost and performance of a ship all largely determined during design these Engineers are on a daily basis dealing with the problems built in to someone else's design. This can be far from satisfying work and tends to create a desire to solve things in the design space, by learning from the lessons of today. But EDMC prevents this and in some cases forces us to repeat the mistakes of the past.

PROPOSED STRATEGY FOR NAVAL SHIP ACQUISITION

The strategy proposed here is focused on the requirements development, options and ship design components of an acquisition. Key changes in these aspects would be expected to flow on to improve the remaining first and second order technical implications described. The approach is entirely consistent with the Naval Shipbuilding Plan [5] and, addressing the challenges identified in this paper, will contribute to the success and long term sustainability of the plan.

In summary the proposed strategy consists of:

- Establishing an effective requirements process and identify real solution independent capability needs and requirements;
- Keeping options sets open and not blindly applying EDMC; and
- Enabling design.

Solution independent needs and requirements

Navy is accountable to Government for the delivery of naval capability. Naval capability is underpinned by the capability of our naval ships and submarines. To ensure that Navy is in a position to deliver on commitments to Government it is essential that solution independent needs and requirements are developed and these are used to evaluate future ship options, and where necessary initiate design activities. Without this there is a risk Navy will be unable to deliver.

Further, because many of the capabilities and functions depend on highly technical solutions where the "devil is in the detail", if only high level requirements are used to evaluate options the technical risks to capability may not be identified until late in a program or even once a ship is in service. Needs and requirements must to be developed in such a way as to enable detail assessment to occur early in a program, regardless of the acquisition approach. As the obligations on Navy to deliver capability effect do not vary with acquisition strategy, neither should the need to establish clear traceability from the high level needs and requirements, down to the detailed elements of the solution. Regardless of approach being one of existing design, modified existing design, or new design, solution independent needs and requirements must be established by Navy. In the past Navy has been accused of 'gold plating', that is setting unrealistic or unnecessarily demanding requirements. Robust requirements development processes with clear traceability, as well as requirements management processes, are the means by which Navy can counter this accusation and demonstrate how and why every requirement is needed.

Once developed requirements must be managed effectively. Requirements creep is one element that can severely impact on the execution of a project. Rigorous requirements processes upfront and strong requirements management processes are a control for this risk. Iterations between design and the requirements should be allowed for, including formal decision making processes to enable trade off decisions.

Requirements that capture the capability need as accurately as possible are essential. Both over specification and under-specification of requirements will cause problems for a project. In the current climate where cost and schedule have such high profile there will continue to be pressure to reduce the requirement to reduce cost and relieve schedule pressure. Over specification of requirements, setting aspirational requirements that are not directly derived from the needs, will unnecessarily drive up cost and make delivery schedules more difficult to achieve. Under specification of requirements, setting requirements below the minimum required to meet the need, will ensure that the capability need is not met.

The relative importance of requirements must also be captured. While some requirements have a more gradual impact on capability, real threshold requirement do exist and should be established as part of the requirements development process. These may relate to compatibility with existing infrastructure, performance against a particular military or environmental threat, Australian legislation or other factors.

Recommendation 1 – Develop solution independent needs and requirements
Recommendation 2 – Use rigorous requirements management processes

Options Open

The principle here is simple. Keep the option set open and evaluate all options on their merits. This is something that is enabled by robust, solution independent, needs and requirements. If there happen to be existing designs that meets the requirement without modification these are likely to be highly competitive. For modified existing designs, the change options to meet the requirement would be presented and the scope and cost of modifications and the associated risk considered as part of evaluation. Where a designer choses to put forward a developmental design, they do so knowing that developmental risk is something that will be taken seriously and factor this into their bid, knowing they are likely to be competing against options with less developmental risk.

For this strategy to be effective it is important that existing options continue to make up part of an option set, but that the development of future capability is not constrained to these past designs. Evaluations must also be designed to account for all types of risk, including the risks associated with a developmental design and the risks associated with a legacy design optimised for a different customer and context. Design changes should also be evaluated and not subject to arbitrary blanket restrictions.

Recommendation 3 – Don't constrain the option set but evaluate all options, be they existing, modified or developmental, on their merits.

Enable Design

A further principle is, regardless of options selected, allow necessary design to occur. Manage it, ensure appropriate processes are in place to mitigate risk associated with design, focus on incremental change, perhaps enable it for dealing with emergent issues only and not on capability requirements creep, but do not implement arbitrary, blanket, no change policies. Removing the ability to consider design change essentially removes the ability to manage technical risk.

Enabling design will also create opportunities and is important for Australian innovation. It would allow the maximum utilisation of Australian industry. A domestic naval ship design capability is also likely to be a prerequisite to any realistic naval shipbuilding export industry.

Some innovation is accessible through design change of an existing design, however to embrace large scale innovation often whole new designs must be pursued starting at the earliest phase. The first submarine would never have been realised by simply adding new technology to an existing surface ship.

Naval ship design is an essential enabler to a sustainable domestic naval shipbuilding industry. In addition to detailed production design, higher level design update and upgrade activities are key features of a continuous build strategy.

Recommendation 4 – Enable design and invest in the engineering to support it
Recommendation 5 - Don't put arbitrary constraints on design but put rigorous processes in place to control design risk

CONCLUSION

Naval ship acquisition is a complex and technically intensive process. It is inherently risky and as such should make use of the full range of risk management mechanisms. The appropriate use of existing designs as a basis for acquisition should be considered, but as an approach - particularly with no change / minimum change - it cannot control all risk. As discussed in this paper EDMC is unable to control all technical risk and, in fact, will reduce a programs ability to manage emergent issues.

It is the author's recommendation that, existing design options continue to make up part of an option set for future capability, but that the development of future capability not be constrained to these past designs. With robust, solution independent, needs and requirements defined, a full range of design options, existing, modified, and developmental, should be able to compete on their merits. Evaluations should be designed to account for all types of risk. Design change should be controlled but not subject to arbitrary blanket restrictions. This in turn would create new opportunities for industry and academia to pursue research and innovation and have these realised in Australian naval ships.

It could be argued that the strategy proposed here is simply a return to the Off-The-Shelf model originally proposed by the Kinnaird Review [1]. It is intended to be entirely consistent with the Naval Shipbuilding Plan [5].

REFERENCES

1. 2003 Defence Procurement Review, *Kinnaird Review*, 15 Aug 2003

2. 2008 Defence Procurement and Sustainment Review, *Mortimer Review,* 18 Sep 2008

3. Brown, Dr A, Salcedo, J, Multiple-Objective Optimization in Naval Ship Design, ASNE 2002

4. Keane, R, Fireman, H, Hough, J, Helgerson, D, Whitcomb, C, Ready to Design a Naval Ship? – Prove It!, SNAME, 2008

5. *Naval Shipbuilding Plan*, Commonwealth of Australia 2017

PACIFIC 2017
International Maritime Conference
International Convention Centre Sydney, Australia 3-5 October 2017

CONFERENCE ABSTRACT

Shipbuilding Lessons Learnt on the Air Warfare Destroyers

Mark Lamarre, Chief Executive Officer, ASC Shipbuilding,
Communications.Department@asc.com.au

In May 2005 ASC Shipbuilding was selected as the Shipbuilder for the Air Warfare Destroyer (AWD) program.

ASC Shipbuilding is responsible for the construction of the AWDs at its dedicated shipyard, ASC South, located at Osborne in South Australia - part of Australia's largest naval shipbuilding hub, Techport Australia.

ASC Shipbuilding is Australia's only builder of complex steel-hulled naval warships.
The Air Warfare Destroyer Alliance, that includes ASC Shipbuilding, will this year deliver the first destroyer, Hobart, to the Government.

Since the launch of Hobart in mid-2015, ASC Shipbuilding has systematically and methodically identified and implemented lessons learnt to the second and third ships; driving significant efficiency and productivity gains.

It has been acknowledged by the customer that since mid-2015 the AWD program has exceeded expectations, achieving all major project milestones on schedule and within the revised budget.

The learning curve achieved by ASC Shipbuilding on the platform build has meant the company is set to achieve international benchmarks for similarly complex naval surface combatants in overseas shipyards.

Valuable input has been received from Alliance participants, the Capability Acquisition and Sustainment Group of the Defence Department and Raytheon Australia, and project partners Navantia and the Royal Australian Navy.

Learnings are outlined and the capability and experience of the ASC Shipbuilding workforce demonstrated, as the company prepares to meet the nation's future naval shipbuilding requirements.

Awaiting the next Revolution in Naval Affairs

Simon Reay Atkinson, University of Sydney/Centre for International Security Studies
simon.reayatkinson@sydney.edu.au
Christopher J. Skinner, Captain RAN (ret.) / NAVSEA FFG-7 T&E Manager,
cjskinner@acslink.net.au
Keith F. Joiner, University of New South Wales/Canberra ADFA, k.joiner@adfa.edu.au
Nicholas HM Caldwell, University of Suffolk/Department of Science and Technology,
N.Caldwell@uos.ac.uk

INTRODUCTION

The *Oliver Hazard Perry* Class (FFG-7) became the benchmark surface warship design against which subsequent designs, be they Russian, British, Chinese, French, or German, drew their inspiration. Criticised at the time for being underarmed and lacking in redundancy, they were not regarded as being part of President Reagan's 500 ship Navy. Nonetheless, its conceptual design space (CDS) created a fundamental break with pre-existing designs. Consequently, it was more representative of the *Information Age* (1970-2015), into which it was conceived and born in the mid-1970s, than the *Industrial Age* (1920-1965) designs that preceded it [1].

This paper argues that subsequent designs, including the Littoral and Global Combat Ships (GCS – the British Type 26 Frigate), are essentially optimised versions of the FFG-7. Concomitantly, they have fallen into the Pugh-Augustine trap [2-5], whereby fleet numbers halve every 20-25 years. By contrast, examination of submarine build programmes subject to regularly refreshed conceptual designs – including modularised build and construction – show submarine Basic Mass Empty (BME) costs [3, 5] have remained below those of other weapon systems; only increasing at or below historic inflation. In simple terms, submarines have become more affordable, not less, and this is reflected in countries like Thailand, Indonesia and Myanmar actively seeking such capabilities [6]. Theoretically, surface warship BME costs should have kept pace with submarines – but they have not. In actuality, frigate and destroyer numbers have often halved over the same period – unlike submarines, surface warships have become less and not more affordable.

Nearing or at the end of the *Information Age*, this paper submits that a reconceptualization of the warship design space; shipyards and build techniques – a Revolution in Naval Affairs – is overdue [7]. Only by restoring political, economic and military affordability of ships and potential losses, will usability be restored [5].

CRITICAL JUNCTURE

This paper has at its core the connective element between science, technology and the social. It is contended that the combination of new science and new technology with the social aggregations of the time leads to revolution. Revolution that has at its core a synthesis of all three: the social, the scientific and the technological. Mario Bunge when addressing the failures of individualism attests [8] 'knowledge is social'. If this is the case, a revolution cannot occur without the human factor. It is human abstract, art, skill, and designs that are used in the formation of science: synthesising the maths to deal with new concepts and ideas; with the technologies derived from them. In their examination of the

British Aircraft Corporation's revolutionary 1960s TSR-2 strike and reconnaissance aircraft programme Law and Callon [9] concluded: '*by following the technologists, we see the kinds of social worlds, institutions, and roles contained in the machinery they create. We also see the diverse objects mobilized to fit these conceptions. The technical thus is social*'. Taking the two precepts together, it is possible to conclude that '*Knowledge is Social and the Technological also*' [10]. The linkage between the social and knowledge may be confirmed by researching previous scientific and technological ages. In their examination of an earlier scientific age (Parson's Turbines), Jarratt and Clarkson [11] and Reay Atkinson [12] determined a Scientific time constant, after Chen and Yi [13], of 45-50 years for change to reach 95% of its steady state value [10]. They also recognised that within the scientific time constant exist technological time constants (driven by the science) of about 15 years. [10] Based on this analysis of Science driving Technology and a Scientific Time Constant of 45-50 years, since the British Industrial Revolution it is contended that there have been five identifiable scientific ages. Suggesting that a new one may be imminent [1, 14]:

Table 1: Different Ages as defined by the Science Time Constant (45-50 years), with the gaps between ages defined by chaotic states as one age dies.

Period	Scientific Age
1770-1815	Steam Age
1820-1865	Locomotive Age
1870-1915	Turbine Age
1920-1965	Industrial Age (as recognised in the literature)
1970-2015	Information Age (as recognised in the literature)
2020-2065	*Synthetical Age* (as posited, Reay Atkinson [14])

The Dreadnought Revolution (1906 [15]) was based on Parson's development of non-compounded steam turbines and, specifically, the introduction of a vacuum (·1900-1904) that quadrupled thermal efficiency. Marder argues that 'at the turn of the [19th] century ideas on naval tactics began to emerge from their chaotic state' [16]. These states of 'successive growth stages of cascading logistic curves; [connecting] natural growth and chaos like states' (see Marchetti [17] and Modis [18]), typically occur at the end of an age when a system comes off-line. Although the Turbine Age had some years to run (with the development of end-tightened blading (1918-1930)), by the beginning of the 20th century it was coming to its end. A new critical juncture was forming with the onset of the Industrial Age leading to mass production, tanks, turbo-prop, jet aircraft, and aircraft carriers. Blitzkrieg and Kantai Kessen (the Imperial Japanese Navy's Decisive Battle Doctrine) were based to a greater or lesser extent on mass produced turbines.

Towards the end of the Industrial Age, in the 1960s similar chaotic states were emerging leading, on the one hand, to the revolutionary designs behind the McDonnell Douglas F-15 Eagle (emerging from the remarkable Skunk Works), nuclear powered attack / deterrence submarines and, on the other, to the Oliver Hazard Perry (FFG-7) class. The FFG-7 was also motivated by the failure of the Israeli Navy in the 6 Day War (1967), and the IDF Navy's determination to be prepared for the next conflict; resulting in the 450 Ton, Sa'ar / Reshef class of missile boats:

> *The Israeli CNO Admiral Yohai Ben Nun placed great emphasis on sophisticated equipment – essentially dividing naval content (weapons, sensors, crewing etc.) from the hull (sometimes considered as the platform). After heated debate, it was decided that the 'boats' (subsequently to be known as Missile Boats) should be based on an existing hull or platform whose operational functionality had already been proven in a [West] European country. It is not clear whether or not Yohai envisioned the vessel in detail. However, his staff made a huge effort to take forward his design thinking. They were aware of the miniaturization process evolving in technology and electronics. They therefore decided to adopt the concept of designing highly sophisticated smaller [missile] 'boats', each capable of working alone or networked, and supporting electronic systems and equipment with the ability for over-the-horizon picture forming and sharing the operational picture to shape the tactical moves and develop firing solutions, in advance.* [19]

Revolutionary designs for the Sa'ar Class, demonstrated in the Battle of Latakia (1973), were scaled successfully into the FFG-07 class; including the addition of Gas Turbines. Details of the revolutionary aspects of FFG-07 will be considered later.

Golding (2011) writing in *Versatile Modular System designs for a Versatile Modular Fleet* [5], concludes that there are peacetime and wartime builds. By this, he means that designs stagnate [11], atrophy and ossify[1] in times of peace and often then fail in subsequent wars, such that new designs are then urgently required. This phenomenon appears to be evident in meta-analyses of modern defeats and victories examined by Biddle [**Error! Reference source not found.**], and concepts like the Revolution in Military Affairs (RMA) [7] when rapid-evolutionary (and revolutionary) technological and organisational changes occur in warfare. Consequently there should be a Darwinistic tendency for nation states to maintain technological adroitness through periods of peace. Reay Atkinson in *Design to Fight* [20] argued that without regular re-fresh and updating of the core design, Fleets incur Defence Cost Inflation (DCI) [3, 21, 22]; halving in size every 25 years. Both the Dreadnought and the FFG-07 Revolutions occurred at the change of scientific ages during 'chaos like states' – potentially also acting as precursors to instabilities leading to WWI, or the Middle East, Vietnam and the Cold War in the 1960s and 1970s.

Through policies such as 'front line first' [21, 23], the formal end of the Cold War in 1992 led to reductions in UK (and other Western Nations) R&D by as much as 85% in real DCI terms between 1979 and 2008 [5, 10]. Investment in Training fell slower than on Procurement and R&D, and recovered more strongly post the 1998 UK Strategic Defence Review (to about 118% its 1979 historical value). Following the unprecedented attack on the World Trade Center in New York (9/11), emphasis was placed on Training (although military education was cut in UK) and investment has continued to grow faster in this sector ever since. Procurement fell rapidly post 1992 before, in 1996, recovering to higher investment levels than for R&D [24] – based on optimisation of existing designs, rather than creating new ones.

[1] See Chris Donnelly on 'War in Peacetime: Coping with today's rapidly changing world' at: http://www.statecraft.org.uk/research/war-peacetime-coping-today%E2%80%99s-rapidly-changing-world, based on Karl Marx 'on War', circa 1861 writing with Frederik Engels on the US Civil War.

The cost imperative of conceptualising and creating new designs removes or significantly reduces DCI from the system. As recognised by Pugh and Augustine, by creating new designs one begins again. In other words, the replacement designs for the UK Type 22 Frigate (itself very similar in design and concept to the FFG-07) were not optimised versions of older designs, such as the Leander Class. Instead, they maintained inflationary adjusted unit costs; designed and conceptualised anew to maintain numbers. Based on the Basic Mass Empty (BME) costs of a 4500 Tonne Batch 1 Type 22 (~£19 per volumetric kg on build), a UK Type 45 Destroyer at 7500 Tonnes would have cost £146M in 1980. Based on historical inflation, the build costs in 2008 would have been about £550M (as opposed to £1 Billion). However, Pugh [3] recognised that frigates, destroyers, main battle tanks and even [nuclear powered] submarines (where R&D was maintained) were attracting a BME DCI of only 6% ('one per cent above Historic Inflation ') between 1950 and 2008 [25]. In actuality, there was downward pressure on the costs of these items. New designs – significantly in commercial shipping – were beginning to drive down build costs. Adjusting for this downward pressure on historic inflation, a new Type 45 design in 2008 may have cost £410M: an equivalent 1980 BME of ~£14 per volumetric kg [5].

Without investment in new designs, concepts and strategies, inadequacies in equipment had to be compensated for by better trained people, and, in conflict, by urgent operational requirements. Cuts to R&D correlated to the failure to invest in a revised Frigate programme (in UK, US and other NATO countries and Australia) in the 1990s, when emphasis was also placed on maintaining status-quo designs. For example, the ANZAC class, the Air Warfare Destroyers and even the Canberra Class Landing Helicopter Docks approved between 2003 and 2004. The 'cause', it is argued, lay in the structural shift between investing in / abstracting new designs (R&D) and optimising existing / status quo ones. The enablers to critical deeper thinking - R&D and education – were drowned out. This was seen by price hysteresis in procuring latter batches of the UK Type 23 Frigates (essentially optimised versions of the Type 22) – indicative of a system exhibiting chaotic states [10]. This potential illusory thinking of saving money and time is most recently evident in the Canadian decision to look for an extant frigate design:

> *Relying on a proven, off-the-shelf warship design from another country takes a lot of the uncertainty out of the planning process ... "We don't know the actual cost per ship yet," she said in an interview. "We're not talking about a custom build anymore. We're talking about existing designs ... and in our view that is likely to have an impact on diminishing all sorts of risks." There would be, however, some modifications to the design to suit unique Canadian requirements. ... Naval buffs will likely mourn the design decision.* [26]

The attractiveness of such arguments early in an acquisition plays well into competitive evaluations and redesigns, that when played out in the fullness of unique requirements and new technological fit-outs (the shopping basket of systems) mean that the off-the-shelf schedule and cost savings are both a fallacy and mirage. An excellent treatment of this fallacy is in the Australian Senate Inquiry into Defence Procurement ([27], Chapters 2 & 12) and there is no better example of it playing out than Australia's Air Warfare Destroyers (AWD), which has so far taken 15 years from requirements to first ship OT&E — much longer than the FFG-7 and ANZAC first-of-classes; and costing three times the original estimates.

The global stagnation of Defence R&D outside of a few key areas in the U.S. and China is clear in reviews like Bitzinger [28], where specific to maritime he covers the USN DDG-1000 program and attempts at new destroyer and cruiser designs (DD-21 & CG-21). He cites Luttwak [29] as concluding that, 'instead of shaping new platforms and weapons configurations to fit today's information technology, communications, sensor and guidance equipment, we are shoving, cramming and moulding such technology to fit the nooks and crannies of 1945-era platforms.' In seeking to explain the reasons for stagnation and re-use rather than innovation, he references Kaldor [30], asserting that 'military bureaucracies, being naturally "conservative" and operating according to "dominant scenarios", are not really comfortable with radical new technologies, since they "pose a risk for organizational survival". Kaldor herself states, "New technologies can only get through the innovation and integration stages if they conform to the requirements of the dominant scenario ... directed towards the improvement in performance of missions that were established nearly 40 years ago...". [30] Such reviews illustrate that while the recent Iraq and Afghanistan wars have been significant for land forces and counter-insurgency or irregular warfare [31, 32], for air and naval platforms, these wars have meant almost no revolutionary design: from a naval and air power design perspective these are largely peacetimes.

This paper argues that the current revolution is based upon the indivisibility of information from the technological, and the social from the IT – hence the *Synthetical Age*, in which 'knowledge is social and the info-technological also' [1]. This suggests we are at a critical juncture when existing designs are no longer affordable and it is necessary to break out of the Pugh [2, 3] and Augustine [4] DCI trap by:

1. **Abstracting, conceptualising and creating new designs**. This is the approach advocated. The revolutionary designs of FFG-7 (and UK Type 42s and Type 22s) were not reconceptualised and maintained in the early 1990s, due to the Peace Dividend and the end of the Cold War. Combined with a new scientific age, a revolution in naval affairs (RNA) is in the offing.
2. **Reconceptualising and redesigning** existing classes and combinations / compositions of systems / capabilities / platforms. For example, HMS OCEAN (L12) designed to commercial standards came in at the same mean unit cost of the *Invincible* Class carriers built 20 years earlier [20].
3. **Maintaining a regular refresh and build rate** – tempo. This requires a programme connecting design and conceptualisation if one is going to maintain change from one generation to the next.
4. **Spending much, much more** (power law increases in budgets) to maintain / preserve existing [obsolescent] design and build capabilities, e.g. the Type 45, the Zumwalt class, or the AWD. This is the risk RAN faces in a premature down-selection for SEA 5000.
5. **Stop and get off**, as the UK Royal Navy appears to have done:

   Even if the GCS were truly an innovative and impressive design, its prospects would be hobbled by the decision of the Cameron government to go back on its plan to buy 13 of them (replacing 19 Type 22 and 23 frigates). Instead of purchasing eight anti-submarine versions and five general purpose versions, the government is now committed to buying just eight ASW frigates. This is fewer than a traditional ship class and that matters because you need to commission and build at least ten vessels to be able to assess their real abilities (to distinguish good, from poor, from average) and make appropriate improvements. [33]

THE REVOLUTION THAT WAS FFG-7

The FFG-7 class frigate was for the United States Navy (USN) a revolutionary design-to-price program designed to compensate for dwindling numbers of anti-submarine warfare escort ships with which to protect convoys to supply NATO or to protect USN amphibious task forces. The FFG-7 class was explicitly not required for escort of carrier task forces for which the more capable (and more expensive) Spruance class DD-963 destroyers were acquired in a similar time frame (from late 70s through early 90's). The cost criterion also extended to the numbers of ships force and air detachment personnel to be accommodated. The corporate mantra was the high mix / low mix approach mandated by Chief of Naval Operations (CNO), Admiral Elmo Zumwalt USN, a charismatic and visionary leader who insisted on new approaches to lingering challenges, especially those central to the Cold War which was then at its height. The FFG-7 design embraced a number of unconventional and even revolutionary required characteristics, including:

- Two high capability ASW helicopters, initially LAMPS-1[2] SH-2D Sea Sprite aircraft and later in the third flight of ships, the much more capable and heavier LAMPS-3 SH-60B Sea Hawk[3] These aircraft also provided over the horizon targetting for anti-ship cruise missiles
- A Recovery Assist, Traverse and Securing [RAST] system based on the Canadian Beartrap to enable operation of heavy aircraft from a relatively small deck up to sea state 6.
- The ASW capability was further enhanced with LAMPS3 by the addition of AN/SQR-19 Tactical Towed Array Sonar System [TACTASS].
- An area, air warfare defence missile system based on the Standard SM-1 missile operated from the proven Guided Missile Launcher System Mark 13 [GMLS Mk13]. These were also used in the DDG-16 later ships of the USS Charles F Adams class as operated by the Australian and German navies.
- A smaller, lighter Guided Missile Fire Control System (Mark 92) based on the Dutch Signaal system with Americanisation of some components and human machine interface (HMI) and the addition of continuous wave illuminator required for target illumination for Standard missiles.
- The GMLS Mk13 also provided magazine and launching capability for the Harpoon Anti-ship cruise missile (ASCM) weapon in widespread use from a variety of launch platforms
- A maximum number of 185 accommodation bunks for all member of the ships force and of the full embarked air detachment for two aircraft
- Ability to operate the embarked aircraft up to Sea State 6 with some limitations and up to Sea State 3 without any restrictions day and night

[2] Light Aircraft Multi-Purpose System (LAMPS).
[3] On which the Australian Navy SH-70 was based.

- Two LM2500 General Electric [GE] marinized gas turbines driving a main reduction gear applying revolutionary frictionless Synchro Self Shifting (SSS) Clutches [34] to a single shaft with controllable pitch propeller.
- Ship service electrical generation by two (later three) ship service diesel generators (SSDG).
- ASW weapons to be delivered by aircraft primarily and to be magazined in an above-waterline space protected by non-metallic armour made from Kevlar proprietary material.
- Close-in defence from anti-ship missiles by mounting of a Close In Weapons System (CIWS) above the dual hangars.
- A secondary general purpose gun system was also provided by a 76mm American built gun to the Oto Melara design; this was mounted amidships at the level of the hangar roof rather than the primary offensive weapon GMLS13 mounted on the forecastle
- Aircraft hangar and lightweight torpedo magazine protected by composite material armour (Kevlar)
- A Land Based Combat System Test Center (LBTS) to provide mock-up of the ship's combat system design options [35].
- The then novel requirement to integrate maintenance planning into the outset of the ship's design and implement maintainability T&E demonstrations during detailed design [36].
- Other requirements were similarly revolutionary in many ways and gave rise to passionate questioning of all design features

The FFG-7 program management was also revolutionary as it was always intended to be run through three shipyards concurrently: Bath Iron Works [BIW] in Bath, Maine; Todd Shipbuilders in two yards – Seattle, Washington, and Long Beach, California. The building tempo was to be such that some eight ships would be delivered in each year. However the US Naval Sea Systems Command did appreciate that many of the features of the design were unproven in any stage of test and evaluation (T&E): developmental (DT&E), operational effectiveness and suitability for fleet employment (OT&E) and production acceptance (PAT&E), and mandated the first of class [4] be constructed in a separate contract with BIW and there be a period of more than a year between the delivery of the lead ship and the first of the 'follow ships' to permit the test and evaluation process to provide the optimum influence on the overall design, while not too far advanced in production [35]. The value of the lessons learned in this deliberate pause are well documented by Stark and Stembel [37].

For each ship there would be a period of shakedown and acceptance trials after delivery, followed by a post-shakedown availability when corrective actions would be programmed. The lead ship FFG-7, the USS Oliver Hazard Perry, would also be subjected to class design evaluations of a formal operational test and evaluation when fully configured[5] and to a ship shock test comprising the deliberate detonation of large underwater explosive charges to verify the integrity of the hull design and the capability of ship systems to

[4] Termed the 'lead ship' of the class.

[5] Conducted later in a ship of the Fiscal Year 1979 [FY79] design baseline when a number of significant changes were introduced including LAMPS-3.

operate effectively after such an explosion. It should be noted that ASW weapons of the era included nuclear-tipped ASROC missiles that could be fired by friendly forces in close proximity to escorts such as FFG-7 class ships.

The other relevant major activities at the time were the ongoing DT&E and OT&E programs for some of the major systems on which FFG-7 design was based. There were concurrent T&E programs for LAMPS3 including RAST, for MFCS Mk92, for Harpoon, SQR-19 TACTASS and for the SQS-56 hull-mounted sonar used in FFG-7. The importance of LBTS in competently managing revolutionary designs is captured by Stark and Stembel ([37], p. 116) as follows and is in stark contrast to the Australian experience of the Collins Class Submarine a decade later, see Joiner & Reay Atkinson [25] and the Landing Helicopter Dock ships only a few years ago:

> *Although costly to design and to build, these two test sites were of inestimable value in accelerating the Lead Ship design and the FFG Program. The Propulsion LBTS permitted ordering and testing of the gas turbine, reduction gear, shaft, propulsion control console, and associated lube oil system more than a year earlier than would otherwise have been the case. Similarly, early development of the Combat System LBTS forced decisions on equipments and arrangements and made data available much earlier than normal. As a result of the two test sites, data for these systems were never a problem in the Lead Ship design, and the successful Acceptance Trials of FFG-7 were attributed in large measure to these Facilities.* [38]

All T&E in the USN, then and now, is based on an approved T&E Master Plan (TEMP) and all of these developmental programs and other test issues were described in the FFG 7 TEMP along with the approaches to be used for all T&E, the responsible agencies and the resources budgets for the T&E to occur. The TEMP required Defense approval and had to be implemented as approved and updated on an annual basis. Quite separately the U.S Congress exercises oversight of all ship programs for the USN through the Board of Inspection and Survey (INSURV) who must be satisfied independently of the merits of USN action.

REVOLUTIONARY TRENDS IN MILITARY RDT&E EFFECTING NEXT FFG DESIGN

To see where the next revolutionary FFG design might come from requires two big steps: first to remove from consideration all recent warfare where naval forces were simply an enabler of majority land conflicts and held largely by one side only; and second, to step away from platform-thinking to systems-thinking, so as to see where technology can go if unconstrained. Taking the first, the last naval engagement between near-peer adversaries, where these engagements seriously shaped the outcome, was the Argentine-UK Falklands war in 1982, 35 years ago [39]. This engagement reaffirmed the importance of aircraft carriers and submarines generally but updated technological advances in the following areas of naval engagement: the timeliness of good intelligence, surveillance and reconnaissance (ISR) by both sides; long-range, remoted submarine sea control (in 1982, the UK had five SSN and one SSK in the South Atlantic); improved at-sea anti-aircraft and surface search radars (ship & airborne); anti-ship cruise missiles, notably Exocet; short-takeoff and Vertical Landing (STOVL) aircraft used by the UK; refuelling aircraft for land-based aircraft projection to sea, or buddy refuelling to extend range and/or time on task – as applied by the Vulcan strategic bomber force; Maritime patrol aircraft operations in reconnaissance for anti-ship strikes; resupply of aviation and ship service fuel, armaments and victuals at sea; and electronic warfare in all its manifestation. Significantly, anti-submarine warfare (ASW) was not effective, and neither were anti-ship missile defences (ASMD). Exactly the lessons that needed [re]-learning.

While it may be harsh to relegate the importance of naval engagement since the Falklands War to enabling projection, deterrence and anti-piracy, the lack of major naval engagement has been the main reason for the lack of impetus for redesign in the warship space. As documented earlier, FFG-7 design arose at the height of the Cold War, itself now 40 years ago; while some other pre-revolutionary developments like the DDG-1000 have occurred since then based on the need for properly envisioned Nation-state naval warfare, but these are the exception. Also, Speller [39] covers the developments of sea-basing as a means for projection and its use by both Russia and the U.S. for such purposes. There are elements of sea-basing in the Australian adaptation of a Spanish amphibious ship design like the Landing Helicopter Dock (LHD) for stabilisation in the South-East Asia and South-West Pacific, however, these are comparatively minor re-use.

Speller ([39], pp. 209-219) looks to the continued development of maritime anti-access/area-denial (A2/AD) weaponry as the greatest effect in naval capabilities, where 'a combination of advanced systems and also the innovative use of cheaper low-technology capabilities might allow a wide range of potential adversaries to deny the seas to even the most powerful navy.' He highlights that *'China has concentrated on developing a range of joint capabilities, across all domains, that are designed to challenge US access and freedom of manoeuvre, aiming to keep US forces away from areas of Chinese interest and to threaten the viability of US bases ...'*. Bitzinger ([28], p. 164) also points to Chinese R&D as the most influential in Defense innovation:

> *... this possible "lull" in disruptive strategic innovation ... may provide a pause or slow-down in the global process of defense technology development that would permit latecomer innovators and "fast-followers" to draw nearer to the state of the art. This is particularly apropos in the case of China. China has ... increasing military expenditures at least five-fold over the past 15 years ... its defense R&D, although classified, probably approaches $6 billion annually ... Certainly, in its pursuit of a fifth-generation fighter aircraft (e.g., the J-20 and the J-31 prototypes), it is poised to overtake Europe in this one particular area.*

Whether from this Chinese investment or because of it, or both, naval spending by the Asia Pacific region is exceeding Europe for the first time in 400 years (Till [40], p. 182). Raska ([41], p. 96) notes the particular attraction of submarines to South-East Asian countries, as was illustrated in the introduction earlier [6]; namely, that they cost-effectively provide sea-denial without requiring an equivalent degree of sea control for sea power projection or trade protection. He explains: 'Stealth attributes of submarines provide strategic advantages over surface ships as the means for sea-based surveillance, particularly in the shallow coastal and littoral waters in the South China Sea and Java Sea.' With these forces at play, returning to the first precondition for revolution and using techniques outlined by Ross [42] for identifying emerging military technologies, Table 2 is a listing of the critical emerging technologies in the Asia Pacific, and especially China, which might force and enable revolutionary naval design.

Table 2: Critical Emerging Technologies effecting naval ship design, especially in Asia-Pacific

Predominately Forcing	Forcing and Enabling
Over-the-horizon Radar (OTHR)	Electronically Scanning Array (ESA) radars
High Frequency Surface Wave Radar (HFSWR)	Ship-borne Sonar-arrays, hull-mounted & towed
Hypersonic missiles, including long-range aircraft launched [43]	Unmanned Underwater Vehicles (UUVs) & Unmanned Aerial Vehicle [UAV]
Cyber-warfare [44]	Cyber-security [45]
Ocean-bottom & tethered remote sensing sonar arrays	Information Dominance (networking+cooperative engagement)
Highly programmable sea mining employed similar to improvised explosive devices	Unmanned Surveillance & Maritime Patrol Aircraft
Submarine endurance from nuclear or air-independent propulsion [41]	Active stealth vice passive reflectance & absorbent designs
Submarine stealth	Swarming of unmanned vehicles or weaponry
Space-based persistent infra-red sensing	Intelligent Integrated Platform Management Systems (IPMS)
Quantum Artificial intelligence (QAI)	Remote Situational Awareness to Hardened Command Centers
Laser weapons	Tectical nuclear weapons (again)
Lightweight armour made of composite materials	Anti-Ship Ballistic Weapons

The existence in multiple countries of strategic surveillance satellites, long-range and reflective fixed radars (i.e. OTHR & HFSWR) and other optical sensors, and fixed or tethered sensing sonar arrays, make it much harder to project ships without detection and so such ships will need to be better defended. To this end, lighter-weight, high-powered ESA radars are promising substantial reductions in the weight and height requirements of ship superstructures, at least to defend against high-speed threats. Unfortunately, the high-speed threats are getting substantially faster, going into the hypersonic regime, where the warning time of advanced strategic sensors like satellites will be effectively taken away as new missiles traverse every 1000 km in around 9 minutes or less ([43], p. 117). Such high-speed threats require phenomenal sensor-fusion and signal-processing rates to defend against. One way to try to hide is active electronic rather than passive stealth, however, if the adversary has multiple sensors operating across many frequency bands, such stealth will be transitory.

The use of remote sensing with 360 degree display to a hardened command room is under development in submarines, enabling all sensor specialists and command elements to see and contribute, moving beyond the individual sensor operators on the deck of the Star Trek Bridge. Such developments may finally offer ship control from networked command spaces rather than the highly vulnerable ship bridges of today, moreover moving those centres to more defensive positions. Heinl ([44], p. 126 & p. 131) points out that:

> *...acquiring offensive or advanced cyber capabilities could seem financially attractive, in particular for less wealthier states in the [South-East Asia] region, relative to the higher costs of other weapons ... defence analysts predict that many South Asian states will undertake state-sponsored cyber programs facilitated by low barriers of*

entry, the availability of large pools of skilled manpower and extensive IT infrastructures.

If future naval warfare comes to a region like South-East Asia it is likely to be hybrid (i.e., include piracy, terrorism & non-state actors with attacks on civil populations and infrastructure as well as military forces) using the comparatively cheap and somewhat indiscriminate and hard to attribute use of cyber warfare and sea mines. As such, future frigate designs must be highly cyber-secure and able to deploy mine countermeasures.

ENVISAGING REVOLUTIONARY WARSHIP DESIGNS

Against a first world contender, 'to survive in the modern battlespace, Fleets will need to be able to afford to take the hits and the losses'. [5] In a non-attritional sense, losses will need to be politically, militarily and economically affordable if ships are going to be used. Therefore, 'Affording to Lose in order to Use; and vice versa', becomes a fundamental design driver.

Examining the likely impact of emerging military technologies on future warship design (by which we mean all surface warships) the following salient features are emerging: Glocal (Global / Local) networked (internal & external) command spaces; enhanced defensive radars and ship defensive missiles to counter hypersonic threats; towed and fixed sonar arrays employed in concert with offboard sensors such as seabed arrays, sonobuoys and seagliders; UUVs to assist with sub-surface threats and associated picture-complication processing; UAVs to augment the maritime patrol of crewed helicopters and aid in defensive awareness especially in congested waterways, and the associated processing within the command spaces; condensated interactive-networking with other ships and other allied assets to enable cooperative engagement and mutual defence; and, Cybersecurity; including cyber-monitoring of all internal databus, external traffic; and, weak-signal RF probing close to land or other vessels, and the associated processing within the command centre [45].

Concepts such as Quantum AI (QAI); condensated interactive networking (CIN); networked command spaces (NCS); and, weak-signal probing, are at the forefront of the Author's research. Considerations set out in this paper, have concomitantly led to the development of a consolidated conceptual design space (CDS), as set out in Table 3.

The Cold War Threat equation considered that 'Threat was equal to Capability plus Intent and Will'. Because Capability could be objectively measured (in terms of numbers of tanks, ships, aircraft etc.), it was. Ultimately, this over-concentration on capability led both to a collective failure to identify the end of the Cold War (on the collapse of Moscow's Intent & Will), and a reductionist concentration on own Capability; including over-reliance on information and technological dominance. To counter this Western-orthodoxy, the Soviet Union designed highly sophisticated, resilient, quantitative, simple (as a form of complex), command doctrines and capabilities, such as Reflexive Control [46] and the AK-47. Similarly, the Pashtun doctrine-d'résistance is based on a simple maxim: 'You the West (NATO, the US, etc.) have your clocks; we have time!' Taking the two concepts together, along with Stalin's maxim on quantity, Capacity to survive, loiter, take the hits, and afford the losses becomes a fundamental warfighting (as differentiated from peacekeeping) design consideration: 'Capacity, Capability, Coherence, Consistency and Continuity[6] have a Quality and Quantity all of their Own'. [32]

Table 3: Consolidated Conceptual Design Space Considerations

Current Position	Assumption	Deduction
Improvements to networking cross-spectrum sensors (beyond CEC), including sonar and EM, and from satellite and cyber tracking (of resources, for example of logistics chains / information / big-data flows.	A Fleet Vessel will be detected at some stage of an operation.	Stealthy hulls are of less value and the expensive premium paid for a quiet hull / tiles etc. (to avoid sonar detection) or for anti-stealth hull forms / RAM etc. (to minimise REA) are less and less effective, when up against a first world contender.
The threat posed by conventional weapon systems - Cruise and Ballistic missiles of the first world contenders is such that not even the largest of US Aircraft Carriers would survive a hit from such a missile. The threat posed by other anti-access systems such as mines, would at least disable most Frigates and Destroyers and, in all probability, sink them.	Such weapon systems are not going to be used singly, but in salvos or fields	Scale counts. Based on Stalin's maxim 'quantity has a quality all of its own', while a Western FF / DD may survive and fight off a first attack – the second, third or fourth waves are going to get through. Either very much larger than current US (and UK) Aircraft Carriers (double-hulled, 100-150,000 Tonne ISO container ships are likely to survive a hit), and / or many more (FFs / DDs etc.) to ensure survival of the whole.
The affordability question becomes key to political, economic and military decision making and taking, for example to contend with a first world power.	The question is not whether or not losses are going to be taken – because they certainly will be – but what price each sector is likely to set.	Political affordability (often tied militarily, as in 'The Invincible Question' during the Falklands War), can become a deciding factor determining operational application. This has economic impact, because it suggests that numbers need to be both affordable (in build) and replicable in a timely way (during conflict) if they are going to have political value in a conflict, and so be used militarily
The application of UAVs, USVs and UUVs is being pursued largely to reduce risk to the up-front operator of systems – this includes smaller aircraft / submarines, without the need for life support systems and, in the case of UCAVs, being able to fling UAVs off much smaller decks at much higher G, G5+	Processing of data for these machines and the number of people necessary to maintain and operate these systems (from distance), actually means that a larger footprint of people and resources is required, than for an equivalent crewed aircraft.	In order to provide real-time processing of data, such processing power may need a) to be closer to the operation and b) more influenced by humans in the real-time loop. Local, as opposed to remote, mobile platforms capable of piloting UAVs and assessing / processing data becomes more important.

A fundamental design error introduced by 1990s reductionist, optimised design space (ODS) thinking was to confuse and conflate scale with numbers – as in numbers of ships and crew sizes. This was further compounded by an accountancy-based predilection conflating capability, with strategy; and ranking (ordering, controlling, tiering etc.), with positioning:

> *In strategy, rank and position are two very different concepts. Ranking does not guarantee victory, but position does. Position is about the ability and agility to adapt and survive and defeat one's opponents. Rank is merely one element of one's position. The mismatch between rank and position is the ultimate strategic nightmare and distinguishes strategy from doctrine and operational art, where Strategy is influenced by wider Political, Social, and Economic considerations.* [47]

In the air power debate in the US, UK, Japan and Australia concerning the F35 Joint Strike Fighter (JSF) [48], the key to Western thinking is in the terminology often used of it "being affordable". In air combat, the loss of single or few F35 aircraft may through high-rate networking disclose the enemy's intentions and even precious electronic warfare codes entirely and enable a quick, almost swarm-like collective response: at least it is an attempt at a simple explanatory theory. Now translate this to the maritime domain and consider the debates about the declining numbers of real ships. Of course, there are some significant down-sides from the F35 analogy.

First, this software intensive aircraft is drowning in difficulties in producing reliable operational software loads [48] and any ship that takes a highly networked design path will

[6] Attributed to Major General John Drewienkiewicz UKA and the First Author, at an ARAG Quesitio, Jun 2006.

experience similar difficulties; although as always, ship designers can learn from the aircraft experience. Second, any networking increases the cyber-attack surface and increases cyber vulnerability and reliance on all cybersecurity measures. Third, the theory is hugely untested in air warfare, and from the obstinate persistence of Western nations, may be the key to whether, or how long, the West remains omnipotent, and whether in the next World War, frigates and land forces must again face the repressive warfare of World War II when an enemy has air and sea superiority for some or all of the time. Consequently, a compromise must occur soon in ship design where singular ships are less defendable but synergistically-networked can be much more dangerous.

This returns to scale, scaling, networking and political, economic, and military affordability. The 1970s revolution was to differentiate between the platform (essentially energy supplies / propulsion, the hull and hotel services); from capability. This probably went too far, as performance-managed accountancy regimes took the platform as being synonymous with the 'ship and crew' to cut crew and ship numbers, accordingly. Furthermore, capability drove strategy and designs: 'capability-driven strategies give the enemy the opportunity to harness and so 'fix' one's own capabilities to their advantage. It confuses weapons and capabilities with political objectives and leads one to define one's enemies in terms of their capabilities rather than their strategy'. [47]

The lack of serious investment in surface warship designs since the 1990s, while on the one hand making them increasingly unaffordable in the numbers needed; is also forcing the Revolution in Naval Affairs that we suggest is coming. Scalability and Composability become critical – which is as important a political and economic consideration, as it is a military one. Such a commitment needs also to be based on re-capitalising surface fleets, which will take political and economic will along with new financial designs, beyond existing models such as PPPs and PFIs. This suggests to us, that if Western Navies are going to bootstrap themselves into the future, they will need to more drastically apply system-modularisation (perhaps[s through "Intermodal Containerization") and enforce not simply the division between platform and capability, but the intermodal interactions between platform and content. This will require Navies to look beyond current platform-orthodoxies, and consider applying perfectly acceptable merchant ship designs (where investment has been maintained) – at scale and numbers – to better fit system-capabilities. "Versatile Modularisation" – a form of agile adaptation – therefore becomes key. [5] The first Aircraft Carrier, HMS ARK ROYAL (II), was laid down in 1914 as a freighter, designed for the coal-grain trade in the Black Sea. More recently, HMS OCEAN applied merchant-marine standards to achieve something of an affordable half-way house, between bastardised RN Engineering and Lloyds Standards, and a fully versatile modularised system. Taken with Army's long-standing criticism of Navies and Air Forces, that 'they man the equipment; rather than equipping the man', this suggests to us four critical design criteria that may in future guide naval systems thinking:

1. Ask first, 'what would we be doing, if we were at war – and, if not, why not?' (the 'ARK ROYAL paradigm');

2. Adopt the *Cult of the Imperfect*[7] – sometimes adapted as 'second best, today': 'give them the third-best to go on; the second-best comes too late, [and] the best never comes'.
3. Enable compositions for 'crewing the ship; rather than shipping the crew';
4. Scale capability-networks for 'fitting the kit; rather than kitting the fit'.

CONCLUSIONS

As we began, this paper submits that a reconceptualization of the warship conceptual design space; shipyards and build techniques – a Revolution in Naval Affairs – is overdue. Only by restoring political, economic and military affordability of ships and potential losses, will usability be restored [5]. We further submit that, for Australia, such an investment programme would restore industrial co-adaptive advantage, at scale and be good for our existing, highly productive, competitive, commercial ship-building yards. In other words, thinking what we would be doing if we were at war... We make five points:

1. Generally there is a Revolution in Naval Affairs underway marking the chaotic state at the end of the Information Age;
2. While many of these features add intelligent, new software-intensive, deployable systems, scaling and composing the size of future fleets through versatile intermodalisation becomes crucial, as a distributed and networked force is more defendable and in peer-warfare, numbers do matter;
3. Rescaling fleets and affordability, are serious design challenge as befits the word revolutionary, but it has like the FFG-7 been done before;
4. Some compromise must occur soon in ship design where singular ships are less defendable but synergistically scaled and composed much more dangerous;
5. This revolution also requires Navies to commit to LBTS that precede ship build and manage system threats and obsolescence through-life. LBTS are particularly key to revolutionary design, since they verify and validate informed decision-making based on iterative T&E and remove the illusionary and ephemeral promise of "off-the-shelf".

We have concluded from our research that there are significant lessons to be taken from the FFG-7 program at the start of the Information Age, and applied now at the start of what may be considered the Synthetical Age.

[7] Sir Robert Alexander Watson-Watt, KCB, FRS, FRAeS (13 April 1892 – 5 December 1973), pioneer of British WWII Radars.

REFERENCES

1. Reay Atkinson S., C., Smallhorn, N.H.M. Caldwell, G. Tolhurst, *Identification and Classification - Designing and Engineering Synthetic Ecologies*, in *System Engineering, Test & Evaluation (SETE) Conference*, E. Australia, Editor, 2016, SETE / INCOSE: Melbourne.
2. Pugh, P.G., *The Cost of Sea Power,* 1986, London: Conway Maritime Press.
3. Pugh, P.G., *Retrospect and Prospect: trends in Cost and their Implications for UK Aerospace.* Defence and Peace Economics, 2007. Vol 18(1), February: pp 25-37.
4. Augustine, N.R., *Augustine's Laws,* 1997, Reston, Virginia, US: American Institute of Aeronautics.
5. Reay Atkinson, S., I. Hassall, N.H.M. Caldwell, M. Romilly, & R. Golding., *Versatile Modular System (VMS™) designs for a Versatile Modular Fleet (VMF™)* in *paper presented at EAWWIV Conference*2011, Old RN College, Greenwich, London.
6. Maas, R., *Indonesian French companies join forces for naval projects*, in *United Press International,* 2017, http://www.upi.com/Defense-News/2017/03/31/Indonesian-French-companies-join-forces-for-naval-projects/9281490979577/: March 31, on-line news article, accessed 26 April 2017.
7. Riekhoff, H., and Gongora, T., *Toward a revolution in military affairs?: defense and security at the dawn of the twenty-first century,* 2000, Westport, Conn.: Greenwood Press.
8. Bunge, M.A., *Ten Modes of Individualism - None of Which Works - And Their Alternatives.* Philosophy of the Social Sciences, 2000. 30(3): pp. 384-406.
9. Law, J., & M., Callon., *Engineering and Sociology in a Military Aircraft Project: A Network Analysis of Technological Change.* Social Problems. Special Issue: The Sociology of Science and Technology., 1988. Vol. 35, No. 3, June.
10. Reay Atkinson, S., *Engineering Design Adaptation Fitness in Complex Adaptive Systems*, in *CUED EDC,* 2012, Cambridge University Engineering Department: Cambridge, UK.
11. Jarratt, J.P., & P.J. Clarkson., *The surge-stagnate model for complex design.* Journal of Engineering Design, 2002. 13 (3): pp.189-196.
12. Reay Atkinson, S., *Returning Science to the Social.* The Shrivenham Papers, UK Defence Academy, 2010. Number 10, July(July).
13. Chen, Y.-H., & C-C. Yi., *Dynamical Properties of Product Life Cycles: implications to the design and operation of industrial processes.* Industrial and Engineering Chemistry Reserach, 2001. Vol. 40: p. pp. 2452-2459.
14. Reay Atkinson, S., *An Age of Evolutionary Step Change*, in *Think-Piece*2012, University of Sydney: CSRG.
15. Fairbanks Jr., C.H., *The Origins of the Dreadnought Revolution: A Historiographical Essay.* The International History Review, 1991. Vol. 13, No. 2 (May, 1991): pp. 246-272.
16. Marder, A.J., *The Anatomy of British Sea Power,* 1940, New York: Alfred A. Knopf.
17. Marchetti, C., *Fifty-Year Pulsation in Human Affairs, Analysis of Some Physical Indicators.* Futures, 1986. Vol. 17 (3): pp. 376-388.
18. Modis, T., *Fractal Aspects of Natural Growth.* Techno-logical Forecasting & Social Change, 1994. Vol. 47: pp. 63-73.
19. Maor, N., *Formation of Navy Part II - The Israeli Defence Force, 1960-1979.* The NAVY Magazine of the Navy League of Australia, 2017. Jan-Mar 2017 Vol. 79, No.1: pp. 22-26.
20. Reay Atkinson, S., *Design to Fight.* Naval Review, 2008. 2, May.

21. UK-MoD., *Defence Cost Study (DCS): Statement on the Defence Estimate 1995.* Presented to House of Commons, 1995. Cm 2800, May.
22. Kirkpatrick, D.L.I., *The Rising Unit Cost of Defence Equipment - The Reasons and the Results, Defence and Peace Economics,* 1995, Volume 6, pp 263-288, GmbH: Harwood Academic Publishers.
23. UK-MoD., *Performance Report:1995/96* in *CM 3448, Nov,* 1996, House of Commons: London.
24. UK-MoD., *Intergenerational Costs of Equipment*, in *MoD DASA-DESA Discussion Paper, 20100208*, Defence-Analytical-Services-Agency, Editor, 2010, MoD: London.
25. Joiner, K.F., and S., Reay Atkinson, *Australia's Future Submarine: shaping early adaptive designs through test and evaluation.* Australian Journal of Multi-Disciplinary Engineering (AJMDE), 2016. Vol 11, No. 2, 16 Nov: Presented at Submarine Institute of Australia 8th Annual Conference, 14-16 Nov.
26. Brewster, M., *Liberals alter course on frigate replacement and rely on foreign design*, in *CBC,* 2016, The Canadian Press: Toronto.
27. AS-Senate, *Senate Inquiry into Defence Procurement*, Australian Senate, Editor, 2012, Australian Parliament House: Canberra.
28. Bitzinger, R.A., *The Future Ain't What it used to be: Strategic Innovation in the Global Defense Industry*, in *Emerging Critical Technologies and Security in the Asia-Pacific, 2016*, R.A. Bitzinger, Editor 2016, Palgrave Macmillan: Hampshire UK. pp. 154-166.
29. Luttwak, E.N., *Breaking the Bank: Why Weapons Cost so Much.* The American Interest, 2007. Autumn.
30. Kaldor, M., *The weapon Succession Process.* World Politics, 1986. July, Vol. 38(4).
31. Kiras, J.D., *Irregular Warfare*, in *Understanding Modern Warfare*, Jordan et al, Editor, 2016, Cambridge, England: Cambridge University Press. pp. 301-378.
32. Reay Atkinson, S., & M Sharma, *Learning and Adapting to Modern Insurgencies*, in *The Comprehensive Approach to Modern Conflicts Conference,*, Marshall-Centre, Editor 2007, UKDA, ARAG: Munich 26-27th March.
33. Foreman, J., *The British Royal Navy - Road to Salvation, Part I.* The NAVY Magazine of the Navy League of Australia, 2017. Apr-Jun, Vol. No.2: pp. 12-17.
34. Blake, R.C., & N., Bellamy, *Before you strike hard and fast: first clutch the engines* (*the adventure of SSS Gears*). The NAVY Magazine of the Navy League of Australia, 2016. Vol. 78, No. 2.
35. Asher, M.W., *Land Based Test Center: A Tool for Design and Construction of FFG 7 Class Frigates.* Naval Engineers Journal, 1978. August: pp. 79-86.
36. Guido, A.J., & S.P., Light, *Ship Design for Maintainability: Experience on FFG 7 Class.* Naval Engineers Journal, 1978. April: pp. 75-84.
37. Stark, R.E., & D.M., Stembel, *Detail Design-FFG 7 Class.* Naval Engineers Journal, 1981. April: pp. 109-119.
38. AS-ANAO, *Test and evaluation of major defence equipment acquisitions*, in *Australian National Audit Office*, ANAO, Editor 2015, ANAO: Canberra.
39. Spellar, I., *Naval Warfare*, in *Understanding Modern Warfare*, Jordan et al, Editor, 2016, Cambridge University Press: Cambridge, England. pp. 159-226.
40. Till, G., *NATO: War Fighting, Naval Diplomacy and Multilateral Cooperation at Sea*, in *Twenty-first Century Sea Power: Cooperation and Conflict at Sea*, P. Dutton, R., Ross, & O., Tunsjo., Editor, 2016, Routledge: London.

41. Raska, M., *Diesel-Electric Submarine Modernization in Asia*, in *Emerging Critical Technologies and Security in the Asia-Pacific*, R.A. Bitzinger, Editor, 2016, Palgrave Macmillan: Hampshire, UK.
42. Ross, A.L., *The Potential Import of New, Emerging, and Over-the-Horizon Technologies*, in *Emerging Critical Technologies and Security in the Asia-Pacific, R.A. Bitzinger, Editor, 2016*, R.A. Bitzinger, Editor 2016, Palgrave Macmillan: Hampshire UK.
43. Kemburi, K.M., *From Subsonic to Hypersonic Cruise Missiles: Revolution or Evolution in Land Attack Capabilities*, in *Emerging Critical Technologies and Security in the Asia-Pacific*, R.A. Bitzinger, Editor, 2016, Palgrave Macmillan: Hampshire UK.
44. Heinl, C.H., *The Potential Military Impact of Emerging Technologies in the Asia-Pacific Region: A focus on cyber capabilities*, in *Emerging Critical Technologies and Security in the Asia-Pacific*, R.A. Bitzinger, Editor, 2016, Palgrave Macmillan: Hampshire, UK.
45. Joiner, K.F., *How Australia can catch up to U.S. cyber resilience by understanding that cyber survivability test and evaluation drives defense investment.* Information Security Journal: A Global Perspective, 2017: pp. 1-11.
46. Lefebvre, V.A., & V.D. Lefebvre., *Reflexive Control. The Soviet Concepts of Influencing an Adversary's Decision Making Process*, SAI-84-024-FSRC-E, Editor 1984, Science Applications Inc (now SAIC): Feb. McLean, VA.
47. Reay Atkinson, S., & A. Goodman., *Network Strategy and Decision Taking.* ARAG Occasional, UK Defence Academy, 2008. 11 / 08.
48. US-DoD, in *Annual Report to Congress on DoD Programs: F35 Joint Strike Fighter*, D.o.O.T.a. Evaluation, Editor. 2016, U.S. Department of Defense: Washington. Accessed at www.dote.osd.mil/annual-report/index.html ,16 January 2017.

Australian Innovation Transforming Modern Naval Shipbuilding

Tim Speer, Austal tims@austal.com

Austal has led the development of modern, modular shipbuilding processes and techniques and has achieved industry-leading efficiencies in naval vessel construction.

At Austal USA, in Mobile Alabama, the company's Module Manufacturing Facility (MMF) includes two production lines that achieve savings in materials handling, fabrication and assembly time in the order of 1:5:10 compared with traditional assembly halls and docks.

One productive hour spent manufacturing modules for the Littoral Combat Ship and Expeditionary Fast Transport in the MMF would take 5 hours in an assembly hall or 10 hours in a dock.

The coordinated handling, movement and consumption of materials in the MMF is saving time and money during construction, reducing program costs and allowing greater investment in research and development into innovative naval technologies.

The fully enclosed MMF allows the potential continuous construction of large vessel modules in a safe and clean environment with production teams, materials and equipment collocated to optimize productivity and inventory management.

The MMF increases Austal's capacity to assemble and outfit unit modules before consolidating them into the full vessel; spreading demands for specific trades more evenly and streamlining the integrated design and production process. Further, by automating component manufacture, including pipe runs from a 3D model, Austal is able to achieve consistent system configurations across vessels, which benefits operators and makes it easier to operate and support naval vessels in service.

Austal's MMF sets a new benchmark for modern shipbuilding efficiency - and an internationally competitive template for future operations around the globe, including Australia.

Human Impact Exposure on board High-Speed Boats
A Multi Agency Study

Carl Magnus Ullman M.Sc.
CEO Ulllman Dynamics - Carl.Ullman@UllmanDynamics.com

Johan Ullman M.D. Assoc.RINA
High Speed Boat Operations Professionals - Johan@HSBO.pro

PURPOSE

The overall purpose of the study is to protect people, working onboard high-speed boats, from injures caused by impacts, resulting from the boat hulls slamming against the water.

BACKGROUND

Human impact exposure on board high-speed boats cause pain, physical fatigue and severe injuries. These problems increase in numbers and severity around the world, along with an increasing number of high-speed boats used in professional operations.

New Knowledge is Needed to Solve the Problems

No scientific data exist regarding what levels of impacts exposure are dangerous, or which characteristics of the impacts cause acute and cumulative injuries.

Counter Productive Standards and Regulations Increase the Problems

Due to lack of relevant knowledge, irrelevant regulations, based on irrelevant standards, based on mean values of vibration, have been adopted and applied to impact exposure, without any support in evidence-based science.
There is a consensus in the scientific community that the currently applied vibration standards and directives have no relevance for quantifying the risks of injury caused by exposure to impact. [1 2 3]
More specifically, an EU-directive[4] is, in some countries applied to set limits for impact exposure. This directive is based on the ISO-standard 2631:5,[5] which sets limits to exposure to continuous vibration, but not to exposure to impacts.
The ISO-standard is it based on mean values of continuous vibration and contains algorithms quantifying the energy transferred to the human body as vibration.[6]
As there is no correlation between mean values of vibration exposure and risk of acute injury[1], this use of the EU directive can increase the risks of injuries caused by impacts.

Laboratory tests can not give any relevant information

Relevant impact exposure cannot be simulated in a lab. Real impact exposure is stochastic and multi-directional, reaching levels up to nearly 20g. Humans can for ethical reasons, not be exposed to relevant exposure levels where injuries may occur.
Lab trials can just test mechanisms, and only in one single axis, and only up to 10g.
Lab trials cannot simulate the human physiologic response with muscular reflexes.
They cannot assess the effect of the human posture, which is crucial for the risk of injury.
Lab trials cannot test the function of mechanisms exposed to lateral forces, which are more dangerous than the pure vertical forces.

Exposure to Repeated Whole Body Impacts has Physiological Effects

A study of human exposure at sea, during 3 hours, showed that, with twice the number of impacts and three times the impacts levels, the physical performance capacity, measured by shuttle run test was reduced by more than 30%.[7]

Vibration standards can not be used for measuring impact exposure

The same study also showed that there was no significant difference between the two teams in exposure to WHB, Whole-Body Vibration, as measured with any of the algorithms used in the standards and regulations in effect today.

A New Way of Quantifying Impact Exposure is Needed

In order to prevent the injuries and the pain, it is necessary to establish a relevant way of quantifying impact exposure.[2]
A new unit must be defined, which quantifies the dangerous forces acting on the human body during impact.[6]

Impacts up to18g can come unexpected, from crossing a wake. Time from take off to impact is still less than ½ a second.

METHOD

The method is developed to quantify actual human impact exposure, measured as whole-body acceleration, on a significant number of high-speed boats while being operated in regular operations in normal sea conditions.
To assess the risk of injury, occurrence and development of pain will be monitored on all personnel serving on board the boats during the trial period.

Whole Body Impact

Whole body impact will be monitored, on two people, on board each boat at all times. This is because the risk of injury - structural failure in anatomical structures - depends on the forces acting on these structures, and these forces depend on the amount of energy transferred to the human body by the impacts.

Pain

Pain is the only physiological parameter, which can be used as an indicator of risk of injury and can be quantified over time and frequently enough to monitor a sufficiently large population of test subjects.

Pain is a physiological function, provided to protect us from injury, by signalling what is potentially harmful. There can be temporary pain without injury, but very rarely a manifest injury without pain.
Pain is also, in itself, a highly relevant symptom, or sometimes a condition, which impairs physical performance and endurance - and even mental capacity.
It should be obvious that exposure causing pain, during demanding operations, must be avoided, or limited as much as ever possible.
This is relevant, not only for liability reasons or the long-term health aspect, but also for enhancing proficiency and effectiveness of the teams operating in threatening conditions.

Multi Agency Study Design
The purpose is to gather sufficient volumes of exposure data and pain data, to identify possible correlations between impact exposure and development of pain.
This can be done by using the same study design, protocols, hardware and software in several agencies and sharing the data.
Several agencies have expressed their interest in finding out how much impact their personnel are subject to and to find out what is relevant to measure.
By joining the efforts and sharing results it should be possible to find the answers to a number of relevant questions.
More than 15 agencies around the world have expressed interest in participating.
The target is to have at least 10 agencies participating with at least 4 boats each wired and monitored over a period of 4-5 months.
By sharing costs, data and results, great synergies can be achieved. The number of subjects and boats can be sufficient to achieve statistically significant results.
The data will anonymous regarding subject identities and stripped of any potentially sensitive information regarding coordinates and whereabouts of operations carried out.

Analysis of data
Data analysis will be done by done at the Royal Institute of Technology in Stockholm Department of Aeronautical and Vehicle Engineering.
Collected data will remain the property of the each participating agency.
Before sharing, data will be cleaned of any potentially sensitive information such as position coordinates, times and dates linked to the missions and identities of the subjects measured for impact exposure and reporting pain experiences.

Sharing of data
Collected data will remain the property of the each participating agency.
Before sharing, data will be cleaned of any potentially sensitive information such as position coordinates, times and dates linked to the missions and identities of the subjects measured for impact exposure and reporting pain experiences.

Dissemination of data
The purpose is to gather the data needed to understand what kind and levels of impact exposure is dangerous. Results are to be published in relevant scientific publications. Scientist from participating agencies/countries are invited to participate in the scientific work. Several different questions can be addressed

DATA COLLECTING EQUIPMENT

Hardware - measuring impacts and vibration

For this study a specific acceleration data logger, the Marec, has been developed. [8]
Marec™ (Marine Acceleration Recorder) continuously logs accelerations data (± 20g) from each hull, in 3 axes and vertical acceleration data from 2 crewmembers, all with a sampling frequency of at 600Hz.

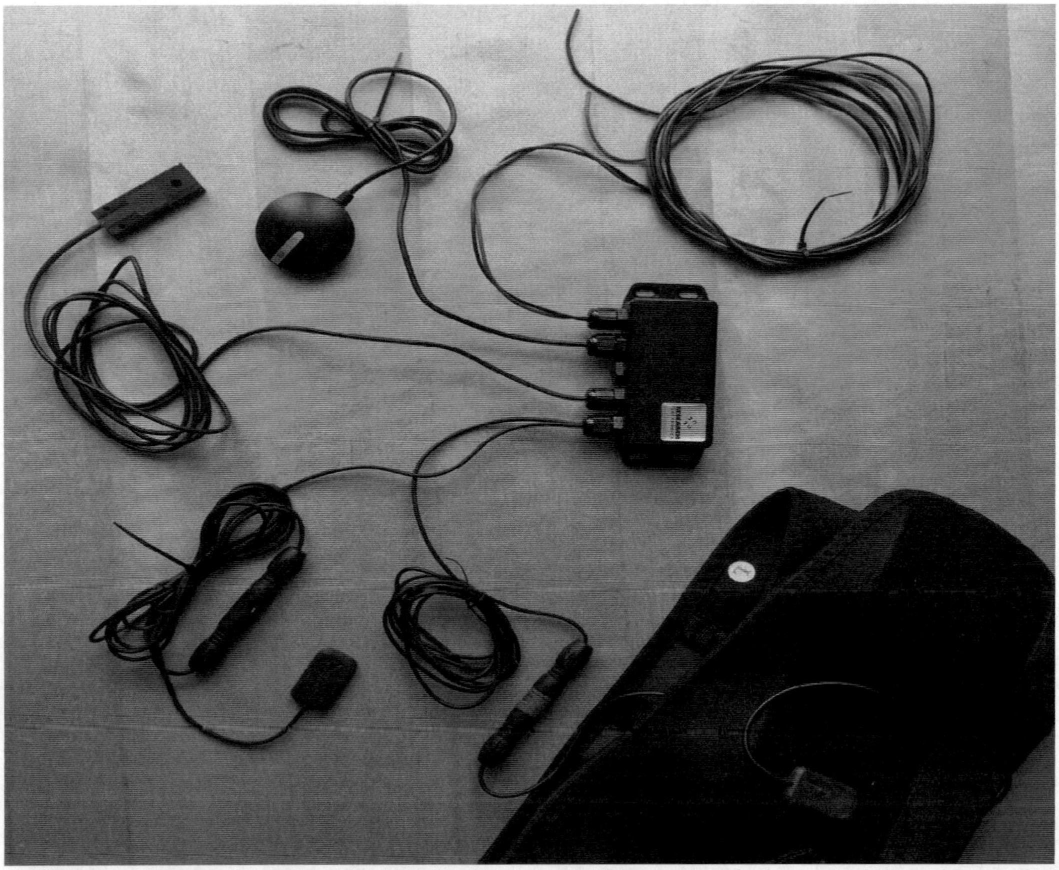

Clockwise from top right: 12 Volt supply cables: One kidney belt with body sensor (red connector): Second persons body sensor (black connector): The 3axis hull sensor: The GPS antenna and in the centre: The Marcec data logger.

Human exposure must be measured on the human body.
2 subjects on each boat will wear kidney-belt mounted accelerometers when operating.
The data recording starts as soon as the boat is moving more than 3kts for more than 60 sec. It gathers data in 30 min files and stops recording when the boat has stopped moving. This means that it can stay connected to power (12 V) for several months.
It's built in GPS tracks GPS time, position, speed and heading.
Data is stored on a 16Gb USB stick, which will suffice for the trial period of 4-5 months.
New transducers developed at Duke University[9] will also be used in some cases. These are wireless, in-ear devices, logging in all degrees freedom, even rotation yaw pitch and roll.

Data Analysis Software

The data analysis software is being developed, which will make it possible to view the collected data on any PC or Mac as view graphs. It will show the number of impacts, the

aptitudes and the characteristics. The software would also make it possible to calculate the energy content of the impacts as well as analysing rise times AKA jerk factor.
For all practical reasons the data can be stored inside the Marec till the end of the trials. Making the results comprehensible even for the test subjects could however hopefully increase the motivation and compliance in the studied populations.

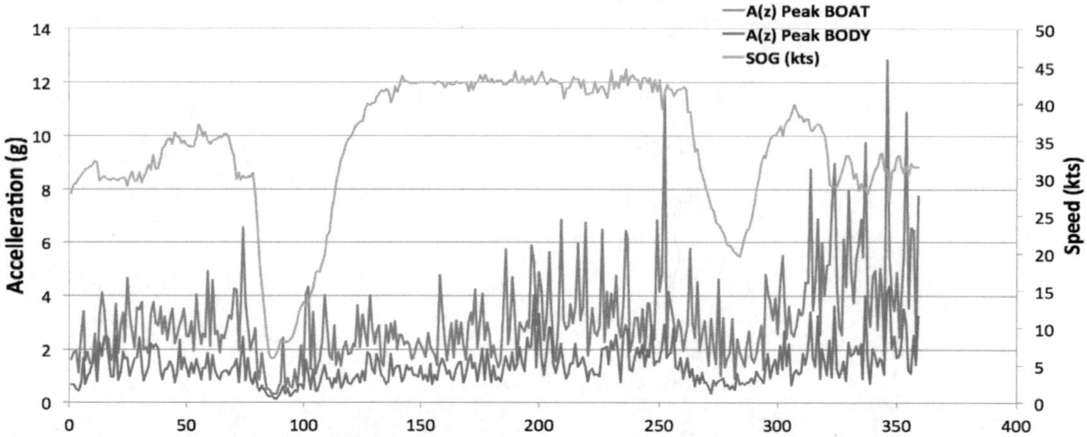

Example of graphical presentation of 400 seconds of data. Blue – impacts measured on the hull; Red impacts measured on the operator in a suspension seat. Green – Speed measured by GPS. It is interesting also to se how impacts on the hull has caused the operator to reduce the speed.

Pain data application

A smartphone app, PainDrawing [10], has been developed for the purpose of collecting information about pain experienced during the trial period.
The app will ping users on their smartphones at predetermined times and prompt them to register any pain, with pain modality, location and intensity. Data will be then be uploaded to a server by the users via Internet.
PainDrawing is based on two scientifically validated accepted methods, the "Nordic Minister Council's Pain drawing form" and a VAS - Visual Analogue Scale.
PainDrawing can be downloaded for free and it runs on OS and Android, even on tablets.

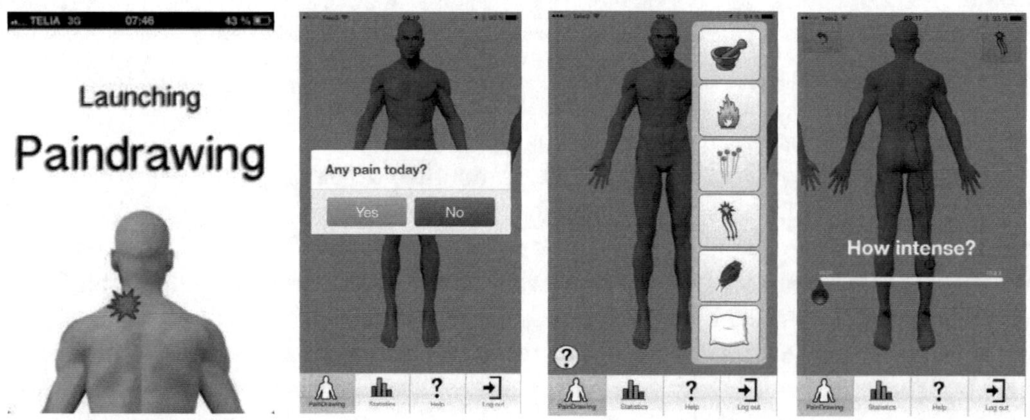

RESULTS and DISCUSSION

If a sufficient number of the participating agencies agree to share their exposure data and pain data, this study should yield sufficient information to establish a scientifically significant connection between human impact exposure and the development of pain. Experience and persistence of pain is used as and indicator of threatening injury.

The data from the study can then be used for various purposes.

1. Give each agency the information they need to assess if personnel is exposed to dangerous impacts
2. Lay the basis for a new relevant exposure standard
3. Provide information for calibrating dashboard mounted impact-indication instruments, showing pilots **Green**, **Yellow** or **Red,** for acceptable, risky or dangerous impacts.

The study will hopefully also help answer many the following questions.

1. What levels of exposure are dangerous?
2. What exposure values can be considered safe?
3. How much impact exposure do people voluntarily expose themselves to?
4. At what exposure levels do people start feeling discomfort - and pain?
5. Which characteristics of impacts are relevant for the risks of injury?
6. Which anatomical structures signal pain related to impact exposure?
7. What is the relevant level of hull impact, for each boat type, not to be exceeded?
8. Will it be sufficient to monitor only hull impacts in the future?
9. Which kinds of hull shapes have the best sea keeping and give the least impact exposure at high speeds in various sea conditions?
10. How does duration of missions affect the wellbeing of subjects?
11. Weighting relative importance of the various parameters defining a shock event, such as Amplitude/Peak acceleration, Rise time, Duration, Energy content, etc.
12. Creating self-teaching software, which will, based on boat type, calculate the exposure at the most exposed position onboard using data from data from one 3-axis sensor in the console.
13. Providing relevant data for calibration of a very low cost instrument, always keeping coxswains aware of when exposure starts getting dangerous.
14. It can also give objective information about differences in operator's driving skills.

SUMMARY

The purpose of the study is to protect people onboard high-speed boats from injuries caused by slamming-induced impacts by letting employers know what levels of impacts their personnel is being exposed to and what levels give risks of injury.
The results can also lay the basis for a new relevant exposure standard - and also for calibrating instruments showing pilots Green, Yellow or Red**,** for acceptable, risky or dangerous impacts. Agencies in more than a dozen countries are interested to join.

References

[1] WBV2017 Abstracts: https://gupea.ub.gu.se/bitstream/2077/52542/4/gupea_2077_52542_4.pdf
[2] Bovenzi, M. Low back pain and exposure to whole body vibration and mechanical shocks /@ WBV2017
[3] Griffin M.J. Predicting discomfort caused by whole-body vibration and mechanical shock /@ WBV2017
[4] Directive 2002/44/EC-Vibration - https://osha.europa.eu/en/legislation/directives/19
[5] ISO-standard 2631:5 - https://www.iso.org/standard/35595.html
[6] Ullman J. Vibration exposure standards are not relevant for impact exposure /@ WBV2017
[7] Dobbins, Meyers et.al, High speed craft motion analysis – Impact Count Index (ICI)
[8] - https://www.researchelectronics.se
[9] - https://scholars.duke.edu/person/dale.bass
[10] - https://www.paindrawing.com

Commercial Ship Management: Strong Risk Management is Good Business

David Parmeter
Teekay Shipping Australia Pty Ltd
david.parmeter@teekay.com

INTRODUCTION

In this paper, I hope to showcase why risk management is good business for all stakeholders:

- Crews aboard the vessels operate in a healthy and sustainable workplace, where work is done effectively and crews get home safely to their families at the end of their tours of duty.
- Customers trust the vessel operators to deliver, to do the right thing, they are outsourcing part of their risk management to the operators so they can 'sleep soundly at night'.
- Our society needs the operators who enable growth in global trade, economic development and delivery of goods around the world, while being good corporate citizens, who implement and support legislation that raises the industry bar that benefits all involved.
- Our environment – "one drop of oil spilt is one drop too many" is one of Teekay's mantras. We want to be a good environmental citizen, to be an industry standards bearer.

Serving our stakeholders is good business. In my presentation, I will draw on the industry as well as Teekay's experiences from building the risk management culture in the organization.

We are all aware that vessel operations involve inherent risks. Traditionally these risks were more related to personal injury, equipment and asset damage or loss, an inability to meet the expectations of stakeholder and customer requirements, financial losses and a loss of reputation.

Since the days of the Titanic the maritime industry has been on a steep learning and improvement journey with regards to safety and risk management. The Safety of Life at Sea (SOLAS) convention, introduced after the Titanic disaster and updated many times since, was one of the first attempts to regulate risk management in the maritime context. More recently, in the last twenty years, the International Safety Management Code has come into effect. Likewise, the technology and standards we use has significantly improved both in efficiency and safety effectiveness. These improvements and regulations create a baseline in the industry.

With growing complexity of the world we live in, influx of disruptive technologies, geopolitical shifts, complexity of shipping risks have been growing exponentially. Therefore, to be successful as a maritime business means not only that effective risk management must be an integral part of the planning and decision making for all commercial and operational decisions. It means the risk management must become one of the pillars of the ecosystem of an organization, evolving and maturing with the organization, its strategy and resource allocation. However, this does not just happen naturally and it doesn't happen quickly. There is variation in risk management focus and implementation across industry segments and between marine companies.

Effective risk management systems and improved culture require not only some improvements to close any existing gaps, but a commitment to this process of building culture as an ongoing organizational journey. As we all know some in the industry will only be driven by regulation, however Teekay believes that taking the lead to have best in industry systems and culture is a sustainable business advantage.

IT STARTS WITH LEADERSHIP

Teekay management identifies that strong and effective risk management is good for business. Positive outcomes include improving people's safety behavior, reducing lost operational time and injury rate frequencies, and better business continuity and productivity through avoidance of incidents, equipment and process failures. These results do have a significant benefit: reduced insurance premiums, higher staff retention, track record supporting tendering for new business, and lead to better all-round economic performance.

It is the people and culture that make all the difference in effective risk management. A company can provide all the risk management tools in the world, but if staff choose to bypass them or have a 'tick the box' mentality, the negative results will speak for themselves. The example must therefore be set at the top of the organization. Leaders must 'walk the talk' - indeed the lowest risk focus on ~~or~~ health and safety standards displayed by leaders and managers, will be the highest standards that employees will achieve! Leaders need to recognize this and realize that their actions will always be seen and have great impact.

It is for this reason in Teekay that our President and CEO along with the senior leadership team are fully committed to our Operational Leadership program. It is often referred to as Teekay`s 'special sauce', and senior managers recognize that it drives significant brand value. The challenge for any organization is how to sustain such risk management focus and ensure consistency of purpose, focus and execution every single day. Risk management focus needs to become a state of mind.

A good example of this comes with the increased focus on hours and work and rest onboard. How many companies still pay lip service to The Maritime Labor Convention and The International Convention on Standards of Training, Certification and Watchkeeping for Seafarers requirements? Does Senior Management demand compliance and provide support to their ships' captains to comply, effectively assess the risks and provide extra resources or stop the ship? Will senior management support a captain and shore team members to achieve compliance, therefore ensuring the crews are well-rested and alert, to be able to focus on working safely? Will they be prepared to defend any associated delays to the customers? And will they do it each and every time such hard decisions to manage fatigue and legislative risk (of a Port State Control detention) are needed? If they do not do this, it sends a message from senior management to their staff that some risk management procedures and tools are just a declaration, and can be ignored. Risk management systems and processes must be backed up by integrity and consistency.

SIGNIFICANCE OF TEAMWORK AND PERSONAL COMMITMENTS

A second critical factor to effective risk management is teamwork and communication, both onboard and ashore. The marine industry is often very traditional, with a 'captain is god' mindset. While the onboard organization structure and captain's authority is extremely important, a healthy teamwork environment with effective leadership will enable the culture where junior team members will feel confident to speak up when there is a need to challenge or stop unsafe behaviors. Leadership and individual team members need to work together to create an engaged team, lead by example, take responsibility, create an environment where anyone from the chief engineer to the cook feel able to speak up if something is wrong. This means leaders who admit their own mistakes and use them as a learning opportunity for the whole team. Naturally, such engaged and self-aware leaders are also more focused on supporting customer needs and delivering a high level of service.

Regular, consultative and consistent communication is the key to creating engaged teams. Teekay's systems and programs formalize the daily communication that needs to occur regarding safety and risk management. In essence, our Operational Leadership program is a communication tool. Our employees are trained and sign a commitment to personal accountability to safety, engagement, communication, when they first join the company. This expectation to the commitments and behaviors gets ongoing reinforcements during regular team meetings and in one-on-one discussions with senior officers when they rejoin a vessel and recommit to Teekay`s six safety commitments.

The focus on taking ownership and responsibility helps to drive increased trust and accountability. Teekay uses a just culture approach to ensure fair, transparent and consistent outcomes. The end result is to create a culture where proactive and generative risk management is simply 'how we do things around here'.

The human element also has a large impact on effective risk management. Individuals have a choice on how they will act. A strong culture will influence the choices they make. However human nature is such that people still make mistakes. A resilient organization recognizes this fact and embeds within systems and processes the barriers for when human error does occur.

Additionally, it is important to help staff recognize their natural behavioral tendencies. How do these impact their ability to lead or to communicate effectively with their teammates? In what circumstances would an individual be naturally reluctant to stop an unsafe act and are they self-aware to recognize own natural tendencies? Developing tools and programs which assist staff with this, and an environment where it is accepted and expected to talk about, helps drive a high-performance risk management culture.

Teekay invests in many programs to drive our Operational Leadership and risk management culture. In any company investments in such programs will often be challenged to demonstrate the "cost vs benefit". However, the cost savings or benefits from an incident that 'did not occur', can be difficult to quantify as they are intangible.

We have established, through measuring the long-term incident trends, that there is a clear correlation between the safety and risk focused programs that are in place, and annualized savings due to reduced incidents, less vessel downtime and reduced costs.

Strong systems and processes, effective, high trust leadership plus engaged employees, result in strong risk management outcomes and better economic results - both good for business.

INTRODUCTION TO TEEKAY

I would like to give some examples how we practically develop the risk management culture at Teekay, through our systems, processes and most importantly – people. But before I go into detail, a couple of introductions to Teekay, who the reader may not be very familiar with.

In the maritime energy business, we are a world leader. In our main markets, which are oil tankers, offshore services and gas carriers, we are the world's largest or are among the top three service providers. Our customers are the world's largest companies. We have financial oversight by the world's biggest investors.

And we play at the global scale. Our vessels service ports and countries in every part of the world. We can serve customers wherever they have requirements for transportation of energy and commodities; wherever they are physically located. And that includes Australia, from ~~form~~ our two locations in Sydney and Perth.

The Teekay culture – and its brand promise – began with its founder, Torben Karlshoej, whose initials form the TK logo. 44 years after the company's founding, and 25 years after Torben's untimely death, all of us take direction from the founder's mentality.

Torben took a keen interest in the lives and practices of his seafarers. It was not the time when Teekay had the scale or business sophistication of today, however our founder knew that right people and the right attitude and culture are the cornerstone to a successful business. He was known to surprise his ships all around the world by showing up at various ports, asking how the food was, if they needed various supplies. These were his people, and he ensured they experienced a culture of excellence and safety from him; so that they could act as the face of the brand for our global stakeholders.

The various Teekay Boards uphold that duty of care. Safety is the first topic discussed at every Board meeting, as it is at every senior leadership team meeting across the Teekay group.

Our safety statistics are ones we are immensely proud of – and that goes right through the organization, all the way to the partner teams building our vessels before our seafarers take command of them.

That culture runs throughout the entire global organization. It is one of our promises to our customers and other constituents. Of course, the culture does not exist outside of the people. It exists because of them. We have the right people – they are always the critical differentiator.

Those people have invested in their careers; and we similarly invest in them. Over eighty percent of our employees work at sea, away from the office and daily in-person interactions with senior management and wider organization. One of the ways of achieving and maintaining the engagement and alignment with our seafarers, is the annual seafarer conferences. They provide a forum for the management and seafarers to have a dialogue rather than a one-way conversation. The management provides an update on business strategies and critical issues, and seafarers are invited to share their feedback, views and concerns, both positive and constructive.

To deliver, we need the right processes, ensuring those core values and principles like safety come alive through the everyday work colleagues undertake. This is about systematic, thoughtful preparedness.

In the background, then, are the tools that supply risk management and audit accountability. They support the activities of our frontline workers.

Tools, processes, people and culture should all fit together like a jigsaw. That brand promise should be transparent, auditable, visible in the everyday interactions and activities of our teams. Anyone who has an interaction with Teekay should be able to acknowledge that culture of excellence at Teekay. Only then will we be certain to be successful – though never content. As the Torben quote says, "Once we have achieved this goal, we shall have to go for the next one – absolute excellence".

So rather than go through every part of the risk management matrix at Teekay, I thought I would share a day in the life of a seafarer – for it is in the actual work at the front line that the rubber hits the road. How do we prepare our people to do the right thing?

This might sound trite, but the first thing we get every employee do – on day one at Teekay – is to sign a pledge to be an operational leader.

As soon as a seafarer arrives onboard, they meet with the captain to discuss various onboarding protocols, to ensure the onboarder understands their various procedural accountabilities. In short, to set the stage for a safe, productive time alongside their team onboard.

In order to enable an environment where personal commitments to safe practices are easier to achieve, we ensure each individual fits into the team. The team dynamic is critical – we need everyone to be engaged, be supportive, be confident and good communicators. But that doesn't mean we need everyone to be the same (although there are some non-negotiables, like a keen focus on safety). We use a program called E-Colours. E-Colours is designed to help individuals learn about their behaviors to improve communication skills, build better relationships and work more efficiently.

Once the onboarding is complete, a new seafarer is ready as an individual. They understand their E-Colours, how they respond and others too. They are ready to perform as a team.

Next up, the daily briefing. This is how the team prepares for the operational risks of the day ahead – it's about clarifying the plans, the roles and the expectations, within a specific work environment. What needs to be checked? Who needs to support whom? How will the work be done in the context of local requirements; in the context of regulations – for work-rest hours discussed above in the paper, for example?

Next up, toolbox talk. We are closer to the action now. Before any work commences, those participating ensure they are following the correct procedures; there is a safety moment. The participants look each other in the eye – they are ready to move ahead, risk managing as they go.

We have thousands of procedures and processes, I won't go into any details. But as an example, we recently underwent a complete review of our navigation program, simplifying the protocols and checklists and updating the information to be as relevant and clear as possible.

Of course, there is a checklist for everything! We have hundreds of checklists to ensure safe, reliable, risk-managed execution of the some extremely dangerous operations across our many fleets.

We discussed culture at the beginning – that culture of safety underscores everything we do. We can train and prepare people as much as we like, but we need there to be an ambiance of togetherness to ensure we hold each other accountable to deliver, and we care for wellbeing and safety of everyone around us.

Culture is as important as operational preparedness. And it means we need to celebrate the wins, together.

Often, it is as simple as enjoying a good meal together, and making sure others around the world see the food and the team togetherness. We share our stories of success with other offices and other ships. We win together.

Our Teekay core values are:

- Safety & Sustainability
- Passion
- Integrity
- Reliability
- Innovation
- Teamwork

Together, they for the acronym SPIRIT.

Many of the vessels are named with the suffix SPIRIT too. We live and breathe the values.

As previously mentioned, the outcome for the hard work and togetherness are long LTI-free milestones. The Goonyella Trader, one of the vessel managed from Australia, completed its entire operating life of 14 years LTI free. Imagine all the millions of man-hours of the crew working safely, managing the risk for themselves, for the company, for our customer, for the environment in which they operate.

Let's switch gears a little here. Our frontline seafarers are doing the hard work; we have the systems in place to maintain a high standard of performance; and the rich Teekay culture has prepared everyone to perform. This is why the customers can see that Teekay will deliver what it says it will do. We have trust, together. The good business outcomes, the financial rewards, follow. We are proud to have partnerships and contracts with some of the world's largest and most particular companies.

One of the challenges and new risks in the industry is commoditization of the marketplace, which often drives pricing wars. Safe and reliable operation can and must be cost-effective, but it cannot be cheap. Our culture will allow us to win the contract because we stand out in the marketplace. Culture is an asset and a barrier to entry, which cannot be quickly and easily replicated.

Legislation creates global level playing field, and the industry standards have been improving over the years. We choose to be the leaders in driving the high industry standard, as integrated risk management is a fair business practice, that benefits all stakeholders. We are gratified to see that it continues to be our differentiator.

SPIRIT ties it all up – we believe in the values. We link our strategy, culture, internal and external interactions to them in many ways. We deliver against them. And our stakeholders see it; they believe it too; we create good business outcomes because of them.

Smart Escape Support System for Passenger Ship : Active Dynamic Signage & Real-time Escape Routing

James Choi, Ph.D., CORNERS Inc., jameschoi@corners.co.kr
Ji-Han Kim, CORNERS Inc., k201a@corners.co.kr
MinJee Son, CORNERS Inc., minjee@corners.co.kr
Chan-Su Yang, Ph.D., Korea Institute of Ocean Science & Tech., yangcs@kiost.ac

INTRODUCTION

It is critical that passengers should be given timely and correct escape or evacuation guidance from captain and crews when there are hazardous situations in a ship. Otherwise the consequences could be disastrous as "SEWOL Ferry" the South Korean passenger ship which sank in southern coastal area on 16th April 2014. Due to the captain's delayed evacuation decision and lack of sufficient number of crews to guide passengers' evacuation, the accident recorded many casualties, most of whom were high school students (302 passengers sank down with the ship while 172 rescued).

Fig. 1 SEWOL FERRY - Early warning and right guidance could have saved many

Building a passenger ship with well-designed physical escape routes is one thing and guiding passengers to those escape routes in real disaster situation is another. Passengers get panic and move to a wrong direction, bottleneck makes situation worse, and even crews get panic also – passive static escape route signage and small number of trained crews might not be enough to take care of them.

SESS (Smart Escape Support System) is being developed sponsored by South Korea Ministry of Ocean and Fisheries starting from 2016 with 4 years of roadmap. SESS comprises multiple active dynamic signage devices which communicate with real-time escape routing server software via LoRa (Long Range) proprietary wireless network.

SESS (SMART ESCAPE SUPPORT SYSTEM) USE CASE SCENARIOS

Fig. 2 shows how SESS works when escape or evacuation situation happens in a passenger ship such as abnormal heeling, fire breakout, and water leakage. SESS can be regarded as a kind of closed loop control system which comprises continuous feedback cycle of "Real-time Sensing", "Precise Decision Making", and "Timely Actions".

Real-time Sensing

There can be 2 ways for gathering crisis event information from sensors – legacy sensor systems integration and new types of sensor network deployment. SESS enables easy-to-install type of new IoT (Internet-of-Things) sensor network deployment by using LPWA LoRa wireless communication technology.

Precise Decision Making

SESS automatically collects and analyzes sensor data streams to identify abnormality within a ship. When SESS detects an event, relevant data and information are visualized for captain's effective decision making whether escape of evacuation should be initiated or not.

Timely Actions

SESS reveals its real value when escape or evacuation procedure initiated by captain. As soon as captain presses "Execute Escape!" button, SERA (Safe Escape Routing Algorithm) runs and generates a safe and fast escape route from every passenger location in a ship to designated emergency assembly stations just within a few seconds. Passengers can easily find their way by simply following the "Active Dynamic Signage – Voice & Arrow".

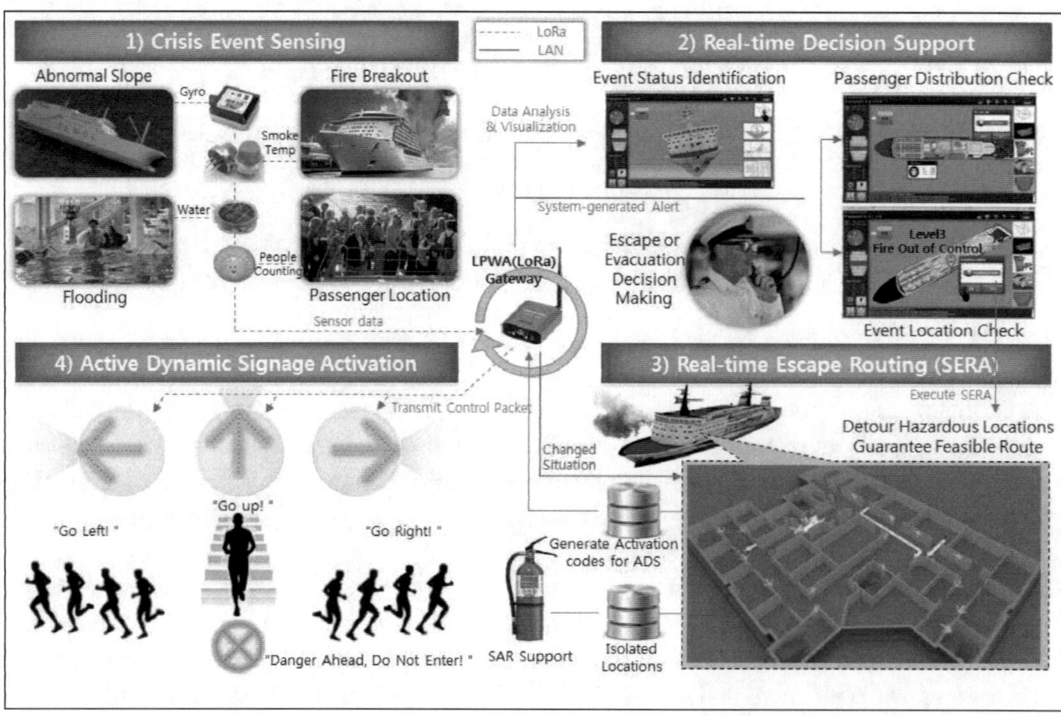

Fig. 2 SESS Overview – Key Components and Closed-loop Control Process (SENSE-DECIDE-ACT)

TECHNICAL BUILDING BLOCKS

SESS's closed-loop control process involves multiple technical components which require hardware, software and communication resources. LPWA LoRa wireless technology has been tested and proved to be suitable for ship area local wireless network. A new type of active dynamic signage device plays critical roles for sensing events and guiding passengers. SESS uses 3D data model of a ship to visualize event status and to interact with captain and crews. SERA utilizes mathematical network problem solving algorithms to find out feasible routing solutions within a few seconds. SESS also provides visual computer-based escape simulation capabilities to analyze and compare the escape performance under various crisis situations within a passenger ship.

LPWA LoRa Wireless Communication Network

LoRa is a kind of LPWA wireless communication technology competing with Sigfox, NB-IoT, etc. Fig. 3 shows the characteristics of LPWA technology in view of power consumption, data rate, and range.

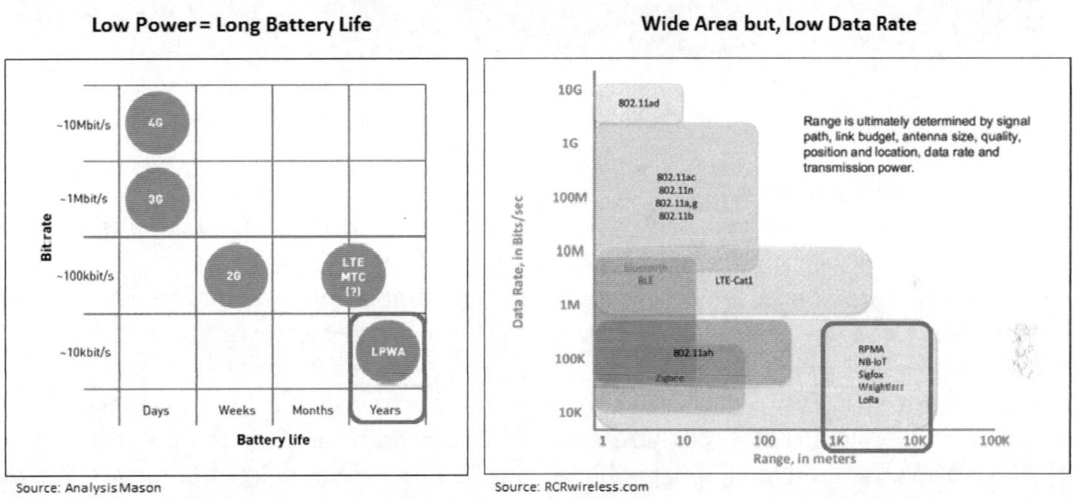

Fig. 3 Characteristics of LPWA compared with other wireless communication technologies

LoRa is a patented chirp spread spectrum modulation technology from Semtech Inc. who provides LoRa chipset to the market, while LoRa Alliance - the global open and non-profit association - publishes protocol specification documents.

LoRa uses unlicensed ISM (Industrial, Scientific, and Medical) frequency band with speed of 3~5kbps. The transmit power ranges from 10mW (10dBm) to 200mW (23dBm) in South Korea. For security concerns, LoRa provides AES-128 encryption. The latest LoRaWAN^{TM} technical specification document is V1.0.2 which published in July 2016.

Fig. 4 shows the actual on-board test for LoRa range performance in 2016 - South Korean passenger ship "Sea Star Cruise" who sails between Mokpo and Cheju island. LoRa test devices were used for checking PER (packet error ratio) among transceiver and receiver, which resulted in good coverage performance even within spaces surrounded by steels.

Fig. 4 "Sea Star Cruise" ship specifications and LoRa performance test results

Active Dynamic Signage

Legacy exit signage and low location lighting is passive and static. They are always turned on and just waiting for passengers to watch them during escape or evacuation – so they are passive. Meanwhile, they are always showing same symbols and same color of light, they cannot change the direction of sign and color of light – so they are static.

Imagining the critical role of crews in real escape or evacuation circumstances will be pictured by "as many crews as possible are dispatched to all over the passenger deck", "the crews are holding megaphone and radio communication device", "they communicate with captain and each other to clarify nearby hazard and find safe ways to assembly station", "guide passengers to follow certain lines and passages". The crews tell and passengers pay attention to them, the crews lead passengers and continuously find and change route to guarantee safe arrival to the assembly station – the crews are active and dynamic.

ADS (Active Dynamic Signage) mimics the crews' active and dynamic guidance activities. ADS has embedded loud speaker system to tell passenger to do what using pre-recorded voice announcement and guidance messages. ADS can show right (currently safe and fast) escape direction (up, down, left, right) with its front panel LED arrow display and edge-side beam light. ADS can also give warning message (do not enter) and visual sign (red "X" front led display) for passengers not to risk any nearby hazard. ADS has embedded temperature sensor and smoke sensor to identify whether current location is safe or not for passengers to get through. Normally ADS get electric power from AC power line and embedded rechargeable 2,600mA battery guarantees more than 1 hour operation under black-out situation. Fig. 5 illustrates LoRa enabled ADS features and design.

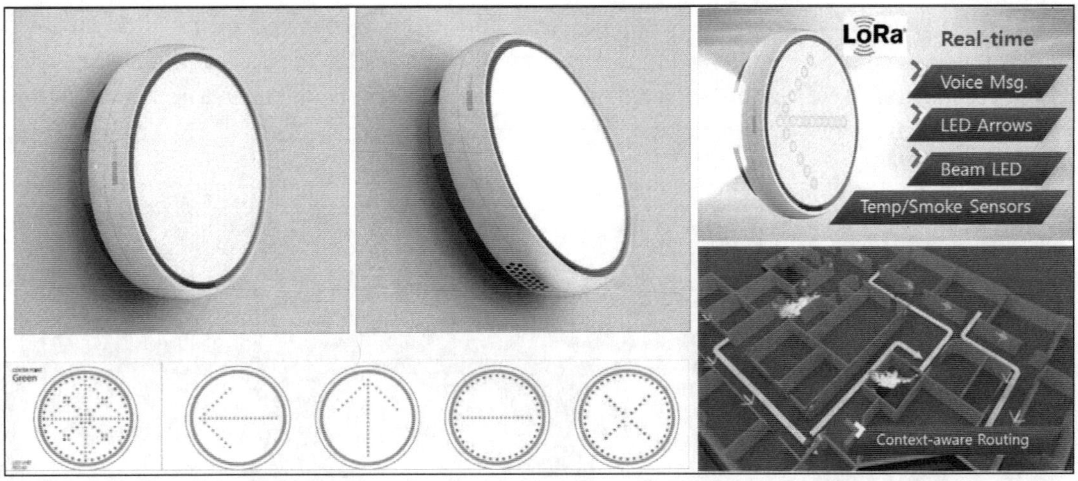

Fig. 5 ADS (Active Dynamic Signage) - mimics active and dynamic role of crews

Escape Route Modelling Tool

In order for SESS to be intelligent enough, SESS should aware the whole escape route structure of a passenger ship as well as the individual location of installed ADS devices. The first step of applying SESS is to prepare a 3D model of a ship and construct a database model of nodes and arcs which represent locations of ADS and escape route respectively. The attributes of escape route include length, width, capacity, and etc. which will be used as input data for SERA (Safe Escape Routing Algorithm). Fig. 6 shows example screen shots from escape route modelling tool.

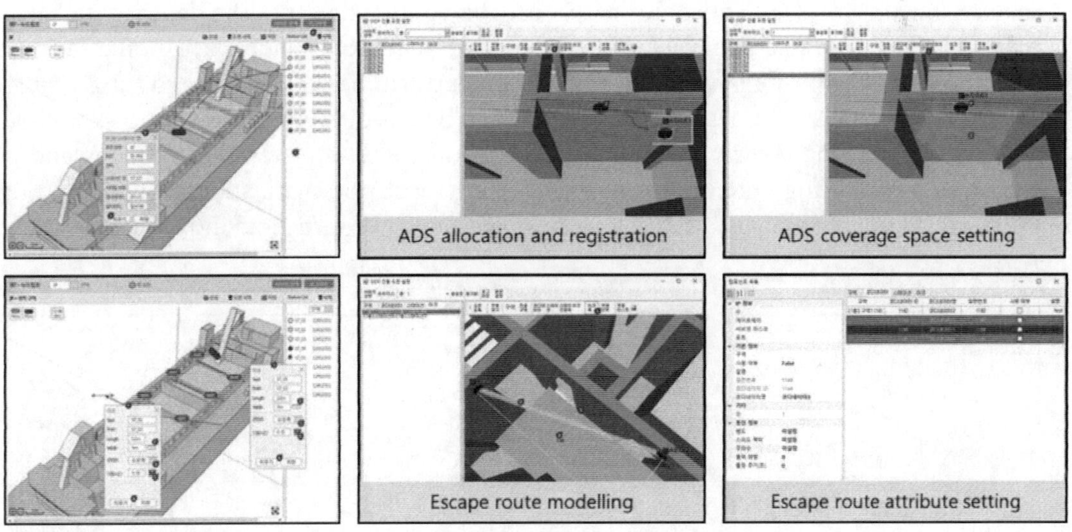

Fig. 6 3D ship model based ADS allocation and escape route modelling

Monitoring and Application Server System

ADS devices normally interact with monitoring and application server software via LoRa gateway which has LAN connection to server system. The server side software system runs MQTT message broker to communicate with sensors, ADS, and legacy fire & gas systems if necessary. The most critical function of server side system is to analyze sensor data streams and to identify abnormal situation event which will be notified to captain and crews. The server side system also runs embedded SERA module and encodes control packets for each ADS. The server system interacts with visual user interface system as well. Fig. 7 explains major functions of server side software system.

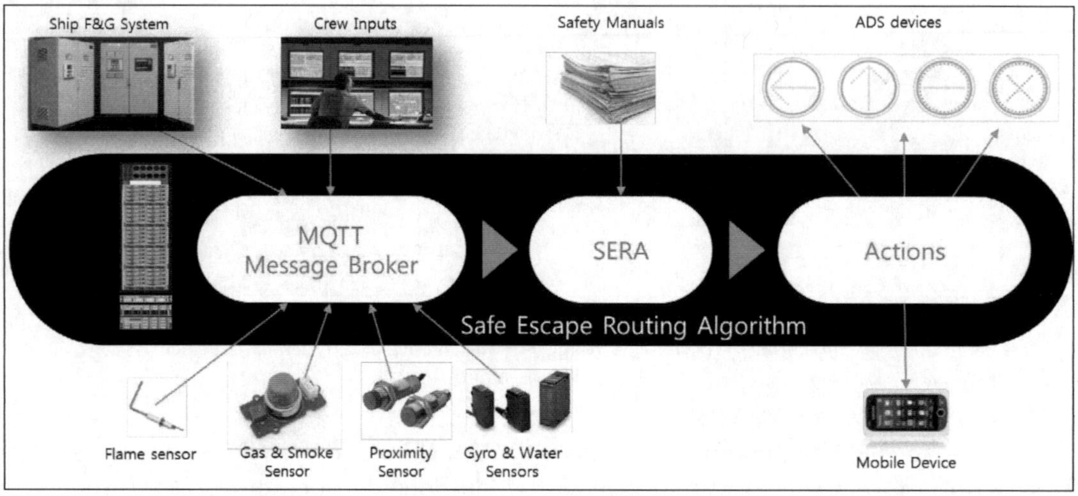

Fig. 7 SESS' server side MQTT message broker and SERA to activate ADS

SERA (Safe Escape Routing Algorithm)

SERA can be described as Time-expanded, Min-Cost, Max-Flow type of heuristic network problem solving algorithm, which maximizes the number of escaping passengers while minimizing the total escape time in the escape route network. The input data include total number of passengers, location of assembly station (final exit), nodes and arcs attributes, and hazardous nodes identified currently. In a given escape routing problem, the travel time of each arc is defined as cost, and flow is defined as the number of moving passengers by using the arc for their escape. Fig. 8 shows an example of nodes and arcs based escape routing network problem. The left most node is a source node, two right most nodes are assembly stations, and the other nodes represents individual location of ADS.

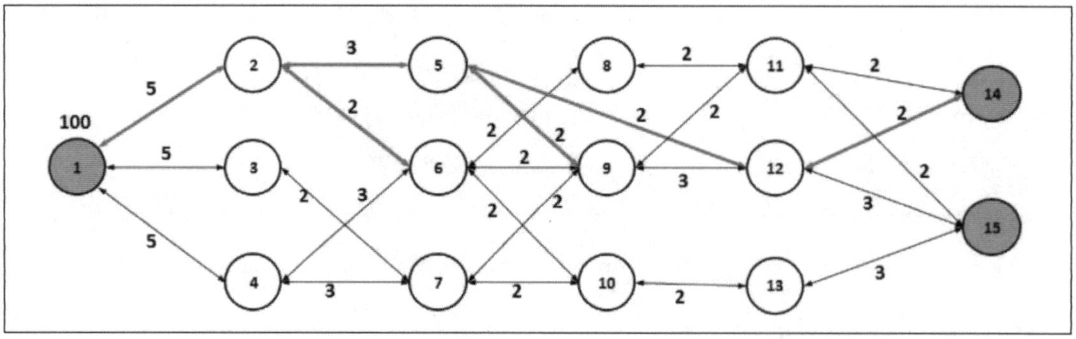

Fig. 8 Sample of nodes and arcs based escape routing network problem

SERA aims to maximize the flow while minimizing the cost in order to move as many people as possible to assembly stations through the available fastest arc if possible. Let's say there are 100 people to be escaped at Node 1, and Nodes 14 and 15 are assembly stations, and the legacy exit signage leads passengers via route 1-2-5-12-14. Since the capacity of the 2-5 path is three people, two out of five people arriving at the Node 2 are bottlenecked, so that only three people can move to the 2-5 path, and two out of the three people arriving at the Node 5 can also move to the 5-12 path without being stuck by a bottleneck situation, and only the two people arriving at the Node 12 will be able to head to assembly station by moving to the 12-14 path without bottleneck. In other words, it can be seen that the number of passengers escaped per unit time and total escape time can't be optimized if only the shortest route to assembly station is guided. Therefore, it can be seen that the guidance for the 2-6 and the 5-9 path is required at the same time, suggesting that the Max flow method that maximizes the available arc capacity at the unit time of the disaster situation should be performed concurrently by crews or ADS.

SERA can also generate bottleneck minimizing escape route by time-expanded modelling technique. Even though there are number of high performance network problem solving engines such as CPLEX and Gurobi, their calculation time (around 10 seconds for 98 nodes and 252 arcs) for finding optimal solution cannot be tolerated during the real disaster situation which requires under second level calculation speed. (SERA: below 0.5 seconds)

Visual Interface System for Captain and Crews

SESS provides 3D ship model based visual user interface software which is able to display relevant alert and escape initiation command to captain and crews. It shows current status of ADS sensor values and device condition as well as user friendly screen navigation functionalities. During the escape stage, it can display current directions each ADS guides and manual status change for specific locations from safe to danger or vice versa is also possible. This user interface software can be used for escape training mode or escape simulation mode also. Fig. 9 highlights key features of it.

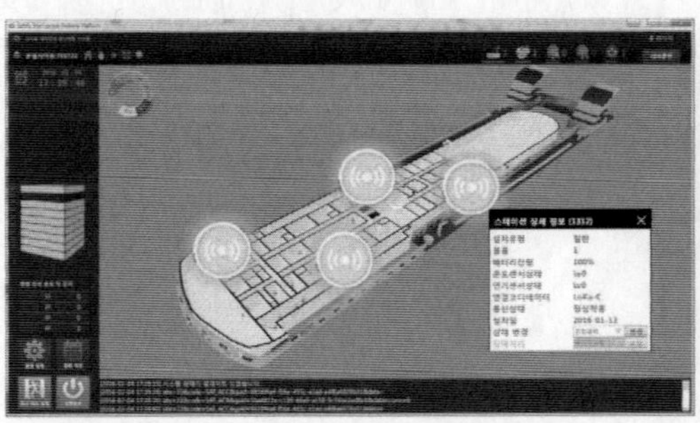

Floor level ADS status monitoring

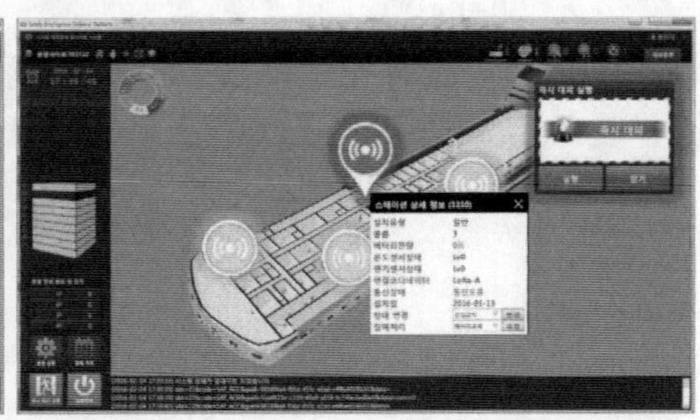

Sensor event alert and escape initiation command window

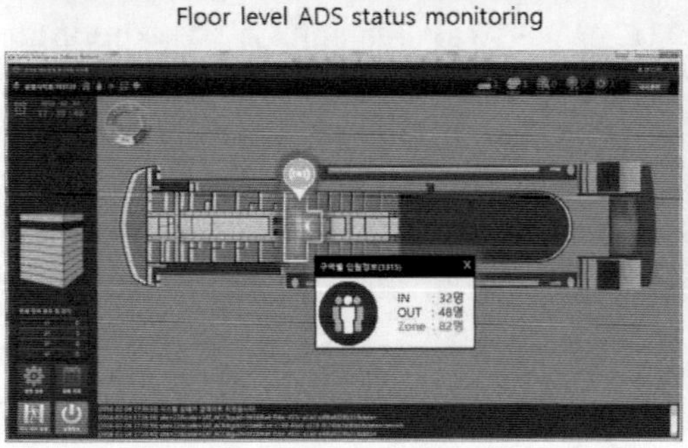

Passenger counting data for major customer deck

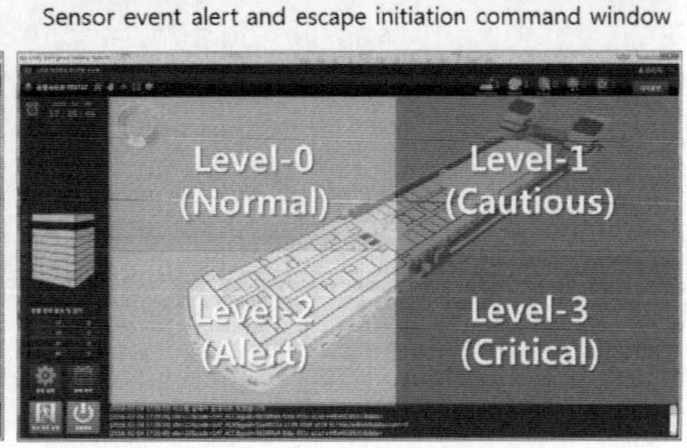

Visualization of escape level event alert

Fig. 9 Key features of visual user interface software

Escape Performance Simulator

There are many kinds of computer-based simulation software for passenger ship escape analysis, which meet MSC. 1/Circ. 1533 requirements for guidelines on evacuation analysis for new and existing passenger ships. But, these simulators do not provide features for SESS type of escape scenarios.

Escape performance by applying SESS can be highlighted for the scenarios such as dynamic situation changes which require escape route change and cause bottleneck. Simulation software for SESS should provide functions which enables applying various dynamic disaster event scenarios even in the middle of a simulation process. Virtual passengers' path finding algorithm should also be adapted to SERA results. Fig. 10 shows some screenshots from SESS simulator.

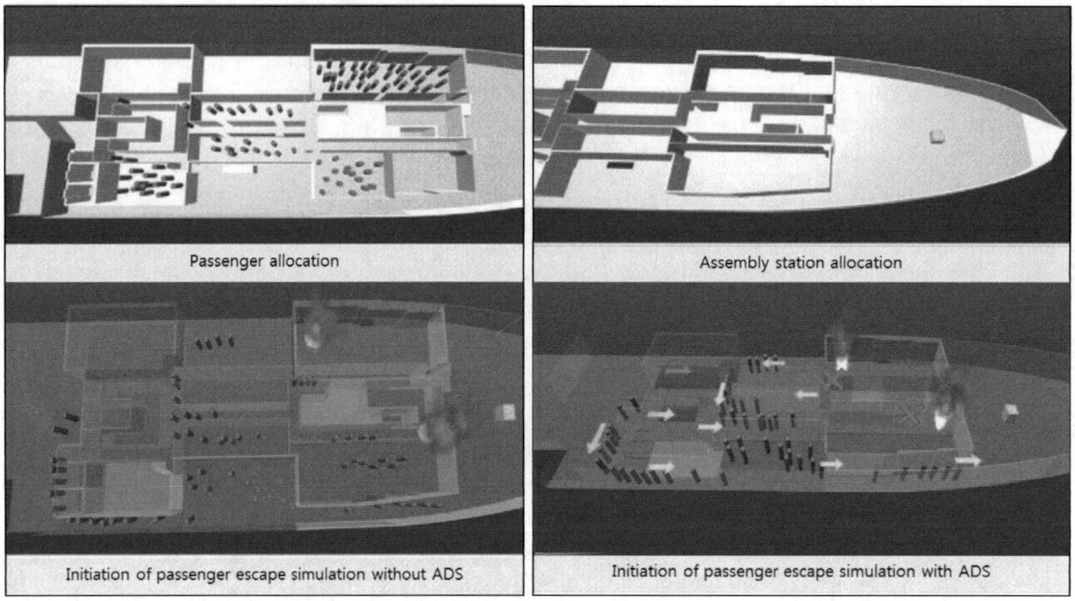

Fig. 10 3D model based passenger ship escape simulator

PILOT ON BOARD EXPERIMENT AND RESULTS

SESS experiment was conducted at "SAEYUDAL" – training ship of Mokpo Maritime University on 4th Oct. 2016 to verify the value of SESS under dynamic emergency escape scenarios. Total 100 participants were divided into two 50 participants groups - comparison group and experiment group (Each group has 42 men, 8 women, 110 days on-board experience). For every escape scenario, individual participant measured his or her own escape completion time with a stopwatch.

Experiment settings are as follows. (Fig. 11)

- 3~8 participants distributed in 12 cabins, which area is surrounded by 4 exit doors
- Normal walking speed when escape from cabins without talking to each other
- Initial escape notification is given via ship area announcement system

- Start stopwatch when each participant starts to move
- Experiment assistants stand outside of exit door blocked participants as scenarios

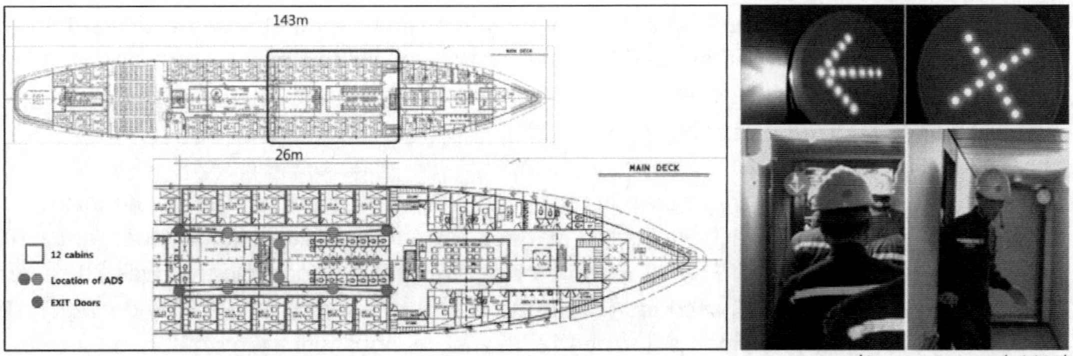

(Source: Ryu et al. 2017)

Fig. 11 Experiment environment setting within “SAEYUDAL” ship

The experiment was conducted for three scenarios which have different settings with number of available (safe) exit doors.

- Scenario 1: All of 4 exit doors are available. Fire detected far away from the experiment area.
- Scenario 2: Only 1 exit door is available. Fire detected across the other 3 exit doors.
- Scenario 3: Number of available exit doors are changed from 2 to 1 after 40 seconds after escape begins. Fire detected far away from the experimental area.

Measures for analyzing escape performance are

- Measure 1: Individual time consumed till escape through an exit door
- Measure 2: Total time consumed till escape completion
- Measure 3: Number of escaped participants for each exit door

From experiment with scenario 1, total time consumed for escape was 64 seconds with comparison group and 40 seconds with experimental group, which shows 38% decrease. (Fig. 12)

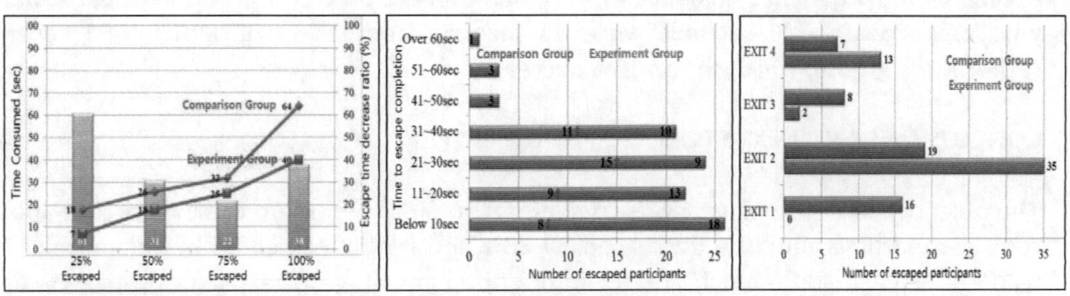

(Source: Ryu et al. 2017)

Fig. 12 Experiment result from Scenario 1

Time duration gap to the 25% participants escape was 7 seconds between comparison group and experiment group, which corresponds to 61% decrease. This result implies the SESS can impact greatly for the early stage of escape. This implication can also be found from the number of escaped participants within 20 seconds was 17 (comparison group) vs. 31 (experiment group).

The results also shows the SESS's bottleneck minimizing capability. 35 participants of the comparison group escaped via EXIT 2 because they preferred to use familiar exit or pre-educated exit before the experiment. Meanwhile, participants of the experiment group used all of 4 EXITS quite evenly.

Scenario 2 emphasizes the capability of SESS which can automatically generate safe escape route which reflects detected danger from the beginning of emergency. This impact can be found from Fig. 13 – completion time to 25% participants was 30 seconds (comparison group) vs. 20 seconds (experiment group), which means 33% decrease. The number of escaped participants during the first 20 seconds were 4 (comparison group) vs. 11 (experiment group), which shows three times more. Total time consumed for escape completion was 89 seconds (comparison group) vs. 69 seconds (experiment group), which converts to 22% decrease impact. The impact will be greater if the environment is bigger and more complex.

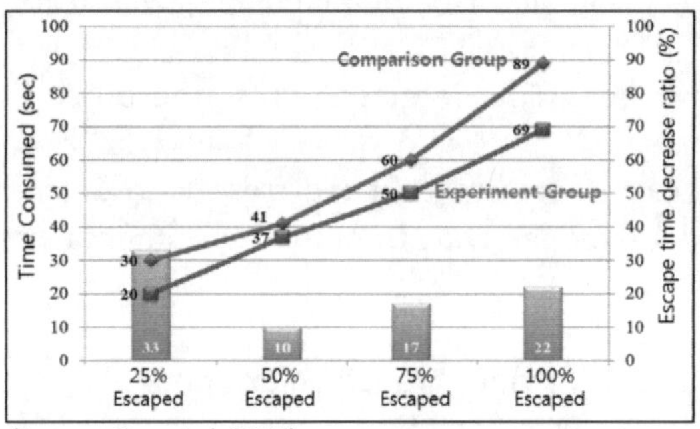

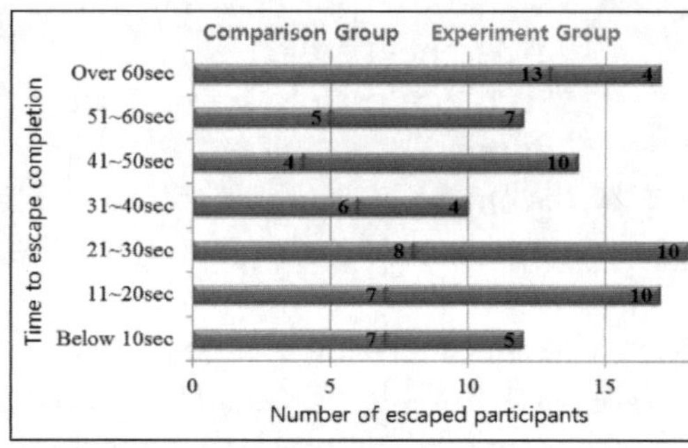

(Source: Ryu et al. 2017)

Fig. 13 Experiment result from Scenario 2 (left) and Scenario 3 (right)

Scenario 3 blocked one of available two exit doors after 40 seconds of escape initiation. The participants of comparison group consumed additional time to find out the only available exit door, meanwhile the participants of experiment group escaped safely and rapidly with the dynamic guidance of ADS. The average time consumption of participants who escaped after 41 seconds were 64 seconds (comparison group) vs. 54 seconds (experiment group), which shows 16% decrease.

CONCLUSIONS AND NEXT STEPS

The objective of developing SESS is to minimize loss of lives. From the experiment above, total escape time could be decreased by 38% and bottleneck could be minimized. The value of SESS could be much greater during the early stage of escape - up to 61% time decrease when crews and passengers prone to get panic. The SESS will be more capable by incorporating multi-type sensor network integration and deployment such as gyro, water, people counting, and etc. The user interface system could be integrated with general ship crisis decision support system. ADS might be required to meet MSC. 1/Circ. 1167-1168 in order to get certification as an alternative escape guidance system to LLL.

REFERENCES

1. IMO(2016), Guidelines for evacuation analysis for new and existing passenger ships, MSC.1/Circ.1533
2. IMO(2005), Functional requirements and performance standards for the assessment of evacuation guidance systems, MSC.1/Circ.1167
3. IMO(2005), Interim guidelines for the testing, approval and maintenance of evacuation guidance systems used as an alternative to low-location lighting systems, MSC.1/Circ.1168
4. SOLAS 2016 amend/ Chapter II-2 /Reg. 13.3.2, Means of escape in passenger ships
5. IMO SSE 4/INF.2 5th Jan. 2017, Information on the technological development of the "Escape Guiding System" for passenger ships.
6. Korea Maritime Safety Tribunal in Ministry of Oceans and Fisheries, Examples and lessons of major maritime accidents and incidents in (2015).
7. Eun-Gyeong Ryu, Chan-Su Yang, James Choi, Fundamental Research for Escape Guidance System Development of Passenger Ship, J. Navig. Port Res. Vol. 41, No. 2 : 39-46, April 2017

CONFERENCE ABSTRACT

The International Maritime Organization Guidelines on Fatigue Mitigation and Management: A Status Quo

Michelle R Grech
Australian Maritime Safety Authority
michelle.grech@amsa.gov.au

Shipping is a 24/7 industry and for seafarers this means long and irregular working hours, sometimes carried out in highly demanding conditions. Several working arrangements are employed on ships, some less ideal than others, with fatigue becoming a hazardous part of the job. Although official figures point to around 20% of accidents as fatigue as a contributory factor, the unofficial percentage may be higher. Despite efforts directed at mitigating the risk of fatigue at sea through the hours of work and rest regulations (Standards of Training Certification and Watchkeeping for Seafarers and the Maritime Labour Convention) this issue still remains a concern. Lack of fatigue management has been identified as a contributory cause in a number of recent accidents.

Australia, considering the view that a global fatigue risk management approach in shipping should be the way forward, commenced the process of submitting a proposed framework for consideration by the International Maritime Organization. Australia's proposal aligned with the IMO's resolutions on the human element strategic plan (A.944(23) and the human element vision, principles and goals (A.947(23)). In 2015 the IMO agreed to include the revision of the *Guidelines on Fatigue Mitigation and Management* as a planned output with a completion date of 2017. Australia has since been leading this work at the IMO by inputting a base document and leading a correspondence group intersessionally. This paper will provide the background, Australia's input, current status quo and way forward on the work being undertaken at the IMO on the revision of the *Guidelines on Fatigue Mitigation and Management*.

Offshore Supply Vessels: A Comparison of Propulsion System Architectures

Cameron Sobie, Christopher Accardi, Eric Richa, and Mike Nicolai
Siemens Industry Software
cameron.sobie@siemens.com

INTRODUCTION

Offshore supply vessels (OSV) are workhorses of offshore industry, carrying equipment, supplies, and personnel to offshore structures as well as performing other duties such as fire-fighting. Their loading profiles are highly characteristic; with a combination between long transits at cruising speed, short journeys between offshore installations, and widely varying electrical loads that can match or exceed the drive power demands. The load profile of each OSV provides a unique opportunity to adapt the electrical and drive power generation to optimize efficiency for its specific anticipated usage. System level simulation is particularly relevant for OSVs where propulsion and electrical loads can vary rapidly, and the interaction between these loads and the propulsion system have a strong effect on the overall fuel efficiency of the vessel. Furthermore, it is advantageous to optimize a vessel not only for a cruise speed and average/maximum electrical load, but for an entire load profile over time to capture the time dependence of certain components such as batteries and dynamic engine responses. A holistic view of the fully connected system is the most natural approach to ship efficiency optimization, where internal combustion engines, electrical machines and loads, vessel resistance, propeller properties and the anticipated load profile all interact to govern the fuel consumption and pollutant emissions.

Concept-level trade-off studies have been well-established in naval architecture for some time, such with the work of Jung [1] who studied possible gearing configurations for steam turbines. The possible topologies for modern ship propulsion systems are multitudinous; low, medium, and high speed reciprocating engines, gas turbines, mechanical couplings/clutches, gearboxes, electric motors/generators, and batteries can all conceivably appear in ship propulsion machinery. Generating and optimizing the feasible combinations of powertrain components is a challenge that has yet to be addressed in an automatic way, and human expertise is the current benchmark for design. At a smaller scale, concept generation has yielded novel and promising ship power solutions such as methanol fuel cells for electrical and thermal energy [2] and combined electric/wind powertrains [3]. In the case of OSVs, diesel engines and electric machines are fundamental to propulsion and power generation, and significant effort has been dedicated to optimizing these configurations in the case of diesel-electric [4, 5] and hybrid electric [6, 7, 8, 9] vessels. Beyond OSVs, hybrid electric power systems have been investigated for warships [10], luxury yachts [11], and ship cranes [12].

In this work, the fuel consumption and NO_x emissions of an offshore supply vessel with four possible propulsion architectures with varying degrees of electrification are compared using vessel simulations at a holistic system level. Constant and rapidly-varying stationary electrical and speed profiles are used to evaluate the architectures under ideal conditions before they are compared using six-day industrial load scenarios representing offshore wind farm and oil field applications.

METHODOLOGY

The simulation models in this work are selected to capture the essential phenomena governing fuel efficiency without building full, high resolution models that could preclude simulations over multi-day load profiles. All simulations are performed using Siemens LMS Imagine.Lab Amesim [13], which enables multi-physics simulation of the mechanical, electrical and control systems for this vessel. By incorporating all factors influencing the vessel propulsion system demands including the anticipated operational profile, end-to-end system optimization is possible with a single tool. This section details the propulsion system architectures, modeling methodology, and control strategies for each architecture. Adaptive numerical analysis methods enable simulations of long vessel trajectories while retaining accuracy during periods with rapidly-varying loads or changes in load profile.

Propulsion System Architectures

Numerous engine and electrical configurations for ship power generation and propulsion have been proposed using combinations of reciprocating engines, gas turbines, and electric machines acting as motors and generators such as combined diesel and diesel (CODAD), combined diesel and gas (CODAG), and combined diesel-electric and gas (CODLAG). As this work focuses on offshore supply vessels, we examine four possible architectures using diesel engines and electric machines that are summarized graphically in Figures 1-4. The vessels are subject to the requirements that they must provide 2 MW of electrical power and sail at a cruise speed of 14 kn. Number of propellers and their size are constant, and hull form is the same in all cases.

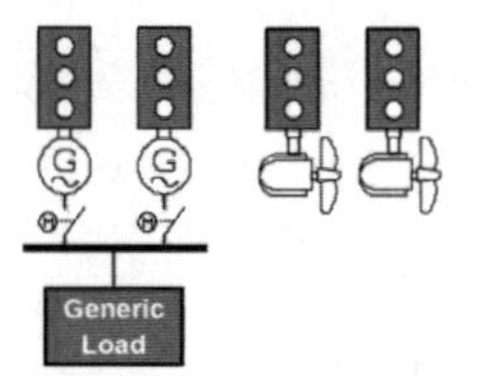

Figure 1. Mechanical configuration: two generators connected to propulsion units, and two electrical generators.

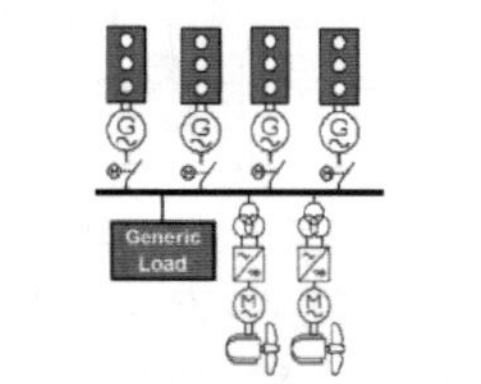

Figure 2. On-demand: the generic load and drive motors are supplied by four diesel generators.

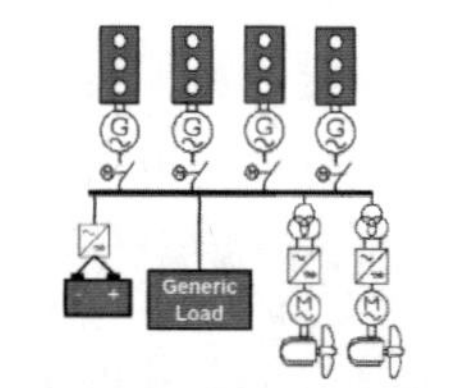

Figure 3. Hybrid: similar to on-demand with the addition of energy storage.

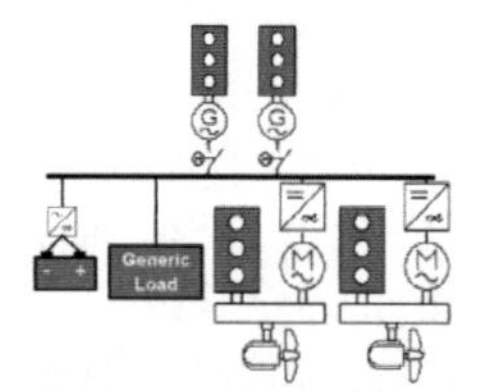

Figure 4. Assist: a combination of mechanical diesel drive and hybrid power systems.

Figure **5** depicts the representation of the hybrid configuration in LMS Amesim. Four generators and their threshold-based controls are present on the left, the electrical storage, losses, and loads in the center, and the drive mechanisms and ship resistance are shown on the right.

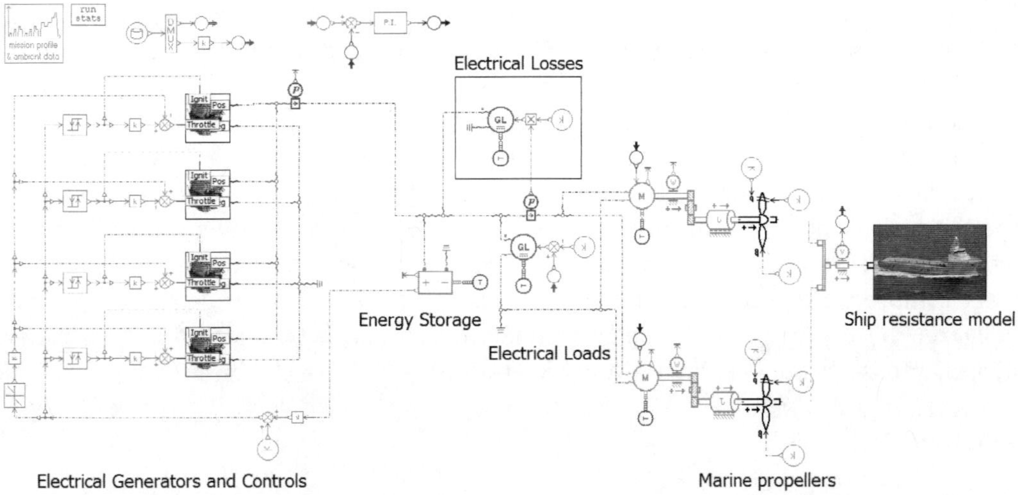

Figure 5. LMS Amesim simulation model of the hybrid vessel topology used in this work.

One of the most common engine configuration used in maritime vessels is a direct drive configuration with heavy fuel oil or diesel engines connected to propellers either directly or through a gearbox. This setup affords the maximum possible drive efficiency with values reaching up to 98% [14]. Electrical power for the ship is provided using separate diesel generators and/or shaft generators. In this work, two drive engines and two engines for electrical generation are used. This configuration is denoted as the *mechanical* configuration.

Diesel-electric power transmission is commonly seen in applications requiring high torque at low rotational speeds with the additional capability of running at high speed. For marine applications, they are also favors in cases where electrical loads are significant compared to thrust loads, as is the case for OSVs and oil platforms. The electrical demands for house loads, industrial machinery, and dynamic positioning while at sea can reach 30% of transit drive load [2], and therefore combining electrical and propulsion power generation can become favourable. The absence of significant electrical energy storage means that the generators operate under a variable load control such that the production matches the requested power by varying the engine speed or torque demanded by the electrical machines. Four diesel motors drive electrical machines to generate electricity, and two 1.4 MW drive motors are used for main propulsion. This configuration is denoted as the *on-demand* configuration, because the electrical energy is provided according to the request.

Hybrid electric vessels are a relatively new concept in industrial applications as the cost and energy density of batteries has been prohibitive to industrial deployment until the recent commercial release of lithium polymer marine batteries. Beyond hybrid vessels, fully electric industrial vessels such as an electric car ferry [15] also now exist; however, the purpose of batteries in this role act as the principal energy supply that are recharged using shore power, whereas for the OSV in this work, batteries are used to smooth the load on the engines and avoid operating internal combustion engines outside of their optimal loading range. The hybrid topology used here is classified as a serial hybrid as it uses diesel engines strictly to provide electrical power. In future sections this configuration is denoted as the *hybrid* configuration.

The final configuration considered in this work is a blend of the *mechanical* and *hybrid* configurations that addresses the weaknesses of the separate configurations. The *mechanical* case suffers when the engine operates away from the cruising speed, whereas the *hybrid* configuration has greater losses while cruising due to losses associated to power conversion and switching. These two propulsion architectures are combined to draw on their strengths: electric motors are used to provide propulsion power up to 9.75kn, after which the diesel motors are activated and used in tandem with the electric motors. The electric motors are used to provide the difference between the torque requested and that delivered by the diesel motors. This configuration can be considered a parallel-series hybrid, and is denoted as the *assist* configuration.

Ship Model

The vessel resistance is modeled using the statistical method of Holtrop and Mennen [16]. The dimensions of the ship used in this study are enumerated in Table 1.

Table 1. OSV model dimensions and coefficients.

Property	Value
Waterline length	76 m
Beam	17 m
Draft	6.5 m
Displacement	5000 MT
Prismatic Coef.	0.75
Waterplane Area Coef.	0.89
Midship Section Coef.	0.969
Block Coef.	0.725

The increased resistance experienced during vessel maneuvering is neglected, and ship dynamics under lateral loading (e.g. during dynamic positioning) are modeled using electrical and thrust demands. An additional 15% increase in resistance is also included to account for fouling [17].

The propeller type selected for all models is a four-bladed Wagenigen B-series propeller [18] with a diameter D=2.2m and a pitch to diameter ratio P/D=1.3 and a blade area ratio of Ae/Ao=1. Two propellers are used in all cases. The dynamic behaviour of the propeller is simulated by generating maps of torque and thrust coefficients K_T and K_Q, and using these to provide the torque load on the engine and force as an input to the differential equation describing the ship motion. The moment of inertia of the propeller and entrained water are calculated to be 295 kg m^2 per propeller using the method presented in [19]. The vessel is assumed to use azimuth thrusters with a loss of 5% of power between the shaft and propeller.

Engine

The engines are modeled using the IFP engine model in Amesim, which uses tabulated values for brake mean effective pressure (BMEP), friction mean effective pressure (FMEP), and pumping mean effective pressure (PMEP). These values can be supplied by the user, or generated using a built-in application starting from a few key details characterizing the engine. Automatic generation is performed using an iterative algorithm to determine pumping losses given the geometric properties, ignition type, and power/torque response of the engine. In this work, these tables are generated using the build-in application for a single four-stroke diesel engines: a 12 cylinder, 48 L engine producing 1320 kW. To model the losses associated to starting the diesel motors, 50% overconsumption of fuel for five seconds is included when the engines are started. All architectures have identical installed power (5280 kW).

Predicting emissions of marine vessels is of ever-increasing importance with stricter emissions regulations such as MARPOL 73/78 Annex VI [20] coming into force. Pollutants such as CO_2 and SO_x are produced proportionally to the fuel consumed with 3.41 grams of CO_2 per gram of fuel and a sulphur content limit of 0.001 grams per gram of fuel for ships in EU waters. The NO_x limit for ships built after January 1, 2016 (Tier III) under MARPOL 73/78 Annex VI defines an NO_x limit of $9n^{-0.2}$ g/kWh (where n is the engine speed in RPM) for diesel engines operating between 130-2000 RPM. Lacking the NO_x emissions maps for specific engines, we adopt this limit as the NO_x production for all engines considered here to provide an upper bound on NO_x production for the engine configurations and load cycles considered.

Electrical System

Electric generators and motors are modeled functionally with a constant efficiency of 96%. A further 5% loss is incorporated into the electrical system to model switching losses for the architectures using batteries. The battery is also modeled using functional models with a constant internal resistance to account for losses. The *hybrid* vessel uses a battery with a capacity of 0.5 MWh and the *assist* variant uses a battery with a capacity of 0.25 MWh.

The electrical loads on the ship are represented with a single generic time-varying load. The power draw is specified using a data file or defined as a function of time before the simulation is launched. The load adapts its internal resistance to satisfy the power draw requirements. The temporal nature of these fluctuations is detailed in each subsection.

Control Strategy

The *mechanical* model uses PID control on drive engine power to match the speed request, and the generators operate in a variable speed mode. Variable engine speed is used to match the electrical load. Once the engine speed and load corresponding to peak efficiency is reached, the next generator is turned on and varied to match the load. If both generators are at their optimal point and more power is required, all engines increase equally to match the load.

The *on-demand* architecture uses variable engine speed to match the electrical load as in the case of the *mechanical* case, with the only difference being a greater number of generators used.

The *hybrid* architecture has the simplest strategy: the engines are turned on and off depending on the state of charge (SOC) of the battery. The generators turn on at 90%, 70%, 50%, and 20% of charge. Below 10%, all generators are changed from optimal efficiency to maximum power production. The generators use an asymmetric start/stop strategy, turning off at a higher SOC than that which activates them so that near one threshold for activation, a generator will not repeatedly start and stop.

The *assist* architecture uses a combination of the *mechanical* and *hybrid* control strategies. Below 9.75 kn, only electrical drive motors are used and the generators are controlled using thresholds on battery state of charge as in the *hybrid* case. Above 9.75 kn, the diesel motors are turned on and provide the maximum possible torque with the electric drives providing the remainder of the requested torque.

Each of the four architectures considered here uses a unique control strategy to maximize its fuel efficiency. The optimization of *hybrid* propulsion systems is a complex problem requiring its own dedicated in depth analysis to achieve optimal performance [21]. While the *mechanical* and *on-demand* architectures have little room for optimization, the threshold-based methods of the *hybrid* and *assist* architectures could possibly be further optimized with more complex control strategies.

RESULTS

The investigation into the performance of the four architectures is split into three subsections. The first elucidates the relative performance of each architecture under constant conditions, the second examines efficiency under rapidly changing thrust and electrical loads (to model dynamic positioning) and the third provides a comparison for three prototypical loading cycles for an offshore supply vessel. Where possible, the *on-demand* case is used as the benchmark as it is widely used in industry, and the performance of the other architectures are reported as a relative quantity.

Quasi-static Conditions

The performance of each architecture is compared under constant speed and electrical loads using a series of 166 simulation for each model. The fuel consumption at constant speed and electrical load is calculated over a range of 0-14 kn and 0-2.0 MW. A simulation length of 1.15 days (10^5 seconds) is used to diminish the influence of the charging/discharging cycle of the battery on the predicted fuel consumption in the *hybrid* and *assist* cases. The simulation run-time for each point is on the order of 10s allowing these maps to be generated quickly.

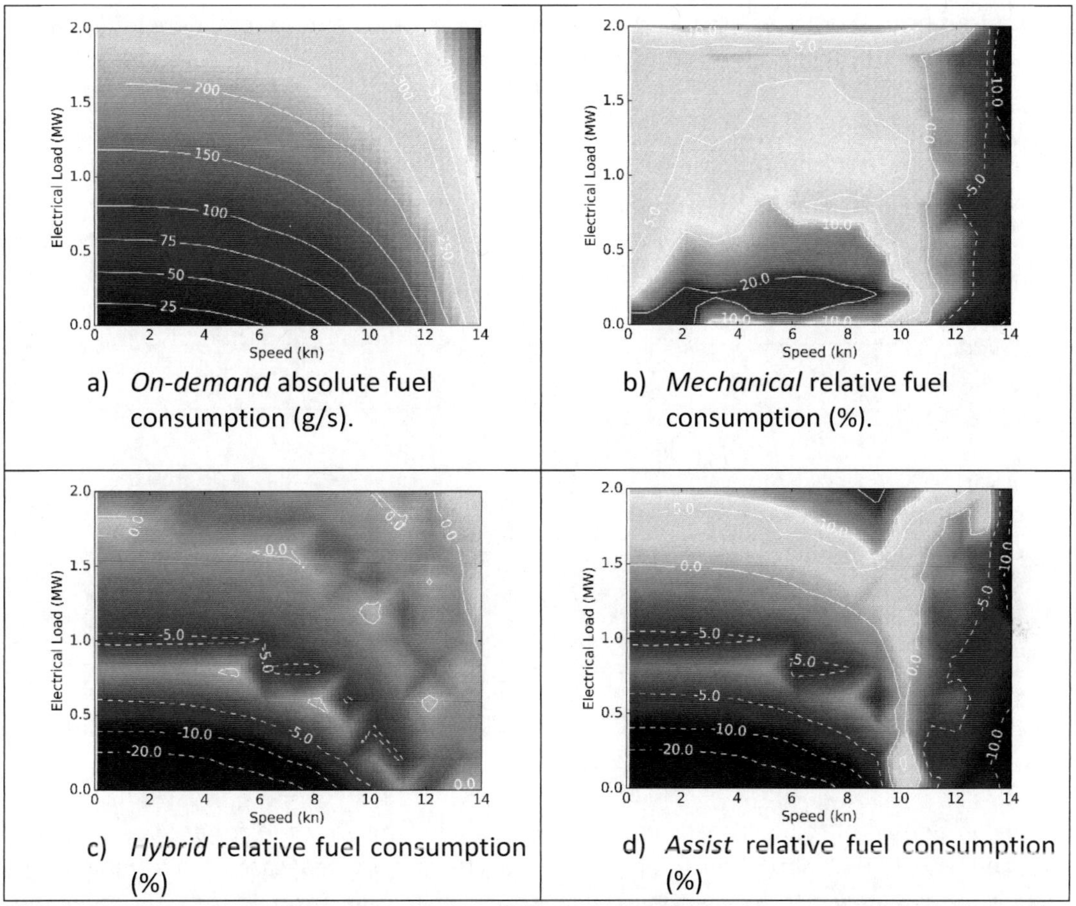

a) *On-demand* absolute fuel consumption (g/s).

b) *Mechanical* relative fuel consumption (%).

c) *Hybrid* relative fuel consumption (%)

d) *Assist* relative fuel consumption (%)

Figure 6. Fuel consumption of the four architectures under constant speed and electrical loads. The *on-demand* case is shown in absolute quantities and the relative consumption of the others is shown to highlight the differences. Dashed lines indicates negative quantities.

The absolute fuel consumption of the *on-demand* case in Figure **6** shows a smooth response as a function of speed and electrical load. While it is difficult to observe in the contour plot, there is a slight stepping effect where the fuel consumption is elevated when a generator is forced to operate outside its optimal RPM range. This effect is clear when comparing the relative fuel efficiency for the *hybrid* and *assist* cases. The *mechanical* case shows complex deviations across the speed range due to the separate electrical generation and propulsion motors contrasting the four generators of the *on-demand* case. The *mechanical* architecture is the most efficient vessel considered at the designed cruise speed of 14 kn. This outcome is expected as the direct mechanical drives do not suffer the electrical losses incurred by the other three configurations at this speed. The *hybrid* architecture fuel consumption is always lower than the *on-demand* case as the generators only operate at the optimum point. These two configurations have relatively high consumption at high speed due to electrical losses; the *assist* case remedies this by providing the most or all of the drive power at high speed with more efficient direct drive diesel motors. This configuration suffers at a combination of high electrical loads and speeds because the generators must operate beyond their optimum point to satisfy the electrical power requirements. From these maps, the *hybrid* or *assist* cases are optimal depending on if the usage is biased to high speeds or high electrical loads.

Next, the fuel efficiency as a function of speed (with zero electrical load) is calculated with a finer discretization to better characterize the propulsion efficiency and NO_x production of the four architecture considered here. Figure **7** shows the efficiency and NO_x production relative to the *on-demand* case, as it is the most commonly used propulsion system for OSVs.

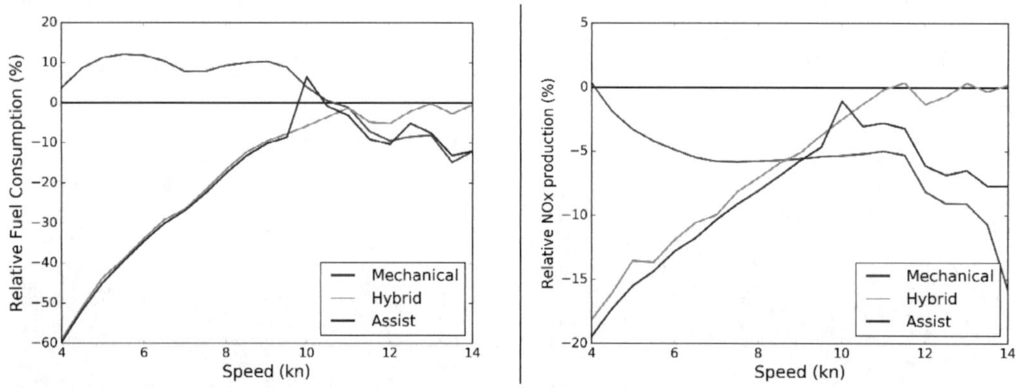

Figure 7. Fuel consumption and emissions relative to an on-demand architecture with constant speed and zero electrical load.

The *on-demand* architecture is an improvement on the *mechanical* architecture at low speeds due to the suboptimal operating point in the *mechanical* case; this trend is reversed at approximately 10 kn as the design cruise speed is approached. The *mechanical* configuration is optimal at the cruise speed, and the *assist* architecture effectively matches it while also achieving optimality across nearly the rest of the speed range. The spike in consumption near 10 kn is attributed to the transition between electrical and diesel power. The slight difference between *hybrid* and *assist* is attributed to the control scheme, battery size, and initial state of charge - these configurations otherwise operate identically at low speed.

The NO_x production in Figure **7** shows strikingly different results with the *on-demand* setup occupying the poorest position across the speed range. The *assist* configuration in almost always the best case with only the *mechanical* configuration being significantly superior at the cruise speed. The contrast between the fuel consumption and NOx production of the *mechanical* and the *on-demand* configurations arises due to the interplay between the engine efficiency and NO_x production limit: the *on-demand* engines extract more power at lower RPM as they operate at the most efficient point, whereas the *mechanical* engines must run at a higher speed to produce the same power. As the assumed MARPOL 73/78 Annex VI [20] NO_x limit decreases with engine speed, the resulting NO_x production is lower. For an engine-specific NO_x production map, this outcome may be significantly different.

Dynamic Conditions

Rapidly changing engine and electrical loading demands are common OSVs when they must maintain a physical position while the forces caused wind and wave action are acting on the vessel. Calculating the power requirements for dynamic positioning requires an accurate characterization of the vessel resistance and dynamics under wind and wave action, which is beyond the scope of this work. Instead, we compare the configurations by assuming rapidly varying velocity and electrical power profiles. These profiles are created by drawing values from a normally distributed random variable every 5 seconds and then averaging these values over 15 seconds to smooth the load. The velocity has a mean of 2 kn and a variance of 0.75 kn, and the electrical load has a mean of 300 kW and a variance of 500 kW. The electrical load distribution is truncated at zero power. The dynamic loading is simulated for two hours, and the resulting fuel consumption and NO_x emission is shown in Figure **8**. These simulations requires roughly 40 seconds to run, which is longer than the static conditions because the variable-step integrator must adapt the integration intervals between the large number of discontinuities in the speed and electrical demands.

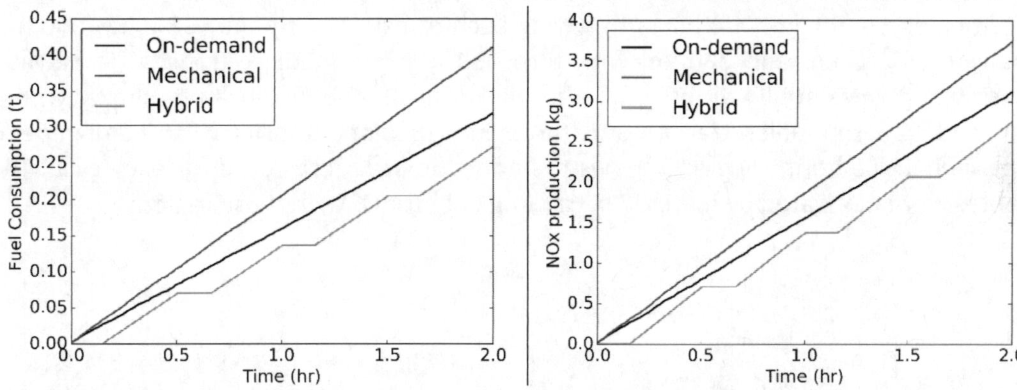

Figure 8. Comparison of the fuel (left) and emissions (right) performance of the on-*demand*, *mechanical*, and *hybrid* architectures under dynamic conditions.

The *hybrid* configuration has a clear advantage compared to the *mechanical* and *on-demand* configurations (the *assist* configuration is not shown here as it is identical to the *hybrid* case at these speeds). The load-smoothing effect of the battery allows the generators to run at optimal speed, resulting in a fuel savings of at least 15% compared to the other two configurations. The NO_x emissions are also comparably lower. In quantitative terms, the *mechanical* and *on-demand* cases consume 50% and 17.1% more fuel and produce 35.9% and 13.3% more NO_x, respectively. In reality, a *mechanical* configuration would not be able to reverse thrust with sufficient speed to follow the velocity target without a variable-pitch propeller and the associated decrease in propeller efficiency would favour the *on-demand* configuration to the *mechanical* one for this application.

Industrial Example

Offshore supply vessels are used heavily in the offshore oil and wind industries, and the operational loading requirements to which they are subject combines longer transits at cruising speed with high electrical loads while working at sea. The first profile represents an OSV based in Denmark providing support to offshore wind farms. The second profile represents a vessel based in Norway supporting oil platforms in the North Sea, and the third for a vessel based in Brazil operating in the Lula oil field. The profiles have different transit distances to the offshore field, as well as between work locations in the offshore field. All profiles requires dynamic positioning while not underway. To model this effect, the electrical loads are increased to reflect the energy used by the bow thrusters, and the velocity profiles have non-zero steady-state values with small random fluctuations to represent the effect of wind.

The electrical loads (separate from the main drive units) are modeled as a normally distributed random variable with a mean of 650 kW and a variance of 450 kW while underway, and a mean of 900 kW with a variance of 500 kW while dynamic positioning. Station-keeping loads include the power demands of the bow thruster power, but the main propulsor power consumption is directly calculated by the simulation independently of these assumed power profiles. In this case, the loads change every twenty minutes (compared to 5 seconds for the previous section) to enable a simulation of a vessel trajectory over six days. To summarize the velocity profiles, the Denmark profile has four major transits of 65-120 km and nine short (approx. 10km) transits with a dynamic positioning time from 1-24 hours following each transit. The Norway case has two major transits of 220 km and 280 km with three 50 km intermediary transits. The dynamic positioning times are 24 or 48 hours after each transit. The Brazil case is similar with 290 km and 350 km major transits and three intermediary transits with waiting periods between 12-48 hours. Figure **9** shows the approximate routes used for each case. These routes were motivated by marine AIS tracking data for OSVs in these regions.

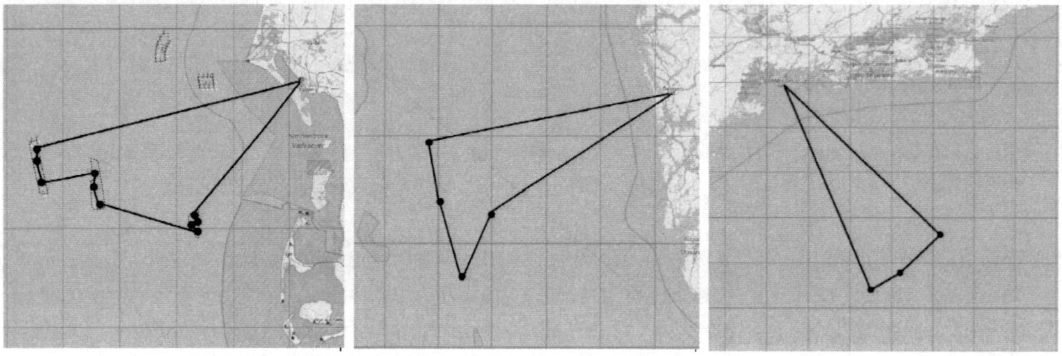

Figure 9. Routes for an OSV supporting wind farms in Denmark (L), oil fields in the North Sea (C), and the Lula oilfields near Brazil (R).

Figure **10** shows the speed profile and fuel consumption for the four architectures in the Denmark case. The time to simulate each load profile is on the order of 100s, so many profiles can be potentially compared with minimal computational requirements.

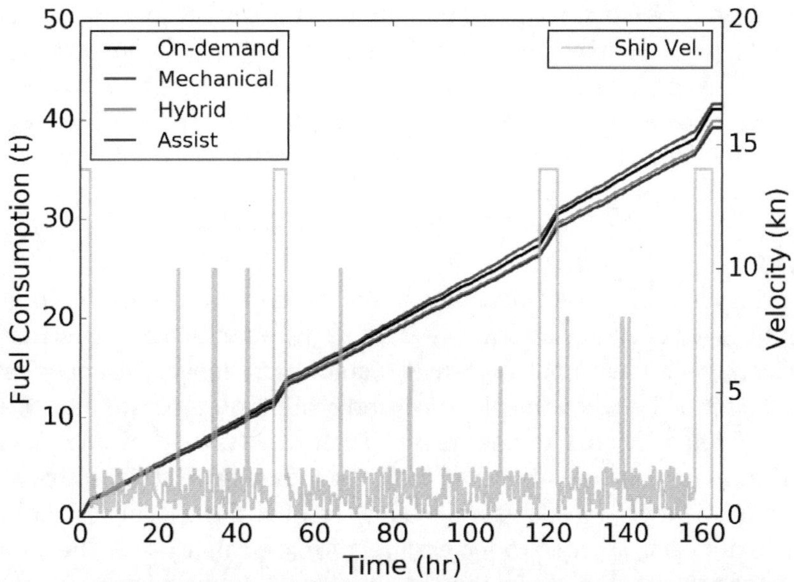

Figure 10. Fuel consumption for the four architectures with the Denmark profile. The majority of the fuel is consumed during dynamic positioning, which favours the two configurations with energy storage to smooth power demands.

While the transits at cruising speed result in the greatest rate of fuel consumption, the majority of the fuel is consumed during dynamic positioning at sea in all cases. The *hybrid* and *assist* cases are therefore superior because they use electric motors to provide thrust at these speeds, and their energy storage smooths the electrical loads to allow the generators to operate at peak efficiency. The *mechanical* and *on-demand* systems suffer because the diesel motors must operate outside of the most efficient RPM range, but the difference is much less clear than in the preceding sections because the loads in the dynamic positioning phases change every 20 minutes rather than every five seconds. The performance of each propulsion system architecture relative to the *on-demand* case is detailed in Figure **11**.

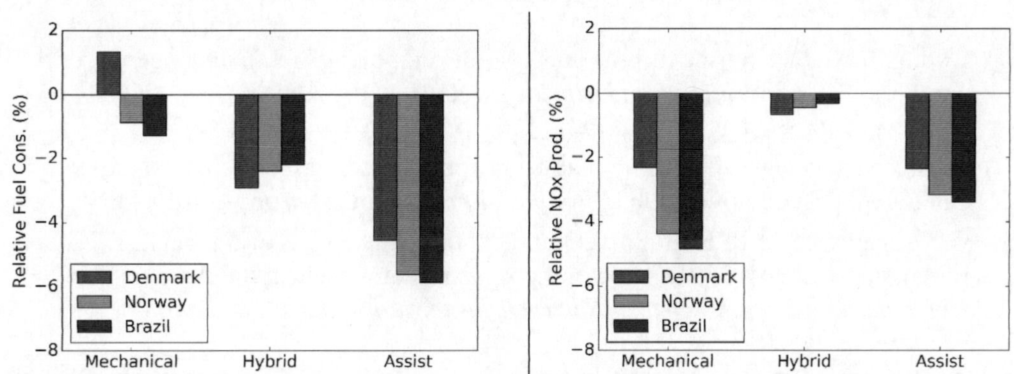

Figure 11. Fuel consumption and NOx emissions for the three load profiles. The assist configuration performs best with regards to NOx emissions because the smaller 880 kW drive motors operate at higher speeds than the generators, resulting in lower NOx emissions.

In all applications, the *assist* configuration has the lowest fuel consumption, and the *hybrid* configuration has slightly greater consumption. The *mechanical* and *on-demand* architectures are comparable in fuel consumption, which is up to 5% more than the other two. Since we assume the MARPOL 73/78 Annex VI [20] limits as the NO_x production of these engines, the higher speed engines of the *mechanical* and *assist* configurations reduces their total emissions output.

CONCLUSION

The four propulsion system configurations presented here represent commonly used or feasible options for offshore supply vessels, as well as for many other varieties of boats and ships. Simulating a vessel from a high level system perspective gives new insights resulting from the interaction between the electrical and propulsion subsystems as well as the speed and electrical requirements of the vessel. Such a system level analysis allows one to optimize the vessel end-to-end, and predict the influence of different configurations and control schemes under realistic vessel load cycles. For the specific case of an OSV, which typically operate using an *on-demand* architecture, we can quantify the improvement that a battery or additional drive diesel motors have on the fuel efficiency of the vessel. Energy storage provides a distinct advantage to reduce fuel consumption and NO_x emissions by load-smoothing and enabling generators to run at optimal efficiency. A holistic, system level approach also enables an economic analysis for the retrofitting a vessel with an *on-demand* propulsion system architecture with energy storage and conversion electronics for in-service OSVs. Furthermore, the calculations can be run quickly so that the system performance with many different parameter combinations and loading cycles can be easily compared. Beyond the work shown here, concept trade-off studies and optimization for marine applications at the vessel system level has the potential to deliver more efficient, purpose-built solutions by creating all feasible propulsion configurations, optimizing parameters and controllers for each case, and simulating vessel performance to select the ideal ship configuration.

References

[1] I. Jung, "Swedish marine turbine and gear development," in *Spring meeting of the Society of Naval Architects and Marine Engineers*, 1966.

[2] M. C. Diaz-de-Baldasano, F. J. Mateos, L. R. Nunez-Rivas and T. J. Leo, "Conceptual design of offshore platform supply vessel based on hybrid diesel generator-fuel cell power plant," *Applied Energy,* vol. 116, pp. 91-100, 2014.

[3] G. Gasparovic and B. Klarin, "Techno-economical analysis of replacing Diesel propulsion with hybrid electric-wind propulsion on ferries in the Adriatic," in *Computer and Energy Science (SpliTech), International Multidisciplinary Conference on*, 2016.

[4] O. C. Nebb, B. Zahedi, J. O. Lindtjorn and L. Norum, "Increased fuel efficiency in ship LVDC power distribution systems," in *Vehicle Power and Propulsion Conference (VPPC), 2012 IEEE*, 2012.

[5] S. Solem, K. Fagerholt, S. O. Erikstad and O. Patricksson, "Optimization of diesel electric machinery system configuration in conceptual ship design," *Journal of Marine Science and Technology,* vol. 20, no. 3, pp. 406-416, 2015.

[6] T. A. Johansen, T. I. Bo, E. Mathiesen, A. Veksler and A. J. Sorensen, "Dynamic positioning system as dynamic energy storage on diesel-electric ships," *IEEE Transactions on Power Systems,* vol. 29, no. 6, pp. 3086-3091, 2014.

[7] F. D. Kanellos, J. M. Prousalidis and G. J. Tsekouras, "Control system for fuel consumption minimization--gas emission limitation of full electric propulsion ship power systems," in *{Proceedings of the Institution of Mechanical Engineers, Part M: Journal of Engineering for the Maritime Environment*, 2014.

[8] H. Lan, S. Wen, Y.-Y. Hong, C. Y. David and L. Zhang, "Optimal sizing of hybrid PV/diesel/battery in ship power system," *Applied Energy,* vol. 158, pp. 26-34, 2015.

[9] J. O. Lindtjorn, F. Wendt, B. Gundersen and J. Hansen, "Demonstrating the benefits of advanced power systems and energy storage for DP vessels," in *Proc. Dyn. Positioning Conf.*, 2014.

[10] S.-Y. Kim, B.-G. Cho and S.-K. Sul, "Feasibility study of integrated power system with battery energy storage system for naval ships," in *Vehicle Power and Propulsion Conference (VPPC), 2012 IEEE*, 2012.

[11] D. Bosich and G. Sulligoi, "Voltage control on a refitted luxury yacht using hybrid electric propulsion and LVDC distribution," in *Ecological Vehicles and Renewable Energies (EVER), 2013 8th International Conference and Exhibition on*, 2013.

[12] E. Ovrum and T. Bergh, "Modelling lithium-ion battery hybrid ship crane operation," *Applied Energy,* vol. 152, pp. 162-172, 2015.

[13] Siemens Industry Software NV, *Siemens LMS Imagine.Lab Amesim (Version 15),* 2016.

[14] E. C. Tupper, Introduction to naval architecture, Butterworth-Heinemann, 2013.

[15] F. Martini, "First electrical car ferry in the world in operation in Norway now," *Siemens, Pictures of the Future,* 2015.

[16] J. Holtrop and G. Mennen, "A statistical power prediction method," *International Shipbuilding Progress,* vol. 25, no. 290, 1978.

[17] A. F. Molland, S. R. Turnock and D. A. Hudson, Ship resistance and propulsion: practical estimation of propulsive power, Cambridge university press, 2011.

[18] W. Van Lammeren, J. Van Manen and M. Oosterveld, "The Wageningen B-screw series," *Nederlands Scheepsbouwkundig Proefstation,* 1969.

[19] R. L. Harrington, Marine engineering, Society of Naval Architects and Marine Engineers, 1971.

[20] International Maritime Organization, *Regulations for the prevention of air pollution from ships and NOx technical code of MARPOL 73/78,* 1998.

[21] J. Vanhuyse, N. Murgovski, M. Nicolai, H. Van der Auweraer and W. Desmet, "Closed-loop design for hybrid powertrains," in *Proceedings of the ASME 2017 International Design Engineering Technical Conferences & Computers and Information in Engineering (accepted)*, 2017.

Mission-centric Design Of Hybrid Propulsion Systems For Multi-purpose Naval Vessels

Pierpaolo Da Fieno
MAN Diesel & Turbo SE
pierpaolo.da-fieno@man.eu

Enrico Chelli
MAN Diesel & Turbo SE
enrico.chelli@man.eu

Bernd Friedrich
MAN Diesel & Turbo SE
bernd.friedrich.a@man.eu

Matthias Schlipf
MAN Diesel & Turbo SE
matthias.schlipf@man.eu

Mark Watson
MAN Diesel & Turbo SE
mark.watson.a@man.eu

ABSTRACT

In the last few decades the design of naval vessels has gone through substantial changes, moving from a rigid hierarchy of highly specialized vessel archetypes, to considerable more flexible platforms designed to cope with a much wider spectrum of tasks, from disaster relief to constabulary and environmental protection operations, from low intensity conflicts up to full-scale maritime interdiction.

These new requirements exposed the clear limits of traditional naval propulsion plants and, driven by the rapid breakthrough in the field of power electronics and automation systems, opened the way to the development and implementation of innovative solutions, which not only fulfil the required flexibility but provide unprecedented levels of efficiency, redundancy and environmental compliance.

This paper presents:

- a brief overview of the most notable hybrid propulsion plants for naval applications;
- an analysis of the typical operational requirement of a modern multi-purpose naval vessel;
- a practical approach to the design of a battery-hybrid propulsion plant optimized on the above mentioned operational requirements and based on the combination of recent and established technologies, including: high-speed diesel engines, variable frequency drives, electric propulsion motors, batteries, exhaust after-treatment systems, and controllable pitch propellers;

- an assessment of the intrinsic advantages of such a design compared to mechanical, diesel-electric or alternative hybrid solutions;
- a concise outlook on future developments.

INTRODUCTION: THE QUEST FOR FLEXIBILITY

Having the right tool for the job is a lesson early learnt by any seafarer. Naval operations are no exception in this regards: being fit for the task, or actually fitter than your opponent, is critical for success.

Over centuries the increasing extent of naval tasks shaped fleets into rigid hierarchies of highly specialized vessels, a pattern so ubiquitous and well-recognized to be reflected still today in the officer ranks. While the technological advancements in marine propulsion introduced since the beginning of the last century, paved the way to new solutions for efficient operation across a broad speed range, such as combined mechanical/electrical plants (CODAD, CODOG, CODELAG, etc.) and integrated electric propulsion (IEP), it was the 21st century, with its well-known changes in geopolitics, to quickly escalate the urge for operational flexibility.

Modern navies are asked to cover the full intensity spectrum in both green and blue waters with fewer vessels, contracted financial resources and reduced crews, thus requiring multi-purpose and dual use platforms.

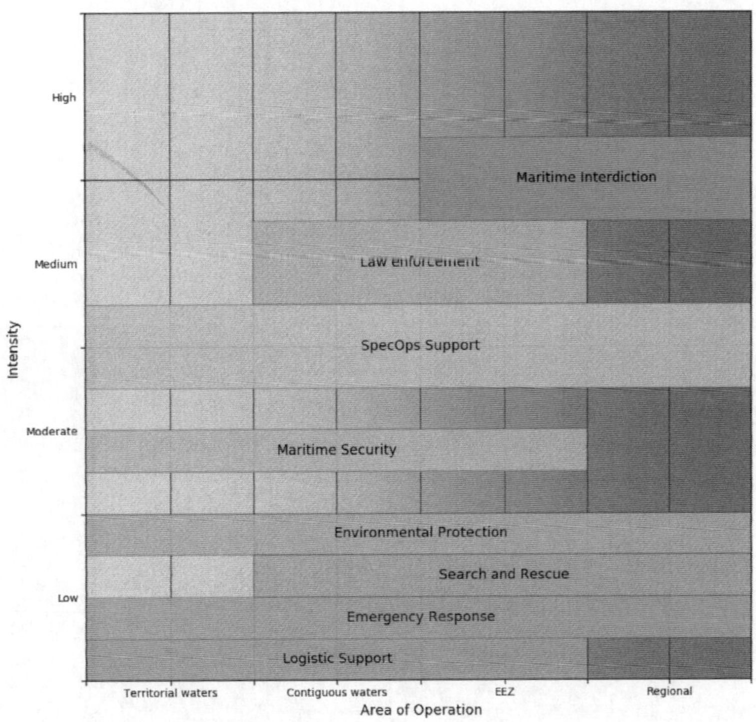

Figure 1 – Operational spectrum of a multi-purpose second line naval vessel

A review of all recent naval designs clearly shows a converging trend toward flexible platforms capable of performing the whole spectrum of tasks traditionally assigned to a variety of second line vessels, such as:

- Maritime Interdiction (deterrence and denial)

- Law enforcement (border protection, fishery surveillance, illegal traffic prevention)
- Maritime Security (intelligence collection, surveillance and protection of assets)
- Marine Environmental protection (prevention, containment, decontamination)
- Special Operations Support (mine countermeasures, diving, hydrography, medical assistance, maritime signaling, etc.)
- Search and Rescue
- Emergency Response (disaster relief, personnel evacuation, salvage)
- Logistic Support

From a propulsion perspective, the immediate issue in dealing with such a broad range of missions is the disproportion of power necessary to achieve the required speed. Even with the latest fuel injection and turbocharging technology a diesel engine (or any other internal combustion prime mover) cannot efficiently operate across such a broad range.

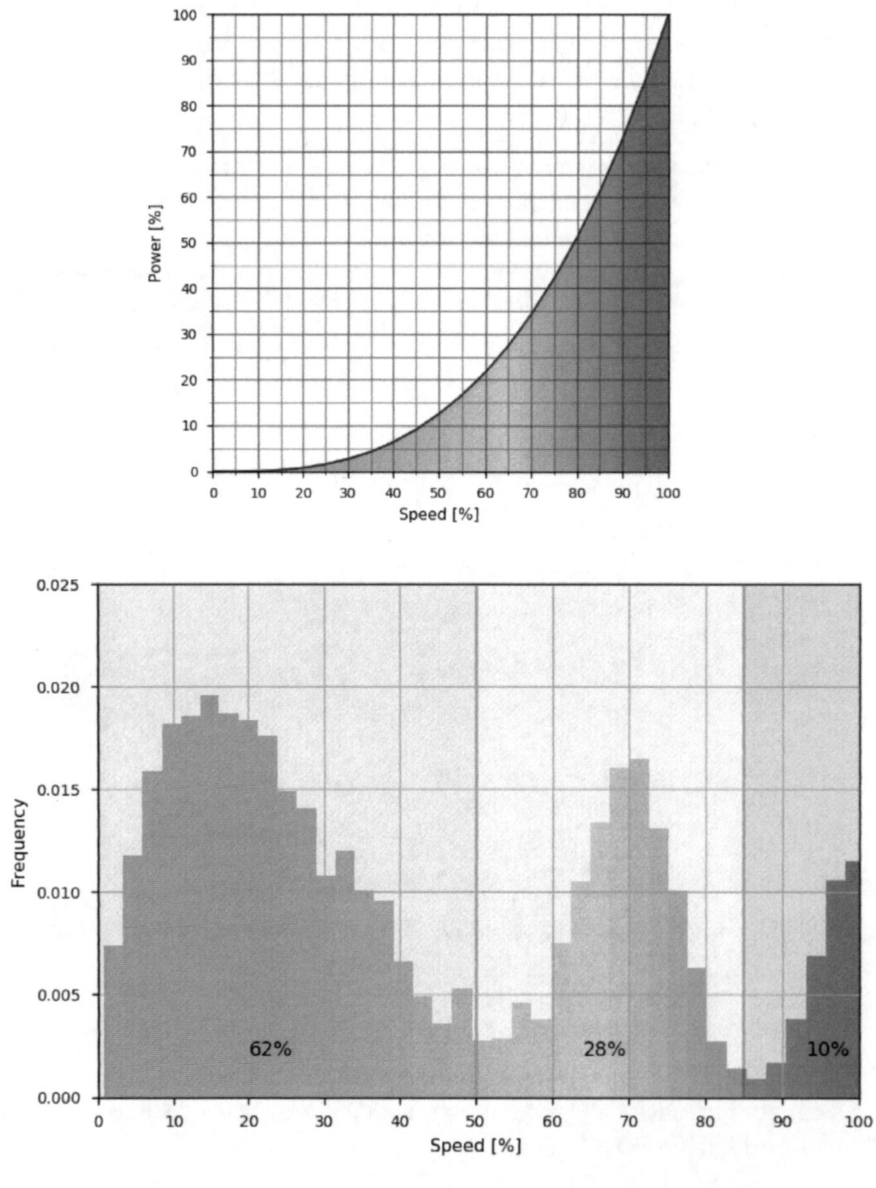

Figure 2 – Exemplary power curve and speed spectrum of a multipurpose vessel

The use of combined mechanical propulsion plants is a classic solution to partially overcome this problem and allow cruise speeds significantly lower than the required maximum speeds (i.e. traditional frigate/destroyer design), but it is still not helpful when an extended loitering mode is required. Electric motors, on the other hand, being able to efficiently operate across an entire power range, represent an optimal solution for such tasks. However, they require significant space as they rely on generating sets and their volume, including drives, is not linear with installed power, making the integrated electric propulsion (IEP) a viable solution only for large vessels.

Since the early 2000s, driven by the considerable advancements in electric equipment, power electronics and automation, hybrid propulsion plants have collected growing interest in both naval and commercial applications[1]. Beside the immediate advantage of combining the operational flexibility of the electric motor with the power density of a conventional prime mover, they offer increased redundancy, higher overall efficiency, faster response and a reduction of the maintenance costs connected to off-design operating points. In addition, due to the rapid development in battery technology the possibilities to include energy storage systems for applications on ships have nowadays occurred. This additional source of energy can be easily integrated in hybrid systems increasing their flexibility. Stored energy might be used to operate the engines and the gensets as close to their design points as possible, to immediately feed the electrical motors for boosting the vessel speed or to be an alternative energy for a certain time reducing noise and exhaust gas emissions.

The potential of battery-hybrid systems is, thus, to provide energy and to control the power delivered by, or to, the shaft machine, in the most efficient way as possible taking all advantages of the electrical and mechanical components. To exploit the potentials of battery-hybrid systems, MAN Diesel & Turbo has developed the advanced and flexible HyProp ECO propulsion system, successfully installed in several types of different marine applications with remarkable benefits in terms of operational flexibilities and operational expenses (OPEX).

In the wake of the trend taking hold of the automotive industry, batteries are getting more and more affordable with reduced overall footprints, weights and costs. However, since the real potential lies in overall optimization, a proper system design requires a detailed knowledge of vessel operating modes and a thorough understanding of each component. Several optimization techniques have been extensively investigated in the technical literature but they all assume a reliable modelling of vessel usage and equipment performance. While the first topic usually resides in a single source (either shipyard or operator), the second one is potentially spread across several solution providers, making a seamless integration of expertise in different fields a key factor for a successful battery-hybrid design.

[1] Even if the first modern Combined Diesel Electric And Gas (CODELAG) plant was commissioned in 1990 on the *HMS Norfolk*, it took almost 20 years for this design to be widely adopted: *USS Makin Island* (CODELOG) was commissioned in 2009, *FS Aquitaine* (CODELOG) and *HNLMS Holland* (CODELOD) in 2012, *ITS Bergamini* (CODELAG) in 2013, *FGS Baden-Württemberg* (CODELAG) in 2016. For a comparison: *Carolyn Dorothy* (tugboat) was commissioned in 2009, *MV Hallaig* (double-ended RoRo ferry) in 2013 and *Prinsesse Benedikte* (RoRo ferry) was retrofitted with a hybrid plant in 2015.

OVERVIEW OF SELECTED TECHNOLOGIES

High speed diesel engines

Traditionally confined to special marine applications (restricted service, sport and leisure crafts, and fast small naval vessels), high speed engines are now largely selected as prime movers for both commercial and governmental applications.

The advancements in fuel injection technology and engine automation, together with the general trend for lower emissions and cleaner marine fuels, have boosted efficiency, flexibility and reliability. Combined with traditional advantages, such as higher power density and reduced footprint, these features tipped the scales making high speed engines a typical choice for a variety of vessels, from harbor tugboats to superyachts, from offshore supply vessels to frigates.

In order to fit such a wide range of applications high speed engines are typically available in several ratings. The same engine architecture can therefore be tuned to light, medium or heavy duty load profiles, with different rated powers, optimized operating ranges and service intervals, providing designers an additional degree of freedom to tailor the propulsion system to specific requirements.

Exhaust after-treatment systems

In the last two decades, following a well established trend in onshore applications, ship emission limits have been significantly reduced. The latest IMO regulation, already enforced in US and Caribbean waters and soon entering into force in the North and Baltic Seas, set a limit which simply exceeds the capability of engine internal measures without compromising efficiency, reliability and service costs.

On the other hand both automotive and stationary applications have already extensively adopted the Selected Catalytic Reduction technology, proving its technical and commercial effectiveness.

Solutions based on this solid experience are currently the best pick for marine plants, providing several advantages:

- usage of a non-hazardous, safe-to-handle, largely commercially available diesel exhaust fluid (DEF) such the automotive AdBlue;
- no interference with engine operation, cancelling any risk of reliability reduction;
- flexibility of operation, making it possible to activate it only in relevant Emission Control Areas (ECAs);
- lower operational expenses, achieved by optimization of fuel and DEF usage.

Controllable pitch propellers

With over a century of history[2], controllable pitch (CP) propellers are among the most mature and reliable products in the marine industry. Latest developments in computational hydrodynamics and continuous refinement in actuator designs allowed a

[2] The first Alpha CPP design was produced in 1902 and patented in 1903.

significant reduction of both the efficiency and maintenance cost gap with fixed pitch propellers, making CP propellers an excellent tradeoff among capital expenses (CAPEX), propulsion efficiency, operational flexibility and operating expenses (OPEX). These benefits can even be increased by using appropriate Efficiency Improving Devices (EID) such as Kappel propeller blades and rudder bulbs. Additionally, CP propellers can be "feathered" when not in operation, significantly reducing the drag and the related negative effects on ship motion of locked/windmilling fixed pitch propeller, therefore providing great operational flexibility.

Electric motors, variable frequency drives and energy storage systems

The concept of electric propulsion is not new, however, with the possibility to control electrical motors by means of frequency converters in a large power and speed range with compact, reliable and cost-competitive electronic solutions, the use of variable speed drives has emerged in electric propulsion since the 1990's.

The following characteristics summarize the main advantages of electric propulsion:

- improved life cycle cost by reduced fuel consumption and maintenance, especially where there is a large variation in load demand;
- reduced vulnerability to single failure and high availability of power in the system;
- possibility to optimize the loading of the prime movers;
- less space consuming and more flexible utilization of onboard space increases the payload of the vessel;
- less propulsion noise and vibrations since rotating shaft lines are shorter, prime movers are running on fixed speed, and using pulling type propellers gives less cavitation due to more uniform water flow.

When integrating an energy storage system (batteries) to an electric propulsion plant a further degree of freedom can be utilized and can be used to optimize the complete system:

- charging and discharging the batteries in such a way that it optimizes the operating point of the running Diesel engines ("Strategic Loading");
- fewer Diesel engines needed online as batteries are the backup for running gensets ("Electrical Spinning Reserve");
- leveling the power seen by the Diesel engines and taking away transient loads ("Peak Shaving");
- instant power in support of running Diesels ("Dynamic Support");
- reduced engine running hours;
- zero Emission operation possible (i.e. in harbor).

CASE STUDY: SECOND-LINE NAVAL VESSEL BASED ON NORCAN 222 ERRSV DESIGN

Due to their inherent flexibility, ship designs developed for Offshore Supply Vessels (OSV) have been commonly used as blueprints for second-line Naval vessels. OSVs have excellent seakeeping capabilities combined with high speed and superior maneuverability, they offer a large deck to accommodate containerized equipment for a wide range of special operations and can efficiently transfer solid and liquid cargo.

Initially used mainly for Coast Guard offshore cutters and governmental auxiliary and research vessels, the designs have recently been selected also for multi-purpose naval vessels. Most notable recent examples are the *Bâtiment multi-mission* (B2M), selected by the French Navy to perform a wide range of tasks in the overseas territories (world's largest EEZ) and the Italian Navy *Multi-purpose Submarine Rescue Ship* (USSP), capable of a large spectrum of operations, including hydro-oceanographic research at high-latitudes and diving support.

Following these recent examples, we selected the NorCan 222 Emergency Response Rescue and Support Vessel (ERRSV), a modern and innovative OSV design, as a reference platform for a Flag Ship case study.

The NorCan 222 ERRSV is a multi-purpose vessel designed to transcend the traditional ocean-going service vessel capabilities. By combining high-speed capabilities, state-of-the-art hybrid propulsion and latest emission control technology, the NorCan 222 ERRSV is able to perform both emergency response/rescue operations and transport of personnel/cargo more effectively and efficiently than traditional OSVs. The high emergency response speed (up to 34 knots), enhanced seakeeping performance, high maneuverability (DP2 certified), capability of operating at extreme latitudes, availability of specialized equipment (e.g. towage, oil spill response, fire fighting, fast rescue crafts) and spacious load deck (360 m^2) are all key features for a second-line multipurpose naval vessels, making the NorCan 222 ERRSV a perfect match for the job.

Figure 3 – NorCan 222 Emergency Response Rescue and Support Vessel.
Courtesy of NorCan Marine Inc.

Length overall	67.7 m (222')
Beam	12 m (40')
Draught	1.76 m (5.8')
Maximum speed	34 knots
Economic speed	27 knots
Bollard pull	80 tons
Fuel capacity	400 m^3
Fresh water capacity	200 m^3
Classification	LR ✠A1 or equivalent

Table 1 – NorCan 222 ERRSV characteristics

The propulsion concept is based on four controllable pitch (CP) propellers turning in a specially designed matched propeller tunnel. The combination of the propeller design and shape of the tunnel, specially designed and optimized for each application results in an unmatched propulsive efficiency.

Figure 4 - Transom view, showing inner propeller feathered for minimum resistance. Courtesy of NorCan Marine Inc. & Servogear AS

The outer shafts, or wing units, deputed to low and cruise speeds are powered by an AKA XeroPoint® Hybrid Propulsion in combination with a CP propeller.

The inner shafts, reserved to high transfer speeds, are powered by a conventional diesel mechanical system with a featherable CP propeller.

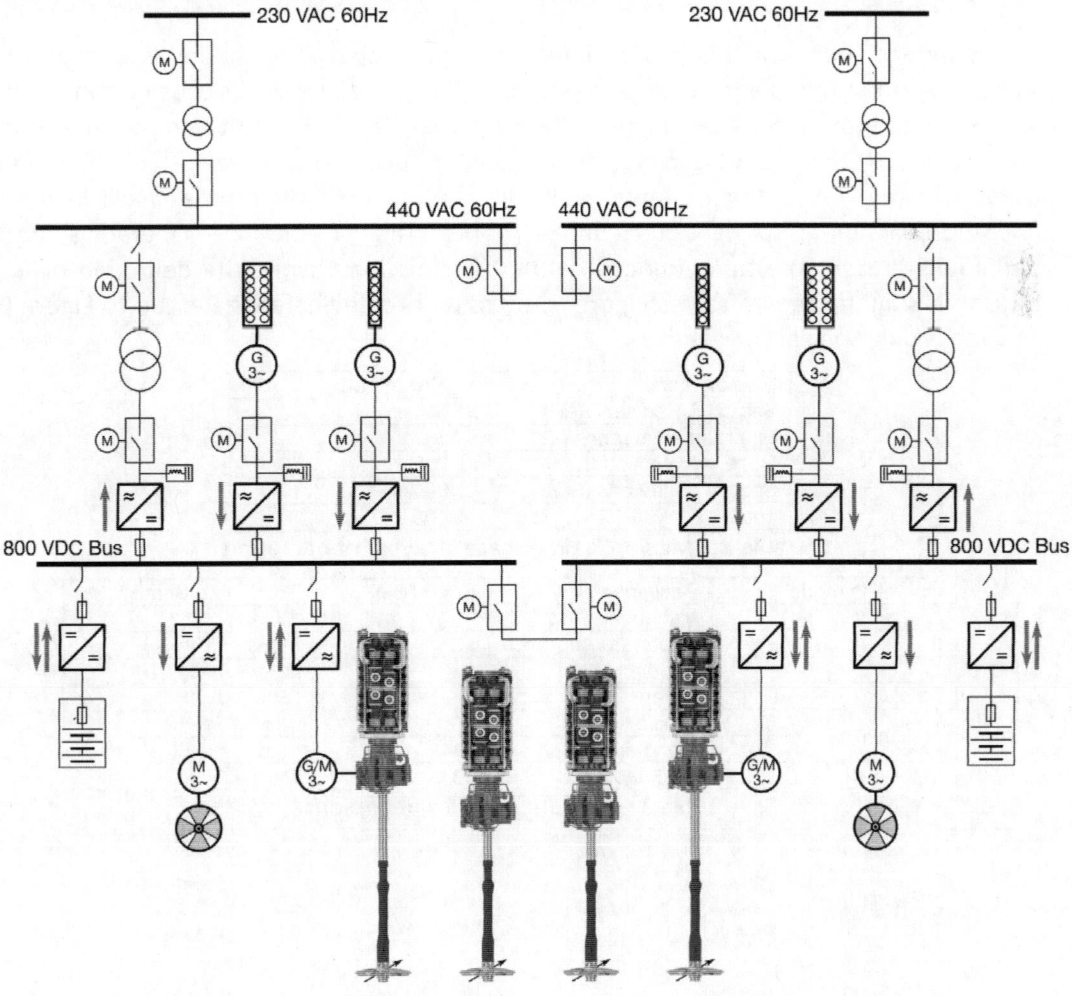

Figure 5 – Conceptual layout of propulsion and power distribution system

This layout has several advantages:

- prime movers are physically independent, providing a high level of redundancy and survivability;
- combining four CP propellers and bow thrusters allows maximum maneuverability;
- electric motors can be operated as shaft generators, reducing the amount of running equipment, increasing the overall fuel efficiency and reducing emission levels;
- the usage of battery banks provides the capability of running in full-electric mode, eliminating both the emission footprint and the risk of running internal combustion engines in dangerous environments;
- the propulsion system has intrinsic reliability during critical operations, without the need for additional sources of uninterruptible power supply (UPS);
- battery banks also act as "peak-shavers" guaranteeing a good load response while avoiding power fluctuations on the diesel engines, therefore improving fuel consumption and eliminating excessive wear;
- feathered CP propellers have minimal impact on ship resistance and impose no operational constraints.

Reference operating modes and profiles.

As mentioned above, a detailed definition of vessel operating modes is critical for a successful propulsion plant design and optimization. In real life this is usually confidential information, provided as an input either by the Naval Authority or by the Prime Contractor. For this study four typical missions of second-line naval vessels have been used. They are presented in Table 2 in combination with their time share in a year operation, assumed to be 2000 running hours. Table Table 3 – Operating modes definitions present the tasks (modes) within the missions, while the definition of each mission is summarized in figures Figure 6 – Mission profile Search & Rescue to Figure 9 – Mission profile Special Operations.

Mission name	Time Share
Search & Rescue (SAR)	50%
Emergency Response (ER)	5%
Fishery protection (FP)	25%
Special operations (SO)	20%
Total	100%

Table 2 – Mission & time share in a year of operation

Operating mode	Description	Speed (kn)	Total Time Share
I	Maneuver	6	1.5%
II	Departure/Approach	12	1.5%
III	Patrol	12	22%
IV	Loiter	5	39%
V	Cruise	16	19%
VI	Transfer	24	10%
VII	Fast deploy	30	7%

Table 3 – Operating modes definitions

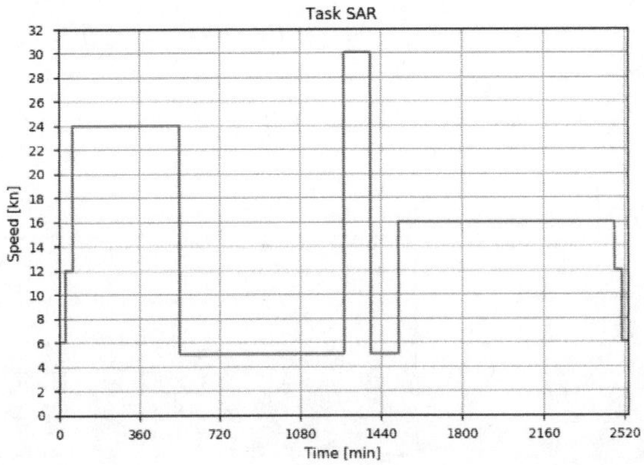

Ident	Mission: Patrol + Search & Rescue	Duration (hrs)	speed (kn)
I	Maneuvering	0,5	6
II	Departure	8,0	12
VI	Transfer	12,3	24
IV	Loitering	2,0	5
VII	Fast Deploy	2,0	30
IV	Loitering	16,0	5
V	Cruise	0,5	16
II	Approach	0,5	12
I	Maneuvering	0,5	6
	Total Mission Time	42,3	

Figure 6 – Mission profile Search & Rescue

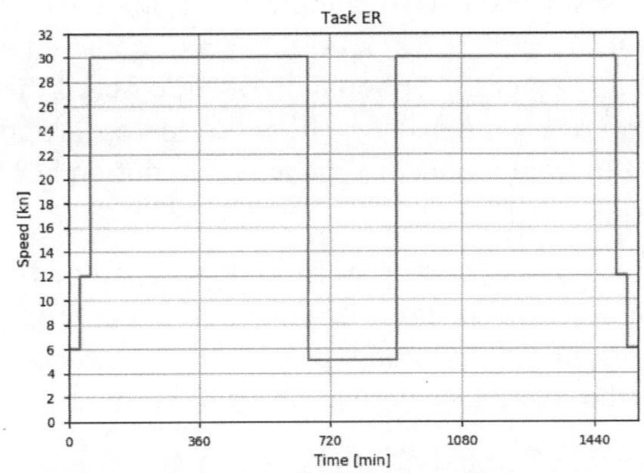

Ident	Mission: Emergency Relief	Duration (hrs)	speed (kn)
I	Maneuvering	0,5	6
II	Departure	0,5	12
VII	Fast Deploy	10,0	30
IV	Loitering	4,0	5
VII	Fast Deploy	10,0	30
II	Approach	0,5	12
I	Maneuvering	0,5	6
	Total Mission Time	26,0	

Figure 7 – Mission profile Emergency Response

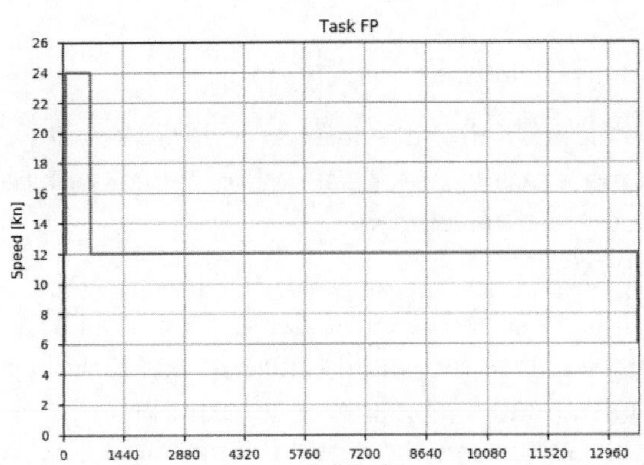

Ident	Mission: Special Operations	Duration (hrs)	speed (kn)
I	Maneuvering	0,5	6
II	Departure	0,5	12
VII	Fast Deploy	10,0	30
IV	Loitering	216,0	5
II	Approach	0,5	12
I	Maneuvering	0,5	6
	Total Mission Time	228	

Figure 8 – Mission profile Fishery Protection

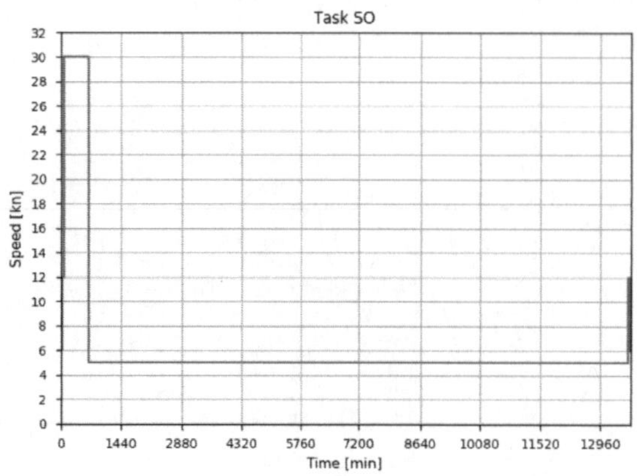

Ident	Mission: Fishery Protection	Duration (hrs)	speed (kn)
I	Maneuvering	0,5	6
II	Departure	0,5	12
VI	Transfer	8,0	24
III	Patrol	216,0	12
II	Approach	0,5	12
I	Maneuvering	0,5	6
	Total Mission Time	226,0	

Figure 9 – Mission profile Special Operations

Even if this deterministic approach has been extensively adopted in traditional design, due to the increased availability of real life operating data, statistical approaches, as described in [1] and [2], have been recently favored. It should anyway be noted that the methodology used to describe the operating profile is uninfluential to the propulsion design optimization process, therefore, the traditional deterministic description has been used in this case study to avoid useless complexity.

In general terms the choice of components of a hybrid propulsion system can be treated as a multi-objective design optimization problem. In contrast with traditional propulsion plants which can only be optimized on single operating points, the combination of elements in a hybrid system can be benchmarked across several criteria to find solutions with the best overall trade-off.

Both the objectives and their relative weights can be defined based on project-specific requirements and are usually complemented with additional constraints in order to limit the search to reasonable solutions.

Constraints translate technical and commercial limitations such as:

- rated power of diesel engines, available only in discrete steps according to ratings;
- power of variable frequency drives and electric motor as well as capacity of battery banks, limited by available space and vessel displacement;
- overall or individual cap on acquisition costs.

The investigation produces a set of optimized solutions. In order to fix a single design a final evaluation round is therefore needed. This step should include additional aspects which are difficult to formally be fixed in the initial phase, such as experience on former plants, technological and commercial risk evaluation for certain components, customer preference and so on.

Extensive and advanced work has already been done in such a field and several examples of design optimization algorithms such as [3], [4] and [5] are available in the technical literature.

The case study here presented is based on a calculation workflow internally established at MAN Diesel & Turbo, with the following exemplary set of objectives:

- minimum number of running engines in each operating mode;

- full-electric loitering (mode III) capability for not less than 30 minutes;
- optimum main engine load, according to rating selection;
- optimum auxiliary engine load (range 50%-100%);
- maximum continuous speed not below 32 kn;
- maximum peak speed not below 33 kn.

Additionally the search was restricted to take into account the following constraints:

- main engines according to rating availability as shown in Table 4 – MAN 175D rating overview;
- auxiliary engines according to market availability;
- power of electric motor not exceeding 1 MW and capacity of battery banks not exceeding 300 kWh, in order to limit the volume of frequency drives and switchboards.
- effect of aging of battery elements.

Rating	Rated Power 12V (kW)	Rated Power 16V (kW)	Optimum load range
Heavy Duty (145 kW/cyl)	1740	2320	60%-100%
Heavy Duty (155 kW/cyl)	1860	2480	60%-100%
Medium Duty (170 kW/cyl)	2040	2720	40%-80%
Medium Duty (185 kW/cyl)	2220	2960	40%-80%

Table 4 – MAN 175D rating overview

RESULT OF CALCULATION

For the above scenario the calculated optimal arrangement is summarized in Table 5 – Configuration of calculated optimal propulsion layout.

Equipment	Qty	Type	Power/Energy	rpm
Main Engine	2	16V175MM, Medium Duty Optimum load range: 50%-80%	2 x 2720 kW	1900
Booster Engines	2	16V175MM, Medium Duty Optimum load range: 50%-80%	2 x 2960 kW	1900
Gear box Main Engines	2	Single reduction, with PTI/PT0, ratio 3.25		585
Gear box Booster Engines	2	Single reduction, ratio 3.435		553
Main Propeller	2	VBS600 Mk5 A – CPP Prop. Diameter: 1650mm	2 x 3600 kW	585
Booster Propeller	2	VBS540 MK5 A – CPP Prop. Diameter: 1650mm	2 x 3000 kW	585
Battery Pack	2	Li-ion, 6C/3C Charge/Discharge rates, water cooled	2x 500 kWh	
PTO/PTI	2	Power factor cos(phi): up to 0,8	2 x 850 kW	Up to 1800
VSD	2	Low Harmonic AFE converter for PTO and PTI	2 x 850 kW	
GenSet	2	MAN D2862 LE322, 12Cyl	2 x 700 kW	1800rpm
GenSet	2	MAN D 2866 LE301, 6Cyl	2 x 280 kW	1800rpm

Table 5 – Configuration of calculated optimal propulsion layout

The benefits of such a configuration can be assessed by briefly analyzing the performance at different operating modes.

Mode I (Maneuver)

As expected from the figures in Table 3 – Operating modes definitions, due to the minimal time share, an optimization in terms of fuel consumption is not significant for this mode. However this mode is mainly performed in harbors, restricted or enclosed waters, where nowadays national maritime and port authorities have engaged in efforts to reduce emissions from all operating vessels within their jurisdiction.

The selected equipment therefore allows maneuvering as well as docking/undocking to be fully conducted in electrical mode meeting even the most stringent regulations.

Additionally the immediate energy availability from the battery packs and the use of variable speed motors provide the vessel with a very high level of maneuverability and agility.

The system operation schematic is presented in Figure 10 – Operation schematics for Mode I.

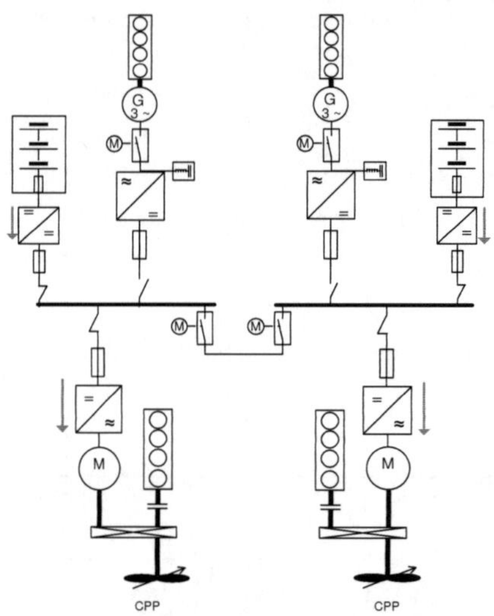

Figure 10 – Operation schematics for Mode I

Mode II (Patrol, Approach & Departing)

A traditional weak spot of conventional propulsion plant, this mode usually requires a single shaft operation to avoid extremely low loads on propulsion engines. This condition has significant drawbacks in terms of maneuverability and response time in case of sudden power request.

The proposed plant overcomes all these limitations, allowing a single diesel propulsion engine to mechanically drive one propeller shaft while providing electric power for both the vessel electric load and the electric motor on the other propeller shaft. This setup, unique to hybrid plants, is conceptually equivalent to a cross-connected gearbox and puts no limitations to vessel capability while minimizing the amount of running equipment and ensuring an efficient engine operating point. Additionally variating the engines speed the propellers can be driven on their efficient combinator curves with evident benefits in term of maneuverability and efficiency.

In this mode batteries can be used as a power reserve, either to compensate wind and current effects, to reach short-term speed peaks or to give a smooth transition to other operating modes.

The system operation schematic is presented in Figure 11 - Operation schematics for Mode II.

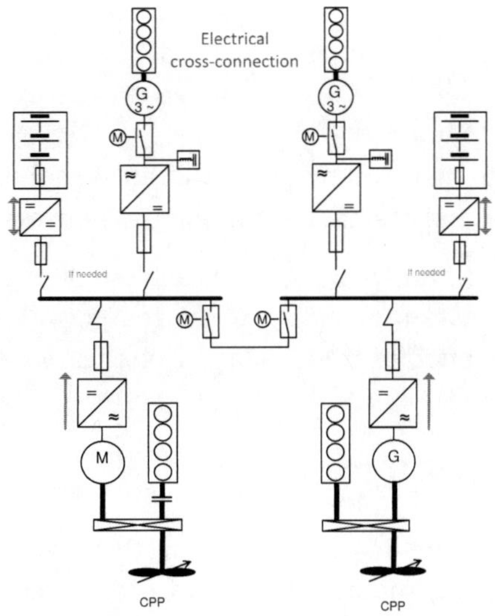

Figure 11 - Operation schematics for Mode II

Mode III (Loiter)

Accounting for almost 40% of the total operating time, performance during loitering is essential for a successful design. With the selected equipment configuration a single genset is used to power to both the electric propulsion motors as well as the power distribution network.

While the advantage over traditional mechanical propulsion plant is evident, this mode also proves itself an excellent example of application on battery-hybrid systems. In loiter mode, due to minimal speed, a major share of power demand is dedicated to vessel own electrical consumption. For a multi-purpose vessel such as the one being considered, this includes also equipment dedicated to cargo handling and special operations, typically highly variable in time. With conventional electric plant, this variability requires several gensets to operate in parallel to avoid the drawbacks of sudden load peaks, such as mechanical wear and reduced reliability with higher risk of blackout, while a battery-hybrid system is simply inherently able to cope with this operation mode without additional effort.

The use of electrical motors allows to move the propellers on their efficient combinator curves with evident benefits in the vessel maneuverability and efficiency.

According to the set objective, full-electric loitering mode with full hotel load is achieved in excess of 30 minutes.

Mode IV (Cruise) and V (Transfer)

Cruise and transfer speeds are achieved by only using the outer diesel engines, while operating the electric motors as shaft generators, thus with no genset running. This outcome is strictly related to the set of objectives as we explicitly searched for solutions to minimize the amount of running equipment.

As discussed above, even if batteries are not directly used they still provide Peak Shaving and Dynamic Support capabilities, with significant performance and reliability boosts in rough sea conditions.

Mode VI (Fast Deploy)

A speed of 30 knots can be reached by using all four diesel engines, with electrical motors operating as shaft generators, therefore keeping the gensets off.

A continuative max speed of above 33 knots is achievable by turning on gensets and using the electric motors as boosters. Even if not perfectly fuel efficient, this layout provides additional flexibility and enables the usage of battery banks as a power reserve, as discussed above.

SUMMARY AND FUTURE DEVELOPMENTS

As shown in the body of the paper, hybrid systems are quickly taking hold in naval designs. By properly selecting and integrating different technologies and including battery banks, it is possible to design solutions optimized for specific requirements.

A practical approach to mission-centric plant design has been shown in the presented exemplary case study and has already been applied to real projects. It is sufficient to observe the quick changes happening in other markets to understand that we are just scratching the surface and the real potential of this technology has still to come.

On the short term DC grids are expected to expand at the expense of traditional AC distribution systems, allowing easier integration of battery banks and usage of variable speed gensets and induction propulsion motors.

In contrast with the traditional fixed speed concept, variable speed gensets can be adjusted for minimum fuel oil consumption according to the system load. The DC grid allows a decoupled operation of the gensets for the consumers and also allows the easy integration of batteries via DC/DC converters. The footprint of such a propulsion plant is up to 25% smaller compared with a classical Diesel-electric propulsion plant based on an AC system.

The DC architecture allows also the usage of induction motors in place of the traditional synchronous machines. Induction motors are cheaper, simpler and more robust machines which can significantly reduce acquisition costs, nowadays the major obstacle to hybrid market diffusion.

Finally DC systems are core technology for leading markets, the automotive one above all, and will therefore benefit the big innovation wave already ongoing, including more sophisticated developments such as autonomous and unmanned operation, currently being investigated also in the marine market.

Even if marine and naval markets have their own specificity, they will definitely be affected by this global trend, therefore battery-hybrid systems can be reasonably considered to stay in the spotlight and gain momentum for the years to come.

BIBLIOGRAPHICAL REFERENCES

1. Greitsch, L.; Eljardt, G. Simulation of Lifetime Operating Conditions as Input Parameters for CFD Calculations and Design Evaluation. *Proc. 12th Numer. Towing Tank Symp. 2009* **2009**, 61–66.

2. Cunningham, A.; Wang, J.; Allanson, D.; Wall, A. Application of Monte Carlo Method in the Marine Environment. *J. Sh. Prod.* **2010**, *26*, 76–87.

3. Sciberras, E.; Grech, A. Optimization of Hybrid Propulsion Systems. *Int. J. Mar. Navig. Saf. Sea Transp.* **2012**, *6*, 539–546.

4. Alwan, S.; Yum, K. K.; Steen, S.; Pedersen, E. Multidisciplinary Process Integration and Design Optimization of a Hybrid Marine Power System Applied to a VLCC. In *16th Conference on Computer and IT Applications in the Maritime Industries*; 2017; pp. 336–350.

5. Chen, C.; Wang, X.; Xiao, J. An energy allocation strategy for hybrid ship DC power system based on genetic algorithm. *IETE J. Res.* **2016**, *62*, 301–306, doi:10.1080/03772063.2015.1093966.

Power and Propulsion System Options for Future Frigate Programmes – Hybrid as an Enabler for Increased Mission Capability

Edward Francis Wright MSc. CEng. MIMechE, Senior Principal Marine Engineer – Propulsion Systems - Rolls-Royce Naval, Edward.Wright@Rolls-Royce.com,
Richard J. Partridge MSc. CEng. FIMechE, Chief of Naval Systems, Richard. Partridge@Rolls-Royce.com (Presenter subject to change).

SYNOPSIS

Building on the paper that Rolls-Royce presented at the Pacific 2015 conference entitled "Modern Power and Propulsion System Options for Future Frigates"[1] which described the Royal Navy's journey from the introduction of gas turbines into the mechanical power and propulsion systems of the 1960s through to the modern single-gas turbine hybrid power and propulsion system of the Type 26 Global Combat Ship, the objective of this paper is to increase the audience's understanding of the advantages and considerations of hybrid systems. Hybrid will be considered as an "Enabler for Increased Operational Capability" in meeting the requirements of today's missions and in providing sufficient flexibility to enable technology insertion and therefore meet future mission requirements throughout the life of the platform, which could be as far into the future as 2060. Reviewed in light of increasing power demand from mission systems driven by top-level requirements such as high-end anti-submarine warfare, an increasing requirement for multi-mission capability and an anticipated increase in unmanned operations - this paper will present the advantages and considerations of hybrid power and propulsion systems in general whilst focusing on the comparison between Combined Diesel eLectric And Gas (CODLAG) and Combined Diesel eLectric Or Gas (CODLOG) which, depending on the platform being considered, are both being evaluated by the Commonwealth for the SEA 5000 program. This discussion will be developed from experience gained by Rolls-Royce Naval having worked on a number of frigate programs worldwide and will make the argument for single-gas turbine CODLOG power and propulsion systems since it provides a highly-capable and robust solution meeting operational, reliability and survivability requirements at the same time as reducing whole of life costs.

Author's biographies

Edward Wright joined Rolls-Royce Marine from Nottingham University in 2009 where he graduated with a first in Mechanical Engineering (Masters). He is now a Senior Principal Marine Engineer with the Rolls-Royce Naval Systems team based in Bristol, UK. Previously, he has been based in Christchurch, New Zealand, as well as in Genoa, Italy, where he worked to support national shipbuilding programmes. Prior to this position in Naval Systems, Edward worked in a number of positions in other areas of Rolls-Royce Marine including offshore propulsion products, marine gas turbines and submarines. His current professional interest is in the impact of unmanned technologies into the marine space.

Richard Partridge is the Chief of Naval Systems at Rolls-Royce PLC, and has worked for the company since 1999. He has worked on a variety of recent naval projects including the Royal Navy's Queen Elizabeth Class Aircraft Carrier, the US Navy's Littoral Combat Ship (Monohull) and the Republic of Korea Navy FFX-II frigate. He worked for the Energy sector of Alstom Gas Turbines between 1996 and 1999, and prior to that spent six and a half years in the Marine Engineering branch in the Royal Navy, including three years serving on an Anti-Submarine Warfare Frigate. He is also a graduate of University College London where he received a Masters degree in marine engineering.

NOMENCLATURE

AER	Aft Engine Room
ASW	Anti-Submarine Warfare
CODAG/CODOG	Combined Diesel And/Or Gas [Turbine]
CODLAG/CODLOG	Combined Diesel Electric And/Or Gas [Turbine]
COGAG/COGOG	Combined Gas [Turbine] And/Or Gas [Turbine]
DEW	Directed Energy Weapons
DG	Diesel Generator
EHM	Equipment Health Monitoring
EM	Electric Motor
EM	Energy Management
FER	Forward Engine Room
GCS	[Type 26] Global Combat Ship
GP	General Purpose
GT	Gas Turbine
HVAC	Heating, Ventilation and Air Conditioning
IAM	Intelligent Asset Management
IFEP	Integrated Full Electric Propulsion
IPMS	Integrated Platform Management System
LDEW	Laser Directed Energy Weapons
MBHS	Mission Bay Handling System
MOTS	Military Off The Shelf
P&P	Power and Propulsion
PG&D	Power Generation and Distribution
Powergen	Power Generation
R&A	Remote and Autonomous
RAN	Royal Australian Navy
RN	[UK] Royal Navy
RoKN	Republic of Korea Navy
SI	Ship Intelligence
UK	United Kingdom
URN	Underwater Radiated Noise
USN	United States Navy
UxV	Unmanned Maritime Vessel
WLC	Whole of Life Costs

INTRODUCTION

At the time of writing, the Royal Australian Navy's (RAN) SEA 5000 programme is, amongst others, considering frigate designs that are powered by single-Gas Turbine (GT) hybrid electro-mechanical Power and Propulsion (P&P) systems. This move to hybrid, from the mechanical P&P systems of the ANZAC class and the Hobart class, represents an exciting opportunity for the RAN in terms of increasing operational capability. As such, this paper will consider such systems in more detail than was previously considered by the paper "Modern P&P System Options for Future Frigates"[1] which was presented by Rolls-Royce at Pacific 2015. At the time, the 2015 paper's objective was to describe the

Royal Navy's (RN) journey from the first introduction of GTs through to the modern single-GT hybrid P&P system of the Type 26 Global Combat Ship (GCS). This 2017 paper will first recap this story in brief before considering hybrid systems in more depth, their nuances and their adoption as an "Enabler for Increased Mission Capability". This will be for meeting the requirements of today's missions such as high-end Anti-Submarine Warfare (ASW) and tomorrow's missions such as the increased adoption of high energy consumers such as Directed Energy Weapons (DEW), additional sensors and unmanned (UxV) technologies. The advantages and considerations of these systems, Rolls-Royce experience from relevant programmes and the contribution of modern propulsion technology to these systems will also be discussed.

A BRIEFER HISTORY OF NAVAL POWER AND PROPULSION

Mechanical Systems

Gas Turbines have been firmly established in RN propulsion systems since they were first adopted on the Type 81 class frigates in the 1960s. Subsequent sea-going experience demonstrated benefits, when compared to steam plants, that included increased reliability and power density, lower maintenance burden, ease of change-out, low load operability and improved supportability. These benefits led to the RN adopting all-GT COmbined Gas Or Gas (COGOG) mechanical propulsion systems for the Type 21, 22 and 42 where the ship's electrical Power Generation and Distribution (PG&D) system was separate from the propulsion system and usually consisted of three or four Diesel Generators (DG) – this is shown in Figure 1.

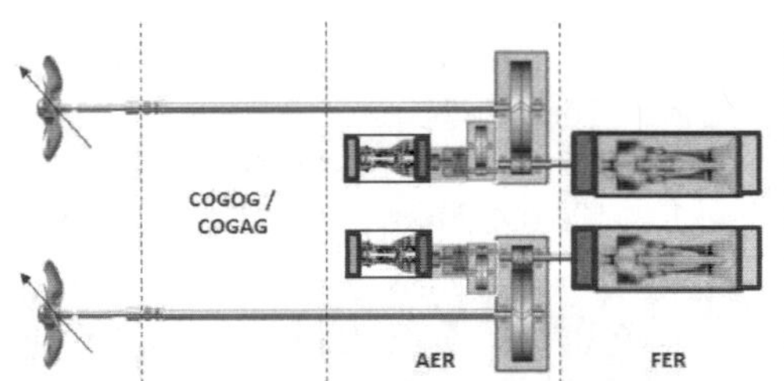

Figure 1: All-GT COGOG propulsion system arrangement as used by the RN. Note that all-GT COmbined Gas And Gas (COGAG) propulsion system arrangements are also referenced in Figure 1 and, today, these are predominantly in-service with the United States Navy (USN). The arrangement does not show the separate PG&D system.

Taken into the context of a typical warship's operational profile which is dominated by low to medium speeds operators would, in practice, run on a single GT. This reduced GT operating hours, however, the low-load operation increased fuel consumption and the trailing shaft resulted in losses up to 20% because of the additional ship resistance[2].

During the same period, many European, Middle Eastern and Asian navies were specifying more affordable COmbined Diesel And/Or Gas (CODAG/CODOG) propulsion systems as shown in Figure 2. Again, the ship's electrical PG&D was separate from the propulsion system and usually consisted of three or four DGs.

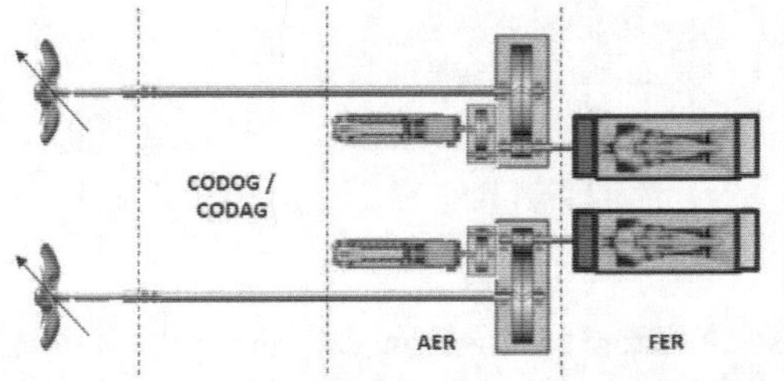

Figure 2: CODAG/CODOG propulsion system arrangement - does not show the separate PG&D system.

Compared to all-GT systems, such systems improved efficiency across the operational profile, namely in the low to medium speed range however, when compared to GTs, the diesels increased the level of on-board maintenance, increased noise and vibration levels leading to higher Underwater Radiated Noise (URN) and reduced plant power density.

More recent single-GT mechanical platforms such as the ANZAC class (CODOG) have refined the application of mechanical propulsion systems by introducing an "eco-mode" as shown by Figure 3. This development improves on traditional mechanical systems such as that shown in Figure 2 by reducing the number of installed prime movers, reducing diesel running hours, reducing diesel maintenance burden and improving on efficiency by eliminating the need for trailing shaft operation. These advances provided important savings in sustainment costs however, operationally, low loading of the main diesel engines during extended low-speed operations remained an issue.

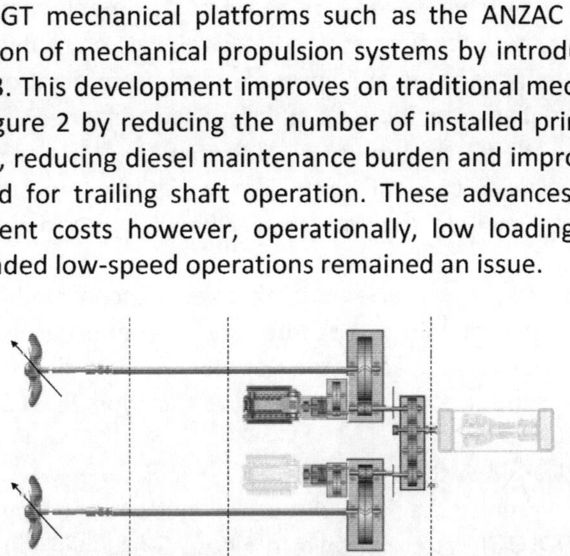
Figure 3: Single-GT CODOG/CODAG propulsion system showing "eco-mode" operation (single diesel/twin shaft). The arrangement does not show the separate PG&D system.

Until recently, mechanical arrangements have been constrained by the market availability of GTs at the time of selection, essentially limited to a unitary power of ≈20MW, mandating boost GT capability via twin- or quad-GT to reach the necessary 30 knots in larger combatants. Single-GT arrangements have been permissible in smaller or <30 knot combatants but have necessitated the use of large diesels and/or an uprated GT.

Regardless of which arrangement was adopted, operational experience with mechanical systems found their high levels of URN to be the limiting factor on the journey toward today's high-end ASW operations. For the RN, this led to the development of hybrid P&P systems for combatants in order to reduce URN during ASW.

Twin Gas Turbine Hybrid Systems

Building on the above operational experience and product developments, the first combatant designed with a hybrid P&P system was the RN Type 23 frigate which used a COmbined Diesel eLectric And Gas (CODLAG) P&P system shown in Figure 4.

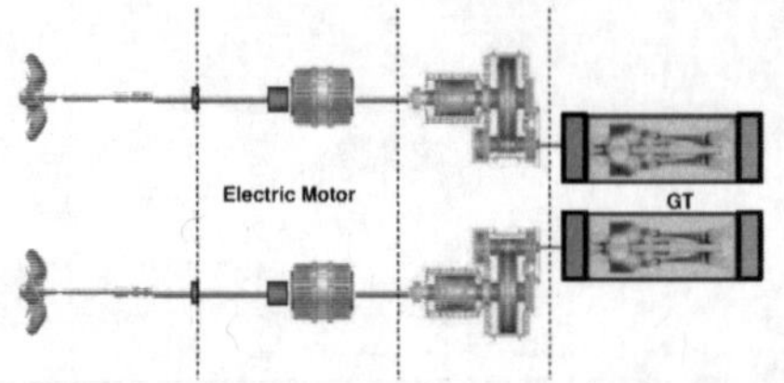

Figure 4: Type 23 CODLAG P&P system arrangement showing shaft-mounted EMs for a cruise speed of 15 knots and twin-boost GTs for speeds up to 28 knots. The arrangement does not show the DGs.

Hybrid systems improved over their mechanical predecessors in the following respects:

- Individual DGs could be brought online to meet the ship's power demand which now included the Electric Motors (EMs) speeds up to the ship's cruise speed - this maintained high individual DG loading to optimize their efficiency and maintenance.
- Increased platform efficiency, particularly in the context of an operational profile dominated by low ship speeds, reduced sustainment costs and extended range yielding military benefits such as a reduced reliance on replenishment ships.
- Reducing the number of installed prime movers from eight on mechanical platforms to six on hybrid platforms reduced maintenance burden.
- For ASW operations, the ability to declutch the EMs from the gearboxes and shutdown gearboxes to significantly reduce URN when compared to mechanical systems. Quiet diesel-electric operation achieved by enclosing and isolating DGs.

The Type 23 hybrid P&P system proved itself able to accommodate successive upgrades which included a change of role to become more General Purpose (GP) and a move to warmer operational climates placing greater demand on the ship's electrical system. These lessons learnt, amongst many others, drove the Type 26 GCS development.

Single Gas Turbine Hybrid Systems

The Type 26 GCS will replace the Type 23 but will move to the single-GT COmbined Diesel eLectric Or Gas (CODLOG) system shown in Figure 5 and will join a growing number of single-GT hybrid platforms including the Franco-Italian FREMM (plus export variants), Italian Navy PPA, German Navy F125 and Republic of Korea Navy (RoKN) FFX-II frigate.

Figure 5: Left: The RN Type 26 GCS CODLOG P&P system showing shaft-mounted EMs for a cruise speed of 18 knots and single boost Rolls-Royce MT30 GT for speeds of 26+ knots. Right: Type 26 GCS with CEA radar for SEA 5000 (Image courtesy of BAE Systems Australia).

For the RN, the Type 26 GCS's CODLOG P&P system improves on the Type 23's CODLAG system because the move to the simpler "or" system reduced complexity (and cost) in the design of the Integrated Platform Management System (IPMS), EM, power converter and gearbox (see later sections). Furthermore, the move to a single-GT further reduces the number of installed prime movers to five which serves to reduce acquisition costs, halves the GT running hours and, overall, reduces the sustainment costs of the platform.

Such single-GT "or" architectures have been enabled by the advent of aero-derivative marine GTs with powers exceeding 32MW per unit, ie: the Rolls-Royce MT30 which can be rated up to 40MW at USN day conditions. In an "or" system, these larger GTs break the relationship between the ship's top speed requirement, EM sizing, availability of the EMs and PG&D system. This reduction in complexity, volume and mass can lead to further optimization opportunities in other areas of the P&P system and a positive design spiral.

HYBRID IN DEPTH – AN ENABLER FOR INCREASED OPERATIONAL CAPABILITY

Most of the platforms being considered for SEA 5000 feature a single-GT hybrid P&P system[3]. Compared to the ANZAC class, this represents a step forward in operational capability for the RAN. The advantages and considerations of this step are explored here since the implications of decisions made during the concept design phase of the candidate platforms will be useful to understand in the selection of the final platform.

Ship Speed - Cruise

Regardless of whether a CODLOG or CODLAG system is chosen, the ship achieves all speeds up to its cruise speed using EMs to deliver power to the propellers where electrical power to the EMs is provided by the DGs. As previously described, these DGs can be brought online individually to meet the ship's power demand which now includes the EMs. At low or loiter speeds, this means that perhaps only two DGs need to be online with their loading maintained at a high level which reduces overall diesel running hours, maintenance burden, etc. As the ship approaches its cruise speed of 15-18 knots, an increased number of DGs will be brought online to meet the higher power requirement. If a vessel has n+1 redundancy specified then three of four DGs will eventually be brought online to meet the cruise speed requirement and ship's service load.

One factor that arises during hybrid P&P system design is the cruise speed that is obtainable when compared to that of legacy mechanical arrangements (eg: CODOG). Modern hybrid platforms tend to have a cruise speed in the range of 15-18 knots whereas mechanical platforms can usual achieve at least 18-20 knots using diesel main propulsion. The challenge that arises with increasing cruise speed on hybrid platforms is in physically and functionally integrated a direct-drive shock rated low noise P&P system as illustrated in Figure 6.

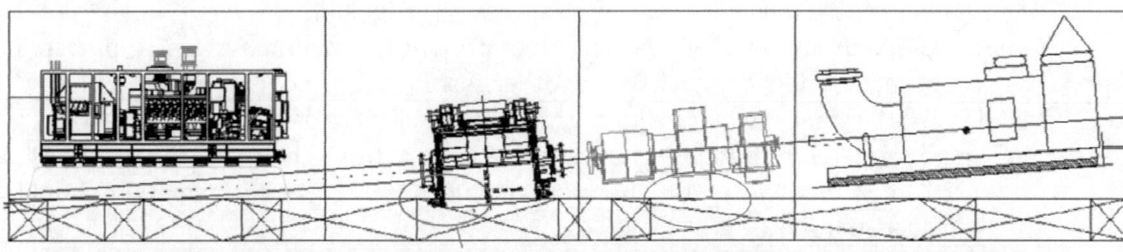

Figure 6: Physical integration issues with a direct drive shock rated propulsion system. The diagram highlights, in red, potential clashes between propulsion machinery and the tank top.

As cruise speed requirement increases; EM, power converter, PG&D size will also increase. This increases the complexity of physically and functionally integrating the P&P system, however, it must be noted that the relationship between EM power and its physical size is not linear as shown in Figure 7 using parametric motor data at various powers. The relationship between EM power, size and cost is also not linear so it is important to carefully trade-off the cruise speed requirement with the wider ship design. Any dependency on the EMs supplementing the GT power in an "and" arrangement for top speed must be avoided to avert a negative design spiral potentially crossing a threshold and invoking High Voltage PG&D. Rather, a higher-rated GT should be specified.

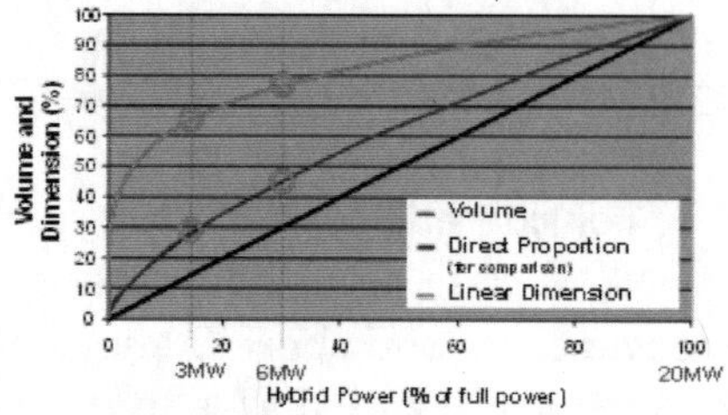

Figure 7: Parametric comparison of hybrid EM size.

Ship Speed - Boost

Considering Figure 5, if a CODLAG system is specified, the ship must achieve its top ship speed through the addition of power from the GT to the power provided by the EMs. Initially, the EMs will be sized according to the power required by the ship's cruise speed. A GT must then be specified to provide sufficient additional power to achieve the ship's top speed. Should the market not be able to provide a suitably powered GT to achieve this top speed requirement, there are cases where projects have investigated using larger EMs to supplement the GT power, however, this risks incurring the problems associated with physically and functionally integrating the P&P system when up-stream effects in the PG&D system are considered.

In comparison, if a CODLOG system is specified, the ship must achieve its top speed using only power from the GT which is specified as such. This means that the EM sizing can be optimized solely for the ship's cruise speed as there is no risk of the EM size being increased as a result of a shortfall in power originating from the GT to reach maximum ship speed. Therefore, the ability for the ship to reach its maximum speed is dependent only on the availability of the GT (and relevant auxiliary systems) leading to a more robust system design ensuring vessel performance.

Ensuring Vessel Performance

One significant challenge in P&P system design is global operation in extremes of temperature without any degradation in performance. A hybrid system is impacted by this sort of requirement as much as any other system and, since the RAN is active in the Middle East, this needs to be considered at the earliest stage of the project to avoid:

- Increased electrical load in hotter climates (eg: from chilled water plants, HVAC, etc.) impacting the cruise speed attainable on EMs. As such, margin should be included in the DGs.
- Specifying the prime movers for operation at lower ambient temperatures or ducting losses that are un-realistic resulting in reduced power.

Furthermore, margin must be included in the P&P design to accommodate for any changes that are made to the platform selected for SEA 5000 at the point of build (eg: the new CEAFAR radar), through life updates and upgrades as well as the inevitable displacement growth of the platform through life.

Anti-Submarine Warfare

When compared to the mechanical P&P systems of the ANZAC class frigate or the Hobart class destroyer, a hybrid-powered SEA 5000 would have significantly reduced URN contributing to a significantly enhanced ASW performance. According to BAE Systems "in excess of five thousand man years of engineering design effort has been applied to the RN Type 26 GCS to make it the world's most potent ASW platform"[4]. This is reflective of the amount of effort that is required to address all ship factors contributing to the platform's overall noise signature and it is understood that this would have included discussions on re-using the Type 45 destroyer hull form as an off the shelf solution for Type 26 GCS, however, "Type 45 was optimized for high speed rather than cruise so it is entirely reasonable that the hull forms would be different[5]". As such, the Type 26 GCS hull form would have also been optimized for low noise during high-end ASW operations.

Again, whilst still a ship-level consideration, the P&P system is a significant contributor to URN and a hybrid system aims to improve URN performance in the following ways:

- The use of EMs enables the use of fixed-pitch propellers which, when compared to controllable pitch propellers typically used on mechanical systems, can be quieter from a hydrodynamic perspective.
- Since the majority of submarine hunting operations are performed at 9-12 knots, a hybrid platform will run on the EMs whilst de-clutched from the gearbox. This removes the gear meshing noise as well as noise from the gearbox's auxiliary systems (eg: the cooling system) that would normally emanate from a mechanical platform running on main propulsion diesels.
- Power for the EMs is provided by enclosed and resiliently mounted DGs. At such a low speed and thus power requirement, it is foreseeable that only two DGs would be required during ASW and these could be physically located higher in the ship to reduce URN. This was the approach that was used by the RN on the Type 23 and, in future, on the Type 26 GCS.

By 2030, it is expected that half of the world's submarines will be in Australia's broader strategic region[6] and so a high-end ASW capability will be required to combat these threats. The Commonwealth Government should look to leverage its partner navy's investments and capabilities in ASW and so establish itself at the forefront in developing the next generation of ASW.

Reliability and Survivability

For single-GT systems, comparing P&P system reliability and survivability must be considered from the perspective of the move from twin to single GT and from the whole system's perspective.

From the perspective of the GT, modern aero-derivative marine GTs offer ultra-high reliability as a consequence of development and legislation within the aerospace industry which has demanded increased unitary power and engine reliability to move from, for example, the quad engine configuration of the Boeing 747 to the twin engine configuration of the 777 – this relationship is shown in Figure 8. Therefore, whilst the redundancy offered by two GTs is perceived to be higher than a single GT, a single-GT based on modern technology with high reliability can, from a system's perspective, be more reliable than a twin installation of legacy GTs with lower reliability.

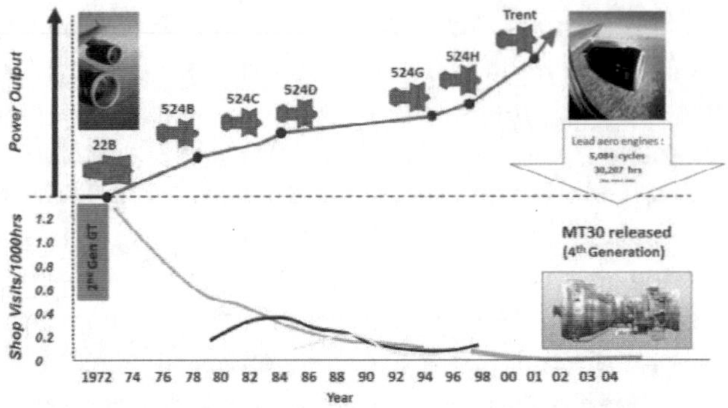

Figure 8: Unitary power output and reliability over time for the Rolls-Royce Trent series of aerospace engines from which the MT30 marine GT is derived.

Since modern warships spend as little as 3% of their seagoing hours at boost speeds[7], a modern marine GT may never need a scheduled or, considering Figure 8, an unscheduled overhaul throughout its life when an annual usage of 500-1000 hours is assumed.

From the system-level perspective, single-GT hybrid with an overall reduction in engines can have excellent survivability when the P&P system is carefully laid out and dependencies/common-mode failures are avoided - this is demonstrated in Figures 9 and 10. Figure 9 shows a legacy twin-GT CODOG system where the GTs are generally co-located in the Forward Engine Room (FER) due to inability to configure a staggered-propulsion layout on a ship with a displacement of less than 8000t. The ship's DGs are located in pairs in auxiliary machinery rooms (eight engines total). The gearboxes are a single point of failure in the Aft Engine Room (AER) where the compartment could be lost due to fire, flood or action damage. By comparison, Figure 10 shows a modern single-GT CODLOG arrangement with just five installed engines, but avoids the single point of failure via careful location of DGs, power converters, transformers, etc.

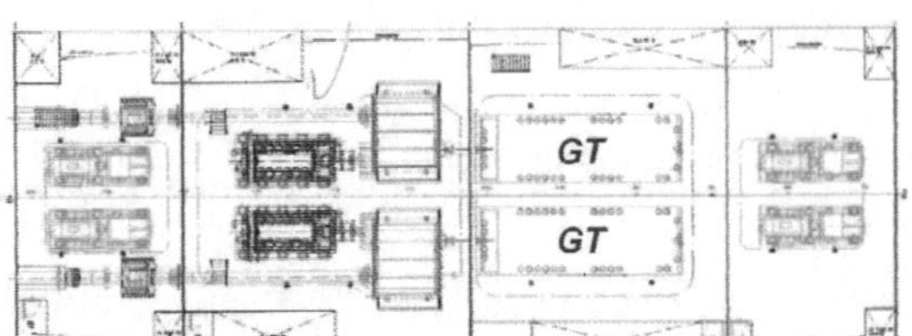

Figure 9: Twin-GT CODOG/CODAG system layout. ANZAC class is the same but with a single GT in the FER.

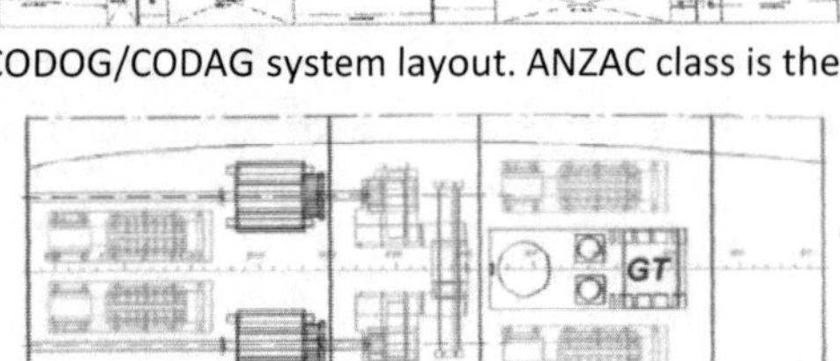

Figure 10: Single-GT CODLOG/CODLAG propulsion system layout showing better redundancy in hybrid systems. No loss of propulsion assumes shaft line components specified to run in a flooded compartment.

Considering Figure 10 in the context of the SEA 5000 programme, it should be noted that modern high-end ASW platforms such as the RN Type 26 GCS have the aft DGs located higher in the ship. This is to attenuate noise and vibration that is permeated into the ship's structure as much as possible to reduce URN and thus improve ASW capability.

Reducing Sustainment Costs

As described, the move from mechanical to single-GT hybrid platforms reduces the number of engines which reduces both acquisition and sustainment costs. Lower engine numbers enables reduced Marine Engineering complement and contribute towards a whole-ship lean manning concept. Whilst this reduction in engine numbers is particularly noticeable in the reduction of diesel cylinder numbers, the move from twin to single-GT also serves to half the number of GT running hours.

Legacy warships were often single-role and were configured around defensive duties such as an escort or for task force protection. The possible reduction in sustainment costs needs to be assessed in the light of new and emerging missions such as anti-piracy operations where the vessel will spend extended periods of time patrolling in diesel-electric mode and fast response will typically be achieved using a high speed interceptor or RHIB[8] rather than GT boost.

Compared to mechanical platforms, hybrid platforms are inherently more future proof as they start life with a higher installed power generation (powergen) capacity which is an advantage in light of wide-spread electrification and the ever-increasing power demand from mission critical weapons and sensors. This reduces the risk of incurring the cost to upgrade the PG&D system or, worse, the platform becoming obsolete.

The RN Type 23 hybrid P&P system demonstrated itself to be sufficiently flexible to accommodate successive upgrades which included a change of role to become more GP and a move to warmer operational climates placing greater demand on the ship's electrical system. These lessons learnt, and others, drove the RN Type 26 GCS development and the move from CODLAG to CODLOG.

CODLOG versus CODLAG

The differences between CODLAG and CODLOG P&P systems have already been touched upon – this section aims to summarize and expand upon this discussion so far.

Using the RN as an example, the Type 26 GCS's CODLOG P&P system improves on the Type 23's CODLAG P&P system because the move from an "and" system to the simpler "or" system reduces complexity (and cost) in the:

- IPMS system since there are now only two main machinery states to control; diesel-electric for all speeds up to and including the cruise speed and GT-only for all speeds up to maximum speed. Whereas, for a CODLAG system, there are three machinery states; diesel-electric to cruise, GT-only for some speed less than maximum ship speed and CODLAG for maximum speed – this introduces an additional control requirement for the gearbox.
- Gearbox since, for CODLOG, only one reduction ratio is required from the GT to the shaft lines. Whereas, for a CODLAG system, a second reduction ratio may be

required to allow both CODLAG mode and, additionally, GT-only mode to ensure good reversionary mode performance.

- EM and power converter since, for CODLOG, they must only be designed to one operational point, ie: max. torque at cruise shaft speed. Whereas, in a CODLAG system, they must be designed to two operational points which can increase EM and power converter size, mass and cost. ie: Max. torque at cruise shaft speed, and, max. EM speed at top ship speed which can dramatically increases the VA rating of the power converter for CODLAG systems.
- DG running hours are reduced since, for CODLOG, power from the DGs to the EMs is not required for the vessel to obtain its top speed, unlike CODLAG.

For the navies of the world who are looking to adopt DEWs and additional sensors in the near future, installed powergen is rapidly becoming a strategic resource. Since a hybrid platform naturally has a higher installed powergen capacity than a mechanical platform, around 10-12MWe for a hybrid versus 4-6MWe for a mechanical, a hybrid platform is naturally more future proof or less at risk of costly PG&D upgrades later in life. Along these lines, platforms such as the RN's Type 26 GCS or RoKN FFX-II frigate who employ CODLOG systems will find that vessel performance can be maintained by the GT, which can provide power for all ship speeds up to maximum speed, whilst the installed powergen can be reserved in its entirety (10-12MWe as above) for current and future high-energy mission systems. By comparison, in a high-threat state where maximum ship speed is important, a CODLAG system can only provide the electrical power which is not being used for propulsion without reducing vessel performance and, therefore, operational capability.

Building on reliability and survivability aspects, if there is an issue as a result of maintenance, failure or battle damage with single installed GT then both the CODLAG and the CODLOG platform are limited to cruise speed – an equal impact. On the other hand, if there is an issue with part of the PG&D system or the EMs then the CODLOG platform is able to reach all ship speeds using GT power along whereas the CODLAG platform will not. CODLOG systems have this luxury as a result of the advent of highly reliable aero-derivative marine GTs with powers exceeding 32MW per unit, ie: the Rolls-Royce MT30 which can be rated up to 40MW at USN day conditions.

Whilst there are benefits in adopting the simpler and less complex CODLOG system both in terms of driving a positive design spiral and in reducing Whole of Life Costs (WLC), the operational aspect is important to remember as a less complex system will be easier to develop and sustain crew expertise and skillsets during peacetime and easier for the crew to manage with during an emergency situation.

IMPACT OF FUTURE TECHNOLOGIES

Power Generation as a Strategic Resource

Whilst it is feasible to design a platform to meet today's mission requirements, it is more difficult to predict tomorrow's requirements. Although envisioned around the concept of an Integrated Full Electric Propulsion (IFEP) platform, the UK Ministry of Defence / RN Marine Engineering Development Strategy[9] and USN Integrated Power System

Roadmaps[10] included an aim to "future-proof" the P&P system so as not to preclude the future retro-fit of increasingly power-hungry mission systems such as DEW and advanced sensors. If their conclusions are taken into the context of a hybrid platform, increased powergen helps this "future-proofing" and presents an opportunity to reduce WLCs in terms of mid-life upgrades to the P&P system. Compared to the ANZAC and Hobart class, the hybrid platforms under consideration for SEA 5000 could offer as much as four times the powergen capacity which will help to "future-proof" the platform in the same way as was demonstrated on the RN Type 23 frigate. Furthermore, having the ability (via higher power output boost GT) to select CODLOG is important as, unlike CODLAG, it avoids the dependency of top ship speed on the availability of all elements of the diesel-electric propulsion system. CODLOG also releases the PG&D system to support the progressively higher ship service load and supports the "incremental acquisition" of mission systems such as Laser DEW as and when they achieve the necessary technology readiness level. These are important considerations since it is conceivable that the platforms procured under the SEA 5000 programme will be in-service as far into the future as 2060.

This paper has already touched upon the need for high-end ASW in Australian waters but it should be added that the best possible ASW platform should be specified today such that, with weapon and sensor upgrades throughout the life of the vessel, the platform remains relevant for as long as possible without a costly replacement. In this respect, it has been argued that a CODLOG platform provides the best possible solution in terms of the availability of redundant electrical power for updated or upgraded mission systems.

Mission Bay Operations

Referring to Figure 11, many modern combatants are being specified with a mission bay and dedicated Mission Bay Handling System (MBHS) that provides a re-configurable and highly adaptable space within the vessel such that a multi-mission role can be adopted. This multi-mission capability may well take the form of the launch and recovery of a UxV toolbox meaning that the vessel will quickly reconfigure to become the central node in a variety of missions such as mine warfare, unmanned ASW, border control, surveillance, search, rescue, surveys, logistics transport, etc.. However, regardless of the mission, the "mother ship" will have to provide power to the systems in the mission bay, loiter (slow speed) and respond quickly (sprint). Considering each of these points in turn:

- For the reasons discussed above, a vessel with a mission bay naturally lends itself toward a hybrid platform due to the higher installed powergen capacity and its ability to accept higher energy consumers onboard via the mission bay.
- If the vessel is forced to loiter or operate at very low speeds for extended periods of time then, again, a hybrid platform is advantageous since DGs can be brought online as required to meet the electric power demand and operate more efficiently rather than, on a mechanical platform, running large main engines at very low MCR levels where fuel efficiency and the maintenance regime is compromised.
- If the platform is forced to sprint to respond to an incident in the UxV area of operations then a CODLOG platform can proceed at the maximum ship's speed on GT power whilst DG power remains reserved for mission systems.

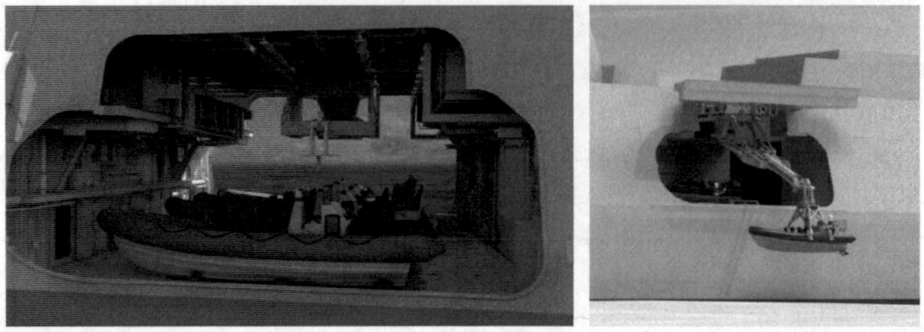

Figure 11: Left: The RN Type 26 GCS mission bay showing the Rolls-Royce MBHS overhead (Image courtesy of BAE systems). Right: MBHS handling a manned boat, alternatively, an UxV asset, ISO container, etc.

During side UxV recovery operations, the ability of the platform to maintain the slowest possible speed whilst maintaining directional control (yaw) and stability (roll) is paramount. For example, some UxVs can only achieve 3-4 knots so they are challenged to keep up with their mother ship that, likewise, is challenged to reduce speed whilst maintaining sufficient stability from the motion control surfaces such that the MBHS operator can quickly "pick" the UxV out of the water.

Ship Intelligence

As organized within Rolls-Royce, Ship Intelligence (SI) breaks down into two categories; Remote & Autonomous (R&A) operations and Intelligent Asset Management (IAM). In the commercial sector, the objectives of SI are to continue to reduce and eventually remove the manned element of shipping so as to reduce costs and improve safety[11]. In the naval sector, small UxV technologies have already been widely adopted and it looks as though larger UxVs are likely to be single role with a dedicated payload[12] or in other governmental ships with a non-combatant role such as harbor tugs[13]. However, for combatants, technologies from the IAM sector such as Energy Management (EM) and Equipment Health Monitoring (EHM) are very likely to play a significant role in coming years in reducing sustainment costs and in improving availability so that a fewer ships can provide the coverage that was previously provided by larger fleets. Simpler, more robust propulsion systems are therefore required which consist of equipment that has longer overhaul periods with EHM to facilitate seamless introduction into service with enhanced reliability, availability and an optimized maintenance regime. This supports unmanned machinery space concepts and reduces the burden on the ship's marine engineering staff.

ROLLS-ROYCE POWER & PROPULSION EXPERIENCE ON RECENT NAVAL PROGRAMMES

The RN Type 26 GCS (Figure 5) will replace the Type 23 frigate and will be a highly capable ASW platform[4] with multi-mission capability. Mechanical, hybrid and IFEP P&P system options were evaluated in detail during the programme's assessment phase and it was concluded that a single-GT CODLOG (Figure 5) would be used. This system consists of a diesel-electric system for ASW and cruise speed operation which will cover approximately 95% of sea-going operation. The single boost GT would then be reserved to provide power up to the ship's top speed. Having been involved in the initial concept and feasibility studies of the Type 26 GCS programme and, later, having been selected to supply the MT30 marine GT and MTU DGs, Rolls-Royce Naval has been associated with the programme for some time and based on this experience, plus that in other similar programmes (eg: RoKN FFX-II frigate, Italian Navy PPA, LHD), the following observations are made on modern warship programmes which are relevant to new builds the implications of the decisions made are important to understand during Military Off The Shelf (MOTS) programmes such as SEA 5000:

- Detailed and iterative analysis is necessary, using various EM ratings and technologies (and including the associated PG&D system components), with consideration of physical attributes (mass, volume, physical integration) to systematically converge on the "sweet spot" for the electrical plant in the context of programme and ship design constraints.
- The P&P system needs to comfortably support prolonged low-speed operation.
- Care is needed to specify a truly resilient PG&D system with built-in redundancy, flexibility, survivability and vulnerability via a carefully configured electrical system with design and growth margins. This should be future-proof in terms of reasonably foreseen upgrades to ship's mission systems.
- DEWs currently in development will one day be available for retrofit to warships with up to 10MW of available electrical capacity. It is likely that when a warship needs to operate its close-in weapons systems it will also need to be highly-maneuverable at top speed. A carefully configured CODLOG arrangement has sufficient mechanical power to achieve top vessel speed whilst at the same time releasing the ship's electrical capacity to support future close-in weapon systems.
- The need for a cost-effective and highly reliable boost GT and gearbox capability, including initial acquisition and sustainment costs.
- High power output boost GTs can be viewed as the enabler for hybrid arrangements in smaller platforms insofar as they permit the co-location of the GT and two of the DGs sets in the FER, avoiding the need for a separate Forward auxiliary machinery room and thus allowing the overall length of the ship to remain within limits.

CONCLUSIONS

This paper has briefly reviewed the history and development of naval propulsion since the advent of marine GTs to set the scene for the modern single GT hybrid platforms that are currently entering service. The paper has discussed the advantages and considerations of the move from the ANZAC's mechanical propulsion system to the hybrid system of the replacement procured under the SEA 5000 programme. Principal conclusions are:

- The replacement could be more "future-proof" by virtue of the inherently larger installed PG&D system, however, margins should also be carefully applied as this will extend the life of the platform – a major affordability consideration given the prohibitively large cost of major upgrades and the need to retain the military relevance of the platform for longer.
- The replacement will re-invigorate the RAN's ASW capability, however, lessons should be taken from other navies where single-purpose platforms have been modified for more GP activities during their service. A carefully specified P&P system will be able to accommodate this change in role, as and when it occurs.
- A platform that is going to be flexible and versatile in its ability to adapt to changing environmental and operational needs should be designed from the keel up with a dedicated mission bay and MBHS.

- Changes to the MOTS platform may increase the vessel's displacement and, thus, increase the powering requirement. A robust P&P system with sufficient margin will help mitigate against the ship being underpowered once in service.
- For global deployment, the P&P system should be specified to operate at extremes of ambient air temperatures to maintain ship performance regardless of location.
- Due to the non-linear relationship between EM power and diameter, cruise speed must be carefully specified to avoid issues with the physical fit and cost of larger diameter EMs and their P&P systems.
- New concepts for supporting equipment should be evaluated to reduce sustainment costs. For example, a 'fit-and-forget' philosophy for the boost capability using a modern GT with high reliability and a long time between overhaul as well as emerging technologies under the SI banner such as EM and EHM.

REFERENCES

[1] Modern Power and Propulsion System Options for Future Frigate Programmes, by Edward Wright and Richard Partridge, International Maritime Conference at Pacific 2015, 6/10/15.

[2] Unclassified Memo from JH Hall & RJ Scrace (DERA Haslar) to RW Tooke (Rolls-Royce PLC) dated 02May1997, subject: Sided-boost powering & trailing shaft losses.

[3] Patrick Durrant. Australian Defence Magazine (ADM). SEA 5000 CEP into the final phase. Available: http://www.australiandefence.com.au/news/sea-5000-cep-into-the-final-phase. Last accessed 21/7/17.

[4] Unclassified Memo from Peter Osbourne (Business Development Manager – Surface Combatants, Strategy and Business Development, BAE Systems Australia) to Edward Wright (Rolls-Royce PLC) dated 6/7/17, subject: Pacific 2017 IMC Program Now Available.

[5] Unclassified Memo from Gavin Rudgely (Chief Engineer, Type 26 GCS Export, BAE Systems Naval Ships, UK) to Richard Partridge (Rolls-Royce PLC) dated 20/7/17, subject: Pacific 2017 IMC Program Now Available.

[6] The Honorable Kevin Andrews, Minister for Defence. (2015). Minister for Defence – Speech – RUSI Submarine Summit – 25 March 2015. Available: http://www.minister.defence.gov.au/2015/03/25/speech-rusi-submarine-summit-25-march-2015/. Last accessed 10/7/15.

[7] Re-writing the propulsion rule book for 21st Century, by Richard Partridge and Ben Thorp, 12th International Naval Engineering Conference and Exhibition 2014, 21/5/15.

[8] Unclassified discussion between E. Wright and S. Elder with crew of HNMLS Groningen, a Dutch Holland-Class OPV, on their anti-piracy operations in the Middle East region during DSEI 2013 ship visit in London.

[9] Future capability through partnership: refreshing the UK MoD's marine systems development strategy through associated international and industrial partnership, by Cdr. PC Carroll, RN and Cdr JC Lowe, RN, INEC 2010.

[10] USN Naval power systems technology development roadmap, by J Kuseian, PMS 320 / NSWCCD-SSES 982.

[11] Ship Intelligence – a new era for warships, by Oskar Levander and Ben Thorp, ASNE Intelligent Ships Symposium 2017, 24th & 25th May 2017.

[12] FedBizOpps.gov, Large Unmanned Surface Vessel (Large USV) Sources Sought/Request for Information, Solicitation Number: N00024-17-6302(2). Available: https://www.fbo.gov/index?s=opportunity&mode=form&id=7e15825c95eade4a1aa0fa59d7864cb4&tab=core&_cview=0. Last accessed 24/7/17.

[13] Rolls-Royce demonstrates world's first remotely operated commercial vessel. Available: http://www.rolls-royce.com/media/press-releases/yr-2017/20-06-2017-rr-demonstrates-worlds-first-remotely-operated-commercial-vessel.aspx. Last accessed 24/7/17.

Optimising High Speed Craft for Hybrid

Robert Hayes
Frazer-Nash Consultancy
r.hayes@fncaustralia.com.au

ABSTRACT

Hybrid technology has made large advances in maturity in the maritime field in the last few years. This paper explores the lessons learned over this period and how hybrid propulsion for small craft can become strategically important in the maritime domain in the coming years.

Operational requirements and flexibility of small boats used by defence are pushing what is achievable through combustion engines alone. With increased emphasis on attributes such as stealth, performance and environmental impact, and subsequently these being key requirements of future small boat platforms, the use of hybrid and innovative propulsion technologies will be essential in enabling vessels to operate effectively. This is compounded by legislation and policy (e.g. the NATO single fuel policy) demanding a change to the current technologies used.

This paper outlines the types of hybrid propulsion available and their respective benefits (environmental, performance, signature). It will also outline the operational and design challenges (space, weight, shock mitigation, hull optimization) hybrid technologies incur and how these challenges can be mitigated.

The paper also briefly touches on what the future may hold for hybrid technology, and how modularity and full electric power trains may completely change how small boats are operated.

INTRODUCTION TO HYBRID

History

Hybrid technology in the marine industry has had quite a challenging past. While it has been influenced by the technology progress in the automotive industry to an extent, the significant difference between vehicles and vessels (e.g. continuous power demands, production volumes, power transmission methods) has meant that a radically different implementation and business drive has been required.

Diesel electric propulsion (without a stored energy source) has been a common propulsion method on large ships for many years. This is due to the flexibility and optimal engine loading provided by this configuration. However, due to many factors, most notably compactness, bringing this technology to High Speed Craft (HSC) has been a lot slower.

The first hybrid implementations started arriving around 2010, with companies such as BAE Hybridrive and the UK based Hybrid Marine developing solutions for large and small vessels respectively. Since then the number of vessels and interest in fitting hybrid technology to (predominantly) passenger craft/ferries has increased several-fold. By 2017 there are nearly 100[1] ships in service worldwide that have some kind of battery or hybrid technology installed. The majority of these vessels have some degree of bespoke hybrid system, making their design and implementation a time consuming and expensive process.

Unfortunately, despite the uptake in hybrid across the board it still remains technically difficult to hybridise very small <10m and very large >80m vessels in a cost-effective manner. Smaller vessels are limited by how compact hybrid systems can be while providing sufficient power and range. Larger vessels face challenges related to the integration of high capacity battery systems, and are limited by cost, space requirements and safety implications of integrating these battery systems.

Typical HSC Suitable For Hybrid

The main candidates for hybrid conversion in the commercial world are ferries and workboats. Ferries are suitable because they often operate a fixed route upon which the system can be optimised. Workboats tend to be suitable as they frequently spend a lot of time on idle while still requiring large engines to achieve transit speeds.

In the defence world, there is no clear cut candidate for conversion with most vessel either fulfilling multiple roles or being highly specialised. However, interdiction type vessels (which need to quickly achieve high speed) may be suitable given the additional boost power that hybrid can provide. Equally, vessels which need to operate stealthily are suitable due to the low noise signature and instant response power offered by hybrid technology. In all cases there remains a case for using the technology to extend theatre operations, which helps achieve the strategic aim of an overmatched capability.

Lessons Learnt

The biggest lesson to date is the high cost of development and re-work. Often this is spurred on by an underestimation of the level of complexity involved. There has been on occasions the expectation that the system integration is a small part of the design; and that once the hybrid system has been decided upon, the vessel designed to accommodate the system and the equipment fitted, the system is most of the way to working[2]. This is rarely the case with significant control system design and development of interfaces the with the Platform Management System needed to get the system working at all, let alone working as intended. Equipment manufacturers tend to emphasise that their system will work with Control Area Network (CAN) bus[1], and while this is true, CAN bus does not necessarily provide the interface or architecture for the control system.

The other lesson has been around ensuring systems are appropriate for the vessel and the conditions, with range requirements not being met because aerodynamic or wave resistance has not been fully considered. This is particularly relevant to smaller vessels where these components may be contribute substantially to total vessel resistance[3].

[1] CAN bus is a protocol designed to allow devices to communicate with each other without requiring a host computer or controller.

The final lesson learnt has been re-training the vessel's crew to operate hybrid vessels and in many cases communicating the impact that small changes in speed have to the range and cost-effectiveness of the hybrid system[4].

TYPES OF HYBRID

Hybrid technology is implemented in one of three ways depending on the vessel and how the system will be used. The ways are Series, Parallel, and Series-Parallel and are indicated in Figure 1 to Figure 3 below. All systems share the following common components:

- A propulsion motor;
- An Internal Combustion Engine (ICE) mounted generator or external power input;
- A battery pack and battery management system (BMS); and
- A control system.

Series Hybrid

With this layout the battery supplies electricity to the propulsion motor(s) and is trickle-charged by a generator connected to an ICE.

Series requires the system to have a motor capable of propelling the vessel to the design speed of the vessel. This makes it suitable for large ships[2] or small displacement ferries operating at slow speeds over short routes. This system is not suitable for high speed craft due to the low power density of electric motors. This system is relatively simple from a control perspective.

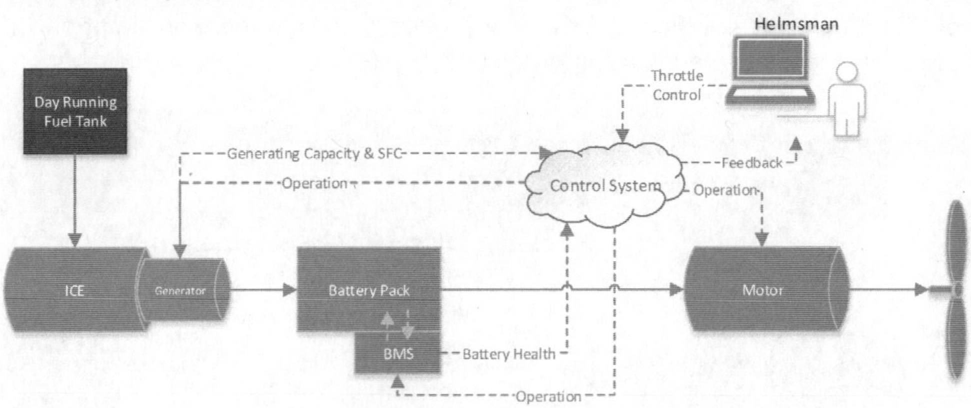

Figure 1: Series Hybrid Layout

Parallel Hybrid

With this layout the propulsion power can be supplied by the ICE or the propulsion motor independently (or in combination) to the propeller.

[2] Diesel Electric drive without a stored energy source (e.g., batteries) is a very common and mature propulsion technology on ships of over 100m.

Parallel requires a special gearbox and clutch that allows for power blending between the two sources. This system is ideally suited to high speed craft operations where there is space for a second generator to trickle-charge batteries or the vessel can be recharged from shore power between operations. The control system is slightly more complex than the series hybrid system, especially in the operation of the power blending gearbox.

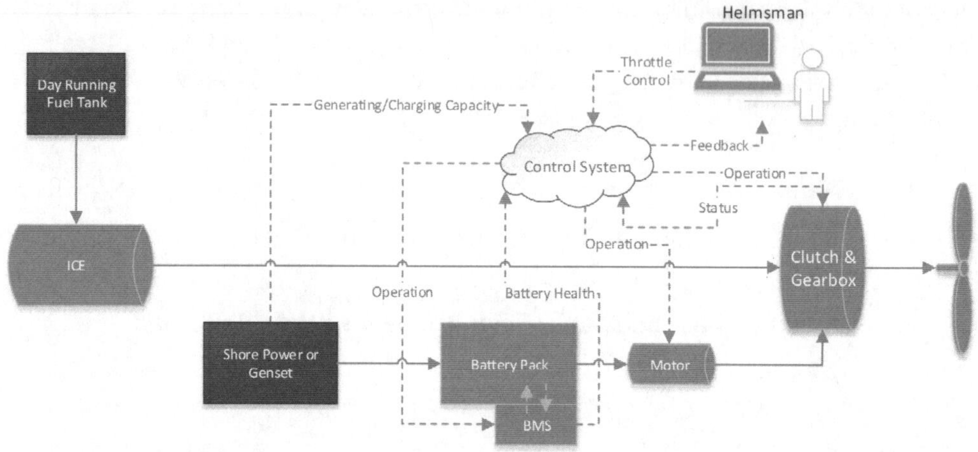

Figure 2: Parallel Hybrid Layout

Series-Parallel Hybrid

This layout is similar to the parallel hybrid but additionally has a power splitter connected to the engine which allows for some power from the ICE to be diverted to a generator as well as being used for propulsion.

This system is complex from a mechanical perspective but allows the ICE to run almost continuously at peak efficiency, thereby reducing specific fuel consumption (SFC). This system also requires the most complex control system.

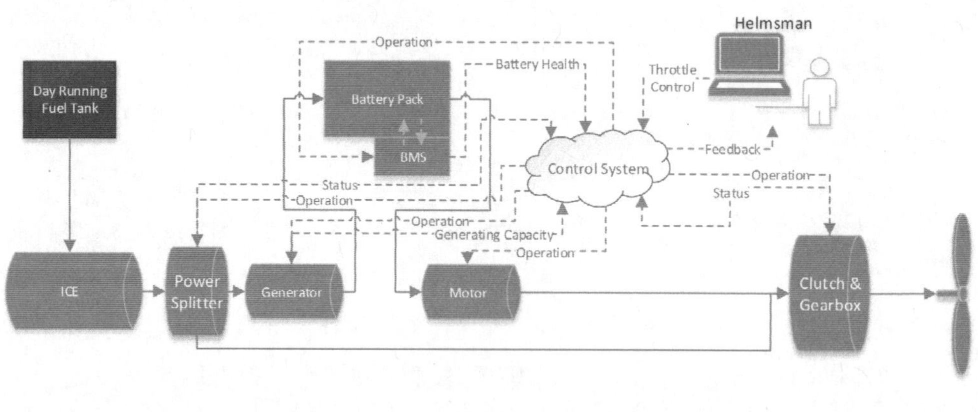

Figure 3: Series-Parallel Hybrid Layout

OPERATIONAL BENEFITS OF HYBRID

Environmental – Lower Emissions

The more time the vessel spends on electric, with the ICE off, the less CO_2/NOx/Particulate emissions produced. While the electricity itself will incur some emissions during generation, if this electricity is sourced from the grid there is a high likelihood that a proportion of this electricity will be obtained from renewables (depending on the country). Since the hybrid system will shift the operation of the ICE to higher vessel speeds, the ICE will run more efficiently and as a result there will be less emissions associated with inefficient running (e.g. CO, unburnt hydrocarbons). As can be seen in Table 1, in a condition where the engines idle for a significant proportion of operations a substantial amount of emissions are produced in a year.

Assumptions:
- CAT 3512C Engine (EPA C1 Tier 3)
- Operates 300 days/yr @ 8hr/day
- Engine idles 60% of operations
- Engine produces 83kW at idle
- Engine emissions are equal to EPA limits

	HC+NOx	PM	CO
EPA Limit (kg/kW-hr)	0.0058	0.00011	0.005
Lim * 83kW	0.48	0.0091	0.42
* 1440hrs	693.2	13.1	597.6

Table 1: Potential Emissions from idling engines[3]

Performance – More Responsive Engines

Diesel engines suffer from lower responsiveness compared with petrol engines. This is due to a combination of engines being turbocharged and having heavier flywheels to spin up. Electric motors can be utilised as part of hybrid systems to provide a boost. This use of electric motors to boost performance at the lower end is becoming commonplace in performance cars (e.g. the Koenigsegg Direct Drive System uses a fixed single gear ratio between the engine and wheels (1:2.73) and the electric motors provide traction at lower speeds).

In particular this is useful for defence applications where there is a need to maintain high performance, responsive engines but also shift to diesel for logistics or safety reasons. Although in practice, the amount of boost power possible will be limited by propeller design and cavitation.

Signature

Especially important for meeting the aims of covert military operations, the silent nature of running on electric motors at slow speeds means that vessels can approach their target while in a powered rather than unpowered state, therefore allowing a quick escape. Additionally, the greater torque afforded by electric motors may mean that larger and slower spinning propellers can be used either further reducing noise, or producing a deliberately misleading signature.

Maintenance

Electric motors and Li-ion batteries have relatively low maintenance needs compared with ICEs. As a result, if the vessel spends a significant proportion of time on electric instead of idling the main engines, the service hours on the engines could significantly drop. Considering most engines need servicing after approximately 1,000 hours this could significantly reduce the cost of operation and increase availability.

[3] Data from Cat 3512C data sheet [10] and EPA emission limits [9].

DESIGN CHALLENGES OF HYBRID

Managing the Challenges

Hybrid technology presents a number of design challenges and complexities which need to be understood and managed throughout the design. Some of the more significant design challenges related to the vessel are discussed in this section, together with how they could be approached and managed.

It is recommended for any project that a systems engineering approach is adopted. This will help to define the scope and functionality of the system; understand the architecture of the system, sub-systems, and interdependencies; provide change management and configuration management; and enable a structured validation and verification process to take place. The outcome of adopting a systems engineering process is that risks are understood and managed, complexity is reduced and the end product provides greater functionality and is more cost-effective.

Space & Compartmentation

Space is at a premium on high speed craft which makes fitting hybrid systems very challenging (especially to existing designs with limited flexibility). The motors, generators and any additional running gear needs to be integrated into the engine room, where an intelligent approach should be taken towards access, noise and vibration, and heat transfer. On smaller vessels (<20m), there may be a need to expand the engine bay or chose the most compact engines to allow appropriate space. In general, use of 3D CAD modelling of the vessel is highly recommended to reduce integration issues as early as possible.

While there is no correct or incorrect location to house batteries, the following points should be considered:

- The battery packs need to be kept in close proximity of the motors to reduce high-voltage cabling runs;
- Good access to the battery packs is essential and the access should be designed so that entire battery modules can be removed from the vessel for maintenance or replacement; and
- On passenger craft batteries should not be placed in or next to passenger compartments.

Consideration should be given to compartmentation of battery packs and where possible they should be placed in a separate compartment. In the event of a battery pack going into thermal runaway, firefighting or flooding the compartment to extinguish the fire should not cause loss of the vessel.

Weight & Centres

The hybrid system, but in particular the batteries, can represent a significant weight increase which will be concentrated in a particular location. Placement of the batteries is crucial to ensure the characteristics and the performance of the vessel is not affected.

Especially for planing craft, the batteries should not be placed too far forward otherwise the ability to plane may be affected.

As discussed later on in structure design challenges, the additional weight will require more structural reinforcement in the local area, these further contribute to weight increases for the vessel.

Resistance & Propulsion

High speed craft are typically designed with a consideration given toward optimising cruise or maximum transit speed. Hull shape is also important, but is usually less flexible as it is driven by operational requirements.

When choosing a design, the maximum speed under full electric needs to be established in the requirements/specification and the design optimised to minimise resistance at this speed. A change of 1 to 2 knots around the 6-10 knots speed range can have significant impact on the power required. This has a compounding effect on the amount of batteries required, the associated additional vessel weight due to these batteries, the increase in vessel resistance brought on by this weight, and ultimately impact on the viability or range of the hybrid system.

Consider Figure 4 which shows speed vs power for a 25m monohull. If the design objective is to have a range of 8 nautical miles, it can be seen that design speed is crucial to achieving this. Going from 6 to 8 knots requires a 77% increase in battery pack capacity (78kWh to 140kWh) to achieve 8 NM of range and going from 8 to 10 knots further increases it by 56% (140kWh to 220kWh). In terms of time, the respective time savings are from 1hr 20 to 1hr and from 1hr to 45 minutes, a relatively small time saving for a significant cost investment.

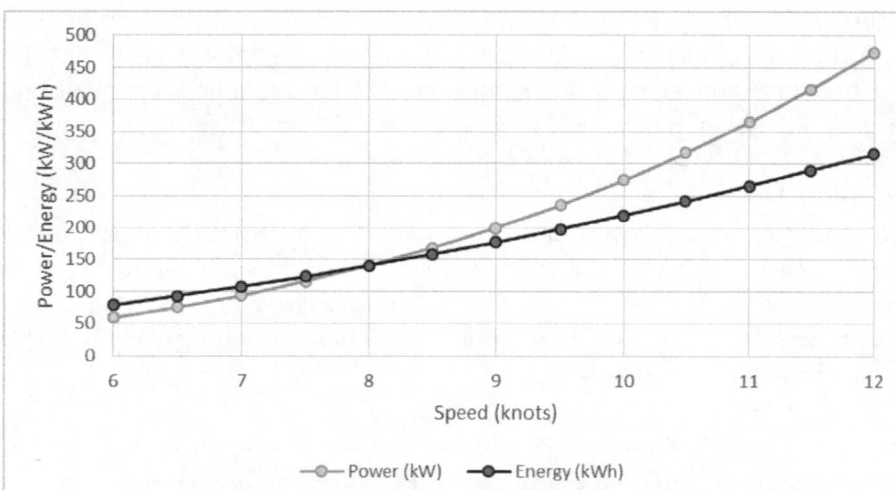

Figure 4: Speed-Power graph for a 25m Monohull[4]

Propeller design and optimisation for hybrid is quite different, with the hybrid motor able to produce 100% of torque from 0 rpm. In some cases it may be possible to fit a larger or

[4] Data from a hull modelled and tested in Maxsurf, figures are indicative rather than absolute.

differently pitched propeller than would be possible with just an ICE in order to gain extra efficiency. Figure 5 shows the propeller-engine curve for a Caterpillar 3512C and twin YASA 750R motors, where it can be seen that the electric motor intersects with the propeller demand curve at below the engine stall speed.

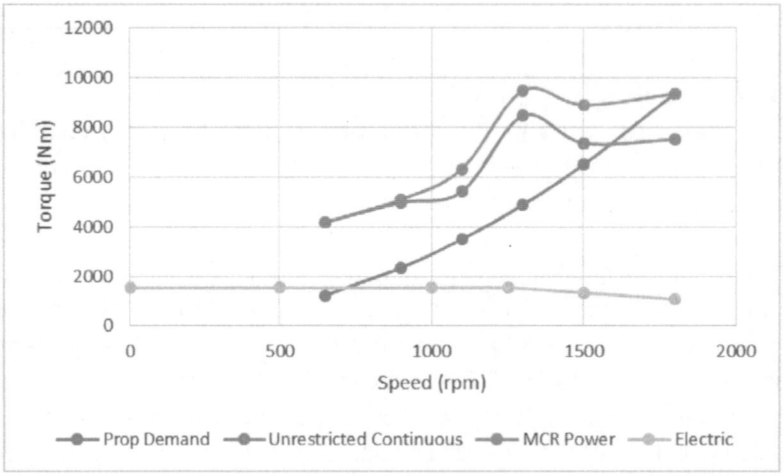

Figure 5: Propeller-Engine matching graph for a Cat 3512C and twin YASA 750R motors[5]

Finally, linking back to lessons learned, aerodynamic drag and wave resistance for smaller vessels must be considered to ensure the hybrid system is not undersized.

Shock Mitigation

High speed craft typically experience high levels of shock, this is particularly important on smaller craft operating in high sea states. From the point of view of a hybrid system, the structural mounting needs to reduce likelihood of slamming loads causing fatigue or structural failure. Batteries that are not designed for maritime environments may not be able to cope with heavy shock loading. In particular, batteries that use capillary cooling may experience reduced cooling performance due to resonant vibrations breaking liquid-vapour meniscuses or mechanical shocks disturbing fluid distribution in the system.

Structure & Class

When modifying an existing vessel design, the vessel structure must be considered. A 200kWh battery pack (typical of what might be on a 30m vessel) would be approximately 2 tonnes of extra mass which is distributed over a $1m^2$ area. From the perspective of structural performance, the local area would need reinforcement suitable to meet classification society requirements.

DNVGL and Lloyd's Register both provide guidance and rules for battery systems on ships, although currently these rules are safety centric.

[5] Data obtained from a combination of Cat 3512C[10] and YASA 750R[11] data sheets.

Integration & Control Systems

As each hybrid implementation is normally different there is a need to architect the control system from the ground up. While this may seem complex, it ultimately means any system integrates properly and will work successfully.

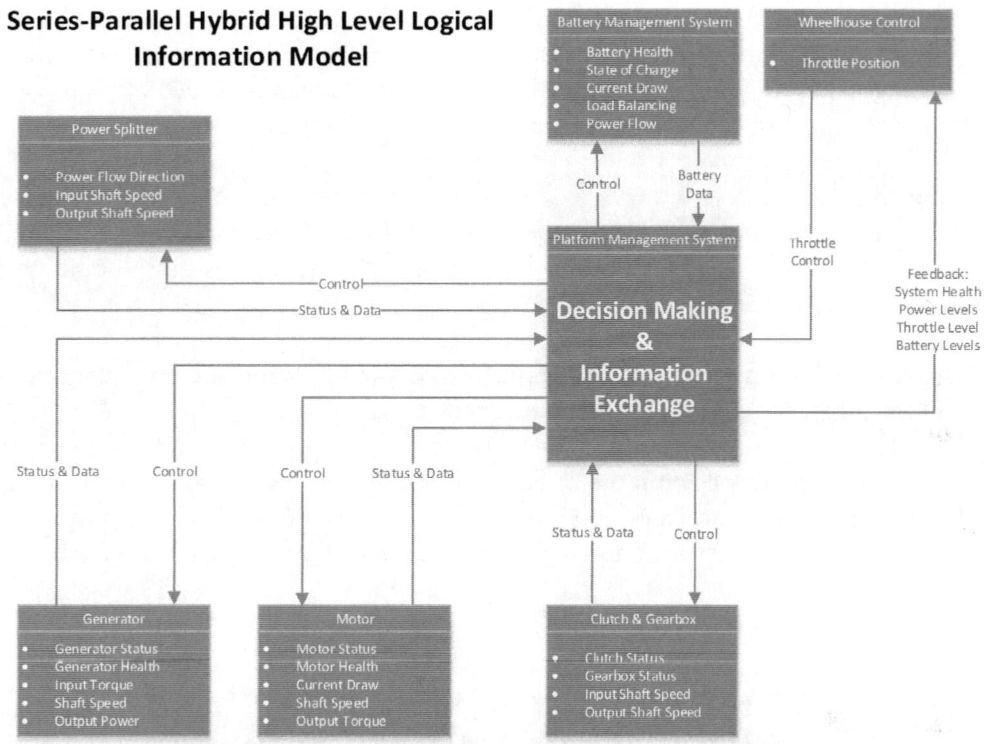

Figure 6: Indicative Data Model for a Series-Parallel Hybrid System

Developing a data model such as in Figure 6 helps to understand what data is present and what data need to be exchanged. From this a physical data model can be developed later in the design which shows how this data is exchanged and the physical connections needed to facilitate this.

Safety

Hybrid systems present a new set of safety considerations to vessel design, manufacture and operation. The classification societies have made good progress in developing rules and guidance advising vessel builders on how to develop systems that are safe[5][6][7].

It is important to note that in most cases installation and maintenance of battery systems will need to be undertaken by specialised high voltage electricians, which may present issues when vessels are built or operated in less economically developed regions.

Training of crew in the safe and economical operation of a hybrid vessel is also an important consideration and in most cases this will need to be done prior to customer sea trials.

FUTURE DIRECTION

Short Term

In the next couple of years it is expected there will be a significant expansion in the number of vessels (especially ferries) that will be built as hybrids, due to hybrid technology reaching a level of maturity where the environmental and economic benefits begin to outweigh the additional cost and engineering effort to implement hybrids. This will be combined with larger vessels being hybridised, for example Ulstein is currently building a 160m hybrid vessel for Color Line to be launched in 2019[8]. This is combined with a strong push in Northern Europe towards greener vessels[1].

Long Term

It is expected in the next 5 to 10 years that all the classification societies will develop regulation for Hybrid vessels and provide guidance on how systems should be designed and implemented in vessels.

There will be a push from battery manufacturers towards more energy dense chemistries and the 200Wh/kg barrier may be exceeded.

It is expected that following the progress in the Automotive industry there will be a push from engine manufacturers to integrate hybrid technology/electric motors into their engines, providing vessel manufacturers with a much better off-the-shelf starting point.

Longer term, it is expected there will be four main changes:

- There will be a better charging infrastructure around busy ports;
- Vessels and hybrid systems will be designed to be modular or be replaced/upgraded as technology progresses;
- Newer forms of propulsion (e.g. Rim Drives) will become more prevalent, paving the way for more flexible and novel propulsion design; and
- There will better progress made for full electric vessels due to technology progress and environmental legislation. However, these are expected to be limited to vessels operating on short routes where the vessel has redundant space (e.g. short ferries or barges). It is unlikely that HSC will become fully electric while battery densities remain at their current order of magnitude.

External Drivers

There are a number of external drivers which are needed to spur on the development and uptake of hybrid technology. Three main ones are:

- Stringent emissions regulations and severe penalties for non-compliance will need to be introduced, thereby making it more economical to have a hybrid vessel;
- Better charging infrastructure at all ports (including minor ones) with an international standard for the charging receptacles; and

- Governments will need to provide further grants and/or tax incentives for smaller builders and/or operators to make it economical to develop small hybrid vessels.

CONCLUSION

At this point it appears that Hybrid technology for HSC could be a serious game changer. This is due to the significant uptake in vessels; the benefits in terms of performance, environmental impact, signature, and operational costs; and the flexibility of propulsion that hybrid technology provides. There are a number of difficult design challenges and significant costs involved in implementing hybrid technology, namely the space, weight, structural impact, power density, and integration into the vessel's platform management system. This means for many vessels hybrid technology is not feasible from purely an economical point of view.

A systems engineering approach is recommended for any project involving hybrid technology. By using this type of approach the requirements can be determined and refined, their inter-dependencies established, and all associated data captured and integrated into the model. As a result complexity is managed, risk is reduced and traceability and verification of requirements is ensured.

It is worth considering that due to the new nature of hybrid technology there is not currently any reliable data to understand what the long term impact of operating a hybrid vessel is; how long hybrid running gear will last, and whether hybrid vessels will need overhauling and refitting sooner or later than those with traditional running gear.

Long term it is still difficult to determine what the impact with be with regards to infrastructure, regulations and incentives but based on the automotive and power generation industries it is expected to be a positive one.

REFERENCES

[1] G. Peters, "Hybrid vessels: here to stay, or fleeting trend?", ship-technology.com, 27 March 2017. [Online]. Available: http://www.ship-technology.com/features/featurehybrid-vessels-here-to-stay-or-fleeting-trend-5769261/. [Accessed 30 May 2017].

[2] M. Downer, "Development of High Technology Hovercraft", in *Next Generation Power and Propulsion Conference*, Southampton, UK, 2016.

[3] C. Dymarski and M. Narewski, "Assessment Of Underpowered Propulsion Machinery In Electrically Driven Small Inland Waterway Passenger Ships From Classification Society Point Of View", *Journal of Polish CIMAC,* 2014.

[4] T. Birr, "New Power and Propulsion Systems Need New Skills Training for Engineers and Integrators", in *Next Generation Power and Propulsion Conference*, Southampton, UK, 2016.

[5] American Bureau of Shipping, "ABS Advisory On Hybrid Electric Power Systems" ABS, 2017.

[6] DNVGL, "In Focus –The Future Is Hybrid – a guide to use of batteries in shipping", DNVGL, 2016.

[7] Lloyd's Register, "Battery Installations - Key hazards to consider and Lloyd's Register's approach to approval", Lloyd's Register, 2016.

[8] Ulstein, "Color Hybrid", 2017. [Online]. Available: https://ulstein.com/references/color-hybrid. [Accessed 30 May 2017].

[9] US Environmental Protection Agency, "Federal Marine Compression-Ignition (CI) Engines: Exhaust Emission Standards", US EPA, 2016.

[10] Caterpillar, "3512C HD Marine Engine Data Sheet", Caterpillar.

[11] YASA Motors, "YASA 750R Data Sheet", YASA Motors, 2016.

Structural Integrity of Maritime Platforms

Bruce Cartwright and Rob Melchers, The University of Newcastle
Bruce.Cartwright@newcastle.edu.au, Rob.Melchers@newcastle.edu.au
Seref Aksu and Teresa Magoga, Defence Science and Technology Group
Seref.Aksu@dst.defence.gov.au, Teresa.Magoga@dst.defence.gov.au
Chong Min Song, The University of New South Wales
c.song@unsw.edu.au
Damian McGuckin and Martin Renilson, Pacific Engineering Systems International
damianm@esi.com.au, martinr@esi.com.au

ABSTRACT

Maritime assets are major investments. Their effective management is a very important factor in not only optimising asset availability, effective lifetime, and functionality, but also reducing downtime for repairs. This paper presents a project that aims to develop an improved understanding of the structural degradation of naval and other vessels. Using an in-service ship as the study platform, structural degradation due to corrosion and fatigue within a representative environment is considered. Seaway loads, based on the operational profile of the ship, are applied to a finite element model to obtain the global and localised structural responses. Thus, stress-critical areas can be identified. Smoothed Particle Hydrodynamics is employed to model the fluid-structure interactions. Further, observational and experimental corrosion data is included to estimate the load-bearing capacity, the remaining safety margin, and the remaining service-life at different stages of the life-cycle of the ship.

NOMENCLATURE

CP	Cathodic Protection	SIoMP	Structural Integrity of Maritime Platforms
FE	Finite Element	SPH	Smoothed Particle Hydrodynamics
MMC	Multi-scale Model Coupling	STL	Standard Tessellation Language

INTRODUCTION

Maritime assets are major investments and their effective management is important to ensuring crew safety, operational availability, and economy. Ideally, asset management should optimise the longevity and functionality of the asset, and reduce the downtime that is needed for its maintenance.

Many concepts and methodologies have been proposed to deal with the structural assessment of ageing maritime assets [1, 2]. Increasingly, mathematical and/or computational models are being used to describe not only an asset's structural response to its operational environment, but also deterioration due to ageing and usage [3]. If sufficiently mature, these models allow the prediction of both the probability of failure in a given lifetime and the expected lifetime (that meets some defined safety or other performance criteria). The quantitative results from these models [4] are often used to support economic analysis and management of assets [3]. However, economic pressures should be balanced with decisions to repair, replace, or retire an asset [1].

As has been shown from a review [1], and from the experience of the authors, the applicability and utility of various structural assessment approaches are dependent on many factors. Over the life of a ship, these factors include: a detailed (historical) knowledge of, and prediction capability for, the climatic and chemical environments; operating conditions; maintenance; the operational profile, including the speed profile and heading distribution, and; the wave environment (that is, the long-term distribution of the wave height, wavelength, and wave frequency).

In order to translate the aforementioned parameters to mathematical models, the parameters must be quantified. Furthermore, for the subsequent predictions to be valid it is necessary to calibrate the models to the actual data. However, currently many of the influencing parameters are ill-defined and have often been estimated from experience of similar structures or ships. Thus, the accurate prediction of the ship's condition at various stages of the life-cycle is subject to the accuracy, detail, and availability of the input parameters. Probabilistic concepts often have been, and are employed, to represent the associated uncertainty. This relatively imprecise approach is considered current best practice and one with which industry appears comfortable at present [1]. It is also broadly consistent with decision-making practices in other fields. For example in finance, probability theory including extreme value analyses [5, 6], is applied widely in risk-assessments. The same theory is applied to high-asset-value, or high-risk, applications such as earthquake risk to nuclear facilities and power stations [7]. It also has been applied to accident insurance for passengers [8].

As noted, the need in industry for the prediction of the structural condition of assets is primarily one of justifying economic decisions, subject to constraints, on matters such as safety. Such decisions typically include the timing, criticality, and extent of maintenance actions. Other options include [9]: repairing or renewing the defective parts of the structural items to enable the asset to continue service at a lower risk of failure; continuing the operation of the deteriorated asset but with a higher risk of failure, and; imposing a service restriction, such as limiting the speed of the vessel and/or operations above a certain sea state, to reduce the risk.

In each case, the economic decision involves comparing the cost of maintenance plus the impact on ongoing productivity against the expected cost of failure and the associated implications. It may also involve developing optimal inspection regimes, including those based on risk-based inspection strategies, or modifying conventional periodic inspection intervals.

In principle, the decision-making process can be enhanced if the true state of the asset can be ascertained. To achieve this, consistent and continuous monitoring of environmental loads and hull condition over the life of a ship has been advocated [10]. However, typically this is still a complex, unreliable and expensive exercise that few vessel owners appear willing to justify. It also requires an understanding of the locations within a vessel where remedial works might be required, together with an assessment of the corresponding priorities.

The present paper outlines a project that aims to develop an integrated risk assessment method for ageing maritime platforms. The paper describes: the overall methodology of the project; the modelling of a specific ship hull using the finite element (FE) method; material modelling and the effects of deterioration, and; the modelling of the hydrodynamic loading on the hull using a mesh-free technique, Smoothed Particle Hydrodynamics (SPH). Finally, the way ahead for the projected is discussed.

PROJECT SIoMP METHODOLOGY

The project Structural Integrity of Maritime Platforms (SIoMP) aims to provide an improved understanding and a predictive capability for the effects on the structural condition of maritime assets subjected to cyclic loads, extreme loading phenomena and structural deterioration.

The SIoMP methodology presented in Figure 1, has three distinct components. Firstly, advanced FE modelling is used to represent the complex structure. Next, the hydrodynamic loads are modelled using the SPH technique to represent complex fluid-structure interactions. Finally, deterioration is modelled such that corrosion, cracking and fatigue damage are incorporated. For the purpose of this project, the data and modelling will be relevant to a specific naval ship referred to henceforth as the *study platform*. One of the important features of this approach is the use of routine hull surveys performed over the life of a vessel to monitor the state and extent of corrosion and cracking. The net result of such a survey is the *hull survey report* containing a documented history of the known state of deterioration of an aged ship.

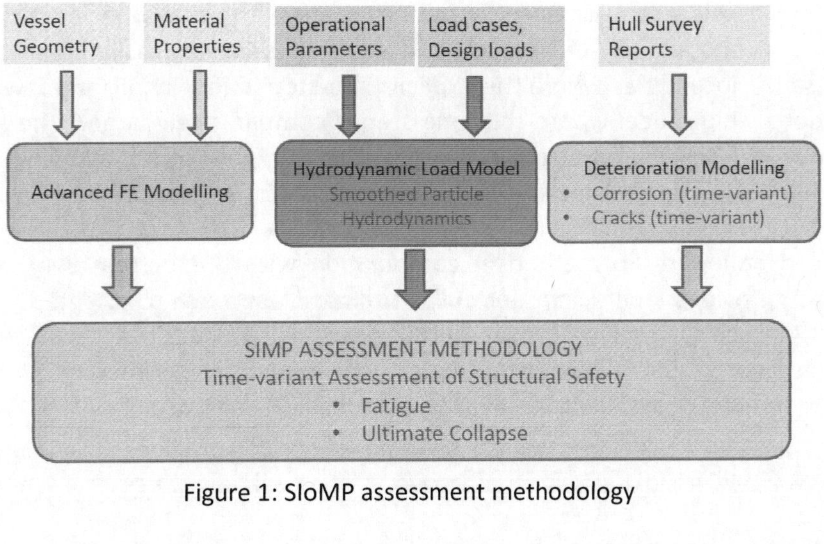

Figure 1: SIoMP assessment methodology

The SIoMP project is a major step forward in the sense that most previous applied research has considered only parts of the overall problem, such as the development of models for wind and wave loads, modelling of fluid-structure interaction, and modelling of fatigue and corrosion in isolation.

Advanced Finite Element Modelling and Analysis

The FE model of a frigate-sized ship, shown in Figure 2, will be modified for the present project. This model consists of shell elements for plating and bulkheads, beam elements for longitudinal members, and point masses for specific items of machinery and other heavy equipment. The FE model has mass and inertia properties representative of the ship in a loaded state, as might be expected on departure from port. Appendages such as rudders, keels, or propellers will not be incorporated in the FE model as these are considered to have negligible effects on the global structural response of the structure. This FE model represents the design condition of the study platform, and contains no defects as might arise from manufacturing, damage or in-service degradation such as corrosion, cracks or dents. The model will be modified by adding local detail to represent specific occurrences of cracking and corrosion as described in the hull survey reports (discussed in more detail below).

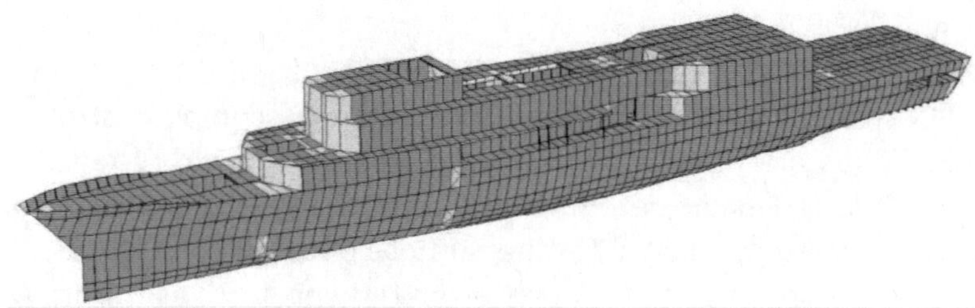

Figure 2: Finite element model of a frigate type ship

An effective way to simultaneously analyse a structure subjected to dynamic loads at the global scale, as well as at the scale of structural details, is through the use of multi-scale model coupling (MMC). MMC [11, 12] is a technique wherein separate explicit FE models of different physical scale seamlessly communicate with each other as if they were one continuous FE model. The technique achieves its efficiency by assigning many processors to analyse the local detail where the explicit time-step is very small, and fewer processors to the global ship model where the time-step is comparatively large. The interaction of these two analyses occurs at the time-step of the larger model, allowing a continuous fully coupled analysis to occur in a computationally economic way.

The scaled boundary FE method in combination with the octree algorithm for mesh generation provides an efficient and fully automatic approach to bridge the difference in scales [13]. Such an analysis can be performed directly using the digital formats of the defect geometry, such as digital images or standard geometry models such as stereolithography or standard tessellation language (STL) models.

Modelling of Corrosion

Corrosion is an inevitable consequence of operating an asset in a marine environment. One of the major issues in modelling of corrosion is to identify the regions in a vessel where corrosion is likely to be the most severe.

Most vessels have either cathodic protection (CP) or protective coatings to attempt to prevent or reduce the incidence of corrosion. CP is often achieved through the use of sacrificial anodes that corrode instead of the hull, but Impressed Current may also be used, either solely or in combination [1, 2]. Neither system is always fully effective. Coatings have a limited life [14] that may be influenced by poor surface preparation and cleanliness prior to application of the protective coating. As well, coatings can degrade in service through their inherent (low) permeability allowing corrosion to commence, or the coatings may be damaged or degraded. Insufficient or ineffective maintenance may contribute to this degradation.

A convenient approach to corrosion modelling is to express the corrosion in terms of a loss of material, or loss of thickness of structural members. In practice, this material loss can occur as a uniform loss over an area, but typically corrosion is uneven and often localised [15]. Furthermore, such loss can occur at varying rates over time [16, 17]. These different types of corrosion will have different consequences on the behaviour of highly stressed structural members. Of most interest for the present project is long-term corrosion and the effect of environmental influences [18 - 20]. The effect of corrosion on a structure was investigated using non-linear FE analyses [21] wherein a uniform loss of material over a structural member was compared with a quasi-random thickness variation to allow for pitting corrosion. Typically, the uniform loss model produced conservative residual structural strength results compared to models employing a quasi-random thickness variation, concluding that local defects had more impact on the loss of strength than any uniform material loss. This finding is made more pertinent when ship survey results indicated that corrosion was more likely to occur in areas of structural detail, at fittings and near weld lines and other structural details that were also known to produce stress concentrations in the structure [10]. This reinforces the notion that specific defect geometry must be modelled accurately to fully understand a loss of strength.

For the prediction of corrosion growth into the future, the project will use an existing model for corrosion pitting of weld zones [22]. This model is based on corrosion fundamentals that consider a number of corrosion phases in sequence. The model has been calibrated using data extending over 33 years for existing steel structures. Critically, the model shows good correlation with corrosion pit depths relevant to different areas of a ship structure, including the parent metal, heat affected zones and the weld zone. It is considered good practice to differentiate between these zones, something that if neglected, may lead to high variabilities (coefficients of variation) in corrosion losses estimated from data [23].

In the present project, corrosion will be modelled by specific topographic representation of the most critical and severe events as revealed by hull survey reports.

Modelling of Cracking

Cracks tend to initiate at stress concentrations that may result from material or fabrication defects, localised damage, corrosion pits, or micro-scale and macro-scale composition changes in the materials [24]. It is not until a crack reaches a critical length or growth-rate that a significant loss of strength or load bearing ability will occur, with the potential to lead to catastrophic failure. Although many cracks are benign and of no structural consequence, the larger cracks, or those that are growing at a significant rate, are of interest for structural capacity. In the present project, crack data from operational vessels (as identified in the hull survey reports) will be used as input data, together with well-established theoretical models.

For crack modelling, particularly in the case of fatigue, structural resistance may depend on the stress resultants local to the crack, and thus on the loads likely to be experienced. In the case of a ship subject to wind and waves, the loads must be predicted by a complex probability function that is dependent on weather forecasts, routes traversed, and the speed of the vessel, that is, the probability that a specific load scenario will be encountered [25].

For cracking, the stress intensity factors will depend on the nature of the topography, and will be demonstrated by small-scale high fidelity finite element models. This will allow a better understanding of the local loads required to propagate cracks. As usual, it will be sufficient to consider only significant cracks so as not to overburden the finite element model for the vessel as a whole with less than critical inputs.

An example of local stresses in a crack is shown in Figure 3, where a finite element model of a tensile specimen predicts a crack prior to ultimate failure. The modelling of the crack propagation can occur through element elimination rules, where elements are removed from the simulation when a failure criterion is met. The load that was carried by the failed element is transferred to adjoining elements over a number of cycles to ensure stable crack growth. The crack can grow to the point that a structural member may ultimately fail, contributing to a larger failure environment.

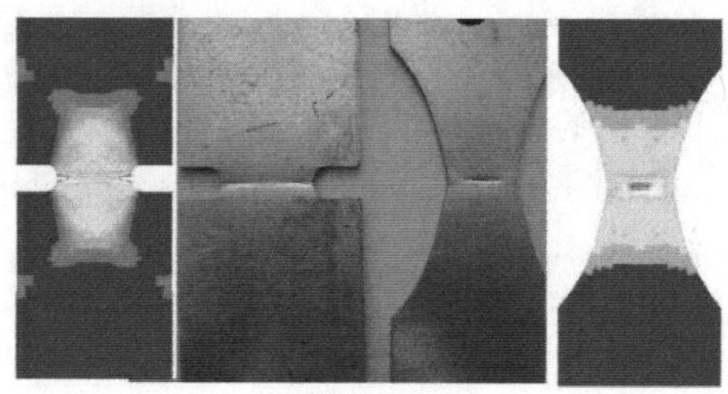

Figure 3: Finite element modelling of tensile coupons, showing crack initiation and propagation

As in the modelling of corrosion, cracks will be modelled by specific topographic representations of the most critical and severe events as revealed in hull survey reports.

Hull Survey Reports

To assess the effect of corrosion and cracking on the structural integrity of the study platform, the location, timeframe, and extent of the degradation at various locations of the hull are required.

The survey of an in-service asset provides valuable information on its condition, and is crucial to smart decision-making regarding maintenance and operations. Examples of defects found during survey of a naval vessel are shown in Figure 4 and Figure 5.

The advantage of hull survey reports is that they are real accounts of the defects that occur. However, the quality and accuracy of the information must be acceptable. Survey reports for the study platform will allow detail in the FE model to be modified locally to represent defects in critical structural components. This facilitates modelling the study platform as it degrades over time.

Figure 4: Typical wastage of a structural member

Figure 5: Vertical fracture in bulkhead plating adjacent to a stiffener weld

Dynamic Loads

The response of a ship to the seaway environment is dependent upon a number of factors such as characteristics of the ship hull, the ship heading, the ship speed relative to the waves, the wave height and length, and the ship's displacement. Consequently these parameters will need to be defined based on the operational limits typical for the study platform.

A novel component of the proposed methodology is the determination and application of the dynamic loads onto the structure. The proposed method is to use the SPH technique to simulate the waves, coupled automatically with the FE model of the ship structure [26, 27]. The current project will use the latest and more accurate SPH technology with implementation of the Reimann-like, diffusive scheme [28].

Two snapshots from a time domain SPH simulation of a rigid model of a generic naval vessel, at forward speed in a seaway, are shown in Figure 6 and Figure 7. The wave height and the vessel speed are chosen such that they represent extreme conditions to generate extreme loadings on the vessel. The dynamic loads calculated from such a simulation are automatically applied to the model of the study platform, through the coupling of the SPH and the FE modelling. In Figure 6, the vessel had traversed a number of waves before encountering the wave shown. The prior waves allowed the vessel to reach a dynamic equilibrium, that is, in terms of response of the vessel. The vessel was free to pitch and heave but held fixed in roll, yaw, sway and surge by means of boundary conditions on the model. Although not considered in this example, roll and sway must also be incorporated.

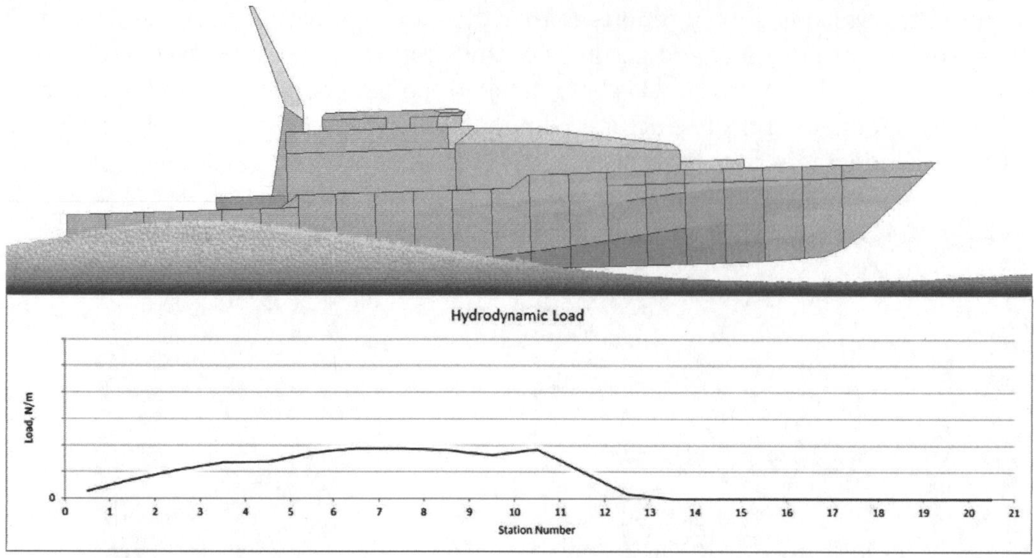

Figure 6: Example SPH model of a generic patrol boat, as forward section exits water

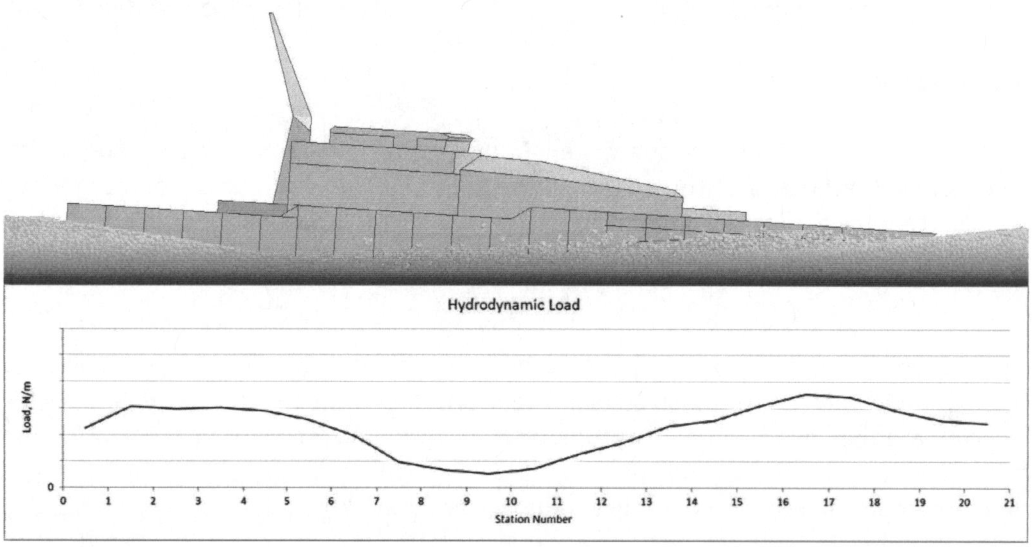

Figure 7: Example SPH model of a generic patrol boat as it re-enters water at wave trough

DISCUSSION

The SIoMP project presents opportunities to insert innovative technical concepts and novel approaches into the life cycle assessment of maritime assets/platforms. Where such assets are subject to potentially severe dynamic loads arising from fluid-structure interactions and also suffer deterioration from corrosion, cracking and fatigue, the technical significance of the research lies in the major improvement in the assessment of the present and future reliability and risk for these assets. It will adapt advanced modelling techniques proven in the structural analysis of other engineering structures and systems, and will develop new numerical tools. It is expected that the algorithms that will be developed within the project will offer clear advances compared to current technology which will potentially have a profound impact on the ability to predict the level of current and future structural safety and reliability of maritime assets.

The present project aims to integrate well established numerical techniques for analysing deterministic, structural behaviour incorporating mesh-free fluid structure interaction and applied dynamic loads with mathematical models for marine and related corrosion. Some of the critical issues involved have been described and resolution of these issues is presently incorporated in the models being developed.

CONCLUDING REMARKS

Outlined herein are the components of a numerical modelling approach to assess and predict the structural integrity of an ageing, deteriorating maritime platform which is exposed to defined sea states. Work is currently underway to refine the models and the inputs required, especially those related to the assumed, experience informed, level of degradation and number of defects.

ACKNOWLEDGMENTS

The authors acknowledge the financial and in-kind support provided by the Australian Research Council, Defence Science and Technology, and Pacific ESI.

REFERENCES

[1] Paik J.K. and Melchers R.E. (Eds), *Condition Assessment of Aged Structures*, Woodhead Publishing Ltd, Cambridge, UK, 2008.

[2] Wang G., Boon B., Brennan F.P., Garbatov Y., Ji, C., Parunov J., Rahman T.A., Rizzo C., Rouhan A., Shin C.H. and Yamamoto N., *Condition assessment of aged ships and offshore structures*, Proceedings 17th International Ship and Offshore Structures Congress, Committee V.6, **2**: pp309-365, Seoul, Korea, August 16-21, 2009.

[3] Ibrahim R.A., *Overview of Structural Life Assessment and Reliability, Part II: Fatigue Life and Reliability Assessment of Naval Ship Structures*, Journal of Ship Production and Design, **31**(2): 100–128, 2015.

[4] Ibrahim R.A., *Handbook of Structural Life Assessment*, John Wiley & Sons Ltd, Chichester, UK, 2017

[5] Galambos J., *The asymptotic theory of extreme order statistics*, Second Edition, Krieger, Malabar, FL, 1987.

[6] Coles S., *An introduction to the modelling of extreme values*, Springer, New York, 2001.

[7] Castillo E., *Extremes in engineering applications*, (in) Extreme value theory and applications, J. Galambos, J. Lechner, E. Simiu and N. Macri Dordecht (Eds), Springer Academic Publishers, pp15-42, 1994.

[8] Charras-Garrido M. and Lezaud P., *Extreme Value Analysis: an Introduction*, Journal de la Societe Francaise de Statisque, Soceite Francaise de Statistque et Societe Mathematique de France, **154** (2): pp66-97, 2014.

[9] INTERTANKO *Guidance Manual for Maintenance of Tanker Structures*, Tanker Structural Co-operative Forum, Witherby, London, 2008.

[10] Phelps B. and Morris B., *Review of Hull Structural Monitoring Systems for Navy Ships*, DSTO-TR-2818, Maritime Platforms Division, Defence Science and Technology Organisation, 2013.

[11] Greve L. and Vlachoutsis S., *Multi-scale and multi-model methods for efficient crash simulation*, International Journal of Crashworthiness, **12**(4): 437-448, 2007.

[12] Croaker P.J., El-Khaldi F., Groenenboom P.H.L., Cartwright B.K., Kamoulakos A., *Structural performance and dynamic response of semi-submersible offshore platforms using a fully coupled FE-SPH approach*, MARINE 2011 - Computational Methods in Marine Engineering IV: 49 – 63, 2011.

[13] Liu Y., Saputra A. and Song Ch., *Automatic polyhedral mesh generation and scaled boundary finite element analysis of STL models*, Computer Methods in Applied Mechanics and Engineering **313**: pp106-132, 2017.

[14] Melchers R.E. and Jiang X., *Estimation of models for durability of epoxy coatings in water ballast tanks*, Journal of Ships and Offshore Structures **1**(1): pp61-70, 2006

[15] Gudze M.T. and Melchers R.E., *Prediction of naval ship ballast tank corrosion using operational profiles*, Transactions of the Royal Institution of Naval Architects Part A: International Journal of Maritime Engineering, 148; 77-86, 2006.

[16] Southwell C.R. and Alexander A.L., *Corrosion of metals in tropical seawaters: Structural Ferrous Metals*, Materials Protection, **9** (1): pp14-23, 1970.

[17] Szklarska-Smialowska Z., *Pitting Corrosion of Metals*, Houston: NACE, 1986.

[18] Blekkenhorst F., Ferrari G.M., van der Wekken C.J. and Ijsseling F.P., *Development of high strength low alloy steels for marine applications, Part 1: Results of long term exposure tests on commercially available and experimental steels*, British Corrosion Journal, **21**(3): 163-176, 1986.

[19] Brossia C.S. and Cragnolino G.A., Effect of environmental variables on localized corrosion of carbon steel, Corrosion (NACE) **56**(5): pp505-514, 2000.

[20] Paik J.K., Thayamballi A.K. and Melchers R.E., *Some recent developments in corrosion assessment and management for steel ships and offshore structures*, Marine Technology **45**(2): pp94-100, 2008.

[21] Mateus A.F. and Wirz J.A., *On the post-buckling of corroded steel plates used in Marine Structures*, Transactions of the Royal Institute of Naval Architects (RINA) **140**: pp165-183, 1998.

[22] Chaves I.A. and Melchers R.E., *Extreme value analysis for assessing structural reliability of welded offshore steel structures*, Structural Safety, Volume 50, pp9-15, Elsevier Publishing, 2014.

[23] Guedes-Soares C., Garbatov Y., Zayed A., Wang G., Melchers R.E., Paik J.K. and Cui W., *Non-linear corrosion model for immersed steel plates accounting for environmental factors*, Transactions of the Society of Naval Architects and Marine Engineers **113**: pp306-329, 2006.

[24] Ship Structures Committee (SSC), *Report 367 - Fatigue Technology Assessment and Strategies for Fatigue Avoidance in Marine Structures*, US Coast Guard 1993.

[25] Beghin D., *Fatigue of Ship Structural Details*, Technical and Research Bulletin **2**-31, Society of Naval Architects and Marine Engineers, 2006.

[26] Groenenboom P.H.L., Cartwright B.K. and McGuckin D., Numerical Simulation of ships in High Seas using a Coupled SPH FEA Approach", RINA conference on Innovation in High Speed Marine Vessels, , Australia, January 28-29, 2009.

[27] Groenenboom P.H.L. and Cartwright B.K., *Hydrodynamics and fluid structure interaction by coupled SPH-FE method*, Journal of Hydraulic Research, **48** (Extra Issue): pp61-73, 2010.

[28] Groenenboom P.H.L. and Cartwright B.K., *Pressure-corrected SPH with innovative particle regularization algorithms and non-uniform initial particle distributions*, Proceedings of the 8th Int. SPHERIC Workshop, Norway, June 4-6, 2013.

Comparative Structural Requirements for High Speed Naval Craft: The 'ROURKE Class' Corvette

John M. Lord,
Independent, Member: RINA (Australian Division) and Engineers Australia
john.lord.310@gmail.com

ABSTRACT

Following on from a paper presented at 'Pacific-IMC-2015' on various concepts for an 'Offshore Combatant Vessel', this paper reviews different structural requirements for High Speed Naval Craft (or Fast Displacement Ships), using a 90 metre monohull Corvette as a representative example. It follows the theme of a Report published in 2005 by the U.S. Ship Structures Committee, SSC-439 'Comparative Structural Requirements for High Speed Crafts', and a paper by the same title published by authors Stone and Novak, that advocated the use of an example to provide a meaningful quantitative comparison.
The work includes an independent review of Commonwealth Defence publication: DEF(AUST)5000 Volume 03 Part 01, Issue 01, 'Surface Ship Structure', 2007. An objective is to incorporate lessons-learned from the engineering analysis of Defence/Navy maritime projects that have experienced structural issues; including vibration and/or cracking, due to a combination of factors such as slamming and hull-whipping, corrosion and/or fatigue. The work considers a variety of methods for the assurance, verification, certification, and acceptance of navy surface ship structural reliability, and also assesses some practical operational limits and constraints, using predictive methods for wind-&-wave forecasting.

1. INTRODUCTION

This paper presents a summary review of professional standards and practices for structural engineering, across domains of civil, structural, marine, and naval engineering, as relevant to naval shipbuilding in Australia. A common theme is that technical solutions fall into one of two categories, either: 'Deemed-to-Satisfy'; or 'Performance Solutions', where compliance with defined performance-based requirements must be demonstrated. For naval shipbuilding projects, all solutions are considered to be 'Performance Solutions', where the method of demonstration of structural reliability is by engineering analysis.

For each of the professional engineering domains, relevant Codes of Practice exist that provide guidance for the theory and application of structural engineering best practice. The Codes of Practice refer to various International and National Standards, and where proven analytical methods, when combined with professional expertise and judgement, can demonstrate compliance with the performance-based requirements of the standard.

The hypothesis is that there is an urgent need for an 'Australian Benchmarking Standard for High Speed Naval Craft', to provide a basis for assurance, certification and acceptance of high-speed naval ships. A suggestion is that the standard could be community-based; typical of a National Standard, similar to the 'National Standard for Commercial Vessels'. The importance and urgency is driven by the Australian Government's Shipbuilding Plan, to provide evidence of 'design acceptance', before the start of ship construction activities.

2. BACKGROUND AND CONTEXT

The Australian Government, via the Defence White Paper 2016 [1], and the National Shipbuilding Plan [2], continues to support the Australian naval shipbuilding sector. VADM Barrett, Chief of Navy (CN), in 'The Navy and The Nation' [3], has addressed topics including: Concepts of sea power, in a contemporary context; and traditions and alliances. The chapter on 'Navy as a national enterprise' included a useful discussion of the benefits of continuous shipbuilding, its implications for innovation, and also the benefits accruing. CN stated: 'Specialist commentators have mentioned the success of the ANZAC Project'; and suggested there could be merit in replicating that project, but with some provisos. In referring to the 'Benefits' of continuous shipbuilding, CN stated [3]:

> *We cannot disconnect from the fact that we will inevitably begin with a mature design. Where we take benefit is in managing the system that supports that design – data recording, real time maintenance management, capability and systems evolution, the exploitation of disruptive technologies. This is essentially about availability. Looking beyond the build, we must design for sustainment.*

The above reference by Chief of Navy to specialist commentators, presumably included the unpublished work by Jacobs Australia: Mega-projects generate their own weather. That report may not be readily available, but an account of the ANZAC Project exists [4].

Shipbuilding comprises modular hull construction and outfitting, set-to-work, and testing. Hull design and construction is the art, science and practice of ship structural engineering. In planning the revitalisation of naval shipbuilding in Australia, and commitment of major capital investment, the need for professional structural engineering practice is essential. Structural integrity and structural reliability provide a basic foundation for navy capability, and represent much more than a safety requirement. Without processes and procedures to assure structural reliability, ships cannot be deployed at sea; the navy cannot operate.

The United States Navy has developed a program to advance the science and technology of ship structures, called 'The ONR Ship Structural Reliability Program', that stated [5]:

> *Ship structural reliability is the assurance that the hull structure will perform its mission for a specified time in a specified environment. The operational environment for a naval ship includes the potential for repeated and extreme seaway and ice loads, and possible combat loading. This program primarily addresses the ability of ship structures to endure wave, ice, and weapons load effects, and methodologies for recognizing these effects.*

Those familiar with the Deming Cycle of Continuous Improvement: 'Plan; Do; Check; Act'; will realise that there is always room for improvement, particularly with 'Requirements'. Australian navy requirements for 'Surface Ship Structure' were first published in 2007 via DEF(AUST)5000, Vol. 03, Part 01 [6], and have not been updated over the past ten years. The policy requirements deferred to Classification Society Rules, but the implementation has lacked consistency. Some shipbuilding projects have opted to 'self-certify', exercising provisions for being 'Deemed-to-Comply'; while other shipbuilding projects, where ships were designed, constructed and maintained to 'Class', have experienced structural issues. Regarding requirements for ship structure, it raises the question: What constitutes a Solution that is ***'Deemed-to-Satisfy'***; versus an alternative ***'Performance Solution'***, where performance can be demonstrated by analysis to comply with defined requirements?

3. HISTORICAL BACKGROUND - ANZAC SHIP PROJECT

The initial stage of the ANZAC Project, then known as the 'New Surface Combatant', involved a complex discussion surrounding the meaning of the term 'existing design'. Similar semantic issues arise for the terms 'mature' or 'proven', in relation to ship design. For the purpose of short-listing candidate designs the project adopted a simple definition: *'An existing design is one which is under contract for construction by December 1986'*.

This definition of an 'existing design' was supported by an engineering policy analysis [7], based on the Australian Frigate project, and its predecessor, U.S. Navy's FFG-7 project. FFG-7 enjoyed a reputation of being the 'best-documented ship acquisition project' [8]. Structural design utilised NAVSEA's Structural Design Manual for Naval Surface Ships [9]. The FFG Program philosophy at the time was: 'If the ship is built to the Drawings, and tested successfully against the Test Procedures, it will comply with the Specification'. This approach represented an example of a technical solution that was **'Deemed-to-Satisfy'**.

Reference [7], Annex A, listed 'Contract or Contract Guidance Drawings', and 'Preliminary Allocated Baseline (PABL) Back-Up Data'. Contract or Contract Guidance Drawings comprised 54 drawings. The Back-Up Data comprised 91 documents other than drawings, and addressed subjects including: Management; Containment/Human Support; Combat Systems; Integrated Logistic Support; Mobility/Systems Support; Systems Engineering; and Test and Evaluation. This exercise drove the development of the Statement of Work and deliverable requirements for Design Development Contracts for the ANZAC Project.

Contract Drawings included the ship specification; lines drawing; midship section, and general arrangement. The Ship Performance Specification (SPS, the equivalent of a Function and Performance Specification - FPS) [10], included a general requirement:

> *The vessel shall be designed and built in accordance with a documented system of established and recognised naval engineering standards and practices, except where otherwise agreed.*

Consistent with the U.S. Navy's Ship Work Breakdown Structure [11], Section 100 of the SPS set out requirements for 'Hull Structure General'. Requirements, as specified, were similar to U.K. and U.S. Navy specifications & design data sheets of the day. For example:

> *The primary stress due to hull girder bending when subjected to a static wave balance at trochoidal form with wave length equal to the waterline length (L_{WL}) of the vessel and wave height equal to 0.6 $\sqrt{L_{WL}}$ (metres). Primary stress in the hull structure under this condition should not exceed 130 MPa (9.43 Tsi) for mild steel or 146 MPa (10.59 Tsi) for High Tensile Steel.*

The structure for the ANZAC ships was designed to German Navy requirements, BV 104-1, 'General Analysis of the Ships Hull (Strength Specification)' [12], and BV-105 'Welded connections (welding specification) Part 1, Surface ships' [13]. To aid in the process of certification, the ANZAC Project engaged Germanischer Lloyd (GL) to provide a 'Certificate-of-Compliance' against the Ship Specification, and the ships were brought into Class with GL - a 'world-first' for a Surface Combatant. The ANZAC Program delivered a **'Performance Solution'** that demonstrated compliance with performance requirements, and where GL concluded: *'This ship is well designed from a structural point of view'* [14]. Lloyd's Register conducted an evaluation of the ANZAC against the Naval Ship Rules [15]: *'The evaluation...shows only a few minor areas where further analysis may be required.'*

4. COMPARISONS OF STRUCTURAL REQUIREMENTS FOR HIGH SPEED NAVAL CRAFT

Structural requirements are set to ensure that naval ships are fit-for-service, or purpose. 'Fitness-for-service' is an executive judgement by the Owner or Naval Technical Authority. There is a risk of a gap between specified structural requirements and fitness-for-service. This leads to a question of whether the risk regime is Caveat Emptor, or Caveat Venditor? In everyday language: 'Buyer beware' or 'Seller beware' where 'Insurance' is a factor [16].

Classification Societies provide a service to all parties, subject to limitation of liability [17]. Classification Society rules and regulations are representative of 'Owners Requirements', but also embody international maritime protocols, and provide guidance to prospective prime contractors, ship builders, and ship designers. Rules are more than a 'Design Code'.

In the United States, the Ship Structures Committee has actively pursued research into ship structures. Topics have included comparative studies of Classification Society Rules for large steel ships [18], and high-speed craft [19], including high-speed naval craft [20].

As an example, the U.S. Coast Guard made a clear statement regarding its requirements for the Offshore Patrol Cutter program at an OPC Industry Day presentation in 2010 [21]:

> *The OPC is to be classed in accordance with American Bureau of Shipping (ABS) Guide for Building and Classing Naval Vessels (NVR) with Coast Guard Appendix, as tailored by the OPC System Specification, and will use rules version January 2010* [22]; and:
> *ABS Class does not replace the Coast Guard's involvement; ABS Class does not address everything – only those aspects addressed in NVR (HM&E); ABS Class does not replace the design development process; ABS Surveyors are not a replacement for a Quality Assurance Department; and the Naval Vessel Rules (NVR) is not a building specification.*

The presentation went on to state that NVR contains unclassified sensitive information that must be protected from unauthorized release. The NVR are governed by: *'Distribution Statement D' – 'Distribution authorized to DoD and DoD contractors only'.* There is a formal NVR request process in order to obtain a copy of the NVR from NAVSEA.

ABS uses a definition for High Speed Craft with a lower threshold than the IMO definition. If a ship's *'speed/length ratio is greater than 1.3' (Knots, Feet)*, it is recommended to use ABS Rules for Building and Classing High Speed Naval Craft (HSNC) [23, refer Fig. 1 below].

From experience with the U.S. Navy Littoral Combat Ship project, Kramer et al compared ABS and DNV Structural Requirements using, as an example: 'structural modifications for converting a commercial, high speed ferry into a high speed military transport capable of unrestricted (i.e., open ocean) operation' [24]. Kramer [24] compared ABS HSNC Rules [23], and the DNV Rules for Classification of High Speed, Light Craft and Naval Surface Craft [25], in order to calculate hull girder, slamming and vehicle deck loads required for the design of a US Navy High Speed craft/combatant. Kramer's approach discusses specific operational limits.

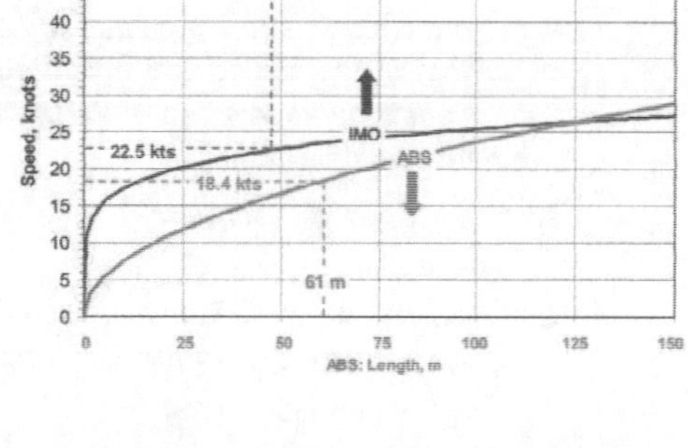

Fig.1. IMO & ABS Definitions of High Speed Craft [24]

5. NATIONAL STANDARDS, INCLUDING THE NATIONAL CONSTRUCTION CODE

5.1 National Construction Code

The National Construction Code (NCC) [26] was introduced effective from 01 May 2016. It addresses: 'Assessment methods; Acceptance of design and construction; Adoption of Standards and other references; Performance requirements; and Verification methods'. Referenced standards include: AS/NZS 1170 Design actions; AS/NZS 4100 Steel structure. The part 'Acceptance of design and construction' refers to:

> *(iii) A certificate from a professional engineer or other appropriately qualified person which*
>
> *(A) certifies that a material, design, or form of construction complies with the requirements of the BCA* (Building Code of Australia)*; and*
>
> *(B) sets out the basis on which it is given and the extent to which relevant specifications, rules, codes of practice or other publications have been relied upon.*

The Verification methods of Section BV1 for Structural reliability require:

> *Compliance with BP1.1 and BP1.2 is verified for the design of structural components and connections when—*
>
> *(a) the calculated annual structural reliability index (β), for each action, is not less than that listed in Table BV1.1; and*
>
> *(b) the annual structural reliability index (β) is calculated in accordance with the following formula (as given in the NCC 2016 Building Code of Australia - Volume One, Page 72).*

The NCC is accompanied by a Structural Reliability Handbook [27], produced by the ABCB:

> *This Handbook is one of a series produced by the ABCB... Handbooks expand on areas of existing regulation or relate to topics which have, for a variety of reasons, been deemed inappropriate for regulation. The aim of the Handbooks is to provide ... advice and guidance on specific topics. Structural reliability has been identified as an issue that requires consistent uniform guidance... The Structural Reliability Handbook has been developed to foster a greater understanding of Verification Methods BV1 and V2.1.1 contained in the National Construction Code (NCC) Volumes One and Two respectively. This information Handbook addresses the methodology in developing the Verification Methods in generic terms, and is not a document that sets out a specific process of using the Verification Methods or an alternative structural reliability process.*

The NCC, and the Structural Reliability Handbook, demonstrate that performance requirements are considered appropriate by the Australian community for civil structures.

The distinction between a compliance solution that is **'Deemed-to-Satisfy'**, and an alternative **'Performance Solution'** perhaps requires some clarification of definitions [28]:

> *The key to the performance based NCC is that there is no obligation to adopt any particular material, component, design factor or construction method. This provides for a choice of compliance pathways. The Performance Requirements can be met using either a Performance Solution (Alternative Solution) or a Deemed-to-Satisfy (DTS) Solution.*
>
> ***Performance Solution**:*
>
> *A Performance Solution is unique for each individual situation. These solutions are often flexible in achieving the outcomes and encouraging innovative design and technology use. A Performance Solution directly addresses the Performance Requirements by using one or more of the Assessment Methods available in the NCC.*
>
> ***Deemed-to-Satisfy Solution:***
>
> *A DTS Solution follows a set recipe of what, when and how to do something. It uses the DTS Solutions from the NCC, which include materials, components, design factors, and construction methods that, if used, are deemed to meet the Performance Requirements.*

5.2 Naval Ship Code

For navy ships operating in Australian and international waters, the global domain of the maritime community, it is reasonable to expect that a similar level of requirements exist. The NATO Standardization Agency, NATO Group MCG-6, published the Naval Ship Code, ANEP-77, Edition E in 2014 [29]. Representatives from the Royal Australian Navy (RAN) contributed to the development of the NSC, and there has been an expectation, partly as the result of a Navy Signal referring to HMAS Choules [30], that the RAN planned to adopt the Naval Ship Code for the fleet, as a matter of naval engineering policy.

The Naval Ship Code (NSC) addresses the topics that might be expected, stating the aim; and setting appropriate goals, performance requirements, solutions, and justification. For example, Chapter II 'Structure', Regulation 0 'Goal', states:

> *For the design life of the ship, the structure shall be designed, constructed and maintained to:*
>
> *1.1 Provide weathertight and watertight integrity;*
>
> *1.2 Carry all loads that may be foreseen;*
>
> *1.3 Permit embarked persons to carry out their duty safely;*
>
> *1.4 Protect the embarked persons and essential safety functions in the event of all foreseeable emergencies and accidents at least until the persons have reached a place of safety or the threat has receded; and*
>
> *1.5 Minimise the risk of loss of the ship.*

Under the heading 'Purpose', the NSC states that: *'Chapter II Structure is written as a "standard for the selection of standards" rather than a standard for direct application in a design office or construction/repair facility. As a consequence, the primary target audience for this Chapter is the Naval Administration and its Recognised Organisations'*.

The NSC goes on to state, under the following sub-heading 'Application':

> *4. In addition to the requirements contained elsewhere in the present regulations, ship structure shall be designed, constructed, and maintained in compliance with the structural requirements of a Classification Society whose rules and procedures are recognised and validated by the Naval Administration as meeting the Goal of this Chapter or with applicable national standards of the Naval Administration which have been demonstrated to provide an equivalent level of safety.*

From the taxpayer's perspective, as the insurer of Commonwealth-owned assets, the priority is one of transparency. The RAN needs to be openly forthcoming about how it intends to address the lack of ship structural standards, and the pre-requisite professional competencies to answer questions of acceptance of ship designs for surface ship projects. What Standard(s) will ship designs be accepted against; how will the acceptance process be conducted; by whom? Who will the Service Chief trust to conduct design acceptance?

If the RAN chooses to manage a ship structural engineering program within 'Naval Class', there is the question: Which Classification Society to use, and then which rule set to use, as being the most appropriate to accept the particular capability that is being acquired? The issue becomes one of strategic engagement between Navy and Class: How to do it?

5.3 National Standard for Commercial Vessels

A National Standard already exists for Commercial Vessels [31], but that standard would need extensive revision, if it was to become applicable to high speed naval surface craft. The question arises: Is it preferable to work within the existing concept of 'Naval Class'; or to start again with a 'first-principles' approach, based on rational ship design methods? Given limited options, a suggestion is to adapt a National Standard for 'Naval Class' [31], and to introduce rational ship design methods to assess optimal structural solutions [32].

5.4 Codes of Practice and National Standards for Naval Ships, also including Naval Class

It is evident that accepted codes of practice, and community or national standards exist for civil buildings, and also for the design, construction and maintenance of naval ships. There is some choice, albeit limited, whether to 'comply with the structural requirements of a Classification Society whose rules and procedures are recognised and validated by the Naval Administration as meeting the Goal of Chapter II'; or to comply with 'applicable national standards of the Naval Administration which have been demonstrated to provide an equivalent level of safety'. There does not appear to be any other option available.

A previous paper [33] noted a lack of clarity surrounding current policy and procedures for the structural design and certification of Navy ships. Existing policy nominally requires the involvement of a Classification Society, but does not mandate such involvement. It has been said that 'Class is not a Capability Requirement'. However, one would expect ship structural integrity to be a Capability Requirement, because a lack of structural integrity (or structural reliability) may render a Maritime Capability 'Unavailable' [34, 35].

In 2016, a no-cost offer was made to assist the RAN with a review of the DEF(AUST)5000 for 'Surface Ship Structure' [36]. The DEF(AUST)5000 from 2007 invokes a requirement for Classification Society involvement. However, there appears to be little appetite within the RAN for the consistent implementation of a policy involving 'Naval Class'. Realistically, since there is no national standard (other than the NCC for buildings) that can claim to provide 'an equivalent level of safety', now there is no viable alternative to 'Naval Class'.

As in other situations where the RAN finds itself in difficulties, a common source of support and technical expertise is the U.S. Navy (USN). In this case, a possible avenue is a request for technical assistance to the USN's Commander Naval Sea Systems Command. However, the USN is unlikely to take on what is essentially a 'Sovereign Risk' for the RAN. The USN is likely to offer support via an agent such as American Bureau of Shipping (ABS). Presumably, technical services that ABS may be able to offer for navy surface ships would be based on the current version of the ABS Guide for Building and Classing Naval Vessels, and similar to the previous example given for the U.S. Coast Guard OPC Program [21].

For vessels with Speed/Length ratio ($V_k/\sqrt{L_{WL}}$: Speed-Knots; Length-Feet) greater than 1.3, the Naval Vessel Rules refer the user to the ABS Rules for High Speed Naval Craft. Kramer [24] has previously provided a worked example of the application of the ABS High Speed Naval Craft Rules, including remarks referring to defined operating limits. [22, 23].

6. THE ROURKE CLASS CORVETTE – PRELIMINARY DESIGN OPTIONS AND SOURCE DATA

The respectful usage of the title 'ROURKE Class' Corvette refers not so much to a product, but a benchmark of high-quality Australian naval engineering and management expertise. Good management expertise includes economic management, and project management. The prospect of continuous shipbuilding is one the late RADM Rourke would have envied. Products that attracted his interest included destroyers, frigates, and also patrol vessels.

The historical basis for the work now entitled the 'ROURKE Class' Corvette was SEA 1427, the Joint Offshore Patrol Combatant that was a competing bid by the Transfield Company for a shipbuilding program involving the Royal Malaysian Navy and Royal Australian Navy. Navy requirements included a maximum speed of 25 knots and unrestricted operations. The project worked with the Classification Society Det Norske Veritas (DNV), using the High Speed Light Craft and Naval Surface Craft (HSLC&NSC) Rules, version July 1993 [25].

Project SEA 1427 OPC did not progress beyond the preliminary project design phase [37]. A review of project documentation identified issues with the ship's structural design, and potential restrictions on operations south of latitude 48 degrees South. Letters from archived Navy files documented those issues [38, 39]. Comments from the time were that to operate in the Southern Ocean (Winter Zone), at latitudes south of 48 degrees South, would require a ship designed and constructed to DNV conventional steel ship Rules [40]. To scope out the issues, it is necessary to summarise aspects addressed in the letter [38].

The examination concentrated on those aspects influenced by the vessel's naval role, such as structural continuity for shock resistance, or design for weapons loads:

a. Inclusion on the Design Load Plan of deck and bulkhead loads arising from weapons reactions, weapons blast pressures and ammunition deck loadings;
b. gun foundations;
c. structural discontinuities, either abrupt endings or abrupt changes, noting that all discontinuities can be a problem in a shock environment;
d. swedge (corrugated) bulkheads used for transverse and longitudinal bulkheads;
e. placement of pillars on soft foundations; and
f. stress concentrations from holes in the stringer plate for uptakes and intake.

The following comments also resulted from the structural review for SEA 1427 OPC [38]:

Every frame, whether it be a shaped hull frame or a straight deck frame (some of which are relatively small), is a fabricated member. This will be an expensive structure to build. Consideration should be given to a wider frame spacing and special steel runs of a limited number of structural Tees. This applies particularly to the stiffeners – angles are not efficient and need to be heavier for this reason and because they are not symmetrical are prone to tripping. More efficient stiffening may allow wider frame spacing with minimal if any stiffening weight increase.

To meet a naval requirement for 'Unrestricted Service', there is a perceived threshold for vessel size, from about 80 to 100 metres length overall, and 3000 Gross Registered Tons. Further work has reviewed ships in this range of length and volume, as typical exemplars. Five options have been reviewed, to consider a range of designs up to length 100 metres. The book 'Fast Displacement Ships' by MARIN includes a case study for a 100m craft [41].

7. PROJECT SEA 1427 OPC, RNZN PROJECT PROTECTOR OPV, & OTHER CASE STUDIES

RAN Project SEA 1427: Design of the Offshore Patrol Combatant progressed in parallel with the ANZAC Frigate, but did not proceed beyond the preliminary design phase. Project deliverables included: Ship specification; hull lines-&-offsets; midship section and structural profile; and general arrangement [37]. Ship design work was performed to the Rules of Det Norske Veritas for High Speed Light Craft and Naval Surface Craft, 1993 [25].

RNZN Project Protector: A seven-ship order from the Royal New Zealand Navy (RNZN) was awarded to Tenix Defence Systems in 2003 that included a Multi-Role Vessel (MRV), four Inshore Patrol Vessels (IPV), and two Offshore Patrol Vessels (OPV). For the OPV, the operational requirement was for a speed of 23 knots. The environmental conditions called for the capability to deploy to the Southern Ocean and Ross Sea, and to operate in ice conditions, and was to conform to Lloyd's Register rules for Special Service Craft, with Class notation: 100A1 SSC Patrol Mono G5 Special Duties: Ice Class 1C, LMC UMS [42, 43].

AMC 91 metre OPV: In 2014, at the Australian Maritime College, four students developed the design for a 91 metre Offshore Patrol Vessel for the subject of 'Ocean Vessel Design'. The students worked on a hypothetical requirement for the Royal New Zealand Navy [44]. Ship design work, including structural design, was performed in accordance with Lloyd's Register Naval Ship Rules, NSR3 [45]. Details of the development of the structural design, and selection of scantlings, is given in the OVD Final Report [46]. LT Marino, developed an Excel spreadsheet to support the structural design and structural weight estimate [47]. The OVD Final Report and Excel spreadsheet offer a guide to application of LR's NSR [45].

University College London 90m Corvette: In 2013, Nicholas Bradbeer, a doctoral student, completed his thesis on the topic: 'Implications for Underwater Shock Response of Adopting Simplified Structural Styles in Warships'. Bradbeer used the Paramarine software tools, originally developed by the MoD UK, to develop a preliminary design and structural concept for a 90 metre Corvette [48, 49]. Design synthesis was performed in accordance with Lloyd's Register Naval Ship Rules, NSR2. The design(s) provided stepping-stones in the process of analysing the structural response to impulsive shock loadings. The alternative structural styles by Bradbeer serve as illustrative examples for this paper.

BMT-Nigel Gee 80 metre Motor Yacht: The paper by Roy et al [50] discussed structural concepts for an 80 metre Motor Yacht, and represented a case study where minimum practical scantlings were developed and discussed. The ship design was for a maximum speed of 16 knots, and utilised Lloyd's Register Rules for Special Service Craft [43].

Ship characteristics are given in Table 1, and four examples are illustrated in Figs. 2a - 2d.

Table 1: Ship design characteristics for five examples:

Option [Ref.]	Ship Design	LOA (m)	LBP (m)	B (Max) (m)	B (WL) (m)	D (m)	T (m)	Δ FL (t)	V (Max) (knots)	Class Society Rule Set
1 [37]	RAN SEA 1427 OPC	81.00	74.93	11.83	11.18	7.20	3.40	1,500	25	DNV HSLC
2 [42]	RNZN OTAGO OPV	85.00	77.60	14.00	12.88	6.80	3.88	1,951	22	LR SSC
3 [46]	AMC Student OPV	91.00	88.00	14.00	13.00	7.40	4.00	2,350	23	LR NSR3
4 [48]	UCL Bradbeer COR	90.00	83.33	11.464	10.85	7.857	2.90	1,800	30	LR NSR2
5 [50]	BMT Motor Yacht	80.00	71.40	14.14	14.00	8.35	3.55	2,000	16	LR SSC

Figures 2a, b, c, d, illustrate the overall similarity between OPC, OPV, & Corvette designs. However, in terms of 'structural style', there are differences, as summarised in Table 2:

Fig. 2a: SEA 1427 Joint OPC (81m)

Fig. 2b: HMNZS OTAGO Class OPV (85m)

Fig. 2c: AMC OPV Design Project (91m)

Fig. 2d: Bradbeer UCL – Corvette (90m)

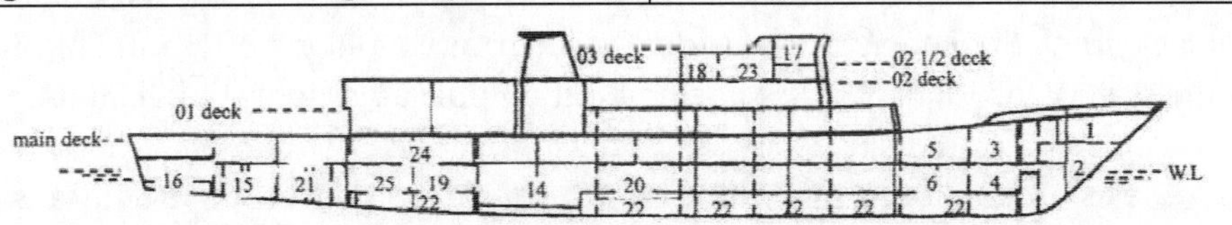

1	cordage room	14	Main engine room	20	1st auxiliary engine room
2	Fore peak tank	15	Main ballast tank	21	Fresh water tank
3	Emergency generator	16	Steering gear room	22	Fuel oil tank
4	Bow thruster room	17	Wheel house	23	Combat direct tower
5	Ammunition chamber	18	Captain's room	24	Engine control room
6	Storage tank	19	Dynamo room	25	2nd auxiliary engine room

Figure 3. Typical profile and geometrical configuration of a 2000-ton patrol-craft [51, 52]

Table 2. Differences between selected examples in terms of the adopted 'Structural Style'

OPTION	Structural Style - Framing System: Transverse, Longitudinal, or Hybrid	RULES
RAN OPC	Longitudinally-framed; Frame spacing – 1,000mm; Stiffener spacing – 500mm	DNV HSLC
OTAGO	Hybrid: Transversely-framed; 1 Deck Longitudinally-framed; Raised Forecastle	LR SSC
AMC	Hybrid: Transversely-framed; 1 Deck Longitudinally-framed	LR NSR3
UCL	Doctoral Thesis reviewed: Transverse-framing; and/or Longitudinal-framing	LR NSR2
BMT	Paper reviewed: Transverse-framing; Longitudinal-framing; or Hybrid	LR SSC

Each of the above examples, whilst sharing similarities in concept and appearance, are different in detail for the structural style and engineering assumptions they incorporate.

For SEA 1427 OPC, the midship section showed a closed section that included 01 Deck. Longitudinal stiffening comprised asymmetric angle sections, prone to failure by tripping. The selected frame spacing was 1,000 mm, with longitudinals spaced at 500 mm apart. This formed a panel aspect ratio of 2:1. The structure was designed for 360 MPa medium strength steel. As mentioned previously, the ship structure was designed to DNV High Speed Light Craft and Naval Surface Craft Rules (DNV HSLC&NSC Rules), July 1993 [25].

For the RNZN OTAGO Class OPV, the raised forecastle deck was carried forward to the bow. The structure was a hybrid style with transverse frames spaced at 500 mm intervals, and where the Main Deck (1 Deck) was longitudinally framed. From unconfirmed reports, the 01 Deck forward was strengthened against buckling, and the Bridge Front was strengthened against wave impact loads. The ship structure was designed to Lloyd's Register Special Service Craft Rules, as current [43]. Note that 'between-deck-height' was 2,800 mm for the OTAGO Class OPV (where other designs were typically about 2,550 mm, or 2,700 mm for the OPV example [59]). As a design variable, 'between-deck-height' has an impact on the beam, since Depth and Beam are related due to stability considerations.

The AMC Student OPV Project, designed to a hypothetical requirement for the RNZN, also selected a hybrid style, where the transverse frame spacing was 1,500 mm, and the Main Deck (1 Deck) and Double Bottom were longitudinally framed. The midship section showed a closed section, also including 01 Deck, 02 Deck, and Bridge Structure at 03 level. To deal with the requirement to operate in ice conditions, the transverse frame spacing for the forward 40% of the ship's length was reduced to 750 mm. The Main Deck was also the weather deck, and was carried forward to the bow. Structural design used 360 MPa medium strength steel. Ship structure was designed to Lloyd's Register Naval Ship Rules, and utilised the provisions for NSR3 [45], including the applicable operational limits.

The design synthesis process used by Bradbeer [48] considered different structural styles: longitudinal framing; transverse framing; or a hybrid style similar to the AMC OPV project. The ship structure was designed to the MOD UK SSCP 23 (the same as Chalmers [53]), and Lloyd's Register Naval Ship Rules, according to the ship type NSR2 and operational limits. As designed, the midship section was an open section, comprising eight panels or strakes. Transverse and longitudinal frames were symmetrical sections, Admiralty Tee Sections, where relative proportions were idealised, using the height (or depth) of the section as the principal dimension, and relative ratios used to determine other section dimensions.

As another example, aspects are included for development of an 80 metre Motor Yacht by Roy et al [50]. This design development reviewed longitudinally framed, transverse framed, and hybrid structural styles, using Lloyd's Register Rules for Special Service Craft. The discussion included consideration of features that contributed to a robust design. Whilst the midship section drawing showed a closed section, related comments suggest:

> *The longitudinal bending moment and shear force should be increased by 25% to account for the inevitable loss of cross-section area due to openings and penetrations in the structure.*

8. ASSURANCE OF ENVIRONMENTAL LOADS, AND STRUCTURAL SCANTLINGS

Bradbeer [48] has explored the application of simplified structural styles in warships, and used the structural design of the NATO Frigate Replacement (NFR 90) as a basis from which to develop the conceptual design for a 90 metre Corvette. Given that the objective related to the design of warships, the focus of the work was on the structure's response to impulsive shock loads. Other researchers who have investigated the response of warship structures to shock loads include Liang & Tai [51], Reid [56], and Szturomski [57].

The alternative structural styles explored by Bradbeer for the 90 metre Corvette included both transversely-framed and longitudinally-framed modules, where the frame-spacing was 1,500mm, and length between watertight bulkheads was 9 metres. Bradbeer's design for the Corvette was performed consistent with Lloyd's Register Naval Ship Rules (NSR), and used requirements for NSR2. Bradbeer explained that the LR NSR were first published in 2000, and evolved from previous work for MoD UK by Chalmers [53]; and originally known as Sea Systems Controllerate Publication 23 (SSCP23). In conducting the work, Bradbeer noted [Sect.3.2, p.59, The Process of Structural Synthesis] that the study did not attempt to draw any comparison between the structures designed to NSR2 and SSCP-23:

> *Since each set of standards uses different methods to estimate the loading while one structure may be heavier than the other, it may also be inherently stronger. Structure designed to the two different standards therefore cannot be considered equivalent.*

The idealised midship section that Bradbeer developed may owe its origins to the NATO Frigate NFR90, but it is also similar to the idealised midship section S-2864 (Fig. 4c), as used to analyse ultimate strength characteristics for a transverse-framed Frigate [54, 55]. This idealised section shape is geometrically similar to the example from Chalmers [53 - as shown in Fig. 6.3 (Idealised Section Shape) and Fig. 7.1 (Dimensions of Example Section)].

Bradbeer explored a number of different options for stiffener geometry for the corvette. Longitudinally stiffened options were examined with nominal longitudinal stiffener spacing of 600 mm and frame spacings of 1,000 mm and 1,500 mm. The 1,500 mm frame spacing option offered a structure requiring significantly less welding than the 1,000 mm option, for a minimal increase in structural weight. The work included modelling a transversely framed structure; and a hybrid system, where the main deck and keel were longitudinally stiffened while the side shell was transversely stiffened. Bradbeer decided to model the longitudinally stiffened structure with a 600 x 1,500 mm frame spacing, and transversely stiffened structure with 1,500 mm frame spacing, as these were considered the most likely practical choices. The chosen spacing addresses the OPC comments [38].

As a part of the research process, Bradbeer developed two corvette structural models: Model 1851 - for the longitudinally stiffened; Model 1852 - for the transversely stiffened:

> *The two corvette models constructed were assigned the numbers 1851 (for the longitudinally-stiffened) and 1852 (for the transversely-stiffened). In both models, all stiffeners were long-stalk tee bars of the standard proportions (previously defined). The same material assumptions were used as in the NFR-90 models: B-Quality crack arrest steel, assumed to behave similarly to mild steel at high strain rates.*

8.1 Indicative Midship Section Drawings - Vessels of Length Overall: 80 - 100 metres

Figures 4a to 4f show indicative midship sections. Figure 4d shows simplified geometry.

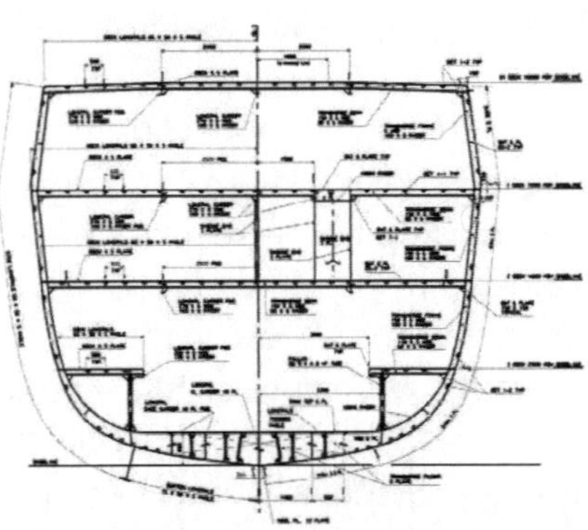

Fig. 4a: SEA 1427 OPC Midship Section [37]

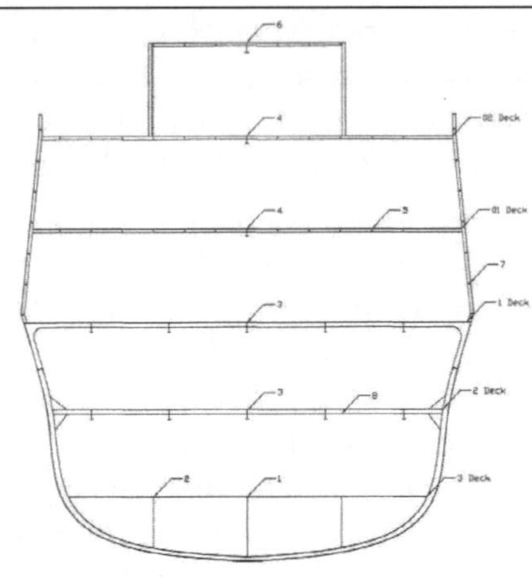

Fig. 4b: AMC - OPV Midship Section [46]

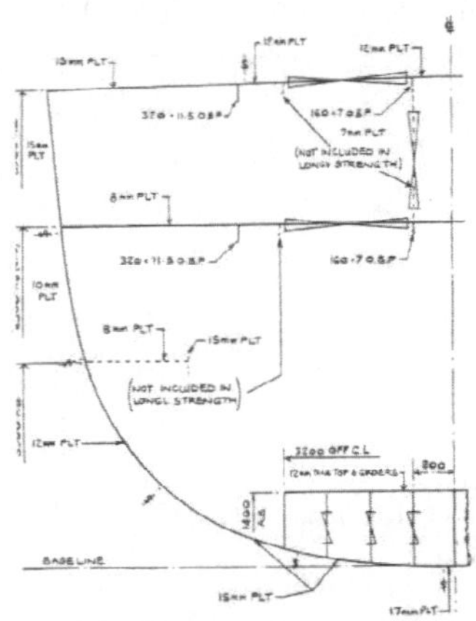

Fig. 4c: Trans. Frame Midship Sect. [54, 55]

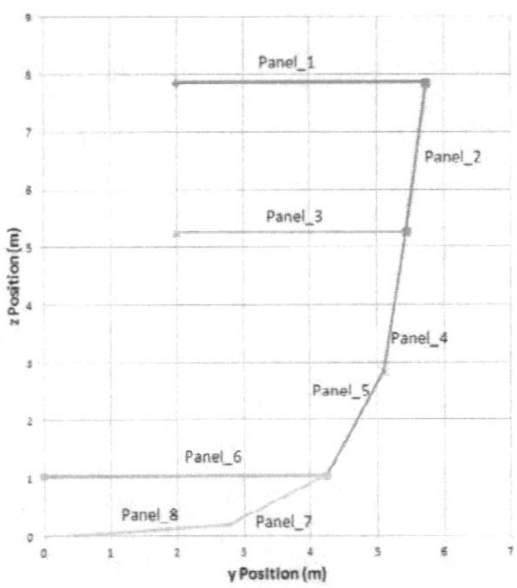

Fig. 4d: UCL Simplified Midship Section [48]

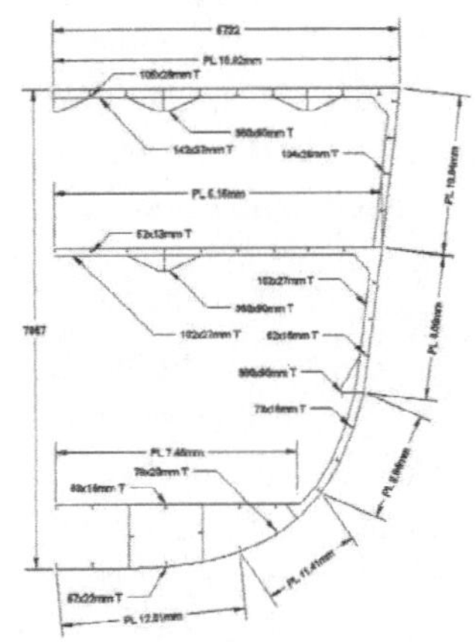

Fig.4e: Model 1851 Longitudinal Frame [48]

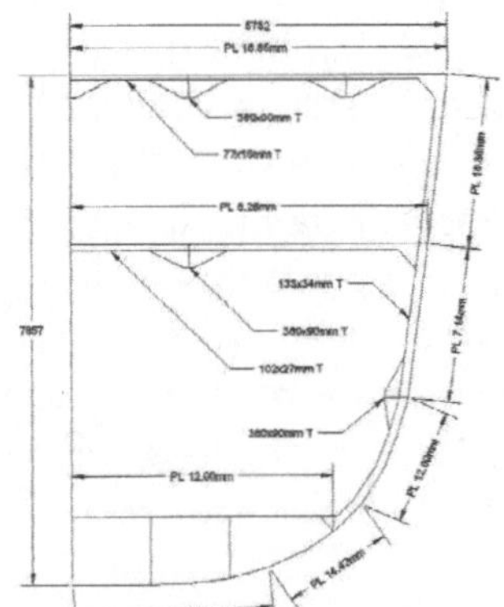

Fig. 4f: Model 1852 Transverse Frame [48]

8.2 Watertight Bulkhead Plating and Bulkhead Stiffening

The structural review of the SEA 1427 OPC included comments or concerns regarding the use of swedge or corrugated bulkheads, in terms of resistance to explosive blast loading. Bradbeer's design synthesis included the watertight bulkhead and stiffener arrangement, as shown in Figure 5a. This follows the style proposed by Chalmers [53, p. 196, Fig.9.1]. According to Bradbeer, [48, page 68, *'3.4 e Assessment of bulkheads'*]:

> *The bulkhead panel was subject to a smaller set of failure modes: Chapter 3 – Modelling Warship Structures. Minimum plate thickness was calculated in accordance with NSR2 Vol 1, Part 6, Chapter 2. Table 3.2.1 (Lloyds Register, 2008). Requirements for stiffener size in terms of required section modulus (Z), second moment of area (I) and web area (Aw) were calculated in accordance with NSR2 Vol 1, Part 6, Chapter 2, Section 2.8.2 (Lloyd's Register, 2008). While in practice it might be more realistic with regard to likely ship structural design practice to allow for lighter plating and stiffeners above the Passing Deck, the plating and stiffeners were kept constant across the whole bulkhead. (This greatly simplified the process of integrating the bulkheads into the surface model later, since the points where the bulkhead webs met their supporting stiffeners were coplanar in the XZ plane.) Unfortunately this precluded the inclusion of a thickened margin plate around the edge of the bulkhead. This is a feature which is included to improve strength and stability under explosive loading (Chalmers, 1993) and so is highly relevant to this research. If possible, a margin plate should be included in any follow-up studies concerned with structural failure mechanics. However, the omission was considered reasonable since the current research was primarily concerned with motions at relatively low stresses (below yield) rather than determining the mechanics of structural failure.*

From Chalmers [53; p.195, Bulkhead Plating; p.196, Fig. 9.1 Explosive Resistant Bulkhead]:

> *Plate thickness over the main part of the bulkhead should be determined to ensure adequate strength under normal or accidental loads as described earlier in this chapter. However, at the edges of the bulkhead where it meets the bottom side of the ship, the bulkhead plating should be thickened to give increased strength and stability under explosive loading. The depth of this thickened strip (measure normal to the hull plating) should be not less than 1.5 times the spacing of the longitudinals (excluding occasional deep members) at the point considered. A possible arrangement of strakes of plating is shown in Figure 9.1. The thickness of this strip should be not less than all of the following: a) 80% of the thickness of the adjacent shell plating. b). The thickness of the web of the nearest longitudinal hull stiffener; and c) 6.5 mm. The normal corrosion allowance should be added to the thickness obtained, as described in Chapter 5.*

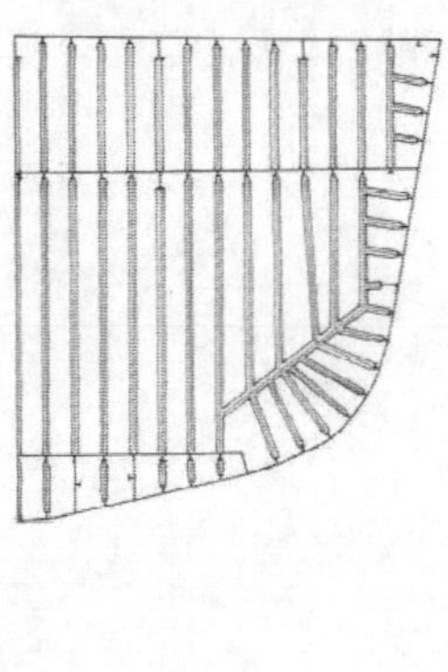

Fig.5a:Watertight Bulkhead [48]

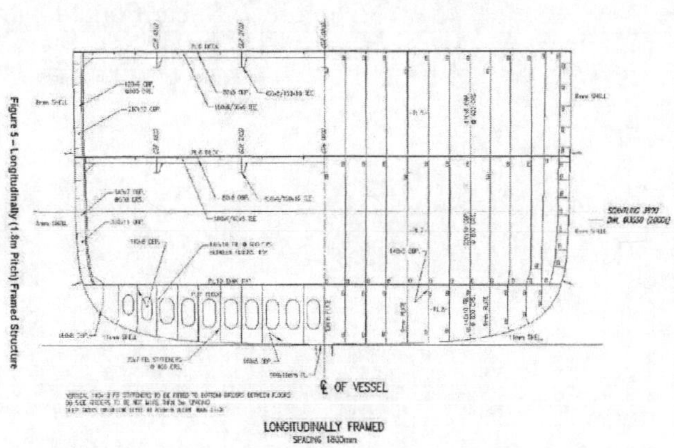

Fig. 5b: 80m Yacht – Midship Section & Bulkhead [50]

8.3. Data - Load Distribution, Shear Force, and Bending Moments, for 90 metre Corvette

Figs. 6, 7, 8, show Load Distribution, Shear Force, & Bending Moments, for 90m Corvette: (derived using Paramarine software by Nicholas Bradbeer University College London [48])

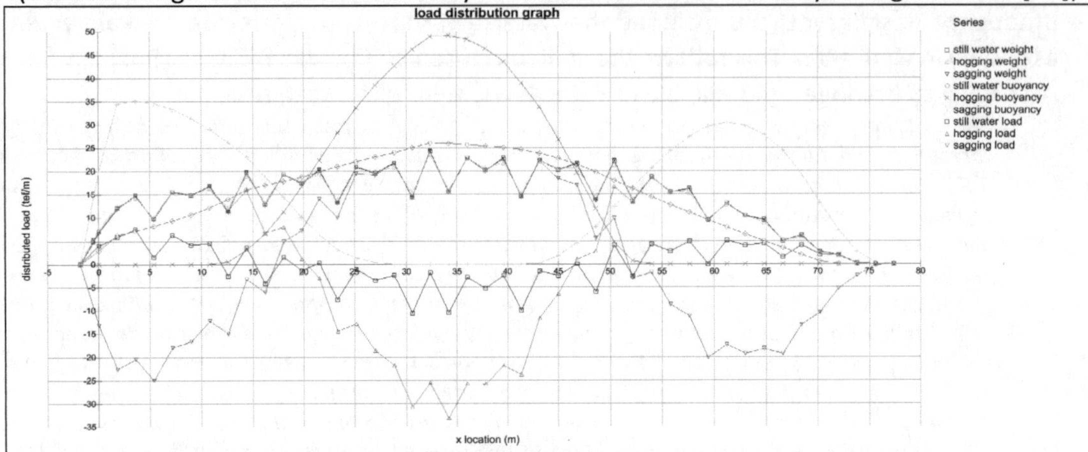

Figure 6: Load Distribution Graph

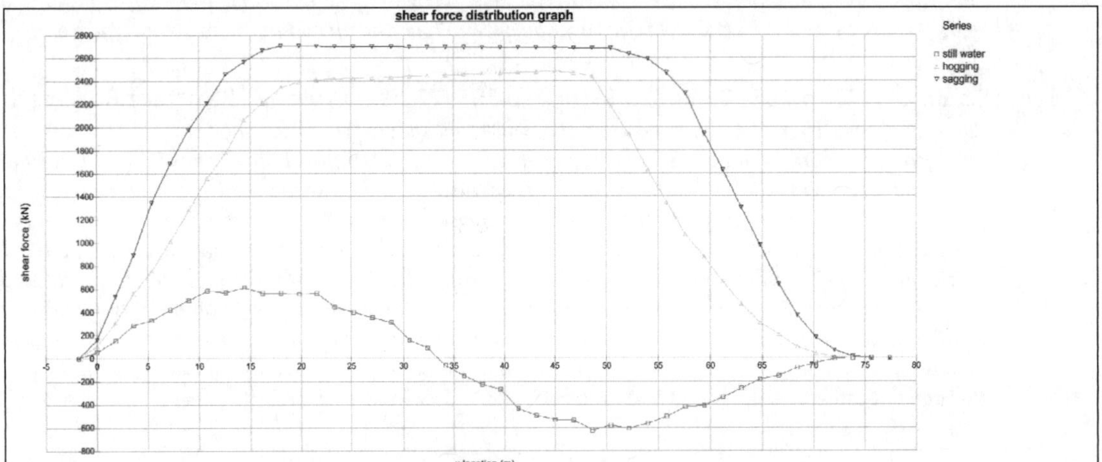

Figure 7: Shear Force Distribution Graph

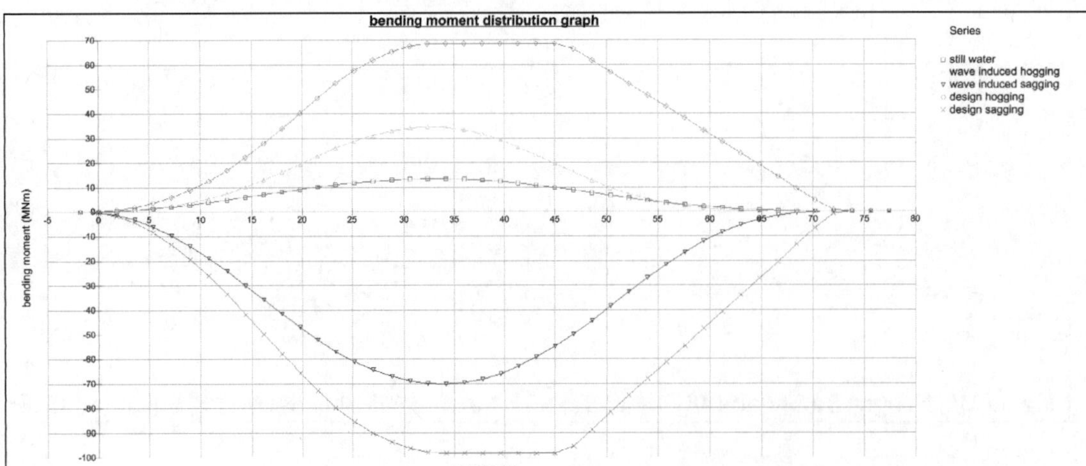

Figure 8: Bending Moment Distribution Graph

8.4. Scantlings - Comparative Shell Plating Thicknesses and Frame Scantlings

Table 3: Comparative Shell Plating Thicknesses and Frame Scantlings in Relation to Fig. 4d.

PANEL	SEA 1427 OPC	OTAGO OPV	AMC OPV	UCL Corvette	BMT M-Yacht
P1 – Plating	4.5	-	11	10.92 mm	6
P1 – Transverse	150x6 Web 80x6 Rider		120x12 Web 50x10 Flange	142x37mm Tee	180 x 6 / 60 x 6 Tee
P1 – L-Girder	270x6 Web 100x6 Rider	-	300x8 Web 100x8 Rider	360x90mm Tee	450x8 / 150x10 Tee
P1 – L-Stiffener	65x50x5 L	-		105x26mm Tee	80 x 5 OBP
P2 – Plating	5	-	7.5	10.84 mm	8
P2 – Transverse	150x6 Web 100x6 Rider	-	120x12 Web 50x10 Flange	142x37mm Tee	260 x 10 OBP @ 1,800 mm
P2 – L-Stiffener	65x50x5 L	-	-	104x26mm Tee	120 x 6 OBP @ 600mm centres
P3 – Plating	4.5	-	8	5.15 mm	6
P3 – Transverse	150x6 Web 80x6 Rider	-	100x8.5 FlatBar	102x27mm Tee	180 x 6 / 60 x 6 Tee
P3 – L-Girder	270x6 Web 100x6 Rider	-	300x6 Web 100x6 Rider	360x90mm Tee	450x8 / 150x10 Tee
P3 – L-Stiffener	65x50x5 L	-		52x13mm Tee	80 x 5 OBP
P4 – Plating	6	-	7.5	8.09 mm	8
P4 – Transverse	150x6 Web 100x6 Rider	-	120x12 Web 50x10 Flange	102x27mm Tee	300 x 11 OBP @ 1,800 mm
P4 – L-Stiffener	65x50x5 L	-	-	62x16mm Tee	140 x 7 OBP @ 500mm centres
P5 – Plating	6	-	9 (Ice Belt 13)	9.84 mm	8
P5 – Transverse	150x6 Web 100x6 Rider	-	120x12 Web 50x10 Flange	102x27mm Tee	300 x 11 OBP @ 1,800 mm
P5 – L-Stiffener	65x50x5 L	-	-	73x18mm Tee	140 x 7 OBP @ 500mm centres
P5 – L-Girder		-	-	360x90mm Tee	Nil
P6 – Plating	8	-	13 mm	7.45 mm	10
P6 – L-Stiffener		-		63x16mm Tee	140 x 8 OBP
P6 – CL-Girder	10	-	12 mm	-	10 mm Plate
P6 – L-Girder(s)	10	-	12 mm	-	9 mm Plate
P7 – Plating	7.5	-	10 mm	11.41 mm	11
P7 – Transverse		-	120x12 Web 50x10 Flange	102x27mm Tee	
P7 – L-Stiffener	75x50x5 L	-	-	78x20mm Tee	160 x 8 OBP
P8 – Plating	7.5	-	11.5 mm	12.01 mm	11 mm
P8 – T/Floors	8	-	12 mm	-	7 mm
P8 – L-Stiffener	75x50x5 L	-	-	87x22mm Tee	160 x 8 OBP
Trans-Bulkhead					
WTB – Plating	8, 6, 3 (swedge)	-	10 mm	10.7	8, 7, 5
WTB – Stiffener	(3mm) Swedge (6) 75x50x5 L (8) 100x75x5 L	-	L Section 100x10 Web 50x8 Flange	Tee Section 181 x 9 Web 90 x 18 Flange	(5) 120x8 OBP (7) 200x10 OBP (8) 140x10 FB

(Note: Above values of Plate Thickness do not include any allowance for corrosion [15]. LR's Naval Ship Rules use corrosion margins based on owner's or designer's requirements. If not specified, typical margins for NSR are: Keel Plate = 2mm; Tank structure = 1mm; Below a line 1 metre above design waterline = 0.5mm; Plate Rolling Tolerance = –0.3mm)

9. GENERAL CONSIDERATIONS FOR CERTIFICATION AND ACCEPTANCE OF STRUCTURES

The RAN's experience associated with the fire on-board Westralia, and the Rizzo Review, has led to the introduction of new processes and procedures for the management of ship Safety and Seaworthiness, including development of a project-related Safety Case, and a Seaworthiness Management Plan. It is not clear if the work relates to structural reliability.

During this task, the question arose: Does the RAN have any structural requirements? The answer is both 'Yes-and-No': 'Yes' in qualitative terms; but 'No' in quantitative terms.

Does the RAN have structural requirements? The answer is Yes! This is evident from the YouTube video [58] of the RNZN ship HMNZS OTAGO operating in the Southern Ocean, and taking green water over the bow in head seas estimated as a height of 20 metres (presumed to be an extreme wave height). The weapons loads must be considered also.

However, the current DEF(AUST)5000 for surface ship structure simply defers to 'Class', and does not specify quantitative performance measures for ship structural reliability; nor is there a method to assess and accept if the performance measures have been achieved.

A decision to progress beyond design, and commit taxpayers' funds to ship construction, implies prior 'design acceptance' by the Customer, Owner, and/or Technical Regulator. Even ordering structural materials, plates and sections, presumes they are fit-for-purpose. Ongoing contract management with a ship designer, builder, or prime contractor may include warranties on materials, workmanship, system- and ship-performance, and also penalties for latent defects and liquidated damages. However, once the ship construction task begins, there is generally no turning back. Various parties must take responsibility.

With regard to the assurance, certification, and acceptance of ship structural reliability, there are a number of other considerations, where opportunity for discussion is limited. Such a discussion may include consideration of the following additional requirements:

- Unrestricted world-wide operation - usually considers operation in ice conditions [59].
- Structural effectiveness and efficiency of superstructures, deckhouses, or helo-decks.
- Plate material properties: For example, the Air Warfare Destroyer project, where use of modern high strength steel grades was the subject of a technical investigation [60].
- Plate thickness: No half millimetres, minimum thickness = 5 mm; and for fairness and robustness a minimum shell thickness of 8 mm on hull sides, & 11 mm on hull bottom; to limit weld distortion a minimum practical plate thickness of 6 mm on decks [50].
- Margins for corrosion in determining plate thickness: Lloyd's Register Naval Ship Rules typically use corrosion margins based on owner's or designer's requirements. If not specified, typical margins for NSR: Keel Plate = 2mm; Tank structure = 1mm; Below a line 1 metre above design waterline = 0.5mm; Plate Rolling Tolerance = – 0.3mm [15].
- Section shapes: Admiralty Tee Section may be an optimal profile. Other profiles are OBP (Holland Profiles); or cut-down from other shapes, or fabricated? Rolled Sections may not be manufactured in Australia. Material grade must be certified by NATA Lab.
- Ballistic protection: Capacity of shell plates to resist attack by small calibre weapons.

10. ASSURANCE, VERIFICATION, CERTIFICATION, AND ACCEPTANCE, OF SURFACE SHIP STRUCTURAL RELIABILITY FOR HIGH SPEED NAVAL CRAFT

In the ten years that have elapsed since the publication of DEF(AUST)5000 Vol. 03, Pt. 01, 'Surface Ship Structure', the regulatory environment governing construction industries, and workplace health and safety, has changed. The National Construction Code, 2016, requires a structural certificate to be issued by a competent professional engineer, and presumably before commencement of construction, to certify 'acceptance' of the design.

The existing provisions of the DEF(AUST) regarding 'Deemed-to-Comply' beg the question: 'Deemed-by-Whom-to-Comply-with-What?', and do not satisfy the test for transparency. The 'Performance Solution' provisions are considered appropriate, and so there is a need to identify mandatory/desirable performance requirements for ship structural reliability. This is not something the principal Owner, Customer, and Technical Regulatory Authority should leave to others to decide. In the context of a major naval shipbuilding program, there is a strong case for clarity to be provided up-front, as to what the requirements are.

Under the circumstances, it is fortunate that the Naval Ship Code has been published [29], and it should be a relatively straight-forward task to ratify that document. There are consequences, because the Naval Ship Code builds on the concept of 'Naval Class', whereby naval surface ships are designed, constructed, and maintained in accordance with selected rules and regulations published by an accredited Classification Society.

Existing naval ship rules and regulations of any one of the major Classification Societies could potentially be utilised. If no decision can be reached quickly, then a fall-back option could be Navy-to-Navy support, and where the U.S. Navy is an obvious alliance-partner. The U.S. Navy would almost certainly direct any task to the American Bureau of Shipping, in which case, the source of the rules would be the Naval Vessel Rules, or, in the case of High Speed Craft (typically where $V_k/\sqrt{L_{WL}} > 1.3$), the ABS Rules for High Speed Naval Craft. Given a selected rule set, and defined structural geometry, a methodology exists that can perform a quantitative assessment of the loads and response, to optimize scantlings [32].

From an analysis of naval surface craft in the range from 80 to 100 metres length overall, a 'community need' has been identified for a national standard for high speed naval craft, to address the key performance requirements for naval surface ship structural reliability. A national standard already exists for commercial vessels [31], but that standard would need extensive revision, if it was to become applicable to high speed naval surface craft. Whilst a range of options should be considered, as a suggested interim 'fall-back option', Australia could adopt the American Bureau of Shipping Rules for High Speed Naval Craft.

The following fall-back option is suggested: 1) National Standard - utilise ABS HSNC Rules; 2) Simplified structural geometry; 3) Load assessment; 4) Initial scantling determination; 5) Scantling optimisation - utilise MAESTRO software methodology outlined by Stone [32].

To embark on a shipbuilding program without appropriate standards would be unwise. Prevarication will delay the identification of suitable performance and acceptance criteria, as needed to support the Government's investment in naval shipbuilding in Australia.

11. DISCUSSION

The following comments resulted from one discussion about the subject of this paper:

- Class Rules are not the measure of 'Seaworthiness'; rather, they are the means by which commercial vessels can achieve (or demonstrate) a level of seaworthiness, and get insurance, both for the vessels, and also including any onboard cargo.
- Commercial vessels have different operating requirements and/or demands of how the vessels are used, and the circumstances in which they operate.
- Taxpayers expect that navy ships will have a minimum level of 'Availability' to undertake the operations that they were required and acquired to perform.
- Possibility exists of a Gap between 'As Designed' versus 'Operational Expectation'; and that translates into 'Risk'.
- For Commercial Vessels, Insurance is predicated on meeting 'Class', and being maintained in 'Class'.
- For Naval Vessels, because the Commonwealth self-insures, there is no process, such as insurance, either initial issuance or renewals, to ensure that the design meets the operational requirements, or is maintained to meet the requirements.
- In the absence of having a DEF(AUST)5000 at the State-of-the-Art, and not ten years out of date, either a large technical development activity is needed urgently, or a fall-back option is to accept the rules or requirements of other navies.
- If you buy MOTS, reliance on parent navy (requirements or) rules is inappropriate, if you have a range of different parent navies: Therefore, there exists a need for a 'Benchmark' – as a set of requirements or rules.
- Benefit is that the capability exists to compare or assess different vessels on offer, qualify where there are differences, and know how to compare against risk profile.
- Having taken that approach going forward, maintain 'Deltas (+/-)'; this will reduce the requirement to establish and maintain a large technical engineering capability.

12. CONCLUSION AND PROPOSED APPROACH

The Navy DEF(AUST)5000 Vol. 03, Pt.01 'Surface Ship Structure' was published in 2007. An offer has been made to assist Navy Engineering to review and revise the publication. This activity is regarded as important and urgent. The proposed approach is as follows:

- Elevate DEF(AUST) publication to a National Australian (Benchmarking) Standard for High Speed Naval Craft, similar to the 'National Standard for Commercial Vessels' published by the Australian Maritime Safety Authority (AMSA).
- As an immediate policy fall-back option (subject to aspects of copyright), adopt the American Bureau of Shipping High Speed Naval Craft Rules, as an interim guideline.
- Establish an informal alliance between UNSW and University College London (UCL, also Qinetiq regarding Paramarine software), to facilitate access to ship design and structural details developed by UCL's Dr Nicholas Bradbeer for a 90 metre Corvette.
- Liaise with the Royal New Zealand Navy Chief Naval Engineer to characterise the extreme environmental parameters, based on available wind/wave statistical data, to define regional environmental parameters applicable to High Speed Naval Craft.
- Liaise with Mr Kevin Stone, DRS Technologies, MAESTRO-Marine, to scope the work needed to optimize ship hull structural scantlings for the example 90 m Corvette.
- Coordinate action with Defence Science & Technology Group and interested parties.

ACKNOWLEDGEMENTS

Thanks to family of William Rourke (RADM RAN 1928-2012) for permission to use the name: The 'ROURKE Class' Corvette. Thanks are also due to Trevor Ruting (RADM RAN Rtd.), David Campbell (Senior Associate Rtd. ACIL ALLEN Consulting), and Andrew Potts (Professor of Practice - Engineering, Monash University), for discussions and comments in preparing this paper, noting that the subject discussed in this paper represents solely the views of the author.

Thanks to Dr Nicholas Bradbeer (University College London) for relevant ship design data, American Bureau of Shipping, Lloyd's Register and DNV-GL (where extracts from relevant reports have been referenced), and Kevin Stone of DRS Technologies (MAESTRO-Marine). Thanks also to the Navy Sea Power Centre Australia for providing the privilege of access by independent researchers to historical information and records held within the Centre, and to Dr Warren Smith, University of NSW, Canberra at ADFA, for supporting this work.

REFERENCES

[1] Department of Defence, '2016 Defence White Paper', Commonwealth of Australia, (web link: http://www.defence.gov.au/whitepaper/docs/2016-defence-white-paper.pdf)

[2] Australian Government, Department of Defence, 'Naval Shipbuilding Plan', 2017, C'wealth of Australia, ISBN: 978-0-9946046-3-7 (Print), ISBN: 978-0-9946046-7-5 (Online).

[3] VADM Tim Barrett, 'The Navy and the Nation', 2017, Melbourne University Press.

[4] Beck, D.J., and Lord, J.M., 'Design and Production of ANZAC Frigates for RAN & RNZN: Progress Towards International Competitiveness', 1998, Journal of Ship Production, Society of Naval Architects and Marine Engineers, New York.

[5] Sielski, R.A., Nahshon, K., Salvino, L.W., Anderson, K., Dow, R., 'The ONR Ship Structural Reliability Program', Naval Engineers Journal, 2013, Vol. 125. No. 4, pp 61-84.

[6] DEF(AUST)5000, Volume 03, Part 01, 'Surface Ship Structure', Iss. No. 1, 04 Nov. 2007.

[7] Lord, J.M., 'Discussion Paper: An Interpretation of what constitutes an Existing Design for the New Surface Combatant', 30 April 1986 (Not published).

[8] MacKenzie, J.U.; 'Documentation of FFG Design', Task 6110-410, Contract N00024-74-C-5482, J.J. Henry Co., Hyattsville, Maryland, USA, File: 6208-15/JUM-2005, 30 May 1975.

[9] NAVSEA, 'Structural Design Manual for Naval Surface Ships', NAVSEA 0900-LP-097-4010, Naval Sea Systems Command (NAVSEA), 15 December 1976.

[10] Commonwealth of Australia and Crown of New Zealand, 'ANZAC Ship Project - Ship Performance Specification', Document No. A013129 (Issue 2), March 1988.

[11] Naval Ship Engineering Center, 'Ship Work Breakdown Structure', Washington D.C., Naval Sea Systems Command, Document No. NAVSEA 0900-LP-039-9010, 1977.

[12] Construction Specifications for Bundeswehr Vessels, BV 104-1, 'Structural Analysis of the Ships Hull (Strength Specification)', Bundesamt fur Wehrtechnik und Beschaffung, Koblenz, Translated from: July 1982 Edition.

[13] Construction Specifications for Bundeswehr Vessels, BV 105-1, 'Welded connections (welding specification)', Pt.1, Surface ships, Bundesamt fur Wehrtechnik und Beschaffung.

[14] Koster, D., Wiechers, H., Asmussen, I., 'Finite Element Strength Calculations for the MEKO 200 ANZAC Frigate', Report S-No. 1907/90/A, Germanischer Lloyd, January 1992.

[15] Bignold, D.J. (Author), White, N.J. (Reviewer), Cheng, Y.F. (Project Manager), 'Appraisal of the Australian ANZAC Class Frigate to the Lloyd's Register Provisional Naval Ship Rules', NSR Report No: NSR/14/R2/GEN, Lloyd's Register of Shipping, Dec. 1999.

[16] Rares, S., 'Striking the Modern Balance Between Freedom of Contract and Consumer Rights', 14th International Association of Consumer Law Conference, Sydney, 2 July 2013.

[17] Miller, M.A., 'Liability of Classification Societies from the perspective of the United States Law', O'Neil, Eichin, Miller, Saporito & Harris, New Orleans, LA 70113, USA.

[18] Bone, C.E., SSC-446, 'Comparative Study of Ship Structure Design Standards', 2007, Ship Structure Committee, Committee chaired by representative of the U.S. Coast Guard.

[19] SSC-439, 'Comparative Structural Requirements for High Speed Crafts', 2005, Ship Structure Committee.

[20] Stone, K., and Novak, D.S., 'Comparative Structural Requirements for High Speed Craft'; sname.org.

[21] Fabling, CAPT, 'Offshore Patrol Cutter Project CG-9322', OPC Industry Day – 04 Nov. 2010, OPCIndustryDay.pdf; http://www.dtic.mil/ndia/2010/USCG/OPCIndustryDay.pdf

[22] 'ABS Guide For Building and Classing Naval Vessels (Naval Vessel Rules)', Jan. 2007.

[23] 'ABS Guide for Building and Classing High Speed Naval Craft', 2017.

[24] Kramer, R.H., Haugan, G. and Fredriksen, A., 'US navy high speed craft - Comparison of ABS and DNV structural requirements', Transactions - Society of Naval Architects and Marine Engineers, Vol 113, pp. 340-366, 2006.

[25] Det Norske Veritas, 'Rules for High Speed Light Craft and Naval Surface Craft', 1993.

[26] 'National Construction Code', 2016, Building Code of Australia - Volume One, The Australian Building Codes Board, Feb. 2016, Print version: 1.0, Release date: May 2016.

[27] 'Structural Reliability Handbook', 2015, Australian Building Codes Board (ABCB), GPO Box 9839, Canberra, ACT, 2601.

[28] National Construction Code, www.abcb.gov.au/NCC/How-it-works

[29] NATO Standardization Agency, NATO Group MCG-6, 'Naval Ship Code', Allied Naval Engineering Publication ANEP-77, Edition E, Version 1, 2014.

[30] CN Signal VZCZCZZZ001, 270050Z September 2011, from CN AUSTRALIA, SUBJ: Establishment of Naval Flag Administrator and Adoption of Naval Ship Code, January 2014, North Atlantic Treaty Organization, NATO Standardization Agency (NSA).

[31] Australian Transport Council, 'National Standard for Commercial Vessels', PART C, DESIGN AND CONSTRUCTION, SECTION 3, CONSTRUCTION, Endorsed by ATC Feb. 2008, Amendment No.1, was approved by the National Marine Safety Regulator 7 June 2016.

[32] Stone, K.; McNatt, T., 'Ship Hull Structural Scantling Optimization', MARSTRUCT 2017, 'Progress in the Analysis and Design of Marine Structures': Proceedings of the 6th International Conference on Marine Structures (MARSTRUCT 2017), May 8-10, 2017, Lisbon, Portugal, Editors Guedes-Soares, Carlos; and Garbatov, Y., CRC Press/Balkema, ISBN 9781138069077, Taylor and Francis Group, London, UK, 2017.

[33] Lord, J.M., Smith, W.F., 'Offshore combatant Vessel: Concepts That Did Not Float', Pacific 2015 International Maritime Conference, Sydney, Australia.

[34] Stewart, C., 'Patrol Boats Crack Up – Asylum Demands Breaking Navy Fleet', page 1, The Australian (Newspaper), Friday 10 August, 2012.

[35] Stewart, C., 'Time's up for navy asylum fleet', The Australian, March 24, 2014.

[36] Letter from J. Lord to DGENG, attaching 'Change Proposal for DEF(AUST)5000, Vol.03, Pt.01, Surface Ship Structure', submitted 24 July 2016, received/actioned 26 August 2016.

[37] Commonwealth of Australia, 'Offshore Patrol Combatant Project, Project Design Phase, Contract C218510', 1995-1997, Canberra, ACT, Australia.

[38] OPC PD Letter OPC T 58/96, 30 August 1996, OPC Project – PDP – Contract C218510 – CDRL 095 Structural Drawings; CDRL 042 Ship Structure Report; Folio 14 (N95/14559 Pt.3)

[39] TDS Letter PV021/97, 01 August 1997, OPC Project PDP – Contract C218510 – OPC Requirements for Macquarie / Heard Island EEZ.

[40] Det Norske Veritas AS, Part 3, Chapter 2, 'Newbuildings Hull and Equipment – Main Class, Hull structural design - Ships with length less than 100 metres', January 2015.

[41] Kapsenberg, G.K., Aalbers, A.B., Koops, A., Blok, J.J., 'Fast Displacement Ships – The MARIN Systematic Series', MARIN (Maritime Research Institute Netherlands), Wageningen, the Netherlands, 2015 (Ch.11 – Case Studies, Fig. 190, 100m Patrol Vessel).

[42] Collier, G., 'Royal New Zealand Navy Project Protector', Brief of 9 November 2009, New Zealand Ministry of Defence.

[43] Lloyd's Register, 'Rules for Special Service Craft' (SSC), editions from 1999 onwards.

[44] Ochtman-Corfe, R.; Project Sponsor, 'Royal New Zealand Navy – Next Generation Offshore Patrol Vessel Operational Concept Document & High Level User Requirements', For Australian Maritime College, 2014, Marked: 'Academic Use Only'.
[Disclaimer: This Operational Concept Document (OCD) is intended for Australian Maritime College (AMC) ocean vessel design students ONLY. It does not represent an actual set of Royal New Zealand Navy (RNZN) strategic intentions, vessel requirements and/or operational concepts.]

[45] Lloyd's Register, 'Rules and Regulations for the Classification of Naval Ships', January 2015, London, UK.

[46] Bhosale, Hubbert, Marino, Jainfeng, 'Ocean Vessel Design Project Final Report', Uni. of Tasmania, Australian Maritime College, 2014, Supervisor: David Harte, UTas Library.

[47] Marino-Gomez, M., 'Developing a structural weight estimation tool for typical frigate hull shape based on Lloyd's Rules and Regulations for Naval Ships'; also including Excel Spreadsheet: Final LR Tool; University of Tasmania, Australian Maritime College, 2014.

[48] Bradbeer, N., 'Implications for Underwater Shock Response of Adopting Simplified Structural Styles in Warships'; PhD Thesis, Dept. of Mech. Eng., UCL, 2013.

[49] Schaffer, R.L. and Kloehn, H.G., 'Design of the NFR-90'. ASNE Naval Engineers Journal, 103(2), pp.29–49. March 1991, NEJ 'www.ingentaconnect.com'.

[50] Roy, J., Munro, B., Walley, S., and Meredith-Hardy, A., 'Longitudinal Vs Transversely Framed Structures For Large Displacement Motor Yachts', BMT Nigel Gee Ltd., HISWA Symposium, 2008.

[51] Cho-Chung Liang; Yuh-Shiou Tai; 'Shock Responses of a Surface Ship Subjected to Non-Contact Underwater Explosion'; Ocean Engineering 33(5-6): 748-772, April 2006.

[52] Xinbei CG127, '2000-ton Class Patrol Vessel'; Refer web link: https://en.wikipedia.org/wiki/Coast_Guard_Administration_(Taiwan)

[53] Chalmers, D.W., 'Design of Ship's Structures', 1993, London, HMSO ISBN: 0117727172.

[54] Report: YM5715/01/A, 'Ultimate Strength of Proposed Midship Section S2864', YARD LTD Consulting Engineers, Glasgow, 18 December 1987.

[55] Smith, M.J., 'Ultimate strength assessment of naval and commercial ships', Defence Research and Development Canada – Atlantic, DRDC - Atlantic TR 2008-059, August 2008.

[56] Reid W. D., 'The Response of Surface Ships to Underwater Explosions', Defence Science and Technology Organisation, Melbourne 1996.

[57] Szturomski, Bogdan, 'The Effect of an Underwater Explosion on a Ship', 015 (LVI) 2 (201) Scientific Journal of Polish Naval Academy, 2015, DOI: 10.5604/0860889X.1172074.

[58] Carney, J., https://www.youtube.com/watch?v=WQUXbkAdZhg New Zealand Navy, 26 May 2016, Video shows a New Zealand Navy ship, HMNZS OTAGO, taking green water over bow in waves more than 20 metres high in the Southern Ocean, Daily Mail Australia, PUBLISHED: 12:19 +10:00, 27 May 2016, UPDATED: 16:59 +10:00, 27 May 2016. http://www.dailymail.co.uk/news/article-3611908/Incredible-video-shows-New-Zealand-navy-ship-smashing-huge-waves-Southern-Ocean.html#ixzz4g6Up8rZl

[59] Braithwaite, R.C., Khan, D., 'Implications of ice class for an offshore patrol vessel', 2014, Journal of Marine Engineering & Technology, 13:3, 17-27, DOI:10.1080/20464177.2014.11658118,

[60] Verdaasdonk, J.P.W., Spyker, R., Sterjovski, Z., 'The use of modern high strength steel grades in the Australian Air Warfare Destroyer (AWD)', Pacific IMC-2012, January 2012.

REFERENCES – ANNEX A

[A1] U.S. Department of Defense, 'Military Standard – Interface Standard for Shipboard Equipment – Section 302A - Weather Environment', dated 29 February 1988.

[A2] Durrant, T., and Greenslade, D., 'Evaluation and implementation of AUSWAVE', CAWCR Technical Report No. 041, September 2011.

[A3] Tolman, H. 2014, 'User manual and system documentation of WAVEWATCH III R°', version 4.18. NOAA / NWS / NCEP / MMAB Technical Note: 316, March 2014.

[A4] Joubert, P.N., 'Some Remarks on the 1998 Sydney-Hobart Race', Proceedings (Transactions) of the Royal Society of Victoria, Volume 117, No. 2, 2005. (copy of article available electronically via the digitised collection of the State Library of Victoria).

[A5] Burston, J., Taylor, D., and Garber, S., 'Contextualising the Return Period of the June 2016 East Coast Low: Waves, Water Levels and Erosion', Baird Australia, Level 14, 309 Kent St., Sydney, NSW, 2000.

[A6] Brown, B.; http://www.abc.net.au/news/2016-06-14/new-map-shows-journey-of-weather-buoy-after-east-coast-low/7508992 New map reveals weather buoy journey after recording monster wave, ABC South East NSW, Updated 14 Jun 2016, 6:17pm.

[A7] Kulmar, M., Manly Hydraulics Laboratory; Email, Date Generated: 08-Mar-2017.

[A8] Lloyd's Register, 'Rules and Regulations for the Classification of Naval Ships', Vol. 1, Part 5, Environmental Loads, January 2015, London, UK.

BIBLIOGRAPHY

[B1] Braidwood, I., Lofthouse, G., 'Development in Bulb Flats and Other Hot Rolled Special Steel Profiles for Advancing Commercial and Naval Shipbuilding in the U.S.', Journal of Ship Production, Vol. 16, No. 4, November 2000, pp. 193-206 (Corus, Skinningrove, U.K.).

[B2] Dey, S., Borvikm. T., Teng, X., Wierzbicki, T., Hopperstad, O.S., 'On the ballistic resistance of double-layered steel plates: An experimental and numerical investigation', International Journal of Solids and Structures, Vol. 44, Issue 20, Oct. 2007, pp. 6701-6723.

[B3] Gehling, R., 'High Speed Craft Codes – Considerations for Naval Operations', 2011, unpublished paper for Head Maritime Systems, Defence Materiel Organisation, Canberra.

[B4] Genco, Giuseppe, 'Lacrosse Building Fire, 673 La Trobe Street, Docklands on 25 November 2014', Municipal Building Surveyor, City of Melbourne, April 2015.

[B5] Hughes, Owen, F., and Paik, Jeom-Kee. 'Ship Structural Analysis and Design', The Society of Naval Architects and Marine Engineers, Jersey City, New Jersey, USA, 2010, ISBN No. 978-0-939773-78-3.

[B6] Kapsenberg, G.K., 'Slamming of Ships: Where are we Now'. Phil. Trans. R. Soc. A (2011) 369, 2892–2919 doi:10.1098/rsta.2011.0118, Ship Hydromechanics and Structures Department, Maritime Research Institute Netherlands, Netherlands Delft University of Technology, Wageningen, The Netherlands.

[B7] Kavanagh, D., 'The Right Stuff: A paper on what it takes to be a Lifecycle Engineer, What they do, Why we need them and How to become one', 2015, The Naval Engineering and Maintenance Bulletin, Issue 1 – Sep. 2015.

[B8] Keane R.G., McNatt, T., Beach, J.E., 'Reducing Total Ownership Cost: Designing Robust Ship Structures', 2017, Naval Engineers Journal, 2017.

[B9] Lloyd's Register Report NSR 13/R1/Gen, 'The Appraisal of the Australian ANZAC Class Frigate to the Rules and Regulations for Special Service Craft', July 1999.

[B10] Lloyd's Register Report NSR/16/R1/Gen, 'Finite Element Analysis of the Australian Anzac Class Frigate Superstructure', Lloyd's Register of Shipping, October, 1999.

[B11] Lloyd's Register Report NSR/19/R1/Gen, 'Anzac Class Review of Structural Details', December 1999.

[B12] Magoga, T., Flockhart, C.; 'Ultimate Strength Calculations of an Aluminium Patrol Boat', January 2012.

[B13] Magoga, T., Aksu, S., Cannon, S., Thomas, G., 'The Need for Fatigue Life Prediction Methods Tailored to High-Speed Craft: A Technical Review', Pacific 2015 International Maritime Conference, Sydney, October 2015.

[B14] Magoga, T., Aksu, S., Cannon, S., Thomas, G., 'Comparison between Fatigue Life values calculated using Standardised and Measured Stress Spectra of a Naval High Speed Craft', Conference paper, September 2016.

[B15] Magoga, T., Aksu, S., Cannon, S., Thomas, G. 'Identification of Slam Events Experienced by a High-Speed Craft', Ocean Engineering, August 2017.

[B16] Michaelson, Robert, W., 'Technical Report - User's Guide for SPECTRA: Version 8.3', Naval Surface Warfare Center Carderock Division, West Bethesda, MD 20817-5700, NSWCCD-65-TR-2000/07 March 2000, Survivability, Structures, and Materials Directorate.

[B17] Morris, Madeleine, 'This is the hidden potential killer in Australian buildings', ABC 7.30 Report, 27 May 2015, Transcript, abc.net.au

[B18] Royal Australian Navy, 'Seaworthiness Management System', March 2017.

[B19] Savitsky, Daniel, 'On the Subject of High Speed Monohulls', Paper presented to the Greek Section of the Society of Naval Architects and Marine Engineers, Athens, Greece, 02 October, 2003.

[B20] Sieve, Michael, W., Kihl, P., Ayyub, Bilal, M., 'Technical Report - Fatigue Design Guidance for Surface Ships', Naval Surface Warfare Center Carderock Division, West Bethesda, MD 20817-5700, NSWCCD-65-TR-2000/25 November 2000, Survivability, Structures, and Materials Directorate.

[B21] Tuitman, J.T., Hoogendoorn, D., 'Fatigue in High-Speed Aluminium Craft: A Design methodology for Predicting the Fatigue Life', Conference paper, May 2014, SSC Symposium - Vessel Safety and Longevity through Ship Structure Research, Baltimore.

[B22] Underwood, James, M., 'Strength Assessment of Damaged Steel Ship Structures', A Thesis Submitted for the Degree of Engineering Doctorate, Fluid Structure Interactions Research Group, Faculty of Engineering and the Environment, University of Southampton, April 2013.

[B23] Zhakarov, Vladimir, E., Badulin, Sergei, I., Hwang, Paul, and Caulliez, Guillemette, 'Universality of Sea Wave growth and its Physical Roots', Journal of Fluid Mechanics, Vol. 780, pp. 503-535, Cambridge University Press, 2015.

[B24] Zhakarov, Valdimir, E., Badulin, S.I., Geogjaev, V.V., Pushkarev, A.N., 'Analytical Theory of Wind-Driven Seas', Lorenz Lecture, AGU Fall Meeting, 14 December 2016, University of Arizona, Tucson, USA.

ANNEX A – LIMITS ON OPERATIONAL ENVELOPE DUE TO ENVIRONMENTAL FACTORS

Issues relating to 'Environmental Factors' are now more prominent. 'Climate Change' is a point of discussion amongst government staff, scientists, engineers, and planners involved in infrastructure development projects. The effects of climate change include extremes of weather (E.g. high winds, sea states) becoming more likely. Consequently, the prediction of extreme weather events is a matter of importance, and it is necessary to identify a source of comprehensive maritime environmental data, to effectively replace MIL-STD-1399 (Navy), Section 302A, Weather Environment, dated 29 February 1988 [A1].

The Australian Bureau of Meteorology has made progress in numerical wave modelling, and the provision of marine wave forecasts. In 2010, the National Meteorological and Oceanographic Centre (NMOC) replaced its Numerical Weather Prediction (NWP) systems with the new Australian Community Climate Earth System Simulator (ACCESS) system:

> *The changes resulted in improved prediction of surface winds, and the wave model needed 're-tuning' to suit the new characteristics of the wind forcing. This led to the introduction of the more widely-used and computationally-efficient WAVEWATCH III® (WW3) model. The implementation of WW3 within the Bureau's forecasting environment is known as AUSWAVE* [A2, A3].

The paper presented at IMC-2015 referred to the 1998 Sydney-to-Hobart yacht race and, subsequent to the event, a paper published by one of the participants, Peter Joubert [A4], to substantiate the claim of an extreme wave of 42.7 metres that occurred following the capsize of the yacht Kingurra, and during the recovery operation of a man-overboard. Discussion at IMC-2015, including with Dr Frans Van Walree from the MARIN Institute, moderated claims of extreme waves to around 20 metres, as representing the maximum that had been simulated by wave tank experiments at the MARIN Institute.

Since 2015, two severe weather events along the NSW east coast have been noteworthy. An event occurred in early June 2016, due to the prevailing wind, a 'Black Nor'-Easter', that caused extensive coastal erosion and damage to beaches and property exposed to the North-East [A5]. Manly Hydraulics Laboratory maintains a series of wave-rider buoys that provides records of data.

Whilst the data is too extensive to include, one aspect of interest was that the conditions were sufficiently severe, such that the wave-rider buoy at Eden broke its mooring, and drifted in the Tasman Sea, and still recording wave data. The journey of the buoy is shown in Figure A1.

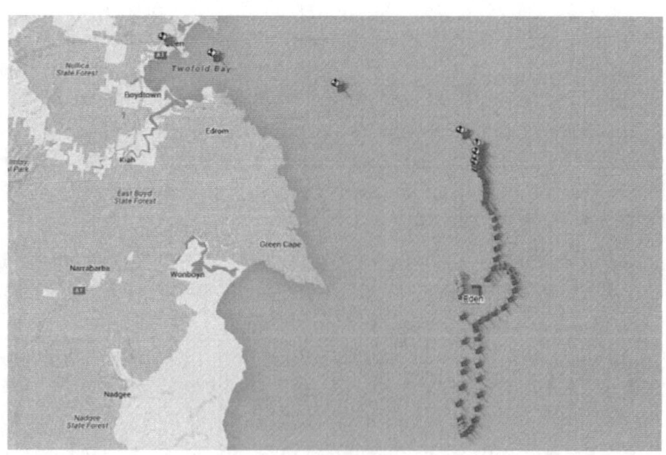

Figure A1: Map reveals weather buoy journey after recording monster wave [A6]

According to the ABC [A6], the above map released by the Manly Hydraulics Laboratory, a business unit of NSW Public Works, shows the path of the buoy after it had broken away from its mooring around midnight on Saturday 4 June 2016. The buoy continued to transmit data, that included the largest-ever wave recorded off NSW, in the early hours of Monday morning. A 17.7-metre wave was recorded in 100-metre-deep water and would have been even higher when it reached the shore. At 17.7 metres, (wave period of 14.1 seconds), the wave exceeded the previous record of 14.9 metres off Sydney in April 2015.

As well as the Eden location, Manly Hydraulics Laboratory has moored buoys off the coast at Byron Bay, Coffs Harbour, Crowdy Head, Sydney, Port Kembla, and Batemans Bay. [A7]

A second severe weather event, also due to an East Coast Low, occurred over the four-day period between Sunday 05 to Wednesday 08 March 2017. Point Perpendicular NSW recorded winds averaging around 20 Knots, with gusts to 30 Knots. [Note: Actual data available from BOM archives: www.bom.gov.au/products/IDN60701/IDN60701.95940.shtm]. This weather event produced the wave data for Crowdy Head NSW shown in Table A1: Hsig of 4.97 metres, mid-range Sea State 6; and a maximum wave height of 9.79 metres. Other details relating to this weather event are included in the following Figures A2 to A4. Figure A4 shows Navy's Sail Training Ship, Young Endeavour, seeking shelter in Jervis Bay.

Fig. A2 – Photo March 2017

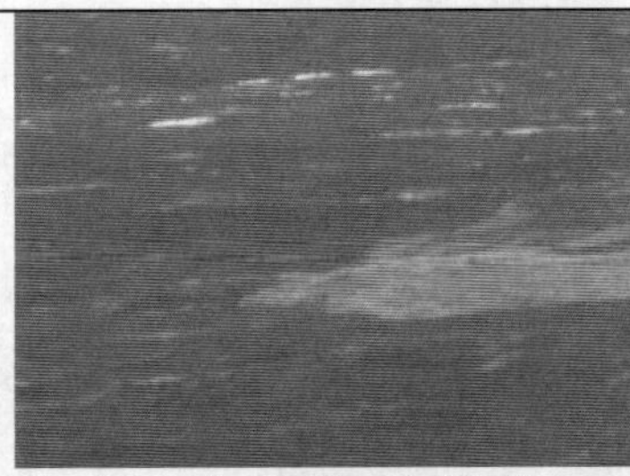

Fig. A3 – Photo March 2017

Fig. A4 HMAS Young Endeavour

Table A1 shows a record of wave data for Crowdy Head from Manly Hydraulics Lab [A7].

Table A1. CROWDY HEAD: Significant and Maximum Wave Heights; and Wave Period

East Coast Low: Sunday 05 to Wednesday 08 March 2017 [A8].

Date/Time	Hsig (m)	Hmax (m)	Tz (s)	Tsig (m)	T P1 (s)	WDIR (deg TN)
7/3/2017 11:00	4.810	7.66	8.40	11.31	12.93	154
7/3/2017 12:00	4.619	7.39	8.78	11.43	12.14	155
7/3/2017 13:00	4.624	9.10	8.46	11.06	12.14	154
7/3/2017 15:00	4.948	8.49	8.17	11.24	12.93	151
7/3/2017 16:00	4.883	9.29	8.86	11.67	13.82	141
7/3/2017 17:00	4.718	9.67	8.74	11.57	13.82	155
7/3/2017 18:00	4.956	7.82	8.75	11.92	12.93	148
7/3/2017 19:00	4.964	7.80	8.99	11.85	12.93	146
7/3/2017 20:00	4.973	8.76	8.86	11.91	12.93	150
7/3/2017 21:00	4.742	9.79	8.53	11.33	12.93	146
7/3/2017 22:00	4.517	8.61	8.04	11.10	12.93	143
7/3/2017 23:00	4.423	8.23	8.50	11.25	13.82	148
8/3/2017 00:00	4.259	9.06	8.44	11.34	13.82	157

The Bureau of Meteorology offers an extensive suite of environmental forecasting data, and extracts for the period 05 to 08 March 2017 are included in Figures A5, A6, and A7. Data includes the wind forecast, significant wave height & direction, and satellite image. The East Coast Low remained relatively stationary for several days, and produced the maximum wave height H_{max}=9.79 metres recorded at Crowdy Head, as shown in Table A1.

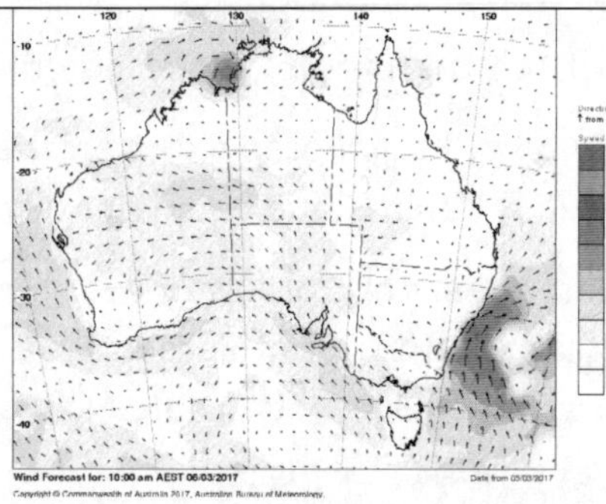

Fig. A5: Wind Forecast: 10:00 on 06/03/2017

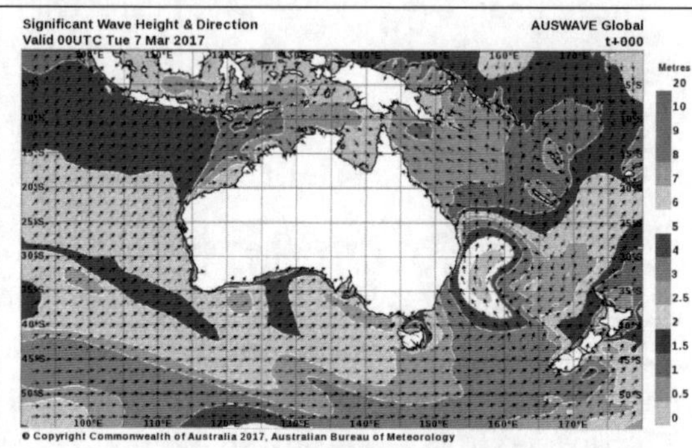

Fig. A6: Significant Wave Height & Direction

Fig. A7: Satellite Image 07/03/17 08:30 UTC

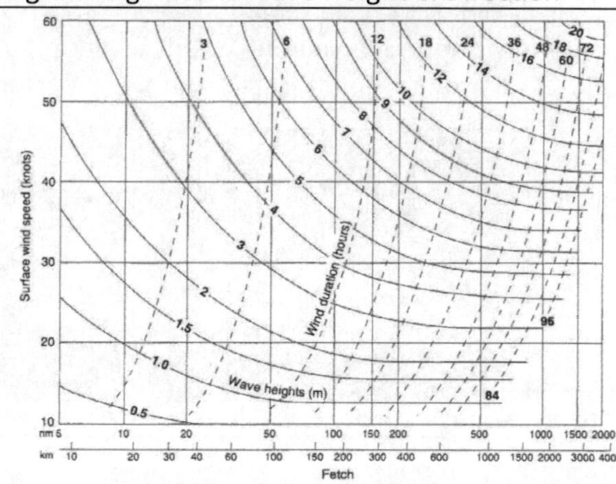

Fig. A8: Wave height as a function of wind speed

From Figure A8, for wave height as a function of wind speed, and unlimited fetch, a wind duration greater than 24 hours, and a wind speed of less than 30 knots, will produce a wave height: 4 to 5.5 metres (Beaufort 7, High wind, moderate gale, near gale, 28-33 Kts).

The above data, on its own, offers little insight into maximum environmental conditions. However, the above data fits reasonably well with Lloyds Register Rules and Regulations for the Classification of Naval Ships, Volume 1 Part 5, Environmental Loads, 2015 [A8]. For sea areas 91, 92, 93; and 99, 100, 101; maximum values are from 17.7 to 20.1 metres, and can be compared with Significant Wave Height Hs from 5.4 to 6.1 metres, and higher.

A question arising from the above review is whether it is more appropriate, statistically, to work with values of Significant Wave Height or Extreme Values, for naval ship design. Depending on the reference, Extreme Values for Wave Height can be 2.1 to 2.35 times greater than the values of Significant Wave Height. A Significant Wave Height of 8 metres could produce an Extreme Wave Height of between 16.8 to 18.8 metres. There is also the question of 'Wave Steepness' and reference is made to work by Zhakarov et al [B23, B24].

Application of Structural Integrity Management Principles Demonstrated in Other Domains to Improve Seaworthiness and Cost of Ownership Outcomes

Michael Houston[1] and Stuart Cannon[2]
QinetiQ Australia, Defence Science and Technology Group
Michael.houston@qinetiq.com.au

INTRODUCTION

The release of the 2016 Defence White Paper provided Australian industry with many challenges associated with the delivery and sustainment of significant Defence capabilities. A requirement is that the whole of life costs need to be stated up front to ensure that each new capability is fully costed. This requirement built on some of the observations made in the 2011 Rizzo review where the primary focus was on acquisition costs rather than sustainment costs. An essential component of such a system is to establish a whole of life approach to managing the structural integrity of our surface ships. This approach entails understanding the basis for the original design and its structural certification, understanding the operational environment with respect to the design assumptions, the use of structural monitoring data and integrating it with structural condition assessments. The outcome of the approach would enable the capability owners to anticipate and mitigate risk to seaworthiness, cost of ownership and capability. Such a closed-loop program, termed a Structural Integrity Program, is applied in other domains and has yielded significant benefits through more informed asset decision making and planning. The ability to recognise a change in operational environment and assess the consequences for structural fatigue, and take action to mitigate risk due to structural degradation is just one example of a structural integrity activity that will deliver great value to operators. This paper describes a Structural Integrity Program in a maritime context and demonstrates the long term benefits to Defence by using relevant case studies.

WHOLE OF LIFE STRUCTURAL INTEGRITY MANAGEMENT

Structural integrity is the capacity of a structure to withstand normal operating loading loads within operational limitations without collapse or unacceptable deformation. It is the fundamental functional requirement of the structural subsystem of any platform, land, sea or air. The regulatory framework within the new Defence Seaworthiness Management System recognizes structural integrity as part of achieving the Exist and Endure goal, one of three goals that make up the Activity and Condition-based Compliance Obligations (ACCO). Structural integrity is ACCO 1.1.

The requirement to demonstrate the structural integrity of a new design during acquisition is captured in the applicable design standards and rules. The rules are involved, reflecting the complex and diverse loads and conditions that a structure must support and the numerous potential modes of failure. Two important, though obvious points are worth acknowledging here as they will have bearing on the subject of this paper:

- The achievement of the structural integrity requirement is intricately intertwined with other platform characteristics and performance requirements such as weight, speed profile, manoeuvrability and payload capacity.
- Changing the structure of a platform affects the loading environment itself, not just the capacity to withstand load.

Done well, design compliance with the rules and demonstration of structural integrity is found during type certification or classification[1]. The platform then enters service with operating, maintenance and logistics instructions and limits that are consistent with a well-established basis for structural integrity and which are required to keep the platform in class.

One aim of the subsequent sustainment program is to maintain structural integrity, and in so doing maintain the certification of the platform, for the entirety of the in-service life. A wide variety of views are held regarding the complexity and magnitude of the effort required to achieve this aim. In the authors' view, this diversity arises due to a similarly varied understanding of the certification basis for structural integrity and of the requirement to keep in class, of the factors that affect structural integrity throughout platform service life, of the way in which degradation or loss of structural integrity develops and of the consequences. These parts of the structural integrity problem are addressed in the following paragraphs.

The structural integrity of a platform is sometimes taken for granted as enduring once established by design, manufacture and certification. Correspondingly, it is assumed that the operating and maintenance instructions and limits, including life of type, will remain valid over the entire service life. However, actions taken over the service life of a platform can impact structural integrity. Such actions include:

- changing the role or role mix of a platform as established in the Operational and Support Intent (OSI),
- changing the usage or usage environment of a platform, and
- modifying the platform.

These actions are rarely taken in isolation. They are typically a product of operational circumstances, part of a package of capability changes or a result of strategic or doctrinal decisions. With reference to a point made earlier in this section, structural integrity is affected because *achievement of the structural integrity requirement is intricately intertwined with other platform characteristics and performance requirements*. However, the relationship between cause and effect with respect to structural integrity is often not straightforward and readily lost.

[1] The terms certification and classification are used interchangeably in this paper as meaning the assurance by various means that the platform design and construction meet a set of predefined, safety related rules and standards. It is acknowledged that aircraft type certification and vessel classification are not strictly equivalent; however, they are analogous and serve the same outcome. The term certification basis refers to the rules and standards specified for classification purposes. Similarly, the phrases maintain certification and keep in class are considered to be equivalent.

These actions impact structural integrity because, if not managed appropriately, they have the potential to invalidate the basis for structural integrity established at certification. For example, modification of a ship may change displacement and weight distribution. The former affects the global response of the ship and therefore loads. The latter may affect longitudinal hull bending moment. In a similar way, modification of an aircraft affects fuselage, wing and control surface loading. In both cases, the consequent loading may exceed certified design limits or, more likely, increase repeated stresses/stress ranges at critical locations sufficiently to increase significantly the rate of fatigue degradation. In effect, the modifications may invalidate the load cases defined by the designer. All modifications must be designed to class and with a full understanding of the OSI to ensure continued validity of the in-service program. The cracking experienced in the bilge keels of the ANZAC Class frigates offers an example of design conducted without an adequately defined OSI.

Introduction of a new weapon, a new countermeasures suite, a new radar (all with associated doctrinal changes), a change in government demands or operational tempo; all will likely affect the usage environment and the frequency of high, medium and lower loads. For a ship, this may come about due to a change in OSI (eg. unlimited operations in the Southern Ocean or a change in home port) leading to operating more frequently in higher sea states and with different speed profiles than were assumed by the designer. Analogous scenarios can be described for other domains and vehicles.

These factors are particularly evident in military endeavours due to the expense of military systems, the long in-service life of military systems typically amounting to several decades, and the evolving specialist roles of military equipment and associated technology development. They emerge with the numerous steps taken over the life of the platform to ensure the capability continues to meet the need.

Simple wear and tear of critical load bearing structure and its repair affect structural integrity over the life of the platform. Individual instances are rarely causes for concern and are readily dealt with via standard maintenance processes. Left unchecked, wear and tear may develop to threaten structural integrity; however, less obvious perhaps is that repeated repair can also threaten structural integrity. Cumulative effects of repair on structure include changes to load paths, peak stresses and susceptibility to new modes of corrosion. The impact develops over time. Consequently, discerning the effect of wear and tear and maintenance action on structural integrity requires analysis of trends in condition and assessment of structural integrity implications over time.

Two other factors are worth recognizing as they arise not from changes to the platform or its use, but from latent certification issues or changes to the sustainment program. Design and design verification always entail approximations of the actual environment and the response of the manufactured structure to that environment. These approximations are almost always conservative with respect to design requirements; however, increasing complexity will often lead to structural performance that is unconservative or structural behaviours that are different to those anticipated. In the worst case, these latent issues appear as surprise failures early in life as seen in the case of the ANZAC Class bilge keels. More typically, they emerge mid to late life as degradation or defects (eg. cracking) that are earlier than expected and/or at slightly different locations and/or of different characteristics. Such latent issues may not be addressed adequately by the maintenance program but, within reason, can be identified via assessment of structural performance over time.

Changes to the sustainment program are made periodically to gain efficiencies or control cost of ownership, address logistics issues such as obsolescence or address organizational changes. The majority of possible changes are inconsequential with respect to structural integrity; however, inspection programs or limits for example, often subject to scrutiny because they are resource intensive or present a capability constraint, may be significant. Again, where an adequate understanding of the certification basis and its reflection in maintenance policy is not a part of the decision making process, apparently small and innocuous changes to the sustainment program can present as structural integrity issues later in life. By the time the issues emerge, options to recover the platform are often limited, expensive and highly detrimental to capability.

The consequences of a degradation of structural integrity are common across all domains and concern safety, cost of ownership and capability. The terms supportability and availability are sometimes substituted for cost of ownership and capability. Where domains vary is in the risk posed by degraded structural integrity to each and the relationships between them. This is perhaps best illustrated by looking at contrasting approaches to regulation. In the aerospace domain, the regulatory framework, known as the Defence Aviation Safety Regulations, is solely concerned with a safe outcome. The aim is not to achieve the safety outcome at the expense of the military outcome; however, the other aims of the capability are not subject to enforceable regulation. From a structural integrity perspective, this approach recognizes both the typically catastrophic consequence (ie. loss of life) of a structural failure and the extreme sensitivity of structural integrity to aircraft design decisions[2]. In the maritime domain, the new DSwMS mandates an integrated view of risk to achieving the desired operational effect whilst minimizing risk to personnel and the environment. From a structural integrity perspective, the importance placed on safety is no less than in the aerospace domain; however, the regulatory approach reflects the lower risk to safety associated with structural integrity due to operational context, structural design parameters and the characteristics of structural failure[3]. Looking forward, while the survivability and damage stability requirements of naval vessels lend the structure redundancy and damage tolerance that underpins this assessment of risk, the emergence of highly integrated and optimized structures points to a potential erosion of this margin and the possibility that this assessment may require review.

The contrast described in the previous paragraph is sometimes emphasized to such a degree that the risks posed by degraded structural integrity to the other two outcomes, cost of ownership and capability, are overlooked. Unanticipated structural integrity degradation in critical load bearing structure is often so significant at the time of

[2] This is due to the emphasis on weight reduction and the relatively high stress levels that typically result.

[3] It should be acknowledged that the contrast described becomes less marked when considering submarines; however, it serves to help focus the consideration of structural integrity appropriately for a given domain.

discovery that the options available to fix the issue are few, costly and time consuming. The need to send eight Armidale Class Patrol Boats (ACPB) to Singapore for repair and to ultimately lease two Cape Class Patrol Boats is, in part, an example of this scenario. The nature of structural integrity degradation and the way it emerges, described in previous paragraphs, means there is a high likelihood of this occurring over the life of a platform. Such issues have a corresponding capability impact, particularly if the options for treatment entail operational limitations (eg. precluding operations in certain conditions or configurations). Ultimately, failure to account for the structural integrity implications of OSI related decisions over the life of maritime platforms may culminate in reduction of the life of type and little time to arrange a replacement or alternate means of meeting the capability need. The late life management of the Fremantle Class fleet offers a case in point. Little time means fewer options for government.

Of course, the argument of the previous paragraph can be expressed in terms of opportunities. Even in the aerospace domain, where the risk is generally framed in terms of safety, structural integrity efforts have generated significant and far reaching benefits for cost of ownership and capability. Examples from this and other domains are provided in the next section. In the maritime domain, meeting ACCO 1.1 imposes a cost burden; however, the magnitude and phasing of that cost over the life of a platform will vary depending on the approach taken to meet the obligation. Establishing a proactive, whole-of-life structural integrity program designed to assess risk and anticipate issues provides the opportunity to control and phase costs. Consistent with the vision of the Defence White Paper, this approach offers a more robust basis for establishing whole-of-life sustainment costs. Moreover, it allows the possibility of a "structural integrity management dividend"; that is, opportunities to gain greater life from the asset or introduce maintenance efficiencies or introduce a capability upgrade because differences between certification basis and design assumptions favour ADF operations.

In summary, establishing an effective approach to maintaining structural integrity over the life of a platform is dependent on:

- understanding the basis for structural integrity established by design, manufacture and certification;
- as part of the above, understanding assumptions made in design about in-service operations;
- understanding how this basis is translated into in-service maintenance and management policies;
- recognising that degradation of structural integrity typically emerges over a long period of time due to the cumulative effects of many in-service actions that in isolation may appear small and benign; and
- recognising that while ensuring static structural limits are not breached is critical, degradation of strength due to fatigue is a more challenging problem due to the mechanisms being less broadly understood, the significance of relatively small changes in load that can be readily overlooked, related design data often not being available, and the cumulative effect of many small changes in configuration enacted through repair or minor modification; and
- understanding how the risks posed by degraded structural integrity vary by domain.

Given the above, a holistic, whole of life approach is required. The approach shown to be effective in other domains (aerospace, offshore oil and gas, nuclear, some infrastructure) is known as the Structural Integrity Program (SIP). The goal of the SIP is to ensure the desired level of structural safety, performance, durability and supportability, with the least possible economic burden throughout the platform's design service life. The objectives of the SIP are to:

- define the structural integrity requirements associated with meeting OSI;
- establish, evaluate, substantiate, and certify the structural integrity of structures;
- acquire, evaluate, and apply usage and condition data to ensure the continued structural integrity of platforms;
- provide quantitative information for decisions on force structure planning, inspection, modification priorities, risk management, expected life cycle costs and related operational and support issues; and
- provide a basis to improve structural criteria and methods of design, evaluation, and substantiation for future systems and modifications.

A SIP typically accomplishes these objectives by conducting the following activities:

1. **Design Requirements and Information.** Definition of design requirements and certification rules related to structural performance of the platform and the maintenance of structural integrity throughout in-service life.
2. **Environmental Characterisation and Design Analysis.** Conduct of analyses and test to characterise the operational environment and analyze the design. This activity establishes design assumptions, material characteristics and load conditions.
3. **Structural Testing and Verification.** Where necessary, conduct of full scale testing and similar physical verification activities to assist in determining the structural adequacy of the design. Structural testing is employed to differing degrees in different domains; however, where used, it typically provides evidence required to certify, life and manage critical structures.
4. **Certification and In-Service Management Policies and Plan.** Develop operating and sustainment policies and procedures (eg. inspections, limits, operating limitations), conduct certification, develop the plan for the in-service structural integrity management program and the data, tools and processes required to support the program. The plan is termed the Structural Integrity Management Plan (SIMP) and the tools/processes include collection and analysis of condition and usage/environmental data (eg. Structural Health Monitoring System), and structural analysis tools including finite element models and fatigue models.
5. **In-Service Management.** Conduct structural integrity management over the in-service life using the resources developed in activity 4. This includes monitoring and assessment of trends in structural condition, platform usage, operational environment, changes in role, modifications etc. always with reference to the established certification basis and design data of activities 1 to 3.

In the aerospace domain, these activities are Tasks I to V specified in United States Air Force (USAF) MIL-STD-1530D (1). They are equivalent to elements of Structural Integrity Management as defined by the off-shore oil and gas industry (2). The activities can also be embodied in an ISO 55000 Asset Management System. This is best illustrated using the Capability Delivery Model described in the Asset Management Body of Knowledge (3). Indeed, closed loop architecture is a fundamental feature of both (Figure 1).

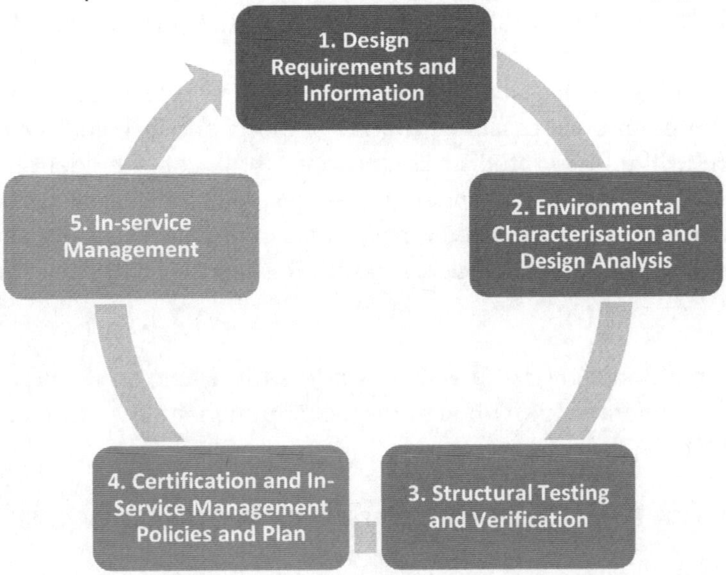

Figure 1: Closed Loop Structural Integrity Management Activities

Activities 1 to 3 occur in the acquisition phase and the degree to which the ADF influences these activities varies by domain; however, of crucial importance is the need to define operating and support intent (ie. the OSI in the maritime domain), to obtain and understand the certification basis, and the data underpinning lifing, limits and maintenance policies (ie. usage assumptions, materials data, critical structures and load conditions, test results). With this information, the ADF is well placed to conduct activities 4 and 5. Indeed, being the in-service element, the vast majority of the duration of a SIP is spent in activity 5. Activity 5 is often represented as shown in Figure 2.

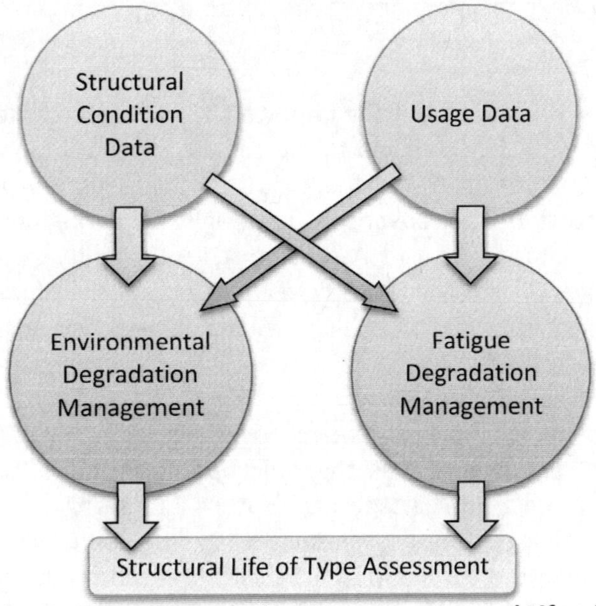

Figure 2: SIP Activity 5 - Data, Degradation Management and Life of Type Assessment

Activity 5 constitutes the closing of the loop through monitoring of lead and lag structural health indicators and assessment against the basis for structural integrity. Activity 5 provides the mechanism to proactively respond to all the factors that affect structural integrity over the life of a platform. Where warranted and to an appropriate degree, activity 5 triggers the revisiting of activities 1 to 4 (Figure 1) to ensure that certification basis, limitations, operations, sustainment arrangements and OSI all remain aligned and, where they don't, the associated risks (or opportunities) are identified, assessed, communicated and managed appropriately. An example of the work being done to improve the effectiveness of this closed loop process in the maritime domain is offered by the United States Navy (USN) Office of Naval Research (ONR) Structural Reliability Program (4).

A program with these objectives and activities offers the whole of life approach required to address the dependencies described earlier and so manage the structural integrity of any platform effectively.

STRUCTURAL INTEGRITY MANAGEMENT EXPERIENCE IN OTHER DOMAINS

SIPs and equivalent programs can be found in a variety of domains, including aviation, nuclear power generation, off-shore oil and gas, and some civil infrastructure such as bridges.

The domain with which the concept of a SIP is most closely associated is aviation. In this domain, the Aircraft Structural Integrity Program (ASIP) was first defined by the USAF as one part of the strategy to recover the B-47 fleet following the loss of five aircraft in 1958 (5). It is sometime suggested that the nascent ASIP was concerned only with safety. On the contrary, in the late 50's and early 60's the US Government was primarily concerned with capability. Nevertheless, over the subsequent decades, the aim of the program evolved to extend beyond safety and capability to also encompass reliability and cost of ownership (1).

For Australia, the ASIP was essentially imported with the purchase of US types and the close relationships with the USAF and USN throughout acquisition and sustainment. The F-111 was the archetype in this respect due to the structural integrity challenges it presented by virtue of its advanced materials and manufacturing technologies. Australia's implementation of the ASIP and ASIP capability matured in the face of numerous challenges posed by various types, particularly Macchi, Nomad, F-111, F/A-18 and C-130. The requirement for an ASIP for each type was mandated by policy and regulation in the early 1990s.

The F-111 is a seminal example that bears closer attention. Withdrawal by the USAF of the F-111 in 1998 left the ADF as the only operator of the aircraft and a Planned Withdrawal Date at least 10 years into the future. Challenges posed by materials and manufacturing technologies were compounded by substantial differences in usage and configuration between the USAF and ADF fleets. When the USAF withdrew the type, the ADF established the F 111 Sole Operator Program. For ASI, the F-111 SOP built on existing

ASI capability with procurement of data, further development of skills, and conduct of test and condition survey activities to baseline our understanding of structural condition and performance. The ADF F-111 fleet continued to deliver critical capability until 2010; however, during this period of sole operation, the capability was threatened by structural test failures and airframe condition findings, yet augmented with new weapons (eg. AGM-142E) and doctrine. Had the ADF not invested over the previous decade in a sovereign ASI capability, this period of operation would have been severely disrupted if not curtailed and possibly marked by loss of life.

In order to ensure that Australia continues to have a strike fighter capability, the service life of F/A-18A/B fleet was required to be extended until 2020 as outlined in the 2009 Defence White Paper. Initially, it was determined that a major structural refurbishment program, including replacement of the centre barrel, was required for the aircraft to reach its planned withdrawal date (PWD). However, a test and engineering program provided the basis for reducing the F/A-18A/B the centre barrel replacement program from 49 to just 10 items, which delivered estimated savings of $443 million. Additionally, significant improvement to aircraft availability levels was achieved during a period of high operational tempo for the platform. Most other operators of F/A-18A/Bs have replaced centre barrels on the vast majority of their fleet.

Elements of the resources industry have implemented structural integrity management frameworks, especially for offshore oil and gas assets and pipelines. The UK and US are prominent in this effort which has been driven by accidents such as loss of the Alexander L. Kielland and by ageing offshore installations. Many installations, several in the North Sea for example, have exceeded their initially specified design life but are required to remain in service. A SIP provides the means to evaluate the structural integrity risks associated with continued operation based on all the principles discussed thus far in this paper: understanding of the basis for structural integrity, collection of data describing the operating environment and condition, and analysis of that data with reference to the basis. One example of such a program is specified by the UK Health and Safety Executive (2) (6), which defines structural integrity as:

the ability of a structure to perform its required function effectively and efficiently over a defined time period whilst protecting health, safety and the environment; and the SIM as,

the means of ensuring that the people, systems, processes and resources that deliver integrity are in place, in use and will perform when required over the whole lifecycle of the structure.

A similar program incorporating quantitative risk assessment into lifing and maintenance policy determination is defined by Bureau Veritas (7).

Programs that implement these principles have been established by BP, Shell and ConocoPhillips amongst others.

While safety is paramount for all the SIP incarnations described here, to suggest that they are solely about safety is incorrect. Ultimately, a SIP aims to maximize the value of an asset whilst minimizing the risk of harm to people and the environment.

THE CASE FOR STRUCTURAL INTEGRITY MANAGEMENT IN THE MARITIME DOMAIN

The issue of structural integrity management is certainly not new to the naval domain. As mentioned above there have been several attempts to address the through life structural management of surface ships. Examples have included research programs carried out by the Ship Structural Committee (SSC) on a US Coast Guard (SSC), an extensive study carried out on the Royal Norwegian Navy (RNoN) Oksoy/Alta- Class mine counter measures vessel the HNOMS OTRA (9) and within Australia, a research program called the Integrated Hull Condition Monitoring Systems program carried out by the Australian Maritime Engineering Cooperative Research Centre along with partners BHP Shipping, IHI Heavy Industries in Japan, Lloyds Register of Shipping and P&O Bulk Shipping. An excellent review of some of the structural monitoring programs that have been embarked upon by navies to understand the loadings placed on warships can be found in Phelps and Morris (10). However as mentioned above many of the structural monitoring and structural life prediction programs tend to focus around the determination of loads and responses to the current condition that the ship is in rather than addressing the through life situation. It is the later problem that is the focus of this paper and we now turn our attention to some of the drivers for this aspect.

A merchant ship is designed for an effective life so that it can make a return on its investment for the ship owners. The process will often involve predicting the balance between the initial purchase costs of the ship, the operating costs including daily running and voyage costs and cargo handling costs with the predicted income from providing the service. However for a warship there is no income to be derived. The benefit arises from the military capability that is offered and its life is limited by factors such obsolescence of the capability or rising costs of maintenance. The rising costs of maintenance will have an impact on the total cost of ownership. The Naval Shipbuilding Plan outlines $90b of new acquisitions including $50b for the future submarine, $35b for the future frigates and $3-4b for the offshore patrol vessels. The acquisition costs account for 30% of the total cost of the investment program leaving another $210b for through life costs for the capability. Structural maintenance and repair form a key part of this cost and it increases as the surface ships become older.

Service Lives

One question that is very simple to ask and yet is proving very difficult to answer is "*what is the life of a naval service ship*"? For a warship there are a number of different lives that can be used to describe the life of type (11). A summary of the different types of lives is shown in table 1.

Type of Live	**Description**
Expected Service Life	Determined by force level planning numbers and is generated by senior defence planners. Within a ship class, all ships are expected to have the service life and this is

	used to determine through life costs.
Actual Service Life	The life of the ship taken from initial commissioning to decommissioning and where appropriate excludes out-of-commission periods where the ship is laid up.
Operational Service Life	The life accounts for the different rate of ship aging between operational years and years out of commission. This age will therefore assume a lower aging rate whilst out of commission.
Assessed Age	A measure that has been agreed by interested parties that accounts for the material condition of the ship. This allows operational flexibility but means that the material condition of the ship needs to be tracked.
Design Service Life	The service life that the ship is designed to achieve. This life, from a structural perspective is similar to the approaches used for most merchant ships.
Economic Life	As discussed above is used by merchant ship owners to determine their return on investment. Economic factors will impact a warships life however the process for defining this is not a transparent process.

Table 1: Life of Types for Naval Vessels

It is extremely important to have a clear understanding of structural load requirements that will be placed upon the ship structural for its lifetime. Many of the standards for design consider the use of the design service life. Typically this may be 30 years with a given operational tempo – say 200 days a year at sea. The assessment of the ship structure will then be completed on the probable loads to ensure compliance. This does not mean that following the 30 year life the ship will fail by some structural means, in fact this is far from the case. However for warships we are seeing a trend that shows the average service life for warships is increasing (12), furthermore the larger the warships the longer the average service life, see table 2.

Ship Type	Average Service Life (Years)
Carriers	40.4
BB-61 Class	48.8
Amphibious Vessels	29.2
Cruisers	26.3
Destroyers	25.4
Frigates	19.8

Table 2: Average Achieved Service Life (after Collette 2011)

To place this life of type question in context with the Royal Australian Navy there are two aspects to consider. Firstly there is a need to determine what will be the expected life of our surface ships. To date, it is probably true to say that the majority of surface ships within the Royal Australian Navy have undergone a Life of Type Extension Program. Examples include the Fremantle class patrol boats and the Adelaide Class Guided Missile Frigates. Both of these classes operated well past their original design life with successful operations. This trend in extending the life of surface ships, beyond their design life is likely to continue. The government's Naval Shipbuilding Plan (13) describes an exciting plan to deliver new Offshore Patrol Vessels and Future Frigates among many other activities under a continuous build program. The plan reports that the *Anzac* class frigates which were built between 1992 and 2006 will not be withdrawn from service until the future frigates enter service in around 2027-30. The first future frigate will realise operational capability by that date, with subsequent ships expected to enter service every two years after that date. Therefore it is evident the design life of 30 years is highly likely to be exceeded. This should not be a major concern with an effective Structural Integrity Program in place. Secondly the data in table 2 suggests that the larger the ship the longer the service life. There are numerous strategic and political reasons that may come to mind when determining the need for current and future naval fleets. These include present and future conflicts that are likely to arise in neighbouring oceans as well as those in the far corners of the globe (14). Additional to this is the level of resources that a nation can provide for the sustainment and future acquisition programs of the country. In 2012 the Australian Government had the lowest budget allocation for defence, in terms of percentage gross national product (GNP) since the Second World War at 1.56% GNP but is improving significantly, however there is also a growing cost with the complexity of modern warships. An analysis in 2012 showed that prior to WW2 the average cost of a surface ship increased by 3.64% per annum. After WW2 this changed to an annual increase of 3.97% for frigates and 4.07% for Destroyers. Therefore it is becoming increasingly likely that the cost for life of type of warships will increase in the future as budgets are constrained and life extensions more likely. Furthermore, as the cost of platforms increase there is a world-wide trend towards modular or flexible platforms. In this approach the surface ship can either be re-rolled by changing its configuration either permanently or for a particular mission. Either of these approaches often means the operational profile for the ship will change and therefore the loads. For Australia it is clear to see the growth in the size of ship types; consider the patrol boat fleet changing from the Attack class to the Fremantle Class and then the Armidale Class to the new Offshore Patrol Vessel – a large increase in size and capability. Similarly the future frigates will be significantly larger than any frigate Australia has operated. Therefore the likelihood that design life will be extended must be underpinned with a sound structural integrity management program throughout the life.

Mid Life Upgrades

The loads and requirements that are placed on the ship structure may also change during the life of the ship due to upgrades. Some upgrades could almost be considered a total redesign to change the operational context whereas others are due to significant upgrades in capability that has arisen as part of an innovation program. An example of a total redesign would be the transformation of the LPA's (HMAS Manoora and Kanimbla) into Australian service from the original USS Saginaw and Fairfax County respectively. This major upgrade included a helicopter hanger amidships and in order to provide enough space for the forward operating helicopter or the carriage of two landing craft the forward end was modified with the removal of the 34m bow ramps and horns. A 70 tonne crane was also installed. In this modification the reporting displacement remained constant (~8534 tons fully loaded) (15). Although these ships were very successful ships in demonstrating the need for an amphibious capability they were subject to some structural management challenges. It was necessary to understand the original design intent for the ships and then to understand the consequences of the modifications for the long service life they endured.

Similar, although smaller, upgrades have occurred on the Anzac class frigates owned and operated by the Australian and New Zealand Navies. These ships are based on the German MEKO 200 concept. Although purchased through a common build program each nation has undergone an upgrade program with different requirements. Naval Technology.com reported that the New Zealand upgrade was "Designed to extend the operational life of Tenix Defence systems-built *Te Kaha* and *Te Mana* frigates for another 15 years, the upgrade programme seeks to replace hardware and software of the combat management systems of the ships and the self-defence missile system." This was completed as a result of some major relocation of equipment on the quarter deck and internal structural changes to support new bulkheads and relocated equipment. Furthermore additional solid ballast has also been added to the ship to address some pre-refit trim issues (16). In a similar fashion the Australian ANZAC frigates have undergone an Anti-Ship Missile Defence (ASMD) system upgrade. In this upgrade the addition of an enclosed mast for the new radar system introduces complexity in terms of evaluating structural life assessments whilst providing the Royal Australian Navy with a tremendous capability upgrade. The consequences of the upgrade system have changed the requirements on the original design of the ships and therefore should be considered in any through life structural management processes. A key question is *how does the capability manager consider these potential upgrades at the start of a program when determining total lifetime costs*? The change in displacement and trim of the ship will have a small effect on the loads and cause changes to lifetime fatigue loads which will need to be managed under a structural integrity program.

Operations

It is not uncommon that the global geo-political environment changes. This may also impose changes on the operational profile to which surface ships operate. Whilst these changes should not impose significant changes to the structural design codes for major surface ships they will cause changes to the loads the ships experience. Through life structural integrity management, as previously mentioned, is concerned with understanding the loads experienced and managing the structure based on usage rather than design. Long periods of operating in extreme sea states to a vessel being laid up or undergoing upgrades changes the lifetime loads on an individual ship. The monitoring or collection of lifetime loads during operations is one of the key components of many of the structural monitoring programs previously mentioned. An excellent example of this type of activity as applied to the Armidale Class Patrol Boat is reported by Gardiner (17). A holistic approach to the tools and methodologies required to assist in the through life structural management of surface ships is given in Eccles (18). In their model they discuss the need for including details of wasted or corroded global and local finite element models that are supported with structural surveys to determine remaining life. This approach is limited in its use since the surface ship is only surveyed at certain times in its life and is historical data rather than based on predictive models. More development and research on predicting future fatigue damage or corrosion is required. An example of the type of prognostic tools that needs further development can be found in Gudze (19). The future operating tempo and environment is crucial to ensuring the sustainability, life of type and total cost of the platform.

THE WAY FORWARD

In order to facilitate a Structural Integrity Program to ensure the successful operational life of surface ships it is apparent that there are a number of contributing elements that must come together. These include having a clear understanding of the original design intent of the ship structure together with assessment margins. This is not sufficient in itself to provide an effective management concept for through life. It is also necessary to be able to determine the potential impacts of any future growth in payloads that might be derived from innovation or capability upgrade programs. This should be managed within the acceptable growth margins derived at the start of the program or might, in the case of the *Anzac* class be increasing the range of the margins. One of the key components is to collect, store and manage data associated with the use and deterioration of the structure whilst it is service. This data needs to be transferred into information by applying it to structural integrity models so that knowledge about the structural system is known and understood. Once this knowledge is clearly understood the concepts of upgrades, life extensions and the determination of when a replacement program should be initiated can be made without undesirable consequences. This overall concept could be defined as a Structural Integrity Program for the surface ship structures.

To be successful it needs ownership of the system from the Royal Australian Navy, the original ship designers, the manufacturer, the ship maintenance contractors as well as regulatory bodies such as classification societies and platform managers. This type of enterprise would fit into the "Ship Zero" concept being developed by Navy. The philosophy of "sustaining where you operate" which exploits the efficiencies of collocating key supporting agencies and training capabilities with industry support centres. For the submarine fleet, "Boat Zero" would also house the land based test site and systems development cell including the finite element and electronic product model. These models support the development of understanding the design intent and the development of training, sustainment and capability assurance requirements (20). The same approach for the surface ship fleet is essential to ensure future success.

Sustainable implementation of a SIP will also be dependent on alignment of the program with the new DSwMS and with emerging asset management systems. The closed loop, through-life and "founded on certification basis" characteristics of a SIP mean that they are aligned and they can be integrated.

Goal 2 of the DSwMS Governance and Management Compliance Obligations (GMCOs) requires systems to align with the OSI. GMCOs 2.1 (Define Operating and Support Intent) to 2.5 (Mission and Support System remain aligned to approved OSI) apply the goal to the different stages of the capability life cycle in the same way as the SIP Activities. Table 3 illustrates the alignment with GMCOs 2 and 3.

CLC Stage	GMCO	SIP Activity
Needs	2.1 Define Operating and Support Intent	1. Design Requirements and Information
Requirements	2.1 Define Operating and Support Intent	1. Design Requirements and Information
Acquisition	2.2 Align systems Design with OSI 2.3 Align Realised Systems with Design	2. Environmental Characterisation and Design Analysis 3. Structural Testing and Verification
Transition	2.4 Transition Systems to and from Service	4. Certification and In-Service Management Policies and Plan
In-Service	2.5 Mission and support Systems remain aligned to aproved OSI, 3.1 Ensure OSI is Understood 3.2 Operate & Support Systems as Intended 3.3 OSI is evolved as required	5. In-Service Management
Disposal		

Table 3: Alignment of Capability Life Cycle, GMCO 2/3, and SIP

All activities of the SIP support the achievement of GMCO 2. However, to fully realise the benefits of this alignment, work is required to develop the roles, responsibilities, processes and tools of a maritime SIP such that the use of their outputs in seaworthiness and fleet planning decision making is clear.

Failure to address a through life approach may lead to unviable life extensions with high cost implications. The enduring success of the national shipbuilding program will rely on each of the different class's ships being designed, built, maintained and managed effectively whilst further programs are being started.

CONCLUSIONS

In this paper the need for an integrated Ship Structural Integrity Program has been discussed. The views expressed regarding the benefits of implementing SIP in the maritime domain are supported by a variety of experiences over time and multiple fleets. Further, an efficient and effective implementation can be achieved due to alignment with the DSwMS and Asset Management systems at both strategic and operational levels. Some of the essential components of this system have been described and what it will take to ensure the success of the future shipbuilding program with respect to ship structures that is outlined in the Naval Shipbuilding Plan.

REFERENCES

1. United States Department of Defense. Aircraft Structural Integrity Program (ASIP), 2016. MIL-STD-1530D.
2. UK Health and Safety Executive Technical Policy Relating to Structural Integrity Management (fixed structures). Health and Safety Executive. [Online] [Cited: August 8, 2017.] http://www.hse.gov.uk/offshore/integritymanagement.htm.

3. Asset Management Council. AMBOK Models. Asset Management Council. [Online] [Cited: August 10, 2017.] http://www.amcouncil.com.au/.
4. Office of Naval Research, United States Navy. Structural Reliability Program. Office of Naval Research, Science and Technology. [Online] [Cited: August 7, 2017.] https://www.onr.navy.mil/en/Media-Center/Fact-Sheets/Structural-Reliability-Program.
5. Negaard, Gordon R. The History of the Aircraft Structural Integrity Program, Aerospace Structures Information and Analysis Center, 1980. 680.1B.
6. UK Health and Safety Executive. Structural Integrity Framework for Fixed Jacketed Structures. 2009. RR684.
7. Bureau Veritas, Risk-Based Structural Integrity Management of Offshore Jacket Structures, NI624, April 2017
8. Karl Stambaugh, Ingo Drummen, Chris Cleary, Rubin Sheinberg, Mirek Kaminski Structural Fatigue Life Assessment and Sustainment Implications for a New Class of US Coast Guard Cutters Ship Structures Committee – US Coast Guard project
9. Erling Torkildsen, H.; Grøvlen, Å.; Skaugen, A.; Wang, G.; Egil Jensen, A.; Pran, K.; Sagvolden, G. (2005) Development and Applications of Full-Scale Ship Hull Health Monitoring Systems for the Royal Norwegian Navy, Recent Developments in Non-Intrusive Measurement Technology for Military Application on Model- and Full-Scale Vehicles (pp. 22-1 – 22-14), Meeting Proceedings RTO-MP-AVT-124, Paper 22. Neuilly-sur-Seine, France: RTO.
10. Phelps, B. and Morris, B. Review of Hull Structure Monitoring Systems for Navy Ships. DSTO-TR-2818 2013.
11. Koenig, P., Nalchaajian, D. and Hootman, J. Ship Service Life and Naval Force Structure. Paper accepted for ASNE symposium Engineering the Total Ship. Tysons Corner, Va, 23-25 Sept 2008.
12. Collette, M. Hull Structure as a system: Supporting Lifecycle Analysis. American Society of Naval Engineers #3 P 45-55. 2011
13. Naval Shipbuilding Plan. Australian Government, Department of Defence 2017.
14. Brown, J.C. A wave induced fatigue strain recorder for surface ships. Society of Naval Architects and Marine Engineers. Proceedings of the Annual meeting 30th September-1st October 1999.
15. Wiki 2017a. https://en.wikipedia.org/wiki/Kanimbla-class_landing_platform_amphibious August 2017
16. Naval Design Committee Report International Ship and Offshore Structures Congress ISSC 2015 7 - 10 September 2015, Portugal
17. Gardiner, C.P., Vincent, P., Wilson, A., Ellery, D. and Armstrong, A. Hull Structural monitoring of the Armidale class Patrol Boat. RINA Innovation in high Speed Marine Vessels. 28-29th January 2009
18. Eccles, T.J., Ashe, G. and Albrecht, S. The Achieving Service Life Program. American Society of Naval Engineers #3 P 103-112. 2011
19. Gudze, M, Cannon, S.M. and Melchers, R.E. Ballast Tank corrosion using Naval Ship Operational Profiles. Pacific2004 Conference. Sydney 3rd-5th Feb. 2004.
20. Barrett, T. Chief of Navy Speeches: Submarine Institute of Australia Future Technology Conference. November 17, 2015.

Trials and Tribulations: Load and Structural Response Measurements of a Naval Semi-Planing Craft

Teresa Magoga
Defence Science and Technology Group
Teresa.Magoga@dst.defence.gov.au

ABSTRACT

In mid-2016 sea trials were undertaken on the Royal Australian Navy Armidale Class Patrol Boat HMAS *Maryborough*. This vessel was instrumented with a hull monitoring system, which included strain gauges, a motion reference unit, a Global Positioning System, and an array of pressure transducers at a forward section of the hull. The objective of the trials was to extend the knowledge base on the loads and structural responses of the vessel, coupled to the wave environment, at specific headings and speeds. The work forms part of a larger effort to improve hydrodynamic, structural, and service life assessment methods for naval craft. The capability to acquire coupled environmental, loads, and response data was demonstrated. This achievement is non-trivial, as acquisition of such data is relatively rare and is used to validate numerical hydrodynamic and structural assessment tools. It was shown that the pressure at locations below and near the mean waterline tended to be harmonic, though the latter featured a one-sided cut-off due to the free surface crossing the hull. Above the waterline, in the bow flare region, the pressure featured a larger degree of non-linearity. The stress and pressure records were correlated, allowing their relationship to be qualitatively examined. The implications of the results on structural assessment procedures are discussed, and recommendations for further analysis are made.

NOMENCLATURE

ACPB	Armidale Class Patrol Boat
AP	Aft Perpendicular
CL	Centreline
DAQ	Data Acquisition System
DNV	Det Norske Veritas
DST	Defence Science and Technology
FEA	Finite Element Analysis
Fn	Froude Number
g	Acceleration due to gravity [m/s^2]
GPS	Global Positioning System
HMAS	Her Majesty's Australian Ship
HMS	Hull Monitoring System
HSLC	High Speed and Light Craft
$H_{1/3}$	Significant wave height [m]
L_{WL}	Waterline length [m]
MAXWB	Micro Air-Launched Expendable Wave Buoy
MRU	Motion Reference Unit
$P_{peak\_1/10}$	Significant tenth of pressure peaks [kPa]
$P_{peak\_1/3}$	Significant third of pressure peaks [kPa]
P_{peak_max}	Maximum of pressure peaks [kPa]
P_{peak_mean}	Mean of pressure peaks [kPa]
STFT	Short-Time Fourier Transform
T_{ave}	Average wave period [s]
USK	Underside of Keel
v	Speed [kn]
$\Delta\sigma_{mean}$	Mean of stress range cycles [MPa]
$\Delta\sigma_{1/3}$	Significant third of stress range cycles [MPa]
$\Delta\sigma_{1/10}$	Significant tenth of stress range cycles [MPa]
$\Delta\sigma_{max}$	Maximum of stress range cycles [MPa]
σ	Stress [MPa]
χ	Ship heading relative to dominant wave direction [°]

INTRODUCTION

The Armidale Class Patrol Boat (ACPB) is a deep-vee, semi-planing monohull, constructed from marine-grade aluminium alloys. The principal particulars are given in Table 1. The ACPBs are operated by the Royal Australian Navy. The vessels were designed, and are maintained, to Det Norske Veritas (DNV) Rules for the Classification of High Speed and Light Craft (HSLC) rules [1] supplemented by navy technical regulatory and safety rules.

Table 1: ACPB principal particulars

Waterline length, L_{WL} [m]	52.1
Beam [m]	9.68
Design draught [m]	2.25
Full scantling displacement [tonne]	336
Maximum speed [kn]	25

Smart decision-making regarding fleet availability and economy requires monitoring of a ship's construction quality, loads, material performance, operational profile, and maintenance quality. These factors can then be incorporated in through-life management tools, such as the prediction of the remaining useful service life of the ship [2, 3]. There is increasing interest in the utilisation of in-service loads and strain data for through-life structural management of both naval and commercial ships. With respect to high-performance warships, the drivers of this trend include [2, 4-7]:

- The increase in vessel size, which has intensified the importance of global loads and dynamic responses.
- The use of weight-optimised structures, which typically feature inserts and cut-outs.
- Maintenance issues (for example, fatigue cracking) and inspection difficulties (for example, the need to remove and re-install insulation to inspect aluminium hulls). Whilst both steel and aluminium alloys are susceptible to fatigue, the latter tends to have a lower capability to withstand cyclic loads due to the loss of strength in the heat-affected zone during welding, greater sensitivity to flaws, a relatively fast crack propagation rate, and more difficulties with fabrication.
- The requirement to operate in extreme environments, leading to the increased likelihood of the vessel sustaining significant wave induced impacts and slamming.

A Hull Monitoring System (HMS) was commissioned on-board an ACPB, HMAS *Maryborough*, in May 2015 by the Defence Science and Technology (DST) Group, Australia. The purpose of the HMS was to gather data on the operational profile, ship motions, and structural responses to global and localised loading. This effort forms part of a larger, on-going research program to improve service life assessment of naval HSLC. Specific sea state information was not available because a wave height measuring device was not incorporated in the HMS. Thus, dedicated sea trials were conducted in mid-2016 to obtain a full-scale dataset related to the wave environment via wave buoy measurements.

Completion of such sea trials is challenging, because the vessel availability is often limited and the sea state cannot be planned in advance. This is in addition to issues with the instrumentation of a vessel which must be overcome, including drilling hull plating to install pressure transducers, accessing the internal structure, and ensuring the functioning of the hardware in the harsh marine environment. Yet, the acquired data enables verification of numerical seakeeping and structural modelling [8], investigation of unexpected behaviours, and improvement of design procedures [9, 10].

This paper describes the capability to acquire loads and response data coupled to the wave environment during dedicated sea trials of a naval semi-planing craft. Preliminary results and recommendations are discussed.

LOADS ON SHIPS

Different load types contribute to the potential for hull failure through overload due to an extreme loading event that exceeds the load carrying capacity of the structure, and structural fatigue resulting from cumulative damage under cyclic loading. Types of loads include [11]:

- Hydrostatic loads.
- Low frequency or "wave" cyclic loads imparted on the hull through the action of the waves in a seaway, including the pressure acting on the hull due to the wave profile and the inertial reactions to the accelerations caused by ship motions.
- Dynamic loads, which can be due to propulsive devices, hull appendages, and unbalanced rotating machinery.
- Operational loads resulting from the ship performing specific activities. For naval vessels, such activities include the landing of helicopters and weapons firing effects.
- Thermal stresses due to a non-uniform temperature distribution.
- Impacts such as slamming and wave slap.

Slamming occurs when a vessel experiences sufficiently large heave and pitch motions such that the bow emerges from and re-enters the water; large dynamic pressures are imparted to the hull structure. The slam force depends on the spatial distribution of the pressure, and the phase shift between the loads (including inertial loads) along the hull. Impact loads that act on the sides of the flared bow zone are termed "bow flare slamming". "Bottom slamming" occurs when the waves impact the bottom forward section of the hull [11, 12]. In comparison to bottom slamming, in general the speed of impact of bow flare slamming is slower, the peak pressure is smaller, and bow emergence need not occur [13].

"Wave slap" describes an impact between the bow-side of a marine structure and a wave system. The large pitch and heave motions associated with both bottom and bow flare slamming need not be present for wave slap to occur [12].

Operational Modes

Hydrodynamic pressure depends on the flow around the hull, and is approximately proportional to the square of the ship speed. In general, the hydrodynamic force effect dominates the buoyancy force above a certain Froude number (Fn). Fn is a function of the ship velocity v, L_{WL}, and the acceleration due to gravity g:

$$Fn = \frac{v \times 0.5144}{\sqrt{gL_{WL}}}$$

Displacement ships operate up to an Fn of 0.4. A vessel that can operate with an Fn greater than 1.0 is called a planing vessel. Vessels operating between these values of Fn are termed semi-displacement or semi-planing [14]. The ACPB is considered to be a semi-planing craft, as it can operate with an Fn of 0.57.

Impacts between the hull and the water, and wave-induced pressure fluctuations, are non-linear phenomena. For displacement ships, the nonlinearities in the ship motions and pressures generally become pronounced in relatively high sea states. However, for HSLC nonlinearities become significant in relatively moderate waves due to the greater susceptibility to slamming and variations in dynamic lift [15].

TRIALS' PROGRAM

20 trials' runs were performed over two days in waters offshore from the Northern Territory, Australia. The runs were defined by vessel speed and heading relative to the predominant wave direction χ. The speeds considered were 10 kn, 15 kn, 20 kn, and 25 kn, and the headings were head seas (χ = 180°), bow quartering seas (χ = 135°), stern quartering seas (χ = 45°), following seas (χ = 0°), and beam seas (χ = 90°). The start of a run was indicated by the establishment of steady speed and ship heading. The Navigator judged the predominant wave direction. During each run the vessel maintained constant speed and heading for duration sufficient to provide adequate statistics for the analysis, that is, a minimum of 100 wave encounters [16]. A wave buoy was also deployed to measure the significant wave height and wave period. A track of the vessel during the 10 kn runs is displayed in Figure 1.

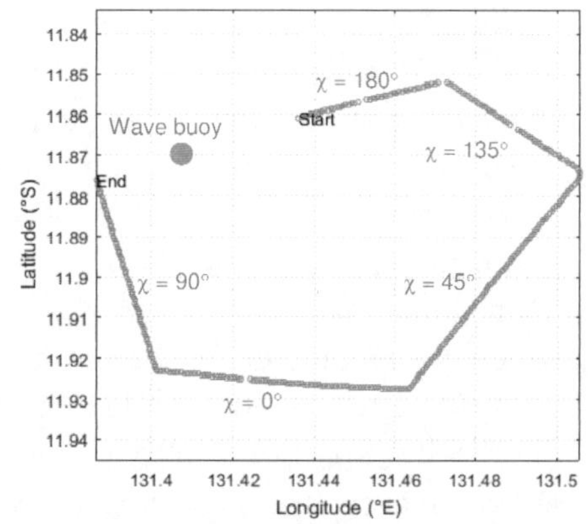

Figure 1: Vessel track during 10 kn runs

The nominal vessel speed, significant wave height $H_{1/3}$, and average wave period T_{ave} during the trials are given in Table 2. In general, the mean of sea state 3 to the bottom of sea state 4 was encountered.

Table 2: Environmental parameters

v nominal [kn]	χ [°]	$H_{1/3}$ [m]	T_{ave} [s]
10	180	1.3	3.8
	135	1.3	3.9
	45	1.3	3.9
	0	1.2	3.9
	90	1.2	3.9
15	180	1.2	3.8
	135	1.3	3.9
	45	1.4	4.0
	0	1.4	3.9
	90	1.3	3.9
20	180	1.1	3.7
	135	1.0	3.9
	45	1.1	3.8
	0	1.1	3.9
	90	1.1	3.9
25	180	1.2	3.9
	135	1.1	4.0
	45	1.2	4.0
	0	1.2	3.9
	90	1.2	3.9

DATA ACQUISITION

The acquisition of data consisted of measurements from the HMS, a high-speed stand-alone Data Acquisition Systems (DAQ), a wave buoy, and manual observations.

The HMS installed on *Maryborough* was based on a network of specialised circuit boards that returned data from several sensors to an embedded personal computer. It was an advanced version of that previously installed on HMAS *Glenelg* [17]. The sensors included foil-based strain gauges, pressure transducers, accelerometers, an external GPS (Global Positioning System) receiver, and a MRU (Motion Reference Unit). Four strain gauges are considered in this paper (refer to Figure 2 for locations). The locations of the pressure transducers were chosen to capture the pressure characteristics at different zones: below the waterline, near the waterline, and the flare of the bow. As three of the transducers were damaged, data from only six of nine transducers is available. The locations of these transducers are shown in Figure 3 and Figure 4. Further information regarding the installation of the transducers is available in [18].

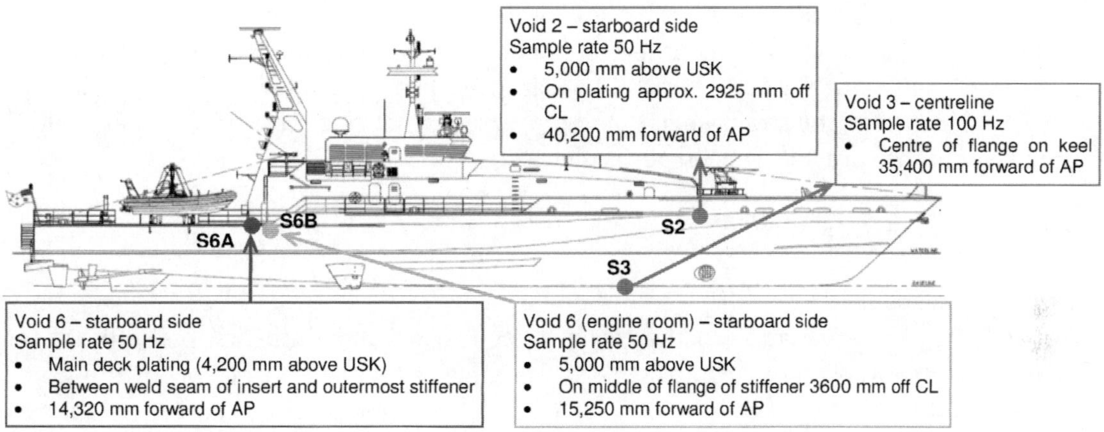

Figure 2: Locations of strain gauges shown on side view of general arrangement (CL = centreline, AP = aft perpendicular, USK = underside of keel)

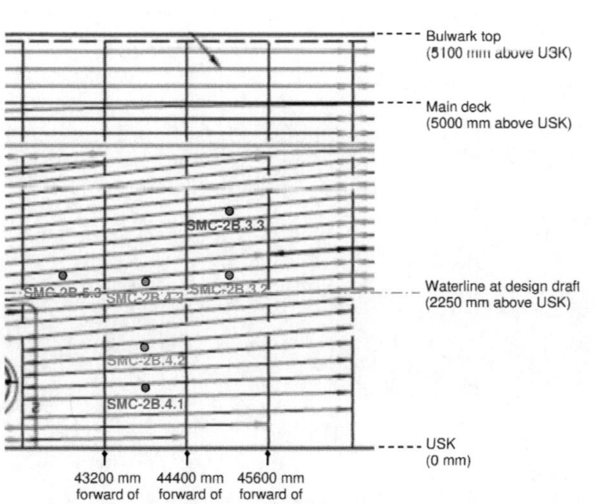

Figure 3: Locations of pressure transducers, shown on side view of sideshell and stiffener plan

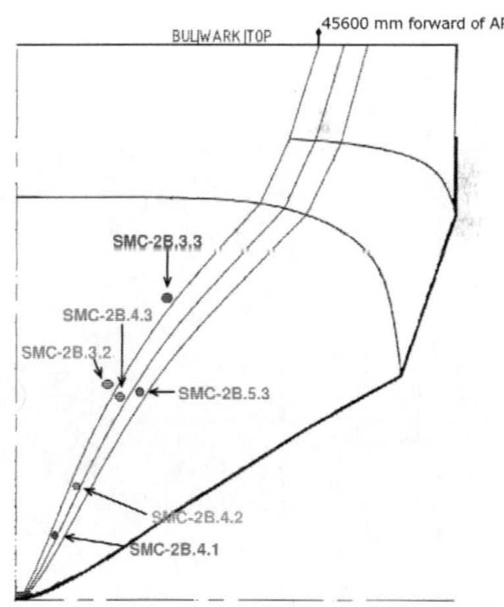

Figure 4: Locations of pressure transducers, shown on sectional view of lines plan

A stand-alone DAQ enabled high-frequency sampling (12.5 kHz) of pressure compared to that capable of the HMS. The DAQ consists of a United Electronic Industries Programmable Automation Controller Cube with serial communication, two high-speed 4-channel analogue acquisition cards, an Inertial Measurement Unit, data storage and a GPS receiver.

A QinetiQ North America MAXWB (Micro Air-Launched Expendable Wave Buoy) was used to measure the non-directional significant wave height and average wave period. To allow reuse of the unit, and more reliable and longer data acquisition, the MAXWB was modified by DST Group to include an oceanic and positional data logging capability, a battery, and increased transmission bandwidth.

DATA PROCESSING

Signal conditioning and data processing was carried out in MATLAB R2015a [19] using tools and methods reported in [20]. Only the longitudinal components of stress are analysed. The stress signal was reduced to cycles using the rainflow counting method proposed by Rychlik [21]. To ensure that statistical values are representative, stress cycles below 3 MPa are excluded from the analysis. In order to handle the large volume of pressure data, the pressure signals were decimated from 12.5 kHz to 2.5 kHz using a low-pass Chebyshev Type I filter. Verification of the strain gauges' and pressure transducers' sensitivities, and analysis of uncertainties, has previously been performed [22].

RESULTS AND DISCUSSION

Structural Responses

Figure 5 shows the variation of the mean $\Delta\sigma_{mean}$, significant third $\Delta\sigma_{1/3}$, significant tenth $\Delta\sigma_{1/10}$, and the maximum $\Delta\sigma_{max}$ of the stress range cycles, as a function of the nominal vessel speed, at strain gauge locations S2, S3, S6A, S6B. Due to a technical issue during the trials, strain gauges S6A and S6B data for the 10 kn runs is unavailable.

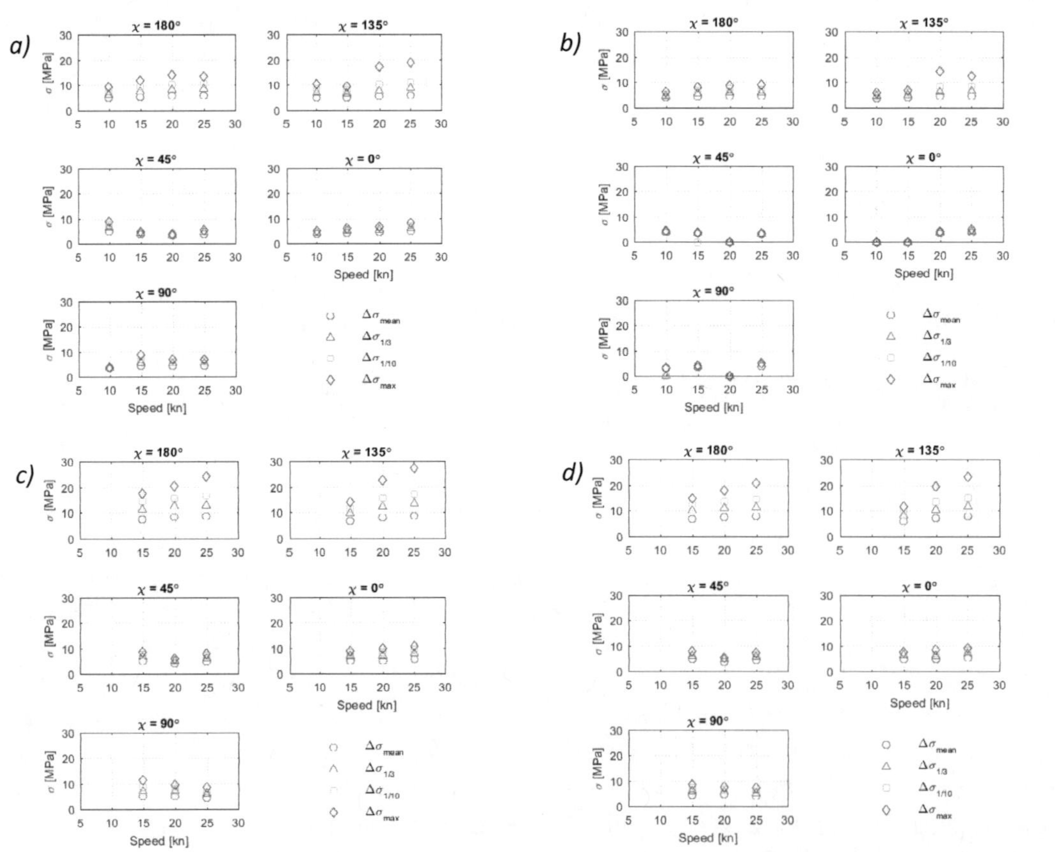

Figure 5: σ_{mean}, $\sigma_{1/3}$, $\sigma_{1/10}$, *and* σ_{max} *as a function nominal speed at a) S2, b) S3, c) S6A, and d) S6B*

The largest stress ranges were measured at strain gauge locations S6A and S6B, followed by S2 and S3. The former two gauges measure strains in the engine room, at the aft end of the superstructure; this location is associated with relatively large deformation due to the significant change in cross-section, and is in proximity to an anti-node of both the two-node and three-node mode shapes (modal analysis of the vessel has been presented in [23]).

The highest values of $\Delta\sigma_{mean}$, $\Delta\sigma_{1/3}$, and $\Delta\sigma_{1/10}$ at the strain gauge locations occurred in head seas and bow quartering seas at 25 kn; the differences in these values for the two headings were generally marginal (ranging between 1 and 15%). Usually, it is assumed that the largest wave loads and subsequent structural responses are generated in head seas. However, larger impact pressures on a Swedish Marine high-speed craft have been measured with χ equal to 150° [24], because the effective deadrise angle (angle between the hull and the water surface at the time of the impact) is smaller than in head seas. Also, on the day that the 20 kn and 25 kn runs were completed a dominant wave direction was less discernible than on the first day of the trials.

Pressures

In this paper, the analysis of pressure is mostly limited to head seas runs. As a consistent significant wave height (~1.2 m) was encountered, the behaviour of the dynamic pressure as a function of speed can be isolated. As a preliminary step, this is achieved through statistical analysis of the peak pressures. The detection of peaks was based on magnitude (set to 1 kPa), extent that the peak stands out relative to other peaks (set to 4), and minimum peak separation (set to half of the average encounter period).

Figure 6a, b, c and d display the mean P_{peak_mean}, significant third $P_{peak\_1/3}$, significant tenth $P_{peak\_1/10}$, and maximum P_{peak_max} of the detected peaks for each pressure record in head seas. As an insufficient number of peaks were detected in the SMC-2B.3.3 signal at 10 kn, the corresponding statistical values are excluded from the plots.

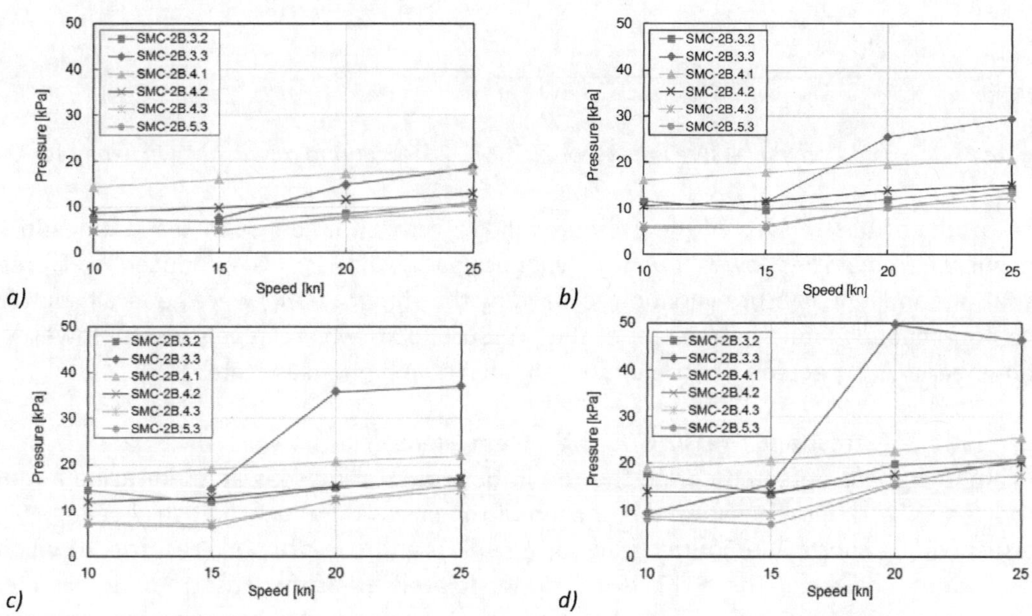

Figure 6: In head seas; variation of a) P_{peak_mean}, b) $P_{peak\_1/3}$, c) $P_{peak\_1/10}$, and d) P_{peak_max}

The following observations are made from Figure 6:

- P_{peak_mean}, $P_{peak\_1/3}$, and $P_{peak\_1/10}$ at locations below the mean waterline varied linearly with speed (SMC-2B.4.1 is located approximately 920 mm above USK, and SMC-2B.4.2 approximately 1470 mm above USK).
- At the locations in proximity to the mean waterline (SMC-2B.3.2, SMC-2B.4.3 and SMC-2B.5.3), P_{peak_mean}, $P_{peak\_1/3}$, and $P_{peak\_1/10}$ varied linearly with speed though there is some variance. This is mainly because $P_{peak\_1/3}$ and $P_{peak\_1/10}$ at 15 kn are smaller than those at 10 kn, and may be due to the significant wave height being slightly smaller during the 15 kn runs (refer to Table 2).
- At SMC-2B.3.3, approximately 3440 mm above USK, P_{peak_mean}, $P_{peak\_1/3}$, and $P_{peak\_1/10}$ varied non-linearly with speed. At this location the applied pressure was characteristic of wave slaps, occurring sporadically, at the lower speeds, as demonstrated in Figure 7. At higher speeds the applied pressure tended to be characteristic of minor water impacts, as highlighted in Figure 8, though were also sporadic.
- The pressures measured at locations near the waterline were generally of harmonic form but distorted by a cut-off on one side due to the crossing of the free surface. The pressures measured at locations closer to the keel varied harmonically, with very sporadic cut-offs when the hull re-entered the water.

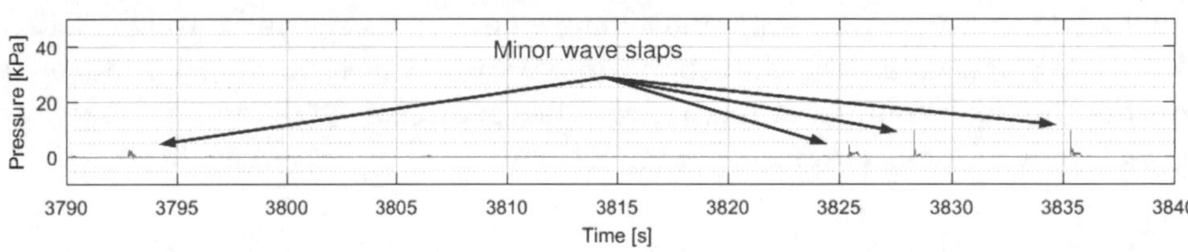

Figure 7: 50 s portion of pressure time record at SMC-2B.3.3 during run at 10 kn in head seas

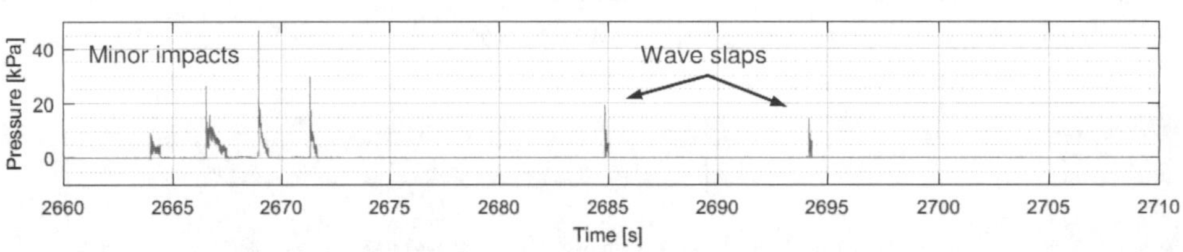

Figure 8: 50 s portion of pressure time record at SMC-2B.3.3 during run at 25 kn in head seas

The results of the analysis of the pressures indicate that for the given sea states and wave encounter frequencies low-frequency dynamic loads (that is, wave-induced hull pressure variations and hull pressure variations caused by the ship motions) were generally dominant. The non-linear dynamic component of the pressure loads was relatively weak. However, the non-linearity is expected to become more significant in higher sea states [15].

Correlation of Stress and Pressure during a Slam Event

A slam event was detected during the run in bow quartering seas at 25 kn using a method reported in [23]. Figure 9 shows the portion of the stress time record at strain gauge S3, the corresponding short-time Fourier transform (STFT), and the pressure records in which the slam event occurred. The STFT is a Fourier-related transform used to determine the frequency and phase content of local sections of a signal as it changes over time. It is a useful technique to visualise and compare the power in the vibrational response, which can result in damage of the hull-girder. The heave and pitch at the MRU location are also displayed in

Figure 9, though with the vertical axis scales removed. The purpose of showing these records is to demonstrate the motion behaviour of the vessel, rather than to focus on the magnitudes. Note that the synchronisation between the motions, stresses and pressures is approximate, but the pressure records are in sync with each other.

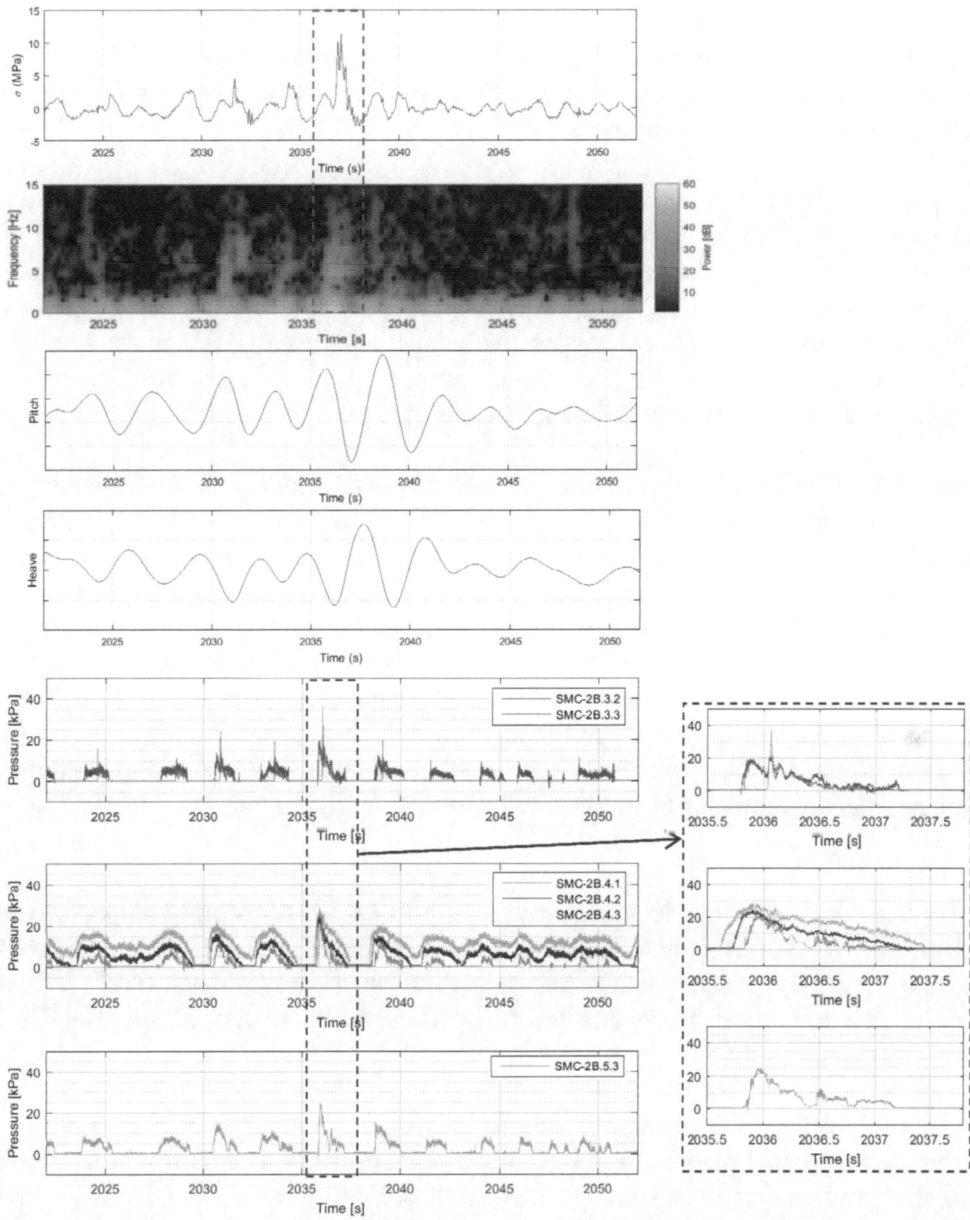

Figure 9: 30 s portion of time records of stress measured by strain gauge S3, corresponding STFT, heave (y-tick label removed) at MRU location, pitch (y-tick label removed) and pressures during bow quatering seas at 25 kn

From Figure 9, it can be seen that the pressure pulse propagated from the keel (SMC-2B.4.1) towards the chine (SMC-2B.3.3 is closest to the chine). The rise time of the pulse was small, and peak pressures were large, relative to those of the proceeding and following wave encounters.

During the slam event, the heave and pitch amplitudes were larger than those of the proceeding and following wave encounters. The water surface profile, timing of the hull and surface motions, and period of the encountered wave were such that the vertical displacement of the vessel was relatively large. Also, the bottom-most pressure sensor (SMC-2B.4.1) exited and re-entered the water prior to and after the impact. The structural response to the slam loading is characterised by a rapid increase during water re-entry followed by high frequency transient vibration, or whipping, which decays rapidly. The slam excited the bending modes of the structure, indicated by the relatively large power in the response at 2.5 Hz and 5 Hz visible in the STFT.

The pulse length of the slam event was 1.03 to 1.85 s, depending on the measurement point. This is 2.6 to 4.6 times the fundamental period of vibration of the ACPB with added mass (the two-node bending vibration occurs at approximately 2.5 Hz, thus the fundamental period of vibration is 0.4 s) [23]. Although only one relatively minor slam event is analysed, this finding may be significant with respect to the idealisation of slam loading for application in Finite Element Analysis (FEA). The pressure impulse, or pressure integrated over a given area or time, is important in structural design [24, 25].

The application of slam loads in a quasi-static manner on FEA models has been investigated by others, via reverse engineering of the loads from full-scale strain data [26], and reconstruction of the pressure distribution from full-scale measurements [27]. Further, in a study of the effect of the impact pressure profile on the failure of ship plating representative of the longitudinal bulkhead of a tanker, the pressure loads were idealised as an impulse; when the duration of the slam loading is three times greater than the natural period of the structure, the impulse can be assumed to behave in the quasi-static domain [25, 28]. Using the trials data discussed in this paper, combined with the continuous monitoring data upon which the operational profile can be derived, it may be possible to expand and validate the quasi-static approach to applying both the wave and slam loads in FEA.

CONCLUSION

In mid-2016 sea trials were conducted on-board the ACPB HMAS *Maryborough* in waters offshore from the Northern Territory, Australia. The objective of the trial was to improve the knowledge base on ship motions, loads, and structural responses of naval HSLC. The work forms part of research to improve and verify hydrodynamic, structural, and service life assessment methods.

The characteristics of the hydrodynamic pressure acting on the hull were illuminated. The mean of sea state 3 to the bottom of sea state 4 was encountered. The harmonic form of the low-frequency wave-induced pressure is distorted near the waterline by free surface crossings and sporadic, slight impacts. Above the mean waterline, in the bow flare zone, water slaps at low speed and infrequent pressure pulses at high speed were measured. Nonetheless, linear motions and loads appeared to be dominant. Preliminary correlation of the stress and pressure records of a slam event was performed, showing the propagation of the pulse from the keel to the chine, relatively large ship motions and structural responses, and excitation of the longitudinal bending modes.

FURTHER WORK

If the loads and structural responses coupled to the wave environment are obtained in higher sea states, valuable work includes:

- Utilisation of the data to evaluate the relevance of applied design loads, and to improve existing semi-empirical design pressure formulae to cover different hull-water interaction aspects for naval ships. For example, Rosén *et al.* [10] showed that the actual panel loads experienced by the Visby Class corvette were distinctly higher than the minimum requirements set by DNV.
- Completion of runs with finer increments in bow seas (that is, χ equal to 165° and 150°).
- Analysis of the pressure impulses, and mapping to FEA models.

ACKNOWLEDGEMENTS

Assistance provided by the following individuals and groups is duly noted:

- Defence Science and Technology Group colleagues that contributed to the HMS, particularly trials' team members Michael Brincat, Peter Graham, and Peter Vincent.
- QinetiQ Australia staff.
- Patrol Boat Systems Program Office.
- *Maryborough* ship's company.

REFERENCES

1. *Rules for Classification of High Speed, Light Craft and Naval Surface Craft*. (1996) Høvik, Norway, Det Norske Veritas.
2. Ferraris, S. and Farinetti, V. (2008) The problem of fatigue on large high-speed craft and the comparison between the behavior of high tensile steel and aluminum alloy structures: The view of an international shipbuilder. *Journal of Ship Production* **24** (2) 116-123.
3. Hess, P., Aksu, S., Blake, J., Boote, D., Caridis, P., Egorov, A., Fjeldstad, A., Hoogeland, M., Murayama, H., Anderson, M. R. and Tammer, M. (2015) Structural Longevity. In: *Proceedings of the 19th International Ship and Offshore Structures Congress*. Cascais, Portugal 211-287.
4. Swartz, R. A., Zimmerman, A. T., Lynch, J. P., Rosario, J., Brady, T., Salvino, L. and Law, K. H. (2012) Hybrid wireless hull monitoring system for naval combat vessels. *Structure and Infrastructure Engineering* **8** (7) 621-638.
5. Soliman, M., Barone, G. and Frangopol, D. M. (2015) Fatigue reliability and service life prediction of aluminum naval ship details based on monitoring data. *Structural Health Monitoring* **14** (1) 3-19.
6. Huang, T., Harbison, M., Kvidahl, L., Niolet, D., Walks, J., Christein, J., Smitherman, M., Phillippi, M., Dong, P., De Can, L., Caccese, V., Blomquist, P., Kihl, D., Wong, R., Sinfield, M., Nappi, N., Gardner, J., Wong, C., Bjornson, M. and Manuel, A. (2016) Reduction of overwelding and distortion for naval surface combatants. Part 2: Weld sizing effects on shear and fatigue performance. *Journal of Ship Production and Design* **32** (1) 21-36.
7. Thomas, G., Davis, M., Holloway, D. and Roberts, T. (2005) The influence of slamming and whipping on the fatigue life of a high-speed catamaran. In: *International Conference on Fast Sea Transportation,* St. Petersburg, Russia
8. Pegg, N. G., Gilroy, L. E. and Kumar, R. (1995) Full scale verification of finite element modelling of a 75 tonne SWATH vessel. *Marine Structures* **8** (3) 1995/01/01/ 211-228.

9. Carrera, G. and Rizzo, C. (2005) Measurements of Motions, Loads and Structural Response on a Fast FRP Pleasure Craft. In: *International Conference on Fast Sea Transportation, FAST'2005,* St. Petersburg, Russia
10. Rosén, A., Garme, K. and Kuttenkeuler, J. (2007) Full-Scale Design Evaluation of the Visby Class Corvette. In: *International Conference on Fast Sea Transportation, Fast'2007,* Shanghai, China
11. Mansour, A. and Liu, D. (2008) *Strength of Ships and Ocean Structures*. Jersey City, United States of America. The Society of Naval Architects and Marine Engineers.
12. Daidola, C. and Mishkevich, V. (1995) *Hydrodynamic Impact of Displacement Ships*. SSC - 385, Washington, United States of America, Ship Structure Committee.
13. *Notes on Ship Slamming*. (1993) Technical and Research Bulletin 2-30, New Jersey, United States of America, The Soceity of Naval Arcitects and Marine Engineers.
14. Faltinsen, O. (2005) *Hydrodynamics of High-Speed Marine Vehicles*. New York, United States of America, Cambridge University Press.
15. Chiu, F.-C., Tiao, W.-C. and Guo, J. (2009) Experimental study on the nonlinear pressure acting on a high-speed vessel in irregular waves. *Journal of Marine Science and Technology* **14** (2) 228-239.
16. *Recommended Procedures and Guidelines: Seakeeping Tests*. (2014) 7.5-02-05-04, Denmark, International Towing Tank Conference.
17. Vincent, P., Gardiner, C., Wilson, A., Ellery, D. and Armstrong, T. (2008) Installation of a sensor network on an RAN Armidale Class Patrol Boat. *Materials Forum* **33** 307-316.
18. Vincent, P., Wilson, A., Brincat, M. and Yiannakopoulos, G. (2015) Design, installation and repair of hull sensors for the measurement of hull slamming pressure loads on Armidale Class Patrol Boats. In: *International Maritime Conference, Pacific 2015,* Sydney, Australia.
19. *MATLAB R2015a* (2015). Natick, United States of America. MathWorks.
20. Dawes-Lynch, J. and Magoga, T. (2017) *HMAS Maryborough Hull Monitoring System: Data Processing and Analysis Methods*. DST Group-GD-0952, Melbourne, Australia, Defence Science and Technology Group.
21. Rychlik, I. (1987) A new definition of the rainflow cycle counting method. *International Journal of Fatigue* **9** (2) 119-121.
22. Magoga, T., Brincat, M. and Dawes-Lynch, J. (2016) *Verification of Pressure Sensors and Strain Gauges Installed Onboard HMAS Maryborough*. DST-Group-TN-1576, Melbourne, Australia, Defence Science and Technology Group.
23. Magoga, T., Aksu, S., Cannon, S., Ojeda, R. and Thomas, G. (2017) Identification of slam events experienced by a high-speed craft. *Ocean Engineering* **140** 309-421.
24. Rosén, A. and Garme, K. (1999) Slamming Studies on High-Speed Planing Craft through Full-Scale Trials and Simulations. In: *International Conference on Fast Sea Transportation (FAST'99),* Seattle, United States of America
25. Paik, J., Lee, J., Shin, Y. and Wang, G. (2004) Design Principles and Criteria for Ship structures under Impact Pressure Loads Arising from Sloshing, Slamming and Green Seas. *Transactions - Society of Naval Architects and Marine Engineers* **112**
26. Amin, W., Davis, M. and Thomas, G. (2008) Evaluation of Finite Element Analysis as a Tool to Predict Sea Loads with the Aid of Trial Data. In: *International Conference on Innovations in High Speed Marine Vessels, HSMV 2008,* Naples, Italy
27. Rosén, A. (2005) Impact pressure distribution reconstruction from discrete point measurements. *International Shipbuilding Progress* **52** (1) 91-107.
28. *Actions and Action Effects*. (1999) Norwegian Standard N003, Norway, NORSOK.

Challenges in Naval Ship Maintenance

Arshad Kunjumon
Abu Dhabi Ship Building
arshadtk@gmail.com

SUMMARY

The challenges associated with the maintenance of a naval ship start right from the conceptual stage of the ship building up until the disposal of the ships. The selection of the most appropriate maintenance strategy for the life cycle of the ship plays a key role in optimizing the maintenance efforts and reducing the overall cost of the ship's maintenance.

Unlike the commercial ships, the range of expertise and skills required for maintaining a naval ship and equipment onboard is very large. The inherent complexities and the risks associated with the naval ship systems make the task of determining the best maintenance strategy for them very effortful. The enhanced operational demands of the naval forces due to the current political scenarios across the world make it difficult for the authorities to make the ship available for the maintenance in accordance with its upkeep plan. Another key factor which makes the naval ship maintenance quite challenging is the varied age of the ships in any leading naval force, which normally ranges from the newly commissioned ships to those which have completed more than 20 years in service.

Maintenance of a naval ship not only involves the design of an efficient Maintenance Management System but also the selection and implementation of the right combination of maintenance strategies. This paper is an attempt to identify the challenges associated with the maintenance of naval ships in order to negotiate them for optimizing the maintenance efforts.

CHARECTRISTICS OF NAVAL SHIPS

A naval ship is a military ship used by a navy. Naval ships are different from commercial ships by construction and purpose. Generally, the naval ships are damage resilient and armed with weapon systems, with the auxiliary and support vessels of the navy being an exception in some cases. Although an average naval ship is smaller as compared to an average commercial ship, a naval ship is quite complex by virtue of the platform and combat systems and equipment onboard, and therefore are more difficult to maintain.

A typical naval ship will have a very high equipment density and consists of a large number of sophisticated systems. The naval ships are normally constructed using high quality materials and are fitted with a large number of high-value, specially designed and customized sensitive equipment which are selected to deliver the envisaged functions and role of the vessel.

The naval ships are classified in to various types such as aircraft carrier, destroyers, frigates, amphibious, submarines etc. based on its design, function and characteristic features.

CONCEPTS AND TERMINOLOGIES

The various common traditional and contemporary concepts and terminologies prevalent in the naval ship maintenance domain and practiced by the world navies are as listed below.

Integrated Logistic Support

Integrated Logistic Support (ILS) is an integrated approach to the management of logistic disciplines in the military with a goal of creating systems that last longer and require less support. ILS is defined as a disciplined approach that influences the product design and develops the support solution to optimise the Life Cycle Cost. It encompasses the technical logistics elements to plan and develop the support requirements for a system. ILS addresses the supportability of the system not only for the acquisition, but also throughout the period of its operational life. [1][2]

Through Life Support

Through Life Support (TLS) involves the maintenance, upkeep and optimisation of the systems and equipment throughout its life cycle. The concept also covers the support required to provide the spares, upgrades, technical documentation, design interfaces, etc. throughout the operational life of the equipment.

Through Life Management

Through Life Management (TLM) involves the life-cycle management of the products, services, and activities required to deliver a fully integrated capability to the customer, while reducing the cost of ownership for the customer.

Through Life Capability Management

Through Life Capability Management (TLCM) is an approach to the acquisition and in-service management of military Capability in which every aspect of new and existing military Capability is planned and managed coherently across all Defence Lines of Development (DLOD) from cradle to grave. The TLCM concept has been evolved based on the partnering with the industry to achieve better outcomes and deliver defence capability, while providing better value for money and greater control of defence acquisition expenditures. [3][4]

Class Management

Class Management involves the integrated management of the TLS of a class of similar ships during the Operational Phase of the life cycle. It is a way to organize the operational part of TLCM for different classes of ships. Class Management brings together the support arrangements for vessels of similar size, complexity, operational roles or vessels with similar support requirements.

TYPES OF MAINTENANCE

A maintenance strategy defines the means by which an organisation sets out to preserve the condition of the ship and systems onboard throughout its service life. Maintenance activities will include the survey and inspection, upkeep, repair and replacement across the entire vessel incorporating equipment, systems and structure.

Typical maintenance strategies practiced in the naval ship maintenance domain are as follows. [5][6]

Corrective Maintenace

Corrective or Reactive maintenance, also known as Break Down Maintenance is based on the policy of 'Run Until It Breaks' or 'Run to Fail'. This is the most avoided kind of maintenance onboard naval ships as it poses a serious safety risk. In this case, no actions are carried out to maintain the equipment and it is expected that the equipment will fulfil its design capability for the duration of its designed life. Reactive maintenance may at times prove to be cost effective type of maintenance through the life cycle of equipment, provided the failure of the equipment does not lead to any major catastrophe. But considering the fact that most of the equipment onboard naval ships are critical, Reactive Maintenance is the least recommended mode of the maintenance of equipment onboard.

Preventive Maintenace

Preventive Maintenance or Scheduled Maintenance is the most common mode of maintenance followed onboard naval ships. In this system the maintenance is carried out based on the Preventive Maintenance Schedule (PMS) and it reduces the malfunctioning of the equipments and thereby reducing the downtime. Preventive maintenance is a calendar based or running hour based maintenance concept which employs various scheduled offline testing methodologies to conduct periodic assessment of the system and thereafter carry out the maintenance activities as prescribed in the schedule. The detailed procedure and schedule for the preventive maintenance routines such as change of filters, insulation tests, lubrication etc. are normally advised by the manufacturer and, if followed correctly, it ensures efficient running of the equipment resulting in reduced life cycle cost.

Predictive Maintenace

In the case of a Predictive Maintenance program the key values of the equipment are measured or monitored and the values obtained are compared with the standard set of values for the respective equipment to identify the deviations if any. The Predictive Maintenance method analyses the causes, symptoms and effect relationship of the equipment's performance to predict the requirement of any corrective actions. The causes of the deviations found are thereafter analysed to identify the maintenance routines to be carried out to normalize the equipment's performance parameters.

Condition Based Maintenace

Condition Based Maintenance (CBM) is carried out based on the knowledge of the condition of equipment obtained from routine or continuous monitoring. Condition Based Maintenance is similar to the Predictive Maintenance but uses both online and offline test data to decide up on the requirement of maintenance to be carried out. In the case of CBM, if the test results fall in the normal or acceptable range, the scheduled Preventive Maintenance may be skipped which shall reduce the maintenance effort and the associated costs.

Application of a customized CBM plans are now gaining wider acceptance in the shipping world with the classification societies coming up with the standards which permit and warrants the inclusion of preventive maintenance methodology in the overall maintenance program of the ships.

MAINTENANCE LEVELS AND ORGANISATION

Typical classification of maintenance levels and maintenance organisations in the naval ship maintenance domain are as follows.

Organisational Level

Organisational level maintenance involves limited pre-planned preventive maintenance, operational checks and adjustments capable of being undertaken by onboard operators utilizing ship held spares, test equipment and facilities. Pre-planned preventive and corrective maintenance and operational checks including diagnostics through Built-In-Test-Equipment (BITE), adjustments and repair by replace; which require onboard spares or tools and test equipment and undertaken onboard ship with the support of Base Maintainers are also considered to be a part of the Organisational level maintenance.

Intermediate Level

Intermediate level maintenance involves corrective, pre-planned preventive maintenance, operation checks, software maintenance, fault finding and repair of defects not reported by systems and adjustments which require either use of specialized base facilities, partial disassembly of equipment or specialist expertise, and base held test equipment and tools. This type of maintenance is normally performed by the specialists from Base Workshops.

Depot Level

Depot level maintenance involves major equipment overhauls and refit work including removal of the equipment to the workshops and the use of specialist personnel, Special Tools and Test Equipment and special facilities as required. This type of maintenance is normally performed by the Shipyards or Original Equipment Manufacturers (OEM) and involves complex maintenance functions.

TYPES OF MAINTENANCE WINDOWS

The upkeep cycle of a naval ship is normally determined as a part of the ILS activities during the engineering design stages of the shipbuilding. The upkeep cycle is designed to accommodate the desired operational profile of the ship and it identifies the windows of opportunity for the maintenance of the ship and equipment onboard. Typical types of maintenance windows that are catered for the upkeep of the naval ships are as follows.

Refits

Refits are extended periods of maintenance during which the ship and the equipment onboard are subjected to several major maintenance activities. Typical activities involve underwater area cleaning and preservation, overhaul, major repairs, additions and alterations etc. Refits are further classified in to various types such as Navigational refits, Major Refits, Mid Life Refits etc. based on the life of the ship and scope of the work involved.

Docking

The ships are normally docked during specified periods of its life cycle in order to undertake the cleaning and preservation of the underwater areas. Typically the dockings are combined with the refit periods in order to optimise the ships availability for operations.

Maintenance Periods

Assisted or Self Maintenance Periods are included in the ship's upkeep cycle to facilitate the conduct of maintenance activities during the operational phase of the ship. During such maintenance periods the ships are kept ready at short notice for the operational requirements. Organisational level maintenance is conducted during the Self Maintenance Periods where as both organisational and intermediate level maintenance activities are undertaken during the Assisted Maintenance Periods.

MAINTENANCE MANAGEMENT SYSTEMS

The maintenance of assets in the naval domain is normally managed using Computerized Maintenance Management Systems (CMMS) or Enterprise Asset Management (EAM) software solutions. CMMS is a software package that maintains a computer database of the information about an organization's maintenance operations. CMMS software solutions used by the navies across the world are SAP EAM-PM, IBM Maximo, InforEAM etc. Use of EAM/CMMS helps to improve the efficiency of maintenance management by integrating the various elements of maintenance such as configuration, technical data and documentation, maintenance plans, maintenance cards, spares information etc. and thereby facilitating the improvement of responsiveness, reliability, quality and safety in maintenance management. [7]

CHALLENGES IN NAVAL SHIP MAINTENANCE

Although most of the challenges associated with the naval ship maintenance surfaces during the operational phase of the ship's life cycle, it can be seen that the cause of many of these challenges are rooted to the life cycle support actions which happens during the conceptual, engineering and construction phases of ship building. It is therefore necessary to carefully plan the life cycle maintenance of the naval ship systems and equipment onboard during the acquisition. Some of the most prominent challenges, or areas and causes which are associated with the emergence of the challenges in the maintenance of the naval ships, are as detailed in the succeeding paragraphs.

ILS Maturity

The timely development and delivery of the ILS elements and their maturity before the commencement of the operational phase of the ship's life cycle is critical for ensuring an effective maintenance program. The Maintenance Planning element of ILS begins early in the acquisition process with development of the maintenance concept. Maintenance Planning is conducted to evolve and establish requirements and tasks to be accomplished for achieving, restoring, and maintaining the operational capability for the life of the system. The Maintenance Schedules, Job Information Cards, Technical Data and Documentation, Spares Catalogue, Onboard and Base Spare Establishment Lists, Special Tools and Test Equipment (STTE), etc. are key ILS elements which are required to support the establishment of an efficient maintenance program for the ship systems and equipment onboard. The availability, accuracy and the maturity of the ILS elements pertaining to the class of ships determines the quality of the Maintenance Management and the availability of the ship systems and equipment.

Quality of Maintenance Plan

The Maintenance Plan is produced before the operational phase of the ship's life cycle as a part of the ILS deliverables. A typical Maintenance Plan will contain the details of the maintenance activity to be carried out for the system/equipment, the periodicity, the spares and consumables required to conduct the maintenance, the activities, tasks and operations to be carried out etc. The Maintenance Plan is normally produced based on the recommendations of the OEM of the equipment and standard engineering practices. The availability of a detailed Maintenance Plan is a key factor for the implementation of an efficient maintenance regime for the ship systems and equipment onboard.

Complexity of Systems and Equipment

The naval ships contain a number of complex platform and combat systems equipment. The maintenance of these systems would require experienced and highly qualified technicians as well as advanced special tools and test equipments. For example, the ships with nuclear and electrical propulsion needs specifically trained specialists to maintain and repair the highly sophisticated systems. Same is the case with the maintenance of the advanced navigation and weapon systems which are key elements in the frontline ships of the world navies.

Choice of Maintenance Strategy

The choice of the right maintenance strategy is normally a trade-off between the various types of maintenance such as Preventive, Corrective, Predictive and Condition Based Maintenance. A typical trade-off pattern between Prevention and Repair costs is as given in Figure 1. [8]

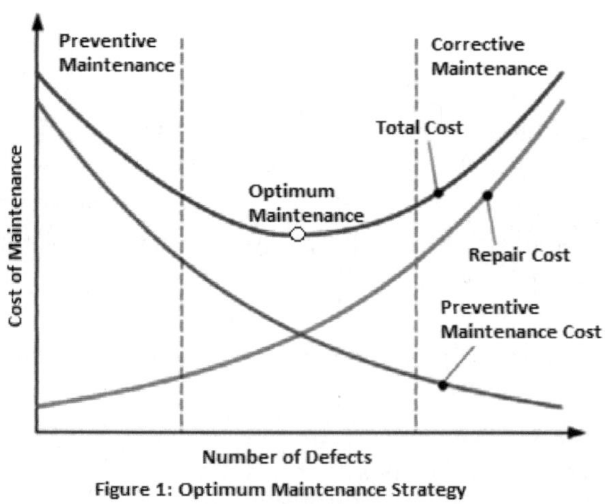

Figure 1: Optimum Maintenance Strategy

However, the choice of the maintenance strategy is not always based on the associated cost factor. There are other factors which influence the selection of the maintenance strategy of systems and equipment onboard a naval ship. For example, in the case of a very critical system, the failure of which may have a severe impact on the safety of the ship, Preventive Maintenance would be the preferred strategy whereas in the case of a non-critical system the preferred maintenance strategy would be based on reactive or run-to-fail. As the selection of the maintenance strategy depends up on several factors and constraints, it is considered as one of the most challenging tasks associated with the maintenance of naval ships.

Adherence to Upkeep Cycle

The upkeep cycle of a naval ship is normally determined during the engineering design and construction phase of the ship building. The upkeep cycle of a ship specifies the periods of operation and maintenance as well as the type of maintenance windows. The demand for increasing the operational availability of the naval ships is at its peak due to the increasing mission commitments across the world. This has forced the naval authorities to increase operational periods and revising the maintenance periods as against the planned upkeep cycle leading to the reduction of preventive maintenance activities recommended for the systems and equipment onboard.

Non availability of the ship for the planned maintenance periods is a major challenge in managing the maintenance of naval ships as it not only leads to backlog of a large number of maintenance work orders but also the deterioration in the performance and productive life span of the systems and equipment onboard.

Data Management & Analysis

The life cycle management of the assets involves the management of large amount of data pertaining to their maintenance and upkeep. This includes the data corresponding to the preventive and corrective maintenance history, configuration changes, spare parts information etc. A systematic recording and utilisation of this data would facilitate the accurate analysis of the asset behaviour and in determining the corrective actions to improve the performance of the asset. For example the failure history of equipment may form the basis for analysing the engineering operational issues associated with it and identifying the requirement to initiate Engineering Change Proposals or modifications in the operation procedure. As the volume of such data is very large, it's timely recording, logical storage and efficient utilisation for analytic purposes is one of the major challenges associated with the maintenance of the naval ships.

Excessive Operation of Equipment

The increase in the operational commitments of the naval ships across the world has resulted in the utilisation of the naval assets beyond their designed profile. The Annual Operating Hours (AOH) specified during the acquisition is one of the major factors for the engineering design and the life cycle logistics planning of the naval ship systems and equipment onboard. The usage of these systems and equipment beyond the design profile leads to the requirement of enhanced maintenance and supply support which is a challenging constraint in the maintenance management of naval ships.

Obsolescence Management

The construction of a naval ship is a very complex evolution and it may take one to several years depending up on the design maturity of the class of the ship being constructed as well as the capability of the yard constructing the ship. Normally the equipment to be outfitted on the ship is decided and ordered during the design stage of the ship building based on the functional and product requirements, and the equipment chosen is more often than not the best and latest models of its type available in the market. However as the construction and delivery of the ship gets prolonged, especially in the case of the First of Class, the technology used on these equipment becomes outdated or it so happens that a better version of the equipment is available in the market with more advanced features. While such issues are contractually negotiated to some extend especially in terms of the software solutions, the negotiation of such challenges arising out of the rapid technological advances in terms of the designed hardware becomes quite challenging. This leads to the early emergence of obsolescence issues associated with the equipment and its spares making the maintenance management of such equipment extremely difficult.

Shortage of Maintenance Funds

The exponential cost of building and maintaining a complex naval ship coupled with the global economic slowdown has constrained even the most powerful governments from venturing in to the programs for fleet expansion and upgrade of the ship systems at a rate proportional to the existing demand and advancement in technology. The non-availability of sufficient funds for enhancing the maintenance capabilities and upgrade of the systems and equipment onboard poses a major challenge in providing efficient maintenance and logistic support for the naval ships. This in turn has resulted in increase in the number of breakdown of the systems and equipment onboard, leading to the reduction in the availability of the naval ship.

Inefficient Management of Maintenance

The effectiveness of the maintenance depends on the manner in which the maintenance is managed. Availability of a very advanced maintenance management system and highly skilled technicians is not a guarantee for achieving the best results in the maintenance of a naval ship if the maintenance management team is not efficient. It is therefore necessary to ensure that the suitably qualified and experienced personnel are positioned at the right places in the maintenance management organisation to overcome this challenge in the maintenance of naval ships.

Availability of Suitably Trained Personnel

The technology behind many of the systems onboard a modern naval ship is normally very advanced and state-of-the-art. As most of the electrical, mechanical, hydraulic and combat systems are automated and interfaced to the Integrated Platform Management Systems (IPMS) or Combat Management Systems (CMS), it becomes challenging for the general technicians to effectively conduct the maintenance of such systems. It is desirable for the technicians maintaining such equipment to have the knowledge of these advanced automation systems at least from the operational and troubleshooting perspective and therefore it should be ensured that the technical team responsible to maintain such systems are suitably trained to undertake the maintenance of these systems in an efficient manner.

OEM Induced Restrictions

Maintenance activities such as depot level maintenance and those requiring advanced troubleshooting skills are normally performed by the OEM representatives due to commercial and technical reasons. However this arrangement becomes challenging at times when the maintenance support is required at short notice and there is no local OEM facility in the region. Moreover due to the complex political equations existing between various countries, it becomes almost impossible for an OEM to get the necessary security clearance from the authorities to provide emergency maintenance support for their systems, especially when the particular naval ship is operating in international waters.

Range of Technology

The current fleet of most of the world navies consist of ships which are newly inducted with state-of-the-art systems as well as legacy class ships which were designed and built decades ago and fitted with equipment based on older technology. Although the upgrade of the naval ship systems and equipment during their major and midlife refits bridges this gap in technology to some extent, the financial constraints make it almost impossible to upgrade all the systems and equipment to the latest technology. This brings to the forefront the challenging requirement to a have a logistics support, support facilities and a maintenance work force with experience in maintaining equipment spread across a wide range of technology.

Configuration Management

The management of configuration of the ships in service is one of the most challenging tasks associated with the naval ship maintenance. During the operational phase of the ships life cycle, there will be requirements to have emergency solutions to the maintenance issues, especially during the breakdown of systems and equipment, to ensure the readiness of the ship for its assigned missions. In such situations, one of the most common approaches taken by the naval ship maintenance authorities is the cannibalisation, wherein the equipment as a whole or in part is removed from one ship and fitted on to another ship. Equipment are also transferred from one ship entering a long refit period to another one which is in operation for optimising the maintenance period. In order to ensure the integrity of the maintenance data, it is necessary to ensure that the configuration information is updated accordingly in the Maintenance Management System and all required processes and procedures are followed for such equipment or part transfers.

Non-availability of Adequate Spares

The analysis to determine the spares required to be stocked onboard and at the bases for the maintenance of the systems and equipment fitted onboard a naval ship is normally carried out as a part of the ILS engineering during the engineering design and construction phase of ship building. Such analysis are carried out by ILS specialists using advance software such as RAM commander and based on the maintenance related specifications and information such as AOH, Mean Time Between Failures (MTBF) etc. provided by the ship designer, OEM and suppliers. Ideally these spares are to be procured and stocked as defined and should be available for maintenance when required. However one of the main challenges which is faced by the maintenance organisation is the non-availability of spares required for the maintenance at the time of the requirement. This is partially due to the inaccuracy of the spares recommendation and more often than not the failure to reorder and stock the recommended spares by the logistics team.

Uncontrolled Engineering Changes

While the engineering changes are strictly controlled through the Engineering Change Proposals (ECP), Change Requests (CR) etc. during the acquisition phase of ship's life cycle, there is a general tendency to undertake engineering changes as a part of the maintenance by the maintenance team during the operational period in order to find urgent solutions to the maintenance problems under emergency situations. Such uncontrolled changes being made in the system or equipment not only leads to the corruption of the configuration data but also poses serious safety risk as no proper study regarding the safety and maintainability is conducted prior to making such changes.

Engineering Design Issues

The design of a ship and the systems onboard are carried out during the acquisition phase of ship's life cycle based on the functional specifications provided by the customer. The maintenance and operational issues due to the flaws in the design and design review process which may surface during the operational phase spawn a big challenge in ensuring the operational availability and maintainability of the ship or the particular system and associated equipment.

Contractual Constraints

Strict contractual constraints imposed by the supplier or OEM may lead to maintenance challenges on some occasions. For example if an equipment is under the warranty period, the supplier or OEM contract prevents the ship's maintenance team in attending to a defect or break down of the equipment until the their representative arrives. While this possible to be practiced for a commercial vessel it becomes almost impossible for a warship as it may not be possible for the OEM representatives to board the ship for providing warranty support while the ship is on a mission and operating in hostile conditions. It is also necessary for all the parties involved to understand the contractual scope, constraints, and the terms and conditions of the contract to facilitate the smooth execution of the maintenance functions.

Stakeholder Relationships and Organisational Synergy

One of the key factors which determine the success of a maintenance organisation and its functioning is the existence of good working relationship and trust among the stakeholders and synergy within the organisation. The most important stakeholders in the maintenance of the naval ship and systems onboard are the ship's crew, base maintainers, base workshops, warehouse and inventory team, the OEMs, suppliers and strategic partners etc. The coordination between stakeholders becomes even more challenging if the maintenance responsibility rests with a contracted non-uniformed organisation. Good working relationship and trust among the stakeholders shall help in the easy negotiation of maintenance challenges which arise due to the highly demanding working conditions and operational requirements. There is also a necessity to have better coordination and synergy with in the various departments of the organisation to promote team work and cooperation for shared success. Each stakeholder is required to carry out their responsibility in a timely and professional manner, especially with respect to the adherence to the processes and procedures associated with the Maintenance Management System.

Poor Communication

Existence of clear and transparent communication between the stakeholders is a must for the success of any program. Poor communication between the stakeholders in the maintenance organisation leads to confusion and undue delay in the completion of maintenance activities, especially in activities which involves the subcontractors. The secrecy and security aspects associated with the operations of a naval ship restricts the transparency in communicating the required information regarding the location of the ship, details of maintenance requirements and the availability of maintenance windows during the operations. However this challenge could be overcome by the implementation of necessary communication procedures and processes, including a clear communication matrix, for optimum communication associated with the maintenance functions, without compromising on the secrecy and security aspects.

Management of Subcontractors

Maintenance of naval ships involves management of several contractors and subcontracts. In the case of a modern naval ship the systems and equipment are sourced from OEMs and suppliers spread across the world. Each contract may have some unique terms and conditions which need to be managed with utmost vigilance. Moreover the support and approach of the subcontractors varies from one subcontractor to another which makes the management of subcontractors a challenging task. It is prudent to establish long term framework agreement with the key suppliers and OEMs for more effective and prompt service and supply support for the systems and equipment supplied by them.

Unlimited Expectations and Gold Plating

Unlimited expectation of the client in a maintenance contract leads to several issues with respect to the contract management. In the case of naval ship maintenance where a non-uniformed organisation is contracted for maintenance support, the contractor employees are sometimes compelled to deliver more than the contracted services to meet the support demands of the ship's crew or client representatives. It becomes quite challenging for the contractor to manage such expectations, especially when the necessary contractual formalities and documentations are not followed and maintained. The contractor obliges to such demands of the client to maintain the working relationship with the client at the cost of extra financial losses and contract management issues.

On the contrary the contractor also sometimes goes out of the way by to deliver service and supply support to the client beyond contractual commitments – also known as gold plating- with an intention to build better relations with the client and secure more business with the client organisation.

Administrative Issues

More often than not the maintenance of naval ships and military equipment are conducted at highly secure areas requiring special permissions and access rights. The issues pertaining to the security clearances and provision of access to the maintenance team to the ship for carrying out the maintenance is a major challenge associated with the naval ship maintenance. This becomes more complex when the maintenance team includes personnel from various nationalities and backgrounds. Inefficient management of this aspect associated with the naval ship maintenance may lead to delay in providing the access to the equipment for maintenance thus resulting in loss of man-hours and decreased availability of the ship.

CONCLUSION

All the challenges mentioned above may not be applicable for the maintenance programs of all the navies as the maintenance management in each part of the world depends on the work culture, professionalism, organizational maturity and contractual terms and conditions of the respective programs. Further, it may be noted that only selected few of the most prominent challenges associated with the maintenance of naval ships are as detailed in the above paragraphs. Negotiating these challenges needs the involvement and application of strategic and technical management skills. This paper has been prepared based on the standard maintenance management practices and the practical challenges identified during the delivery of Through Life Support programs.

REFERENCES

1. Arshad Thaikoottam Kunjumon – Life Cycle Management Matrix – Optimising Life Cycle Maintenance of Naval Ships.
2. Ministry of Defence - UK, 'Integrated Logistics Support for MOD Projects', Web-British Defence Standard 00-600, dated 11 November 2011.
3. Jacques S. Gansler, William Lucyshyn, and Lisa H. Harrington – An Analysis of Through Life Support – capability Management at UK's MoD.
4. The People in Systems TLCM Hand Book - "A guide to the consideration of People Factors within Through Life Capability Management".
5. Operation and Maintenance Best Practices Guide Release – Chapter 5 Types of Maintenance Programs.
6. NA Tomlinson, IEng MIET MIMarEST, BMT Defence Services, UK - What is the ideal maintenance strategy? A look at both MoD and commercial shipping best practice.
7. Christopher Wenz, 'Maintenance Life Cycle Planning – An Introduction'.
8. An Roinn Cosanta, Óglaigh na hÉireann, Defence Forces – Naval Service Vessel Maintenance.

ACKNOWLEDGEMENTS

The author gratefully acknowledges the support of the Management of Abu Dhabi Ship Building for their guidance and encouragement towards the participation in the PACIFIC 2017 International Maritime Conference.

AUTHOR BIOGRAPHY

Arshad Kunjumon holds the current position of the Head of TLS Operations at Abu Dhabi Ship Building, UAE. He is responsible for the Management of the Maintenance and the development of Through Life Support solutions for the Naval Ships. His previous experience includes tenure as Project Manager with Converteam Power Conversion and as an Officer with the Indian Navy.

3D LASER SCANNING APPLICATIONS FOR SHIPBUILDING AND THROUGHLIFE SUPPORT

Mr. Chris Deegan
Mr. Joe Daley
Mr. Keith Harper
Mr. Tom Schubert
Gibbs & Cox, Inc.
cdeegan@gibbscox.com

ABSTRACT

3D Laser Scanning is now affordable for broad application in engineering and design. Software tools have increased the speed and reduced the effort required to integrate multiple scans into a single point cloud file, as well as integrate point cloud data and associated photographic images into 3D CAD models. This paper discusses these 3D scanning topics and their application to new construction and in-service naval and commercial ship programs. The authors use recent work on the SS UNITED STATES as a case study to illustrate the application of 3D laser scanning in an engineering and design environment.

Gibbs & Cox, Inc. designed the SS UNITED STATES, one of the last great "ships of state", for high-speed transatlantic luxury service. SS UNITED STATES is the largest ocean liner ever constructed entirely in the United States, and could accommodate nearly 2000 passengers. The U.S. Government financed a portion of the project, and the ship was designed so that it could be rapidly converted to a troop transport that could carry 15,000 troops 10,000 miles without refuelling. Put into service in 1952, she set the record for fastest transatlantic crossing (east and westbound) on her maiden voyage. While other ships in have since broken the eastbound crossing, SS UNITED STATES still holds the Blue Riband record for the fastest westbound crossing for a liner in passenger service, achieving an average speed of 34.51 knots over the 3 day, 12 hour, and 12 minute crossing. Competition from increasingly cost effective and reliable transatlantic air travel resulted in the retirement of the UNITED STATES from service in 1969. Over the past two years, Gibbs & Cox, Inc. has worked with the SS UNITED STATES Conservancy (the current owner of the ship) and several developers interested in future projects centred around the ship. The use of 3D laser scanning has been invaluable in capturing digital representations of the ship to support and inform these feasibility and conversion studies.

INTRODUCTION

Ships have been designed using computer-aided drafting techniques and tools since the 1980s. Digital scanning capabilities, initially developed for large-scale surveying and mapping applications, were applied sparingly to engineering and construction projects (including shipbuilding and repair) because of the high cost of the scanning equipment. In the last few years, the prices of the scanning hardware and associated software have decreased dramatically, making it much more cost-effective to employ scanning technology on construction and shipbuilding projects. This paper provides a brief overview of current scanning technologies applicable to shipbuilding and ship repair projects. The authors describe the tools and techniques used to integrate scan data with other computer-based engineering and design data, and highlight potential applications of scanning technology in new construction, in-service and modernization projects. The authors use their recent experience on the SS UNITED STATES as a case study to demonstrate potential applications of scanning in a variety of shipbuilding project applications.

3D LASER SCANNING EQUIPMENT

Equipment for 3D laser scanning is available in three general configurations: tripod-mounted scanners, hand-held scanners, and Unmanned Aerial Vehicle (UAV) mounted scanners. Each type has its distinct advantages and disadvantages, and is best suited for particular types of applications. Tripod-mounted scanners are capable of providing extremely high resolution/accuracy (3mm at 40m) over a large area, and are therefore well suited to scanning large, complex areas. In scanning a ship compartment (or the exterior of a hull), multiple scans are taken from around the area to ensure that a full 360 degree view has been captured. Hand-held scanners, as their name implies, are extremely convenient and portable, and able to provide good resolution over smaller areas. UAV mounted scanners are valuable for capturing imagery to support condition assessment of large tanks or cargo holds, reducing or eliminating the need for staging. For capturing geometry to support design, we typically employ tripod-mounted scanners, using hand-held equipment for either adding localized detail or in regions where there is insufficient access for a tripod-mounted unit. Typical output data generated during the scanning process includes raw scan data (3D point cloud files with correlated images (photographs) from each scan location. The scan data files are then downloaded from the scanner to a personal computer and post-processed using software tools (typically per the scanner manufacturer's recommendation) to make them accessible and more useful in the overall design process, as described in the following paragraphs. A key element in successfully post-processing scan data files is a computer with sufficient processing, memory and graphics capability. Our experience has shown that a computer with an 8-core processor, 64 GB of RAM, a 4 GB video card, and a 512 GB solid-state drive, running Microsoft Windows 7 or higher operating system is ideal for processing and registering data.

SCANNING SOFTWARE TOOLS AND INTEGRATION WITH DESIGN TOOLS

Scanning software tools used with 3D scan data generally fall into three categories: image processing tools, CAD interface tools, and visualization tools. Image processing tools (such as SCENE by Faro) combine multiple scan data files into an integrated, geometrically consistent and accurate digital representation of a large area. By identifying common points in each scan file, those files can be "registered" or joined together into synchronized point cloud files. Photographic images can then be overlaid in the registered file to provide a cleaner view of the associated area. For our applications, we typically would register scans of a compartment or deck together into a single file. From the image processing tool, the registered files can then be exported into formats that allow the user to integrate the scan data with CAD information in a design tool, or read and interact with the scan data through a visualization tool. It is important to note that software for digital imaging and design integration is evolving rapidly – with new tools frequently released. The tools noted within this paper are examples of tools used by Gibbs & Cox recently, and are representative of the full suite of tools that are available.

Visualization tools (such as Navisworks Freedom by Autodesk) provide inexpensive, highly accessible visualization capabilities that do not require the user to have the supporting CAD modeling or image processing software. These tools are extremely useful in conducting 3D design reviews with clients, as the files are compact enough to be easily viewed in a web meeting environment. Navisworks can also be used to allow the user to "fly through" the point cloud data – a powerful capability for developing demonstrations and promotional materials. Other software, such as Clear Edge's Verity, takes that a step further and allows the user to compare point cloud data against a production 3D CAD model to identify differences between the design and the actual construction. Intergraph and other software manufacturers make similar tools that allow for querying and visualization of scan data with design models.

In addition, there are numerous tools on the market that interface with standard CAD tools (such as CloudWorx by Leica Geosystems) to allow these CAD applications (AutoCAD, Rhino, etc.) to read and use scan data files. These interface applications allow a designer to directly model "over" scan data in 2D or 3D – using the commands and techniques he or she is accustomed to in the native CAD tool. Other tools (such as ClearEdge-Edgewise, Faro Pointsense, Intergraph S3D) are able to recognize collections of points as components. Once the association is made between the point field and a component definition, the software then can identify all of the instances of similar collections of points and classify them as the same components. This transforms the scan data into a useful design model – which can then be updated as the design changes in the future.

APPLICATIONS FOR SCANNING IN NEW CONSTRUCTION PROGRAMS

Many applications of 3D scanning to new construction programs involve the comparison of the as-constructed ship (or system) with the design model. For example, 3D scanning can be used to help resolve interferences between systems that are identified during construction. By scanning the area in question (using a hand-held scanner for convenience), the as-built configurations of the systems can be directly compared against the 3D model, and any dimensional or other installation differences quickly identified. The waterfront engineer can then determine the best course of action to clear the interference, implement the change in the design model and extract new production information for the production trades to accomplish the necessary changes. In a larger sense, 3D scanning can be highly effective in identifying block and module misalignments before unit erection. Being able to identify and resolve any alignment issues prior to executing the lifts can be a tremendous cost-saver for the shipbuilder.

During construction, 3D scanning can be used by the owner or the shipbuilder as a means of objectively documenting construction progress, in support of compartment closeout or earned value reporting. Similarly, during inclining of the ship, a digital scan can be an affordable way to document the exact condition of the ship at the time of the experiment that can easily be compared to the design model to identify systems and equipment that are not yet installed or construction tools and material that are aboard, but not part of ship's configuration. 3D scanning results can reduce the effort required for preparation of the final inclining test report and associated weight and stability reports, by providing CAD tools to determine the volume and geometric center of construction material left aboard at the time of inclining.

Once the ship is complete, 3D scanning can be used to precisely and affordably document the as-delivered configuration. This digital record is useful in many ways – from a contractual evidence perspective, as well as in planning for future modifications and upgrades. Another potentially useful application for 3D scanning data of the completed ship would be for training and situational awareness, particularly for first responders in the event of an on-board emergency. Being able to quickly visualize access routes to the areas of concern is essential in minimizing the response time to a casualty.

An operational enhancement for ships at delivery is to combine engineering and technical data for ship systems and equipment into the 3D model. Examples include ship's information books, equipment and system technical manuals, troubleshooting guides, maintenance procedures and records, and spare parts information – all of these digital publications can be linked to the equipment in the model – so that a simple click on the image of the equipment sends you to the associated engineering and technical data. The data can be accessed from the ship's computer network, or even downloaded to a tablet so that the operators and maintainers have access to the information right where they need it for diagnosing issues or performing routine maintenance.

For commercial passenger ships, the 3D scan with images overlaid can be used to generate high resolution virtual "fly throughs" of the ship, to be used in promotional materials by the owner/operator.

APPLICATIONS FOR SCANNING IN IN-SERVICE OR MODERNIZATION PROGRAMS

All of the applications identified above for new construction ships can also be directly applied during maintenance or modernization activities for in-service ships. In addition, scanned data of the ship's existing configuration can be an important input to the maintenance or modernization planning process. Rather than sending large teams of personnel to "walk the ship" with tape measures and red pencils to mark up the selected record drawings by hand to create a configuration baseline to develop the upgrade or maintenance design packages from, a relatively small team can scan the entire ship, accurately recording every dimension the first time. The scan data can then be integrated into the design model as described above, allowing seamless, accurate adoption of the survey data without requiring any manual data recording or translating of dimensional and arrangement information. For complex changes supported by a manual shipcheck process, it is frequently necessary to conduct two shipchecks: and initial shipcheck to collect the data to support developing the alteration and a validation shipcheck to confirm that the alteration as designed will work. Because the data collected in a laser scan shipcheck is more accurate and comprehensive, it is often possible to verify the alteration design in the model, eliminating the need for the second shipcheck. Finally, because of the cost of conducting a manual shipcheck, the team will almost always capture only the data need to support the immediate change under development. By its nature, a 3D laser scan collects complete data on the regions scanned making it quite possible that the scan files can be used to support the next change affecting the same area of the ship.

At a more detailed system/equipment level, scan data can be used to visualize the arrangement and fit-up of new or replacement equipment. After deleting the removed equipment from the scan data, equipment models from the original equipment manufacturer (OEM) can be overlaid in the scan data to quickly identify any interferences with structure or distributed systems and verify the fit before the development of the ripout and installation packages – thereby ensuring that there will be no surprises during the execution of the industrial work.

Another potential application of scanning technology to in-service ships is to periodically scan the exterior of the ship, and compare the data to the as-delivered data. This simple digital comparison process will clearly identify any changes to or deformation of structure over time – which can then be correlated with more advanced engineering analysis tools to understand the failure mechanisms and predict structural health of the ship.

A special, and particularly important case of this would be battle damage assessment, where a highly accurate scan of the damaged condition would not only facilitate rapid review and repair planning at headquarters (regardless of where the ship was damaged), but would also be valuable for validating weapons effects models.

The most important and valuable aspect of scanning for in-service ships (or facilities, for that matter) is the fact that the resulting digital data is an exact replica of the ship's configuration – whether the ship was originally designed using CAD or not. This creates a significant opportunity for an owner or ship manager. For older ships that were not originally designed using CAD, scanning can be used to capture the current configuration, and support the design of future upgrades or enhancements using today's CAD tools. The same would apply for ships that were originally designed using 3D CAD, but the original design model is either not available to the owner, or has not been kept up to date over the ship's life. Similar advantages exist for new construction ships: If the ship is designed using a 3D product model, that model can be maintained throughout the ship's life; or, the ship can be scanned at delivery and then any future modifications or upgrades designed within the original design tool, or in any other design tool. The ability to change design tools perhaps multiple times over the ship's life can be an important economic advantage – as technology and tools change over the long life of a ship, the owner or manager is not forced to continue to use a particular tool. Rather, the best design tool for the job at hand can be used by the designer – whether that is the original design tool or something else.

CASE STUDY: SCANNING APPLICATIONS FOR SS UNITED STATES REDEVELOPMENT

Gibbs & Cox, Inc. designed the SS UNITED STATES, one of the last great "ships of state", for high-speed transatlantic luxury service. SS UNITED STATES is the largest ocean liner ever constructed entirely in the United States, and could accommodate nearly 2000 passengers. The U.S. Government financed a portion of the project, and the ship was designed so that it could be rapidly converted to a troop transport that could carry 15,000 troops 10,000 miles without refuelling. Put into service in 1952, she set the record for fastest transatlantic crossing (east and westbound) on her maiden voyage. While other ships in have since broken the eastbound crossing, SS UNITED STATES still holds the Blue Riband record for the fastest westbound crossing for a liner in passenger service, achieving an average speed of 34.51 knots over the 3 day, 12 hour, and 12 minute crossing. Competition from increasingly cost effective and reliable transatlantic air travel resulted in the retirement of the UNITED STATES from service in 1969.

To provide a sense of scale of the project, the table below highlights the principal characteristics of the SS UNITED STATES.

Table 1. SS UNITED STATES Principal Characteristics

Principal Characteristics:	SS UNITED STATES[1]
Displacement:	46,129 MT
Length Overall	302m
Beam (max)	30.9m
Draft (design)	9.53m
Depth	23m
Propulsion:	4 screw steam turbine propulsion, 8 boilers, 180 MW total installed power
Passengers:	1928
Crew	900

Recently, Gibbs & Cox supported the SS UNITED STATES Conservancy and two potential redevelopers of the ship by scanning the full interior and exterior of the ship. These scans, resulting in over 1.3 TB of high-resolution image data, have been invaluable during the feasibility stages of the redevelopment projects. In addition to providing a highly accurate, easily accessible dataset for developers to base their studies on, the images have been valuable to the Conservancy in giving news media and potential donors a view of exactly what the ship looks like today.

The figures below demonstrate the application of scanning technology and the capabilities related to CAD model interface and visualization using one area on the SS UNITED STATES – a view of structure in the Ballroom on the Promenade Deck.

Figure 1 shows a single scan of the structure. Figure 2 shows the result of registering multiple scans of the same area, resulting in a more complete view. Figure 3 shows the same registered point cloud that has been colorized based on the overlaid photographs. This view is the best representation of the existing condition developed using the scanning software and is ready for integration with CAD models or visualization tools.

For CAD model integration, Figure 4 (in the lower right corner) shows structure that has been modeled in AutoCAD "over" the point cloud data. Figure 5 shows the modeled structure without the point cloud data. From this point, additional details or design changes could be easily incorporated by a designer using native CAD software.

For visualization tools, Figure 6 shows a registered point cloud with overlaid photos that has been converted to a webshare using SCENE from FARO – for local visualization or

[1] Downloaded from https://en.wikipedia.org/wiki/SS_United_States on June 18, 2017.

sharing with users in remote locations, including customers and partners. Figure 7 shows the same webshare, with dimensions and annotations added. A fly-through video, while incompatible with the format of this paper, would have visual characteristics similar to a webshare.

Each of the development projects used the scan data and related data files to develop their own visualizations of the post-conversion ship – a powerful visual tool in presenting the concepts to potential investors.

In one case, the developer used a model based on scan data to track equipment or systems installed on the ship that would require environmental remediation. In the model, each component that required remediation was assigned an attribute (using the annotation process shown above) that described the hazard, the remediation required, and provided a place to track the completion of each step in the remediation process. This provided an effective visual roadmap of the work required to bring the ship into environmental compliance and the progress achieved at any given point in the process.

In addition to the work by the potential developers, the Conservancy plans to use the scan data to support the development of future exhibits and educational materials about the ship. A fly-through of the Promenade Deck was also used by the Shipbuilders' Council of America and America's Natural Gas Alliance in a television commercial about the shipbuilding and repair industry in the United States.

Figures 8 and 9 are examples of colorized scans of large areas of the ship, to give a sense of perspective of the clarity of the images available. Figure 10, taken in the Aft Generator Room, again is a single scan – but this time taken in zero light conditions. Note the clarity of the scan data despite the challenging environment. This highlights the tremendous capability that scanning can offer in collecting accurate information that otherwise would be impossible to capture manually. Similarly, Figure 11, taken on D deck, illustrates the ability of scan technology to accurately capture complex geometry and dense system and equipment layouts.

SUMMARY

3D Laser scanning is affordable for application on a variety of shipbuilding projects – from new construction through in-service maintenance and modernization. The scanning process is far more efficient, accurate and fast than a traditional ship survey, with far less opportunity for human error in recording dimensions. This is particularly significant for ships that are stationed far from the "home office". The use of scanning provides an opportunity to create digital design information for ships that were not originally designed using CAD, or for ships where the original design models are not available or not current. The scan data can be integrated with other CAD and digital engineering information, and can also be converted to generate high resolution images and videos for visualization, walk-throughs, and demonstrations. Gibbs & Cox has demonstrated the utility of scanning in a number of shipbuilding for new construction and in-service shipbuilding programs, including our support to the SS UNITED STATES Conservancy.

Figure 1: Raw scan data from a single scan.

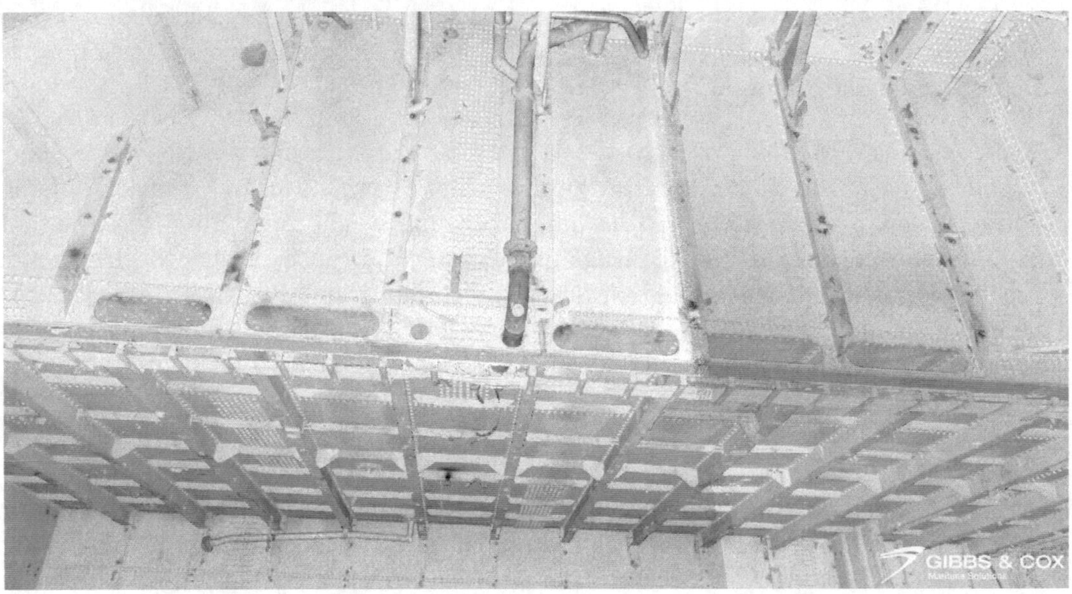

Figure 2: Multiple scans registered to a single point cloud

Figure 3: Registered scans point cloud data colorized

Figure 4: Structure modelled over point cloud data (point cloud overlay)

Figure 5: Structure modelled from point cloud only

Figure 6: Webshare without dimensions

Figure 7: Webshare with dimensions and annotations added

Figure 8: SS UNITED STATES, colorized registered scan data, bow looking forward

Figure 9: SS UNITED STATES, colorized registered scan data, bow looking aft

Figure 10: SS UNITED STATES, Aft Diesel Generator Room, registered scan data taken in zero light conditions

Figure 11: SS UNITED STATES, D Deck, registered scan data taken in zero light conditions

High Entropy Brasses & Bronzes – New Marine Alloys

Kevin J Laws, School of Materials Science & Engineering, UNSW Australia,
k.laws@unsw.edu.au
Warren McKenzie, Advanced Alloy Holdings,
Warren.McKenzie@advancedalloyholdings.com

Research into high entropy alloys is an emerging field of metallurgy that has discovered many new alloy systems with unique properties relative to commercially available alloys of a similar composition.

We present a range of new marine brasses and bronzes containing copper, nickel, iron, manganese, aluminum and tin in various quantities that exhibit superior, mechanical, corrosion and wear resistance properties than current state-of-the art marine alloys.

Typical marine brasses, bronzes and copper-nickels consist primarily of a copper-rich base with alloying additions of nickel (copper-nickels), zinc (brasses), tin, silicon or aluminium (bronzes). High entropy brasses and bronzes utilise the broad mutual solid solubility of the copper-nickel-manganese alloys system which makes for a high performance base alloy with properties comparable to current heavy copper-nickel marine alloys at a considerably lower price (due to the very low cost of manganese). Here we present the vast range of properties of these novel new materials in marine environments, and discuss their applications in ships, submarines and for the offshore resource industry.

Low distortion welding processes for naval shipbuilding

Bruno Levieil[1], University of Wollongong, brunol@uow.edu.au
Stephen van Duin[1], University of Wollongong, svanduin@uow.edu.au
François Cortial[1], Naval Group, francois.cortial@naval-group.com
Lyndon Edwards[1], ANSTO, lyndon.edwards@ansto.gov.au
Zoran Sterjovski[1], DST, zoran.sterjovski@dsto.defence.gov.au
[1]Defence Material Technology Centre

The drive to minimise ship weight, and hence reduce fuel consumption, has stimulated research into the effects of reducing the plate thickness of the hull structure. However, thinner plate structures significantly increase the risk of major distortion, which can lead to substantial post-welding rework. A number of novel welding techniques have shown potential to reduce welding induced distortion and its associated cost.

Two novel welding processes are of particular interest: Tandem-Gas Metal Arc Welding (T-GMAW) and, *so called*, GMAW with Dynamically Controlled-Low Stress No Distortion (DC-LSND). Although, both techniques modify the thermal heat fields associated with the welding process, they achieve this aim in fundamentally different ways. The T-GMAW process uses a second welding electrode to introduce a second heat source whilst DC-LSND technique uses a trailing CO_2 heat sink nozzle to accelerate the cooling of the GMA weldment. Both techniques have the potential to influence the stresses introduced by the welding process and, thus, the distortion.

In this study, these two processes are applied to both butt weld and fillet welds specimens made of 5mm DH36 steel. A comparison of the measured out-of-plane distortion for both processes is presented and compared to traditionally used processes, such as Submerged Arc Welding or Gas Metal Arc Welding. After considering other influencing factors such as weld mechanical properties and productivity, an assessment is presented of the suitability of both the T-GMAW and DC-LSND processes for use in naval shipbuilding.

INTRODUCTION

Reducing weld induced distortion in thin welded plates has been a challenge for industry for at least 40 years [1]. TWI Ltd. estimates the cost of correcting weld distortion to be as high as 30% of the total fabrication cost of a ship hull [2]. In addition, with a trend towards using higher strength thin section materials to achieve lower material costs and lighter weight structures, post weld distortion becomes a more significant issue [3]–[5].

Welding induced distortion control can be achieved through either process-independent techniques, such as clamping [6] and sequencing optimization [7]–[11], or alternatively, by changing the welding process itself. Naval shipbuilding currently uses a range of traditional welding processes specific to the joining of naval steels. A number of emerging welding technologies may provide the industry with alternative processes, potentially having better control of heat input than standard single wire Gas Metal Arc Welding (GMAW) and therefore reducing the amount of distortion in the final structure. Some of these include hybrid-laser GMAW [12]–[14], Double Electrode-Gas Metal Arc Welding

[15], Tandem-Gas Metal Arc Welding (T-GMAW) [16]–[18] and active cooling in-process methods such as Dynamically Controlled-Low Stress No Distortion (DC-LSND) [19]–[21].

This paper focuses on T-GMAW and DC-LSND, because of their relatively low cost and commercial availability. Experiments have been carried out to provide a broad assessment of the two process' suitability for naval shipbuilding, with each compared to classical GMAW for weld performance of 5mm DH36 for both butt and fillet welds. The targeted application is stiffened panels typically used in shipbuilding.

Tandem Gas Metal Arc Welding is a relatively new process, which combines higher deposition rates at lower heat inputs compared with more conventional welding processes than single GMAW. In T-GMAW, two wires are fed into the same weld pool through electrically isolated contact tips, which allow different welding parameters to be set for each wire. Most T-GMAW systems achieve best performance in the anti-phase synchronized pulsed mode, avoiding an overlap of peak pulse current between the two arcs [16], [22]. T-GMAW has previously been reported to provide improved penetration and productivity [23],[24], reduced distortion and minimal changes in the metallurgical structure when compared to Submerged Arc Welding (SAW) process for the butt welding of ship panels [25]. Here, the effectiveness of T-GMAW in reducing distortion is compared to similarly optimised GMAW for both butt welds and T-fillet joints.

A second set of welding experiments were used to compare classic GMAW with process assisted GMAW + DC-LSND. Using a jet of CO_2 cryogen gas as a trailing heat sink, a rapid thermal cycle can be introduced which reduces distortion. Although reported to be experimentally successful for butt welds of DH36 weldments [26], the violent nature of the CO_2 snow gas stream and the complex interaction with the welding gas shield makes the process difficult to apply. Shielding the CO_2 snow gas [19], [20], [27] or applying the coolant on the other side of the weld [28] are solutions that have been demonstrated to be effective. The option explored in the current work was to apply intermittent DC-LSND. It consists in dividing the total weld into a series of stitches. A stitch was welded before immediately applying the coolant. The next stich was then subsequently welded and again followed by the coolant. This alternating sequence was continued for the full length of the joint. This paper assesses the effectiveness of this technique on butt and fillet welds. For that purpose, a series of experiments that compare welds with similar bead geometry and mechanical properties from each process are reported. The resultant distortion from the process is then compared to both GMAW and T-GMAW.

In addition, a two-pass procedure, using a metal-cored wire, is tested on butt-welded joint. This procedure uses a relatively low speed, inducing a high heat input and deposited metal volume, similar to SAW. This wire is also tested on fillet welds using a speed similar to the other GMAW procedure with a lower wire feed rate, inducing a lower heat input.

The present paper is organized as follows: Section 2 introduces the experimental set-up for the tests and measurements. It also provides an overview of the various welding process parameters selected. Section 3 provides a detailed evaluation of bead geometry and basic mechanical properties. Section 4 provides the results of distortion measurements. Conclusions are presented in Section 5, followed by acknowledgments.

EXPERIMENTAL

Specimen material and geometry

The base material used for the welding experiments was made of 5mm thick DH36 steel, widely used in naval shipbuilding. The butt welds were made by joining two 1000mm x 200mm plates. In order to ensure a steady-state weld in the main plate, run-on and run-off plates were tacked top the specimen using GMAW. As shown in Figure 1, butt welds run-off plates were larger to mitigate against any opening effect during welding.

For the T-joint specimens, the one-metre long fillet welds were made with a 400mm wide base plate and a 100mm height T-joint stiffener. All plates were 5mm thick. 25mm long tacks welds were placed every 250mm on one side of the joint, and 100mm wide run-on and run off plates were used as shown in Figure 2.

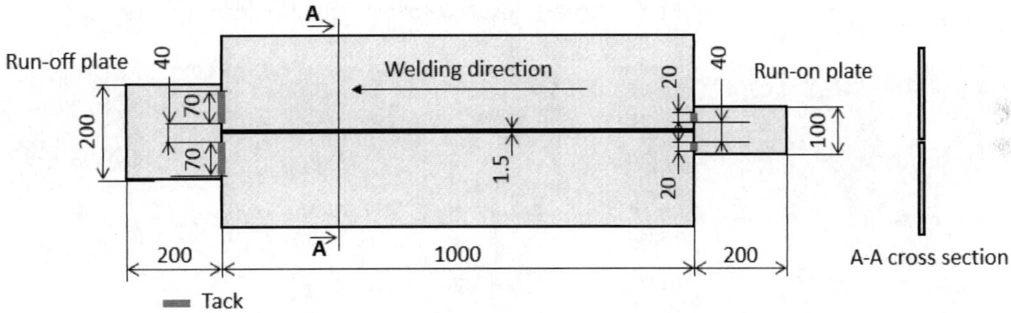

Figure 1: Geometry of the butt weld samples

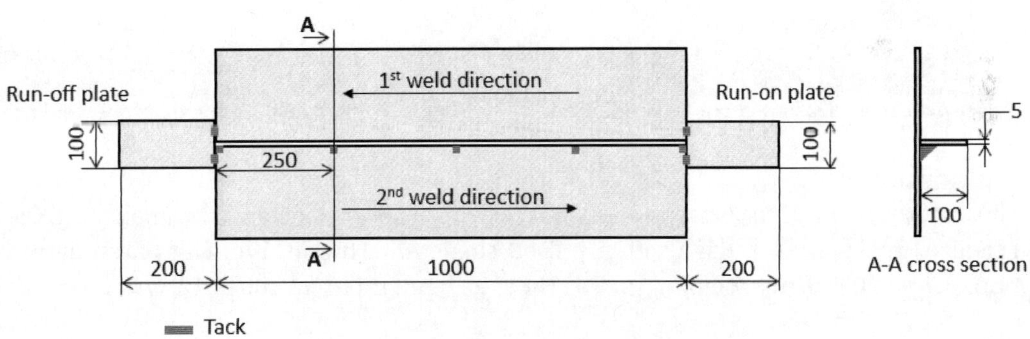

Figure 2: Geometry of the fillet weld specimen

Welding set-up

The welds were carried out in the Manufacturing Robotic Laboratory of the University of Wollongong using an ABB IRB 4400 robot. The welds were carried out on a 1000mm x 450mm x 30mm steel table. In the butt weld configuration, a copper bar was placed in the centre with a slot to host an eight-millimetre groove porous ceramic backing. The copper bar was replaced by a solid steel bar of 50mm x 30mm section in the fillet weld configuration to obtain a homogeneous heat sink.

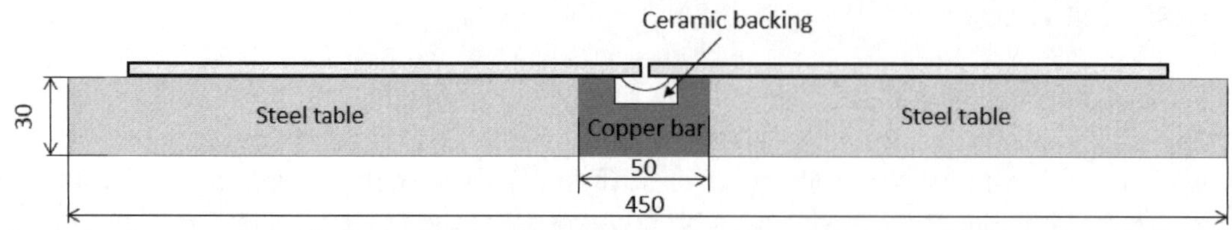

Figure 3: Cross-section of the welding table used to complete the butt welds. In the fillet weld configuration, the copper bar is replaced by a full steel bar of 30mm x 50mm cross section

Clamping

Plates were clamped on the edges 200 mm away from the weld centreline. Four clamps were used on each side. The clamping was the same for butt welds and fillet welds as the widths and lengths of the specimen were the same. Photographs of the butt welds and fillet welds before welding can be seen respectively in Figure 4 and Figure 5. The clamps were released after each weld had cooled down to the ambient temperature.

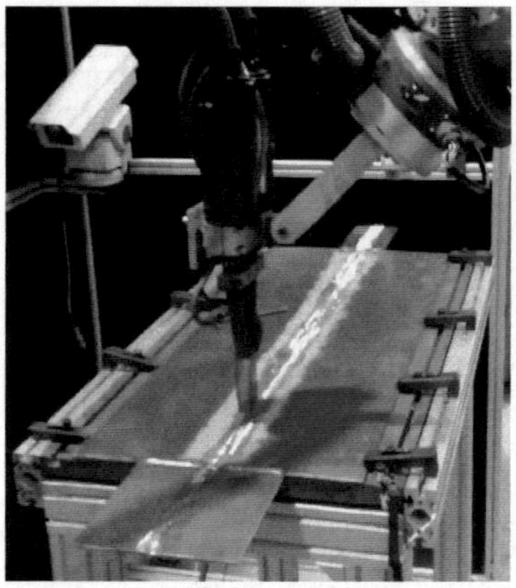

Figure 4: Butt weld clamped before welding

Figure 5: Fillet weld clamped before welding

Welding processes

The reference process was pulsed-GMAW with 1.2mm Lincoln Ultramag® ER70S-6. A Fronius TimeTwin® set was used in pulsed single wire mode. Welding parameters were optimised at 0.6m/min welding speed. The wire-feed rate was set at 12m/min.

The second weld process was carried out with metal-cored arc welding wire at a travel speed of 0.12m/min with constant current and voltage. The filler wire was a Böhler HL51T-MC filler wire. The welded plates were then unclamped and scanned before re-clamping and proceeding to the second weld pass. During the second pass, the travel speed was increased to 0.15m/min, while keeping the same wire feed rate.

The third process used was the tandem-gas metal arc welding process (T-GMAW). This process has two independently controlled and insulated electrodes using a Fronius TimeTwin® equipment. The total deposition rate was the same as in the reference pulsed-GMAW process. Contact-tip-wire-distance (CTWD), wire and gas mixture were also kept constant. This commercial set uses an arc spacing of 4 to 6 mm, depending on the CTWD. This spacing is relatively small in comparison to other commercial sets and requires the use of antiphase synchronisation of the electrodes pulses [16]. The ratio of the wire feed rate of the trailing electrode over the wire feed rate of the leading electrode has been shown to be an essential parameter of the T-GMAW process [17]. Ueyama et al. have shown that reducing this ratio will lead to higher possible welding speed [29] without undercut or humping problem. Larkin et al. [25] have shown that increasing this ratio reduces heat input and therefore distortion, but it is incompatible with very high welding speeds. A compromise between welding speed and heat input therefore seems necessary. Reducing the welding speed allows for a higher wire feed rate ratio and potentially less distortion than a fast welding process with a lower wire feed rate ratio. Here, an optimised welding speed of 1.2 m/min with a wire feed rate ratio (leading:trailing) of 1.5 was chosen.

The fourth welding process used was the intermittent DC-LSND GMAW process. CO_2 snow was used for the coolant as it has shown its efficiency in previous work [20], [27], [30]. Two Linspray® TSF 30 nozzles delivering each 430g/min of CO_2 from a liquid cylinder were used. The set-up is illustrated in Figure 6. The weld was divided into stiches of 110mm long with each stich overlapping the previous one by 10mm. A cooling pass was applied between each stich welding operation. Including the time for the robot to change position, the cooling pass was applied one second after the arc extinguished inducing an arc to cooling time of 12 seconds. Using intermittent DC-LSND, the cooling speed can be independently varied from the welding speed. Here, the cooling pass was applied at the same speed as the welding pass (0.6m/min). The welding parameters were set identical as those of the GMAW reference weld.

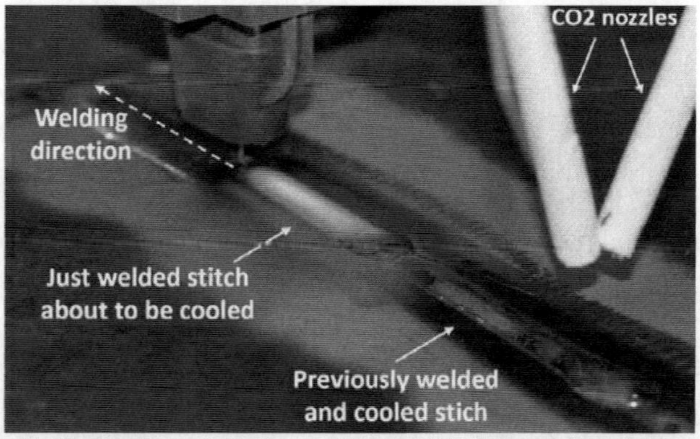

Figure 6: Illustration of the intermittent DC-LSND process on butt weld

All the welding operations were monitored through an independent current and voltage measurement at a 2kHz frequency. The trace of the two electrodes during the T-GMAW process are illustrated in Figure 7. The heat input induced by each welding process are calculated at 100% efficiency and summarized in Table 1.

	Pulsed GMAW	Metal-cored GMAW		Tandem-GMAW		GMAW Intermittent DC-LSND
		1st pass	2nd pass	Lead	Trail	
Welding speed (m/min)	0.6	0.12	0.15	1.2		0.6
Wire feed rate (m/min)	12	4.5	4.5	9.5	14.5	12
Heat input (kJ/mm)	0.85	2.3	1.8	0.41	0.51	1.3

Table 1: Main parameters for the four butt welds trials. Heat inputs are calculated at 100% efficiency.

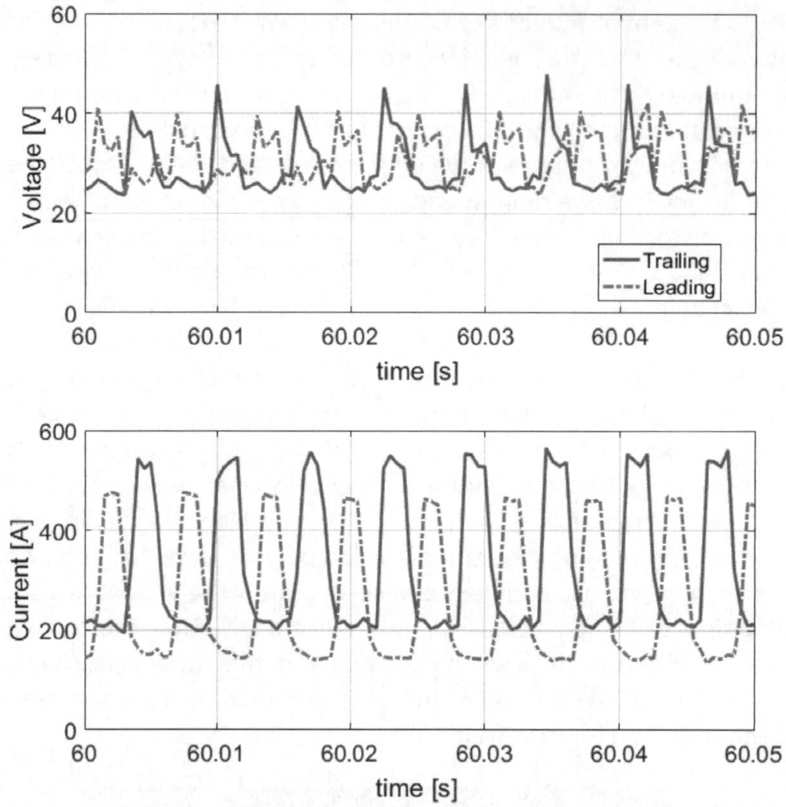

Figure 7: Current and voltage trace from the T-GMAW process for the butt weld

All four processes were also used to produce fillet welds. However, some adjustments in the welding parameters were required. The main settings are summarised in Table 2. For the fillet welds, the torch was inclined at 45°. The plates were welded first on the opposite side of the tacks. After complete cooling, the plate was unclamped, rotated and clamped again to weld the second side. Reversed-sided welding was applied. After complete cooling, the plates were unclamped and scanned to measure plate distortion resulting from welding.

	Pulsed GMAW	Metal-cored GMAW	Tandem-GMAW		GMAW Intermittent DC-LSND
			Lead	Trail	
Welding speed (m/min)	0.6	0.59	0.9		0.6
Wire feed rate (m/min)	12	8.5	7	11	12
Heat input (kJ/mm)	1.2	0.83	0.24	0.34	1.3

Table 2: Main parameters for the four fillet welds trials. Heat inputs are calculated at 100% efficiency.

RESULTS

Macrographs

Sections were cut-out, polished and etched to produce macrographs of the cross-section of the different butt welds (Figure 8 a-d) and fillet welds (Figure 9 a-d). The metal-cored GMAW procedure using two passes shows a very large weld area and fusion zone (FZ). The heat-affected zone (HAZ) was almost twice as large as the other processes. The macrographs were similar between the tandem and the single GMAW processes. The intermittent DC-LSND HAZ was slightly larger than the GMAW reference one, which reflects the higher heat input that may be caused by the numerous arc ignitions linked to the intermittent process. In addition, the DC-LSND FZ exhibited coarse structure, due to the rapid cooling. Closer observation of the microstructure shows inhomogeneous austenite areas in this sample with a very small grain size, in comparison to the other structures. However, no martensite has been observed.

Although full penetration is not required for fillet welds, it is interesting to observe that the pulsed GMAW procedure achieved full penetration whereas the others exhibit only partial penetration. Although the deposition rate was the same, the T-GMAW has a smaller penetration than the pulsed GMAW. This can be explained by the smaller heat input of the tandem process. Both these processes induced a twisted buckling mode, which implies high distortion. The metal-cored GMAW procedure, which had a heat input similar to the pulsed GMAW despite the lower deposition rate, has a smaller throat thickness than all the other processes. The DC-LSND macrographs exhibit a very large throat thickness as it has been cut in a section where the stitches overlap. The intermittent DC-LSND process induces a double heating/cooling cycle in that region, which can be seen in the macrograph.

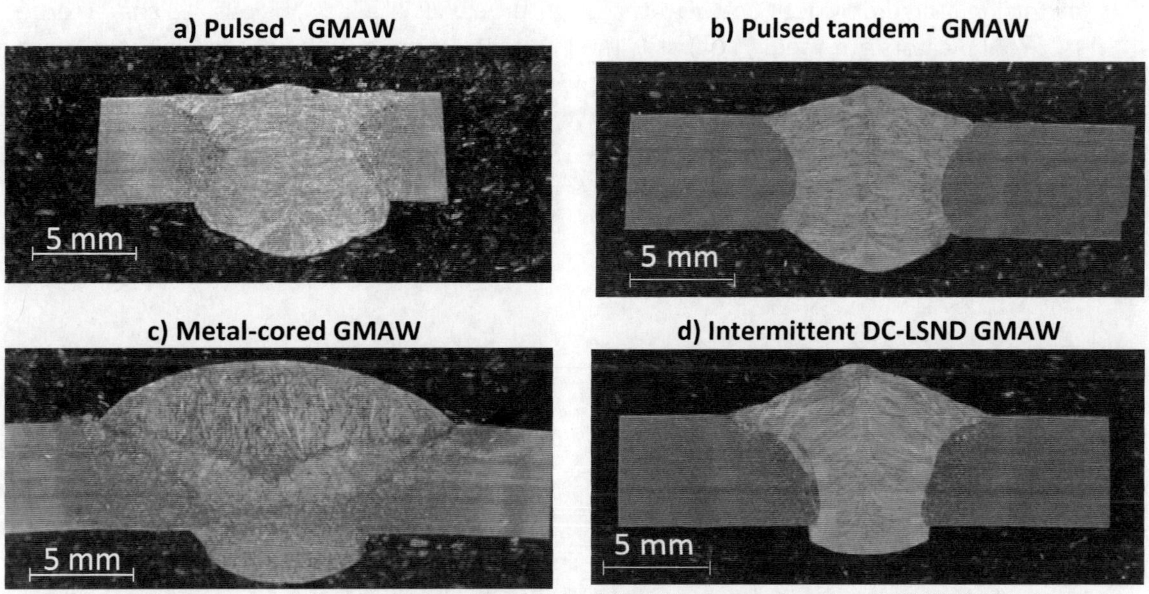

Figure 8: Macrographs of the four butt welds: a) pulsed GMAW, b) pulsed tandem GMAW, c) Metal-cored GMAW Intermittent, d) DC-LSND pulsed GMAW

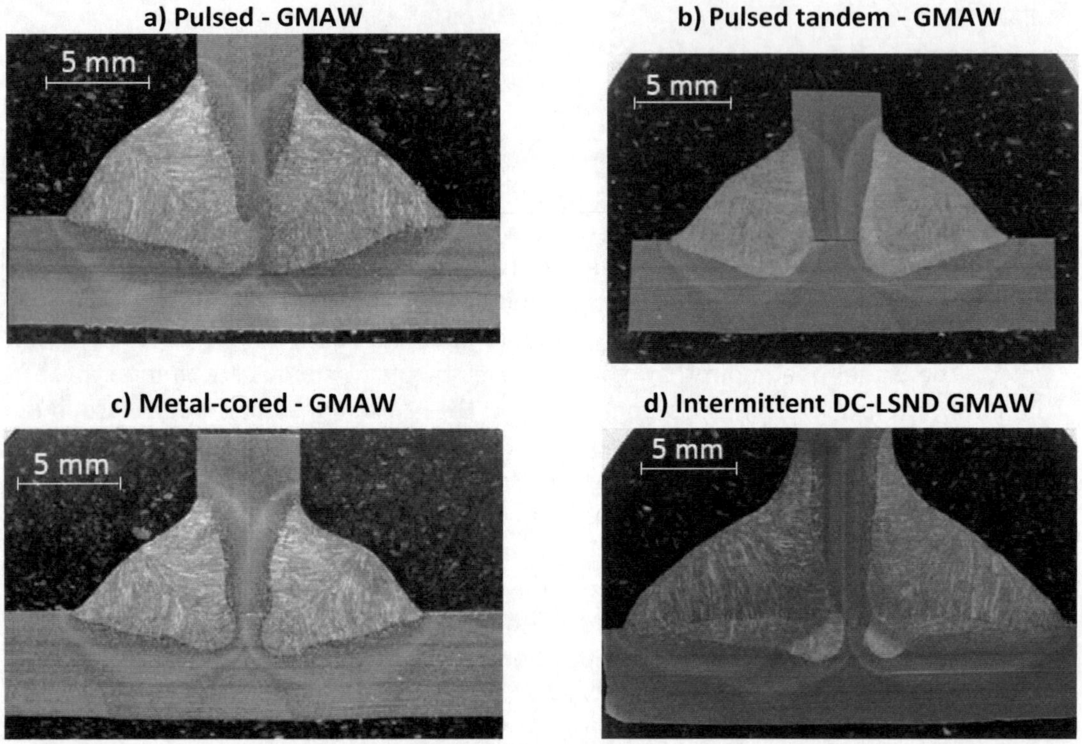

Figure 9: Macrographs of the four double-sided fillet welds: a) pulsed GMAW, b) pulsed tandem GMAW, c) Metal-cored GMAW Intermittent, d) DC-LSND pulsed GMAW.

Distortion

All samples were scanned using a Micro-Epsilon® laser scanner attached to the ABB 4400 robot at a 2m/min speed. The butt welds scans are shown in Figure 10 a-d. The metal-cored welding procedure using two passes clearly exhibits the most out-of-plane distortion. The differences between the other processes are less obvious. To characterize it, two indicators are used. The first is the peak distortion, and the second is the average angular distortion, respectively visible in Figure 11 and Figure 12. Although the peak distortion is similar between pulsed GMAW, pulsed tandem GMAW and intermittent DC-LSND, the last two processes offer a 15% reduction of the angular distortion in comparison to the reference pulsed GMAW process.

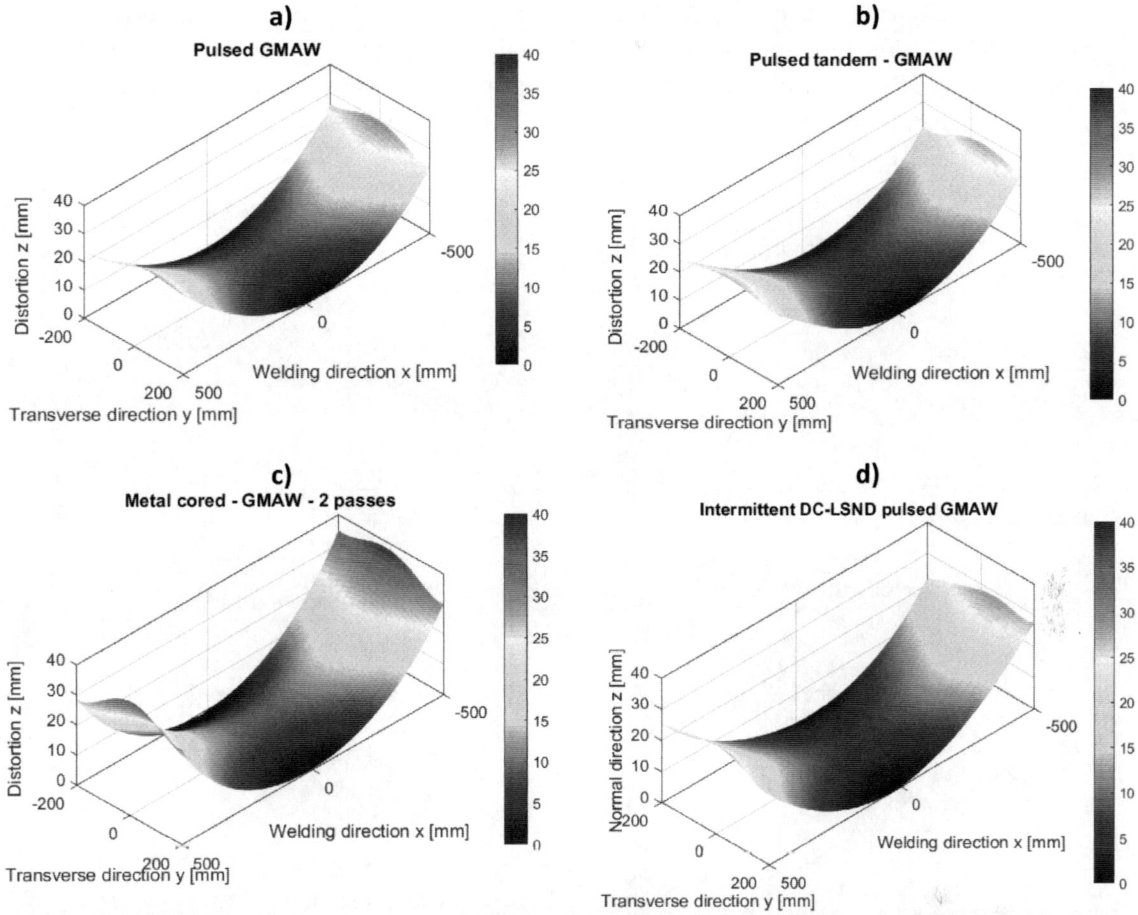

Figure 10: Scanned butt welds. a) pulsed GMAW, b) pulsed tandem GMAW, c) Metal-cored GMAW Intermittent, d) DC-LSND pulsed GMAW. The out-of-plane distortion is exaggerated by a factor 10.

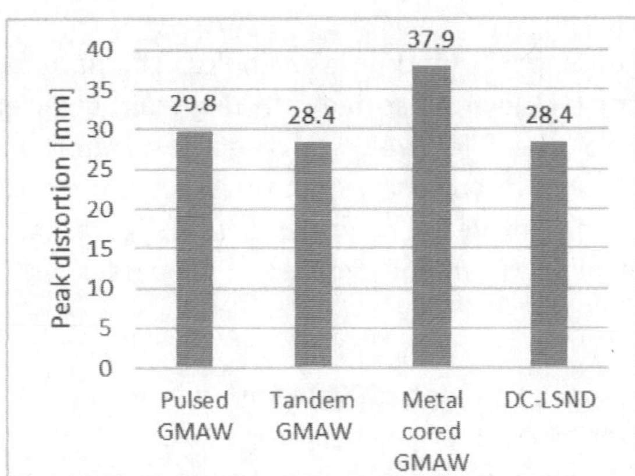

Figure 11: Peak distortion of the butt weld samples

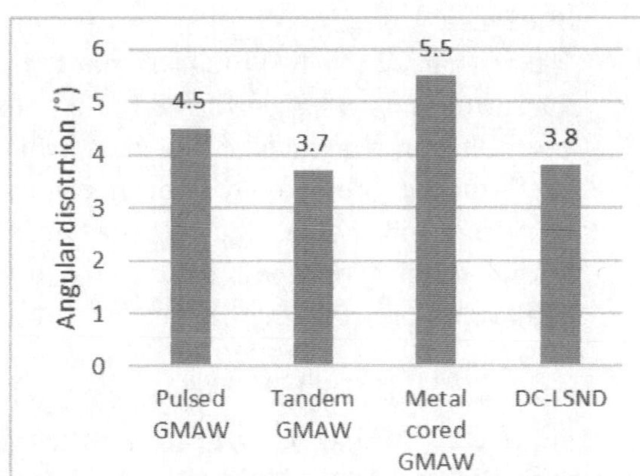

Figure 12: Angular distortion of the butt weld samples

Fillet welds specimen have also been scanned. For the sake of clarity, only the base panel plate is plotted. The indicators used for the butt welds distortion analysis are not pertinent for the fillet welds as the distortion mode differs between the processes. Pulsed GMAW and pulsed Tandem GMAW exhibit a similar distortion mode with a twisting of the base plate along the weld axis. This is believed to be linked to a buckling distortion mode of the plate induced by the reversed welding sequence. The metal-cored GMAW and the intermittent DC-LSND welded plate did not buckled but presented instead mostly a slight angular distortion (respectively -0.8° and -1.6 °).

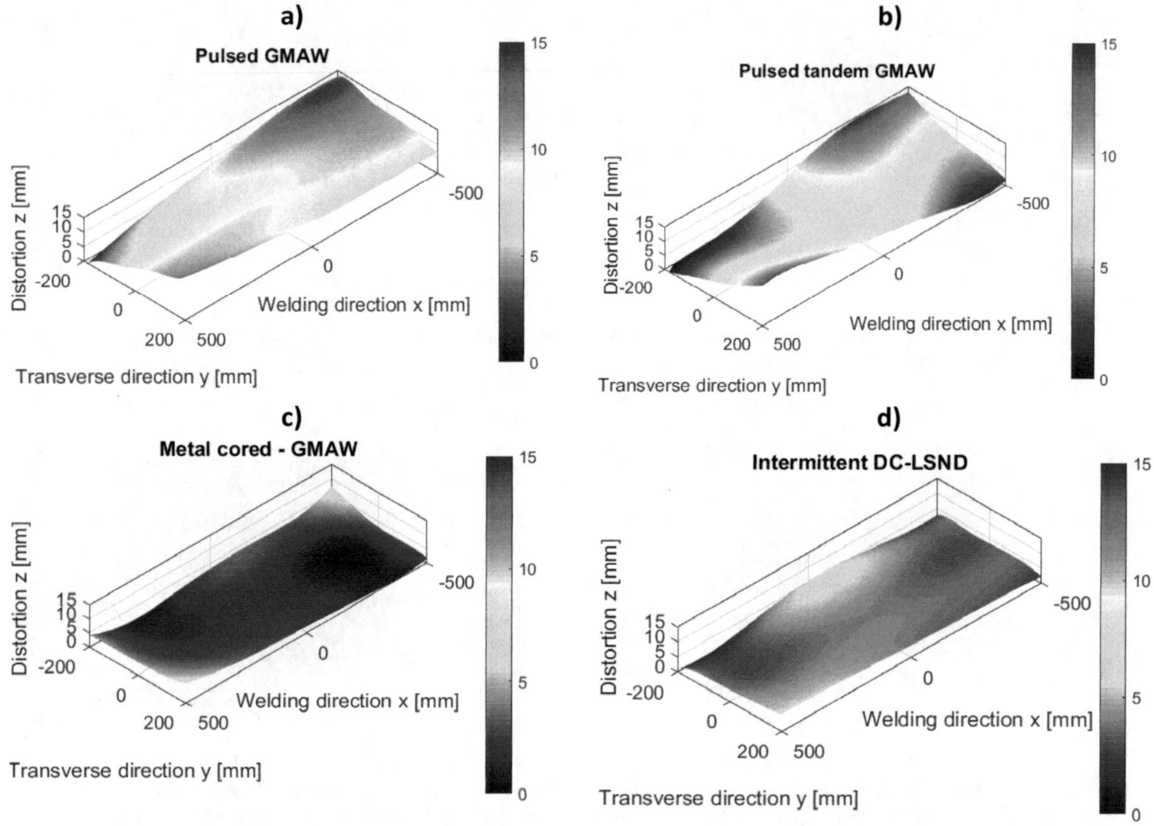

Figure 13: Scanned fillet welds. a) pulsed GMAW, b) pulsed tandem GMAW, c) Metal-cored GMAW, d) intermittent DC-LSND pulsed GMAW. The out-of-plane distortion is exaggerated by a factor 10

Hardness

Vickers hardness (HV_1) measurements has been performed on all samples. The hardness measurements were performed one millimetre below the surface starting from the weld centreline in the case of the butt welds. For the fillet welds, a diagonal filiation was performed in order to cross each grain size zone. In both cases, hardness was measured using one millimetre spacing between each indent. Results can be seen in Figure 14. In the fillet welds, hardness values are quite similar between the different processes with a peak value of 200-230kgf/mm^2.

For the butt welds, DC-LSND exhibits a high peak value of 250kgf/mm^2, whereas GMAW and T-GMAW peak value is around 200kgf/mm^2. The metal-cored procedure, which has a very large fusion zone area, presents hardness values that are close to the parent metal hardness.

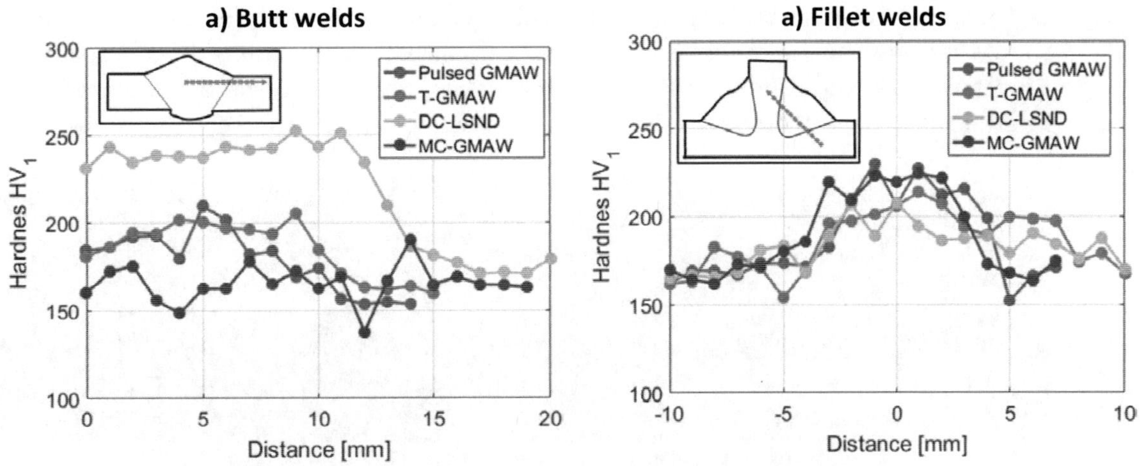

Figure 14: Hardness measurement on butt welds (a) and fillet welds (b)

DISCUSSION ON INDUSTRIAL SUITABILITY AND FURTHER EXPERIMENTS

Under equivalent deposition rates, tandem wire welding has shown a 15% reduction of the average angular plate distortion in butt welds in comparison to single wire welding but has not shown any improvement in the peak distortion. As the specimens exhibit large distortion due to a buckling mode of the reversed-sided fillet welds stiffeners, no significant gain in distortion has been shown between single and tandem wire welding on fillet welds. This observation corroborates with Thompson et al. conclusions that tandem welding is useful in reducing distortion when buckling is not the main distortion mode [16]. The tandem welding procedure does however allow a doubling in travel speed in comparison to the GMAW process.

A metal-cored procedure using two passes on butt welds has also been tested. As the literature suggests [31], the high heat input and deposition rate induce a very high distortion in the butt welded specimen. Furthermore, due to the very low speed required for the procedure, the welding time was increased by 9 times in comparison to the GMAW process, without taking into account the inter-pass cooling time. A faster metal-cored welding procedure was used on the fillet welds with a relatively low heat input and deposition rate. This procedure was shown to be very efficient in reducing distortion as no buckling was observed on those specimens.

The last process of this study was the intermittent DC-LSND. This intermittent process avoided the shielding issue of the conventional in-process DC-LSND procedures and gave an industrial capability for these reasons. The main drawbacks were that the total welding time is multiplied by two and the presence of numerous restarts along the weld, as shown in Figure 15. Here, an effective stitch length of 100mm has been used to obtain an arc to cooling time of 12 seconds. Keeping the arc to cooling time constant, the stitch length could be increased by combining intermittent DC-LSND with much faster tandem welding. However, surprisingly, the distortion on butt welds was not significantly improved by the intermittent DC-LSND process. On fillet welds, the distortion was reduced by the intermittent DC-LSD process, as these welds did not exhibit strong buckling. Further investigation is required to determine whether the intermittent welding process or the cooling effect is responsible. In addition, the microstructure of the intermittent DC-LSND welds were shown to be very different with a relatively high hardness. Further testing should include the comparison of the mechanical properties induced by the different processes. Validation should also include particular attention to the overlap area in the intermittent DC-LSND process, as it experiences two full welding and cooling cycles.

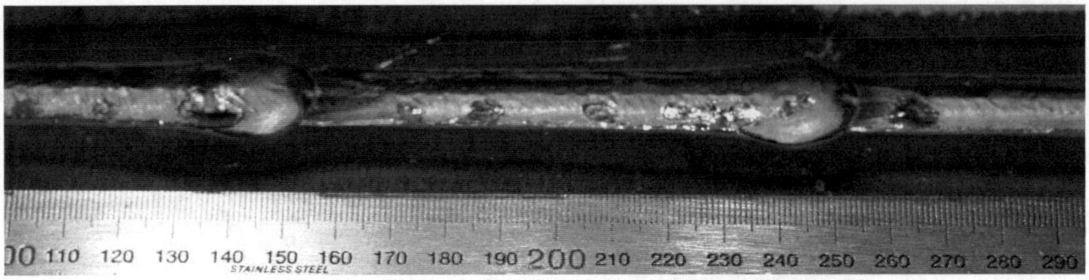

Figure 15: Visual aspect of the intermittent DC-LSND butt weld

CONCLUSION

The principal objective of this on-going collaborative project is to evaluate innovative welding processes capability in order to reduce the distortion induced by welding in shipbuilding. The main benefits would be to improve hull geometry while reducing the costs associated with rework. These recently developed welding processes also have the potential to increase productivity. Four welding processes have been assessed with particular procedures: GMAW, tandem pulsed GMAW, metal cored-GMAW and DC-LSND GMAW. The four welding procedures have been tested on one-metre long butt welds and fillet welds. The following conclusions have been made:

- For butt welds, the tandem-GMAW process exhibits the least peak and angular distortion, and in particular less than the GMAW procedure with similar deposition rate. The tandem-GMAW procedure was also carried out at twice the speed of the single wire GMAW procedure offering a significant productivity increase.

- A metal-cored butt weld procedure using two passes has been tested and exhibits large distortion. This can be attributed to the high heat input induced by the high volume of filler metal deposited. However, the same wire used with a low heat input has shown very little distortion for fillet welds.

- Intermittent DC-LSND did not produce significant distortion reduction as expected on butt welds, in comparison to the equivalent GMAW procedure. More investigations are needed to understand the reasons why such result was obtained on butt-welded sections whereas it allows a measurable reduction in distortion on fillet welded T-sections. In both cases, the intermittent DC-LSND is two times slower than the equivalent GMAW reference procedure.

ACKNOWLEDGEMENTS

The authors acknowledge the support of the Defence Materials Technology Centre, which was established and is supported by the Australian Government's Defence Future Capability Technology Centre (DFCTC) initiative. This is an on-going collaborative project between University of Wollongong, Naval Group, Australian Nuclear Science Technology Organisation and DST. In addition, acknowledgment is given to DMTC partner NAVAL Group for the supply of plate and consumables used in this work and BOC for the supply of Linde LINSPRAY® CO_2 cooling nozzles.

BIBLIOGRAPHY

[1] K. Masubuchi, Ed., "Analysis of Welded Structures," in *Analysis of Welded Structures*, Pergamon, 1980, p. ii.

[2] TWI, "Hybrid Laser Arc Welding at TWI," 2017. [Online]. Available: http://www.twi-global.com/capabilities/joining-technologies/lasers/hybrid-laser-arc-welding/. [Accessed: 14-Feb-2017].

[3] Marine Corps Armored Vehicles, and Commercial Applications, "Control of Weld Distortion and Accuracy Benefits Navy Ships," *EWI Insights- Mater. Join. News Lett.*, vol. 17, p. 3, 2004.

[4] P. Michaleris, Ed., *Minimization of Welding Distortion and Buckling*. Woodhead Publishing, 2011.

[5] T. Gray, D. Camilleri, and N. McPherson, *Control of Welding Distortion in Thin-Plate Fabrication: Design Support Exploiting Computational Simulation*. Elsevier, 2014.

[6] T. Schenk, "Modelling of welding distortion; the influence of clamping and sequencing," 2011.

[7] G. Fu, M. I. Lourenço, M. Duan, and S. F. Estefen, "Influence of the welding sequence on residual stress and distortion of fillet welded structures," *Mar. Struct.*, vol. 46, pp. 30–55, Mar. 2016.

[8] L. Gannon, Y. Liu, N. Pegg, and M. J. Smith, "Effect of three-dimensional welding-induced residual stress and distortion fields on strength and behaviour of flat-bar stiffened panels," *Ships Offshore Struct.*, vol. 8, no. 5, pp. 565–578, Oct. 2013.

[9] C. L. Tsai, S. C. Park, and W. T. Cheng, "Welding distortion of a thin-plate panel structure," *Weld. J.-N. Y.-*, vol. 78, p. 156, 1999.

[10] N. Syahroni and M. I. P. Hidayat, "3D Finite Element Simulation of T-Joint Fillet Weld: Effect of Various Welding Sequences on the Residual Stresses and Distortions," 2012.

[11] C. Seyyedian, M. Haghpanahi, and M. Sedighi, "Prediction of welding-induced angular distortions in thin butt-welded plates using artificial neural networks," *Comput. Mater. Sci.*, vol. 62, pp. 152–159, 2012.

[12] S. M. Kelly, S. W. Brown, J. F. Tressler, R. P. Martukanitz, and M. J. Ludwig, "Using hybrid laser-arc welding to reduce distortion in ship panels," *Weld. J.*, vol. 88, no. 3, pp. 32–36, 2009.

[13] C. M. Allen, C. H. J. Gerritsen, Y. Zhang, and J. Mawella, "Hybrid laser-MAG welding procedures and weld properties in 4mm, 6mm and 8mm thickness C-Mn steels," in *IIW Commission IV / XII*, Vigo, Spain, 2007.

[14] C. H. J. Gerritsen, C. M. Allen, and J. Mawella, "Development and evaluation of CO2 laser-MAG hybrid welding for DH36 shipbuilding steel," presented at the 11th

CF/DRDC International Meeting on Naval Applications of Materials Technology, Halifax, 2005.

[15] Y. Lu *et al.*, "Double-electrode arc welding process: Principle, variants, control and developments," *J. Manuf. Process.*, vol. 16, no. 1, pp. 93–108, Jan. 2014.

[16] A. M. Thompson *et al.*, "Improving the competitiveness of the European steel fabrication industry using synchronised tandem wire welding technology." European Commission - Reserach Fund for Coal and Steel, 2008.

[17] Z. Sterjovski, J. Donato, and H. Li, "The effect of travel speed and CTWD on the bead profile and microstructure of tandem GMA steel welds," *Australas. Weld. J.*, vol. 56, no. 1, pp. 40–46, 2011.

[18] Z. Pan *et al.*, "Evaluation of tandem gas metal arc welding for low distortion butt-welds in naval shipbuilding," *Australas. Weld. J.*, vol. 58, no. 2, pp. 35–41, 2013.

[19] R. O'Brien, W. Veldsman, and D. Baglee, "The development of an industrial robotic Low Stress No Distortion (LSND) welding system," presented at the IIW International Conference High Strength Materials - Challenges and Applications, Helsinki, 2015.

[20] Q. Guan, C. X. Zhang, and D. L. Guo, "Dynamic control of welding distortion by moving spot heat sink," *Weld. World Soudage Dans Monde*, vol. 33, no. 4, pp. 308–312, 1994.

[21] S. Okano, M. Mochizuki, M. Toyoda, and T. Ueyama, "Effect of welding conditions on reduction in angular distortion by welding with trailing heat sink," *Sci. Technol. Weld. Join.*, vol. 17, no. 4, pp. 264–268, May 2012.

[22] U. Tomoyuki, O. Toshio, Y. Kei, T. Manabu, U. Masao, and N. Kazuhiro, "High-Speed Welding of Steel Sheets by the Tandem Pulsed Gas Metal Arc Welding System," *Pulse*, vol. 34, 2004.

[23] V. Fersini and S. Matera, "The synchronized tandem wire welding process applied to the welding of structural plates," *Weld. Int.*, vol. 23, pp. 597–605, 2009.

[24] G. Sproesser, A. Pittner, and M. Rethmeier, "Increasing Performance and Energy Efficiency of Gas Metal Arc Welding by a High Power Tandem Process," *Procedia CIRP*, vol. 40, pp. 642–647, Jan. 2016.

[25] N. Larkin, Z. Pan, S. van Duin, M. Callaghan, H. Li, and J. Norrish, "Tandem gas metal arc welding for low distortion butt welds," *Adv. Mater. Res.*, vol. 337, pp. 511–516, Jan. 2011.

[26] R. Holder, "Development of a Cryogenic Cooling Process for Weld Distortion Control in Shipbuilding," Master's thesis, University of Wollongong, 2011.

[27] E. M. Van der Aa, "Local Cooling during Welding: Prediction and Control of Residual Stresses and Buckling Distortion," 2007.

[28] D. Bertaso, "Advanced Manufacturing Using Low Stress No Distortion (LSND) Techniques IIW DOC," 2004.

[29] T. Ueyama, T. Uezono, T. Era, M. Tanaka, and K. Nakata, "Solution to problems of arc interruption and arc length control in tandem pulsed gas metal arc welding," *Sci. Technol. Weld. Join.*, vol. 14, no. 4, pp. 305–314, May 2009.

[30] R. Holder, N. Larkin, H. Li, L. Kuzmikova, Z. Pan, and J. Norrish, "Development of a DC-LSND welding process for GMAW on DH-36 Steel," 2011.

[31] P. Colegrove *et al.*, "Welding process impact on residual stress and distortion," *Sci. Technol. Weld. Join.*, vol. 14, no. 8, pp. 717–725, Nov. 2009.

Enhancement of safety and operational reliability of modern submarines through the involvement of a Classification Society

Christian von Oldershausen

DNV GL

christian.von-oldershausen@dnvgl.com

THE ROLE OF CLASSIFICATION

Classification societies develop, establish and subsequently audit rules and technical standards for commercial and naval ships to achieve accepted safety requirements on board.

The origin of the first classification societies can be traced to the 18th century. Seaborn trade was flourishing but it was not without danger and many commercial ships lost their cargo or drowned with all hands. In many cases, the cause was material fatigue or insufficiently equipped ships. Therefore, underwriters and ship owners were looking for an objective and reliable means to assess the vessels they were writing policies for, or entrusting with their seamen and cargo. They were seeking information regarding the type of vessel, its construction, outfitting, age and condition. Information which could only be provided through a systematically applied survey by a qualified, competent and, most of all, impartial party. The categorisation of ships according to their quality, laid the foundation for classification societies. The first such register of ships and their condition, the Lloyd's Register Book, was published in 1764 in London by a group of underwriters who had established themselves at Lloyd's Coffee House. The concept of classification of ships quickly found its acceptance across the shipping community, with national shipping societies springing up across the globe in quick succession. Bureau Veritas was founded in 1828 in Antwerp, Lloyd's Register of British and Foreign Shipping (LR) had been active since the 1760s and established itself as independent classification society in 1834. Various other societies who are still very active today followed: Registro Italiano Navale (RINA) 1861, American Bureau of Shipping (ABS) 1862, Det Norske Veritas (DNV) 1864, Germanischer Lloyd (GL) 1867, Nippon Kaiji Kyokai (NK) 1899 to name a few.

Classification societies originated from the need of the maritime industry for independent safety requirements which considered economic viability, resulting in standards that reduce risk down to an as low as reasonably practicable (ALARP) level. While classification societies originally focussed on analysing and documenting the seaworthiness of vessels, using their own technical rules as shipping entered the modern age, they have added the preventive safety as part of their responsibility. And for many years, classification societies have taken the responsibility of verifying compliance with the regulations of the International Maritime Organisation (IMO) for the safety of life at sea and the maritime environment (SOLAS and MARPOL). But even as classification societies have continued to develop and adapted their role to change and grow, their independence, knowledge and competence has never been called into doubt.

Thus, classification has established itself as a fundamental instrument to ensure ship safety and statutory authorities, Flag Authorities, in many countries have recognised this. These statutory authorities have authorised classification societies to undertake surveys and inspections on their behalf by granting them the status of a Recognised Organisation (RO). Initially classification societies operated independently, developing and implementing their individual set of rules. To increase their influence regarding the implementation of technical safety on board a coordinating umbrella organisation, the International Association of Classification Societies (IACS) was founded in 1968, comprising 10 classification societies. The IACS statutes define minimum technical and organizational standards for its members. Since its inception IACS has held an observer status at the IMO.

By applying a stringent and mandatory classification process, commercial maritime industry stakeholders are assured that a ship complies with a set of rules or other safety and technical standards that have been demonstrated to be appropriate for its function. This process is achieved through independent audit of the design, equipment and material supply and construction, after which a certificate is issued, followed by regular surveys of the through-life maintenance of the vessel.

NAVAL VESSEL SAFETY ASSURANCE

In terms of naval applications, shipbuilders and navies find themselves faced with many challenges throughout the whole project lifecycle from the conceptual phase of a navy vessel or submarine until the end of an often-extended operational lifetime. Typical constraints include: budget limitations, a critical dilution of in-house expertise for shipbuilding and design, a shrinking pool of qualified personnel and finally, increasing demands from the public for safety and environmental protection. At the same time requirements, such as sufficient capability and reliable function of naval vessels and their systems, an expanded and flexible operational profile, exploitation of latest technology, high safety levels for the crew on board and compliance with environmental standards, must be met. Ensuring that these objectives are achieved requires a robust and well-proven assurance process. Consequently, classification is increasingly being used by navies as a powerful process and assurance tool, bearing in mind however, that unlike in commercial shipping this is an optional process depending on the requirements of the respective Naval Administrations. In some cases, additional technical standards need to be considered to achieve a broader level of assurance. This might be enhanced through the application of a barrier management risk-based safety case assessment approach to the platform and/or its systems.

Simply applying the existing commercial classification regime to naval vessels has proved not to be feasible due to the fundamentally different operational requirements of naval ships and their complex procurement processes. This has led to DNV GL developing a more tailored classification approach for naval ships which provides for a more differentiated assurance process depending on the requirements of the involved navy. It is therefore imperative for naval ships that the scope of classification matches the desired functional and operational requirements of the planned vessel, often described in a concept of operations (CONOPS). The classification process captures these requirements by determining a set of "class notations" that describe solutions to address the functional and operational requirements and associated risks. Naval classification seeks to provide assurance to the naval stakeholders that the vessel represents an appropriate level of safety when being operated as intended.

DNV GL's Naval Rules have been specifically developed to meet the challenges designers and yards, procurement organisations, naval flag authorities, and navies are facing implementing a naval ship or submarine project.

It should be noted that the process of classification can also be applied to existing naval ships, subject to certain criteria. There are numerous examples of naval vessels having been brought into class and that the assessment of the platform and systems against Rule requirements has had a significant positive impact on the level of safety and quality, for example the three HDW-Class 209/1400mod submarines for the South African Navy.

Figure 1: South African Navy submarine SAS Charlotte Maxeke, Type 209/1400mod

NAVAL SUBMARINE RULES

DNV GL first undertook Classification and Certification of submarines in the 1960's and thus can look back over 50 years of experience in this field. Over the years DNV GL has developed classification rules for the construction of naval submarines and other underwater technology. Today DNV GL is the only classification society in the world with a full set of class rules for submarines specifically adapted to naval standards. Reflecting the German Military Standards (Bauvorschriften für Schiffe, or BV in short) and the Manned Submersible Rules of Germanischer Lloyd, DNV GL's Submarine Classification Rules combine national and international standards and regulations, specific technical requirements (where suitable standards are not available), best practices based on the experience gained from numerous submarine newbuilding, conversion and retrofit projects. The rules are regularly reviewed and updated to include experience gained with new designs and materials, changing regulatory requirements and developments in technology.

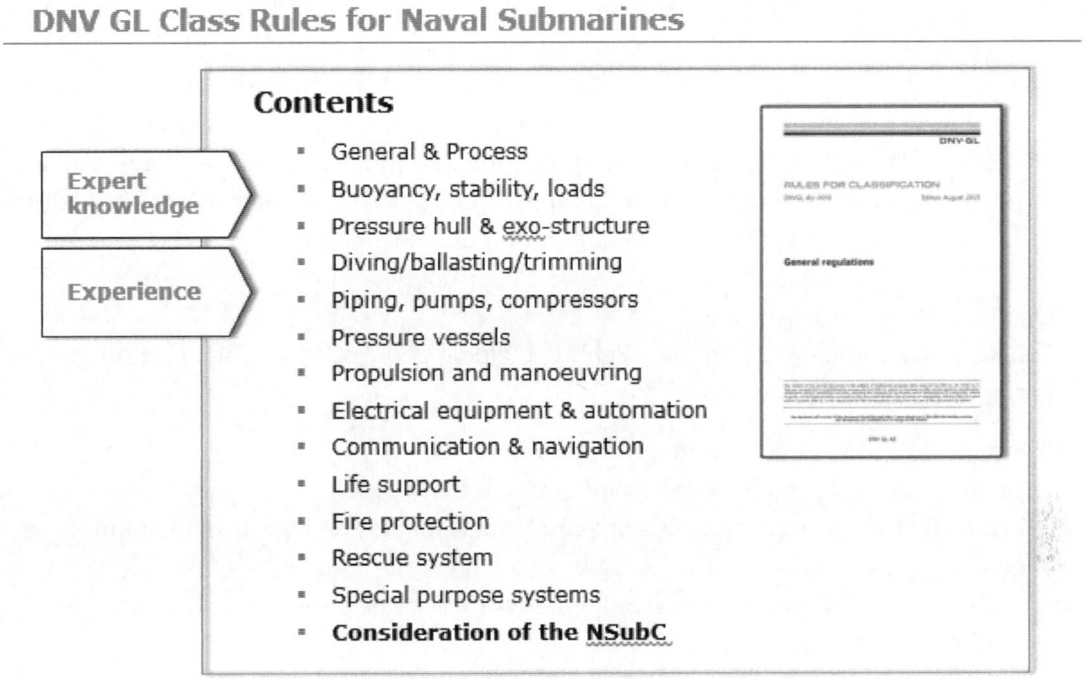

Figure 2: Directory of DNV GL Submarine Rules

SUBMARINE SAFETY

On April 10, 1963, while engaged in a deep test dive, approximately 200 miles off the north-eastern coast of the United States, the USS THRESHER (SSN-593) was lost at sea with all persons aboard - 112 naval personnel and 17 civilians. Launched in 1960 and the first ship of her class, the THRESHER was the leading edge of US submarine technology, combining nuclear power with a modern hull design. She was fast, quiet and deep diving. The loss of THRESHER and her crew was a devastating event which brought submarine safety into focus in the most tragic way. Other more recent accidents as the fire suffered by the Dutch submarine WALRUS during construction, the cracking suffered in steel used in the US submarine SEAWOLF, the fire suffered underway by USS BONEFISH, the sinking alongside of USS GUITARRO, the collision, sinking and subsequent escape from the Peruvian submarine BAP PACOCHA, and the collision of British nuclear submarine HMS AMBUSH with a tanker while submerged off Gibraltar, and other cases show that the enhancement of safety and operational reliability of modern submarines has becomes of paramount importance and priority.

Figure 3: Peruvian submarine BAP PACOCHA breaking the surface on 23 July 1989, eleven months after she was sunk

Submarine Safety is a complex, multi-dimensional task covering all technical areas of the submarine platform. The environment on board must be sustained by careful control of the atmospheric conditions, uncontained fire and flooding can threaten both the crew and the vessel, and finally the ability of a distressed vessel to surface complicates rescue and recovery operations. The importance of managing safety is therefore imperative at every major phase throughout the lifecycle of the submarine - from design to construction, through to testing, commissioning for service, and the operation of the submarine.

Consequently, the complex nature of designing, building and testing a submarine requires a robust platform and safety framework to:

a) identify any safety issues during the concept phase
b) ensure that safety can be ascertained at the design phase
c) assure that the manufacturing work carried out meets the high quality standard required
d) ensure the safety requirements and emergency systems are vigorously tested to provide maximum assurance.

Concept Phase

During the concept phase the functional/capability requirements of a submarine are determined, taking account of technological, commercial and market parameters. The concept phase will typically include the selection of the regulatory regimes, class rules, standards, codes and certifications to be used in the design and construction of the submarine, resulting in a binding schedule of certificates listing all performance and quality records to be provided. DNV GL can support the project during this phase by advising on submarine safety considerations through the provision of a robust safety assurance process in line with the functional/capability requirements of the navy.

Design Phase

The subsequent design phase requires that a determined number of plans and documents are submitted for technical appraisal against the agreed standards and class rules. Typical aspects to be considered when assessing submarine safety include the structural integrity of the platform, machinery, electrical equipment and safety systems. The objective of these design reviews and the plan approval is to ensure that the platform and systems design meet the contractually agreed standards for safety and quality. The design assessment covers many aspects including: ensuring that the material selected is suitable, a review of risk assessments (e.g. FMEA), and an appraisal of emergency procedures. Experienced engineers and former Navy submariners provide expert guidance based on their broad knowledge and long-standing involvement in the submersible industry.

Construction Phase

The involvement of classification includes not only the survey of construction at the shipyard but also the inspection of material, equipment and systems at the suppliers' premises. It might also involve a welding certification by DNV GL prior to the commencement of manufacturing with a mandatory workshop approval at the shipyard and of sub-contractors. This includes certification of welders, approval of materials and welding consumables, and the implementation of a Welding Procedure Specification (WPS) and a Welding Procedure Qualification Record (WPQR).

The classification process ensures the material and components used meet the technical specification and classification requirements, and furthermore a confirmation that the construction is in accordance with the design.

Testing Phase

During the testing phase, which includes Factory Acceptance Tests (FATs), Shore Base Integration Tests (SBITs), Platform Integration Tests (PITs), Harbour Acceptance Test (HAT) and Sea Trials with diving tests, DNV GL will witness the functioning of all safety relevant systems, assuring that the mitigating measures for safety risks identified during the design and construction phase have been implemented. DNV GL submarine experts will also participate in the deep diving tests of the submarine.

Operational Phase

Through the submarine's life, periodic inspections are conducted by DNV GL's experienced submarine surveyors, assuring that compliance with the standards agreed during the previous phases are maintained. The inspections vary according to the position of the submarine within the classification cycle, which typically is five years. The length of the classification cycle can also be adjusted to match overhaul and/or maintenance periods, for example the three South African HDW-Class 209/1400mod submarines follow a three-year classification cycle.

There are different types of periodical inspections:

- Annual surveys
- Intermediate survey after the 2nd year
- Renewal survey after completion of the "classification cycle", which includes a dry docking to inspect the hull and appendices

- Special survey, for example due to a damage repair, and which requires the engagement of the classification society to assess the situation and provide advice on the repair to maintain classification requirements.

The results of the inspections are documented in survey reports, which detail any defects and/or conditions of class which must be rectified within a certain period.

The process of classification covers every phase of the submarine's life and oversees all safety aspects of the platform. Specific military aspects such as the combat systems are outside the scope of DNV GL's involvement. However, interfaces, for example the hydraulics, electrical supply, cooling system, etc. are included.

THE NAVAL SUBMARINE CODE (NSubC)

DNV GL's Submarine Classification Rules, as part of DNV GL's RU-NAVAL Rule Set, also fulfil the requirements of the Naval Submarine Code, an international goal-based standard for managing risks, which has been developed with the participation of DNV GL.

The new Naval Submarine Code, following in the footsteps of the Naval Ship Code (NSC) which was first released as NATO ANEP 77 in 2008 (Allied Naval Engineering Publication), was discussed and approved at the 2014 International Naval Safety Association's (INSA) annual general meeting. The Naval Submarine Code is a goal-based standard, which sets a minimum level of safety for submarines and provides a framework for further work and development in this field. The benefits of this goal-based standard are that it considers the ultimate safety objectives and allows for the consideration of alternative designs and technical solutions that meet the objective. The Naval Submarine Code has been forwarded to NATO for ratification and publication as a new ANEP.

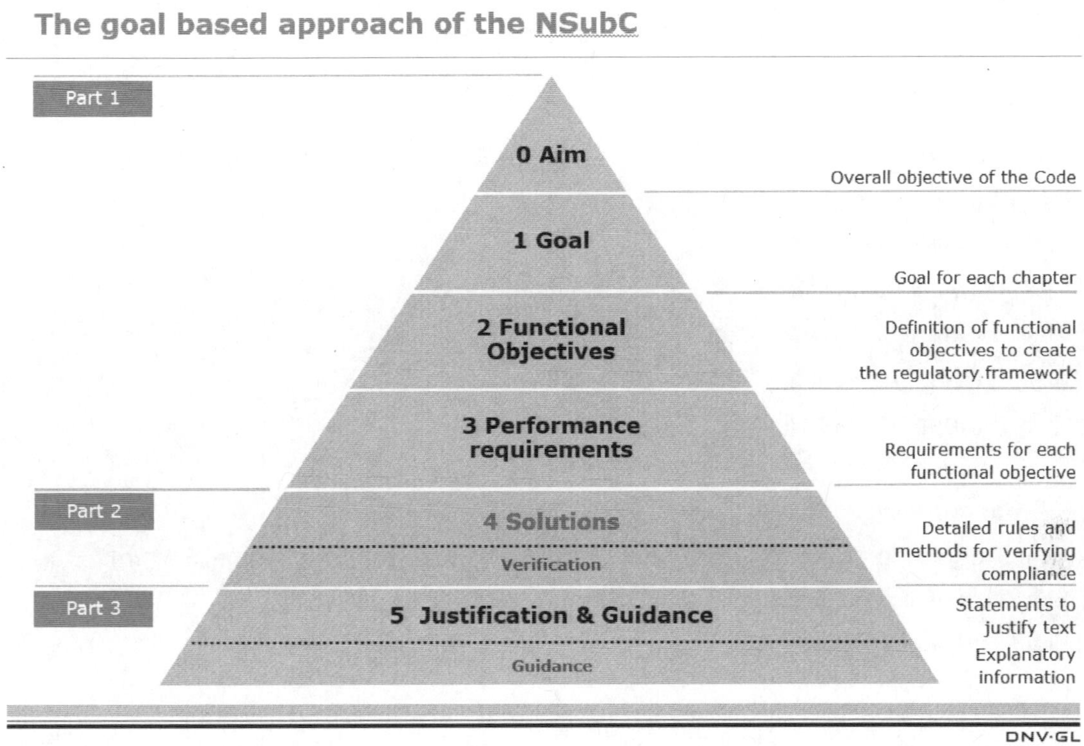

Figure 4: Structure of NSubC

Members of INSA, a group of navies and classification societies, express the hope that the Naval Submarine Code will gain acceptance in a similar way to the Naval Ship Code, which has now reached a level of maturity through the support of several countries who have implemented the NSC on their existing naval fleets and new vessels.

Directory of the Naval Submarine Code: goal based requirements for all safety-relevant areas

I – General Provisions
II – Structure
III – Buoyancy, Stability and Controllability
IV – Engineering Systems
V – Seamanship Systems
VI – Fire Safety
VII – Submarine Escape, Rescue, Abandonment and Survival
VIII – Communications
IX – Navigation
X – Dangerous Goods
XI – Integration of Platform, Combat, Navigation and Communication Systems
XII – Atmosphere Control
XIII – Nuclear *(not issued)*

DNV·GL

Figure 5: NSubC Directory

CONCLUSION

Using the naval classification services of DNV GL provides:

- an efficient way of ensuring a safe, reliable and ready for duty submarine platform in compliance with all relevant standards and regulation
- transparency evidence of an adequate process for control and the achieved level of safety
- the application of continuously maintained and updated classification rule set
- a safe and reliable design in compliance with all relevant national & international standards
- assurance of high quality of materials and components and a high fabrication quality incl. welding procedures etc.
- enhances the ability of naval authorities and/or procurement organisations to focus on the military aspects of the submarine

- the benefit of an organisation used to making decisions on acceptance limits and the proper interpretation of formal requirements as an independent and impartial Recognised Organisation
- a cost-effective approach to safety and quality through clearly agreed standards, procedures, requirements and specifications

ACKNOWLEDGEMENT

The author wishes to thank Simon Adams, Communications Manager at DNV GL Maritime, Hamburg for the excellent work of proof reading and editing.

REFERENCES

/1/ Rules for Classification Naval Vessels, Part 4 Sub-surface ships, Chapter 1 Submarines, DNV GL, Edition December 2015

/2/ Rules for Classification and Construction, I Ship Technology, 5 Underwater Technology, 2 Manned Submersibles, Germanischer Lloyd AG, 1 November 2000

/3/ Bauvorschriften für Schiffe (BV), Federal Office for Defence Technology and Procurement (BAAINBw), Germany

/4/ Unterseebootbau, Ulrich Gabler, Bernhard & Graefe Verlag, 1997

/5/ Leiser, Tiefer, Schneller – Innovation im Deutschen U-Boot-Bau, Jürgen Rohwedder / Peter Neumann, E.S. Mittler & Sohn, 2015

/6/ Die neuen deutschen U-Boote, Eberhard Rössler, Bernhard & Graefe in der Mönch Verlagsgesellschaft mbH, 2009

/7/ Naval ship safety assurance – Guidance for navies and shipbuilders, Lloyd's Register Group Limited, 2014

/8/ Approach to Deliver a Safe Type 218SG Submarine, Chee Yang Terence Ong, Li Koon Ong, Yu Ming Chee, Naval Systems, Defence Science Technology Agency, Republic of Singapore, 2017

/9/ The Navy and use of DNV Class, Albert Fredriksen, Det Norske Veritas, 2010

/10/ Using the Naval Submarine Code and DNV GL Class Rules for Assuring the Technical Safety of Submarines, Daniel Wilhelm, DNV GL, 2017

PACIFIC 2017
International Maritime Conference
International Convention Centre Sydney, Australia 3-5 October 2017

CONFERENCE ABSTRACT

Submarine Sovereignty – ASC's Collins Experience

Martin Edwards – General Manager Future Submarines, ASC Pty Ltd
Communications.Department@asc.com.au

Australia has a sovereign submarine capability and intends to retain it in coming decades.

ASC and the Australian Submarine Enterprise have in the past five years succeeded in significantly improving the availability of Collins Class submarines to the Royal Australian Navy to the point it that submarine sustainment was labelled an "exemplar program" by the Federal Government in late 2016.

ASC's experience in delivering sovereign submarine sustainment as part of the Submarine Enterprise is outlined with particular reference to:

- Supply chain management: ASC has been responsible since 2012 for managing more than 1000 suppliers, increasing the percentage of Australian suppliers to more than 90 per cent (by value), compared to about 72 per cent during the Collins build phase;
- Innovation, with reference to fostering Australian-led innovations at ASC, a key driver of the improved deep maintenance performance by ASC since the Coles Review of 2012;
- Enterprise relationships, including with the Royal Australian Navy and Capability Acquisition and Sustainment Group (CASG) of the Defence Department. The Enterprise structure provides clear lines of responsibility for different aspects of submarine sustainment, while engendering a team ethos aimed at maximizing overall submarine availability.
- Australian capability: ASC has built its capability in submarine platforms by strategic investments in its more than 1000 personnel and facilities. ASC's capability is the nation's capability, given the company is 100 per cent owned by the Australian Government.
- Training, with reference to ASC's management of the Submarine Training School at Fleet Base West, trains future submarine crews and leaders in a range of disciplines, including with the use of a submarine simulator.
- Building Australian expertise, with reference to ASC's involvement in Masters level university instruction (Masters of Marine Engineering – University of Adelaide), research and development projects involving university post graduates and researchers as well as direct support for undergraduate scholarships and high-school STEM programs (eg Subs in Schools).

Integrated Performance Modelling of Submarines to Support Strategic Decision Making for the Future Submarine Program

Hamid Diab, Karl Slater and Simon Harrison
Defence Science and Technology Group, Australia
hamid.diab@dst.defence.gov.au

1. INTRODUCTION

The intricate design, complex roles and criticality of many Defence assets make it essential to consider platform performance in an integrated and systematic manner. Otherwise, it would not be possible to assess whether complex platforms, such as Australia's future submarines, meet extensive technical and operational requirements.

Prior to 2009, the Maritime Division of the Defence Science and Technology (DST) Group – Department of Defence – conducted analysis of platform performance using a variety of methods, one of which was using empirical based equations in an Excel spreadsheet, aimed at supporting early stage submarine design trade-off studies. Whilst this spreadsheet was able to undertake various tasks, it was not a robust method for providing decision makers with objective quality evidence, nor was it repeatable or simple to integrate other tools; and as with all large spreadsheets was prone to user error.

Considering the complex and classified nature of submarine operations; coupled with the extensive requirements stipulated by Australia's Future Submarine Program (FSP) it became apparent to DST that a capability was required that could:

- understand top-level requirements and their impacts on submarine platform designs
- assess the overall performance of various platform concepts/designs whilst simulating the influences of different mission scenarios
- integrate the capability of sub-systems/technologies to assess the impacts on platform performance parameters, enabling trade-off studies to be completed
- highlight key design and performance drivers
- validate manufacturer's claims
- support the maturation of performance and operational requirements for the FSP.
- ensure any assessment was auditable and repeatable; and was conducted using a 'level playing field'.

The above challenges required a suite of system modelling capabilities; however there were no suitable tools available to capture and integrate all the diverse requirements. Although the body of knowledge on using model driven techniques to support trade-off studies in design had begun to amass [1], there was very little literature on the use of these techniques in the submarine domain. The integrated nature of submarines means that the design is tightly coupled with the desired mission scenarios, which is highly sensitive, and hence modelling complexity is increased and access to data is difficult.

Compounding the above challenge was that the required modelling capabilities would be required in the early years of the procurement phase of the FSP, where any outcomes would provide the most benefit, as illustrated in Figure 1.

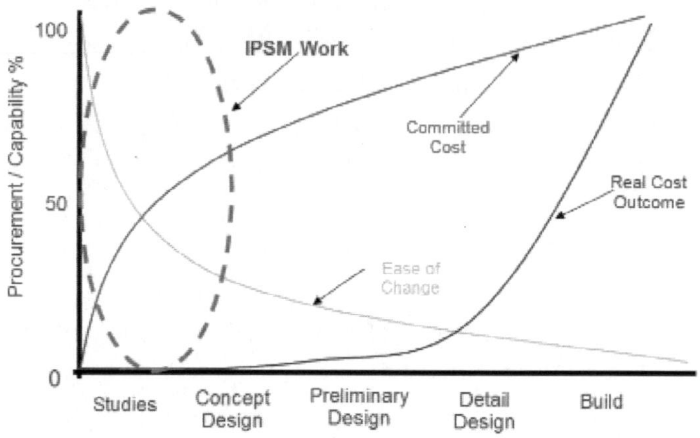

Figure 1: IPSM's Value in the Life-cycle

As such DST, in support of the Commonwealth of Australia, embarked on a task to develop an in-house software tool known as the 'Integrated Platform System Model' (IPSM). The IPSM is a platform systems analysis tool that has been and continues to be developed; and has become a critical modelling tool to support whole of platform performance studies during the FSP. This paper provides an overview of DST's development of the IPSM tool.

For details on the genesis, theories and methodologies adopted in developing this tool, the reader is referred to [2]. It should also be noted that due to the sensitive nature of the FSP, specific outcomes of DST's work in the development, use and validation of the IPSM cannot be provided. However, open literature details on the IPSM are available in [3], [4] and [5].

2. SOFTWARE AND MODEL OVERVIEW

2.1 The Core of the IPSM

The Platform Systems Analysis team of DST's Maritime Division have applied the following principles in the development of the IPSM over the past decade:

- continual software development with continual feedback from stakeholders to ascertain current and future needs
- continual improvement of processes for software development, numerical modelling and simulation
- adoption of 'SysML' (systems modelling language) principles without creating constraints/techniques imposed by propriety software
- platform system models are to be validated - validation efforts should be considered upfront
- quality control and pre/post configuration tools are essential.

A conceptual framework showing functional inputs, system models; and various infrastructure and support tools of the IPSM, is shown in Figure 2. The blocks inside the box labelled 'Modelica Environment' represents the core of the IPSM, consisting of various 'system models', systems requirements and design rules, all of which are written in an object oriented modelling code Modelica (discussed in more detail in Section 2.2).

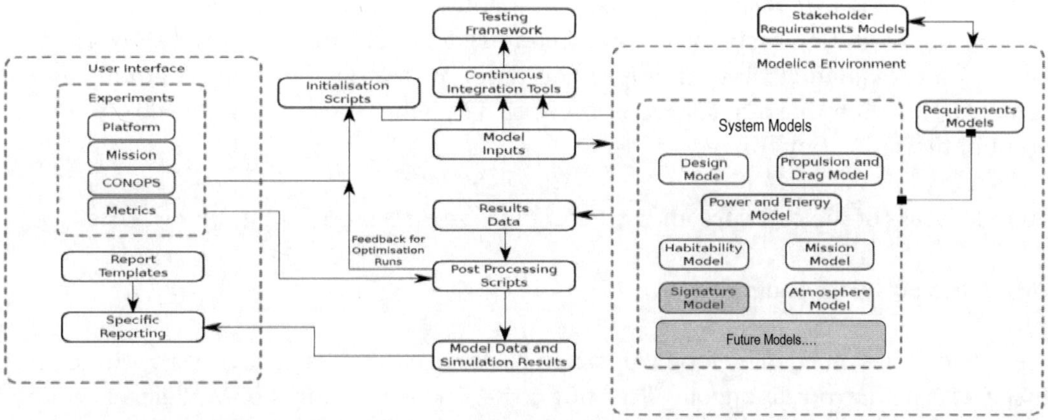

Figure 2: The 'IPSM Framework' (shaded blocks are planned for future software builds)

The 'Modelica Environment' is capable of modelling the integrated effects of the 'system models' on mission performance, enabling an assessment on whether particular measures of success or requirements are being met. The algorithms and equations underpinning several of these platform system models are derived from research, development and analyses conducted by DST subject matter experts over several decades, in addition to external sources such as [6].

An idealised representation of an example submarine mission modelled for analysis in the IPSM is shown in Figure 3, highlighting attributes of the Operational Area (OA). The OA encompasses the movement of a submarine through a patrol area, transit to a safe area for battery charging using diesel engines (snort box), and returns to the patrol area. The ability to model this level of mission detail capability has become a feature of IPSM Phase 3.0, as will be highlighted in Section 3.

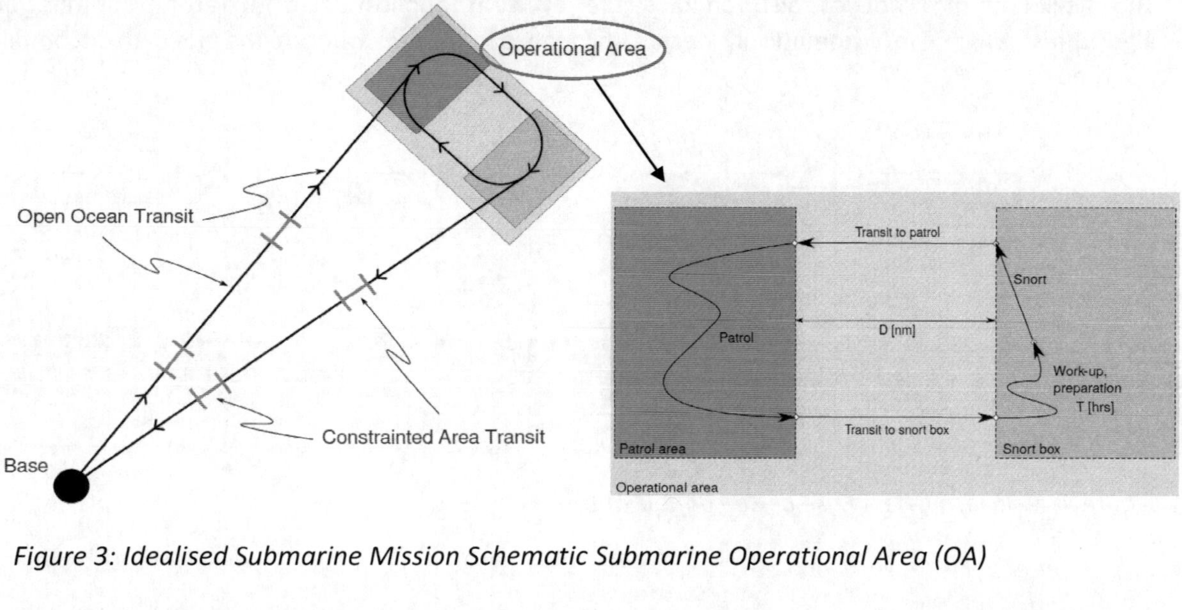

Figure 3: Idealised Submarine Mission Schematic Submarine Operational Area (OA)

The remaining blocks in the 'IPSM Framework' shown in Figure 2 represents components and tools of the software infrastructure, as well links to the stakeholder requirements and data products. These products (visual, tabular and textual results) are derived from performing model simulations. If IPSM simulations are conducted via single experiment, parametric studies (i.e. multiple simulations are performed across a range of values); and in the most recent version - Phase 3.1 - optimisation techniques have been used to determine the overall mission time and transit speed combination that maximises the time on task. Optimisation methods are currently implemented via the Dakota software package [7], which allows the flexibility of exercising different optimisation options.

Additional details of the software infrastructure will be discussed further in this paper.

2.2 Modelica Software Language

The IPSM tool has been developed using the Modelica language, a systems-based modelling language with no upfront or ongoing licencing costs. The key features of Modelica include:

- object-oriented language with replaceable classes
- equation-based modelling (also referred to as a 'system dynamic simulation')
- capable of modelling complex physical systems such as those contained in a submarine
- widely used in the modelling community, as demonstrated recently in [8].

The main advantage with an equation based modelling approach is that it gives more flexibility to re-configure and optimise the system (Figure 4), in comparison to a traditional time-step approach. Modelica implements system models internally as a nonlinear, constrained system of equations which are solved during simulation execution, given a set of initial parameters. This enables parametric analysis to be conducted, by altering these input parameters over several simulation runs, allowing the IPSM to determine the sensitivity of performance to the parameters of interest.

The flexibility of Modelica also enables the efficient application of external optimisation algorithms, which can automatically iterate through parameters without the need to recompile the model.

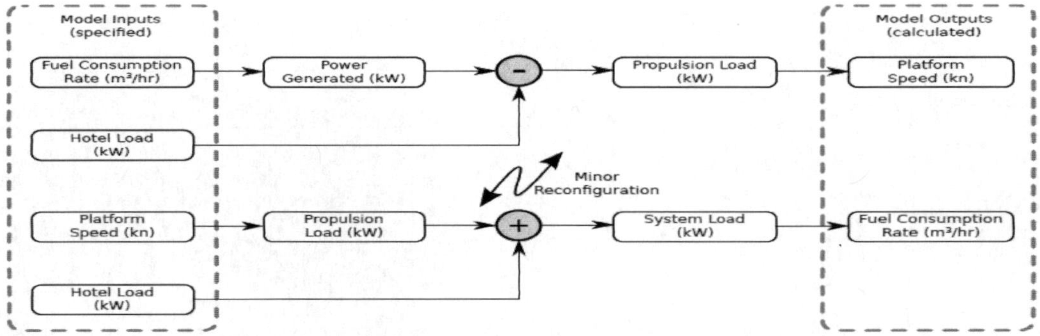

Figure 4: Modelica Offers Ease of Re-configuration

3. PHASED DEVELOPMENT APPROACH

The IPSM has been developed in a phased approach over several years as shown in Figure 5. The IPSM development team has generally comprised of 3 - 5 staff, with details of the various phases are provided in the following sections.

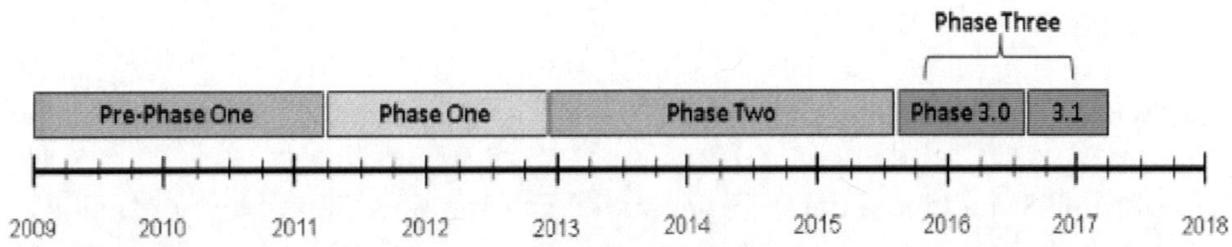

Figure 5: IPSM Development Timeline (future phases will be developed from 2017 onwards)

3.1 Pre-phase One

Prior to the IPSM, modelling was undertaken in the group using custom software written by DST staff, mainly investigating the influence of scaling submarine displacement with performance. This approach was very limited in its capabilities and wasn't suited for overall systems analysis of different submarine options. Later, a spreadsheet using empirical based equations, as previously mentioned, was developed for analysing the alternate submarine options. This approach resulted in a large cumbersome spreadsheet that was difficult to use, manage and would not allow robust or repeatable results.

The experiences learned from these early attempts demonstrated that a different approach and method for analysing platforms systems and concepts were needed. Investigations were undertaken to find a more suitable method that was robust but flexible; repeatable but always auditable, and capable of system based modelling, hence Modelica was adopted.

3.2 Phase One

Development of the IPSM Phase One started in 2011 with a view of providing a platform systems analysis capability that could evaluate whole of platform performance of future submarine concepts. The initial focus was on developing power and energy modules, as power and energy mediums were considered to have a significant influence on the relative performance of concept submarine designs.

A visual-based version of Modelica – *ModelicaML* – was used to develop IPSM Phase One. However, upon release of the IPSM Phase One, it was deemed that any future enhancements would be done using the language-based version of Modelica. The visual based package wasn't powerful or flexible enough and as such modellers devoted significant effort editing the code generated by the in built code generator.

As part of the exploratory process involved in developing the IPSM Phase One, a collaborative program was established between DST and QinetiQ GRC (UK based) to link the IPSM with Paramarine, with a view to assess the trade-space between platform performance and naval architecture attributes [4]. The intent was not to create a submarine concept design tool, but to use the advantages of each tool to complement each other to assess potential design configurations. This exercise was deemed successful as not only did it demonstrate the success of IPSM to interrogate further aspects of the design space, it also showed the flexibility of IPSM to interact with other modelling packages. As this activity was conducted as a demonstrator, no additional efforts were expended in linking Paramarine with IPSM.

Phase One also included a basic Graphical User Interface (GUI) to demonstrate its functionality, and was complete in mid-2012.

3.3 Phase Two

Phase Two of the IPSM added a parametric modelling capability (instead of only single experiment simulations), submarine atmosphere models; and waste and victualling models. It also included a significant redesign of the GUI to make it more intuitive and user friendly (Figure 6). Phase Two was delivered in late-2013.

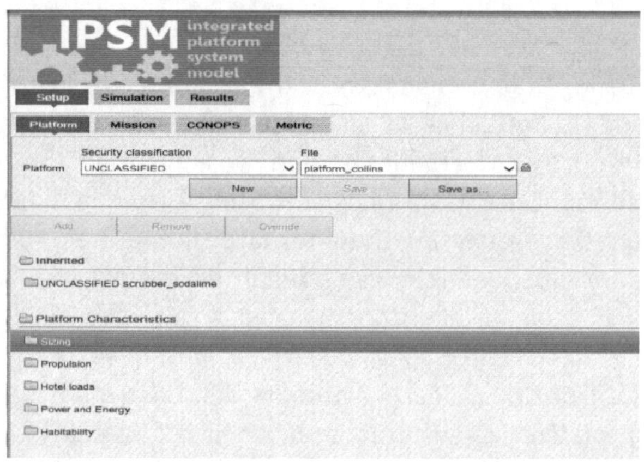

Figure 6: IPSM GUI Screenshot

As part of this phase, the development team was modelling the impact of habitability aspects of submarine design. New modules were developed to investigate aspects such as fresh, waste, grey and black water; victualling parameters (cold, cool and dry stores) and atmosphere control consumables (CO_2 canisters and oxygen candles).

Phase Two was used to support analyses of data provided to DST in relation to initial military off the shelf designs considered by the FSP in 2012. The use of the IPSM and its outcomes were received very well by the FSP stakeholders and hence further investment into the IPSM's development was made.

3.4 Phase Three

Phase Three (includes 3.0 and 3.1 – see Figure 6) was developed and used to support the FSP Competitive Evaluation Process (CEP) in 2015/16 and most recently, to provide support and evidence for developing and maturing effective requirements in the Preliminary Function and Performance Specification (PFPS). As part of the development of this phase, extensive validation activities were undertaken to validate the IPSM against 'real-world' Collins Class data. Upon successful validation, Phase 3.1 was deployed to staff embedded in the FSP, to allow them to run their own simulations.

Phase Three added additional functionality that included an enhanced battery model and an increased mission management module. This functionality allowed the user to model missions of any format so long as they can be decomposed for representation by discrete phases. This enhancement has been especially useful as it allows for significantly more flexibility in the way in which missions are modelled, for example the ability to add any number of choke points and mission leg lengths.

During the development of IPSM Phase Three the functionality of the IPSM grew significantly and as such maintaining the GUI to remain 'in-step' with the IPSM became a resource drain. Consequently, due to the prioritisation of work, the GUI is obsolete for Phase Three. Rather, the IPSM development team have developed fit for purpose training guides to assist any new users to access the IPSM via command lines in UNIX, providing greater functionality and flexibility to build platform models and run simulations. Additionally, using command lines allows users to interrogate the tool, providing an insight on how the model has been developed.

4 VERIFICATION AND VALIDATION APPROACH

The purpose of model verification is to verify that the model is consistent with its specification and fulfils its intention. This differs from model 'validation', which is primarily concerned with determining the degree to which the model is an accurate representation of the real world, from the perspective of the intended use of the model.

Software testing is conducted in the IPSM using Continuous Integration practices, which allows isolated IPSM code changes to be immediately tested and reported on when added to the main code base. Continuous Integration provides rapid feedback so that if a defect is introduced into the model, it can be identified and corrected as soon as possible. The IPSM development team have created a testing framework that has been integrated into the Jenkins software, which is used for implementing the Continuous Integration process.

For IPSM Phase One and Phase Two validation, unclassified open literature data were used. Additionally, 'brochure data' from prospective manufacturers were also used to support the validation of IPSM Phase Two.

For Phase Three, the validation activities were much more rigorous due to the availability of data. Firstly, the ISPM was cross-validated against various outputs from various power and energy models internal to DST and the FSP. Additional validation activities were then conducted using 'real-world' Collins Class submarine fleet data. The validation results showed that the outputs of the IPSM provided a close representation of the real world data, and as such the IPSM has been deployed for use by staff embedded in the FSP – a major achievement for the IPSM program and its developers. Due to the purposes of this paper and the classified nature of the validation work, validation results cannot be discussed in detail.

5. OVERVIEW OF IPSM SUPPORT AND INFRASTRUCTURE

Supporting the IPSM requires hardware, software and supporting processes. Beyond model development and execution and technical support, it is essential to ensure the outputs of the model are accurate, robust, repeatable, and traceable; and in a form that serves the needs of the intended audience. As such, standards or best practice guides in systems engineering (e.g. IEEE 1220-2005), quality management (e.g. ISO9001) and software development (e.g. ISO/IEC 12207:2008) were used to guide the design and development of the IPSM; however the strict adoption of these standards would've introduced far too much overheads for the IPSM's size.

5.1 Hardware Overview

Early in the development of the tool it was decided that it would be advantageous to have the tool available on a dedicated Defence server rather than on a stand-alone platform. This would allow multiple users to be able to use the tool from different premises across Defence and would allow the developers to keep tight control of the versions being used.

Subsequently, the IPSM is hosted on blade servers located within Defence premises. Multiple servers are used due to the different security classifications of data used in IPSM simulation runs. As the IPSM is hosted on multiple servers, an integration manager has been established that ensures the IPSM version that exist on each server are compatible, where possible – this is achieved via rigorous software testing, configuration management and version control.

5.2 Software Suite

The IPSM servers are configured with a Ubuntu Linux based operating system, however putty or Xwin-32 is used for remote display of Unix from Windows workstations. The operating system uses several software and support tools to enable the model to be developed, compiled, tested, executed and managed, such as:

- *Modelica*: IPSM is implemented in the Modelica modelling language and is compiled using the Open Modelica Compiler (OMC).
- *Git Distributed Version Control System*: In the IPSM Git is used for version control of model code, data, build scripts, test files and some documentation.
- *Jenkins*: an open sourced Continuous Integration Server that provides continuous integration services for software development. This includes building and testing (i.e. verification) software projects continuously, and monitoring the execution of externally-run jobs. In the IPSM, various Jenkins jobs are created to monitor the IPSM test suites. When the developer makes a commit to a code repository it can trigger a build of the IPSM code base, a build and execution of the IPSM unit test suite and a build and execution of the IPSM system test suite.
- *TracWiki*: an open sourced, web-based project management and bug tracking system hosted on a Wiki page. In the IPSM, Trac is used to manage and track all IPSM issues, from future enhancements to existing 'bugs' and, provides wiki facility to store supporting material for IPSM user and developers, including background information.
- *Make*: a build automation tool that automatically builds executable programs and libraries from source code by reading files called Makefiles. Make is used to build executables for the IPSM.

- *Perl*: used for building scripts in IPSM for various purposes.
- *Gnuplot*: used in IPSM to generate output plots.

5.3 Configuration Management

The Configuration Management (CM) approach for the IPSM is as light weight as possible, whilst still retaining the ability to adequately control and manage any changes to the model code, the server software and hardware configuration, and any associated documentation. The goal is to give the IPSM developers the ability to identify (and return to) a known state at any point in time.

The IPSM consists of the model code itself, the model infrastructure, formal documentation, and numerous other project artefacts that are produced as part of the development process. Each of these items is placed under configuration control and baselined, as this information is important to reproducing any previously known state of the IPSM. The following is a list of configuration items identified for control in the IPSM:

- model source code
- model input and data, including reports
- requirements specifications
- test plans
- review evidence (e.g. validation data)
- design documents
- test procedures and procedures (unit tests, system tests and acceptance tests)
- verification and validation reports
- user guides
- security documents
- software release notes
- version description documents.

The configuration management tools used in the IPSM consist of the Git Distributed Version Control System, TracWiki and internal defence archiving systems.

6. EXAMPLE ANALYSIS

As demonstrated by the analysis conducted in [9], the IPSM is capable of supporting a broad range of analysis tasks, ranging from evaluating design alternatives (to determine what capability is to be built), to developing concept of operations for the capability being introduced (to determine how the capability is to be deployed).

However, for demonstration purposes, an example analysis from using the IPSM is presented below to provide a high level overview of the IPSM's capability. This example also demonstrates how the model can be used to generate insights into the domain. The aim of this example analysis is to investigate the sensitivity of operational area (OA) performance to variation of operational parameters.

As discussed previously, a representation of a platform and of the scenario of interest is required. The platform representation at a minimum requires descriptions of the size of the platform (displacement) and parameters describing the power systems included in the design.

The subsystem used for representing scenarios or missions in the IPSM is flexible with the only requirement being that the desired scenario can be decomposed into discrete segments. Within the IPSM, the scenario is described as a state machine where each state represents a segment of the scenario. Each state has a configuration applied describing the submarine's behaviour for the present state, this includes: submarine speed, battery use profile, and relative priorities and availabilities of the installed power systems.

The example scenario used in this analysis is an idealised representation of a submarine area of operations and is shown in Figure 3. This scenario models the movement of a submarine through a patrol area, transit to a safe area for battery charging using diesel engines (snorting), and returns the patrol area.

In this analysis the patrol box in Figure 3 represents the area of interest to the submarine; as such a measure of performance is defined as the OA effectiveness: the proportion of time spent in the patrol box to the total time, Equation 1.

$$OA_{effectiven\ ess} = \frac{Time_{patrol}}{Time_{total}}$$

Equation 1: OA Effectiveness

The ability of the IPSM to conduct parametric analysis allows the IPSM to determine the sensitivity of performance to the parameters of interest. This analysis will investigate the performance variation with regard to the distance between patrol and snorting boxes (in the OA), investigated by means of a parametric sequential search through a specified range. The resulting operational area effectiveness will be recorded for each run of the scenario.

Figure 7 shows the results of the analysis with respect to the OA effectiveness. The key points of the results are that the OA effectiveness reduces linearly with the increasing distance between the patrol box and the snort box (see Figure 3). Where the distance between the two boxes is 0 nautical miles (nm) the OA effectiveness is large ~95%, however even at a modest 25 nm standoff distance the OA effectiveness has been significantly reduced to ~40%.

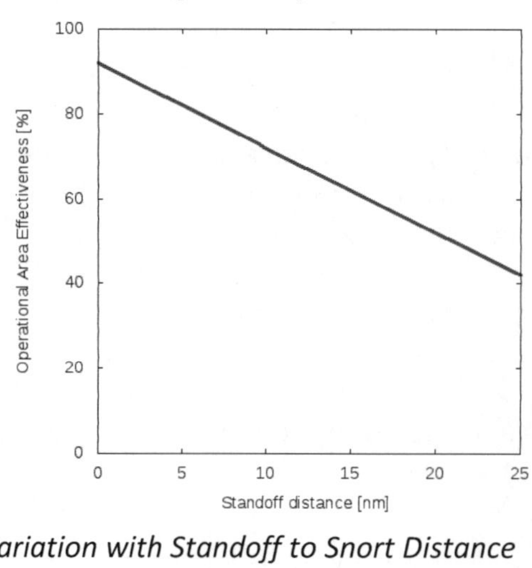

Figure 7: OA Effectiveness Variation with Standoff to Snort Distance

In the context of a study conducted to support maturation of requirements for a future platform, the outcomes of the example analysis could include:

- a requirement that specifies the minimum OA effectiveness for the submarine
- specification of a minimum distance between the patrol and snort boxes based on expected platform signature levels and expected threat environments.

7. OBSERVATIONS AND LESSONS LEARNT

Throughout the development of the IPSM tool a number of lessons have been learnt and observations made by the development team. A summary of these are listed below.

- Platform system modelling tools are highly effective and useful when representative mission scenarios and operational parameters are being considered.
    - Subsequently it is paramount that a continuous feedback loop is established between modellers and the submarine user community.
- The time required to undertake verification and validation tasks are often underestimated and not planned for sufficiently in the inception stages.
- GUIs designed for conducting analysis can be of limited utility as they tend to constrain the type of analysis that can be performed.
    - The IPSM development team spent a lot of time developing the GUI for use in Phase Two, but ultimately it became underutilised as the IPSM's capability grew.
- Effective visualisation of results is critical for communicating findings to a broad audience.
- Enabling access of the IPSM using a remote login ensures that all users are using the most updated software build of the tool.
- Invest effort in reducing the run-time of the model - a single experiment simulation of the IPSM is in the order of a few seconds, making it highly attractive from a usability perspective.
- The end goal of the model cannot be forgotten i.e. it is not necessary to have a perfect or fully developed model to generate effective results.
- Analytical tools do not follow a linear environment where 'development' is followed by 'use'. For example:
    - tools can highlight limitations in the analysis being conducted, requiring further development to address these limitations.
    - introduction of new technologies need to be considered, requiring new system models to be developed and integrated into the overarching model.
    - different mission scenarios or platform tactics need to be evaluated, requiring a different approach or new models to be developed for operational analysis.
- A software development team must be diverse in their skills and not all focussed on the technical modelling aspects.
- Having multiple models with overlapping capability, that are independently developed, promotes trust in the development, use and validation of models.
- The capability provided by the IPSM provides a useful test-bed for novel concepts of operation as demonstrated in [9].

- Reconstructing performance estimates provided by prospective tenders has enabled the impact of uncertainty in sub-system performance to be quantified; and all tenders evaluated fairly with a common set of assumptions.

8. SUMMARY OF IPSM KEY FEATURES

In summary the IPSM provides the following features and capabilities:

- Platform level performance parameters are calculated based on integrating multiple system models, as opposed to a coarse single-equipment approach.
- Post processing of simulations to produce graphical, tabular and textual results to analyse and compare results of multiple platform concepts against the same criteria.
- The execution of single experiment simulations, parametric simulations and optimised simulations to generate performance and operational attributes.
- Effective management and segmentation of data, coupled with strict access control, to ensure varying levels of security classification restrictions are handled appropriately.
- The ability to undertake platform related sensitivity analysis of performance for existing, alternative and new technologies.
- Ease of re-configuration between inputs and outputs; and reuse of various parameters.
- Ability to accept user defined metrics to be applied in order to test a series of conditions on simulated results (i.e. for verification and validation activities).
- An automated and flexible reporting function that is highly effective to the FSP.
- Continuous and automated software testing (at unit level and system-level); and effective version control, quality management and configuration management.
- Extensive logs of simulations and data products for tracking and auditing purposes.
- A series of secure wiki pages used for information sharing, which provides up to date user guides; and other supporting material and requirements for users and developers.

9. IPSM EXPANSION PLANS

The success in generating auditable, repeatable, clear, meaningful and reliable results using the IPSM has shaped and matured various platform requirements for the FSP. Subsequently, the IPSM has become an exemplar modelling tool and in the short term the IPSM will continue to be deployed during various phases of the FSP, as more platform details emerge. The IPSM will continue to be maintained, verified and validated as more modelling and in-service Collins Class data becomes available. Additionally, IPSM outputs will be used to support the validation of deep-dive models.

The flexibility, adaptability and effectiveness of the IPSM, enables the tool to be used for in service applications. As such, there is interest from the Collins Class submarine community in using the IPSM to support pre/post-mission planning analysis and activities.

DST, with support from key FSP stakeholders, is also interested in expanding the functionality of the IPSM to directly support assessments of measures of effectiveness. The stealthy nature of submarine operations means that platform performance parameters strongly interface with the threat domain. As such the IPSM development team are in the inception stages of developing frameworks to understand the trade-off space between platform performance, mission effectiveness and susceptibility to exposure; and are aiming to provide a demonstration of this capability over the next year.

The case study provided in [9] showed that the IPSM can be adapted to assess the benefits of particular technology concepts. As such, the IPSM could be used to assess the capability impacts of various technology demonstrators considered in the FSP. Additionally, this assessment could also consider any integration impacts that this technology has on overall performance.

10. CONCLUDING REMARKS

The IPSM has been developed incrementally over the past ten years using DST's knowledge on submarine systems, coupled with their diverse experience in systems modelling and simulation. Subsequently, the IPSM has evolved from an Excel based exploratory tool to a highly powerful, integrated, sophisticated and validated integrated platform systems modelling tool.

The DST IPSM tool is a highly effective and respected tool for providing analysis and advice on platform performance parameters and trade-off studies for Australia's FSP; one of Defence's flag ship programs over the next few decades.

ACKNOWLEDGEMENTS

It is acknowledged that the content in this paper was first published by the Royal Institution of Naval Architects (RINA) as proceedings for 'Warship 2017 - Naval Submarines and UUVs', 14-15 June 2017, UK.

The authors acknowledge the following personnel that have contributed to the development of the IPSM: Mr Alex Bulach, Mr Simon Demediuk, Mr Jon Hornstein, Mr Rob Jarvis, Mr Josh Rodgers, Mr Mario Selvestrel, Mr Andrew Tynan; and in particular, Dr. John Wharington.

REFERENCES

1. KASSEL, B., COOPER, S. and MacKenna, A., 'Rebuilding the NAVSEA Early Stage Ship Design Environment', ASNE Day - Engineering the Affordable Global Navy Through Innovation, USA - Virginia, 2010.
2. RODGERS, J. et al., 'A Framework for Integrated Platform System Modelling', Maritime / Air Systems and Technologies (MAST) Conference, Europe - Malmö, 2012.
3. HARRISON, S. et al., 'Analysis of Platform Configurations Using the Integrated Platform Systems Model', International Maritime Conference – Pacific 2012, Australia - Sydney, 2012.
4. HEPWORTH, G. et al., 'High Level Capability Analysis of Submarine Platforms', RINA Submarine Conference, UK - Bath, 2014
5. HARRISON S., Wharington, J. and Slater, K., 'An Investigation into the Penalties Incurred due to Operational Constraints Imposed on an SSK-type Submarine in an Operational Area', International Maritime Conference – Pacific 2015, Australia - Sydney, 2015.
6. BURCHER, R. and RYDILL, L., 'Concepts in Submarine Design', Cambridge University Press, 1995
7. ADAMS, B.M. et al., 'Dakota, A Multilevel Parallel Object-Oriented Framework for Design Optimization, Parameter Estimation, Uncertainty Quantification, and Sensitivity Analysis: Version 6.5 User's Manual', Sandia Technical Report SAND2014-4633, updated November 2016.
8. SUTHERLAND, J. et al., 'System-Level Design Trade Studies by Multi Objective Decision Analysis (MODA) utilizing Modelica', Proceedings of the 1st Japanese Modelica Conference, Japan - Tokyo, 2016.
9. HARRISON, S., et al., 'The Operational Benefits of Covertly Refuelling a Conventional Submarine whilst at Sea', International Maritime Conference – Pacific 2017, Australia - Sydney, 2017.

The Operational Benefits of Covertly Refuelling a Conventional Submarine whilst at Sea

Simon Harrison[1], Hamid Diab[1], Karl Slater[1] and Martin Renilson[2]
[1] Defence Science and Technology Group, Australia
[2] University of Tasmania, Australia
simon.harrison@dst.defence.gov.au

1. INTRODUCTION

The Defence Science and Technology (DST) Group, which forms part of the Australian Department of Defence, have been applying integrated systems modelling techniques to provide whole of platform advice to the Australian Future Submarine Program (FSP). Since 2010 DST has been developing a tool known as the Integrated Platform System Model (IPSM) to model, simulate and analyse platform performance parameters [1-4]. This capability forms part of a wider DST initiative to provide Defence stakeholders methods for clear and robust assessment of top level capability objectives, platform characteristics, and system requirements.

The IPSM enables trade-off studies to be conducted between platform performance; platform concepts and configurations; and operational parameters. The IPSM integrates system models of power and energy systems (including main storage batteries, diesel and air independent propulsion systems), propulsion and habitability systems (including atmosphere management, consumable food, waste stowage and discharge; and accommodation spaces), and can compare the platform's overall performance against various capability and performance requirements and constraints.

In this paper, a demonstration of the IPSM's adaptability to assess the impacts of novel technology concepts is presented. The IPSM has been used to consider the operational benefits of refuelling a conventional diesel-electric submarine at sea. The intent of replenishment at sea (RAS) of diesel is to increase the level of capability achievable by existing submarines without significantly compromising the stealthy nature of their operations.

The technical feasibility of the refuelling concept is not the subject of this paper. However, the operational benefits and impacts are investigated with the use of DST's IPSM. A brief overview of the technology concept is given prior to the analysis and provides an excellent case study for the IPSM to examine a novel technology with low technology readiness level.

The advantage of using the IPSM at the conceptual phase is that it can be re-configured relatively quickly to assess operational impacts resulting from different technologies, concepts of operations, and scenarios. Therefore the IPSM has the potential to provide valuable inputs to business cases considering whether or not to develop the technology further. This makes the IPSM a powerful tool that could be used to support trade-off studies for future technology concepts considered as part of the FSP.

2. TECHNOLOGY CONCEPT

A concept for refuelling conventional submarines at sea is presented in [5]. This is based on the use of submerged fuel tanks that could be statically located at strategic positions at sea. For routine operations the fuel tanks could be transported to the designated area in advance by surface ships, and then moored to a float in the desired location. The engineering issues associated with transport of tanks into theatre are outside of the scope of this study. Figure 1 shows a diagram of the concept as described by Renilson [5].

It is proposed that the submarine will be equipped to allow it to approach a fuel tank which is moored in place and capture it. The connections between the submarine and the fuel tank may be similar to what is proposed in [6].

The submarine will deploy its catching equipment when approaching the moored fuel tank arrangement at a depth of around 50-60 metres (m). At this depth motions imposed by the surface waves will be minimal, and the submarine will be able to be controlled easily [7].

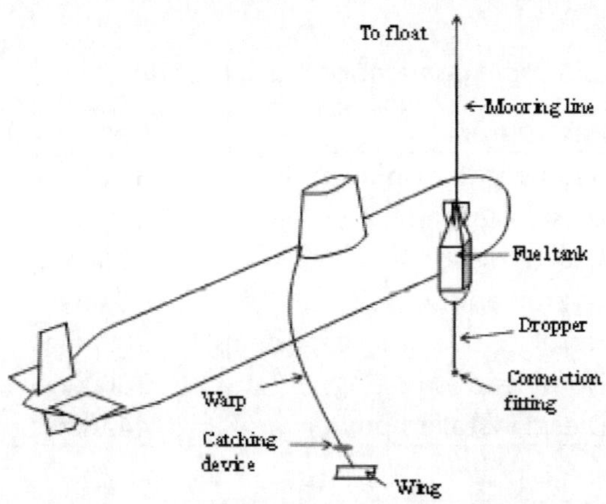

Figure 1: Schematic of Submarine Approaching Fuel Tank and Equipment (not to scale)

Once the fuel tank has been captured and appropriate connections between the submarine and fuel tank have been made refuelling will occur. Equipment will be required to ensure that the fuel tank can remain neutrally buoyant. Likewise the submarine will need to take appropriate action to maintain the desired buoyancy during the refuelling process.

It is acknowledged that this novel concept is far from reality at present and that there are many engineering difficulties to be overcome before being able to refuel a submarine at sea. However, assessing the technical feasibility of the technology concept is not the intent or priority of this paper. Rather, the refuelling concept provides a case study for DST to demonstrate the ability of IPSM to assess operational impacts of various technology concepts, irrespective of its technology readiness level.

3. ANALYSIS

In order to determine the potential benefits of RAS of diesel fuel for a conventional diesel-electric submarine, it is useful to demonstrate the conditions under which conventional submarines may become constrained in operation by the quantity of embarked diesel fuel. Constraining factors for conventional submarines can include: overall endurance, platform speed, power, energy storage, internal atmosphere, victuals, and diesel capacity. The constraining factor may vary from mission to mission for any given submarine design, but the overall range and endurance for a single deployment is usually constrained by either the submarine's overall endurance limit or by the quantity of usable diesel fuel. Typically, overall endurance is characterised by the endurance of the submarine related to the embarked quantity of food and endurance of the crew.

To demonstrate these constraints an example submarine and mission are modelled within DST's IPSM and described by the parameters in Table 1. These values are based on open literature sources [8-9] for Australia's Collins Class submarine. Figure 2 shows the format of the example mission, where the measure of success is time spent in the operational area, and the transit range to the operational area has been set at 2500 nautical miles (nm).

Table 1: Parameters of an Example Conventional Submarine

Parameter	Value
Submerged displacement	3353 tonnes
Diesel fuel capacity	268 m^3
Usable diesel	90% of capacity
Overall endurance	70 days
Hotel load submerged (deep)	200 kilowatts (kW)
Hotel load (snorting)	200 kW
Diesel installed power	4420 kW

3.1 Determining when the Submarine is Diesel Limited

In order to show how a submarine can become diesel constrained, a parametric analysis was conducted to show the variation of time spent in the operational area and diesel consumed with transit speed. In order to reduce the degrees of freedom for this scenario the speed whilst snorting was fixed at 10 knots. This was done for the purposes of this study only; the IPSM supports the specification of speed in snort and submerged deep states. Variation of transit speed was achieved through variation of the submarine's speed whilst submerged deep using energy from the main storage battery.

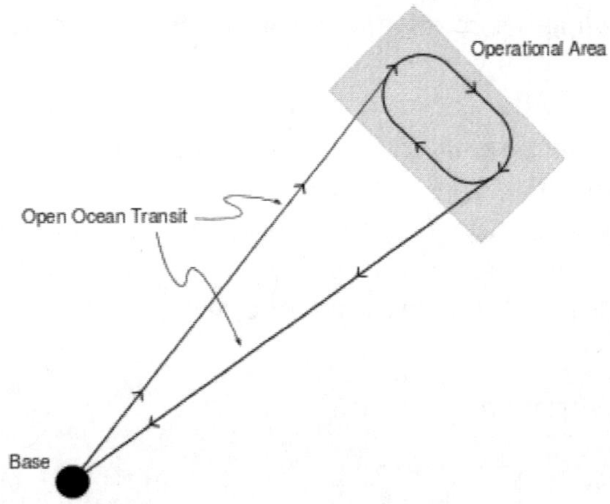

Figure 2: Example Mission

Figure 3 shows the results of a parametric variation of transit speed, reported as speed of advance (SoA), and the resulting percentage of the diesel fuel and overall endurance resources consumed. The shading on the plot indicates the proportion of resource used in transit and in the operations area (OA). The figure demonstrates that the SoA to achieve the maximum time in the operations area occurs at approximately 8.5 knots and that at this SoA the diesel fuel and overall endurance resources are exhausted coincidentally. At a SoA less than 8.5 knots the entirety of the available overall endurance is exhausted whilst leaving some portion of diesel fuel unused. Conversely, at SoA higher than 8.5 knots all available diesel fuel is consumed whilst a portion of the overall endurance remains unused. The later indicates the case where the submarine is constrained by the quantity of embarked diesel fuel and this is the focus of this study.

The simulations conducted for this study placed no additional constraints on the submarine whilst in transit. In real world submarine operations it is likely that various constraining factors will interact and result in constraining the speed of the submarine below the optimal speed for maximum OA endurance. Examples of such factors include varying threat levels and prevailing weather conditions. Therefore it is not unreasonable to assume that a submarine conducting a mission similar to that shown in Figure 2 may have difficulty in maintaining transit speed to attain the maximum OA endurance associated with the submarine design.

The SoA required to achieve the maximum OA endurance is a function of the submarine design and the mission, specifically the range to the OA that is being conducted. Longer transit distances result in a reduced speed for maximum OA endurance. This in turn means that a larger range of transit speeds result in a diesel constrained operation. This is analogous to reducing the embarked quantity of diesel fuel, which may be the case for different submarine designs. Hence the degree to which a submarine can become diesel constrained increases with range to the OA and reduced diesel capacity. Therefore it is expected that the benefit of RAS of diesel would be less for a submarine with large diesel capacity conducting a short range mission and greater for a submarine with a smaller quantity of diesel conducting a long range mission.

The key message is that if the submarine conducts a transit at any speed greater than the SoA to achieve maximum OA endurance, it will be constrained by the quantity of diesel fuel. This is clearly shown in Figure 3. Refuelling the conventional submarine mid-deployment is a possible strategy to relieve this effect.

The following sections will now investigate the potential benefits of RAS of diesel fuel. The hypothesis of the analysis is that RAS of diesel fuel will allow the submarine to operate at a transit SoA that would otherwise result in diesel constrained operation. With the key factor of the hypothesis being that faster transit reduces the proportion of overall endurance spent in transit legs and increasing the proportion spent in the operations area.

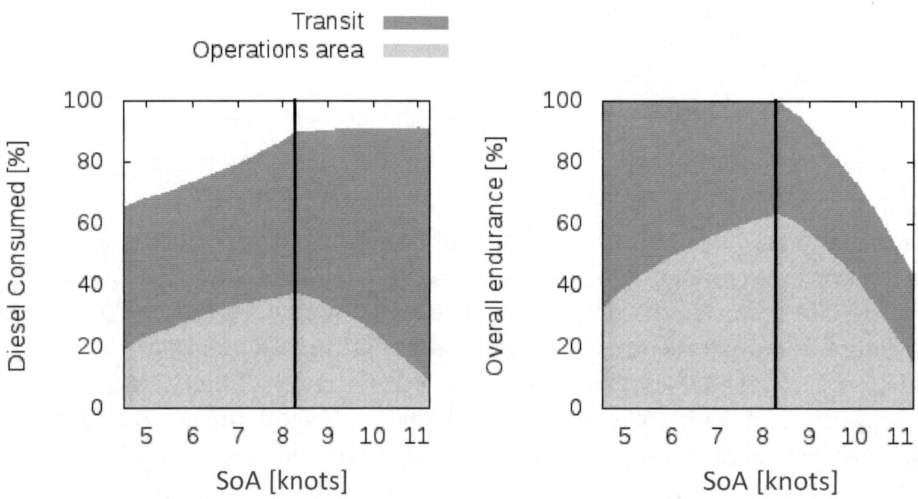

Figure 3: Diesel Use and Overall Endurance vs SoA (constant snort speed, range to OA of 2500 nm and final 10% of diesel is held in reserve)

3.2 Maximising OA Endurance

Given that a faster transit will result in greater operational area endurance if the issue of increased diesel consumption can be resolved through RAS, then the maximum transit SoA of the submarine needs to be determined. To achieve this, a parametric simulation of snort cycle transit was setup and run in the IPSM to show the resulting transit SoA for different submerged deep and snorting speed combinations. As previously, the snorting speed was held constant at 10 knots, usable diesel capacity was capped at 90% of total embarked diesel fuel, and the simulation was terminated at the exhaustion of the usable diesel capacity.

The results of the parametric simulation are shown in Figure 4 and illustrate the variation of specific diesel consumption with SoA. SoA was varied by increasing the speed of the submarine when submerged deep under battery power whilst the snort speed was held constant at 10 knots.

There are two points of interest, “A” and “B”, marked on Figure 4. Point A on the figure indicates the minimum with respect to specific diesel consumption and occurs at a SoA of approximately 5.5 knots, which corresponds to the speed for achieving maximum range. Thus any increase or reduction in speed will result in a reduced possible range.

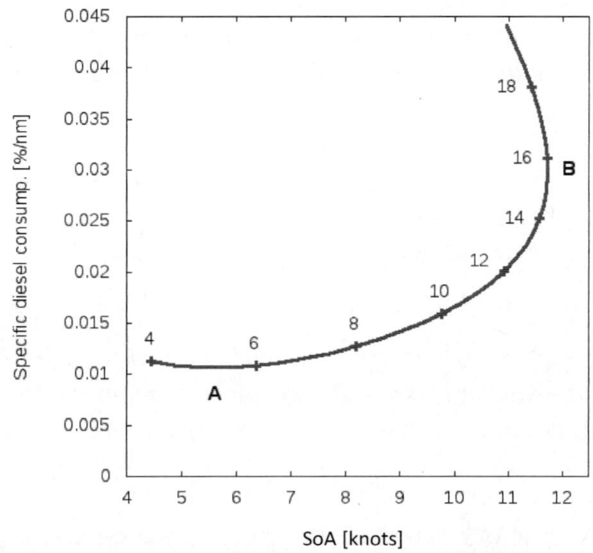

Figure 4: Specific Diesel Consumption Variation with SoA (labels on the figure show the submerged deep speed and snorting speed was held constant at 10 knots)

Point B in the figure shows the maximum achievable SoA of approximately 11.75 knots. Increases to the speed whilst deep beyond 11.75 knots will reduce SoA. This result is somewhat counter intuitive as it shows that increased speed whilst deep utilising battery power can result in a reduction of SoA. This effect occurs because as the speed of the submarine increases the endurance of a single battery discharge decreases. Eventually the submarine spends a greater amount of time snorting than submerged at depth thus resulting in the lower SoA.

The maximum achievable SoA of the submarine occurs at point B in Figure 4. This therefore represents the fastest possible transit SoA the submarine can achieve. If we assume that the submarine has enough diesel fuel to transit at the maximum SoA whether through RAS or otherwise, the maximum possible OA endurance can be calculated. The reader should recall that for this study the speed during snorting has been held at a constant 10 knots for all simulations. A faster maximum SoA than what is shown in Figure 4 may be possible with variation of the snorting speed.

Equation 1 shows the maximum OA endurance as a function of overall submarine endurance, transit speed, and range to the operational area. Where E is endurance given in days, R is range in nm, and S is SoA given in knots.

$$E_{oa,\max} = E_{overall} - \left(\frac{R_{oa}}{12\, S_{\max}} \right)$$

Equation 1: Maximum OA Endurance

If the submarine can transit at the maximum transit SoA, this will result in the greatest amount of time spent in the operations area. The increase in OA endurance will need to be traded off against the steep increase in diesel consumption. It is possible that some short range missions could be conducted with transit at maximum SoA but the achievable transit distances are likely to be very short. Similarly it may be possible for submarines with large diesel capacity relative to their displacement, to complete missions at maximum SoA.

In the presence of the overall endurance constraint, the maximum endurance in an operations area occurs when transit is conducted at the maximum achievable SoA. As previously mentioned, this is not an achievable goal due to the high diesel consumption at maximum SoA. Furthermore, the previous section demonstrated that the optimal transit SoA to maximise OA endurance is significantly lower than the maximum SoA which corresponds to a reduced OA endurance. Refuelling at sea provides one possible solution to allow the submarine to transit at maximum SoA and therefore achieve the maximum OA endurance.

3.3 Timing of RAS

Given that transit at maximum SoA requires a specific fuel consumption that very likely cannot be supported by the embarked quantity of diesel, it is important to investigate the possible timings for RAS of diesel.

Consider the case where a single RAS event is conducted for the duration of a deployment at the completion of the first transit section prior to entry to the OA. If we assume that this event refuels the submarine to maximum diesel capacity, the usable diesel quantity must be sufficient for both the required OA endurance and the return journey transit. Thus when the transit range to the OA becomes large enough such that the return journey and OA patrol together require more than the usable quantity of diesel embarked, two RAS events will be required. Nominally this could be one before and one after patrol in the OA.

Given that for the example submarine in this study (see Table 1) the diesel consumed in a 2500 nm transit is ≈75%. The return journey is assumed to require the same quantity of diesel but may vary in reality according to tasking, prevailing weather, and other factors. As such, if a single RAS is conducted immediately prior to entering the OA, 75% of diesel must be reserved for the return transit. This leaves 15 % (allowing for 10 % reserve) for patrol in the OA, approximately 17 days of patrol. This brings the total duration of the deployment to 35 days, indicating that the submarine is still constrained by diesel availability.

This also shows that the required capacity of the refuelling tank is dependent on the quantity of fuel used in transit. The sizing of the refuelling tank is investigated in the following section.

If two RAS events were conducted, each refuelling the submarine to maximum capacity one before entry and one after exiting the OA, diesel sufficient to exhaust overall endurance is available (assuming refuelling back to maximum capacity) whilst still allowing for transit at maximum SoA. In this case the OA endurance will be limited to 52 days by the overall endurance and maximum transit SoA constraints.

There is a significant risk for the submarine when operating in a manner as just described. Allowing the available diesel to reduce to a point where the submarine may not have enough diesel to transit to a friendly port presents a situation with significantly increased risk. This risk needs to be traded against the operational benefit of the increased OA endurance. This trade-off is beyond the scope of this study and is a potential avenue of future work.

3.4 Sizing Refuelling Tanks for Maximum OA Endurance

The conditions under which diesel constrained operation have been shown, and the illustration of the maximum transit SoA that limits the maximum possible operational area endurance has been discussed. The benefits from refuelling a submarine mid-deployment are now investigated, specifically with respect to the mission shown in Figure 2.

Table 2 shows the increase in the achievable operational area endurance for three set points of transit SoA. The benefit to OA endurance from refuelling is significant. However, there is a significant increase in the quantity of diesel consumed and the values shown represent what is possible if the submarine can be refuelled to maximum diesel capacity. Additionally, in order to achieve high transit speed the indiscretion ratio of the submarine is greatly increased. This should be traded along with other considerations such as size of refuelling tanks, cost, logistics, stealth, and maintenance of a RAS system against the increased operational effect achieved by the increased OA endurance. Furthermore, the signature profile and risk associated with RAS of diesel fuel is unknown and is a key factor in determining the difference in operational effect of the submarine. However this is beyond the scope of this study.

Table 2: OA Endurance at Selected Transit SoA (assumes refuelling to maximum diesel capacity)

Transit SoA [knots]	OA Endurance [days]
5.5 (min diesel consumption)	32
8.5 (no RAS, baseline)	44
11.75 (max SoA with RAS)	52

Table 3 shows a comparison of the OA endurance possible with a number of different combinations of transit SoA and refuelling tank capacities.

Table 3: OA Endurance for Different Transit SoA and Refuel Tank Capacities

Transit SoA [knots]	Refuelling Tank(s) Diesel Capacity [tonnes]	OA Endurance [days]
8.5 (baseline)	Nil	44
9.5	34	48
11.75 (max SoA)	34	0
11.75 (max SoA)	237	52

The first row of Table 3 shows the baseline case where no refuelling occurs. The best OA endurance possible in this case was illustrated in Figure 3 and is 44 days using a transit SoA of 8.5 knots.

The second and third rows in Table 3 show the original concept from [5] with a refuelling tank with capacity of approximately 34 tonnes of diesel. A refuelling station with 34 tonnes capacity would be sufficient to refuel only 15 % of the submarine's diesel tank and would not be sufficient to allow any patrol of the OA when using the maximum transit SoA, 11.75 knots. However, an increase in OA endurance is achievable using a transit SoA of 9.5 knots. Further work to investigate different transit speeds across a range of mission profiles and the resulting possible benefit over the baseline case should be done as follow-on work to this study.

A possible upper limit for the refuelling diesel capacity is shown in the final row of Table 4. The capacity in this case is sufficient to allow transit at maximum SoA and OA endurance such that the overall endurance is exhausted, 52 days in this case. The capacity of the required diesel for refuelling to satisfy this is 237 tonnes of diesel and will need to be delivered across two refuelling events. The required diesel in this case is very high due to the steep increase in specific fuel consumption when using maximum transit SoA and the very long transit conducted for the example mission. This case will allow for the full use of overall endurance and maximum transit SoA and will exhaust diesel and overall endurance simultaneously. Whilst shorter transit distances will require a reduced capacity of diesel for refuelling, the reduction of time spent in transit achieved by using maximum SoA is also reduced, thus reducing the overall benefit to time in the OA.

To envisage the physical dimensions of tank sizes required, Table 4 shows the length and diameter for each of the tank capacity options shown in Table 3 allowing only for the volume of diesel required per refuelling event. The tank sizes are given as a length of 2 and 3 metre diameter cylinders. The values are very large indicating that refuelling tanks sized to maximise OA endurance and allow for the use of maximum transit speeds are infeasible. Additionally, towing of tanks this size by the submarine during the refuelling process is likely to be problematic.

Table 4: Example Refuelling Tank Sizing

Refuelling Tank Diesel Capacity [tonnes]	**2-metre dia. Cylinder Length [m]**	**3-metre dia. Cylinder Length [m]**
34 (concept)	12.7	5.7
119 (maximum)	44.4	19.7

Large benefits from refuelling the submarine can only be realised when the submarine transits at maximum SoA and the transit distance is significant. A possible maximum tank size has been presented but is significantly larger than proposed in the original concept and sensitive to the quantity of fuel used in transit. Alternatively, a succession of refuelling events from smaller tanks may be able to provide the required diesel quantity. This presents a number of feasibility issues that would need to be resolved for refuelling submarines at sea from tanks such as those presented in [5] to become worthwhile. Furthermore the duration of the refuelling event, which has not been investigated in this study, will reduce the benefit of refuelling significantly if the process is slow. This would be compounded in a concept where multiple refuelling occurs from a succession of smaller tanks. This may occur in order to embark the desired quantity of diesel whilst limiting refuel tank sizes to feasible dimensions. The technical aspects of this are beyond the scope of this study and should be investigated as future work.

4. CONCLUSIONS AND FUTURE WORK

In summary, platform performance impacts of the ability to refuel a conventional submarine whilst at sea have been investigated as a case study for the DST Group's IPSM. The IPSM has been designed with the intention of providing a capability to conduct trade-off studies across different technologies, concepts of operations, and scenarios in a robust and validated manner. This study provides an example of how the IPSM can be used as a test-bed to assess novel technology options in order to assist decision makers and stakeholders to determine which options may be worth pursuing further.

With regard to the specific findings of the case study, the following conclusions have been drawn:

- An optimal speed for transit exists specific to the mission and submarine performing the mission, where the measure of success is endurance in the operations area.
- A physical upper limit exists for transit SoA, placing a lower limit on the time transit stages can be completed.
- When operating at the maximum transit SoA it is very likely that a conventional submarine will become constrained by its diesel capacity. Refuelling the submarine offers a potential relief to this effect if the required capacity of diesel can be embarked.
- A maximum refuelling tank size has been discussed that allows diesel quantity sufficient to exhaust overall endurance and allow transit at maximum SoA. However, the tank size is very large.
- The quantity of diesel required for refuelling is sensitive to the submarine design and the transit range to the OA.
- The largest benefits from refuelling occur for submarines with small diesel capacity and in missions with long transit distances.
- If RAS of diesel can be shown to be practical it may have an impact on submarine design. This may result in designs with reduced diesel capacity, and allow for increased food stores.

Whilst there are some clear benefits to the ability to refuel the submarine at sea, a range of areas for future work and other considerations have been identified:

- Further investigation and optimisation of transit speed across a range of mission types is required in order to fully determine the potential benefits of refuelling to OA endurance.
- Required refuelling diesel capacity is large when transit range is large. This presents a significant engineering and logistical problem.
- The signature aspects and risk associated with refuelling at sea should be further investigated.
- The benefit to OA endurance from refuelling is significant. However there is a significant increase in the quantity of diesel consumed in order to achieve this increase and as a result the quantity of diesel for refuelling is very large. This should be traded along with cost, logistics, stealth, and maintenance of a RAS system against the increased operational effect achieved by the increased OA endurance.
- The risk associated with allowing the submarine to use the majority of embarked diesel whilst operating far from friendly ports should be assessed in terms of the operational effect that may be gained.

- If self compensating fuel tanks are used in the submarine, refuelling will require diesel contaminated seawater to be discharged from the submarine during refuelling. The potential environmental and legal impacts of this should be investigated further.

5. ACKNOWLEDGEMENTS

It is acknowledged that the content in this paper was first published by the Royal Institution of Naval Architects (RINA), and presented at 'Warship 2017 - Naval Submarines and UUVs', 14-15 June 2017, UK.

The authors acknowledge the following personnel that have contributed to the development of the Integrated Platform System used in this case study: Mr. Alex Bulach, Mr. Simon Demediuk, Mr. Jon Hornstein, Mr. Rob Jarvis, Mr. Josh Rodgers, Mr. Mario Selvestrel, Mr. Andrew Tynan; and in particular, Dr. John Wharington.

6. REFERENCES

1. DIAB, H. et al., 'Integrated Performance Modelling of Submarines to Support Strategic Decision Making for the Future Submarine Program', International Maritime Conference – Pacific 2017, Australia, Sydney, 2017.
2. HARRISON, S. et al., 'Analysis of Platform Configurations Using the Integrated Platform Systems Model', International Maritime Conference – Pacific 2012, Australia, Sydney, 2012.
3. HEPWORTH, G. et al., 'High Level Capability Analysis of Submarine Platforms', RINA Submarine Conference, UK, Bath, 2014.
4. HARRISON S., Wharington, J. and Slater, K., 'An Investigation into the Penalties Incurred due to Operational Constraints Imposed on an SSK-type Submarine in an Operational Area', International Maritime Conference – Pacific 2015, Australia, Sydney, 2015.
5. RENILSON, M., 'Extending the Range of a Conventional Submarine by Autonomous Covert Refuelling when on Patrol', International Maritime Conference – Pacific 2015, Australia, Sydney, 2015.
6. RENILSON, M., 'A Simplified Concept for Recovering a UUV to a Submarine', Underwater Technology, Vol. 32, No. 3, pp. 193-197, 2014.
7. RENILSON, M., 'Submarine Hydrodynamics', Springer Briefs in Applied Sciences and Technology, ISSN: 2191-530X, ISBN: 978-3-319-16183-9.
8. GRAZEBROOK, A., 'Collins Class Comes Up Down Under', Jane's Navy International, Janes Information Group, 1998.
9. 'Jane's Fighting Ships 2012-2013', Jane's Information Group, 2013.

The Influence of Appendages and their Stall on Submarine Hydrodynamic Loads

Gregory J. Seil[1,4] , gregory.seil@dst.defence.gov.au
David Pook[2,4], david.pook@dst.defence.gov.au
Mai Chi Nguyen [1,4] , mai-chi.nguyen@dst.defence.gov.au
Zhi Quan Leong [3,4] , zhi.leong@utas.edu.au

[1] Advanced VTOL Technologies Pty Ltd
[2] School of Engineering, RMIT University
[3]Australian Maritime College, University of Tasmania
[4] Maritime Division, Defence Science and Technology Group

ABSTRACT

Global hydrodynamic loads as a function of drift angle are presented for the Defence Science and Technology Group (DST) generic submarine geometry BB2. Loads are calculated for a model-scale Reynolds number using Computational Fluid Dynamics (CFD) at angles of drift. Abrupt change in the global loads are attributed to appendage stall. Surprisingly, CFD-predicted aft-control-surface (ACS) and bridge fin stall are not easily identified from the sway force. Instead, secondary load components, such as the heave force, must be used to identify that an appendage has stalled. Identifying which ACS has stalled is not possible from global load data. Appendage pressure loads and surface flow topology are required to confirm appendage stall. The results presented will aid the interpretation of experimental load tests currently being conducted by DST.

INTRODUCTION

When a submarine manoeuvres and is at an angle-of-incidence to the oncoming flow, the global forces and moments may exhibit both linear and non-linear variation with incidence angle. For example, at an angle-of-drift (AoD), the sway force, roll moment and yaw moment acting on the submarine exhibit (near) linear variation for small AoD changes. The heave force and pitch moment exhibit a non-linear (quadratic) variation with AoD. Another form of non-linearity is the abrupt load changes that can occur as individual appendages stall.

Appendage stall has implications for the prediction of manoeuvring performance and the design of control systems. It makes the prediction of manoeuvring performance difficult for two reasons: the first is the need for the manoeuvring model to account for abrupt changes in the hydrodynamic forces and moments; the second is reduced confidence in the force and moment prediction due to difficulties predicting stall and post stall loads accurately. Post stall, hydrodynamic loads may vary significantly as a function of time due to unsteadiness in regions of separated flow. Reynolds number effects may also be significant. For stable operation of the boat and the design of its control system, it is therefore essential to be able to predict incidence angles when abrupt changes in global forces and moments occur.

Prediction of appendage stall with semi-empirical methods is often not reliable or possible. Determination of appendage stall from experiment can also be difficult. For example, from the DARPA Suboff load data in Roddy [1], bridge fin stall can be inferred for a drift angle of approximately 16°. This is done by examining loads data for the Suboff tested without the aft control surfaces (ACS), and identifying the abrupt change in the sway force that must be attributed to bridge fin stall. However, due to the effect of the bridge fin wake and trailing vortex when at an AoD, the stall of the ACS cannot be found experimentally by removing the bridge fin. This can make determination of a given appendage stall, or attributing abrupt changes in global loads to the stall of a given appendage, impossible without some other measurement quantity (e.g. surface flow topology).

The data from Mackay [2] for the `Standard Submarine Model' geometry, which tested the same submarine geometry in multiple facilities to assess load dependency on facility and experimental procedures, again identified bridge fin stall by tests without the ACS. However, bridge fin stall varied from approximately 12° AoD to in excess of 20° depending on the test conditions. Stall of the hull and ACS only configuration was reported as possibly visible in only one data set, for an AoD of approximately 20°. This result suggests that appendage stall might be difficult to discern from experiments that measure only global loads.

The Defence Science and Technology Group (DST) generic submarine geometry BB2 is currently undergoing experimental testing to determine loads. A priori knowledge of appendage stall, and how it manifests in global load measurements, will aid the interpretation of experimental results. Computational Fluid Dynamics (CFD) potentially offers the ability to predict appendage stall, and possibly even boat loads post appendage stall. However, experimental validation of CFD-predicted appendage stall, and how it manifests in the measured global loads, will be required.

In this paper, the non-linear changes in the force and moments on the BB2 submarine with AoD are predicted using CFD. Abrupt, non-linear changes in loads predicted by CFD are related to the stall of appendages. Stall is determined by examining the pressure loads and surface flow topology on individual appendages. The results presented, when combined with future BB2 experimental data, will be used to validate CFD prediction of submarine appendage stall and increase confidence in submarine hydrodynamics and manoeuvring based on CFD predictions.

DST GROUP EVOLVED GENERIC SUBMARINE GEOMETRY (BB2)

The DST submarine geometry (BB2) is shown in Figure 1. The full-scale BB2 has a submerged displacement of 4343 m^3, an overall length of 70.2 m, and a maximum hull diameter of 9.6 m with a 10.6 m distance from the keel to the top of the casing. The bridge fin has a span of 5.6 m and a chord of 11.0 m.

The BB2 features an X-form arrangement of the ACS and forward hydroplanes mounted on the bridge fin. The X-form ACS arrangement extends to the "bounding box" [3] defined by the width and depth of the hull. The BB2 has been tested to verify that it has straight line stability (Overpelt et al. [4]).

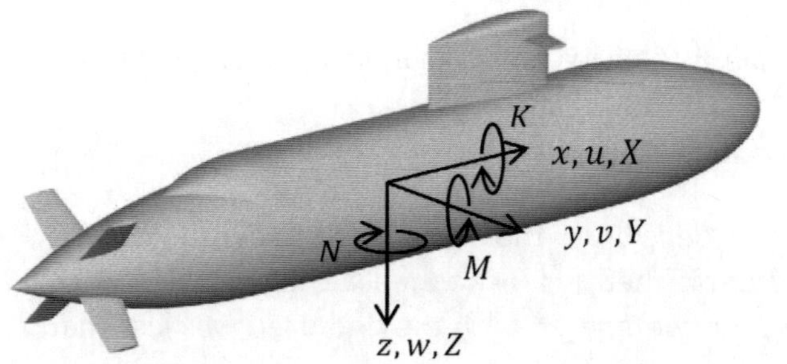

Figure 1 DST Group Evolved Generic Submarine Geometry (BB2) and the body-fixed coordinate system with its origin located amidships.

AXIS SYSTEM

The body-fixed Cartesian coordinate system (x, y, z) used to define the positive direction of the forces (X, Y, Z), moments (K, M, N), and linear velocity (u, v, w) is shown in Figure 1. The origin of this axis system is located amidships. Moments are reported about amidships.

COMPUTATIONAL FLUID DYNAMICS MODELLING

ANSYS Fluent 14.5 was used for the CFD simulations, with closure of the Reynolds Averaged Navier-Stokes (RANS) equations provided by the shear-stress transport (SST) $k - \omega$ turbulence model [5]. Second-order upwind differencing was used for the convective terms in the governing momentum equations, and the turbulence quantities k and ω. Diffusion terms used a second-order central difference. The SIMPLE algorithm was used for pressure-velocity coupling.

The submarine was modelled at a depth absent of free-surface effects using a rectangular domain of length $6.0L$ and lateral dimensions $3.0\,L \times 3.0\,L$. The submarine was placed on the centreline of the domain, $2.0\,L$ from the upstream boundary condition. A fully structured grid of 60.9 million cells was generated for the domain surrounding the submarine using Pointwise V17. Figure 2 shows the surface grid on the submarine, as well as the grid on the centre plane of the flow domain.

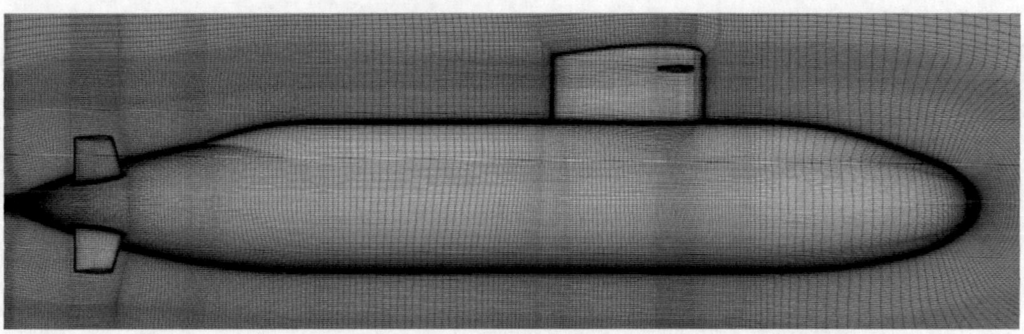

Figure 2 Surface grid of BB2 CFD model and grid on the domain centre plane.

All sides of the rectangular domain were specified as uniform velocity-inlet boundary conditions. An AoD was created by changing the direction of the flow at the velocity-inlet boundary conditions. A turbulence intensity of 1%, and ratio of turbulent viscosity (μ_t) to dynamic viscosity (μ) of 10 was specified at the velocity-inlet boundary conditions. All surfaces of the submarine were specified as non-slip walls.

The freestream speed, density and dynamic viscosity were set to achieve the desired Reynolds number given by,

$$\mathrm{Re} = \frac{\rho U L}{\mu}, \tag{1}$$

where ρ is the fluid density, U the freestream speed and L the length of the boat. All calculations were performed for a model-scale Reynolds number of 5.2×10^6. For this Reynolds number, the average y^+ with the grid used was less than 1.

CFD modelling was performed for a range of drift angles (β) from 0° to 20°. The drift angle is defined as,

$$\beta = \tan^{-1}\left(\frac{-v}{u}\right). \tag{2}$$

A positive drift angle implies the flow is from the port side.

Grid Sensitivity

A comprehensive grid sensitivity was undertaken for the current grid by Seil et al. [6], for drift angles of 0°, 10° and 15°. Five grids with total cell counts of 12.9, 29.9, 60.9 (baseline), 122.5 and 250.2 million cells were used in the refinement study. Each successively finer grid had approximately double the number of cells. The methodology of Eça and Hoekstra [7] was used to assess the numerical uncertainty associated with the baseline grid. It was found that for 10° AoD, the forces calculated on the 60.9 million cell grid were within 2% of the Richardson extrapolated grid independent values, and moments within 1.2%. At this AoD, the surface flow on the boat is predominantly attached. For 15° AoD, it was difficult to assess the numerical uncertainty because the forces and moments did not exhibit asymptotic convergence. However, the dominant sway force and yaw moment on the 60.9 million cell grid were within 1.1% of the values obtained on the finest grid. The greatest difference was for the pitch moment, which was within 8.1% of the value on the finest grid. It was thus concluded that for AoD ≤10°, the 60.9 million cell baseline grid produces a solution which is relatively insensitive to further grid refinement. This grid was considered suitable for the study presented in this paper.

RESULTS

The variation of the global forces and moments acting on the boat at an AoD are shown in Figure 3. The forces and moments have been non-dimensionalised according to,

$$X' = \frac{X}{\frac{1}{2}\rho U^2 L^2}, \quad Y' = \frac{Y}{\frac{1}{2}\rho U^2 L^2}, \quad Z' = \frac{Z}{\frac{1}{2}\rho U^2 L^2}, \tag{3}$$

$$K' = \frac{K}{\frac{1}{2}\rho U^2 L^3}, \quad M' = \frac{M}{\frac{1}{2}\rho U^2 L^3}, \quad N' = \frac{N}{\frac{1}{2}\rho U^2 L^3}, \tag{4}$$

where X, Y, Z is the surge, sway and heave forces, and K, M, N is the roll, pitch and yaw moments. The variation of the non-dimensional pressure force magnitude (F'_P) acting on

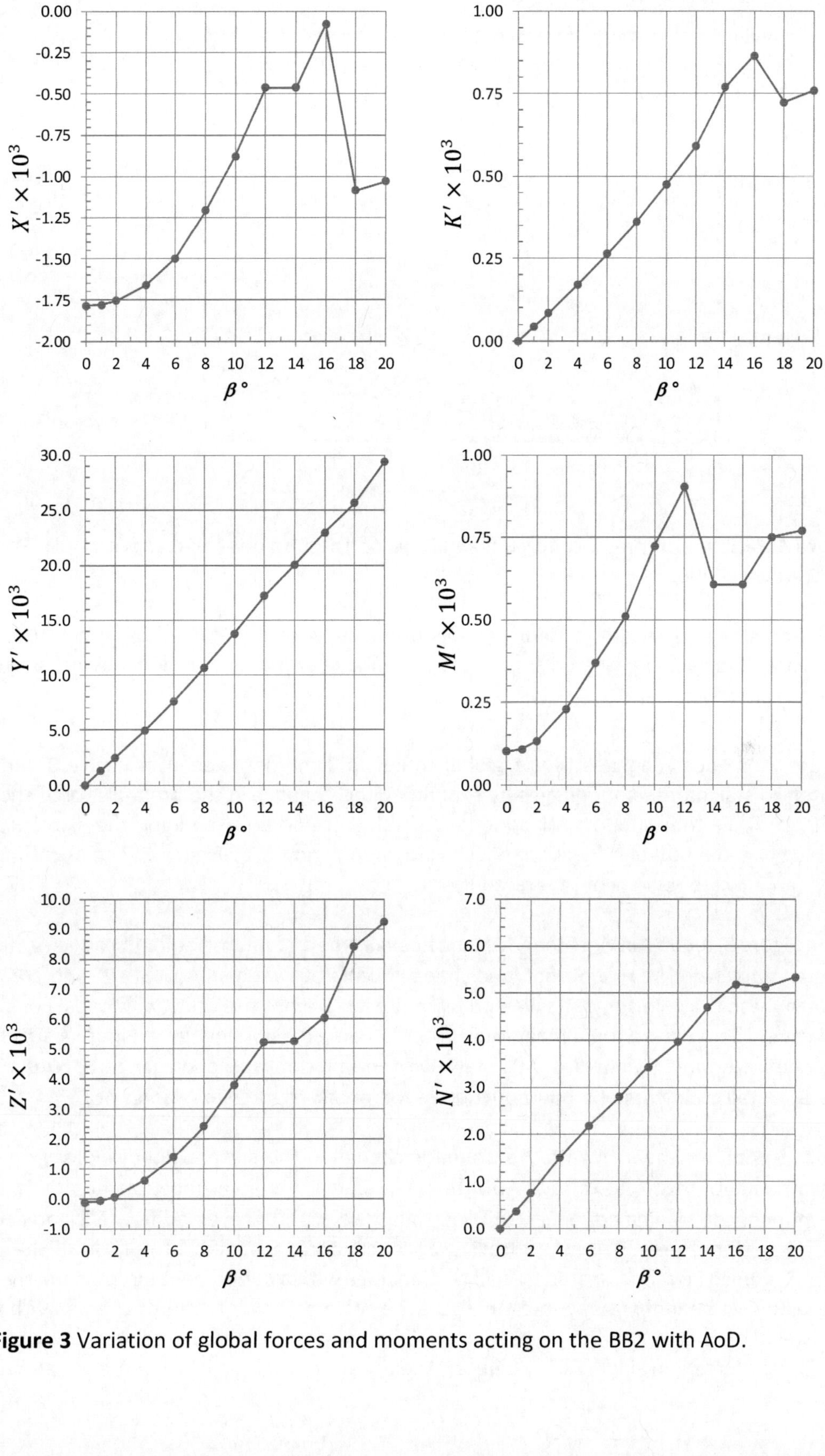

Figure 3 Variation of global forces and moments acting on the BB2 with AoD.

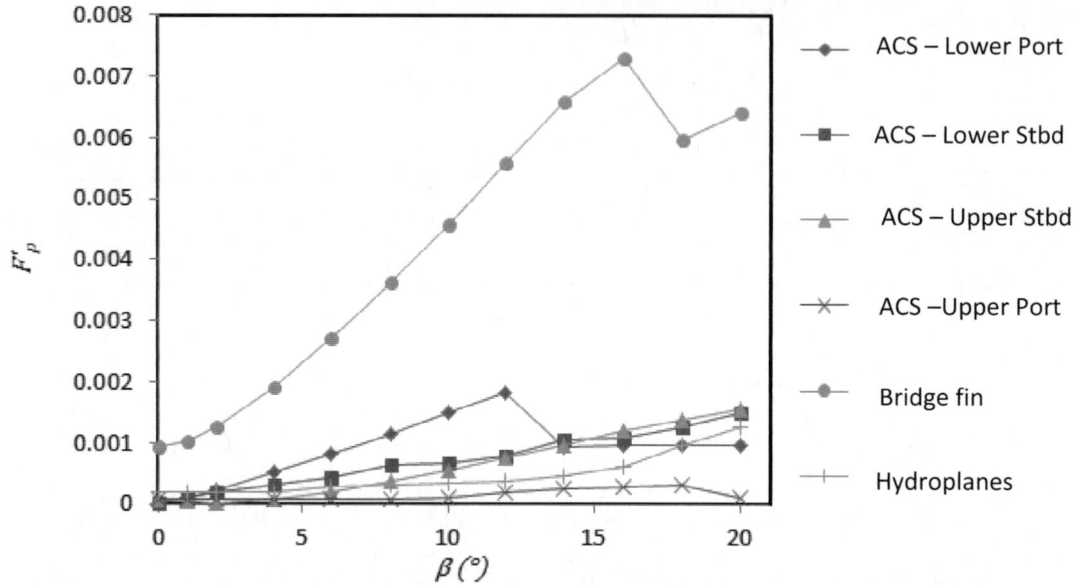

Figure 4 Variation of pressure force magnitude on the BB2 appendages with AoD. Port is the windward side.

each appendage is shown in Figure 4. This pressure force is obtained by integrating the static pressure over the wetted area of the appendage and then non-dimensionalising this force by $(\frac{1}{2}\rho U^2 L^2)$, consistent with (3).

Abrupt, non-linear changes in the global forces and moments seen in Figure 3 can be related to stall on the appendages by examining the changes in the pressure force shown in Figure 4. Further support for appendage stall is found by examining the skin friction topology on the hull and its appendages, shown in Figure 5 to Figure 7. The skin friction lines can indicate regions of separated flow.

From Figure 3, for an AoD less than 12°, the heave force and pitch moment produce a stern-dipping tendency. The first abrupt load change occurs between an AoD of 12° and 14° and is clearly visible in the surge, heave and pitch (which are not the dominant loads, i.e. sway and yaw). The pressure force magnitude variation on each appendage in Figure 4 suggests that the lower port (windward) ACS stalls with the first abrupt load change between 12° and 14°. Stall is indicated by the reduction of the pressure force at a larger AoD.

Figure 5 confirms the stall with the dramatic change in the surface flow topology on the leeward side of this ACS. For 12° AoD, the flow is attached over most of the ACS chord, except near the trailing edge. For 14° AoD, a large region of reversed flow is clearly seen extending from the ACS root and trailing edge, covering a significant portion of the ACS span. Reversed flow is indicated by skin friction lines with a component directed upstream. Interestingly, the global sway force in Figure 3 does not show an abrupt change with this ACS stall. This is attributed to the large sway load created by the hull-casing relative to the small sway force of a single ACS. However, the abrupt change is visible in the global heave force as the ACS produces a relatively significant heave force compared to the hull when the boat is at an AoD.

The pressure magnitude curves in Figure 4 show the next appendage to stall is the bridge fin, between 16° and 18° AoD. Figure 6 shows the dramatic change in flow topology on the leeward side of the bridge fin between these drift angles. For 16° AoD, a complex flow surface flow topology exists, due in part to the interaction with the bridge fin hydroplane which results in flow separation on the underside of the hydroplane and the formation of two vortices at the indicated foci. The flow around the leading edge of the bridge fin remains attached. For 18° AoD, the flow separates immediately downstream of the bridge fin leading edge and a large region of reversed flow covers most of the leeward side. The reversed flow region also extends onto the adjacent casing.

From Figure 3, the bridge fin stall is accompanied by abrupt changes in all global load components except the sway force. Again, this attributed to the sway force of the hull being significantly greater than that of the bridge fin.

The pressure force magnitude curve for the lower starboard (leeward) ACS in Figure 4 is quite non-linear but no clear indication of stall is seen. Plateaus in the pressure force are seen between 8° and 10° AoD, and 14° and 16° AoD. Surface flow topology for this ACS is shown in Figure 7. Between 8° and 10° AoD, a corner separation develops.

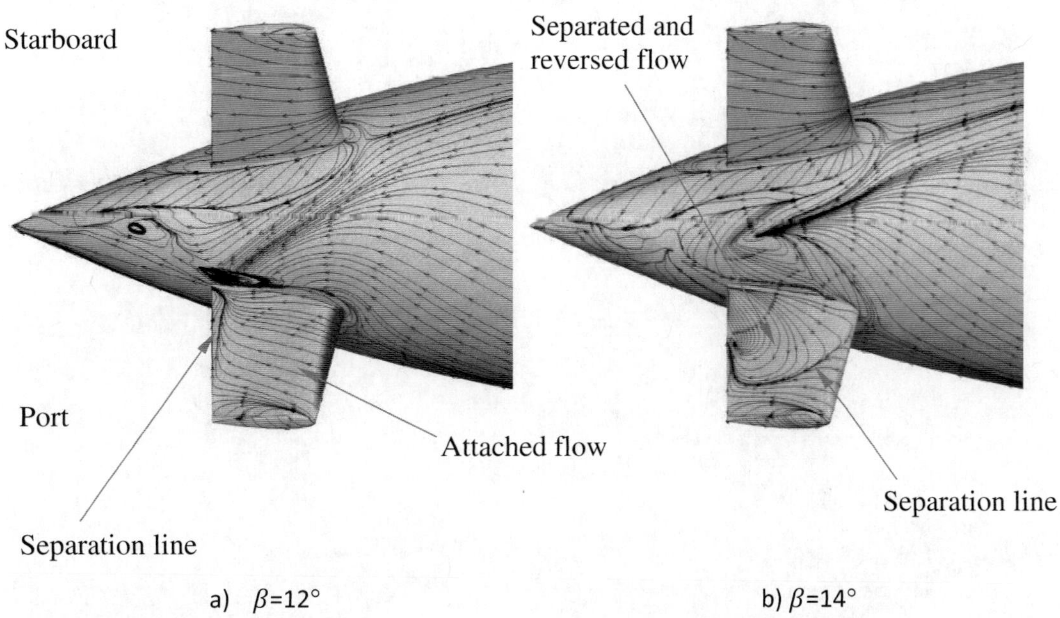

Figure 5 Surface skin friction lines showing flow topology in the vicinity of the lower port (windward) ACS, view from the underside of the hull. Contours are static pressure (red high, blue low, values not shown)

Between 10° and 16° AoD, the corner flow separation grows in size, and by 16° AoD it reaches the full span of the ACS, but only along the trailing edge. However, the larger pressure force magnitude with greater AoD indicates the ACS is not stalled.

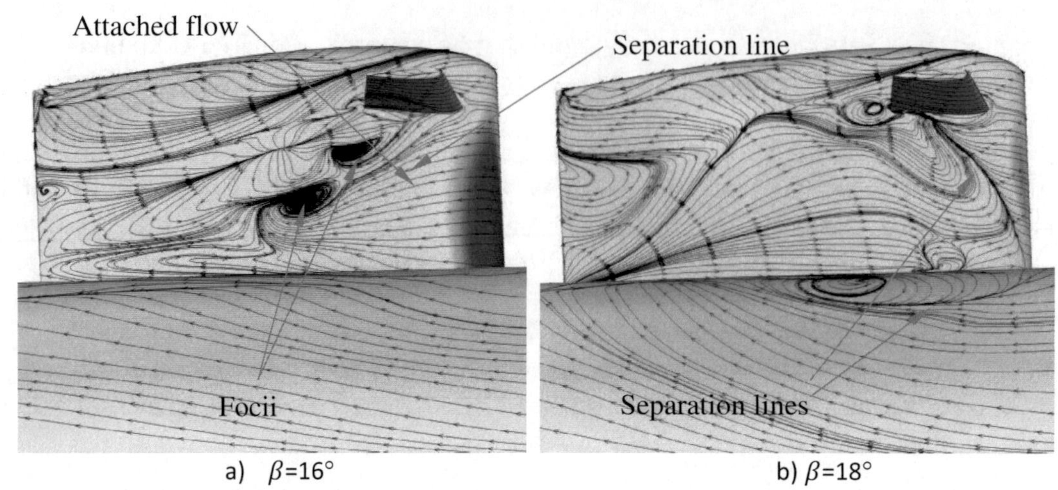

Figure 6 Surface skin friction lines showing surface flow topology on the leeward side of the bridge fin. Contours are static pressure (red high, blue low, values not shown).

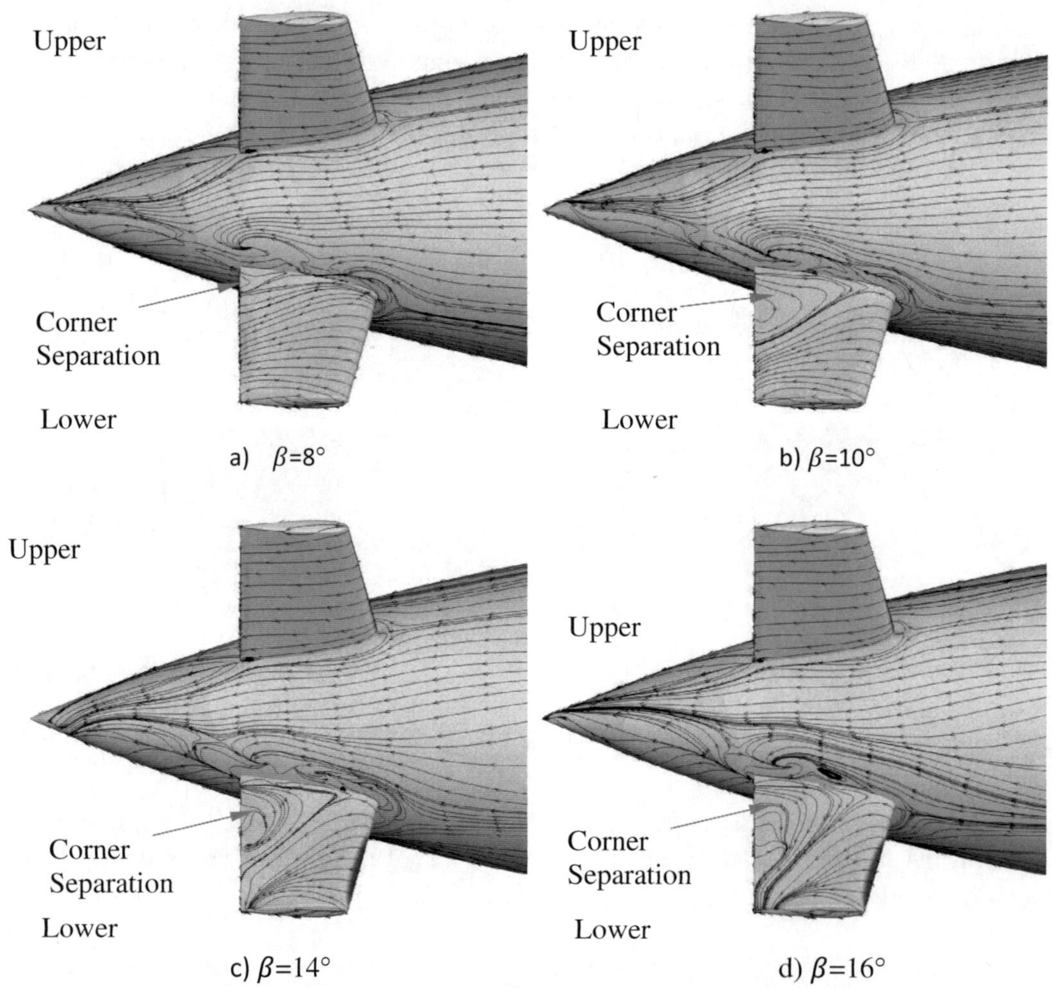

Figure 7 Surface skin friction lines showing flow topology in the vicinity of the lower starboard (leeward) ACS, view from starboard. Contours are static pressure (red high, blue low, values not shown).

The pressure force magnitude of the upper port ACS is seen in Figure 4 to reduce significantly between 18° and 20° AoD. However, surface flow topology (not shown) again does not indicate a stall

DISCUSSION

The results presented demonstrate the abrupt changes in global loads are due to appendage stall. The CFD predicted order of appendage stall in the AoD range studied is:

1) Lower port (windward) ACS: $12° \leq \beta < 14°$,
2) Bridge fin: $16° \leq \beta < 18°$.

The lower port ACS stalls at the lowest AoD due to the crossflow over the hull and the tip vortex of the bridge fin interacting with the ACS. The blockage effect of the hull accelerates the crossflow on the windward side. This increases the incidence seen by both port (windward) ACS If the bridge fin and casing were absent, then the incidence increase on the upper port ACS would be identical. However, trailing flow structures from the bridge fin (e.g. tip vortex) induce circulation around the hull and alter the incidence seen by the upper and lower port ACS. The bridge fin wake structures create an anticlockwise, when viewed from stern, circulation around the hull which decreases the incidence on the upper port ACS and increases the incidence of the lower port ACS. Hence, the stall of the lower port ACS is at a lower AoD than the bridge fin.

Stall of the bridge fin will be influenced not only by its design (e.g. hydrofoil cross-section) but also by the increased incidence due to the hull blockage effect. Again, hull-blockage accelerates the crossflow past the bridge fin, effectively increasing the incidence. The stall of the bridge fin at a higher AoD than the lower port (windward) ACS can be attributed to its lower aspect ratio (approximately 0.5 based on exposed area) compared to the aspect ratio of a single ACS (closer to 2), as well as the bridge fin's wake increasing the lower port (windward) ACS incidence.

It is interesting that, despite the use of ACS to generate sway and yaw for stability or control, stall of the lower port ACS is not readily visible in the global sway and yaw curves. The abrupt change in the global heave force is the clear indicator of stall. However, which ACS had stalled is not identifiable from global loads. In an experiment, supplementary measurements would be required, e.g. individual appendage loads or surface flow topology. Even more surprising is that the stall of the bridge fin is not clearly visible in the global sway force. Reliable identification of stall from experimental loads will likely require CFD to confirm individual appendage stall, assuming validation of CFD stall predictions can be obtained.

The pressure force magnitude change with AoD for each ACS varies significantly and is attributable to hull and bridge fin vortex influences as previously discussed. However, the force changes for each ACS, particularly the lower starboard ACS, do not vary `smoothly' with AoD. This appears to be attributable, at least in part, to the changing corner flow separation predicted by the CFD. If the CFD is accurate, then semi-empirical determination of ACS loads and stall is not likely to be accurate. However, the corner flow separation—a wing-body junction flow—cannot be predicted accurately by eddy-viscosity RANS models (e.g. SST $k - \omega$) [8]. The loads predicted when significant separated flow is present in this CFD study, must be interpreted with caution. However, since the hull and casing are the dominant sources of loads, then large errors in ACS load prediction may not significantly affect global loads. The grid refinement study conducted by [6] for the grid used in this paper, and the minimal separated flow for AoD less than approximately 12°, does provide some confidence in predicted loads at least up to 12° AoD.

CONCLUSION

The CFD modelling in this paper provides an understanding of the flow around the BB2 submarine. The results are relevant to manoeuvring and control system design. The results presented will allow more targeted testing in load experiments currently being conducted on the BB2 geometry at DST.

Over a range of drift angles ($0° \leq \beta \leq 20°$), CFD predicts the BB2 will create highly non-linear global forces and moments. Abrupt load changes are related to appendage stall. Surprisingly, appendage stall was not easily discernible in the dominant sway force. Secondary loads such as the heave force in the case of the ACS, provided a clearer indication of stall. However, identifying which appendage had stalled required the examination of individual ACS loads and surface topology. In an experimental program, global loads would not be sufficient to reliably detect and identify which ACS has stalled.

REFERENCES

1. Roddy, R.F. (1990), "Investigation of the Stability and Control Characteristics of Several Configurations of the DARPA SUBOFF Model (DTRC Model 5470) from Captive-Model Experiments", *David Taylor Research Center*, DTRC/SHD-1298-08
2. Mackay, M. (2003), "The Standard Submarine Model: A Survey of Static Hydrodynamic Experiments and Semiempirical Predictions", *DRDC Atlantic*, TR-2003-079
3. Renilson, M.R. (2015), "Submarine Hydrodynamics", *Springer Briefs in Applied Sciences and Technology*.
4. Overpelt, B., Nienhuis, B. and Anderson, B. (2015), "Free Running Manoeuvring Model Test on a Modern SSK Class Submarine (BB2)", *Proc. Pacific 2016 International Maritime Conference*, Sydney, 6-8 Oct. 2015.
5. *ANSYS Fluent User's Guide*, ANSYS Inc.
6. Seil, G. J., Nguyen, M.C., Pook, D. and Leong, Z. Q. (2017), "CFD Study of the Evolved DST Group Generic submarine (BB2)", *DST Group Technical Report*, Under Review.
7. Eça, L. and Hoekstra, M. (2009), "Evaluation of Numerical Error Estimation based on Grid Refinement Studies with the Method of Manufactured Solutions", *Computers & Fluids*, Vol. 38, pp. 1580-1591.
8. Gand, F., Brunet, V. and Deck, S. (2010), "A Combined Experimental, RANS and LES Investigation of a Wing Body Junction Flow", *40th Fluid Dynamics Conference and Exhibit*, Chicago, 28 Jun-1 Jul.

Sensitivity of a Submarine Manoeuvring Model to Parameter Uncertainty

Siobhan Giles[1] and Francis Valentinis[2]
[1]BMT Design & Technology, [2]Defence Science and Technology Group
sgiles@bmtdt.com.au

ABSTRACT

Submarine manoeuvring models are used to predict performance, identify safe manoeuvring limits and provide simulation models for training platforms. A manoeuvring model for a submarine typically has many parameters, however not all of these need to be characterised with a high level of accuracy to guarantee good simulation outcomes. In order to present some guidance for the preparation of these models, the manoeuvring model of the DST generic BB1-X submarine has been analysed for parametric sensitivity.

Parameters of the well-known Gertler-Hagen based manoeuvring model were varied individually and ranked based on the simulation model's sensitivity to their variation. A frequency domain analysis was used to indicate which modes of motion would be most affected and consequently, how the stability characteristics of the model would change. The intent of this study was to compare parameters based on the extent to which a prescribed level of uncertainty influences the accuracy of a submarine manoeuvring simulation. The results highlight that precision may be sacrificed, or uncertainty allowed to increase, in certain parameters while not significantly affecting the motion. This can reduce the characterisation workload and save development time and cost. It is shown that the linear damping terms had the greatest influence on the dynamic characteristics in terms of overall fidelity and stability. However, some other terms were also found to be quite significant. The findings enable future modelling studies to concentrate on characterising the critical parameters while still ensuring accurate simulations.

1. INTRODUCTION

The process of producing a parametric manoeuvring model of a submarine must always trade off desired model fidelity against available resources and time. Depending on the purpose of a particular model, fidelity may be sacrificed in certain areas in the interests of reducing computational time or data input requirements. However, before making this reduction it is important to understand how various simplifications will affect the overall dynamic manoeuvring characteristics. If this is well understood, models can be created in a timely way while still serving their intended purpose.

The parametric data for a manoeuvring model is often produced through experimental work or computational fluid dynamic (CFD) simulations. A certain amount of force and moment data can provide enough information to calculate a set of hydrodynamic coefficients to populate a manoeuvring model. The higher the data requirements, the longer these processes take. When reducing the extent of parametric data required, it is desirable to reduce detail in a manner that has minimal effect on the behaviour of the manoeuvring model. Thus, it is important to know against which hydrodynamic coefficients the manoeuvring behaviour is most sensitive. Ray et al. (2010) provides an analysis of a submarine simulation in a vertical zigzag manoeuvre with a view to understand its sensitivity to variation of characteristic hydrodynamic coefficients. This analysis achieves some insight into which parameters are most important for this particular manoeuvre. The analysis is limited however, as it focuses only on eight coefficients and is restricted to the time domain and a vertical plane manoeuvre.

The purpose of this study is to determine which data is most crucial to the stability and accuracy of a manoeuvring model, specifically for that of the Defence Science and Technology (DST) Group BB1-X submarine geometry. An analysis is carried out to determine how the manoeuvring model is affected by variations in parameters, in this instance, obtained through CFD simulations. The results can then be used to inform future CFD or experimental work, so that the extent of these studies and resources can be reduced while obtaining meaningful manoeuvring and stability results. A frequency domain approach was used for this analysis in order to give a more complete picture of the submarine dynamics and identify natural modes of motion. The natural modes identified in this study are limited to those present during straight and level motion. While some of these modes are likely to be present during other manoeuvring scenarios (such as steady turning or diving), there will be some variation.

2. NOMECLATURE

u	body-fixed surge (forward) velocity [m/s]
v	body-fixed sway velocity [m/s]
w	body-fixed heave velocity [m/s]
p	body-fixed roll rate [rad/s]
q	body-fixed pitch rate [rad/s]
r	body-fixed yaw rate [rad/s]
x	earth-fixed axial position [m]
y	earth-fixed lateral position [m]
z	earth-fixed vertical position [m]
φ	roll angle [rad]
ϑ	pitch angle [rad]
ψ	yaw angle [rad]
m	mass [kg]
g	gravity [m/s^2]
l	body length [m]
x_g	centre of gravity x coordinate [m]
y_g	centre of gravity y coordinate [m]
z_g	centre of gravity z coordinate [m]
x_b	centre of buoyancy x coordinate [m]
y_b	centre of buoyancy y coordinate [m]
z_b	centre of buoyancy z coordinate [m]
I_{xx}	moment of inertia about the x axis [kg/m^2]
I_{yy}	moment of inertia about the y axis [kg/m^2]
I_{zz}	moment of inertia about the z axis [kg/m^2]
ρ	density [kg/m^3]
X	body-fixed external axial force [N]
Y	body-fixed external lateral force [N]
Z	body-fixed external heave force [N]
K	body-fixed external roll moment [N m]
M	body-fixed external pitch moment [N m]
N	body-fixed external yaw moment [N m]
$F_{\dot{\alpha}}$	added mass coefficient of the force or moment F, due to an acceleration $\dot{\alpha}$
F_{α}	hydrodynamic coefficient of force or moment F, due to velocities u and α
$F_{\alpha\beta}$	hydrodynamic coefficient of force F, due to velocities α and β

3. THE BB1-X SUBMARINE

The BB1-X submarine is a generic submarine geometry outlined in Joubert (2004) and shown in Figure 1. The submarine geometry consists of an axisymmetric hull form with a tangential casing, sail, four aft control surfaces in an 'X' configuration and two hydroplanes mounted on the sail.

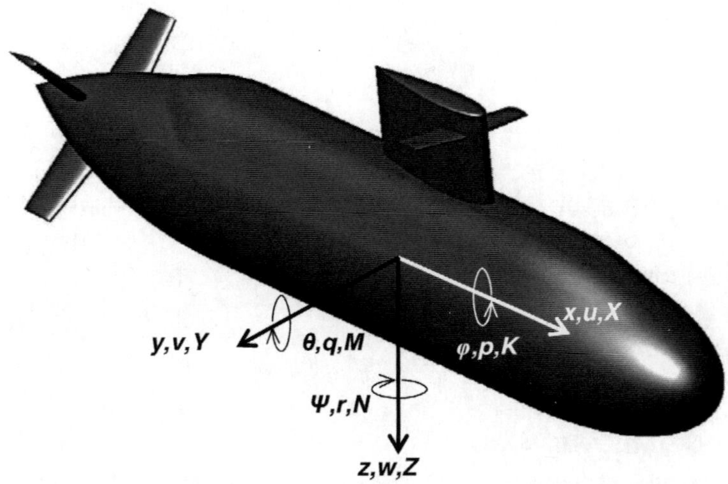

Figure 1 The BB1-X Submarine geometry

Figure 1 shows the coordinate system located at midships with three axes in the *x*, *y* and *z* directions as well as the rotations, φ, ϑ and Ψ, about each of these axes. The x axis is located at the centre of the cylindrical section of the hull (ie. *D/2* above the keel). The forces and moments in each direction are denoted by *X*, *Y* and *Z* and *K*, *M* and *N* respectively. Also shown are the translational velocities, *u*, *v* and *w*, and the rotational velocities *p*, *q* and *r*.

The unclassified BB1-X geometry was used in this study due to the availability of existing CFD data. Although, at the time of writing, not all the data required for the manoeuvring model was available, the data that did exist was easily accessible and the missing data easily estimated. Multiple sets of data from Seil and Anderson (2013) and Valentinis and Kofmansky (2013a, 2013b) were used to create a full set of data, in order to create a fully populated parametric model.

A summary of the vehicle parameters as used in the manoeuvring analysis of this study is presented in Table 1. The mass was calculated using the body volume and fluid density of the CFD simulations. The moments of inertia were estimated by comparing the submarine to an underwater vehicle of similar size and mass. The centre of volume was determined numerically from a 3D CAD model while the centre of gravity (CG) was approximated by scaling (by diameter) from a larger known vehicle. The centre of buoyancy (CB) and centre of gravity vectors are defined relative to amidships, with the centre of buoyancy being coincident with the centre of volume.

Table 1 BB1-X Parameters

Parameter	Value
Length l [m]	70.307
Volume V [m^3]	4331
Mass m [kg]	4.3223E6
Moment of inertia about x axis [kg.m^2] I_{xx}	35E6
Moment of inertia about y axis [kg.m^2] I_{yy}	1.5E9
Moment of inertia about z axis [kg.m^2] I_{zz}	1.5E9
Diameter D [m]	8.0
CG vector (x_g, y_g, z_g) [m]	(0, 0, 0.085)
CB vector (x_g, y_g, z_g) [m]	(0, 0, 0)

4. MANOEUVRING SIMULATION MODEL

A BB1-X manoeuvring simulation model was created using the six degree of freedom equations described in Fossen (1994). The model is a Gertler-Hagen based parametric model (Gertler and Hagen, 1967), which was characterised using CFD data from sources outlined in Section 3. In such a model, the hydrodynamic forces are calculated by summing the contribution due to the velocity and acceleration in each direction. A coefficient is used for each type of motion causing a force or moment in each direction. These coefficients are denoted by F_α where F is the force or moment and α is a velocity which results in the force or moment. Also included were the added mass terms, $F_{\dot{\alpha}}$ where $\dot{\alpha}$ is an acceleration causing the force or moment F.

5. FREQUENCY ANALYSIS

For this study, the authors have chosen to focus on the analysis in the frequency domain. As described in Stevens and Lewis (2003), frequency domain approaches are widely used in the aircraft flight dynamics community, but have not often been adopted for submarine dynamics applications. The higher coupling between modes that is present in submarine dynamics makes clear and useful identification of modes more difficult, however, the work presented in this study shows that it is possible.

While clear and useful results can be obtained by studying the time domain (for example Ray et al. (2010)), a frequency domain analysis will give a more complete picture of the submarine dynamics, as it will capture the behavioural trends of all modes of interest.

Using the six degree of freedom equations of motion described in Fossen (1994) together with the state vector of the submarine in trim (that is at some steady state of equilibrium), the dynamics were scrutinised by means of an Eigenanalysis. This type of analysis gives the natural modes of a dynamic system, a description of the dynamic characteristics that will influence and govern the behaviour of the system. The connection between natural modes and dynamic behaviour is often utilised in aircraft autopilot design to indicate handling quality of an aircraft (Stevens and Lewis 2003).

The two states of equilibrium used were straight and level forward motion at 5m/s and 10m/s. Although this study is limited to the straight and level case, some modes will be common across multiple conditions. As the condition moves further from the straight and level case it is expected the modes will change more drastically. Extreme manoeuvres at high pitch or yaw rates would be considered far from the straight and level case however a considerable amount of time is spent outside of these extreme conditions. The straight and level case will reveal sensitivities that are relevant to a significant portion of manoeuvring operations.

While the submarine model used is a fully submerged model with no free surface interaction effects, the magnitudes and balance of the Gertler-Hagen manoeuvring coefficients across each axis will be similar to that of a near-surface model. Although ocean waves will have less of an effect at depth, the effect of wave disturbance to the vehicles stability was still considered in this analysis. This analysis was still considered because response of the near-surface and submerged models will be similar and likely to exhibit similar trends.

Any variation to a model will change the response of the system. Although different hull form geometries will have a significant influence on the model's sensitivities to various hydrodynamic coefficients (McCormick 1995, Ray et al. 2010), many trends will remain. This study aims to give a general indication of the most sensitive aspects of a fully submerged submarine model.

5.1 Eigenanalysis

The equations of motion described in Fossen (1994) were linearised to allow the motion to be represented in state space form. An eigenanalysis of the resulting equations was then used to analyse the eigenmodes of the linearised dynamics. The eigenmodes give insight into the dynamic characteristics and behaviour of a system. They are scrutinised here to give an indication of changes in dynamic behaviour of a submarine due to a parameter change. A change in one or more of the natural modes present suggests a change in the characteristic behaviour of the submarine therefore a parameter that causes a change in natural mode will also cause a change in manoeuvring behaviour.

Modes which are lightly damped and have a high natural frequency will play a strong part in the transient behaviour of the submarine. When a mode has an extremely low natural frequency, the dynamic behaviour associated with the mode approaches a steady state. If these slow modes are shown to be sensitive to certain factors, it is likely that the steady motion of the submarine will be sensitive to the same factors.

6. SUBMARINE DYNAMIC MODES

Using a trim state for straight and level motion, an eigenanalysis was carried out to give the natural modes for the submarine. The submarine is considered `in a trim state' when all velocities are constant and the system is in equilibrium. The trim state was determined by running a simulation until the desired motion converged to a steady state. Each eigenvalue represents a natural mode. From the eigenvalue it is possible to deduce the associated natural frequency, period and damping ratio. The results of this process are listed in Table 2. The remaining eigenvalues, equal to zero, were omitted. Each eigenvalue has an associated eigenvector and these give the natural modes of the system. Each non-zero element of an eigenvector indicates the extent of which that state will participate in the mode (Stevens and Lewis 2003). In Table 2, each natural mode has been labelled by the states which are most involved in the mode (ie. the largest elements of the eigenvector). Of the modes listed, those which were deemed more significant have been highlighted and are discussed below.

Table 2 Eigen modes for straight and level motion at 5m/s

States	Eigen Value	Frequency (rad|s)	Period (s)	Damping Ratio
v p y ϕ	-0. 51 + 0.12 i	0. 53	11. 9	0. 97
w z	-0. 19 + 0. 18 i	0. 26	24.	0. 72
y ϕ	-0. 22	0. 22	27. 9	1
x y	-0. 014	0. 014	464	1
y	-0. 0086	0. 0086	727	1
z	-0. 0021	0. 0021	2941	1

Modes of particular interest to the transient behaviour of the submarine are those that have a lower damping ratio and shorter periods. The two modes of greatest significance in the above results are the sway roll mode and the heave mode. These modes have periods between 11 and 24 secs and are not critically damped like the other modes (they have a damping ratio of less than 1). Eigenvector plots of these two modes are given in Figure 2 to show the mode shapes giving the relative amplitudes of oscillation for each state.

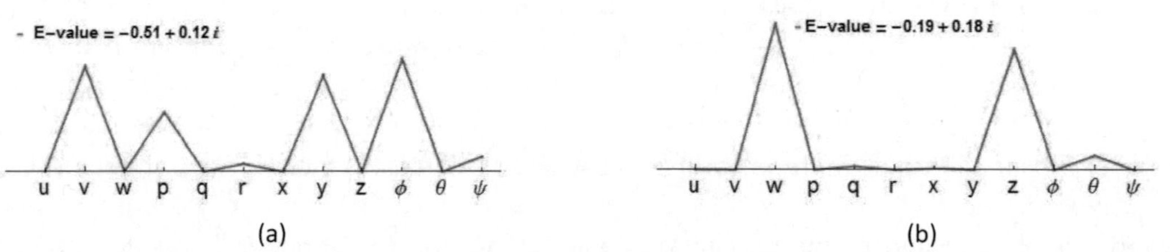

Figure 2 Eigenvectors for straight and level motion (a) Sway Roll Mode (paired) and (b) Heave Mode (paired)

A wave disturbance is likely to affect either the sway, roll and heave states, or surge, pitch and heave states, depending on the submarine's orientation to the waves. The length of the submarine along the x-axis acts as a filter in this direction and makes wave disturbance less of a concern for the pitch and surge motions.

The sway roll mode, having a period of around 12 seconds, is likely to be excited by wave motion. The frequency and period of this mode fits within a typical spectrum of regular ocean waves, caused by wind patterns (Munk 1950). This will only be of concern in surface and periscope depth manoeuvres and, due to the high damping ratio, should have little effect during forward motion.

The heave mode also has a shorter period of 24 seconds (relative to the last three listed modes). This makes it more likely to be excited by wave disturbances during straight and level motion, however the high damping ratio will still restrict the oscillations. This will minimise the effect of this mode on the system's dynamics.

The y roll mode has an entirely real eigenvalue. This implies that the damping ratio will be equal to one and the mode will therefore have minimal effect on the submarine's transient dynamics. However, if the eigenvalue was to become complex and thus a damping ratio of less than one, the mode is of a low enough period (less than 30 secs) that the effects of this oscillation may be noticeable and/or susceptible to wave disturbances. This observation will be important while considering eigenvalue sensitivity in Section 7.

Other modes are of very long period and are therefore not significant to the transient behaviour. The eigenvalues would have to alter by multiple orders of magnitude before their effects would be significant to the transient motion. While unlikely to be excited by environmental disturbances, these modes are more closely related to the steady state behaviour of the submarine. The same force balance which keeps these modes stable, allows the submarine to achieve steady-state motion.

Table 3 provides a summary of the modes discussed above, with the label `t' assigned to modes which have more effect on the transient behaviour of the submarine and `s' assigned to those which have more effect on the quasi-steady behaviour. Modes listed as damped have a damping ratio of 1, while others have a damping ratio of less than 1. The final column shows if that mode is likely to be excited by ocean waves based on the natural frequency and characteristic states.

Table 3 Dynamic mode summary, straight and level motion. Modes are classified by relevance to transient or steady motion, whether they are critically damped and whether they are likely to be excited by ocean waves.

Mode	Trans/Steady	Damped	Wave Excited
Sway Roll	t	×	✓
Heave	t	×	✓
y Roll	t	✓	✓
x y	s	✓	×
y	s	✓	×
z	s	✓	×

7. COEFFICIENT SENSITIVITY

In order to assess which input parameters are most important to the manoeuvring model, a sensitivity analysis was carried out. This analysis again focused on the frequency domain and tested which coefficients caused the greatest changes in the submarine response as their magnitudes were varied.

7.1 Eigenanalysis

Using the six degree of freedom model, each coefficient was progressively varied in 10% increments from 50% of the original value to 150%. This was used to identify the change in eigenvalues as each coefficient was varied. For each eigenvalue that became unstable (indicated by a positive real part), the point at which it became unstable was interpolated and given as a percentage change in the coefficient that resulted in instability. For eigenvalues that changed significantly but did not destabilise, the phase and magnitude change was determined in order to identify their influence on each mode. Each eigenvalue shift of more than 10% in magnitude or 10° in phase is shown in Figure 3. For coefficients with an initial value of zero, the analysis was carried out by assigning and varying a value which corresponded to the smallest similar non-zero coefficient. Coefficients were considered similar if they were dimensionalised using the same formula. Two trim states were considered for the eigenanalysis. Both were straight and level trim states, however with two different forward speeds of 5m/s and 10m/s.

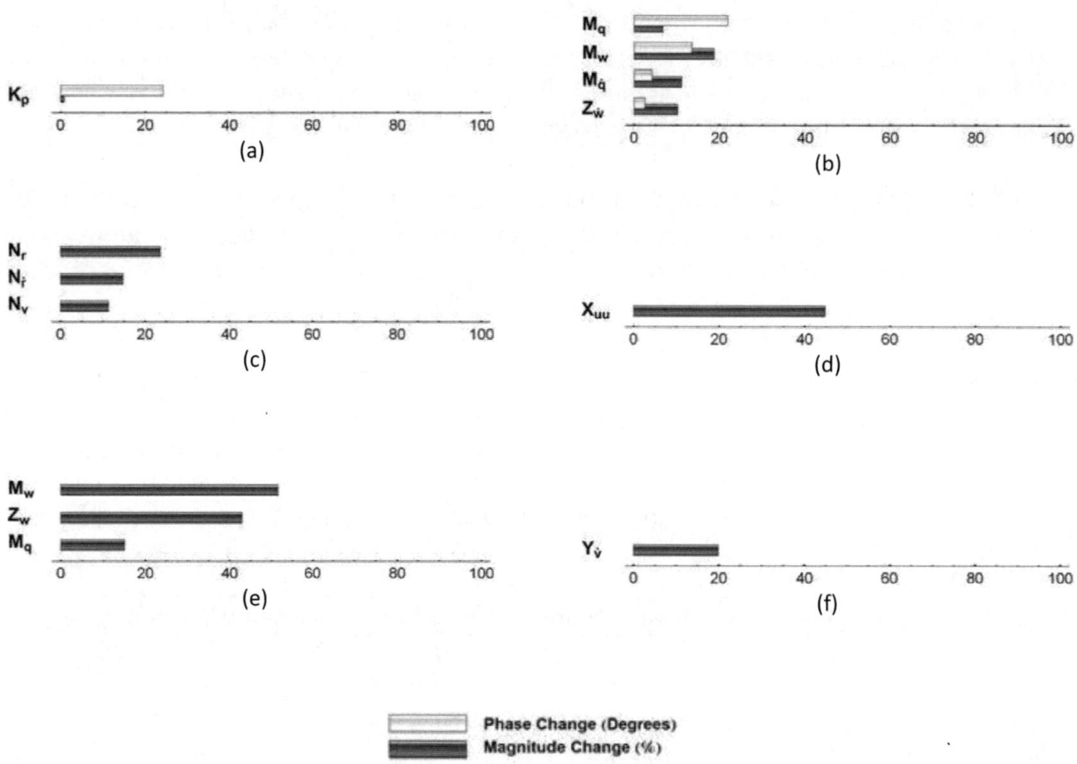

Figure 3 Coefficient Sensitivity for Straight and Level Motion u=5m/s (a) Sway Roll Mode 1 and 2; (b) Heave Mode 1 and 2; (c) Sway Mode; (d) x Mode (e) z Mode (f) y Mode.

When altering the value of the hydrodynamic coefficients, instability exists when the real part of any eigenvalue becomes positive. Instances where this occurred for the straight and level trim state at the two forward speeds are indicated in Table 4 and Table 5 respectively. The percentage increase or decrease which caused instability for each of the relevant coefficients is listed along with the mode which became unstable. It was noted that the *y* mode became unstable for both cases. As seen by comparing Table 4 and Table

5, an increase in axial speed had no effect on the margin of stability or the coefficients causing instability. The latter were identified as the damping terms Y_v, Y_r, N_v and N_r.

Table 4 Change in hydrodynamic coefficients causing instability, u=5 m/s

Coefficient	Percentage Change	Unstable Mode
Y_v	-7.4	y
Y_r	-18.1	y
N_v	+8.1	y
N_r	-7.4	y

Table 5 Change in hydrodynamic coefficients causing instability, u=10 m/s

Coefficient	Percentage Change	Unstable Mode
Y_v	-7.4	y
Y_r	-18.1	y
N_v	+8.1	y
N_r	-7.4	y

Although only the results for the 5 m/s case are included, results of both speeds are discussed below. Coefficients of particular significance for straight and level transient motion are K_p, which had significant effect on the sway roll mode, M_q which had a strong impact on the heave mode and N_r which had a strong impact on the sway mode. The modes with a low natural frequency were sensitive to $X_{u|u|}$ (*x* mode), M_w and Z_w (*z* mode). Some changes to these sensitivities were observed at the higher speed (10m/s). In particular, terms relating to the sway roll mode became significant.

At lower speed the sway roll mode had a lower damping ratio. The change in phase due to the variation of coefficient K_p caused a shift in damping ratio. As K_p decreased, so did the damping ratio. When K_p was equal to 50% of the original value, the damping was reduced to 0.44 for the 10m/s case and 0.20 for the 5m/s case. Only a slight increase in K_p was required to critically dampen the system at the higher speed, while at the lower speed a K_p of 150% of the original value resulted in a damping ratio of 0.82. It is therefore important to obtain an accurate estimate of the coefficient K_p. Any significant inaccuracies could lead to a model that exhibits a distinctly different dynamics and disturbance rejection behaviour in the sway roll mode.

The heave mode exhibited a phase shift with variation in M_q, M_w, $M_{\dot{q}}$ and $Z_{\dot{w}}$. This shift was more significant for coefficients M_q and M_w. The damping ratio increased as M_q increased and M_w decreased. Once again, inaccuracies in these coefficients may lead to the model exhibiting more or less susceptibility to heave disturbance than is realistic.

While the sway mode had an entirely real eigenvalue at low speeds (and therefore a damping ratio of one), at higher speeds it became a *y* roll mode and some coefficients induced a phase shift in the eigenvalue. The phase shift caused the system to become more lightly damped and for this reason these particular coefficients are more important than those that didn't induce a phase shift. For small values of N_r (less than 60% of the original value) and for large values of K_v (greater than 140% of the original value) a phase change in the eigenvalue meant the eigenmode was no longer critically damped. As the damping ratio decreases, the system is more susceptible to *y* and roll disturbance.

At the higher speed, the *x* mode became an *x y* mode and more sensitive to Y_r. The *z* mode sensitivities stayed the same, while the *y* mode became more sensitive to Y_p and N_p. No phase change was observed for any of these modes, thus they all remained critically damped. The disturbance rejection properties of these modes, due to their low natural frequency and high damping, will remain largely the same. The sensitivities demonstrated by these modes will more likely affect the slow and quasi-steady motions of the submarine.

7.2 Coefficient Sensitivity Summary

Coefficient sensitivities are summarised in Figure 4. The red grid shows the coefficients affecting the transient modes. The blue grid shows the coefficients affecting the slow, quasi-steady modes. The model's sensitivity to each coefficient is represented by the shading surrounding it. Coefficients shaded darker cause larger changes in eigenvalues. A coefficient causing instability is identified by a black outline. These grids have been condensed to show only relevant coefficients, thus coefficients that caused no change were omitted. The exact changes these coefficients induced are shown in Figure 3.

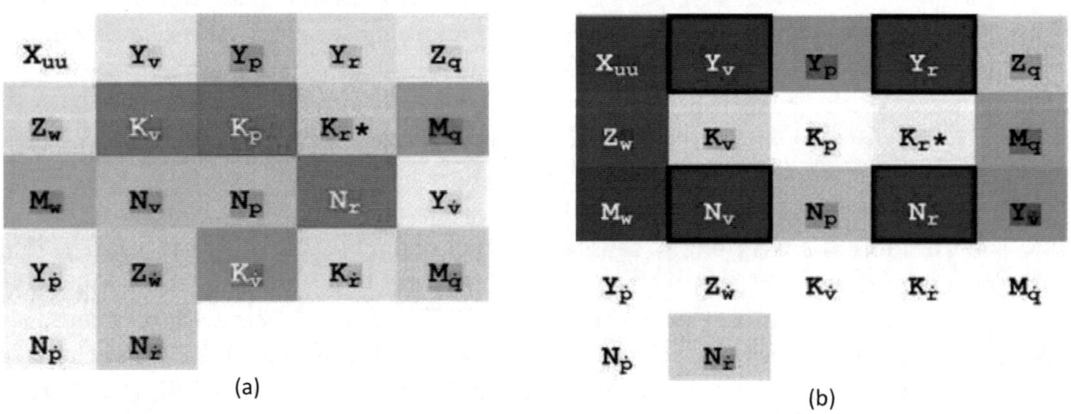

Figure 4 Coefficient sensitivity in straight and level motion: (a) Transient modes; (b) Slow modes

8. CONCLUSION AND FUTURE WORK

A frequency domain analysis was used to identify the main characteristics of submarine motion. Using a parametric submarine manoeuvring model, the natural modes were identified and categorised based on their impact on the overall dynamics. Parameters of the manoeuvring model were then varied to identify which input data caused the largest changes to the manoeuvring characteristics.

The damping terms Y_v, Y_r, N_v and N_r are the most significant coefficients to the system's stability. Changes to these coefficients cause instability. This is consistent with commonly used stability proofs for the horizontal plane (Roddy, 1990). Coefficients which didn't cause instability but did cause significant changes in eigenvalue included K_p, K_v, N_r and $K_{\dot{v}}$ for transient motion and X_{uu}, M_w and Z_w for quasi-steady motion. Other coefficients that had significant impact varied across different speed ranges. These sensitivities are relevant to the straight and level motion. Other sensitivities may exist for other manoeuvring conditions.

These results do not corroborate with those of Ray et al. (2010), but offer additional insight. Ray highlights the control surface coefficients as being very significant. This study only focuses on the systems natural response and neglects the control surfaces. Of the coefficients governing the natural response in the vertical plane, Ray emphasises the role of Mq and Mw. While these coefficients are not among the most significant in this study, the results do also show that these coefficients are important. The authors believe that if the analysis in this report was extended to other equilibrium conditions, such as a vertical pitch condition, the resulting sensitivities would agree with those of Ray. This is, however, deferred to further work.

When producing data to be used in manoeuvring models, the coefficients identified by this study should be determined with high accuracy, as small inaccuracies will cause significant changes in the results predicted by the manoeuvring model. Although many other coefficients have very little effect on the overall dynamics of the submarine, it does not follow that they can be completely neglected. A change of ±50% may have little effect from a sensitivity perspective; however, to neglect a term completely may change the force balance significantly.

This study has neglected some areas. The first is that it only considered the linearised system dynamics about the straight and level state. Using linear models that consider perturbations about this state is very common. Linear stability derivatives, for example, are determined using this assumption. Linearisations in other conditions (such as turns and climbs) would, however yield additional insight and provide an opportunity for a more in-depth comparison with the work of Ray et al. (2010). The second area is an examination of the sensitivity of the model to the accurate characterisation of the viscous component of the coupled hydrodynamic coefficients. While these values have been varied to assess model sensitivity, a more accurate starting point (including the viscous component) for these coefficients would provide a deeper understanding of their importance. This would be an important investigation for future work, as some of these terms are neglected by practitioners. The third is the study of how the sensitivities are likely to vary as the geometric features of the submarine change. The authors believe that the variation is likely to be small, however, further work is required to explore this. No body parameter variation was considered in this study such as mass or CG and CB location, this may also alter sensitivities and could be further investigated.

9. REFERENCES

Fossen, T. I. (1994). Guidance and control of ocean vehicles, John Wiley & Sons Inc.

Gertler, M. and Hagen, G. R. (1967). Standard equations of motion for submarine simulation (No. NSRDC-2510). David W Taylor Naval Ship Research and Development Center Bethesda MD.

Joubert, P. N. (2004). Some Aspects of Submarine Design. Part 1. Hydrodynamics, Defence Science and Technology Organisation.

McCormick, B. W. (1995). Aerodynamics, aeronautics, and flight mechanics, Wiley New York.

Munk, W. H. (1950). Origin and Generation of Waves. 1st International Conference on Coastal Engineering, Long Beach, California.

Ray, A., Sen, D., Singh, S. and Seshadri, V. (2010). "Quantification of Uncertainty in Manoeuvring Characteristics for Design of Underwater Vehicles." International Journal of Maritime Engineering 152: A71-A84.

Roddy, R. F. (1990). Investigation of the stability and control characteristics of several configurations of the DARPA SUBOFF model (DTRC Model 5470) from captive-model experiments (No. DTRC/SHD-1298-08). David Taylor Research Center Bethesda MD Ship Hydromechanics Dept.

Seil, G. J. and Anderson, B. (2013). Calculation of linear stability and wake distortion of of fully-appended BB1 and BB1-X Geometry, Defence Science and Technology Organisation.

Stevens, B. and Lewis, F. (2003). Aircraft Control and Simulation, 2nd Ed.,Wiley Interscience

Valentinis, F. and Kofmansky, T. (2013a). A Comparative Analysis of the DSTO Generic Joubert Submarine with Different Fins, Defence Science and Technology Organisation.

Valentinis, F. and Kofmansky, T. (2013b). A Dynamic Stability Analysis of the DSTO Generic Joubert Submarine with Two Aft Control Surface Designs, Defence Science and Technology Organisation.

The Effect of Hull Form on the Optimum L/D ratio for Minimum Resistance for Submerged Bodies

Pierre-André Crété[1], Zhi Quan Leong[2,3], Dev Ranmuthugala[2,3], Martin Renilson[3]
[1]ENSTA Bretagne, France
[2]DST Group, Australia
[3]Australian Maritime College/University of Tasmania, Australia
zhi.leong@utas.edu.au

ABSTRACT

For minimum resistance, the optimum slenderness (*L/D*) ratio for an underwater vehicle hull based on a near optimal 'teardrop' hull shape is generally considered to be around six. However, in practice the optimal hull shape is rarely achieved, with most modern underwater vehicles such as submarines and autonomous underwater vehicles (AUV) having parallel mid-body sections. Thus, the optimum *L/D* ratio is different from that established on the basis of a teardrop shape.

This study involves the use of Reynolds-Averaged Navier-Stokes (RANS) Computational Fluid Dynamic (CFD) simulations to quantify the resistance characteristics of three submarine hull shapes with different *L/D* ratios. The results were used to identify the optimum *L/D* ratio for minimum resistance for each of the hull forms. The hull forms include the DST generic conventional Joubert submarine geometry, the DARPA standard SUBOFF hull form, and the Teardrop shape that is optimal in terms of hydrodynamic resistance. The CFD simulations were validated against published measurements from captive model experiments. The results presented provide a comparison of the resistance for the shapes and the optimum values of *L/D* for each different shape are given.

The results are also used to establish an approximation technique which can be used to predict the resistance of a hull shape for different *L/D* ratios. The value of the approximation technique is at the concept stage of a submarine or underwater vehicle design, where it is necessary to have a prediction for the resistance prior to the commissioning of any model tests or CFD simulations, both of which are relatively expensive and time consuming.

1. INTRODUCTION

The resistance of a deeply submerged submarine is primarily made up of two components: pressure resistance and friction resistance. For a given volume, the pressure resistance is smaller for larger values of slenderness ratio (i.e. larger *L/D* ratios), while the friction resistance is larger for larger values of *L/D* (Renilson, 2015). This is illustrated in Figure 1 for a deeply submerged Teardrop hull form, with the resistance non-dimensionalised based on the velocity and the volume of the hull. The notation and non-dimensionalisation schemes are provided in *section 7*.

Consequently, from a resistance perspective, there is an optimum *L/D* value for a deeply submerged submarine. Although influenced slightly by the Reynolds Number (R_e), the approximate optimum L/D value is essentially similar for different speeds. Earlier studies based on the results of model tests (Burcher & Rydill, 1994; Gertler, 1950;) have shown that this optimum L/D is approximately six, and this is now generally accepted as the optimum ratio. However, as the pressure resistance is influenced by the shape of the hull, the optimum L/D value will depend on the hull form. For a given L/D value, a hydrodynamically optimized hull form, i.e. the classic 'teardrop' shaped hull, will generate a lower pressure resistance and a higher friction resistance for a given volume than a conventional submarine hull form with a large parallel middle body length.

In practice the optimal hull shape or the optimum L/D ratio are rarely achieved, with most modern underwater vehicles such as submarines and autonomous underwater vehicles (AUVs) having parallel mid-body sections and significantly larger L/D ratios than six (Renilson et al., 2011). Current hull forms are influenced by many factors other than the resistance characteristics, such as payload, machinery, sensors, manoeuvrability, deck space, and draft limitations that affect the hull form and its dimensions (Joubert, 2004; Renilson, 2015).

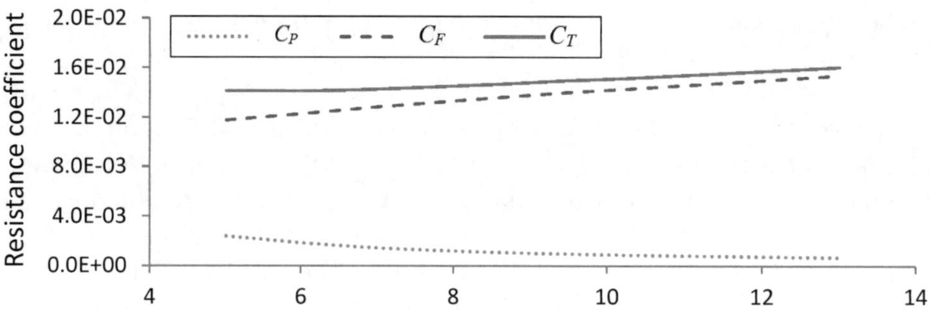

Figure 1: Components of the volume-based resistance coefficient (pressure, C_p, friction, C_f, and total, C_T, coefficients) on a deeply submerged Teardrop shaped submarine hull form at a Reynolds number, R_e, of 4.8×10^8

This study uses computational fluid dynamics (CFD) simulations to investigate the resistance of hull forms as functions of the L/D ratio and how much the optimum L/D varies for the hull forms. The results are also used to establish an approximation technique which can be used to predict the resistance of a hull shape for different L/D ratios.

2. INVESTIGATION PROGRAMME

The effects of different slenderness (L/D) ratios on the resistance of a submarine hull were investigated for three different bare hull forms (see Figure 2): the optimal hydrodynamic Teardrop hull form (Joubert, 2004); the DST Joubert generic hull form (Joubert, 2006); and the DARPA standard SUBOFF hull form (Groves et al., 1989).

The variables investigated included the total resistance, pressure resistance, and friction resistance coefficients defined in *section 7*. With the exception of the validation study, all results presented in this paper are at full-scale with an operating speed of 7.2 ms^{-1} (length based R_e = 4.8×10^8, 14 knots). The hull forms were scaled to achieve the desired slenderness ratio while keeping the overall length constant at 60 m to maintain the same R_e.

Table 1 summarises the investigation parameters and vehicle dimensions. Table 2 outlines the volumes of the hull forms at the different *L/D* ratios.

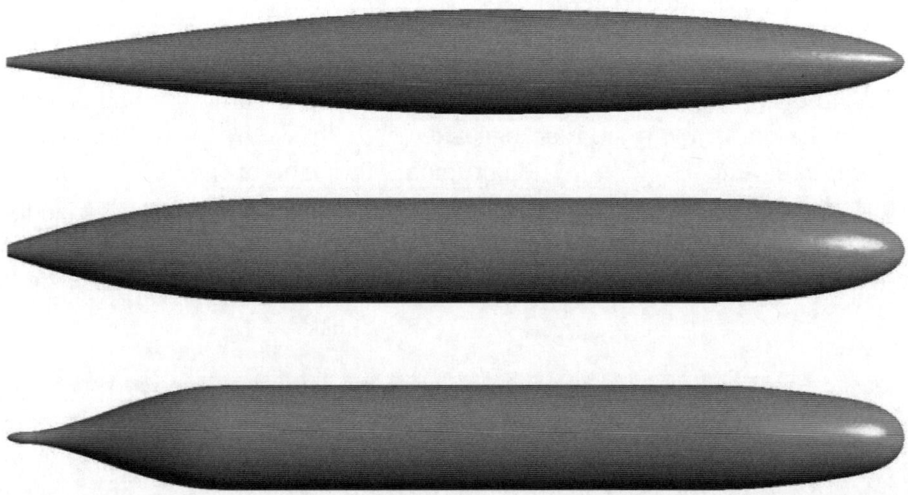

Figure 2: Bare hull forms (top to bottom): Teardrop (Joubert, 2004); Joubert (Joubert, 2006); and SUBOFF (Groves et al., 1989).

Table 1: Hull form dimensions and investigation parameters.

Parameter	Value	
Length, L	60	[m]
Diameter, D	12.0, 10.0, 8.6, 7.0, 6.3, 5.9, 4.6	[m]
Slenderness ratio, L/D	5.0, 6.0, 7.0, 8.6, 9.5, 10.2, 13.0	
Forward speed, U_0	7.2	[ms^{-1}]
Fluid density, ρ	997	[kgm^{-3}]
Fluid dynamic viscosity, μ	1.028×10^{-3}	[$kgm^{-1}s^{-1}$]

Table 2: Hull form volume and surface area parameters for different *L/D ratios*.

L/D	Volume [m^3]			Surface Area [m^2]		
	Teardrop	**Joubert**	**SUBOFF**	**Teardrop**	**Joubert**	**SUBOFF**
5.0	3904	5414	5374	1630	2006	1983
6.0	2711	3760	3732	1352	1660	1639
7.0	1992	2762	2742	1155	1417	1398
8.6	1328	1841	1827	941	1152	1136
9.5	1082	1500	1489	848	1038	1023
10.2	931	1291	1281	786	962	948
13.0	578	801	795	618	756	745

3. SIMULATION SETUP

The commercial CFD code, ANSYS-CFX 17, was used to model the flow around the hulls undergoing the straight line manoeuvre using Reynolds-Averaged Navier-Stokes (RANS) based steady state simulations with double precision and the high resolution advection scheme.

Figure 3 shows the computational domain and coordinate system with the following boundary settings: hull as a no-slip wall; top, side and bottom boundaries as free-slip walls; and boundary behind the hull as an outlet opening with zero relative pressure. The boundary in front of the hull was defined as an inlet, with a prescribed flow speed of U_0. The coordinate system origin was located at the bow of the hull. The outlet boundary distance to the coordinate system origin was 6*L*, while the inlet and far field boundaries were located a distance of 3*L* to ensure the pressure and wake fields generated by the hull are sufficiently resolved within the domain with no blockage effects, as shown by the authors (Leong et al., 2015). A symmetry boundary along the longitudinal mid-plane of the hull (i.e. *xz*-plane) was used to reduce the computational effort by effectively halving the domain (see Figure 3). The simplification of using the symmetry boundary was verified by comparing the resistance prediction from a full domain simulation with differences in the results found to be less than 0.001%.

The RANS-based Baseline Reynolds stress turbulence model was utilised in this analysis as it was found to more accurately predict the forces and moments acting on torpedo-shaped bodies compared to the RANS-based eddy-viscosity models within CFX (Leong et al., 2015; Toxopeus et al., 2012). The Baseline Reynolds stress model is more comprehensive in modelling of flows that are strongly anisotropic, thus offering better fidelity in capturing rotational flow, flow separation, and secondary flows commonly encountered around a torpedo-shaped body due to its streamlined stern region.

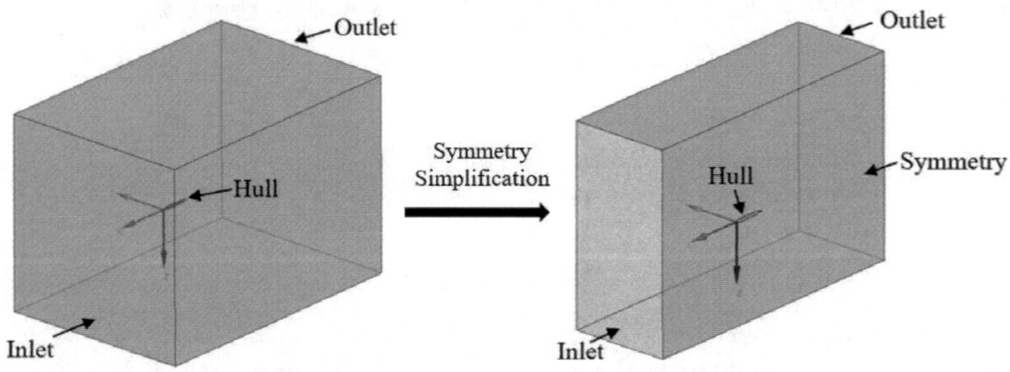

Figure 3: Computational domain and coordinate system (at bow of hull) for straight-line manoeuvre: full domain (left) and symmetry domain (right).

3.1 MESH MODEL

The computational mesh was generated with ANSYS Meshing and is composed of triangular prismatic layers around the hull to capture the boundary layer and unstructured tetrahedrons in the far field (see Figure 4). An unstructured mesh approach, although requiring a higher mesh density, was established to offer the same degree of accuracy in comparison to a structured mesh (Duda et al., 2011).

A non-dimensional wall distance (y^+, see *section 7*) of 1 is used for the first prismatic layer height in combination with the low-R_e wall treatment model which is required to accurately capture flow separation at the streamlined stern region of a torpedo-shaped body, as shown by the authors in Leong et al. (2015). The overall mesh cell count for the symmetry domain used in this study was five million, which was deemed sufficient for the purpose of this study as it provided an uncertainty in resistance of less than 3% when compared to further mesh refinements of up to 16 million.

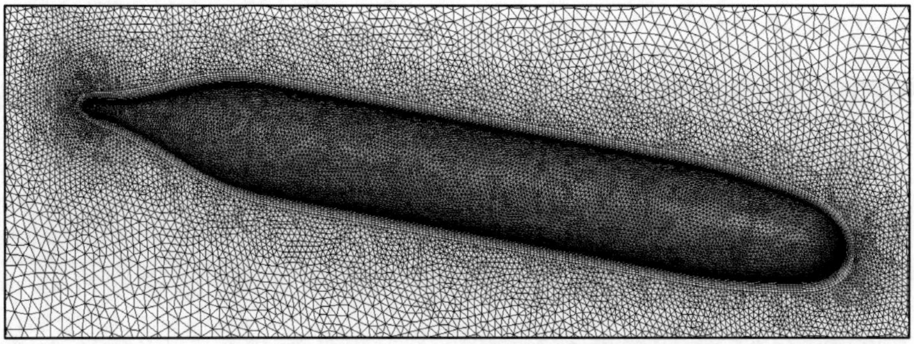

Figure 4: Mesh model of the SUBOFF hull form

4. CFD VALIDATION

As there was no available experimental resistance data for an axisymmetrical bare hull at the full-scale conditions investigated in this study, validation was carried out at model-scale against experimental data for the SUBOFF at $R_e = 1.4 \times 10^7$ by Huang et al. (1992) and at $R_e = 2.1 \times 10^6$ by Ranmuthugala et al. (2012), comparing loads and resistance distribution along the length of the hull. The latter was considered essential as the following calculations are based on accurate predictions of the resistance components.

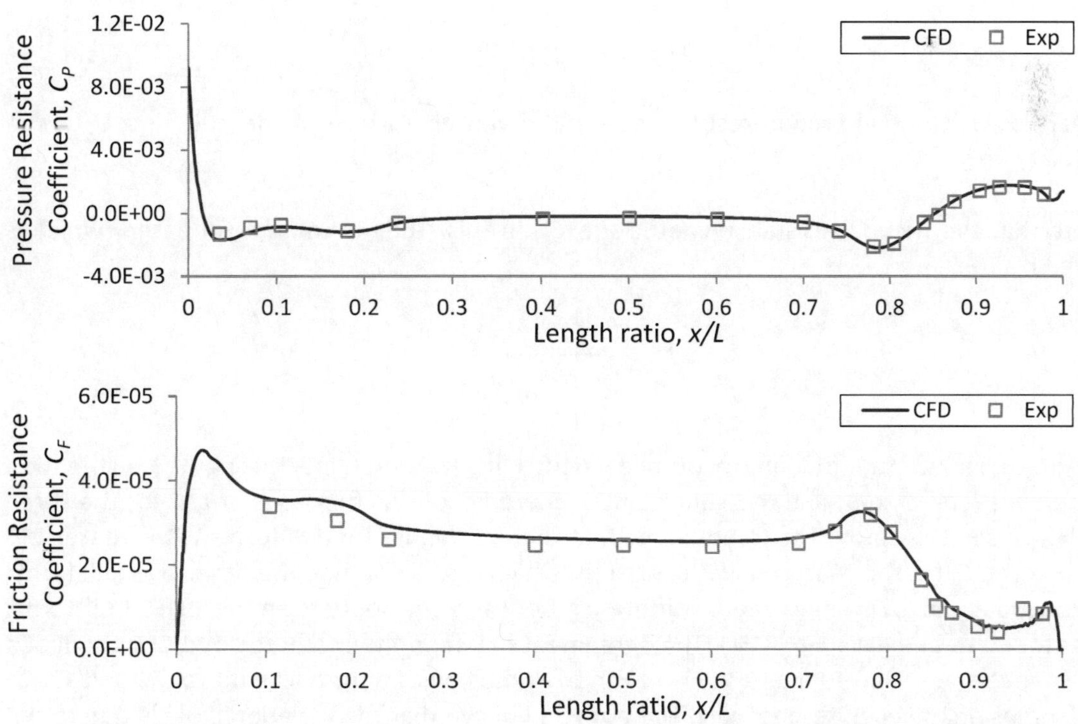

Figure 5: CFD and experimental (Huang et al., 1992) results of the resistance distribution along the unappended SUBOFF hull form at Re = 1.4×10^7: C_P (top) and C_F (bottom).

Figure 5 shows the resistance of the SUBOFF bare hull in terms of: (1) pressure distribution and (2) friction distribution, along the length of the hull. The experimental and CFD model-scale results are in good agreement, with the latter well within the experimental uncertainty of 10%, thus providing sufficient validation for the CFD model to be extended to the higher R_e full-scale condition investigated in this study.

5 RESULTS AND DISCUSSION

The following discussion focuses on the resistance of the three full scale hull forms for the range of slenderness ratios (*L/D*) outlined in Table 1. An approximation technique which can be used to predict the pressure resistance of a hull shape for different *L/D* ratios is also presented.

5.1 Friction Resistance

The friction resistance for an axisymmetrical hull can be estimated using the ITTC-57 correlation line (ITTC, 2008), as given in Eq. 1.

$$C_{F*ITTC-57} = \frac{0.075}{(log_{10}R_e - 2)^2} \tag{1}$$

where $C_{F*ITTC-57}$ is the non-dimensional friction resistance including the frictional-form resistance component and is based on the hull wetted area, as defined by Eq. 2.

$$R_F = C_{F*ITTC-57}(\frac{1}{2}\rho S {U_0}^2) \tag{2}$$

where R_F is the hull friction resistance, *S* is the wetted surface of the hull, and U_0 is the velocity.

Note that the friction resistance coefficient (C_F) in this study is volume (*V*) based, which is given by Eq. 3.

$$C_F = \frac{R_F}{\frac{1}{2}\rho V^{2/3} {U_0}^2} \tag{3}$$

Figure 6 shows the friction resistance coefficient (C_F) obtained using the estimation equations (Eqs. 1 to 3) and the equivalent CFD predictions as functions of *L/D* for the three hull forms in this study. The trends show that the C_F is higher for hull forms with greater *L/D* ratios. The C_F for the Teardrop was consistently higher (2 - 3%) than the other two hull forms throughout the *L/D* range. The C_F difference between the Joubert and SUBOFF hull forms was found to be negligible (<1%). The agreement between the estimated and CFD predicted resistance was found to increase at higher *L/D* ratios, with differences up to 3% and 6% at *L/D* ratios of 13 and 5, respectively. The authors believe that the variation is likely due to the accuracy of the ITTC-57 correlation line diminishing at the lower *L/D*, however, further work is ongoing to explore this. Nevertheless, these differences are considered small and highlight the value of the estimation technique compared to CFD that is significantly more computationally expensive.

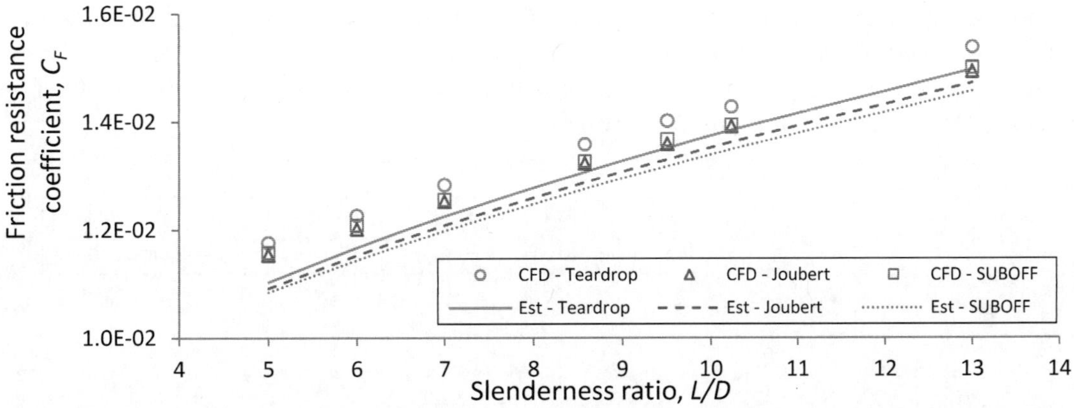

Figure 6: C_F as a function of L/D for CFD predicted values and estimated values using Eq. 1 to 3 for the three hull forms: Teardrop, Joubert and SUBOFF.

Note that the estimation was for a hydrodynamically smooth axisymmetric hull. Additional allowance may need to be made for imperfections of a submarine hull, ranging from general unevenness and fouling on the hull, to vent holes for the ballast tanks (Renilson, 2015; Joubert, 2004).

5.2 Pressure Resistance

The pressure resistance is a smaller component of resistance than the friction resistance on a typical submarine hull (see Figure 1). The pressure resistance coefficient (C_P) can be obtained as a function of C_F using Eq. 4 (Renilson, 2015).

$$C_P = K_P C_F \tag{4}$$

The value of K_P represents the ratio of pressure resistance to friction resistance for a hull form. K_P changes as a function of the L/D ratio, which can be rather expensive to obtain over a range of L/D using CFD and/or experimental measurements. Using power law regression analysis of the Teardrop and SUBOFF data within this study, Eq. 5 (Renilson, 2015) is derived to provide an estimation of K_P as a function of the L/D ratio,

$$K_P = \xi_{hull}(\frac{L}{D})^{-1.7} \tag{5}$$

where ξ_{hull} is the hull form factor for the shape of the hull and is assumed constant for a small change in R_e. If ξ_{hull} for a hull form can be determined by calculating the K_P at one L/D ratio using CFD or experiments, then Eq. 5 can be used to estimate K_P for other L/D values. For example, the ξ_{hull} for the Teardrop, Joubert and SUBOFF hull forms were calculated to be 3, 5 and 6 respectively using the CFD results at L/D = 8.57, as outlined in Table 3. Figure 7 shows the K_P obtained using the estimation equation (Eq. 5) and CFD as functions of L/D for the three hull forms. As can be seen in the figure, K_P for the SUBOFF shape is larger than for the Joubert shape, with the Teardrop shape, as expected, having the lowest K_P. The K_P for the hull forms decreased as *L/D* increases. The agreement between the estimated and the CFD predicted K_P is good, with differences less than 6%.

The same level of agreement was observed for C_P as functions of *L/D* in Figure 8 for the CFD predicted and calculated values using the K_P estimates from Eq. 4. The C_P for the hull forms decreased as *L/D* increases. It is also noted that the SUBOFF has the highest C_P, followed by the Joubert hull form and then the Teardrop hull form.

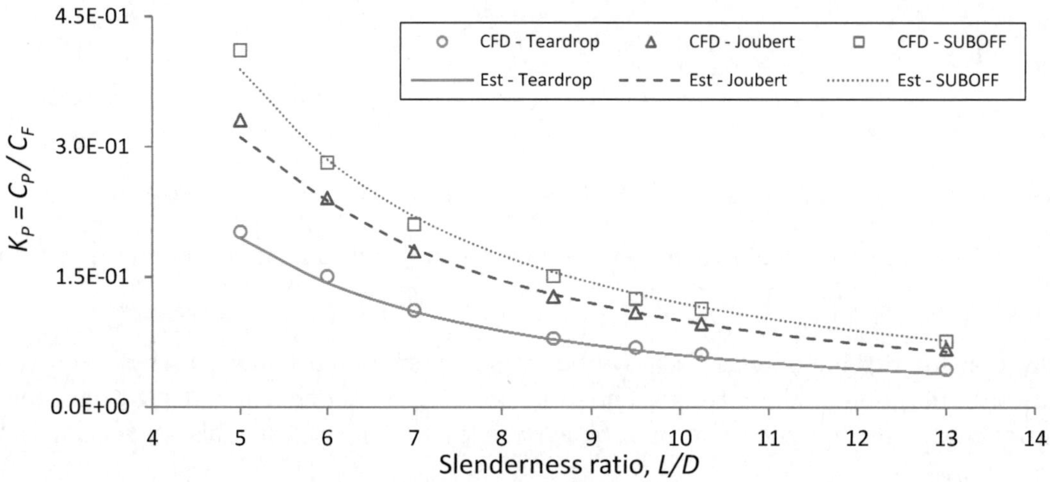

Figure 7: K_P as a function of *L/D* for CFD predicted values and estimated values using Eq. 5 for the three hull forms: Teardrop, Joubert and SUBOFF.

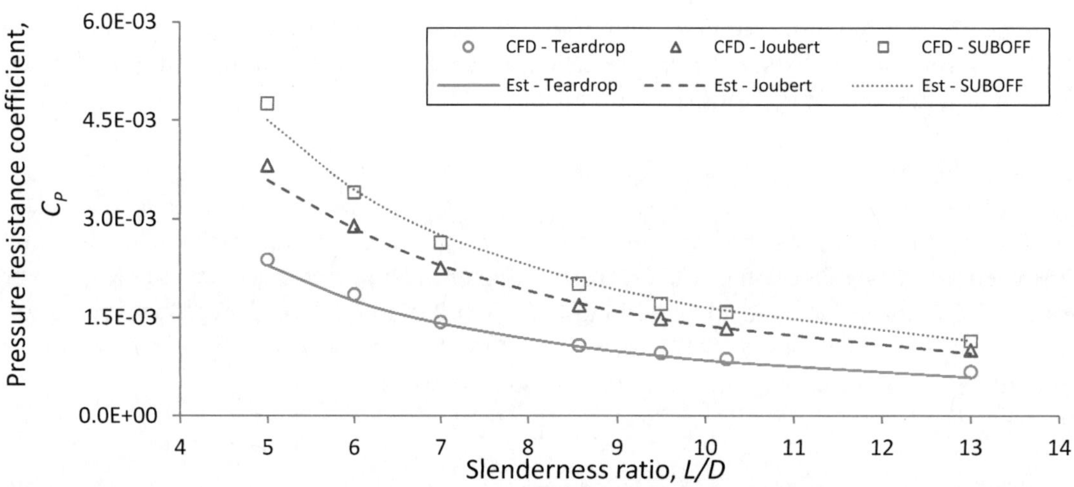

Figure 8: C_P as a function of *L/D* for CFD predicted values and calculated values from the K_P estimates using Eq. 4 for the three hull forms: Teardrop, Joubert and SUBOFF.

Table 3: Calculated hull form factor, ξ_{hull}, using Eq. 5 with K_P values obtained from CFD for the three hull forms at *L/D* = 8.57

Hull form	K_P at *L/D* = 8.57	Hull form factor, ξ_{hull}
Teardrop	0.079	3
Joubert	0.127	5
SUBOFF	0.151	6

5.3 Total Resistance

Figure 9 shows the total resistance coefficient (C_T) obtained using the estimation equations (combining both friction and pressure, i.e. Eqs. 1 to 5) and CFD predictions as a function of *L/D* for the three hull forms. The optimum *L/D* ratio for Teardrop, Joubert and SUBOFF hull forms are approximately 6, 7, and 8 respectively. This indicates that optimum *L/D* increases as the hull form becomes 'fuller', i.e. an increasing prismatic coefficient (ratio between the volume of the hull and the product of the length and the maximum cross-sectional area).

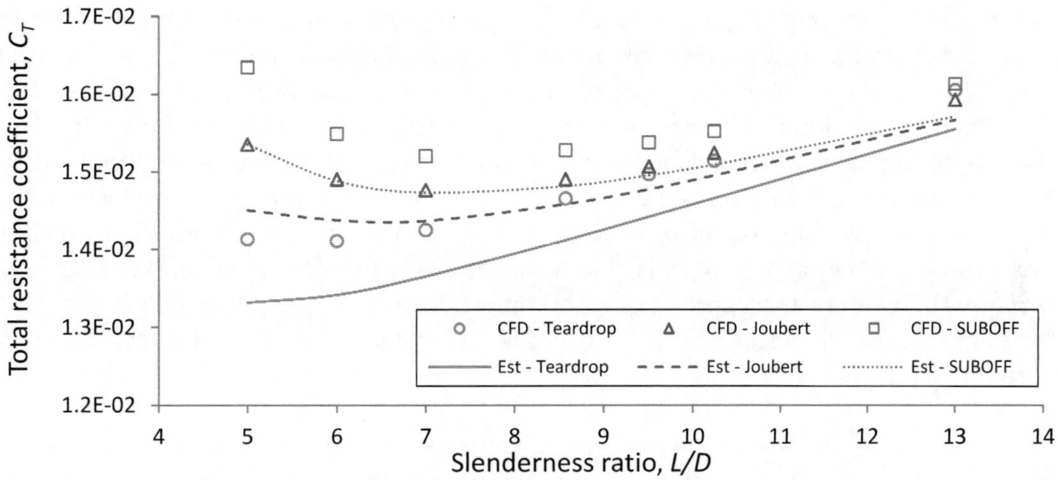

Figure 9: C_T as a function of *L/D* for CFD predicted values and estimates using Eq. 1 to 5 for the three hull forms: Teardrop, Joubert and SUBOFF.

The agreement between the estimated and the CFD predicted values of C_T was good, with differences less than 6%. These differences are considered small given that the estimation technique is significantly computationally cheaper in comparison to CFD predictions. Although the estimation technique provides a similar trend in total resistance and can estimate the approximate K_P, it tends to under predict the friction resistance values (especially at low *L/D* ratios within 6%). However, the estimation technique can be used to predict the approximate optimum *L/D* for minimum total resistance for the hull forms within 10%, thus provide a good indicator for preliminary studies.

6 CONCLUDING REMARKS

This paper presented a study conducted to quantify the resistance of three unappended submarine hull forms (i.e. Teardrop, Joubert and SUBOFF) as functions of the *L/D* ratio. The investigation utilised CFD techniques to quantify the friction, pressure, and total resistance acting on the hulls at full-scale conditions operating deeply submerged at a speed of 14 knots. The results were used to examine the optimum *L/D* ratio of hull forms for minimum resistance and to establish an approximation technique which can be used to predict the resistance of a hull shape for different *L/D* ratios.

Previously published experimental data was used to validate the capability of the CFD model to predict the resistance of the SUBOFF at model-scale. The experimental and CFD results showed good agreement, thus validating the CFD approach.

The results showed that the pressure resistance coefficient decreased and the friction resistance coefficient increased with greater *L/D* ratios. Throughout the *L/D* range investigated, the Teardrop hull form had the highest volume-based friction resistance coefficient. The Joubert and SUBOFF had very similar friction resistance coefficients (with differences <1%). The SUBOFF hull form had the highest pressure resistance coefficient, followed by the Joubert and Teardrop hull forms.

The optimum *L/D* ratio for minimum total resistance was found to increase as the hull form became fuller (i.e. an increasing prismatic coefficient). The Teardrop hull form has the lowest value of optimum *L/D*, followed by the Joubert and the SUBOFF hullforms.

The estimation technique presented in this paper was found to provide friction and pressure resistance values which had similar trends as functions of *L/D* to the CFD predictions and within 6% (under predicted) for the three hull forms throughout the *L/D* range investigated. The under prediction is mainly due to the errors in the friction resistance estimation. However, the estimation technique is able to satisfactorily predict the K_P values for different *L/D* ratios and capture the approximate optimum *L/D* ratio for the three hull forms within 10%. Further work is ongoing to improve the estimation of the friction resistance for different hull forms.

The value of the estimation technique is at the concept stage of a submarine or underwater vehicle design, where it is necessary to have a prediction for the resistance prior to the commissioning of a significantly more expensive and time-consuming model test and/or CFD program.

7. NOMENCLATURE

C_F	Friction resistance coefficient; $R_F/(0.5\rho V^{2/3}U_0^2)$	R_P	Pressure resistance [N]
$C_{F*ITTC-57}$	ITTC-57 non-dimensional friction resistance; $0.075/(\log_{10}R_e-2)^2$	R_T	Total resistance [N]
C_P	Pressure resistance coefficient; $R_P/(0.5\rho V^{2/3}U_0^2)$	S	Hull wetted surface area [m^2]
C_T	Total resistance coefficient; $R_T/(0.5\rho V^{2/3}U_0^2)$	U_0	Forward speed of the hull relative to fluid [ms^{-1}]
D	Diameter [m]	u_*	Shear velocity [ms^{-1}]
K_P	C_P/C_F	y_{wall}	Mesh node distance to wall [m]
L	Overall length [m]	y^+	Non-dimensional wall distance; $(u_* y_{wall})/v$
L/D	Slenderness ratio; L/D	ρ	Fluid density [kgm^{-3}]
R_e	Reynolds number; $(\rho U_0 L)/\mu$	μ	Fluid dynamic viscosity [kgm^{-1}s^{-1}]
R_F	Friction resistance [N]	V	Volume of displacement [m^3]
		v	Fluid kinematic viscosity [m^{2}s^{-1}]; μ/ρ
		ξ_{hull}	Hull form factor

8. REFERENCES

1. Burcher, R. and Rydill, L., *Concepts in Submarine Design*, Ocean Technology Series, Cambridge University Press, 1994

2. Duda, B., Menter, F., Hansen, T., and Esteve, M-J. 2011, Scale-adaptive simulation of a hot jet in cross flow, *Journal of Physics: Conference Series*, vol. 318, pp. 1-6
3. Groves, N., Huang, T., and Chang, M., *Geometric Characteristics of DARPA SUBOFF Models (DTRC Models Nos. 5470 and 5471)*, David Taylor Research Center, Maryland, 1989.
4. Huang, T., Liu, H., Groves, N., Forlini, J., Blanton, J., and Gowing, S., *Measurements of Flows Over an Axisymmetric Body with Various Appendages in a Wind Tunnel: the DARPA SUBOFF Experimental Program*, David Taylor Model Basin, USA, 1992.
5. ITTC, Recommended Procedures and Guidelines: Testing and Extrapolation Methods: Propulsion, Performance Propulsion Test, *ITTC Recommended Procedures and Guidelines, Procedure 7.5-02-03-01.1*, 2008.
6. Gertler, M., *Resistance Experiments on a Systematic Series of Streamlined Bodies of Revolution – For Application to the Design of High-Speed Submarines*, The David W. Taylor Model Basin, Washington D.C, 1950.
7. Joubert, P., *Some Aspects of Submarine Design: Part 1 Hydrodynamics*, DSTO Platforms Science Authority, Melbourne, 2004.
8. Joubert, P., *Some Aspects of Submarine Design: Part 2 Shape of a submarine 2026*, DSTO Platforms Science Authority, Melbourne, 2006.
9. Leong, Z.Q., Ranmuthugala, D., Penesis, I., and Nguyen, H., RANS-based CFD Prediction of the Hydrodynamic Coefficients of DARPA SUBOFF Geometry in Straight-Line and Rotating Arm Manoeuvres, *RINA International Journal of Maritime Engineering*, vol. 157, pp. 41-51, 2015.
10. Ranmuthugala, D., Fletcher, D., Neulist, D., Renilson, M., and Anderson, B., The effect of length to diameter ratio on the hydrodynamic coefficients of a submarine operating near the surface in calm water, *Proceedings of the International Maritime Conference 2012*, 30 January - 2 February 2012, Darling Harbour, Australia, pp. 617-629, 2012.
11. Renilson, M., Ranmuthugala, D., Dawson, E., Anderson, B., van Steel, S., and Wilson-Haffenden, S., Hydrodynamic design implications for a submarine operating near the surface, *Proceedings of Warship 2011: Naval Submarines and UUVs*, Royal Institution of Naval Architects, 2011.
12. Renilson, M., Submarine Hydrodynamics, Springer Briefs in Applied Sciences and Technology, 2015, ISSN: 2191-530X, ISBN: 978-3-319-16183-9.
13. Toxopeus, S., Atsavapranee, P., Wolf, E., Daum, S., Pattenden, R., Widjaja, R., Zhang, J.T. and Gerner, A., Collaborative CFD Exercise for a Submarine in a Steady Turn, *Proceedings of OMAE ASME 31st International Conference on Ocean, Offshore and Arctic Engineering*, Rio de Janeiro, Brazil, 2012.

Integrated Survivability Analysis of Naval Platforms in High Threat Environments

John Brett, Grant Gamble, Warren Reid, Sue Smith,
Anthony Woolley and George Yiannakopoulos
Defence Science & Technology Group
[anthony.woolley@dst.defence.gov.au]

ABSTRACT

Survivability of a naval platform is critical to mission success in a high threat environment. Survivability is the ability to avoid, withstand, contain and recover from weapon threats and damage events. Platform survivability is characterised by the susceptibility, vulnerability and recoverability of the platform; and the temporal permutation of these characteristics is referred to as Integrated Survivability. The ability to model and assess platform Integrated Survivability will give mission planners assurance that the platform can survive exposure to high threat environments. Unfortunately, domain specific survivability models have generally been used in isolation, without communication across the domains of susceptibility, vulnerability and recoverability. Recent commercial software models provide some degree of integrated analysis; however, they do not consider the holistic nature of Integrated Survivability. The ability to provide Integrated Survivability analysis would enable acquisition programs to perform trade-off studies to ensure acquisition of a platform best suited to survive threats and threat effects. Integrated Survivability analysis would also facilitate assessment of platform configuration changes and changes to Standard Operating Procedures to ensure there is no detrimental effect to crew safety and mission success. This paper reviews Integrated Survivability definitions; how to quantify Integrated Survivability; and overviews the current state of Integrated Survivability analysis technology.

1. INTRODUCTION

The ability of a naval platform to survive in a high threat environment is generalised as combinations of the platform's ability to: avoid detection, classification and successful targeting by an adversary; withstand damage effects; and recover from damage effects. The combination of these factors defines the platform's survivability. Understanding naval platform survivability is necessary to make informed tactical and strategic decisions. It is also essential for the design, construction and acquisition of new platforms. Whilst the benefits of survivable platforms are undeniable, it is recognised that the pursuit of survivability must be compatible with other operational requirements and budget constraints. To justify survivability, it is necessary to quantify survivability and establish operational and cost implications. This will enable the provision of platform survivability assessment in high threat environments; inform improvements to platform survivability; identify how to best achieve survivability aims within budget constraints; and identify how survivability affects platform capability and the task force to which it is assigned.

However, survivability is generally conceptualised as a static idea. Consequently, the concept of 'Integrated Survivability' enhances the notion of survivability with the inclusion of factors that occur during a mission. Primarily, this is the temporal feedback between susceptibility, vulnerability and recoverability that develops during an incident, during engagement with hostile forces, and during the interaction of a platform with other platforms of a task force.

Integrated Survivability is not a static concept and, subsequently, the evaluation of Integrated Survivability becomes a complex task. Individually quantifying susceptibility, vulnerability and recoverability has unique difficulties, and there are significant technical challenges to overcome. Integrating susceptibility, vulnerability and recoverability to quantify platform Integrated Survivability presents additional challenges when simulating a naval mission. Whilst susceptibility, vulnerability and recoverability have been studied individually for some time, only recently have there been attempts to integrate these components to provide an overall assessment of Integrated Survivability for an individual platform. This paper presents a review of Integrated Survivability assessment for naval platforms, and discusses how it can be quantified and applied. The paper reviews the definition of survivability (the static domain) and identifies how that relates to Integrated Survivability (the temporal domain).

2. DEFINING INTEGRATED SURVIVABILITY

Quantifying Integrated Survivability is understandably a complex process and a prerequisite is a clear definition of terminology. Due in part to the evolution in the understanding and application of the concept of survivability of naval platforms, there is variation in terminology in the literature. Simplistically, survivability is commonly defined as a combination of susceptibility, vulnerability, and recoverability. However, there is some inconsistency within this accepted terminology. Susceptibility and vulnerability describe situations that are detrimental to a platform's condition whereas recoverability describes an operative condition. The alternative terms 'stealthiness' and 'sustainability' would make the definitions consistent but these are not the accepted usage.

One of the earliest uses of the term survivability for naval platforms is defined with respect to enhancing readiness and combat sustainability by addressing fundamental design considerations and associated protection level requirements, as found in the United States (US) Navy OPNAV Instruction (OPNAVINST) 9070.1 [1]. This definition was given in the context of survivability policy for surface platforms of the US Navy at the instigation of US Congress. Three levels of survivability were defined: Level I (low); Level II (medium); and Level III (high). Each level, from least severe to most severe, reflects the level of damage that could be sustained by a naval platform whilst maintaining mission capability. However, these definitions did not include recoverability, nor do they form a definition of survivability in terms of the three subordinate components: susceptibility, vulnerability and recoverability. However, with the development of the engineering management discipline, *Total Ship Survivability,* utilised by NAVSEA, survivability for naval platforms was then considered in terms of the subordinate components [2]. OPNAVINST 9070.1A [3] revised the definitions of survivability and its components, defining each term as a *measure of capability* in relation to combat and non-combat incidents. Whilst many other authors have defined, and continue to define survivability, in terms of an ability or inability to achieve a goal, consideration of recoverability is sometimes omitted from the definitions [4, 5, 6, 7, 8, 9]. The definitions found in OPNAVINST 9070.1A most accurately reflect the need for quantifying survivability, encompassing susceptibility, vulnerability and recoverability [3]:

1. *Survivability is a measure of both the capability of the ship, mission critical systems, and crew to perform assigned warfare missions, and of the protection provided to the crew to prevent serious injury of death. The three principal disciplines of survivability are susceptibility, vulnerability and recoverability.*

2. *Susceptibility is a measure of the capability of the ship, mission critical systems, and crew to avoid and or defeat an attack and is a function of operational tactics, signature reduction, countermeasure, and self-defense system effectiveness.*
3. *Vulnerability is a measure of the capability of the ship, mission critical systems, and crew to withstand the initial damage effects from conventional, Chemical, Biological and Radiological (CBR) or asymmetric threat weapons, or accidents, and to continue to perform assigned primary warfare missions, and protect the crew from serious injury or death.*
4. *Recoverability is a measure of the capability of the ship and crew, after initial damage effects, whatever the cause, to take emergency action to contain and control damage, prevent loss of a damaged ship, minimize personnel casualties, and restore and sustain primary mission capabilities.*

Inherent in these definitions is the implicit acknowledgement that survivability is a temporal concept. However, this does not explicitly recognise Integrated Survivability. The distinction between the terms 'survivability' and 'Integrated Survivability' is important. Integrated Survivability is the explicit acknowledgment that platform and mission survival is dependent on the ongoing situation. An incident involving any form of threat encounter may result in primary and secondary damage. Primary damage is a result of the initial incident (such as an explosion event), and secondary damage results from the ongoing effects initiated by the primary event (such as fire or flood). Primary and secondary events affect susceptibility, vulnerability and recoverability in the temporal domain, and many authors do not acknowledge this when quantifying survivability. Therefore, (perhaps, unwittingly) survivability becomes defined as a static concept and, in many instances, only considers susceptibility and vulnerability to quantify survivability (as a probability of kill) from the primary damage event. So, taking the definition of survivability one step further:

Integrated Survivability is the temporal measure of survivability (as defined), recognising that susceptibility, vulnerability and recoverability are interdependent, that they change with respect to time, and are dependent on the threat environment and mission requirements.

3. QUANTIFYING INTEGRATED SURVIVABILITY

The literature predominantly quantifies survivability using probability measures and, as can be expected, this quantification is from a static point of view. Therefore, to quantify Integrated Survivability, survivability can initially be considered as it is generally defined in terms of susceptibility and vulnerability, where the probability of an event occurring lies between the interval (0, 1) [10, 11]. So, the probability of platform survivability, $P(S)$, can be expressed as the complement of the probability of a mission kill, $P(K)$:

$$P(S) = P(\overline{K}) = 1 - P(K) \quad (1)$$

Now, susceptibility is the inability to avoid a threat, quantified by the probability of being hit by a weapon threat, $P(H)$; and vulnerability is quantified by the conditional probability that a platform is 'killed' given that it is hit by a threat, $P(K|H)$. The probability of kill is the product of the probability of the platform being hit, $P(H)$, and the platform's vulnerability, $P(K|H)$:

$$P(K) = P(H) \times P(K|H) \quad (2)$$

Thus, survivability is defined as:

$$P(S) = 1 - (P(H) \times P(K|H)) \quad (3)$$

and it can be reformulated to include other subordinate elements of susceptibility:

$$P(S) = 1 - (P(E_C) \times P(E_N) \times P(H) \times P(K)) \quad (4)$$

where

$P(E_C)$ = the probability of being encountered;
$P(E_N)$ = the probability of being engaged;
$P(H)$ = the probability of being hit; and
$P(K)$ = the probability of being killed.

Survivability equation (4) considers susceptibility (represented by $P(E_C)$, $P(E_N)$ *and* $P(H)$) and vulnerability (represented by $P(K)$), as mutually exclusive. However, in reality, there may be interaction between variables contributing to the probabilities – a damage event may affect ongoing platform susceptibility. Furthermore, recoverability is not considered. Platform recoverability is dependent on the degree of inflicted damage, and the ability to recover from that damage. Consequently, equation (4) is insufficient, and a different approach is required to incorporate recoverability into the equation of probability of survivability. Consider, then, Figure 1.

Figure 1 presents a survivability conditional probability tree for the ability of a platform to continue a mission, or not continue a mission. In Figure 1, the platform has a probability of being hit, $P(H)$, and a probability of not being hit, $P(\bar{H})$. If the platform is not hit, then it can continue the mission. If the platform is hit, then it has a probability of sustaining damage, $P(V|H)$; and, likewise, there is a probability of not sustaining damage, $P(\bar{V}|H)$, and the platform can continue the mission. Finally, if the platform sustains damage, there is a probability of being able to recover from the damage $P(R|V)$; and, there is a probability of not being able to recover from the damage $P(\bar{R}|V)$ and, therefore, being unable to continue the mission. So, assume the platform did encounter a threat, and damage was sustained, and the crew were able to recover from the damage and continue the mission, the probability of survivability can be expressed as:

$$P(H) \times P(V|H) \times P(R|V)$$

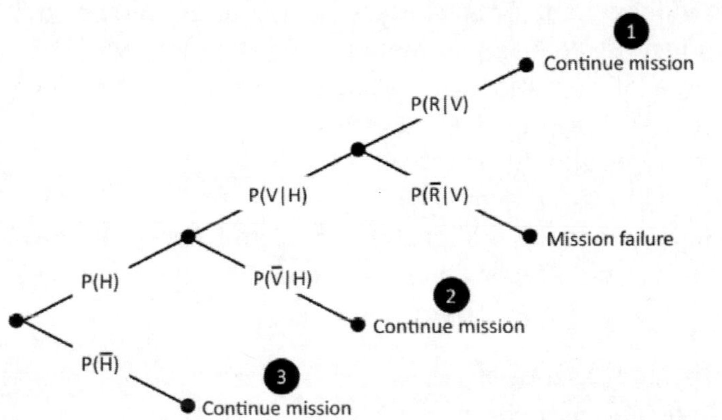

Figure 1: Probability of survivability (and, therefore, continuing the mission) as a conditional probability function of: probability of being hit, P(H); probability of sustaining damage, P(V); probability of recoverability, P(R); and the respective complementary probabilities

This represents location ❶ in Figure 1. However, this is not the complete representation of mission survivability. The platform has two other possible options for continuing the mission: the platform being hit but not sustaining damage, location ❷ in Figure 1; and the platform not being hit, location ❸ in Figure 1. Locations ❷ and ❸ can be represented by:

❷ $P(H) \times P(\bar{V}|H)$; and

❸ $P(\bar{H})$.

Hence, the probability of survivability can be represented as:

$$P(S) = P(H) \times P(V|H) \times P(R|V) + P(H) \times P(\bar{V}|H) + P(\bar{H}) \tag{5}$$

From Figure 1, the following can be deduced:

$$P(\bar{V}|H) = 1 - P(V|H)$$
$$P(\bar{H}) = 1 - P(H)$$

Substituting these into equation (5) gives,

$$P(S) = P(H) \times P(V|H) \times P(R|V) + P(H) \times (1 - P(V|H)) + (1 - P(H))$$
$$\therefore P(S) = 1 - P(H) \times P(V|H) \times [1 - P(R|V)] \tag{6}$$

Equation (6) concurs with Manly [12], where the probability of being damaged, *P(V),* replaces the probability of kill, *P(K).* This implicitly takes into consideration platform functionality to achieve the mission requirements, and expresses it in terms of reduced capability. However, Integrated Survivability stresses the interaction between susceptibility, vulnerability and recoverability in a mission context. Therefore, equation (6) is still insufficient for the study of Integrated Survivability because, as discussed by Coulthard and Hardy [13]:

(i) it applies to a single encounter, whereas a mission may involve multiple encounters with differing values of *P(H)*, *P(V)* and *P(R)*; and

(ii) *P(H)*, *P(V)* and *P(R)* may change during an encounter if a threat eventuates and causes damage.

Furthermore, equation (6) is scenario independent. That is, there is no consideration of the mission or threat type. The mission affects system requirements – specific systems will be required for specific missions. The area of operations for each mission will expose the platform to differing threats. Each threat will have a different effect on the platform and its systems. Therefore, *k* can refer to threat type; and the variable *t* is the time.

Now, the probability of being hit, *P(H)*, becomes the probability of susceptibility to threat *k.* Similarly, the probability of sustaining damage affecting the mission requirement, and the probability of recoverability also need to consider the threat, *k*. That is:

k = threat types expected for the mission scenario;

$P(H) \Rightarrow P(H_k)$ = probability of susceptibility to threat *k;*

$P(V) \Rightarrow P(V_k)$ = probability of sustaining damage by threat k that affects the mission requirement; and

$P(R) \Rightarrow P(R_k)$ = probability of ability to recover mission systems damaged by threat k.

Also, damage effects may take time to occur, and time is required to recover from that damage. That is,

t_{damage} = time for damage to occur where some damage will be instantaneous (for example, resulting from weapon explosion), and some might be ongoing (for example, resulting from ongoing fire or flood); and

$t_{recover}$ = time to recover damage systems;

such that, t_{damage} and $t_{recover}$ occur in the range $t = t_0 \rightarrow t_x$, *where*

t_0 = start time for the scenario; and

t_x = time at which the scenario concludes.

Therefore the probability of Integrated Survivability is,

$$P(Integrated\ Survivability) = 1 - P(H_k) \times P(V_k|H_k) \times [1 - P(R_k|V_k)] \quad \forall k;\ t = t_0 \rightarrow t_x \quad (7)$$

Finally, each of $P(H_k)$, $P(V_k)$ and $P(R_k)$ can be defined by subordinate component probabilities, such as $P(E_C)$ and $P(E_N)$ for $P(H_k)$ in equation (4). As a result, equation (7) becomes a complex equation, requiring Monte Carlo simulation to provide a solution. An example of subordinate component probability analysis for recoverability, $P(R|V)$, in relation to the ability to control a fire near a platform's magazine in order to prevent cook-off of munitions, has previously been described [14]. The advantage of equation (7) is the ability to calculate the probability of survivability with respect to time. This will enable Integrated Survivability analysis to identify where changes might be made to the platform, thereby producing a more favourable outcome (for example, improving mission success, platform safety and crew safety). Equation (7) also supports the development of on-board decision support tools to facilitate decision-making in high threat environments and damage control situations. Integrated Survivability analysis and related decision support tools will be further enhanced with the use of physics based simulations and mathematical modelling as described in Sections 4 and 7.

Equation (7) relates specifically to the platform. Further complexity arises when the concept of Integrated Survivability is extended to the mission success of a multi-platform taskforce. Coulthard and Hardy [13] characterise interaction between susceptibility, vulnerability and recoverability in terms of immediate intra-engagement feedbacks and less immediate inter-engagement feedbacks. Intra-engagement feedbacks occur within a single multiple threat engagement, for example damage to a platform's hard kill system (such as close-in weapon systems) will increase platform susceptibility to ongoing threats. An example of an inter-engagement feedback is a platform damaged in an initial engagement that requires assistance from other platforms within the task force during recoverability operations. This becomes a force divider, and generates a larger signature for subsequent engagements. A force divider will affect the probability of mission success (for the task force). However, while an important consideration, inter-engagement feedback complicates the Integrated Survivability equation considerably. For now, only survivability effects on a single platform and the platform's ability to continue its designated mission, without consideration of the effect on a task force mission requirement, is assumed.

4. APPROACHES TO QUANTIFYING INTEGRATED SURVIVABILTY

Integrated Survivability may be derived experimentally or by application of Modelling and Simulation (M&S). Experimental determination will generally be cost prohibitive. This is because susceptibility must be measured in various environmental conditions, and vulnerability and recoverability require destructive testing. M&S is a cost beneficial solution, though requiring experimental results for the development and validation of suitable models. A conceptual overview of an Integrated Survivability M&S analysis framework recognises that the simulation commences with susceptibility followed by vulnerability and then recoverability. The framework must be connected such that it forms a loop connecting susceptibility, vulnerability and recoverability. Subsequent iterations of the simulation must then include feedback and interaction between susceptibility, vulnerability and recoverability. Inputs to the model would include platform descriptions and threat environment, and a description of the mission scenario. The framework concept may be constructed using different techniques; three of these are outlined.

Concept 1: Perform susceptibility, vulnerability and recoverability assessments contained within a single Integrated Survivability model. A scenario description would be defined for the survivability model, the model executed, and results obtained. The advantage is that descriptions of the platform and the mission scenario are input once, and no manipulation of data between separate models of susceptibility, vulnerability and recoverability is required. The disadvantage is that a large model must be developed using common techniques, incurring a large development cost. This approach is illustrated in Figure 2.

Concept 2: Perform susceptibility, vulnerability and recoverability assessments in three separate models and communicate results from one model to another. For example, results for damage from the vulnerability model would become the input for the recoverability model. A higher-level control program would combine the values of susceptibility, vulnerability and recoverably into an Integrated Survivability assessment within the context of the mission loop. The advantage of this would be that smaller models are developed, which do not require common development techniques, incurring a lower development cost. However, while smaller, the individual models would still be extensive and would incur significant development and integration costs. Additional disadvantages are that manipulation of data would still be required between models, and platform descriptions may be required in different formats. This approach is illustrated in Figure 3.

Concept 3: Perform the assessment by the assembly of results from software models that address specific elements of susceptibility, vulnerability and recoverability. For example, the recoverability assessment would be performed by several software models such as one each for fire growth, flooding, and damage control. Interactions between the software models, and the generation of the final overall value of susceptibility, vulnerability or recoverability would be controlled by an interaction and assembly control process. As with Concept 2, a higher-level control program would combine the values of susceptibility, vulnerability and recoverability into an Integrated Survivability assessment within the context of the mission loop. The advantage of this approach is that existing models could be used, lowering development costs. The disadvantage is that automatic manipulation of data would be required between the software models, and platform descriptions may need to be defined in different formats as required by the software codes. This approach is illustrated in Figure 4.

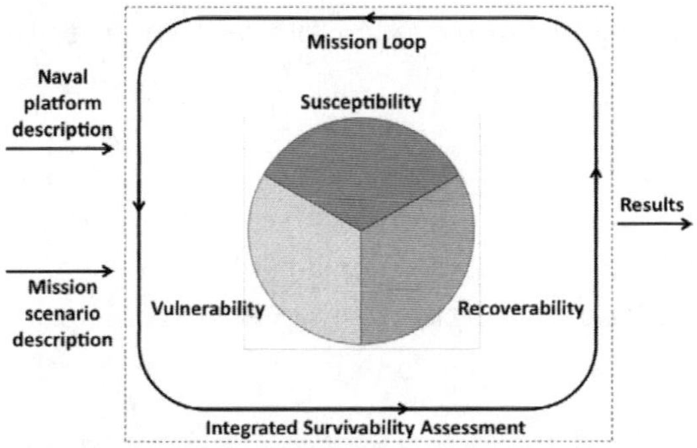

Figure 2: Single Integrated Survivability model

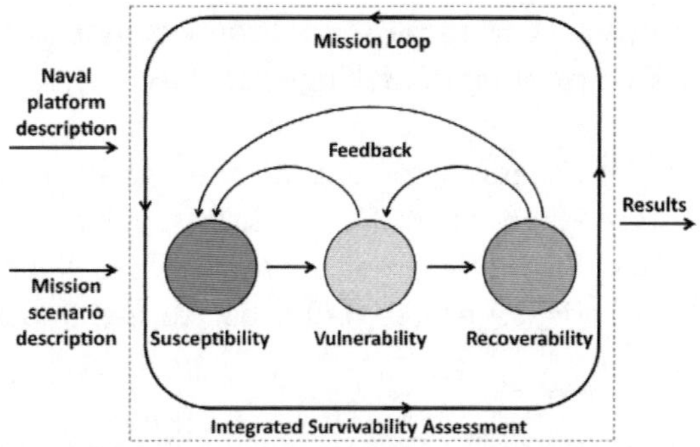

Figure 3: Integrated Survivability component model

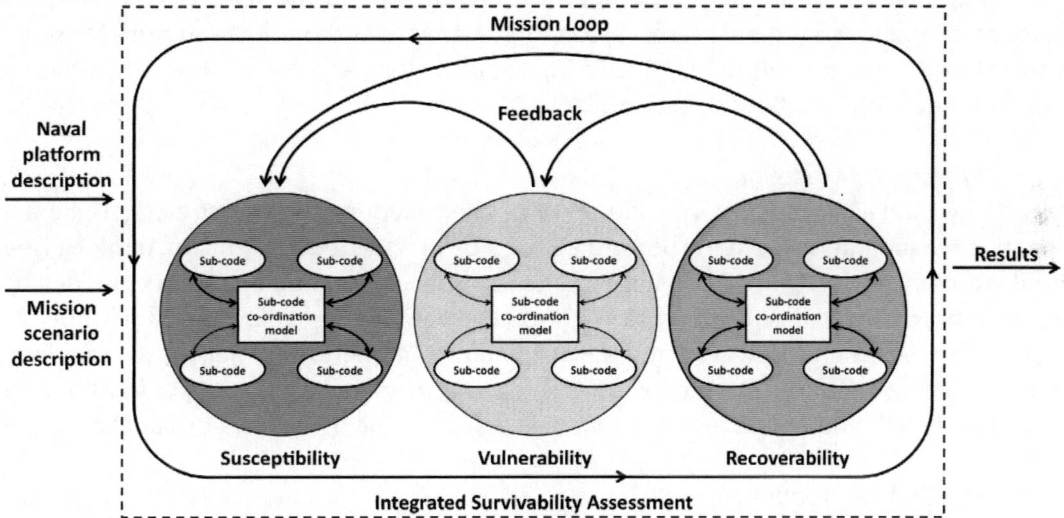

Figure 4: Integrated Survivability model using individual software models

5. ANALYSIS CONSIDERATIONS FOR QUANTIFYING INTEGRATED SURVIVABILITY

Quantification of Integrated Survivability requires coordination of many inputs. It is possible to conceptualise methods for quantifying Integrated Survivability; however, issues can be identified that will require attention if an Integrated Survivability M&S assessment capability is to be realised.

5.1 Susceptibility Modelling

Assessment of platform susceptibility must include the sequence of events that occur during the lead up to engagement between the threat and the platform. The susceptibility analysis must then include target acquisition and subsequent interaction between the threat and the platform. Realistic simulation of this sequence will require the inclusion of target-seeking capabilities of hostile surveillance platforms and the threat itself, the signature of the platform, the countermeasure capability of the platform, and the influence of environmental factors. Military tactics must also be considered since this clearly affects the efficiency of a weapon system. If the platform is part of a task force, the assessment may also include analysis of these features for every other platform within the task force, and analysis of the interaction between the platforms.

The starting point for this assessment is to consider the threat, and its search and tracking capabilities. In this regard, it is necessary to assess details of its sensor capability such as wavelength coverage and sensitivity. This will of course vary with each threat. For example, a torpedo will require an acoustic modelling capability, whilst a missile will require an infrared or radar modelling capability.

It is also necessary to consider the signature characteristics of the platform. For example, the platform's emissivity for passive seeking threats, and its reflectivity for active seeking threats. A wide range of signature types will need to be simulated to encapsulate the sensor characteristics of the potential threats, including radar, infrared, magnetic, visible and acoustic.

The effects of the environment must be considered as an overlay to the surveillance characteristics of the threat and the signature characteristics of the platform. The physical conditions of the environment cause absorption of emissions from both the platform and the threat and generate natural emissions. This leads to degradation of threat surveillance capabilities, masking the intended target platform.

5.2 Vulnerability Modelling

Vulnerability modelling requires consideration of the response of structures, systems and humans to weapon attack. The response needs to be considered in terms of the localised, immediate damage and remote damage that may also occur. Empirical based modelling, using experimental testing results, has traditionally been used to develop algorithms for vulnerability assessment. However, the advancement of numerical modelling capabilities is beginning to surpass these techniques. Detail relating to modelling capabilities will not be discussed due to the number of software models that could be utilised for this application.

5.3 Recoverability Modelling

Models for dynamic physical phenomena (for example, fire, smoke and flood) and structural damage are available for aspects related to recoverability modelling. Human behaviour, however, is especially difficult to model but critical for recoverability analysis. Additionally, these models must interact at each time step within the recoverability simulation. To illustrate the complexity of recoverability modelling, some of the required modelling capability is described.

<u>Fire and Smoke</u>: Simulating the spread of fire and smoke requires a platform compartment model. Smoke spread is influenced by the openings between compartments and the ventilation system. Fire behaviour is influenced by, for example, fuel load, fuel type, compartment size, compartment boundary materials, and ventilation. It is usually depicted as a temperature-time curve with attributes, such as rate of growth and fire burn time, dependent on the characteristics of the compartment in which the fire occurs [14].

<u>Flooding</u>: Simulating the ingress of water requires a platform compartment model [15]. Flooding progresses through the platform via internal openings and the difference in water pressure between compartments. A flooding model may also consider environmental conditions (such as, sea state) and the stability of the platform.

<u>Structural Damage</u>: The loss of platform structural strength due to damage resulting from a threat encounter, grounding, or collision can be quantified [15]. The platform state

(direction and speed) and environmental conditions (sea state) affect whether the residual strength of the platform structure is able to withstand the loads applied to it.

Access and Evacuation: Civilian evacuation models have been applied to naval platforms to meet safety requirements [16, 17]. However, differences between naval and commercial platforms must be considered. Evacuation models simulate movement of individuals based on attributes such as age, gender and group behaviour [16, 17, 18]. Models of platform damage, including structural, fire, smoke, and flooding would need to be considered due to blocking of access and evacuation routes. An evacuation model would also serve as an access model, that is, to simulate the movement of personnel throughout a platform to perform various tasks. Alternate routes could be found using network routing algorithms.

Damage Control and Fire Fighting: The resources required to perform damage control would need to be included in any model of damage control. The simulation of personnel movement to perform damage control tasks is necessary, and would be achieved using the access and evacuation model. The realistic evaluation of workload and fatigue levels of personnel during damage control activities, and resulting effectiveness, would be a necessary inclusion in damage control modelling. The application of Cognitive Work Analysis techniques to understand crew action and behaviour during damage control [19, 20, 21] will facilitate development of crew models for damage control and recoverability analysis.

6. APPLICATIONS FOR INTEGRATED SURVIVABILITY ASSESSMENT

The most obvious application for Integrated Survivability analysis is establishing baseline information for the survivability of existing platforms in a variety of high threat environments. This information has clear application in the development of naval tactics and missions, including the aspect of force composition. The survivability of a task force engaged in combat operations, must include factors such as: the offensive and defensive capabilities of the task force elements; and the consequences of any successful counteractions by the opposing force. Whilst simulation and analysis techniques can be used to assist in such decisions, Integrated Survivability analysis can add an additional perspective. Incorporating Integrated Survivability analysis into these studies, it is possible to investigate, for example, the way in which a particular platform contributes to the overall Integrated Survivability of a task force. It would also be possible to measure the degree to which other platforms can substitute their capabilities during a threat encounter. This could have immediate application in the selection of available platforms for specific mission objectives.

Another application for Integrated Survivability analysis is in the acquisition process for new platforms. As described by Manley [12], the United Kingdom (UK) has incorporated Integrated Survivability into the Royal Navy shipbuilding and acquisition process. Similarly, Said [2] describes the Total Ship Survivability management program, utilised in the US, incorporated throughout the life-cycle of the platform. Several authors highlight the need for survivability, and Integrated Survivability, analysis to be incorporated during the early design phase of capability acquisition to ensure suitable trade-off analysis and to avoid costly redesign, or unachievable design, during later design phases [5, 22, 23, 24, 25, 26]. Historically, Royal Australian Navy (RAN) capability acquisition has primarily been Military-off-the-Shelf (MOTS), where the parent navy or the original equipment manufacturer made platform design decisions. This leaves the RAN with design decisions that may not reflect

RAN capability and Integrated Survivability requirements. Even though this capability acquisition trend for the RAN is changing, Integrated Survivability analysis applicable to the evaluation of competing platform designs will still be required. Additionally, Integrated Survivability analysis will allow for the assessment of technology insertion during platform upgrades. Furthermore, because Integrated Survivability analysis incorporates mission level analysis, it could provide the RAN with the opportunity to select the platform having maximum survivability when engaged in missions with objectives, environments and taskforce compositions specific to RAN requirements. Integrated Survivability analysis will, therefore, facilitate risk analysis and risk mitigation strategies for MOTS designs.

An Integrated Survivability framework also offers a method for improving the value of training. Integrated Survivability highlights important feedback between susceptibility, vulnerability and recoverability that can develop within a mission. Therefore, by means of Integrated Survivability analysis, it is possible to more closely approximate the complexity of conflict in the real world and this will lead to a better trained fighting force.

Finally, an Integrated Survivability analysis capability may be utilised as a decision support tool aboard the platform. Input parameters defining the situation would be supplied by the crew and platform sensors, and the Integrated Survivability analysis capability would simulate the encounter, and predict the most favourable course of action. The capability would thereby allow the crew to assess their available options.

7. SURVIVABILITY ANALYSIS MODELS

Several software tools are currently available that provide some degree of Integrated Survivability analysis capability, and there are a number of software tools that have integrated some aspects of survivability. Many of these tools have evolved by linking existing software capabilities that analyse elements of susceptibility, vulnerability and recoverability. These integrated models generally follow a form of Concept 3 (Section 4) for coupling component models of Integrated Survivability.

7.1 Measure of Total Integrated System Survivability

Measure of Total Integrated System Survivability (MOTISS) is described as a vulnerability-recoverability analysis software tool [27], which also 'allows prediction of signatures' [28]. However, weapon impact locations within MOTISS are defined by the normal distribution in the platform's longitudinal direction, and a Weibull distribution in the hull's vertical direction [29, 30]. As such, MOTISS provides a probabilistic assessment of weapon impact location. MOTISS is also described as using 'first order physics' routines coupled with empirical data to reduce computational time and enable identification of events requiring higher precision consideration [28]. MOTISS is accredited by the US Navy to support naval platform design [27, 31], and is used by the Republic of Korea Navy 'as the only acceptable methodology for performing survivability evaluations' [32].

7.2 Advanced Survivability Analysis Program

The Advanced Survivability Analysis Program (ASAP) is primarily a vulnerability analysis software tool to simulate blast effects on large naval platforms [33, 34]. ASAP is designed to be modular to allow for integration of software to model primary and secondary weapon effects. As such, ASAP is able to model structural response to air and underwater explosion blast effects [34], and to model the spread of fire and smoke, and flood.

7.3 SURVIVE

SURVIVE evolved from several vulnerability analysis software tools designed to predict system failure and crew casualty, and determine the damage spread resulting from blast and fragmentation [5, 35]. Crew and equipment movement modelling within SURVIVE enable systems recoverability analysis, and examination of control and recovery from ongoing fire and flood. SURVIVE does not include a susceptibility analysis component; however, demonstrated coupling with the Spectre radar cross-section software, utilising an Integrated Survivability Interface Module (ISIM), has been achieved [5].

7.4 Maritime Integrated Survivability Simulation

Maritime Integrated Survivability Simulation (MISSION) is an Integrated Survivability capability for analysing mission survivability for an individual platform and also for a task force. This is a mission level simulation tool for predicting the probability of mission success. MISSION incorporates various susceptibility analysis tools, and includes SURVIVE for vulnerability and recoverability analysis [36].

7.5 Integrated Recoverability Model

Integrated Recoverability Model (IRM) is a recoverability modelling software tool integrating systems analysis (for estimating platform capability, and system vulnerability and recovery in relation to an ongoing damage event); fire modelling (using Fire and Smoke Simulator (FSSIM) [37]); flood modelling (using Flooding Model Advanced Stability Algorithms (FLMASA)); with crew movement and behaviour for modelling system repair and mission recovery. IRM is described as 'a network connectivity time based simulation used to evaluate the complex interaction of systems and identify survivability (vulnerability and recoverability) design issues' [38].

7.6 Survivability Manager Application

Survivability Manager Application (SURMA) is described as a 'software tool for comprehensive navy combat survivability assessment' [39]. Development of SURMA commenced with a vulnerability capability for the Finnish Navy; however, being designed over a number of phases, SURMA now facilitates coupling of software tools for broader Integrated Survivability analysis. Specifically, radar cross-section susceptibility analysis is enabled by linking to CAST [40]; vulnerability analysis is facilitated with the use of NAPA software tools [41]; and recoverability analysis is enabled by linking to NAPA tools for flooding analysis, Fire Dynamics Simulator (FDS) [42] for modelling fire and smoke spread, and to IRM (Section 7.5).

7.7 Naval Damage Integrated Recoverability Toolset

Naval Damage Integrated Recoverability Toolset (NavDIRecT) is a software M&S environment and toolset in development to enable integration of commercial-off-the-shelf (COTS) and bespoke naval recoverability software in one common framework. The intent is to enable beginning to end vulnerability and recoverability modelling and analysis [9, 20]. The design philosophy for NavDIRecT utilises an open architecture that will enable other software tools (such as those described in this section, as well as individual models of, vulnerability and recoverability) to connect to the framework. In the longer term, this will enable the development of a holistic analysis capability known as the *Integrated Ship Survivability Analysis Capability* (ISSAC). ISSAC is being designed for analysing a threat encounter, and repeat encounters, understanding survivability measures, and the effects on mission requirements [19]. Currently, the NavDIRecT framework utilises COTS software technology: SMARTFIRE for fire and smoke spread modelling and analysis [43, 44]; maritimeEXODUS for crew movement [16, 17], to be enhanced with crew behaviour models for recoverability modelling [19, 45]; and the UNITY simulation engine [46] to temporally couple the models.

8. CONCLUSION

Integrated Survivability is a temporal concept dependent on the threat environment, mission requirements and the naval platform's susceptibility, vulnerability and recoverability capabilities. A review of the literature identified inconsistency in definitions of platform survivability and, generally, recoverability was ignored as part of the survivability equation. Integrated Survivability has been shown to be a conditional probability equation that enables platform survivability assessment to be evaluated in the temporal domain. This is in contrast to a static assessment as described in much of the literature. A temporal assessment provides an analysis capability to support the acquisition program to perform trade-off studies between competing platform designs, and trade-off between susceptibility, vulnerability, recoverability and capability for a specific platform. Integrated Survivability analysis provides a powerful decision support tool that could be deployed aboard platforms to enable crew to analyse their options to identify the most favourable solution to their threat encounter.

9. ACKNOWLEDGEMENTS

The authors acknowledge Director Navy Technical Bureau, Royal Australian Navy, as sponsor of this work. The authors thank Mr. Tom Whitehouse, DST Group, for identifying relevant survivability software models. The contribution of Dr. Moya Tyndall, DST Group, in developing Figure 1 and related equations, and associated discussion is greatly appreciated.

10. REFERENCES

1. US Department of Defense (1988) *OPNAV Instruction 9070.1 Survivability Policy for Surface Ships of the US Navy*. Office of the Chief of Naval Operations, Department of the Navy.
2. Said, M. O. (1995) Theory and Practice of Total Ship Survivability for Ship Design. *Naval Engineers Journal,* 107(**4**), pp. 191-203.
3. US Department of Defense (2012) *OPNAV Instruction 9070.1A Survivability Policy and Standards for Surface Ships and Craft of the U.S. Navy*. Office of the Chief of Naval Operations, Department of the Navy.
4. Keuning, P. J. and Smit, C. (2002) Survivability in Warship Design. *NATO Symposium on Combat Survivability of Air, Sea and Land Vehicles.* Aalborg, Denmark.
5. Turner, S., Horstmann, P. and Bain, G. (2006) Warship Survivability. *Warship 2006: Future Surface Ship.* 20 – 21 June 2006, London, UK.
6. Vilchez, F. and Tejedor, J. (2002) Survivability enhancing Measures in modern frigate designs. *NATO Symposium on Combat Survivability of Air, Sea and Land Vehicles.* Aalborg, Denmark.

7. Annati, M. and van Erkel, A. G. (2003) What if...? A ship's survivability depends on many factors. *Naval Forces*, 24(**3**), pp. 110.
8. de Yong, L. (2008) Integrated Ship Survivability for the Royal Australian Navy. *Pacific International Maritime Conference 2008*. 29 – 31 January 2008, Sydney, Australia.
9. Woolley, A., Ewer, J., Lawrence, P., Travers, A., Deere, S., Whitehouse, T. and Galea, E. (2016) A Modelling and Simulation Framework to Assess Naval Platform Integrated Survivability. *MAST Europe 2016*. 21 – 23 June 2016, Amsterdam, Netherlands.
10. Bush, R. S. and Cimiluca, A. E. (1996) *Tradeoff Analysis Model for Arsenal Ship Survivability and Sustainability*. Monterey, CA, Naval Postgraduate School.
11. Reese, R., Calvano, C. and Hopkins, T. (1998) Operationally oriented vulnerability requirements in the ship design process. *Naval Engineers Journal*, 110(**1**), pp. 19-34.
12. Manley, D. (2002) The development of "Smart" requirements for ship survivability. *NATO Symposium on Combat Survivability of Air, Sea and Land Vehicles.* Aalborg, Denmark.
13. Coulthard, R. and Hardy, J. (2005) Maritime integrated survivability – the modelling approach. *Journal of Defence Science*, 10(**2**), pp. 73-77.
14. Gamble, G., Suendermann, B., Mathys, Z. and Woolley, A. (2014) Fire Modelling and Event Tree Analysis for Naval Platform Fire Incidents. *Fire at Sea*. 26 – 27 March 2014, London, UK.
15. Turner, T. and Cannon, S. (2007) Onboard rapid assessment tools for damaged naval vessels. *MAST 2007: 2nd Annual Maritime Systems and Technology Global Conference and Trade Show*. 14 – 15 November 2007, Genoa, Italy.
16. Galea, E. R., Lawrence, P. J., Filippidis, L., Blackshields, D., Cooney, D. and Sharp, G. (2016) *maritimeEXODUS v5.1 Theory Manual, Revision 1.1*. University of Greenwich, UK.
17. Galea, E. R., Deere, S., Brown, R., and Filippidis, L. (2013) An Experimental Validation of an Evacuation Model using Data Sets Generated from Two Large Passenger Ships. *Journal of Ship Research*, 57(**3**), pp. 155-170.
18. van Hooren, C. and Noordegraaf, M. (2002) Improving Ship Design with Passenger Flow Simulation. *Interferry Conference 2002*. Brisbane, Australia.
19. Woolley, A. and Gould, R. (2015) An Integrated Capability to Assess RAN Platform Recoverability after a Damage Event. *Pacific 2015 International Maritime Conference*. 6 – 8 October 2015, Sydney, Australia.
20. Liu, J. and Woolley, A. (2015) Holistic Analysis of Crew Cognition and Interaction During Naval Damage Control. *Damage Ship III*. 25 – 26 March 2015, London, UK.
21. Torenvliet, G., Jamieson, G. and Cournoyer, L. (2006) *Functional Modelling, Scenario Development, and Options Analysis to Support Optimized Crewing for Damage Control. Phase 1: Functional Modelling*. DRDC Toronto CR 2006-090.
22. Piperakis, A. S. (2013) *An Integrated Approach to Naval Ship Survivability in Preliminary Ship Design*. Doctor of Philosophy thesis, Department Mechanical Engineering, University College London.
23. Pawling, R. J., Harmer, S., Campbell, J. D. C., Launchbury, C. (2015) Survivability: The Human Element. *Warship 2015: Future Surface Vessel*. 10 – 11 June 2015, Bath, UK.
24. Martin, A. A. (2007) Survivability and the Affordable Warship. *Warship 2007: The Affordable Warship*. 20 – 21 June 2007, Bath, UK.
25. Schofield, J. S., Horstmannd, P. O., Robb, M. D. and Lewis, D. H. (2012) The Influence of Survivability in the Design of a Capable Affordable Surface Combatant. *Warship 2012: The Affordable Warship*. 20 – 21 June 2012, Bath, UK.

26. Liwang, H. and Jonsson, H. (2015) Comparison Between Different Survivability Measures on a Generic Frigate. *International Journal of Maritime Engineering,* 157**,** pp. A125-A134.
27. Alion Technologies (2017) *Naval Architecture and Marine Engineering,* http://www.alionscience.com/Technologies/Naval-Architecture-and-Marine-Engineering/MOTISS (Accessed 9 May 2017).
28. Sajdak, J. A. W. (2015) Putting a Number on It. *Maritime Reporter & Engineering News 2015 World Yearbook*, 77(**6**), pp. 18-21.
29. Kim, K. S., Hwang, S. Y. and Lee, J. H. (2014) Naval ship's susceptibility assessment by the probabilistic density function. *Journal of Computational Design and Engineering*, 1(**4**) p266-271.
30. Walker, J. C. (2016) *Multi-Attribute Tradespace Exploration for US Navy Surface Ship Survivability: A Framework for Balancing Capability, Survivability, and Affordability*. Master of Science thesis, Department of Mechanical Engineering, Massachusetts Institute of Technology.
31. Shafiepour, R. and Sajdak, J. W. *Total Ship Survivability System (TS3) Integrating MOTISS and BDCS to Provide Predictive Awareness Capabilities*. White Paper, L-3 Communications MAPPS, Inc., Canada.
32. Sajdak, J. A. W. and Cowardin, W. H. (2013) The use of Integrated Barrier Assessments to Improve Safety and Reduce Risk in Offshore Oil and Gas Exploration and Production. *Proceedings of the 18th Offshore Symposium.* 7 February 2013, Houston, Texas.
33. SURVICE, *PMC Operation Fact Sheet – SURVICE Engineering Company. http://www.survice.com/uploads/file/SURVICE_PMC.pdf* (Accessed 28 June 2017).
34. Freitas, C. J. (2012) Application of V&V 20 to the Simulation of Propagating Blast Loads in Multi-compartment Structures. *ASME Verification and Validation Symposium*. 2 – 4 May 2012, Las Vegas, Nevada.
35. QinetiQ (2017) *Naval Architecture*. https://www.qinetiq.com.au/services/maritime/naval-architecture.aspx (Accessed 31 May 2017).
36. Thornton, J. S., Coulthard, R. and Hardy, J. N. (2006) Modelling Maritime Survivability. *Warship 2006*: *Future Surface Ships*. 20 – 21 June 2006, London, UK.
37. Floyd, J., Hunt, S., Williams, F. and Tatem, P. (2005) A Network Fire Model for the Simulation of Fire Growth and Smoke Spread in Multiple Compartments with Complex Ventilation. *Journal of Fire Protection Engineering*, 15(**3**), pp. 199-229.
38. Test & Evaluation Solutions (2017) *Integrated Recoverability Model (IRM)*. http://www.irm-software.org/ (Accessed 9 May 2017).
39. SURMA. (2016) *SURMA Technical Description Software for Comprehensive Navy Combat Survivability Analysis*. White paper, SURMA Technical Description, SURMA-M-101-G.
40. VTT (2017) *CAST*. http://virtual.vtt.fi/virtual/proj2/cast/index.html (Accessed 9 May 2017).
41. NAPA (2017) *Solutions for Design and Operation of Ships*. https://www.napa.fi/ (Accessed 9 May 2017).
42. FDS-SMV (2017) *FDS-SMV*. https://pages.nist.gov/fds-smv/ (Accessed 9 May 2017).
43. Ewer, J., Jia, F., Grandison, A., Frost, I., Galea, E. and Patel, M. (2013) *Technical Reference Manual and User Guide for SMARTFIRE 4.3*. University of Greenwich, UK.
44. Wang, Z., Jia F., Galea, E. and Ewer, J. (2013) A CFD Simulation of a Post-Crash Aircraft Fire Test. *Journal of Aircraft*, 50(**1**) pp. 164-175.
45. Woolley, A. and Liu, J. (2016) Cognitive Modelling of Crew Response During Naval Damage Control Operations. *Human Factors in Ship Design and Operation*. 28 – 29 September 2016, London, UK.
46. Unity, *Unity.* (2017) https://unity3d.com/ (Accessed 9 May 2017).

Medium-Scale Experiments for the Validation of Naval Platform Vulnerability Simulations

G. Yiannakopoulos[1]; C. Flockhart[1]; J. M. Brett[1]; W.D. Reid[1]

[1]Defence Science and Technology Group

George.Yiannakopoulos@dst.defence.gov.au

ABSTRACT

The vulnerability of Royal Australian Navy (RAN) vessels to an underwater explosion (UNDEX) is a major consideration during the acquisition process and the platform's operational life. Numerical simulation procedures, using finite element analysis, are currently being developed with the aim of predicting the vulnerability of hull structures and equipment to UNDEX induced shock and bubble pulse loading. Numerical techniques to simulate UNDEX fluid-structural interaction and structural response are highly non-linear and complex, and require validation through comparison with experimental data. While full-scale testing remains an essential element in assuring minimum levels of integrated survivability are achieved, such tests are generally limited to low shock levels and are a prohibitively expensive means of gathering the detailed data needed for validation of simulation codes. Testing of small and medium-scale model structures, including destructive tests, offers a cost effective means of code validation enabling detailed investigations of underlying physical phenomena. Small-scale tests typically involve very simple structural geometries and idealized representations of equipment. Medium-scale tests allow the investigation structural forms and equipment which are more representative of the full-scale. This paper outlines an underwater explosive experimental program that has been conducted at the medium-scale for the purpose of validating finite element modelling techniques used for vulnerability analysis. The test program includes a range of equipment and structural models subjected to UNDEX shockwave and bubble pulse whipping loads. The experiments were extensively instrumented and yielded comprehensive data sets. This has permitted closer scrutiny of detailed finite element models and numerical techniques.

INTRODUCTION

DST is conducting a research program to develop a computational vulnerability prediction capability to support RAN operations and acquisition programs. A critical aspect is predicting the structural and equipment response to the shock and bubble pulse loading from underwater explosions. This includes the response of the hull, internal structures such as bulkheads and tanks, and also equipment and systems mounted on the hull or on shock isolated rafts.

Validation data for comparison with computational vulnerability modelling predictions is not readily available in the open literature. DST group can draw on measurements gathered from full-scale UNDEX shock testing, including the first of class inshore Minehunter in 1987 [1] [2], on drone boats for mine countermeasures in 1989 and 1991 and two full scale tests on the Minehunter Coastal in 1999 and the Collins Class Submarine in 2003. These trials involved low shock levels relative to structural and equipment design limits. Other trials on decommissioned vessels [3, 4] and critical equipment [5] have investigated severe loading conditions resulting in structural failure. Due to high cost, opportunities for full-scale testing are rare. Furthermore, it is often difficult to deploy the number and type of sensors during full-scale trials desired for detailed numerical validation studies. Thus, to achieve the ultimate aim of validating numerical procedures applicable to full-scale naval platforms, it is necessary to supplement full-scale trials data by conducting tests on simplified, scaled model systems which can provide detailed data on the underlying physical processes of loading and response, including failure.

UNDEX research programs involving testing of simple, small-scale model structures, including submerged plates [6, 7] and cylinders [8-10], have been undertaken by DST over a number of years. These studies have focused on simulation of the interaction of a structure with the shock wave and explosive bubble, and the resulting elasto-plastic deformation and fracture of the structure.

Medium-scale tests allow for the inclusion of hull geometries, equipment and mounting arrangements which are more representative of full-scale structures. Currently there is a lack of experimental data at the medium-scale to help fill the gap between small and full-scale test data. To help fill this gap, a number of medium-scale UNDEX experiments have been recently conducted by DST. These experiments, conducted in support of RAN acquisition and operational programs, have initially focused on elastic response to UNDEX loading. This paper summarises two medium-scale experiments reflecting the range of investigations undertaken into early-time response to the initial shock wave. An overview is also provided of current efforts to numerically model the experiments. Global bending response of submarine structures due to the relatively late-time bubble oscillation phase of UNDEX loading, known as whipping, is also of vital interest in vulnerability assessments. Details of a recent medium-scale whipping experiment can be found in reference [11].

SUMMARY OF EXPERIMENTS

The UNDEX experiments described in this paper were conducted at the DST underwater explosive test facility located in Melbourne. The first experiment investigated the shock response of equipment hard mounted to hull plating. The second experiment focused on the shock response of an analogue submarine model and the transmission of the shock wave within it to both hard mounted and resiliently mounted rafted equipment.

Experiment 1 – Hull Mounted Equipment

Equipment that is hard mounted to the hull of a submarine is exposed to the most severe UNDEX shock environment. The performance of equipment such as hull valves is critical to the function and safety of the pressure hull. An experimental investigation of the response of hull mounted equipment was conducted with a specially designed test rig as shown in Figure 1. The test rig incorporated three air-backed rectangular chests, each with a bolted top test plate. The test plates were of thickness, width and material strength representative of typical submarine hull plating between stiffeners. Equipment items were hard mounted to the inside of the test plates. A series of explosive events were conducted with hull valves of bore size 50mm, 65mm, 100mm and 125 mm at high shock intensity levels corresponding to the certified design shock survival level of the valve. Additional tests were performed with lumped mass representations of equipment in the form of solid cylindrical steel discs of mass 12 and 40 kg. Pressure was recorded in the water adjacent to the test plate. Acceleration and strain were recorded on the test plate and on the item of equipment.

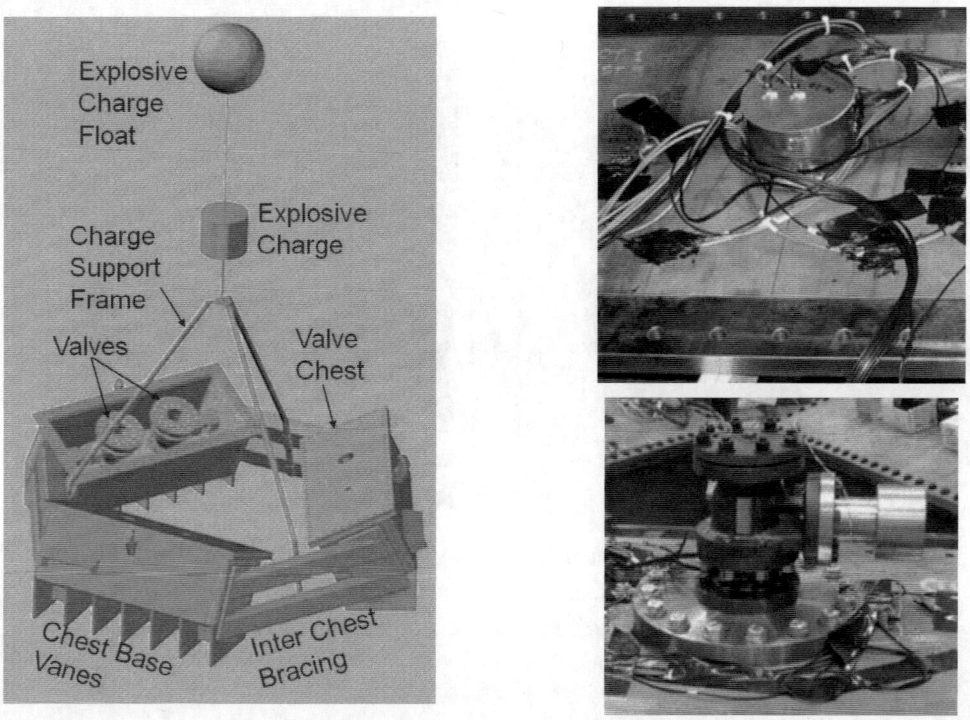

Figure 1 : Experiment 1. Shows over-all view of the test rig comprised of three air-backed chests, left, and two views of an instrumented test plate fitted with lumped steel cylinder mass, top right, and a hull valve, bottom right.

Experiment 2 - Shock Response of an Analogue Submarine Structure

A cylindrical test structure representative of a submarine (analogue submarine) was constructed with a length, diameter and mass of approximately 4.4 m, 1.2 m and 6 tonnes respectively. The analogue submarine consisted of a circumferentially stiffened centre section of length ~ 2 m and two elliptically domed end sections of length ~ 1.2 m. The centre section was fitted with two transverse rafts placed longitudinally end-to-end. Support plates at the four corners of each raft could be either hard mounted (bolted) to the hull or resiliently mounted by means of rubber shock isolators placed between the hull and raft supports. The end sections were fitted with hull penetrator ports to which full-scale 65 mm hull valves were mounted. Figure 2 shows overall views of the analogue submarine test structure, the centre section and the method of deployment for a fully submerged test. A series of twenty UNDEX events with varying charge size, stand-off distance and orientation were performed with the rafts either hard or resiliently mounted and with the test structure either floating or submerged to a depth of 5 m. For each event, sensors were deployed to measure; (a) incident pressure in the water and reflected pressure on the wet surface of the hull, (b) acceleration, velocity and strain of the hull, (c) acceleration of the hull valves and (d) acceleration and velocity of the rafts. This data set captured the shock-induced local and global (whole-body) motion response of the hull structure, shock transmission to the raft supports and the attenuation offered by resilient mounts.

Figure 2: Set-up for Experiment 2. Shows the analogue submarine test structure, top right, the centre section with rafts and instrumentation, bottom right, and overall view of deployment of test structure submerged to a depth of 5m, left.

NUMERICAL MODELLING OF EXPERIMENTS

The development of validated numerical modelling procedures is currently being carried out at DST primarily using the codes LS-DYNA [12], USA/LS-DYNA [13] and Trident FEA [14]. LS-DYNA is well known as a general purpose, Finite Element (FE) based, multi-physics explicit code with extensive capabilities for modelling highly non-linear transient events including far-field and close-in UNDEX. The USA (Underwater Shock Analysis) code is a Boundary Element (BE) code, based on the Doubly Asymptotic Approximation (DAA), for modelling a linear-acoustic fluid domain surrounding a submerged or floating structure. When coupled with LS-DYNA as the structural solver, the USA/LS-DYNA code is applicable to the analysis of far-field UNDEX problems. The Trident FEA code can similarly be coupled to the USA code. Additionally, Trident FEA includes a range of analytical and semi-analytical tools that are suited to rapid assessment of UNDEX response during early stages of design. Simulation results presented in this paper will be limited to investigations using LS-DYNA.

Experiment 1

In general the environment experienced by a submerged structure subjected to an underwater explosive loading is complex but particular challenges involved in the simulation of Experiment 1 are the severity of the shock loading, the likely occurrence of cavitation on the test rig chest top plate and the additional loading arising from reflections from the bottom of the underwater test site on which the test rig sits. Effective simulation techniques are still in development; here we present our progress to date.

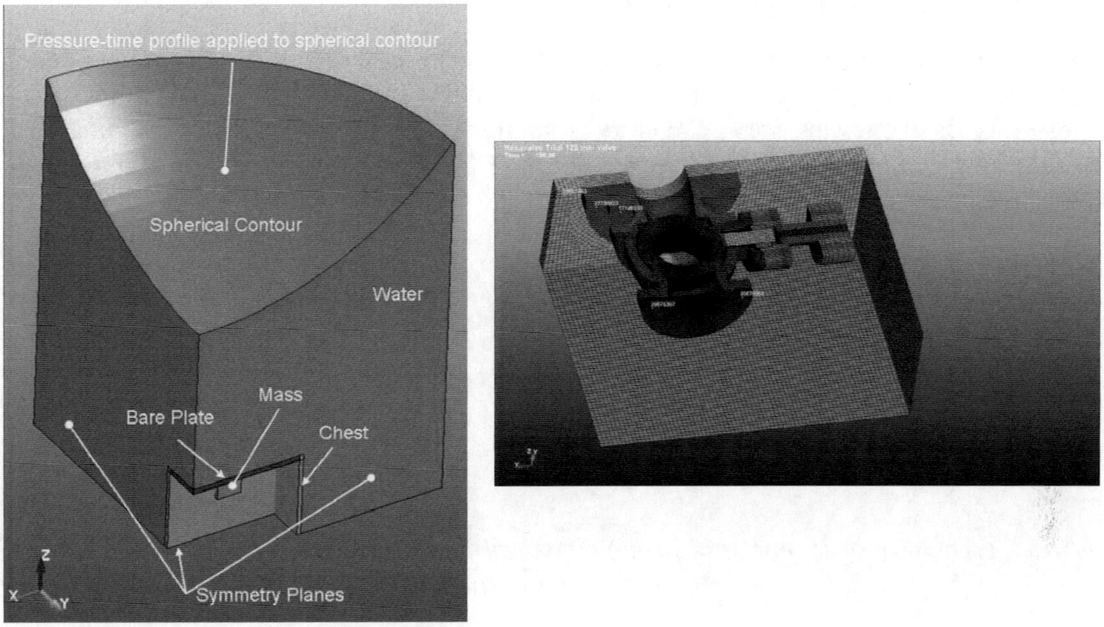

Figure 3: Finite Element models for Experiment 1. Overall features of the model (with a simple cylindrical equipment mass) are shown at left. A close-up view of the test rig chest structure, incorporating a 125 mm hull valve is shown at right

The FE simulations were conducted with LS-DYNA using a Lagrangian element formulation for both the structure and external fluid. The model is illustrated in Figure 3. The symmetry of the experiment permits the model to be limited to one half of a test rig chest, thereby reducing the required number of nodes, and enabling higher levels of mesh resolution. Non-reflective boundaries were applied on the outer faces of the water domain. Both the water domain and the chest were constrained at the bottom in the vertical axis (consistent with the test rig being positioned on the bottom of the test site).

Sufficient water was included to incorporate the fluid-structure interaction. The water domain had a spherical contour at the top, coincident with the expanded shock wave at some distance from the point of detonation. The applied pressure-time profile was derived from empirical similitude equations guided by the experimental measured peak pressure. The water domain was initialised to the appropriate hydrostatic depth and an atmospheric pressure load was applied on the interior of the chest and top plate.

The chest was modelled using shell elements (Hughes-Liu formulation), whilst the top plate, attached masses and water were modelled with brick elements. An elastic-plastic material model (with appropriate constants) was used for the test rig and attached masses. The LS-DYNA null material model with a Gruneisen EOS was used for the water.

A thorough validation of the modelling methodology for this experiment will require an extensive comparison to measurements of acceleration, strain and pressure. For the purpose of this report we present a comparison of two particular cases – the velocity response at the inboard end of the equipment (i.e. furthest from the explosion) for both a 12 kg cylindrical mass and a 125 mm hull valve.

The comparison of experimental and predicted velocity at these points is presented in Figure 4. This shows good agreement for the initial velocity peak. The match of subsequent velocity is less successful showing some similarity in features but with increased amplitude. This overly elastic response is possibly a consequence of the simplified constraints of the model and/or the adopted material properties.

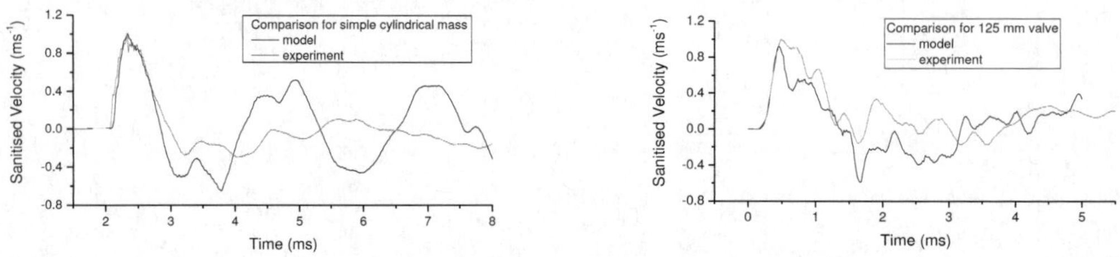

Figure 4: Experiment 1. Comparison of experimental and predicted initial velocities on the inboard edge of a simple cylindrical mass (left) and a 125mm hull valve (right).

Modelling of a hull valve can be expected to be more problematic, due to its structure which consists of multiple parts bolted together, including a side attached actuator. A simplified valve model was developed for this initial comparison, as shown in Figure 3. It incorporates the main sections of the valve but these are joined by merging of the interface nodes without the joining bolts. As shown in Figure 4, the comparison to experimental data for this model is less successful than that achieved for the simple cylindrical mass, but there are some encouraging aspects. Peak velocity is matched to within 10%. The subsequent response shows similarities of form but appears to be smaller in magnitude and occurs more rapidly. A likely explanation for this is the absence of bolts in the model. It is expected that the concentration of the required restraint into the bolts rather than in the overall body of the valve would give rise to a more exaggerated and longer period response. Future models that incorporate bolts will be able to test this possibility.

Experiment 2

Finite element simulations were performed with the aim of modelling both the local and global response of the hull and the rafts. Details of the LS-DYNA FE simulation model for Experiment 2 are presented in Figure 7. The hull and raft structure were modelled primarily with Lagrangian shell elements, however solid brick elements were used to model the support fixtures at the corners of the rafts. Bolted joints at the raft supports were modelled by means of rigid nodal constraints around bolt holes. The structural model was surrounded by an annular domain of solid Lagrangian fluid elements, the extent of the domain being sufficient to capture the shock-structure interaction. Shock loading of the model is achieved by applying an experimentally measured free-field pressure-time history to the external surface of the fluid. Similarly to the modelling of Experiment 1, propagation of the spherical shock wave front into the fluid domain is achieved by contouring the outer surface of the fluid domain to match the radius from the charge location. It is also necessary to apply non-reflecting boundary conditions to the fluid outer surface. This modelling arrangement is known as the fluid island method.

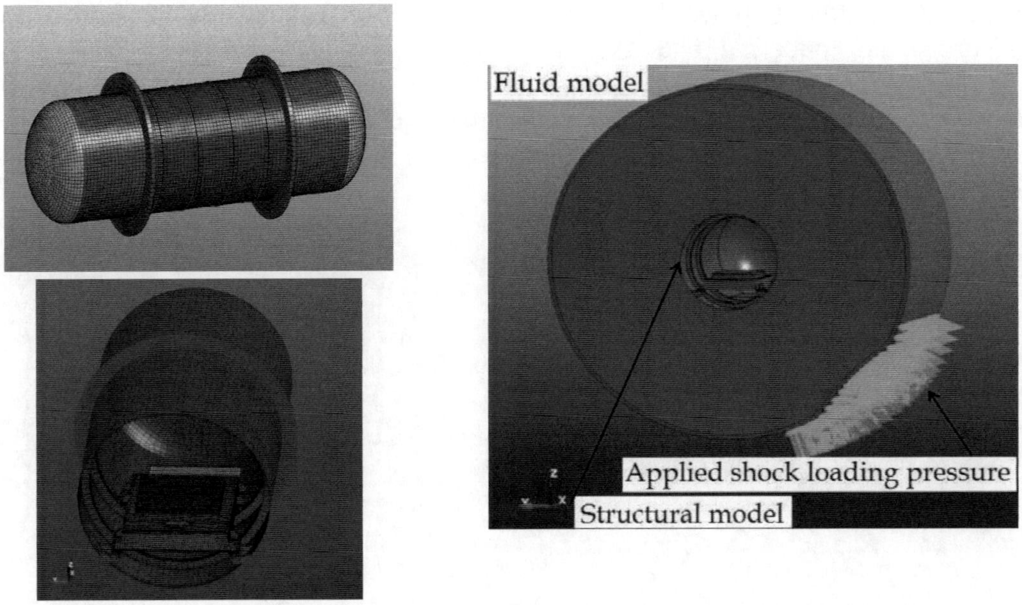

Figure 5: LS-DYNA simulation model of Experiment 2. Shows an overall view of the finite element model of the hull, top left, and an internal view (half model) of the hard mounted raft, bottom left. Shock loading pressures are introduced through the external boundary of the fluid element domain surrounding the cylinder, right.

Figure 6 compares the hull wet surface pressure (reflected pressure) recorded at the point on the hull closest to the charge. The FE analysis can be seen to under-predict the peak pressure and over-predict the specific impulse (area under the pressure-time curve) and rise-time of loading pulse. This was due to a loss of high frequency components and broadening of the shock front as the shock wave propagates through the fluid domain towards the structure. This can be greatly improved in future simulations by increasing fluid mesh density.

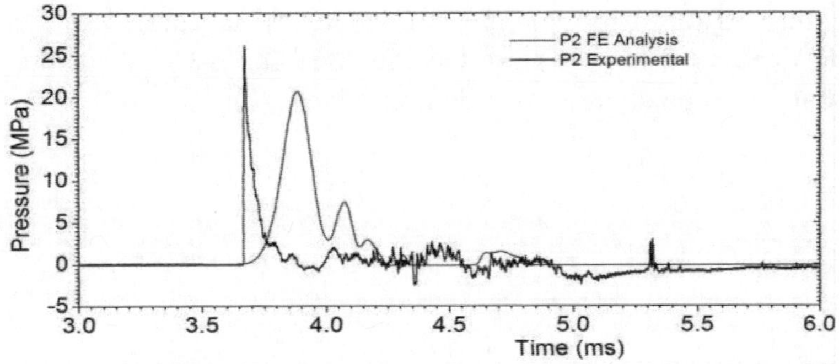

Figure 6: Experiment 2. Comparison of predicted and measured reflected pressure loading on the hull at the closest point to the charge

Simulation of local strain response on the hull at the closest point to the charge (maximum strain point) showed a reduced frequency content with respect to the experimental data, Figure 7. The measured early-time, high frequency strain peaks over the initial ~3 ms of response were not accurately modelled, with strain being over predicted and under predicted at different times during the early response. Beyond 3-4 ms from the initial response, the modelling and experiment were in good agreement. The modelling deficiencies for the initial strain response are thought to be caused primarily by the deficiencies in the shock pressure loading predictions described above. Thus the strain results can also be expected to be improved by increasing the fluid mesh density used in future simulations.

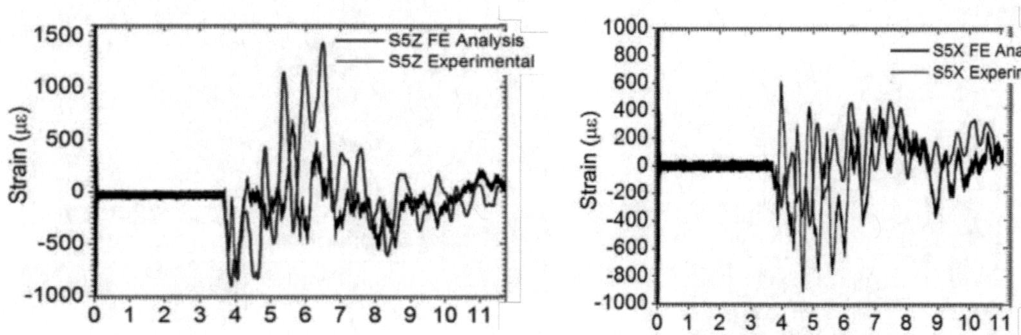

Figure 7: Experiment 2. Comparison of predicted and measured hoop strain, left, and axial strain, right, on hull near to the closest point to the charge. Tensile strains are negative.

The kick-off velocity is defined as the initial peak velocity response at a particular location. For a relatively rigid structure, such as the present test structure, a similar kick-off velocity is expected at different locations around the hull, this velocity being an indication of the global, whole-body velocity of the structure. A comparison of simulated and measured velocity responses on the hull is shown in Figure 8. The comparisons include a location on the hull closest to the charge and on the distal side of the hull from the charge. Both the initial peak velocity and the frequency of the subsequent response were quite well represented by the modelling at the two locations. The kick-off velocity in response to UNDEX is generally observed to depend primarily on the total impulse imparted and be less sensitive to the high frequency peak content of the shock wave pressure loading. The level of agreement between the predicted and experimental hull velocities indicates, despite over-prediction of local specific impulse of the pressure loading discussed earlier, that the total impulse imparted to the hull was modelled reasonable well.

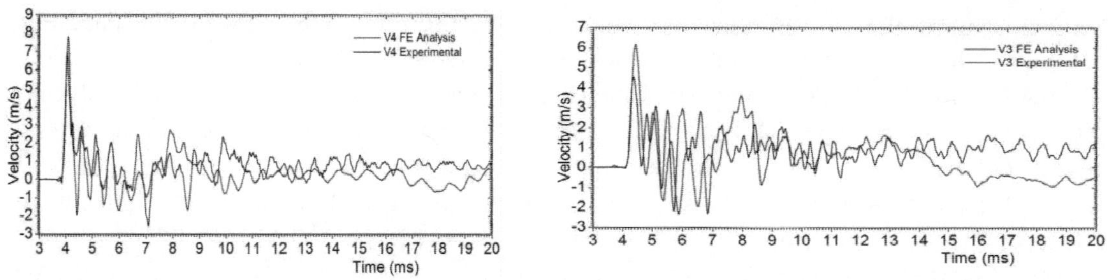

Figure 8: Experiment 2. Comparison of predicted and measured velocity on the hull. Comparison of vertical velocity at the bottom of the hull close to the charge, left, and velocity on the distal side of the hull in line with the charge, right.

Figure 9 compares the experimental velocity response of the hard mounted raft, measured at the raft centre in the direction normal to the top of the raft, against the modelling prediction. The predicted and measured initial peak velocity response showed quite good agreement at ~2.6 m/s*Figure 9*, indicating that transmission of the shock impulse through the hard mounting supports to the raft was well modelled. Following the initial peak velocity, the measured velocity data showed a more rapid decrease compared to the simulation. Beyond approximately 10 ms, the measured and predicted velocity responses are out-of-phase, however both responses indicate a dominant modal frequency of ~120 Hz, corresponding to the fixed base natural frequency of the hard mounted raft. The differences observed between the predicted and measured response are thought to be the result of overly simplified modelling of the contact interfaces between the top surface of the raft and plates placed on the raft representing equipment mass. Further work is required to investigate and rectify this issue.

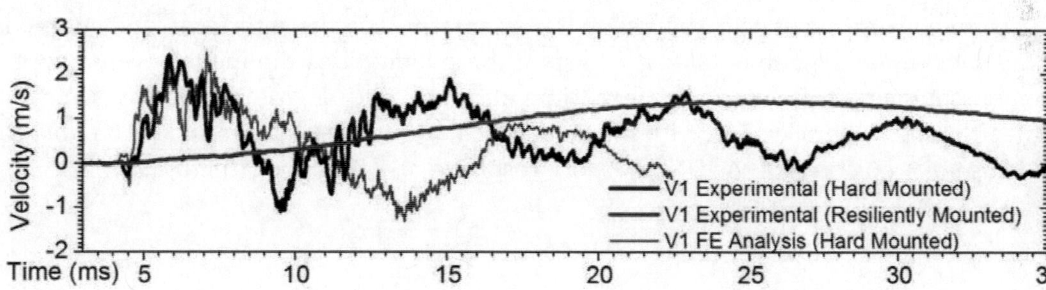

Figure 9: Experiment 2. Comparison of predicted and measured vertical velocity at the centre top of a hard mounted raft. Also shown is the velocity response of a resiliently mounted raft.

Also shown in Figure 9 is the measured velocity response of the raft when supported on rubber resilient mounts. The same charge size and position were used in testing of the hard mounted and resliently mounted rafts. In comparison with the hard mounted raft, the effect of the resilient mounts is to reduce in the initial peak velocity from 2.6 m/s to 1.5 m/s and increase the rise time from ~1.5 ms to ~20 ms, thereby isolating the the raft from high frequency shock compents transmitted to the raft supports.

Modelling the response of the resiliently mounted raft requires material property data for the rubber mounts at large strains and high strain-rates suitable for shock analysis. Such data must be obtained from a range of material tests and calibrated (curve fitted) to obtain the parameters needed by models describing the hyperelastic, viscoelastic and cyclic stress softening behaviour of material. Reliable material data for the rubber mounts used in Experiment 2 are currently not available. Testing to obtain such data will be carried out in the near future.

CONCLUSION

A program of medium scale experiments have been performed to provide essential validation data for the development of an FE based UNDEX vulnerability prediction capability. In this report we have provided illustrative results from two particular investigations.

Experiment 1 studied the response of steel plates fitted with hull valves and cylindrical masses, to high shock load levels. The results will support the development of a predictive capability for the UNDEX performance of hull mounted equipment which is critical to the function and safety of the pressure hull of a submarine.

Experiment 2 investigated hull whole-body motion and the response of hard mounted and resiliently mounted rafts under relatively low levels of shock loading. The data will enable a better understanding of shock transmission through the hull and shock isolation achieved by resilient mounting.

FE modelling of these experiments was performed with the LS-DYNA code. Some simplifications were needed to represent the geometry, contact interfaces and material specifications of the experimental tests. Despite these limitations the models were able to provide encouraging matches to the experimental data. The degree of success achieved indicates that more detailed FEA with more realistic material parameters should be able to provide useful predictions of the UNDEX shock response of submarine structures.

REFERENCES

1. MacIntyre I., Hatt D. and Elischer P. P., MRL in Support of the Mine Hunter Inshore Project, MRL video report No. VT 1622, 1987.

2. Campanella G. and Elischer P.P., “Establishing an Australian Capability for Shock Testing the RAN Minesweeping Catamaran”, ABCA-7, Passive Ship Defence, UK (Bath), March 1988.

3 Walsh B.E., Burman N M and Buckland M E (1996), “Ship Survivability Enhancement Program (SSEP) - Manual of Conventional Weapons Data”, DSTO-GD-0105, September 1996, (Limited Release).

4 Walsh B.E. and Yiannakopoulos G., 2001, “Structural arrangements to provide resistance to air-blast and underwater shock in surface warships”, DSTO-TR-1142 (Limited Release), Melbourne, AMRL, 2001.

5 Shock Integrity of NAB Hull Valves with Alternate Fasteners: Part 2 The Shock Test, DSTO-CR-2014-0001, (Confidential), Brett J., Yiannakopoulos G., Krelle A., Edwards D.

6. Yiannakopoulos G., Brett J. M. and Krelle A. (2008), "A Preliminary Investigation of Plate Penetration by an Underwater Explosion", *79th Shock & Vibration Symposium,* October 26-30, 2008, Orlando, FL.

7. Brett J. M. and Yiannakopoulos G. (2009), "An experimental study of the detonation of an explosive device in close proximity to a submerged structure", 8th Asia-Pacific Conference on Shock & Impact Loads On Structures, 7–9 December 2005, Perth, W. Australia.

8. Yiannakopoulos G. (2007), "Fluid and Structure Interaction Studies on a Submerged Platformed Cylinder near an Underwater Explosion", *78th Shock & Vibration Symposium*, November 4-8, 2007, Philadelphia, PA, USA.

9. Brett J.M. and Yiannakopoulos G., (2008) "A study of explosive effects in close proximity to a submerged cylinder", *The International Journal of Impact Engineering*, 35, pp. 206-225.

10. Yiannakopoulos G. (2004), "Lagrangian/Eulerian Finite Element Studies For Undersea Vessels Subjected To Explosives", UDT Conference, Hawaii 2004, 19-21 October 2004.

11. De Candia S., Ojeda, R., Reid, W., Ratcliffe, M. and Binns, J. (2017), "The Whipping Response of a Submerged Free-Free Cylinder Due to Underwater Explosions " Pacific 2017, Sydney Australia, October 2017.

12. LSTC. LS-DYNA R9.0 Keyword Manual. Livermore, CA: Livermore Software Technology Corporation, 2016.

13. USA LS-DYNA Users Manual - USA Release 7.5. Livermore Software Corporation, CA, 2014.

14. Trident FEA, Applied Technology Group, Lloyd's Register Group Ltd

Rapid Vulnerability Assessment of Naval Structures subjected to Localised Blast

Luiz Bortolan Neto[1,3], Michael Saleh[1,3],
Vanessa Pickerd[2,3], George Yiannakopoulos[2,3],
Zenka Mathys[2,3], Warren Reid[2,3]

[1]Australian Nuclear Science and Technology Organisation,
Locked Bag 2001, Kirrawee DC, NSW 2232, Australia

[2]Defence Science & Technology Group, 506 Lorimer St,
Fishermans Bend, Melbourne, VIC 3207, Australia

[3]Defence Materials Technology Centre,
Level 2, 24 Wakefield St., Hawthorn, VIC 3122, Australia

luiz.bortolanneto@ansto.gov.au

ABSTRACT

The development of modern naval vessels is driven by the optimum balance between operational performance, technology restrictions and the costs of ownership. These factors impose limitations on all features of surface ships, including weaponry, structural materials, radar systems, and propulsors. Strategies must be set to identify design features and materials that can enhance the vessels protection in the event of shock loadings *e.g.* air blast and underwater explosions. Assessment of design solutions is a complicated task due to the large number of unknowns involved. Appropriate computational models and experimental tests can give insights into the expected mechanical behaviour to support the design process. The authors are developing a framework for vulnerability assessment, which includes experimental tests and appropriate finite element (FE) models of representative structural parts subjected to blast loading. This combined approach provides a comprehensive analysis tool but its complexity prevents the quick assessment of the vessel structural vulnerability when various design features and a range of materials are to be considered. To overcome this hurdle, a machine learning model based on Artificial Neural Networks is proposed to identify patterns in numerical and experimental data, yielding timely conclusions about the structural response.

Keywords: vulnerability assessment, artificial neural networks, finite element analysis, high strain rate, blast loading

1 INTRODUCTION

In naval vessels, the blast resistance of hull structures is usually analysed by means of single-degree-of-freedom (SDOF) models [1]. While these models may be suitable for some simple applications, they lack the ability to accurately describe the localised failure of bulkhead structures subjected to internal blast loading [2-4]. Currently advanced numerical tools are available and can deliver reasonable predictions [5-10]. In some situations, however, engineers and naval architects are interested in rapid assessments of structural hull vulnerability to meet tight development schedules or to ascertain overall platform vulnerability. In these cases, the long lead times inherent to numerical simulations can pose a significant issue. Machine learning techniques have proven to be viable in the development of fast running models for steel columns [11], hypervelocity impact events [12] and fatigue behaviour [13, 14], provided an adequate number of training data is available. Building on this idea, an Artificial Neural Network (ANN) model is proposed herein and its ability to rapidly yield simplified but clear conclusions, with respect to the response of solid plates to localised blast loadings, is evaluated. A finite element analysis (FEA) model is also developed to give insights on the parameters needed for creating more reliable ANN models.

2 ARTIFICIAL NEURAL NETWORKS

An artificial neural network (ANN) is a computational approach inspired by early models of sensory processing of biological brains. In their seminal work, McCulloch and Pitts [15] postulated that neurons can operate as a switch, which can be activated or deactivated depending on the input signals received. From this idea an algorithm for outputting a binary classifier, known as the perceptron, was introduced [16].

The inability of the perceptron to handle nonlinear classifications and operations resulted in a decline in activity in this field until the development of the error back-propagation method [17]. This method has the ability of combining several perceptrons into a self-organising multi-layered network that mimics the processes of biological neural networks. This analogy led to the Artificial Neural Network moniker for the multilayer perceptron architecture depicted in Figure 1. More details about the underpinning theory of the ANN can be found in Refs. [17, 18].

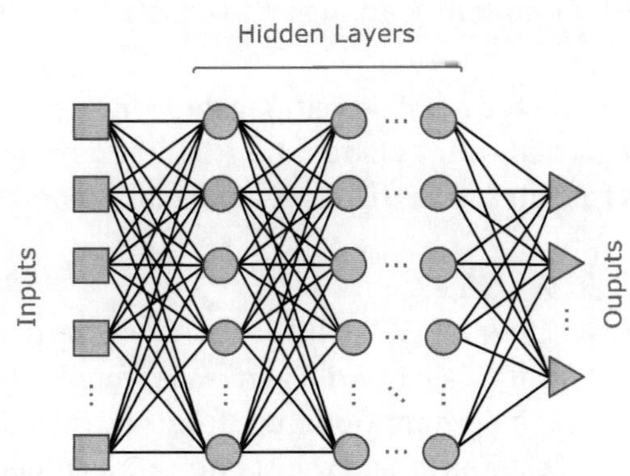

Figure 1. Schematic of the multilayer perceptron, also known as the feed-forward network. In each layer the perceptrons are arranged in parallel. The layers are processed in series.

In recent history ANNs have been applied for modelling numerous problems in engineering, science and medicine; including speech and image recognition, data mining and structural reliability and damage evaluation. In the current study an ANN model is used in the predictive analyses of structural behaviour to blast loadings. The resilient propagation (RProp) [19] and Levenberg–Marquardt (LM) [20] algorithms have been chosen for this study due to robustness and reliability and were computationally implemented via the C programming language.

3 ARCHITECTURE AND EVALUATION OF THE ANN MODELS

The ANN architecture chosen has three layers: an input layer, a hidden layer and an output layer. The number of perceptrons in the hidden layer must be carefully selected to avoid poor predictive models because of either under-fitting or over-fitting of the training data. The data for training the ANN models come from Jacob et al. [21] experiments for quadrangular plates under localised blast loading for various charge configurations and plate geometries. The tests were performed on standard cold-rolled mild steel (density of 7700 kg/m³) plates. The impulsive load was generated by cylindrical shaped plastic explosive (PE4) with varying density (ρ) and placed at the centre of the plate with a standoff distance of 12 mm.

Seven inputs and four outputs for the ANN were identified from the experimental results of Jacob et al. [21]. The identified inputs are: the plate thickness, width and length, the static yield stress, and the charge diameter, height and mass (m). These experiments measured (outputs): the shock impulse (I), the mid-plate deflection (δ), the burnt diameter on the plate (d) and the plate's failure mode (f) as per Table 1. These measurements are the four identified outputs used in the ANN. An ANN model for each output was developed as this proved to deliver results closer to the experimental data in comparison to a single model with four outputs.

Table 1. Modes of failure [21]

Mode	f	Details
0	0.00	Elastic deformation
I	0.25	Large inelastic deformation
Itc	0.50	Large inelastic response with thinning in the central area
II*c	0.75	Partial tearing in the central area
IIc	1.00	Complete tearing at the central area (capping)

Other mechanical properties, such as elastic parameters, were found to be identical for all specimens and can be disregarded as inputs for the ANN models, limiting the application of the developed ANN models to plates with the same material properties as those described by Jacob et al. [21].

A total population of 148 experiments was considered. Two-thirds of this set was used for training the model and one-third for validation. Each experimental datum was normalised to its respective maximum value. The number of perceptrons in the hidden layer was limited to 10 for the burnt diameter, shock impulse and mid-plate displacement. For the failure mode, 15 neurons were employed in the hidden layer, which is just greater than the optimum limit of 14, calculated as per Ref. [13]. Each model was trained with the RProp algorithm and then refined with the LM algorithm.

The ANN models revealed trends similar to Jacob et al.'s [21] for the impulse correlation to the mid-plate displacement, charge mass and charge density. These trends are illustrated by Figures 2 and 3. Figure 2 depicts the linear growth of the impulse with increasing explosive mass as well as impulse decay with charge compaction, as predicted by the ANN model. Figure 3 demonstrates the capability of the ANN model to behave akin to the experiments. Figure 3-a shows the mid-plate displacement versus shock impulse from the experimental data of Jacob et. al. [21] and Figure 3-b shows the results from the ANN model. Note that Figures 2 and 3 were produced using the complete dataset available.

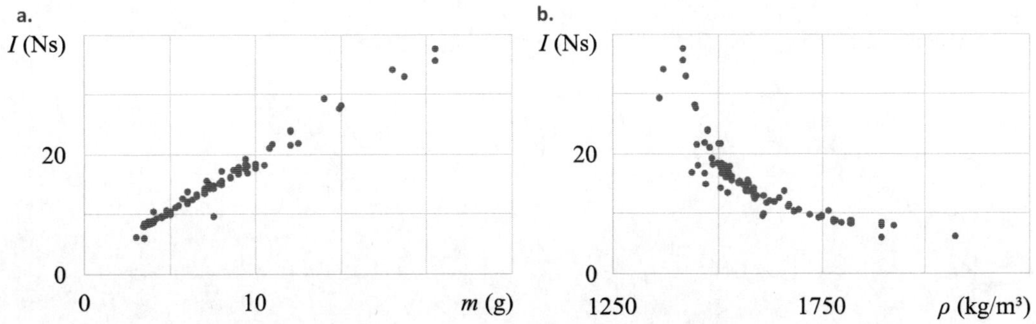

Figure 2. Impulse versus charge mass (a) and impulse versus charge density (b) predicted by the ANN model for the complete (training and validation) dataset.

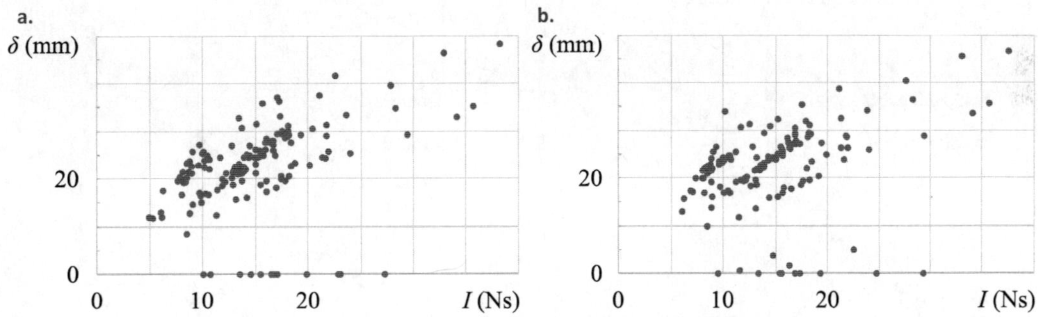

Figure 3. Mid-plate displacement versus impulse from Jacob et al.'s [21] experiments (a) and from the ANN model (b). These graphs were produced with the complete (training and validation) dataset.

Other pertinent comparisons between the ANN predictions and Jacob et al.'s [21] experiments are summarised in Figures 4 and 5. Figure 4 presents straightforward comparisons of the ANN predictions to the normalised experimental data. Figure 5 presents the percentage of results within the ±10% error band. These figures show good correlation between the ANN results and the experimental data for the charge impulse and for the burnt diameter on the plate. However, the figures also illustrate the poor predictive behaviour of the ANN models towards the mid-plate displacement and failure modes. This fact is highlighted further by comparing the training and validation root mean square errors (RMSE) shown in Table 2, where the validation RMSE for δ and f are an order of magnitude higher than the training RMSE.

Table 2. Root mean square errors (RMSE) of the normalised ANN predictions

	I	d	δ	f
Training	0.01169	0.01693	0.02570	0.03997
Validation	0.03221	0.05061	0.12900	0.24795

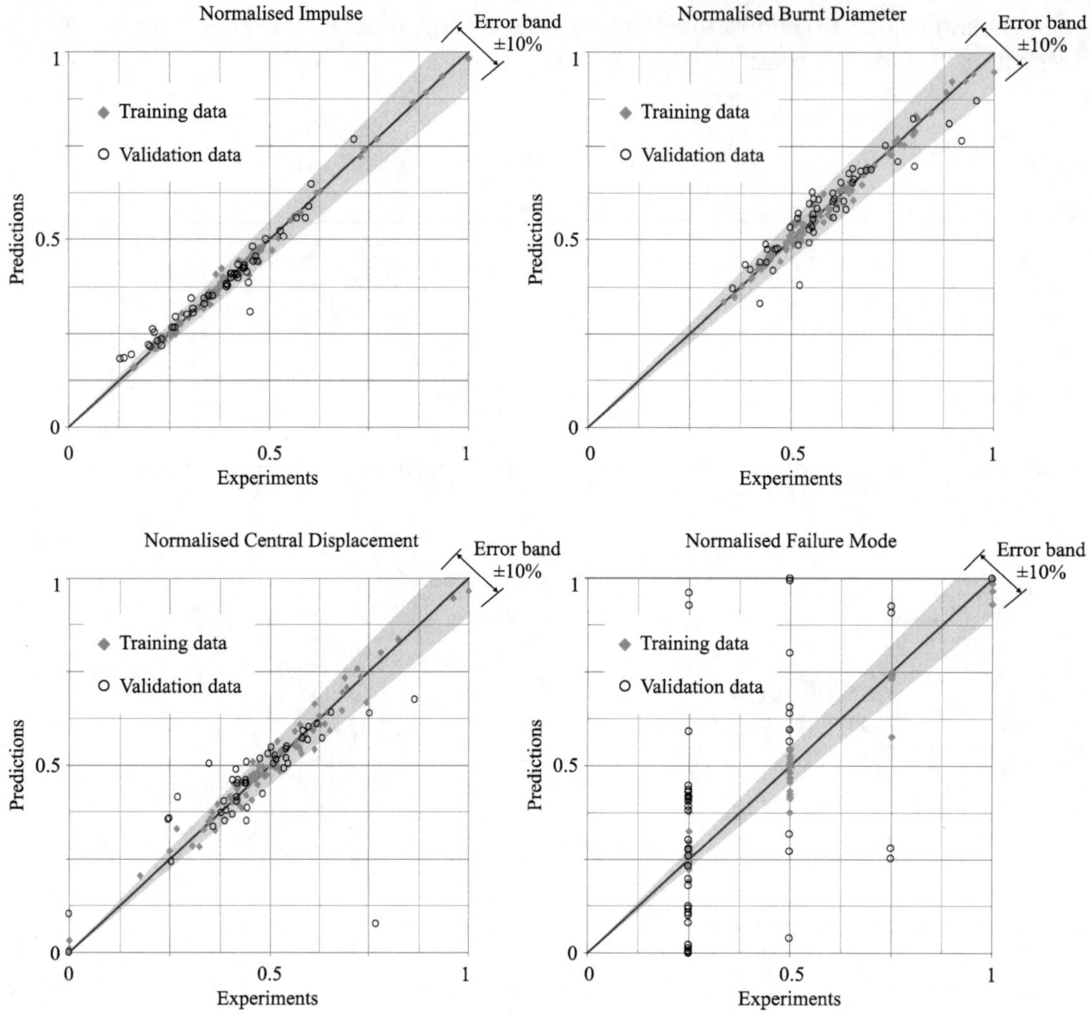

Figure 4. Comparison between normalised experimental values and the ANN predictions.

The absence of viscoplastic parameters of the plate as inputs to the ANN model is the major contributor to the poor prediction of the ANN models as these correlate directly to the plastic deformation and failure behaviours of the steel plate. Such parameters would have given the models the means to incorporate strain hardening, strain rate dependence and temperature influence on the material behaviour, yielding more accurate results.

Additionally, the experimental dataset utilised in the development of the ANN models provided a limited number of outputs. The use of strain gauges and digital image correlation (DIC) systems can be effective in obtaining displacement and strain profiles of the specimens [3, 22] to enhance the number of available output metrics. However, the long lead times and high costs associated with experimental set-ups can restrict the data available for developing ANN models. This issue can be minimised with the utilisation of numerical models that could supply a number of relevant parameters. A FEA simulation potentially capable of providing additional parameters for ANN models is discussed in the next section.

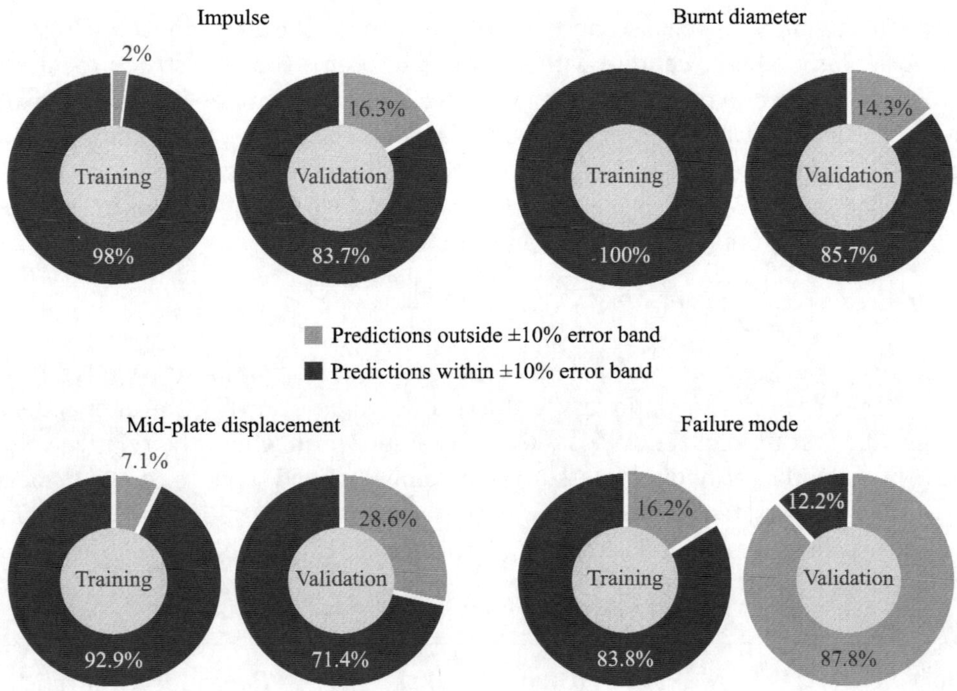

Figure 5. Percentages of the ANN results within and outside the error band for both training and validation sets.

4 FINITE ELEMENT SIMULATION

The discrepancies between the experimental and ANN results for the mid-plate displacement and failure modes stem from the lack of strain rate dependent properties in the input data. In the experiments of Jacob et al. [21] the material yield point is the only descriptor for the plastic properties of the steel. In practice each charge/plate configuration will induce a spatial distribution of stresses across the plate's face. Since mild steel is known to exhibit sensitivities to strain rate and strain hardening [23] the authors elected to simulate the experimental work of Jacob et al [21] to better elucidate the deformation mechanisms. The blast simulations were carried out on a 3 mm thick plate (ℓ = 290 mm, b = 145 mm) with a cylindrical 9 g (Ø 29x9 mm) PE4 charge at a stand-off distance of 12 mm in LS-DYNA [24] to compute the spatial and temporal variations in the plate's deformation; interrogate the complex strain rate behaviour of the steel under blast load; and evaluate the resultant stress-strain contours.

To compute the high strain rate deformation of the steel plates a reduced version of the Johnson-Cook (JC) constitutive model [25] was used to compute the high strain rate (HSR) flow stress. The reduced JC relation is as follows:

$$\sigma_{eq} = [A + B\varepsilon^n][1 + Cln\dot{\varepsilon}^*] \quad (1)$$

Where σ_{eq} is the flow stress, ε is the effective plastic strain, $\dot{\varepsilon}^* = \dot{\varepsilon}/\dot{\varepsilon}_0$ is the dimensionless plastic strain rate and $A, B, n,$ and C are material constants. This phenomenological model accounts for (i) the increase in yield stress due to strain hardening (first bracket), and (ii) the increase in yield and tensile strength with increased strain rate (second bracket). The constants are reproduced from the work of Vedantm et al. [23] for mild steel.

Jacob et al. [21] applied the blast load as a pressure by considering the burnt diameter, the burn time and the delivered experimental impulse (measured through a pendulum apparatus). The simulations focused on the steel's response (displacement/bulging/ failure) which the ANN was unable to capture with sufficient accuracy. The fluid structure interaction and the resultant structural response were evaluated by explicitly modelling the air, explosive and the constrained mild steel plate. The solid explosive was modelled using a *MAT_HIGH_EXPLOSIVE_BURN material model previously employed for the analysis of explosives in soil [26-29]. The explosive product was modelled using the Jones, Wilkins and Lee (JWL) equation of state (EOS) with the pressure term given by:

$$P = A\left(1 - \frac{\omega}{R_1 V}\right)\exp(-R_1 V) + B\left(1 - \frac{\omega}{R_2 V}\right)\exp(-R_2 V) + \frac{\omega E}{V} \quad (2)$$

Where, A, B, R_1, R_2, ω are material constants usually derived from cylindrical explosive testing or thermo-chemical kinetic simulations. V is the relative volume and E is the energy per initial unit volume. The first term of the JWL is the high pressure term where the relative volume is close to 1. The second intermediate pressure term applies when V is close to 2. As V becomes larger in the expanded state the EOS reduces to the last term, $\omega E/V$ [30]. The JWL parameters are analogous to those used in Ref. [31] for PE4 and are shown in Table 3.

Table 3. JWL EOS parameters for PE4

ρ_0 (kg/m^3)	D (m/s)	PCJ (GPa)	A (GPa)	B (GPa)	R_1	R_2	ω	E_m (MJ/kg)
1601	8193	28	309.8	12.95	4.50	1.40	0.25	5.621

The Air is modelled using a *MAT_NULL material model specifying only density and a linear polynomial EOS with the pressure term given by:

$$P = C_0 + C_1\mu + C_2\mu^2 + C_3\mu^3 + (C_4 + C_5\mu + C_6\mu^2)E \quad (3)$$

Where, $C_0, C_1, C_2, C_3, C_4, C_5, C_6$ are material parameters; $\mu = 1/V - 1$; V is the relative volume; and E is the energy per initial unit volume.

The cut-off pressure and viscosity were set to zero as viscosity was found to be negligible and to ensure the pressure stays positive. The polynomial EOS can be reduced to an ideal gas type relation by setting $C_0 = C_1 = C_2 = C_3 = 0$ and $C_4 = C_5 = \gamma - 1 = 0.40$. To initialise the atmospheric pressure of 101 kPa, the relation below can be used:

$$P = (\gamma - 1)\frac{\rho}{\rho_0}E \quad (4)$$

Where P is the atmospheric pressure, γ is the specific heat ratio, ρ and ρ_0 are the current and initial densities ($\rho_0 =$ 1.25 kg/m^3), respectively, and E is the energy per unit volume (E = 253.2 kPa).

To reduce the computational cost a quarter-model (Figure 6) was used, owing to the symmetry of the experimental study, for the assessment of the blast. Arbitrary Lagrangian-Eulerian (ALE) elements were used to compute the fluid structure interaction between the explosive/air and the steel plate. The plate is constrained around its periphery as per the experimental work. The steel plate is modelled using solid elements with reduced integration and hour-glassing controls. The ALE domain was meshed using 8 nodes, 1 point integration ALE multi-material elements. Since the computational domain is finite, a non-reflecting boundary condition was applied to the top and periphery of the cubic domain; the bottom face of the plate was fully constrained with no inflow/outflow conditions. The non-reflecting boundary ensures the stress waves and the advected fluid propagating through the container and the air do not reflect back into the computational domain. This implementation is partially correct due to the limitation of the artificial damping imposed on the boundaries which are only applicable to elastic stress waves.

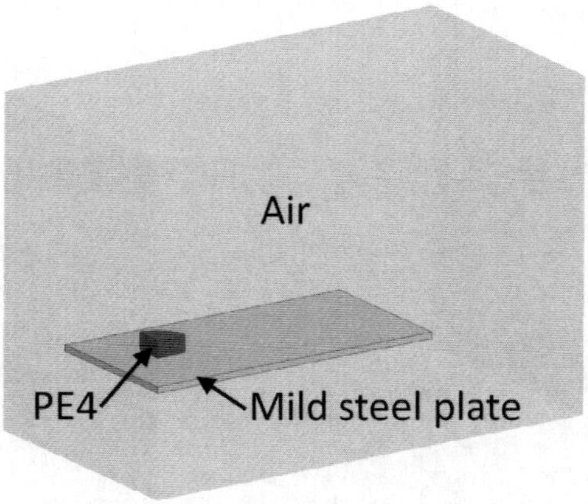

Figure 6. FE setup of the blast model.

The resultant mid-plate displacement is 27 mm for this computational assessment and is shown in Figure 7, and this is in good agreement with the reported experimental result of 28.7 mm (an error margin of 5%). Experimentally this particular test configuration is characterised as having large inelastic response with thinning in the central area, "*Mode Itc*". This was also observed in the current simulations with a 46% thickness reduction at the centre point of the plate, as illustrated by Figure 8. The iso-surfaces of the explosive products are shown in Figure 9, highlighting the interactions of the blast products with the flat plate and subsequent plate velocity. The propagation of the blast products beyond the plate's edge and beyond the experimentally prescribed burnt diameter is also shown. This particular propagation pattern has ramifications for more complex geometries where the interaction with neighbouring structures could lead to pressure wave reflections and secondary pressure loads. The back face velocity of the plate is in excess of 500 m/s with the displacement being largely confined to the centre region below the charge.

The von-Mises stress contours are shown in Figure 10, exhibiting a range of 200–400 MPa under load, with the stresses being confined to the unclamped region. The regions exhibiting the largest plastic strain are those directly below the charge and at the edge of the clamped region closest to the charge. This area forms a plastic hinge and is only apparent along one of the clamped edges. The presence of the plastic hinge would indicate that edge effects would likely have affected the plate's maximum displacement.

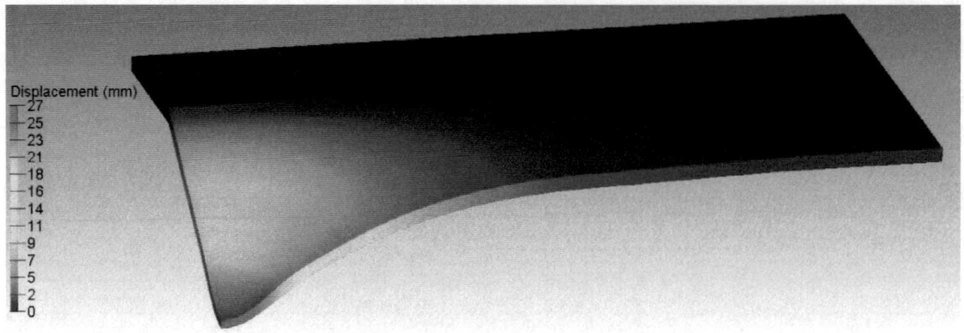

Figure 7. Final displacement (mm) contours of the plate at t=300 μs.

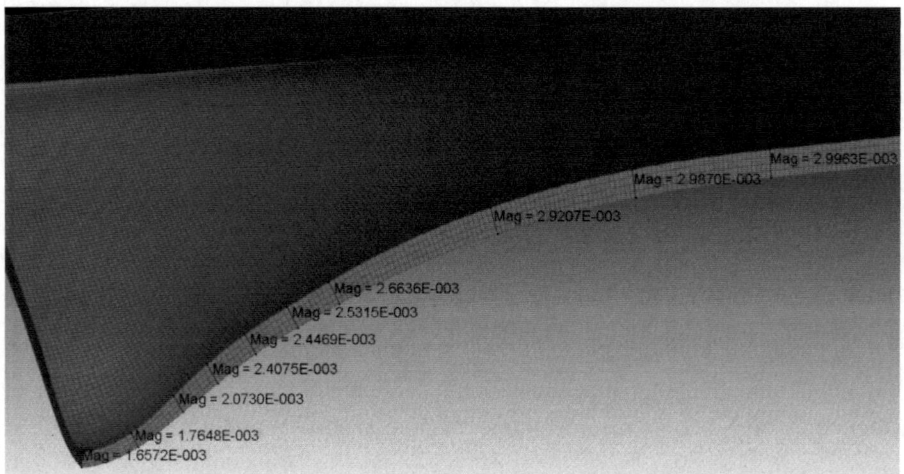

Figure 8. Thickness variations along the plate's centreline (m).

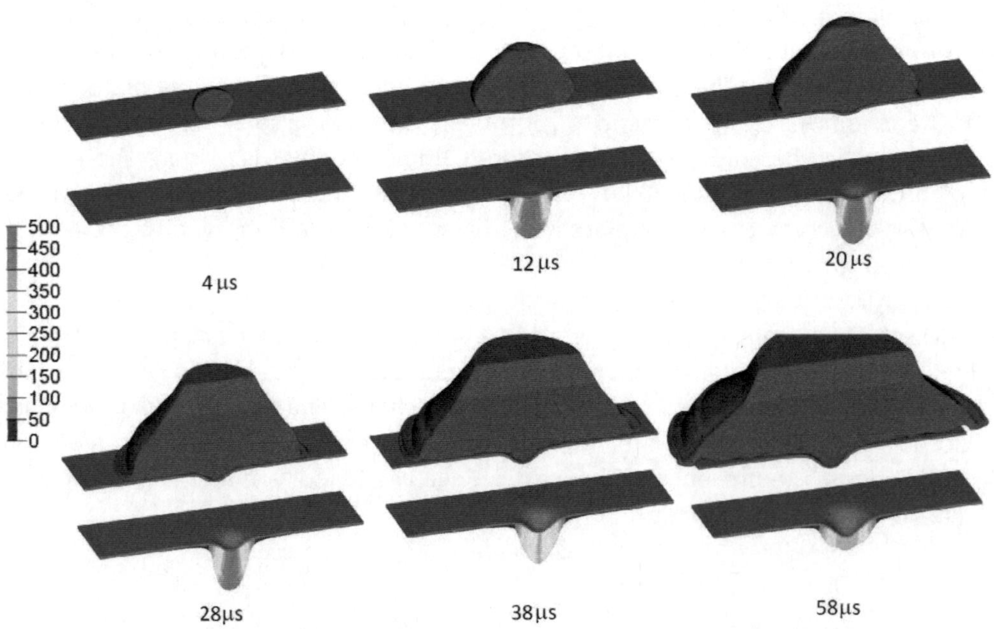

Figure 9. Temporal variation of the back face velocity (m/s) of the steel plate alongside the iso-surface of the blast products. Note that the maximum velocity is attained at 28 μs. (NB: the model is reflected for clarity).

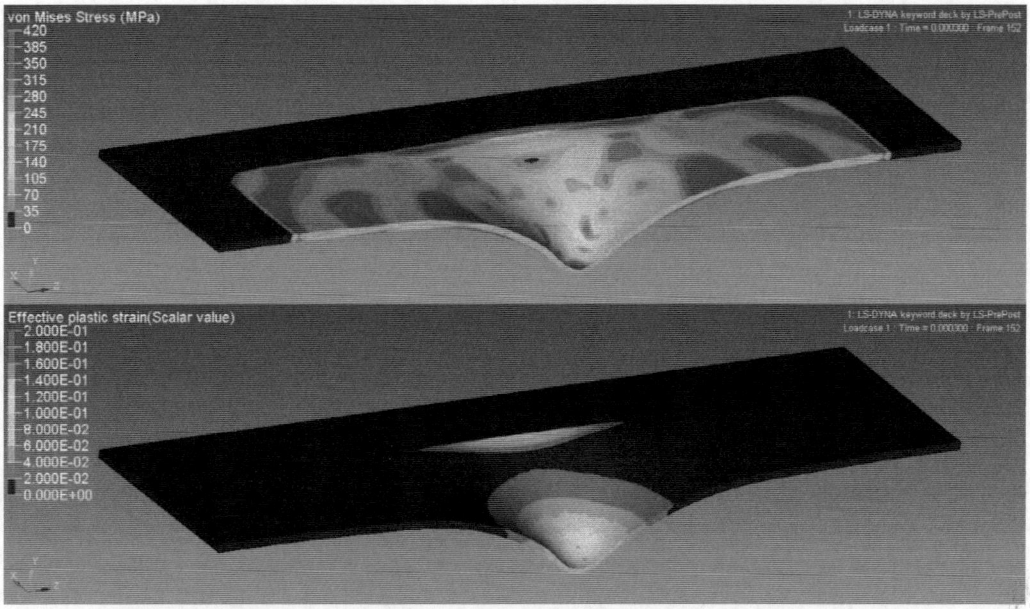

Figure 10. von Mises stress and plastic strain of the plate at t = 300 µs (NB: the model is reflected for clarity).

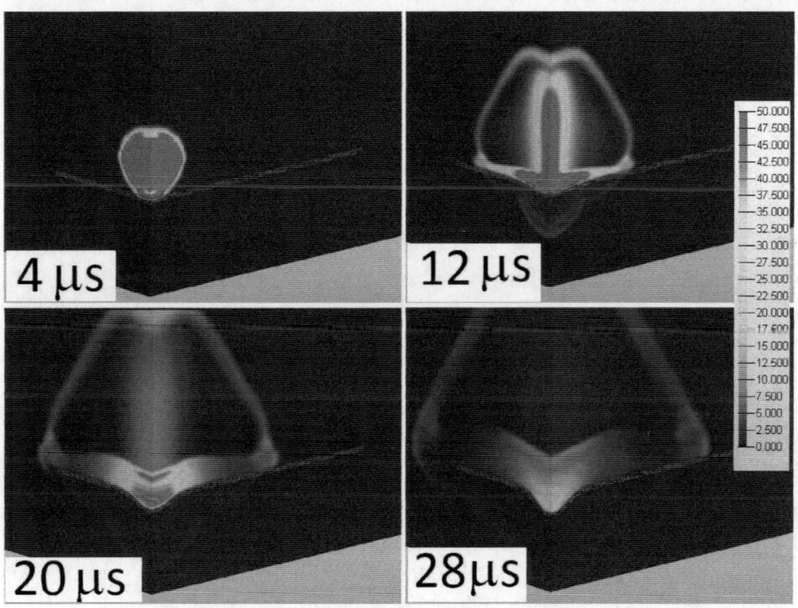

Figure 11. Temporal variations of the pressure contours (MPa).

The plastic deformation of the plate is driven by the pressure imparted on the top face by the PE4 charge. The contours of the pressure in Figure 11 show a pronounced focusing effect of the pressure at $t = 12$ µs with a vertical zone of high pressures positioned above the plate's centreline. This is characteristic of the asymmetric propagation of cylindrical charges and is experimentally corroborated by the localised bulging and cap rupture of the experimental plates. The pressure drops significantly between $t = 20$ µs and $t = 28$ µs with the latter time corresponding to the maximum back face velocity of the plate.

The use of finite element modelling in this example demonstrates an ability to obtain viscoplastic strain data from numerical modelling, which could be used to supplement experimental data in ANN models.

5 CONCLUSIONS

An Artificial Neural Network (ANN) model designed to deliver rapid predictions regarding the blast load response and subsequent vulnerability of naval vessels was implemented. This model applied the Resilient Propagation and Levenberg–Marquardt learning algorithms on the experimental data of Jacob et al. [21]. The ANN replicated the trends observed on the experiments reasonably well. However, the limited number of material properties used as inputs reduced the model's ability to accurately describe the response of solid plates to localised blast loadings. The Finite Element analysis performed replicated the experimental results extremely well, yielding useful insights about the influence of viscoplastic parameters on the plate response to impact loadings, albeit at a much larger computational cost. The authors envision that the ANN model can be further refined by incorporating simulation outputs into its training program. Experimental blast testing is inherently a low data environment with many attributes poorly quantified due to resource restrictions and the harsh blast testing environment. Rigorous simulations have the capacity to provide the required data to supplement the ANN training and improve the validity and confidence in the machine learning algorithms.

ACKNOWLEDGEMENTS

The authors would like to acknowledge the support and funding of the Defence Materials Technology Centre (DMTC). The DMTC was established and is supported under the Australian Government's Defence Future Capability Technology Centres Program. The authors also wish to thank the IT department at ANSTO for the provision of HPC.

REFERENCES

1. Lotovonen, H., *Beyond Biggs - Review, Implementation and Comparison of Modern SDOF Analysis*, 2014, Surma Ltd: Helsinki, Finland. p. 12.
2. Burman, N.M., et al. *Deformation and Fracture of Compartments Subjected to Internal Blast Loading*. in *5th Australian Aeronautical Conference*. 1993. Institution of Engineers Australia.
3. Pickerd, V., et al., *Analysis of the structural response and failure of containers subjected to internal blast loading.* International Journal of Impact Engineering, 2016. **95**: p. 40-53.
4. Pickerd, V.L. and P.W. McCarthy. *Damage of welded structures from internal explosive loading*. in *3rd International Conference on Protective Structures (ICPS3)*. 2015. International Association of Protective Structures.
5. Langdon, G.S., et al., *Performance of mild steel perforated plates as a blast wave mitigation technique: Experimental and numerical investigation.* International Journal of Impact Engineering, 2010. **37**: p. 1021-1036.
6. Longère, P., A.-G. Geffroy, and B. Leblé. *Experimental and numerical aspects of ship structure failure under airblast loading*. in *DYMAT 2012 - 10th International Conference on the Mechanical and Physical Behaviour of Materials under Dynamic Loading*. 2012. EDP Sciences.
7. Longère, P., et al., *Ship structure steel plate failure under near-field air-blast loading: Numerical simulations vs experiment.* International Journal of Impact Engineering, 2013. **62**: p. 88-98.
8. Saleh, M., L. Edwards, and I.G. Crouch, *Chapter 9: Numerical modelling and computer simulations*, in *The science of armour materials*, I.G. Crouch, Editor. 2016, Woodhead Publishing: Cambridge, England. p. 483–579.

9. Tyler-Street, M. and J.M. Luyten. *Developing failure criteria for application to ship structures subjected to explosive blast loadings*. in *7th European LS-DYNA Conference*. 2009. DYNAmore.
10. Tyler-Street, M. and J.M. Luyten. *Failure criteria applied to ship structures subjected to explosive blast loadings*. in *10th International Conference on Computational Plasticity - COMPLAS X*. 2009. International Center for Numerical Methods in Engineering (CIMNE).
11. Stewart, L.K., *Testing and analysis of structural steel columns subjected to blast loads*, 2010, University of California: San Diego, California. p. 491.
12. Ryan, S., S. Thaler, and S. Kandanaarachchi, *Machine learning methods for predicting the outcome of hypervelocity impact events.* Expert Systems with Applications, 2016. **45**: p. 23-39.
13. Haque, M.E. and K.V. Sudhakar, *Prediction of corrosion–fatigue behavior of DP steel through artificial neural network.* International Journal of Fatigue, 2001. **23**: p. 1-4.
14. Iacoviello, F., D. Iacoviello, and M. Cavallini, *Analysis of stress ratio effects on fatigue propagation in a sintered duplex steel by experimentation and artificial neural network approaches.* International Journal of Fatigue, 2004. **26**: p. 819-828.
15. McCulloch, W.S. and W. Pitts, *A logical calculus of the ideas immanent in nervous activity.* The bulletin of mathematical biophysics, 1943. **5**: p. 115-133.
16. Rosenblatt, F., *The perceptron: A probabilistic model for information storage and organization in the brain.* Psychological Review, 1958. **65**: p. 386-408.
17. Krogh, A., *What are artificial neural networks?* Nature Biotechnology, 2008. **26**: p. 195-197.
18. Bishop, C.M., *Neural Networks for Pattern Recognition*. 1995, Oxford, England: Clarendon Press. 498.
19. Riedmiller, M. and H. Braun. *A Direct Adaptive Method for Faster Backpropagation Learning: The RPROP Algorithm*. in *IEEE International Conference on Neural Networks*. 1993.
20. Hagan, M.T. and M.B. Menhaj, *Training feedforward networks with the Marquardt algorithm.* IEEE Transactions on Neural Networks, 1994. **5**: p. 989-993.
21. Jacob, N., et al., *Scaling aspects of quadrangular plates subjected to localised blast loads—experiments and predictions.* International Journal of Impact Engineering, 2004. **30**: p. 1179-1208.
22. Pickerd, V.L., *Optimisation and Validation of the ARAMIS Digital Image Correlation System for use in Large-scale High-strain-rate Events*, 2013, Defence Science and Technology Organisation: Melbourne, Australia. p. 31.
23. Vedantam, K., et al., *Johnson - Cook Strength Models for Mild and DP 590 Steels.* AIP Conference Proceedings, 2006. **845**(1): p. 775-778.
24. LSTC, *LS-DYNA KEYWORD USER'S MANUAL R8.0*, 2015, LIVERMORE SOFTWARE TECHNOLOGY CORPORATION (LSTC): California.
25. Johnson, G.R. and W.H. Cook. *A constitutive model and data for metals subjected to large strains, high strain rates and high temperatures*. in *Proc. 7th Int. Symp. on Ballistics*. 1983. The Netherlands.
26. Saleh, M. and L. Edwards. *Evaluation of a hydrocode in modelling NATO threats against steel armour*. in *Proceedings of the 25th International Symposium on Ballistics*. 2010.
27. Saleh, M. and L. Edwards. *Application of a soil model in the numerical analysis of landmine interaction with protective structures*. in *26th International Symposium, USA*. 2011. DEStech Publications, Inc.
28. Saleh, M. and L. Edwards, *Evaluation of soil and fluid structure interaction in blast modelling of the flying plate test.* Computers & Structures, 2015. **151**: p. 96-114.
29. Saleh, M., et al. *Numerical FE modelling of occupant injury in soil-vehicle blast interaction*. in *28th International Symposium on Ballistics, USA*. 2014.
30. Zukas, J.A., W. Walters, and W.P. Walters, *Explosive Effects and Applications*. 2002: Springer New York.
31. RIGBY, S.E. and P.W. SIELICKI, *An Investigation of TNT Equivalence of Hemispherical PE4 Charges.* 2015, 2015. **62**(4).

Rapid System Vulnerability Assessment Using Fault Trees with System Location Attribution

Serap Aksu
Defence Science and Technology Group
serap.aksu@dst.defence.gov.au

ABSTRACT

A ship's vulnerability to weapon attack, and its retained operational capability after sustaining damage, is a major concern for the ship's survivability and / or mission success. The operational capability is based on the functionality of a ship's systems, sub-systems and components that may be dispersed across the ship in multiple compartments. Fault Tree tools are often used to identify a system's retained functionality after sustaining damage. However, Fault Trees contain no physical location information regarding the systems, sub-systems and components that provide system functionality. This is important, because to determine the effect of blast damage in a specific compartment requires the development of a geometrical ship model, and this is a time consuming process. This paper presents a new approach to determine the effect on system functionality, from the effect of blast damage to specific ship compartments, using Fault Trees and without the need to develop geometrical ship models. Fault Trees are created with the functionality to assign systems, sub-systems and components to specific ship compartments. Compartment(s) that are considered damaged can then be selected in the fault tree, which will emulate the global effects of blast damage within that compartment. Consequently system functionality and retained operational capability can be predicted. This approach will significantly reduce the time to perform damage assessments particularly during a ship's procurement phase, where designs change rapidly, and hence the development of a geometrical ship model would not require.

INTRODUCTION

A warship's vulnerability to weapon attack, and retaining the vessels operational capability after sustaining damage, is a critical measure when assessing survivability and mission success. Vulnerability measures may be used for survivability trade-off analysis when comparing competing warship designs. Some of the current software suites for vulnerability assessment, such as PREVENT [1], SURMA [2] and XVAM/CVAM [3], use probabilistic Fault Trees (FT) to estimate a warship's retained operational capability after damage. The FT is used to represent functional dependencies, and relationships, between systems, sub-systems and components that contribute to the war fighting capability. Consequently, FTs are an integral component of the aforementioned vulnerability assessment software suites. However, there are two issues concerning the current use of FTs. The first issue is that FTs have limited modelling capability. This is mainly due to the FTs within the survivability software suites not being updated regularly to keep up with the most up-to-date modelling features of traditional FTs [4, 5]. The second issue is the FT dependency on a ship geometry model for identifying the locations of the systems, sub-systems and components. FTs embedded within the vulnerability software tools do not contain location information for the systems, sub-systems and components. Instead, the software tools require a three-dimensional ship geometry model to correlate the necessary location information. Identifying the location of systems, sub-systems and components is important for predicting a ship's retained capability and survivability after sustaining blast damage to specific compartments. However, the creation of a ship geometry model requires time, which could be a crucial constraint during the procurement phase if various ship designs need to be rapidly assessed.

This paper proposes an alternative FT modelling technique to overcome the limited modelling and dependency issues. The proposed modelling technique "FTs with System Location Attribution" (FTwSLA) takes advantage of the most up-to-date modelling features of traditional FTs to enable more accurate representation of relationships between systems, sub-systems and components. In addition, and more importantly, the FTwSLA modelling technique introduces a novel approach for location attribution of systems, sub-systems and components to be recorded directly within the FT structure. This novel approach enhances FT analysis and makes the technique self-sufficient, enabling faster ship vulnerability assessments without the need for ship geometry models.

PROPERTIES OF FAULT TREES WITH SYSTEM LOCATION ATTRIBUTION

FT modelling has traditionally been used for the purpose of probabilistic quantification of risk and reliability of systems when assessing undesired events [6-20]. Traditional FT modelling has a top down approach to identify the potential causes leading to the undesired event, where the undesired event is known as the 'top event'. FT modelling uses a graphical tree structure starting with the top event, and branches off downwards with the potential causes that potentially contribute to the occurrence of the top event until a desired level of resolution is reached. The events at the desired level of resolution are called 'basic events' and represent potential root causes of the top event. Basic events are, therefore, the most essential and fundamental part of traditional FTs. The branches of traditional FT structures must terminate with basic events to allow for analytical and algebraic assessments. To perform the assessment, the relationship between the top event, sub-events and basic events are defined in the traditional FTs using logic gates. Consequently, probability of occurrence of the top event is calculated using the logic gates based on logic propagation through the sub-events from the basic events.

The early, traditional FTs used simple logic gates, AND Gate and OR Gate, which model the static mode of operation for the systems. The AND Gate simply assumes that the output event occurs when all input events occur. The output event of an OR Gate occurs when at least one of the input events occurs. The traditional FT technique was first introduced in the 1960s and, since then, many types of logic gates have been proposed to make the traditional FTs more capable of modelling the dynamic behaviour of today's modern systems [21-24]. For example, traditional FTs were extended with additional logical gates such as a Voting (k out of n) Gate and Priority-AND (PAND) to define the dynamic nature of modern systems with multiple redundancies, spares, safety arrangements and compensating functions. The Voting (k out of n) Gate assumes the output event occurs if and only if k or more input events occurs. This gate is useful when a system is composed of a number of redundant elements, such that the system can provide full or degraded functionality when only some of the sub-systems and/or components are operational. The PAND Gate requires that all input events (sub-events or basic events) occurring in a defined order (typically left to right) to be TRUE. Symbols of the above-mentioned logical gates and the basic events are given in Table 1.

Table 1: Legend of building blocks of Traditional FTs.

Building Blocks of Traditional FTs	
Logical Gates	
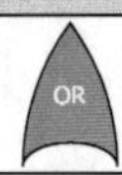	OR Gate
	AND Gate
	Voting (n out of k) Gate
	Priority AND Gate
Basic events (Root Causes)	
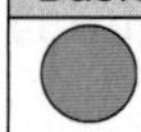 or	Circle-shape Basic Event and House- shape Basic Event_

A typical, traditional FT diagram incorporating the building blocks of top event, sub-events, and basic events connected to each other via logical gates is given in Figure 1. In this example, the top event, 'Loss of functionality of System X' occurs when any of Sub_Event 1, Sub_Event 2 or Sub_Event 3 occur. Sub_Event 1, 'Loss of sub-system X', will only occur if both 'Component A fails' and 'Component B fails' occur. Sub_Event 2, 'Loss of sub-system Y', occurs only if three basic events occur in a particular, user defined, order. Assume that order is left to right. Therefore, Sub_Event 2 occurs only when 'Primary Component C fails' occurs first, followed by the occurrence of 'Stand-by Component C1 fails' and then 'Stand-by Component C2 fails'. Finally, Sub_Event 3, 'Loss of sub-system Z', occurs when at least two out the three basic events (represented by 2/3 in the voting gate) occur. That is, if 'Component D fails' and 'Component E fails' occur, then Sub_Event 3 occurs; or, if 'Component D fails' and 'Component F fails' occur, then Sub_Event 3 occurs. Similarly if 'Component E fails' and 'Component F fails' occur, or if 'Component D fails' and 'Component E fails' and 'Component F fails' occur, then Sub_Event 3 will also occur.

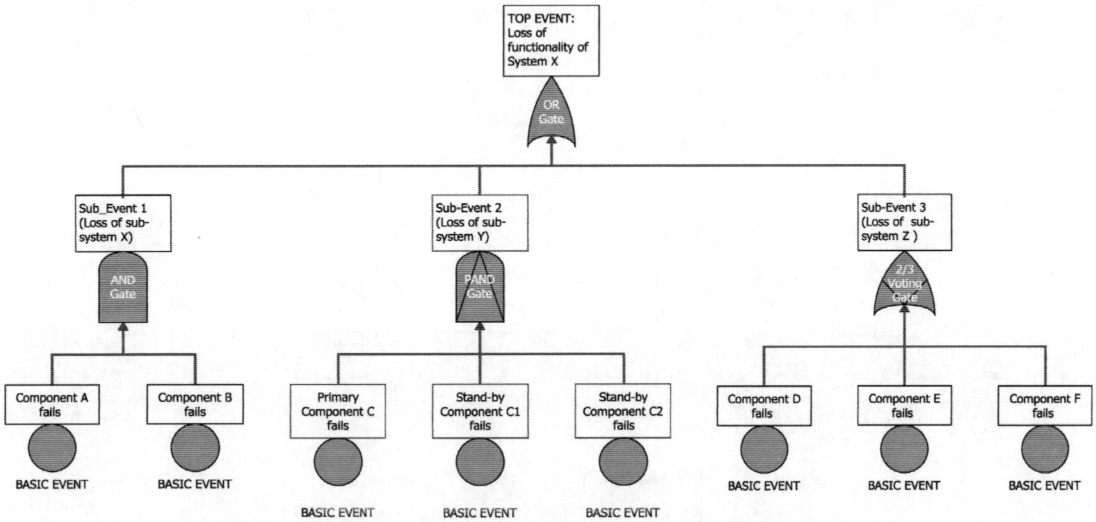

Figure1: Building blocks of a traditional FT diagram

The FTwSLA modelling technique proposed in this paper utilises the structural properties of traditional FTs, where complex logic gates may be used as required. However, the exception occurs when modelling non-functionality of a component which is assumed to occur due to damage resulting from a weapon attack. In order for a FTwSLA model to be aligned with weapon-caused damage, non-functionality of a component is linked to only two, uniquely defined, basic events via the OR Gate. These two basic events are:

- ***'Local damage to a Component'*** which is used to represent any possible event that may cause the loss, or non-functionality, of an individual component. Such damage may result from fragmentation (shrapnel) or a bullet. This basic event is represented in the FTwSLA by a circle shape. This basic event enables the FTwSLA to simulate and analyse component level damage, which would affect the retained capability of the warship; and
- ***'Global damage to a compartment'*** which is used to represent weapon-induced fire or explosion that may likely destroy an entire compartment, including the component of interest. This basic event is represented by a house-like shape in the FTwSLA. This basic event definition is a state-of-the-art application of the FTwSLA modelling technique. The compartment definition details the components within the compartment and, if the compartment is damaged, the components contained within it are also damaged. This makes the FTwSLA technique self-sufficient enabling the analysis of compartment level damage without a need for ship geometry model.

Figure 2 presents an example implementation of the FTwSLA. The two basic events that begin with, 'Local damage to Component X' and 'Global damage to Compartment containing Component X'; where 'X' represents components A, B, C and D, and are shown circled in Figure 2. These are shown to be connected to sub-events representing non-functionality of components via an OR Gate. Using different shapes for the two basic events (circle and house-like shape) in the FTwSLA model is only for demonstration purposes, it does not have any influence on the algebraic manipulations.

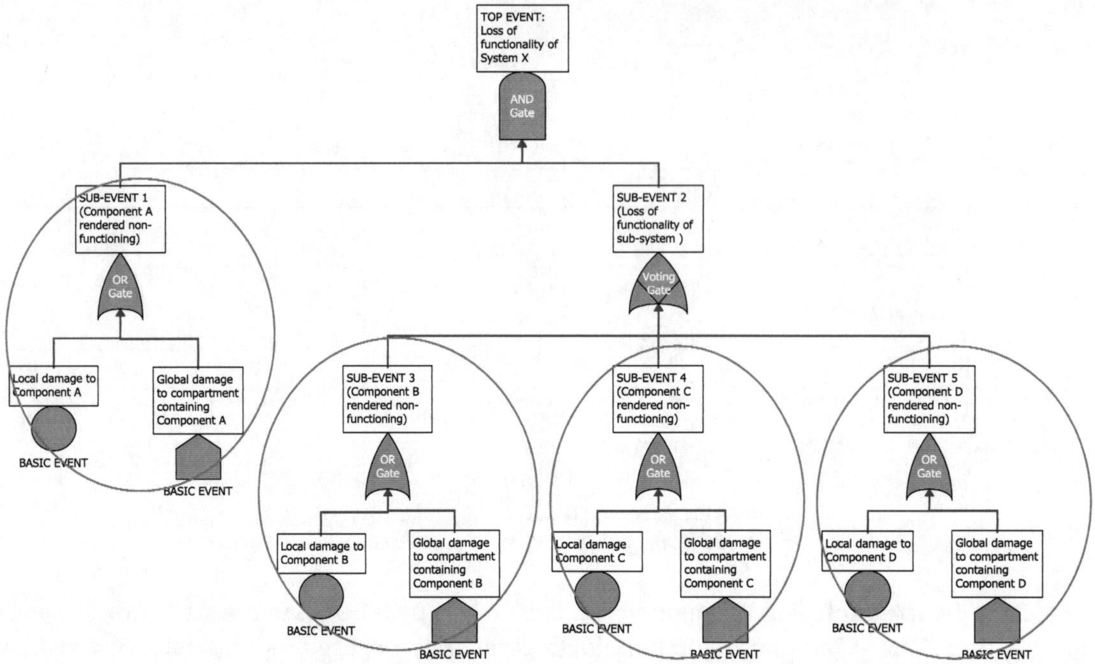

Figure 2: A sample FTwSLA diagram showing modelling of the basic events

CASE STUDY: MODELLING A GENERIC POWER GENERATION AND DISTRIBUTION SYSTEM USING FTwSLA

In this case study, a FTwSLA model was developed for a simple, generic form of a warship's Power Generation and Distribution (PG&D) System. The PG&D System is considered as one of the most critical functions that affects the operation of almost all of a warship's capabilities.

Functional Description of PG&D System

The generic PG&D System consists of: sub-systems; power generation; high voltage power distribution; and auxiliary systems of the power generation system. The components of the sub-systems are located in multiple compartments over three decks of the generic warship. Each compartment is named in accordance with the associated deck. The configuration of the systems, sub-systems and components throughout the various compartments is shown in Figure 3.

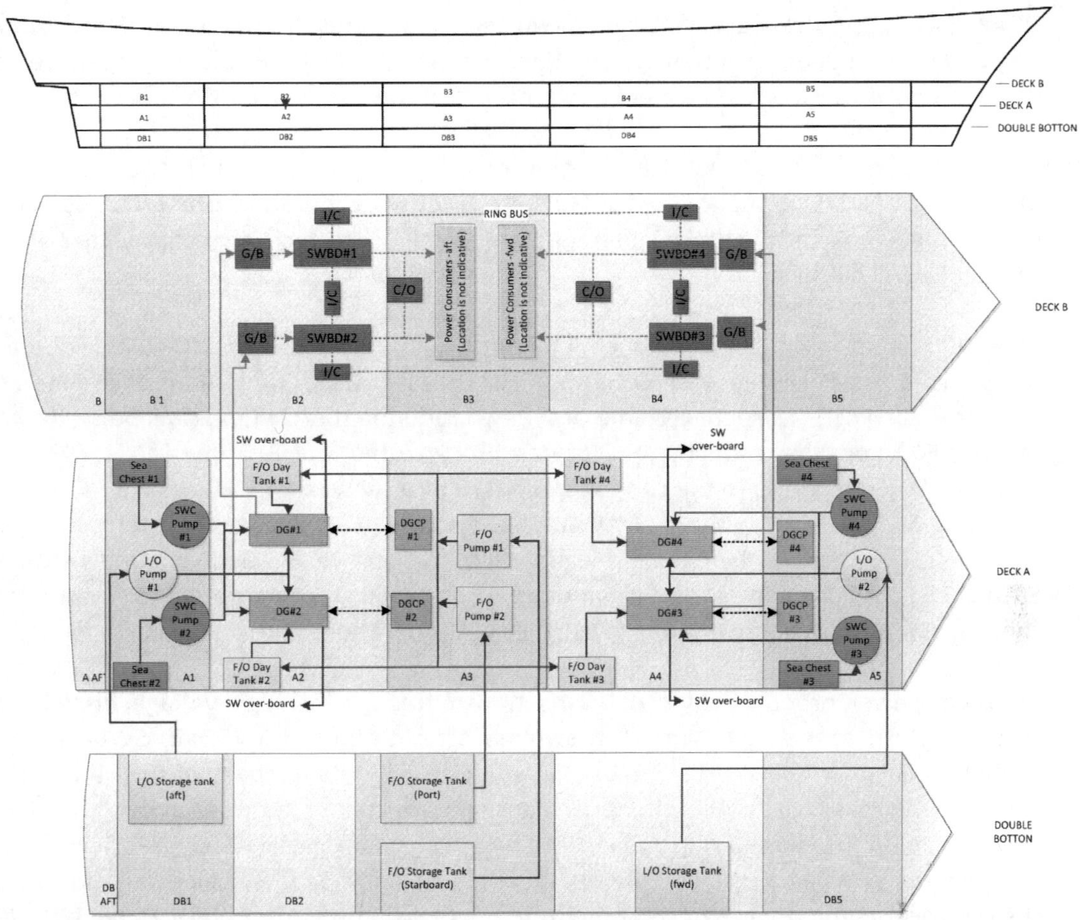

Figure 3: General arrangement of the generic PG&D system

Key to Figure 3:

- Components of Lubricating Oil (L/O) System
- Components of Fuel Oil (F/O) System
- Components of Sea Water Cooling (SWC) System
- Components of Power Generation System
- Components of Power Distribution System

In this generic design, the electrical power is generated by four sets of identical Diesel Generators (DG). Each DG set consists of an alternator, diesel engine, and other internal systems such as cooling and lubrication. DG sets are located in compartments A2 and A4 on Deck A, as follows:

Diesel Generator 1 (DG#1) - in Compartment A2
Diesel Generator 2 (DG#2) - in Compartment A2
Diesel Generator 3 (DG#3) - in Compartment A4
Diesel Generator 4 (DG#4) - in Compartment A4

The operation of the DGs relies on continuous support from the auxiliary systems: L/O System; F/O System; and SWC System. One SWC system is located forward (fwd), and is dedicated to DG#3 and DG#4; and the other is located aft, and is dedicated to DG#1 and

DG#2. Each SWC System also consists of two Sea Chests; and two redundant SWC Pumps, one of which remains on standby. Each SWC Pump takes suction from its respective Sea Chest to cool down the fresh water circulated within the respective DGs.

The L/O System has two lubrication pumps: L/O Pump #1, which lubricates the aft generators; and L/O Pump #2, which lubricates the fwd generators. The L/O is stored in L/O Storage tank (aft) and L/O Storage tank (fwd) with each tank connected to its respective L/O Pump.

The F/O System has two redundant F/O Pumps: F/O Pump #1 and F/O Pump #2, each takes suction from respective F/O Storage Tank; F/O Storage Tank (Starboard) and F/O Storage Tank (Port), respectively. One of the two redundant F/O Pumps will be on duty to transfer F/O from the respective F/O Storage Tank to the four F/O Day Tanks. F/O from each F/O Day Tank flows to the corresponding DG by gravity.

The control of each DG is managed remotely by respective DG Control Panels (DGCP). Each DGCP provides control and monitoring of essential DG parameters to ensure they remain within an acceptable range for efficient DG operation.

Electrical power produced by the DGs is transmitted to the high voltage distribution system, which consists of four main switchboards: SWBD#1; SWBD#2; SWBD#3; and SWBD#4. Each switchboard's busbar is energised by its respective DG through a Generator Breaker (G/B). All switchboard busbars are connected to each other via Inter Connectors (I/C) to make a ring busbar. The ring busbar configuration creates flexibility to energise the switchboards from any operational DG. Additionally, the power supplied to the electrical components (consumers) from the aft switchboards, SWBD#1 and SWBD#2, are provided with a Change-Over (C/O) facility. This provides the flexibility for the power supplied to the consumers to be transferred from one switchboard to another in case one has failed. A similar C/O facility is also available for power supplied to the consumers of the fwd switchboards to be supplied from either SWBD#3 or SWBD#4.

Assumptions

The assumptions used in the FTwSLA model in this case study include:

- all components are initially fully functional;
- total power requirements for full ship services and capabilities can be met with any two operational DGs;
- initially, busbars of all switchboards are connected to form a ring busbar configuration;
- power loss to the electrical components (consumers) will only occur if both fwd and aft power distribution systems are lost;
- redundant components have identical capabilities;
- redundant components can alternate between each other automatically without a time delay;
- system conduits, such as pipes and cables, are not modelled; and
- initially, each DG is running, and energises its respective switchboard via the respective DG breaker in its closed state.

Power Generation and Distribution System FTwSLA Model

The FTwSLA model was designed to identify the possible disruptions and/or events that could occur and could cause the PG&D System to have the undesired top event of 'Loss of power to consumers'. The hierarchy of the FTwSLA model captures the failure events which could potentially originate from interruption of the functional sequence between the power distribution system, the power generation system and the auxiliary services to generators, at the basic events level. Due to the large tree structure of the FTwSLA model, only a small section of it is presented in Figure 4.

The file folder icons at the base of the FTwSLA in Figure 4 indicate that there is more structure to the FTwSLA model.

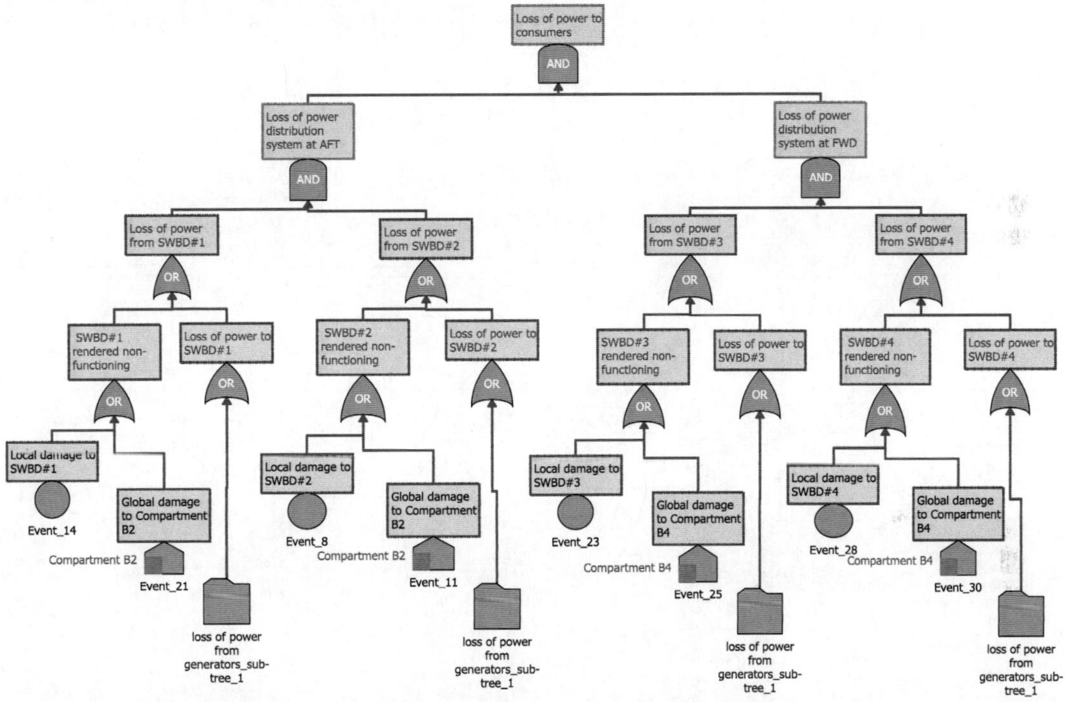

Figure 4: Section of the FTwSLA

Damage Scenario

The compartment level damage scenario was simulated for Compartment A1 by setting the occurrence of the basic event 'Global damage to Compartment A1' to damage probability of 100% as shown in Figure 5. That is, the occurrence of the basic event is TRUE and signifies a detonating warhead destroying the whole compartment. To identify the effect on the top event, 'Loss of power to consumers', of solely losing 'Compartment A1', all other basic events of the FTwSLA model were assigned a damage probability of 0%. That is, the occurrence of all other basic events was FALSE, indicating that no damage occurred.

The Compartment A1 is a common location of the following components:

- L/O Pump #1;
- SWC Pump #1;
- SWC Pump #2;

- Sea Chest #1; and
- Sea Chest #2,

Hence, this basic event 'Global damage to Compartment A1' becomes a common cause for non-functioning of all these components. Consequently, the global damage to Compartment A1, as shown by crossed out basic events in Figure 5 are a cause of non-functionality of all these components. These basic events are the basic events of sub-trees constructed for the loss of auxiliary services to DG#1 and DG#2, as shown in Figure 5. Auxiliary services for DG#3 and DG#4 would have similar sub-trees but are not shown due to space restrictions. Noting that loss of Compartment A1 does not impact auxiliary services to DG#3 and DG#4.

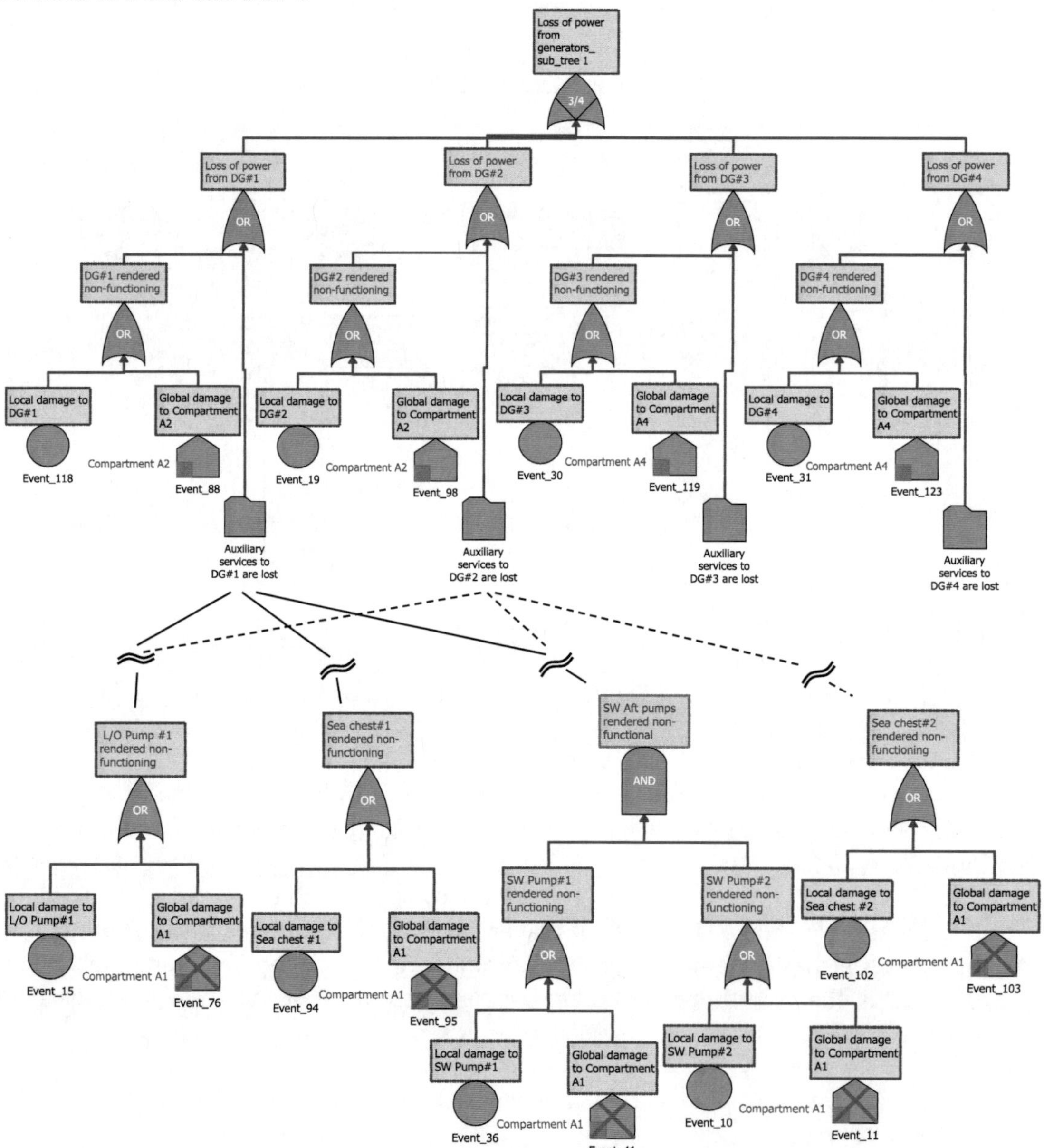

Figure 5: Sections of FTwSLA where the damage was simulated. X denotes 100% probability of failure

The cascading effects of non-functioning components are easily assessed by running the FTwSLA model. For example, the TRUE outcome of the basic event 'L/O Pump #1 rendered non-functioning' has resulted in both Sub-events of 'Auxiliary services to DG#1 are lost' and 'Auxiliary services to DG#2 are lost' are TRUE. This is because L/O Pump#1 is a dedicated pump for lubricating the DG#1 and DG#2. Furthermore, because 'SW Pump#1 rendered non-functioning' and 'SW Pump#2 rendered non-functioning' are TRUE, and 'Sea Chest #1 rendered non-functioning' and 'Sea Chest #2 rendered non-functioning' are TRUE, the SWC services to DG#1 and DG#2 failed. Since these auxiliary services are essential for operation of these DGs, disruption of these services caused 'Loss of power from DG#1' and 'Loss of power from DG#2', to be TRUE. However, the design principle of this generic system stipulates that the ship's power requirements can be achieved using only two DGs. Then, the FTwSLA will predict the probability of sub-event 'Loss of power from generators' as 0%. That is, the occurrence of 'Loss of power from generators' is FALSE. This indicates that sufficient power will successfully pass to all switchboards connected on the power distribution system through the ring bus configuration. Consequently this power will be available to all consumers on this distribution system. That is, the occurrence of 'Loss of power to consumers' will be FALSE and the warship will be able to continue its mission.

In this example damage scenario, the damage is totally contained within Compartment A1; however, FTwSLA is capable of modelling damage in multiple compartments simultaneously. In the event the warship sustains damage to multiple compartments or, even localised damage to multiple components, the occurrence 'Loss of power to consumers' may eventuate. That is, the warship will be without power to conduct its mission. If there was global damage to multiple compartments, FTwSLA analysis will enable identification of the root causes, and identify the compartments contributing the root causes without the need for a ship geometry model. Identification of root cause compartments and components will then enable ship designers to redesign the warship so that critical events affecting survivability, and the mission, do not occur. The redesigned warship may include multiple redundancies for critical systems and components; or for the inclusion of armour or strengthening of bulkheads to be applied to critical compartments. Since FTwSLA does not require a warship geometry model, the redesigned warship may be rapidly assessed using FTwSLA to determine if the design changes are effective.

CONCLUSION

FTwSLA is a novel modelling technique that eliminates the need for platform geometry models. The use of two basic event types, 'Local damage to a Component X' and 'Global damage to a Compartment containing Component X'; combined with an OR Gate at the basic event level in FTwSLA, provides an adaption of the FT modelling technique for warship vulnerability survivability analysis. Furthermore, the basic event type 'Global damage to a Compartment' contains information relating to systems, sub-systems and components within the compartment. Consequently, the FTwSLA facilitates rapid capability and survivability analysis resulting from the effects of blast damage to ship compartments. The FTwSLA will, therefore, contribute to understanding survivability of ship functions and capability, and predicting mission success. FTwSLA analysis will enable rapid re-assessment of potential design changes identified utilising the FTwSLA technique. The case study has shown that FTwSLA is a self-sufficient modelling technique that will enable rapid assessment of ship designs during the requirements phase of acquisition.

ACKNOWLEDGEMENTS

The author would like to acknowledge Director Navy Technical Bureau, Royal Australian Navy, as sponsor of this work. The author would like to thank Mr. Anthony Woolley, Mr. Tom Whitehouse and Mr. Warren Roberts for their comments on an earlier version of this paper.

REFERENCES

1. Heywood, M. (2006) PREVENT - *A Tool to Reduce Vulnerability Early in the Design.* In: RINA, Royal Institution of Naval Architects - International Conference Warship 2006 - Future Surface Ships, The Royal Institution of Naval Architects
2. Holmberg, T. and Kotiranta, R. (2012) *Bringing efficiency in the naval ship design process with cots software*. In: RINA, Royal Institution of Naval Architects - International Conference Warship 2012: *The Affordable Warship*, UK, Elsevier
3. Skeen, M. and Pickerd, V. (2011) *CVAM v2.1- User Manual.* DSTO-GD-0627, [General Report] Defence Science and Technology Organisation
4. Ni, J., Tang, W. and Xing, Y. (2013) *A Simple Algebra for Fault Tree Analysis of Static and Dynamic Systems*. IEEE Transactions on Reliability 62 (4) 846-861
5. Yuge, T. and Yanagi, S. (2013) *Dynamic Fault Tree Analysis for Systems with Nonexponential Failure Components.* IEICE Transactions on Fundamentals of Electronics, Communications and Computer Sciences E96.A (8) 1730-1736
6. Vesely, W. E. (1971) *Reliability and Fault Tree Applications at the NRTS*. IEEE Transactions on Nuclear Science, Nuclear Science, IEEE Transactions on, IEEE Trans. Nucl. Sci. (1) 472
7. Ramamoorthy, C. V., Ho, G. S. and Han, Y. W. (1977) *Fault tree analysis of computer systems.* AFIPS Joint Computer Conferences 06/13/ 13
8. Locks, M. O. (1980) T*he Fail-Safe Feature of the Lapp & Powers Fault Tree*. IEEE Transactions on Reliability, Reliability, IEEE Transactions on, IEEE Trans. Rel. (1) 10
9. Georgiannis, V. C., Caragiannis, I. E. and Tsoukarellas, M. A. (1997) *A general framework for applying safety analysis to safety critical real-time applications using fault trees.* In: Proceedings. Ninth Euromicro Workshop on Real Time Systems (Cat. No. 97TB100144), Place of Publication: Los Alamitos, CA, USA; Toledo, Spain. Country of Publication: USA.: 1997/01/01/, IEEE Comput. Soc. Spanish Gov. Board for Sci. & Technol. (CICYT) Spanish Gov. Board for Sci. & Technol. (CICYT)
10. Szabó, G. and Gáspár, P. (1998) *Probabilistic dependability analysis of adaptive functions*: A fault-tree based approach and its applications in transportation. Periodica Polytechnica Transportation Engineering 26 (1-2) 01 / 01 / 187-200
11. Aljassmi, H. and Sangwon, H. (2013) *Analysis of causes of construction defects using fault trees and risk importance measures*. Journal of Construction Engineering & Management 139 (7) 870-880
12. Anthony, M., et al. (2013) *Reliability Analysis for Power to Fire Pump Using Fault Tree and RBD*. IEEE Transactions on Industry Applications 49 (2) 997-1003

13. Ayav, T. and Sözer, H. (2016) *Identifying critical architectural components with spectral analysis of fault trees.* Applied Soft Computing 49 12// 1270-1282
14. Chen, J. M., Bi, T. S. and Sun, M. (2016) *Study on quality risk assessment for power transformer based on fault tree analysis*. In: 2016 China International Conference on Electricity Distribution (CICED): 10-13 Aug. 2016
15. Chiremsel, Z., Nait Said, R. and Chiremsel, R. (2016) *Probabilistic Fault Diagnosis of Safety Instrumented Systems based on Fault Tree Analysis and Bayesian Network.* Journal of Failure Analysis and Prevention 16 (5) 2016// 747-760
16. In-Hwan, C. and Daejun, C. (2016) *Reliability and availability assessment of seabed storage tanks using fault tree analysis*. Ocean Engineering 120 1-14
17. Jin, W., et al. (2016) *Fault-tree-based instantaneous risk computing core in nuclear power plant risk monitor.* Annals of Nuclear Energy 95 35-41
18. Yifeng, G., et al. (2016) *Fault tree analysis of fire and explosion accidents for dual fuel (diesel/natural gas) ship engine rooms*. Journal of Marine Science and Application 15 (3) 331-335
19. Yingjie, L. V., et al. (2016) *Damage assessment analysis of reconnaissance satellite based on the multi-state fault tree*. In: 2016 IEEE International Conference on Aircraft Utility Systems (AUS), IEEE
20. Kabir, S. (2017) *An overview of fault tree analysis and its application in model based dependability analysis*. Expert Systems with Applications 77 7/1/ 114-135
21. Andrews, J. D. and Moss, T. R. (2002) *Reliability and Risk Assessment.* Wilthshire, UK, Professional Engineering Publishing
22. Rausand, M. and Hoyland, A. (2004) *System Reliability Theory*. Wiley-Interscience. John Wiley
23. Limnios, N. (2010) *Multi-Component Systems*. In: Fault Trees. ISTE 33-48
24. Munro, R., et al. (2013) *Fault Tree Analysis (FTA)*: Concepts and Applications. In., American Society for Quality (ASQ)

Experiments investigating the Mass Moment of Inertia of Fluids in Roll

Darris Fortescue[1], Martin Grimm[2], and Warren Smith[3]

[1] CASG, Department of Defence, Australia. darris.fortescue@defence.gov.au
[2] NTB, Department of Defence, Australia. martin.grimm@defence.gov.au
[3] UNSW Canberra at ADFA, Australia. w.smith@adfa.edu.au

Seakeeping characteristics of any vessel are important, influencing the ability of a vessel to remain at sea. During design analysis, it follows that it is important to have a full understanding of the influence of cargo while calculating ship motions. It has been common practice while estimating ship motion parameters, such as radius of gyration, to treat all bodies within a vessel as solids. In some cases this can include large bodies of fluids which make up a significant portion of the total displacement, such as in tankers and replenishment ships.

To determine how a fluid's inertia can differ from a solid, laboratory experiments and computational fluid dynamic (CFD) simulations were conducted. The laboratory experiments used a purpose-built pendulum apparatus to test the inertia of water and a solid of the same mass and geometry in various configurations. The results showed that the inertia of the water was only a fraction of the inertia of the solid and in some cases as low as 3%. Physical visualisation using tethered ribbons and photography and CFD simulations showed a difference in relative velocity between the container and the fluid. The results that were found using these methods all indicated that fluids do not behave the same as a solid. The results are also shown to be generally consistent with previous numerical analysis [1]. Therefore, for improved accuracy while calculating ship motion, the fluids need to be treated on a case-by-case basis, depending on tank geometry.

NOMENCLATURE

CFD	=	Computational Fluid Dynamics	g	=	gravity [m/s2]
CG	=	Centre of Gravity	H	=	box (tank) height [m]
CR	=	Centre of Rotation	I	=	mass moment of inertia [kg.m2]
DNPS	=	Director Navy Platform Systems	m	=	mass [kg]
UDF	=	User Defined Function	m_b	=	mass of steel block [kg]
A	=	box (tank) width [m]	T	=	time [s]
E	=	eccentric distance [m]	x	=	distance [m]

INTRODUCTION

By having a full understanding of a ship's characteristics while calculating ship motion, the seaworthiness can be confidently assessed such that the ship will operate safely in all conditions. There are a number of characteristics that a ship needs to exhibit as a seaworthy vessel, such as stability, strength, maneuverability, endurance seakeeping and related phenomena [2]. A ship is considered to have six degrees of freedom, these include the three rotations; roll, pitch and yaw and the three translations; heave, surge and sway. The rotational motion characteristics of a ship and more specifically the roll motion is central to this paper. The natural frequency of a ship's rolling motion is dependent upon three key parameters; the ship's displacement, the metacentric height and the mass moment of inertia.

The past practice while calculating the motion characteristics of ships with large bodies of fluid such as the Australian Navy's replenishment ships HMAS Sirius and HMAS Success has been to treat these bodies as a solid. Ship motion characteristics are commonly calculated using techniques such as strip theory, as implemented in programs like SHIPMO7. Parameters, including inertia information are entered to determine responses such as natural roll period and roll amplitude of the vessel.

Through an analytical approach, Lee [1] has shown that the roll inertia of a fluid can be expressed as a percentage of that of a solid with equivalent geometry and is dependent on the shape and aspect ratio of the container. The ratio of inertia between the effective and solid moments of inertia for roll (C_r) of a fully filled tank is a function of the aspect ratio A/H as described in [1] [3] [4].

To determine experimentally how the inertia of a body of fluid compares to the inertia of an equivalent solid, a simple pendulum apparatus was designed and fabricated at UNSW Canberra. The pendulum consisted of a lightweight aluminium frame held by two low friction bearings that were connected to a heavy, rigid steel support. The aluminium frame supported both the test container and the pendulum counter weight, which sat 269mm below the centre of rotation (CR). The container held either water or a steel block depending on what was being tested. The steel support, repurposed from other experiments, served as a convenient solid foundation for the current experiments.

EXPERIMENTAL DESIGN

The experiments were setup so that a container of fixed volume could be placed at varying heights on an aluminium frame pendulum allowing the container's centre of gravity (CG) to be varied relative to the CR. The aluminium frame had a 5kg pendulum mass located 269mm below the CR. The container was capable of having a variety of baffles placed within it. A single baffle or twin baffles were used. Container dimensions, offset positions and baffle locations that were tested are shown in Figure 1 to Figure 6.

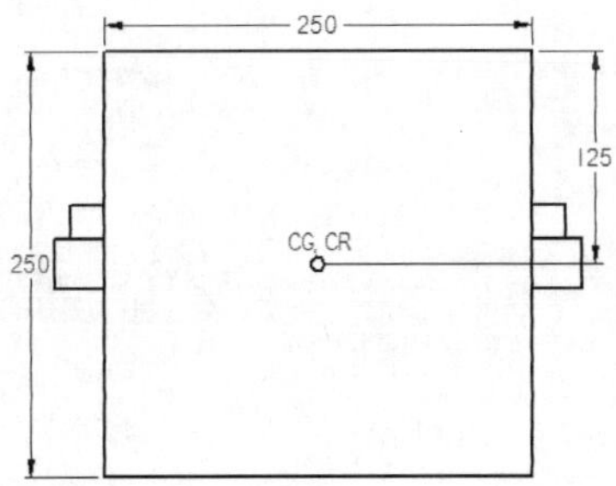

Figure 1. CG aligned with CR

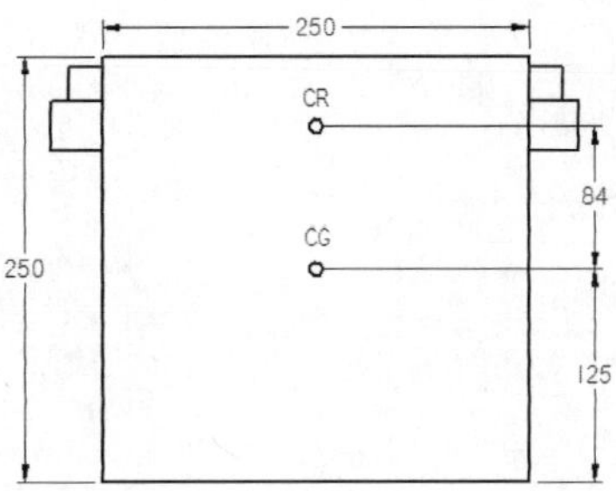

Figure 2. CG 84mm below CR

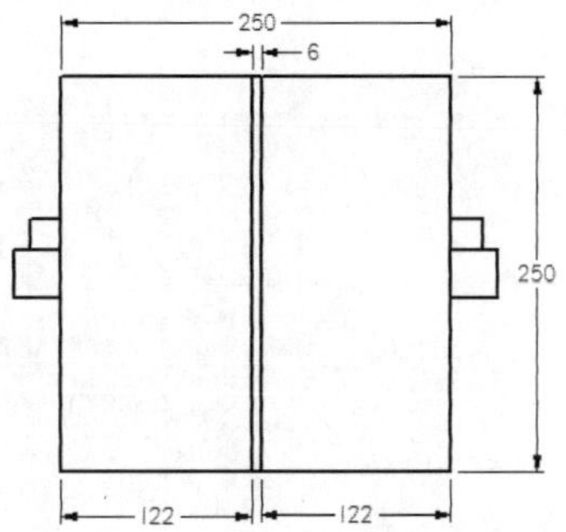

Figure 3. Centre Baffle 50|50

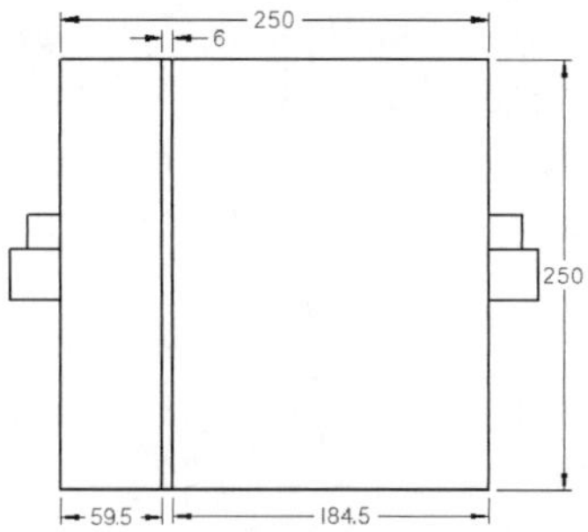

Figure 4. Baffle 25|75

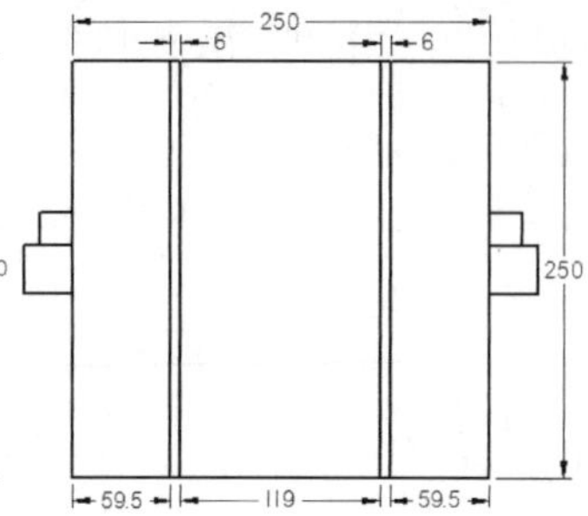

Figure 5. Baffle 25|50|25

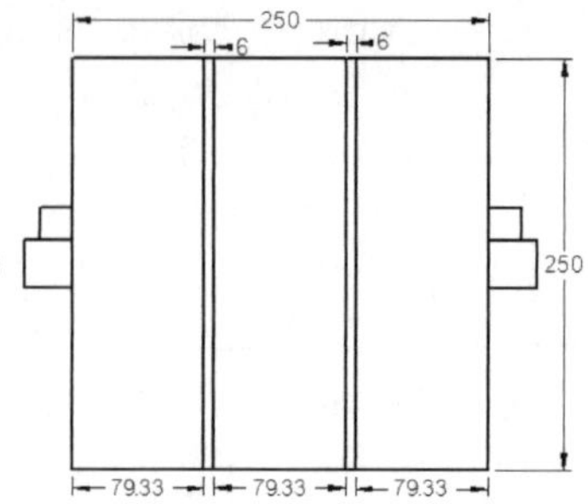

Figure 6. Baffle 1/3|1/3|1/3

For each location of the container, experiments were conducted with the container empty and with the container pressed full of water. This allowed the inertia of the frame, container and pendulum masses to be subtracted from the system's inertia and leave only the inertia of the fluid. As a comparison, a steel block with the same cross sectional area was also tested. The depth of the block in the x (longitudinal) direction was cut to length so that it had the same mass as the water (± 0.66%). Along with comparing the experimental inertia of the fluid to the experimental inertia of the solid, the theoretical inertia was also calculated, initially for the solid case and subsequently also for the fluid cases using the approach proposed by Lee [1].

The period was timed with a gate meter which used a highly accurate electronic beam breaking system. The gate was secured in a stationary position placed underneath the pendulum. The beam was broken by a small tab hanging from the base of the pendulum weight and created a time stamp every time the beam was broken.

The pendulum was tested with the container empty, full of water and containing the steel block. For each of these cases, the pendulum was allowed to oscillate between 30 and 80 full cycles depending on the setup and the test was repeated three to five times to get an average time period between oscillations. A lower number of oscillations had to be accepted for some of the test conditions, mainly water filled, where the pendulum's amplitude decayed very quickly. Outlined in the Summary of Results section is a sample of the 24 locations, 121 scenarios and over 400 samples tested during the experiments. More detailed results for the experiments can be made available through contacting the authors. Images of the pendulum design in CATIA and the final experimental rig respectively are shown in Figure 7 and Figure 8.

Figure 7. CATIA Design of Pendulum

Figure 8. Container Holding Steel Block

The inertia that was calculated theoretically, for the solid case, and from the results of the experiments is as defined in equations 1, 2 and 3.

Theoretical: $$I = \frac{m_b.(A^2+H^2)}{12} + m_b.E^2 \quad (1)$$

where I is the inertia, m_b is the mass of the block, A is the width in the y direction, H is the height in the z direction, the x coordinate is parallel with axis of rotation and E is the distance between the centre of the block and the CR. The container measured 250mm by 250mm.

Experimental: $$I = \left(\frac{T}{2.\pi}\right)^2 g.m.x \quad (2)$$

$$x = \frac{\sum_{i=1}^{N} m_i x_i}{\sum_{i=1}^{N} m} \quad (3)$$

where I is the inertia, T is the oscillation period, g is gravity (9.81m/s^2), m is the total mass and x is the distance of the centre of gravity for the summation of all bodies with respect to a datum point.

SUMMARY OF RESULTS

Centre of Gravity Aligned with Centre of Rotation

For this experiment, the CG of the container was aligned as close as possible with the pendulum's CR. It was expected that at this location the water would have the lowest inertial contribution and therefore, have the largest difference between the solid's inertia and fluid's inertia. Along with having no baffle in the container, this experimental rotation point also related to cases with four longitudinal baffle locations as shown in Figure 3 to Figure 6. To test whether the two transverse faces of the container had any effect on the movement of the water through extra boundary layer effects, transverse baffles were also included in tested configurations. One case was with a transverse baffle placed in the centre (50|50 transverse) and another was with two baffles splitting the tank in thirds (1/3|1/3|1/3 transverse). The idea of this test configuration was that if the inertia increased with these baffles in place then it would demonstrate that end effects contribute to how the fluid reacts. This would also show whether the chosen length of the container was significant. Shown in Table 1 and Figure 9 are the results for these configurations.

Table 1: Inertia Results for when Centre of Gravity is Aligned with Centre of Rotation

	Inertia Calculation when the Centre of Rotation is located at the Centre of Gravity						
		Longitudinal Baffle				Transvers Baffle	
	No Baffle	25|50 Baffle	50|50 Baffle	25|50|25 Baffle	1/3|1/3|1/3 Baffle	50|50 Baffle	1/3|1/3|1/3 Baffle
Frame (T_0)	1.25416	1.25624	1.25602	1.25771	1.25592	1.25802	1.26125
Water (T_1)	1.25850	1.32297	1.34705	1.36579	1.37182	1.26104	1.26635
Steel (T_3)	1.40017	1.40017	1.40017	1.40017	1.40017	1.40017	1.40017
Water (I_1)	0.0040	0.0633	0.0872	0.1043	0.1120	0.0028	0.0047
Steel (I_2)	0.1426	0.1426	0.1426	0.1426	0.1426	0.1426	0.1426
I_1/I_2	2.8%	44.4%	61.10%	73.2%	78.6%	2.0%	3.3%

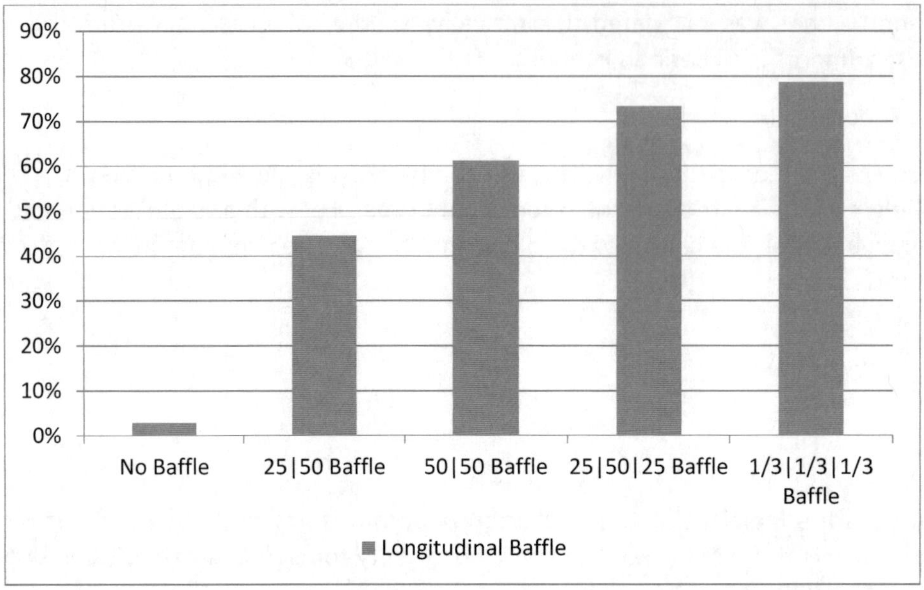

Figure 9. Fluid Inertial Results as % of Solid for Centre of Rotation Located at the Centre of Gravity

The highlighted results in Table 1 are the percentage of inertia for the water from the experiment with respect to the inertia of the steel also shown in Figure 9 and will be the main point of discussion. As anticipated the inertia for the water with no baffling is very low, getting down to less than 3% of the solid's inertia. Having either one or two extra baffles placed transversely across the face did not change the inertia of the water significantly, noting the baffles were smooth. The differences between the inertias with and without transverse baffles can be considered negligible or within experimental error. Small differences in time do impact the calculated inertia, even though this was minimised by taking several samples and using the average time period.

Where a single longitudinal baffle is placed significantly influences the total inertia. Fluid inertia was measured as 61% of the solid when the single baffle was placed in the centre. The inertia continues to increase when a second baffle is placed vertically in the container, with 78% of the solid case when equally spaced (1/3|1/3|1/3).

Out of the five different container setups having two baffles divide the container in thirds had the greatest influence in increasing the inertia for the system. By having an even spacing and the smallest central sized box the fluid inertia has increased the most for this experiment.

Centre of Gravity 85mm below Centre of Rotation

For this experiment the container was lowered so that the CG was now 85mm below the CR, this was the largest vertical distance that was practically possible for this particular container in the z direction. As mentioned above in 'Centre of Gravity Aligned with Centre of Rotation' cases these experiments were done with the same pendulum weight and again repeated for an empty container, water filled and steel mass with and without the baffling. The results obtained from these tests are shown in Table 2 and Figure 10.

Table 2: Inertia Results for Centre of Gravity 85mm below Centre of Rotation

	Inertia Calculation for the Centre of Gravity 85mm Lower then Centre of Rotation				
		Longitudinal Baffle			
	No Baffle	25|50 Baffle	50|50 Baffle	25|50|25 Baffle	1/3|1/3|1/3 Baffle
Frame (T_0)	1.156755	1.150551	1.14773	1.148493	1.146517
Water (T_1)	0.971981	1.011036	1.027569	1.036560	1.042628
Steel (T_3)	1.061399	1.061399	1.061399	1.061399	1.061399
Frame (I_0)	0.5788	0.5863	0.5835	0.5917	0.5897
Water (I_1)	0.1416	0.2062	0.2353	0.2507	0.2627
Steel (I_2)	0.2828	0.2828	0.2828	0.2828	0.2828
I_1/I_2	50.1%	72.9%	83.20	88.7%	92.9%

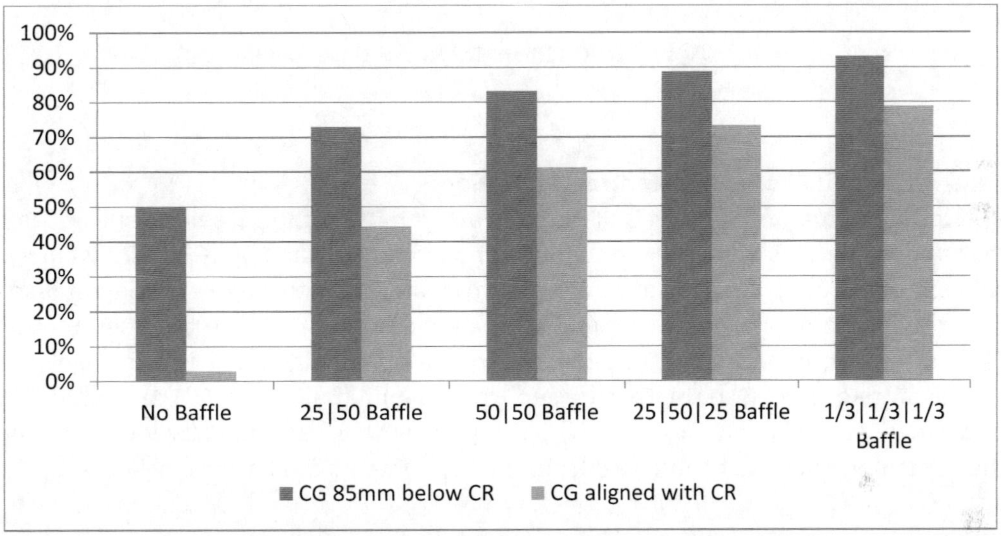

Figure 10. Centre of Gravity 85mm Lower then Centre of Rotation

The results show that the inertia has increased with the offset of the CG from the CR, (see Figure 10). However, a similar trend as before is evident in that the inertia increases from no baffle to a single baffle case and continues to increase with a second baffle, which as before is dependent on the location of the baffling. As before the highest inertia measured is when the container had two baffles placed equally apart (1/3|1/3|1/3).

Centre of Gravity 125mm Transverse Offset to Centre of Rotation

Tests were also performed with a transverse offset of the CG from the CR of 125mm, as shown in Figure 11. At this offset, it was planned to carry out all previous baffle configurations. However, due to safety concerns with large counter weight loads and inaccuracies from small deflections only non-baffled tests were conducted. The results are plotted in Figure 12 as the right hand set of points where CG to CR is 125mm.

The result of this experiment follows the expected trend in that the inertia of the fluid continued to increase with the increase in distance between the CG and CR. As this was the largest difference between the CG and CR the results have the highest inertia for the water with no baffling, being 77% of an equivalent solid case. The results showed that even when the CR was entirely outside the body of interest that the inertia for the water was still not the same as if it were solid.

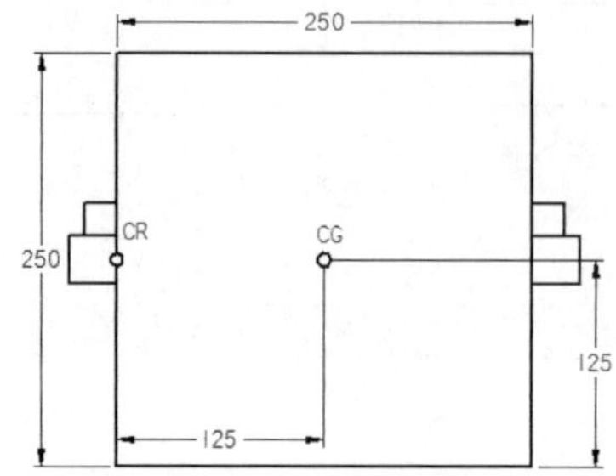

Figure 11. CG 125mm Transverse Offset to CR

Centre of Gravity Varying to Centre of Rotation

Due to the inertia of the water still not approaching that of the steel case, it was decided that a series of tests of inertias should be conducted with varying CG to CR distances. The concept of the experiments was developed to clearly show through experimental analysis how the inertia of the water changed with an increase in distance between the CG and CR and to note at what point, the water's inertia essentially matched the solid's inertia. The distance of the CG to the CR changed from -84mm (below the CR) to 0mm (CR at CG) by 6mm increments; this was done by placing 6mm MDF packers between the frame and the container. The results obtained from the experiments are plotted in Figure 12.

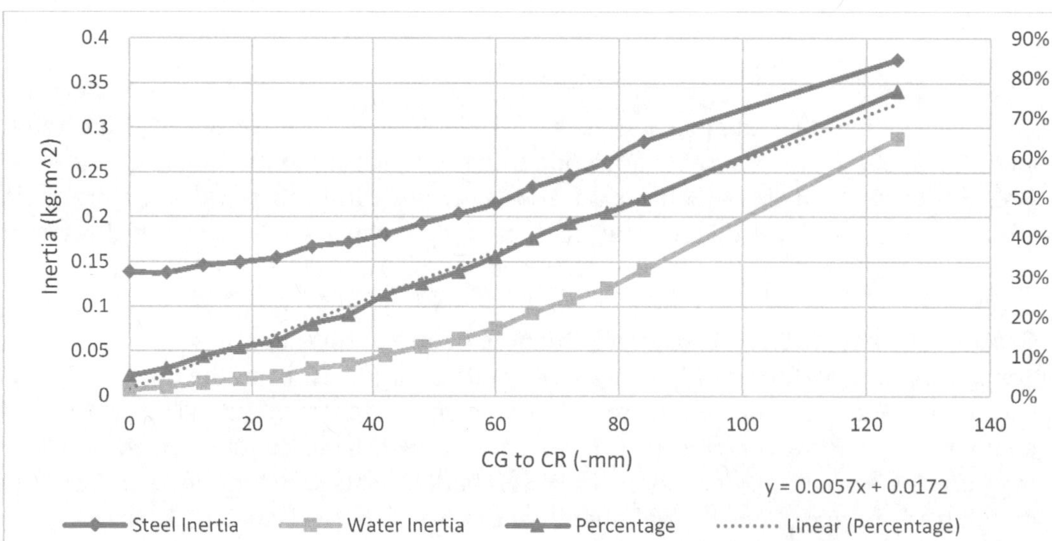

Figure 12. Inertia as a Function of Distance between CG and CR

The results from this set of experiments provide a clear picture of how the water's inertia changes with distance from the CR. While the fluid inertia approaches that of the solid equivalent, theoretical considerations outlined in the following section suggest that the inertia will never entirely match the equivalent solid, no matter how large the CR to CG separation.

Comparison between Experimental Results and Theoretical

A comparison was made between some of the results obtained during the experiments and Lee's available analytical equation [1]. However, results were adjusted to account for roll centres being away from the CG [4]. This comparison is shown in Figure 13.

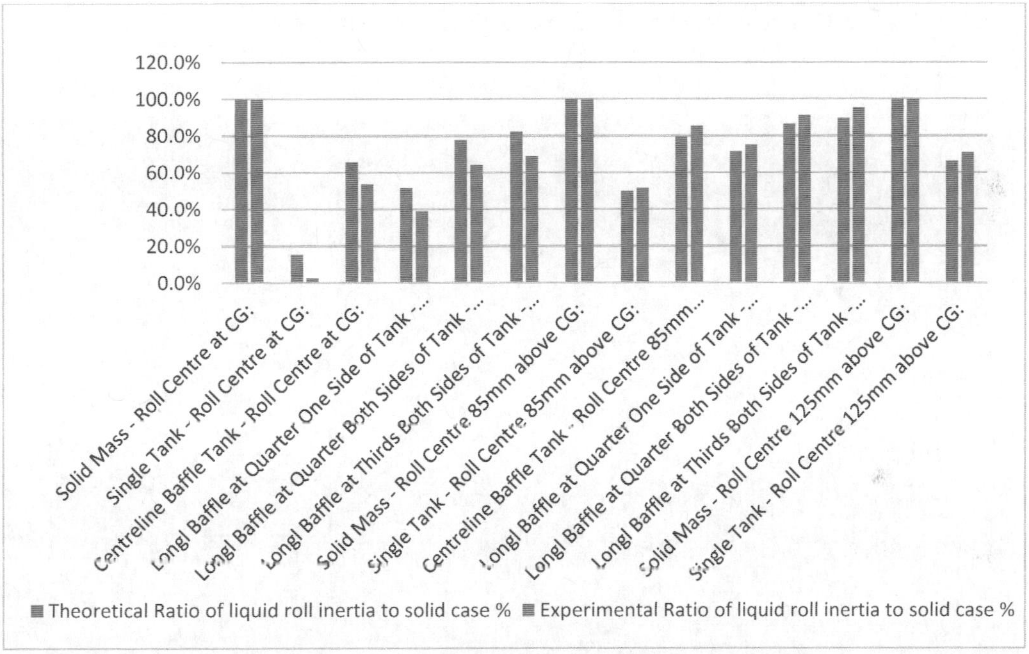

Figure 13. Compare Theoretical vs. Experimental Roll Inertia Results

Good correlation between most of the results is demonstrated and the discrepancies can probably be attributed to either experimental error or the viscous effects that are not included in the analytical method of Lee [1].

VISUAL ANALYSIS

As a way to try to visualise what was happening in the container during the roll oscillations, multiple approaches were trialled to determine an appropriate method. The best and simplest solution was by attaching ribbons on a mesh grid within the fluid. The idea behind this was that if the ribbon does not move/sway then the water is stationary relative to the container and it is moving with the container thus contributing to inertia. If the ribbons move/sway, then the water is moving relative to the container and not moving with it and therefore not contributing to the inertia of the system. The visual analysis was conducted by filming and conducting a photo burst at 20 frames per second for the container rotating with its CG at the same location as the CR without a baffle. Some images for this are presented in Figure 14.

The nine images (annotated in sequence) in Figure 14 are from a half period oscillation starting from very little to no amplitude of the ribbons through maximum and back to little amplitude. As the images show; the amplitude increases as the relative velocity increases between the fluid and the container. Following the ribbons from the outer edge to the centre the maximum amplitude decrease. This does not illustrate that the outer edge is not contributing to the inertia while the centre is; it is simply due to a lower maximum fluid velocity near the centre of rotation. In a very basic form this shows that the majority of the fluid is moving in relation to the tank which corresponds with the data collected from the experiments. Note; the camera was mounted to the tank for this test and the CG and CR are in the centre of each of the nine images.

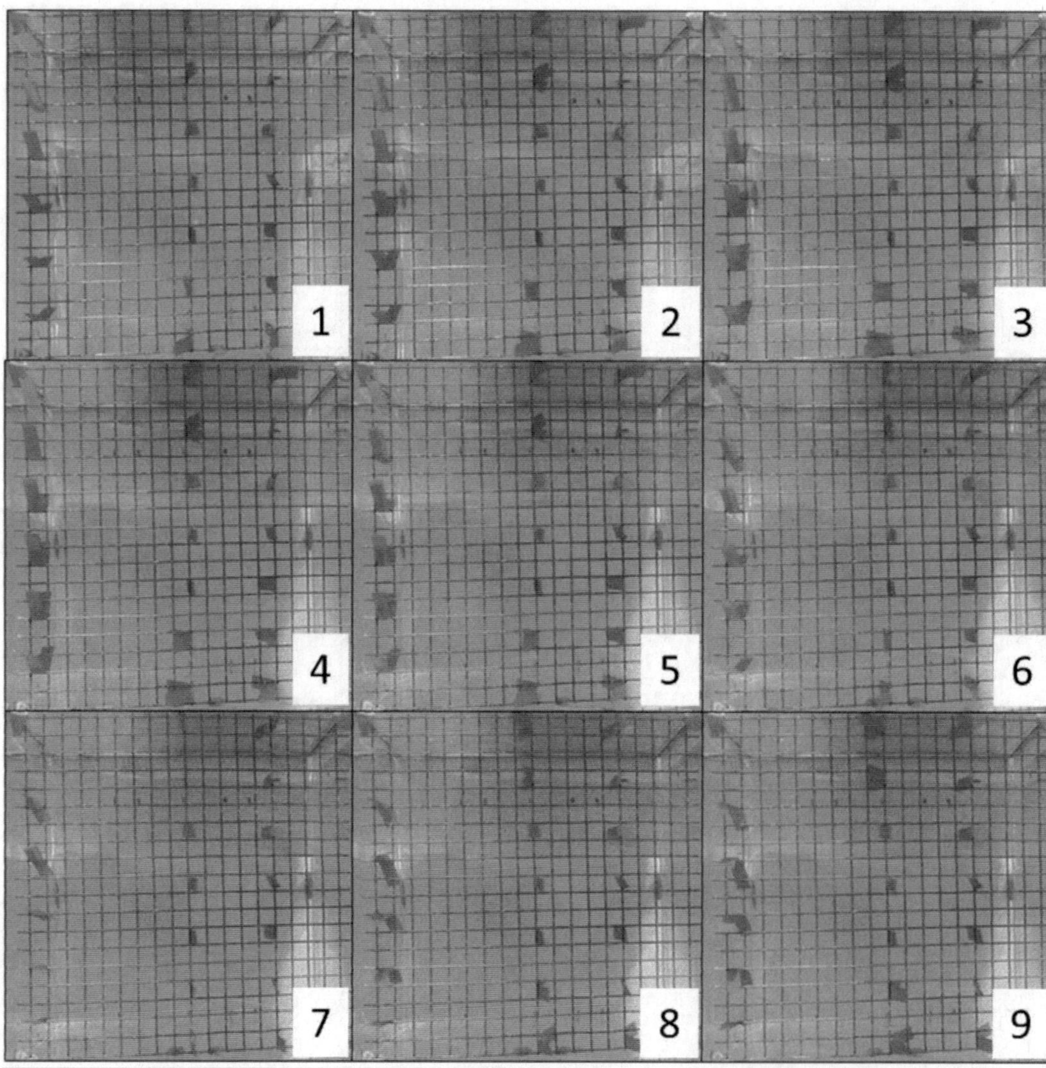

Figure 14. Amplitude of ribbon motion during oscillation – Visual Sequencing 1-9

COMPUTATIONAL FLUID DYNAMICS SIMULATION

Some of the experiments that were discussed earlier were also modeled in the Computation Fluid Dynamics (CFD) software package, ANSYS Fluent. The purpose of these simulations was not to compare the inertias of the fluid with a solid, but rather to visualise the fluid flow noting the velocity profiles of the fluid. A secondary motivation was that replenishment ships usually do not carry large quantities of water but rather oil and diesel and using the CFD code a comparison between water and diesel could be made.

The simulation was modeled in 2D running a User Defined Function (UDF) code to simulate the oscillation of the container. Simulations were performed for scenarios with the CG located at the CR with no baffling, a baffle in the centre, two baffles in the centre in a cross pattern and two baffles separated by thirds for both water and diesel. All simulations conducted used the same UDF and a period that matched the centre un-baffled experiment of 1.2585 seconds, the simulations ran at 0.01-time step intervals and images were captured every 5-time step intervals. A simulation was also done using the geometry for HMAS Sirius, where the oscillation time was based on natural roll period calculations obtained from SHIPMO7. A different UDF with a frequency of 0.1 and an angle of 12 degrees was used. The time step intervals remained the same for this simulation.

The fluid properties for the water were selected from the library within ANSYS, with; density equal to $998.2 kg/m^3$ and dynamic viscosity equal to $0.001003 kg/m.s$. There are several different fuels that are typically carried in large volumes on many of the replenishment ships. This includes F76 which is a diesel fuel for ship propulsion and power generation and F44 which is a fuel used by helicopters. Though the more common fuel is the F76, F44 was used during the CFD analysis as it had a higher viscosity [5]. By using the higher viscosity, the simulation was conducted as a worst-case scenario to determine the legitimacy of the laboratory experiments and their use of water and a small scale model. Two sources used for the properties of the fluid did not contain data at 20°C but at −20°C and 40°C, so a simple extrapolation was used to determine the approximate properties at the desired temperature [6]. The following properties were used for the simulation [7];

- Density: $\rho = 840 kg/m^3$
- Dynamic viscosity: $\mu = 0.003528 kg/m.s \approx 20°C$
- Kinematic viscosity: $\nu = 4.2 \times 10^{-6} m^2/s \approx 20°C$

The velocity profile shown in Figure 15 is for a solid. It changes linearly from the centre outwards increasing in velocity and is used as reference. The profile was based on data stated above and it used the time period for the water with no baffling. Each colour stamp was based on the radius at the midpoint for that particular colour.

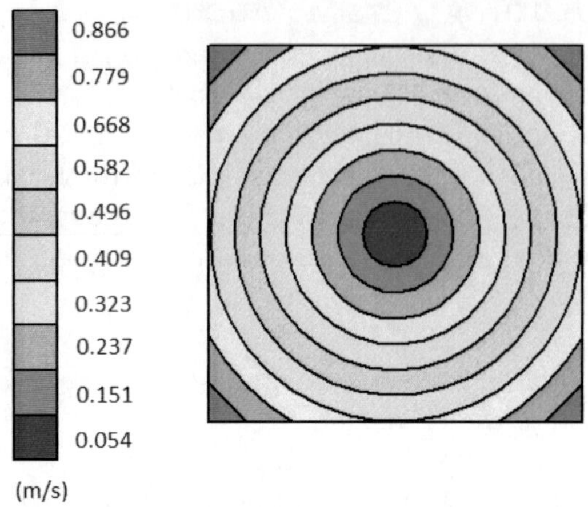

Figure 15. Velocity profile for a solid mass

The first two simulations that were conducted had the CG at the CR with no baffling; the first was filled with water, the second with F44 fuel. Both simulations ran for the same time. Shown in Figure 16 and Figure 17. are images captured during the simulation taken at identical times. The images both start at 0.3 seconds and finish at 1.5 seconds recorded at 0.15 second intervals. This started from one extreme amplitude and finished just before one full cycle. The velocity profiles appeared to go through a cycle of having a large area with very little to no velocity, to a higher velocity and back to little velocity again.

As Figure 16 and Figure 17 show, there is little difference between the two fluids that were assessed. Only in the last three images, in the pair of figures, can a noticeable difference in the velocity fields be seen. Though there is a difference in the colour map this is only a very small change and therefore is considered to be negligible. Though several different experiments were conducted, they all showed similar results in that there were very little differences between the fuel and the water. Due to having very small differences it can be assumed that the laboratory results are a valid representation for other common fluids in larger ships tanks.

ERROR ANALYSIS

An error analysis was conducted for one of the experiments. This was to assess what degree of uncertainty there was in the laboratory experiments. The error analysis is not detailed here, however considered the errors of the three types of measurements used throughout the experiments:

- Mass ± 0.00005kg
- Distance ±0.0005m
- Time ±0.0000005s

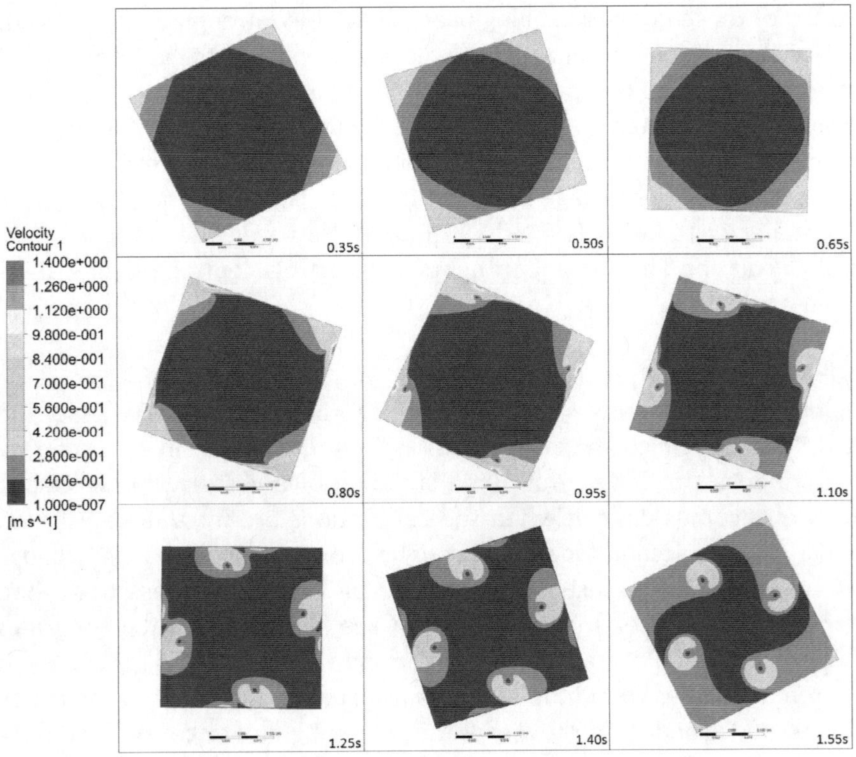

Figure 16. CFD simulation of water filled container

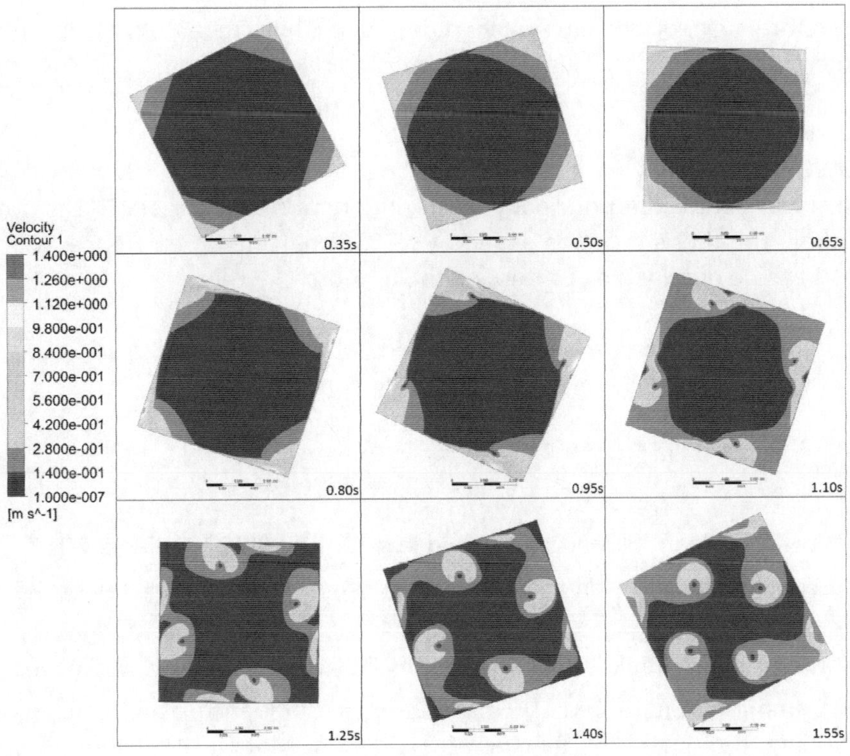

Figure 17. CFD simulation of fuel filled container

Due to each of these quantities being used repeatedly throughout several equations and having very little error, the final error was also expected to be very low. This did not account for the average times used in each individual test, but the error was calculated for the averages between the tests. The calculation was done for when the CG was 85mm lower than CR with no baffling for both the solid and water so that the final error was in respect of the percentage between the two. The error analysis suggested that the experimental error to be $I_1/I_2 = 50.1\% \pm 0.18$. However, the comparison between the analytical calculation and the experiments suggest this could be higher though they do follow the same trend.

CONCLUSION

Seakeeping and seaworthiness are important characteristics of any vessel. These are the features of a vessel which influence its ability to remain at sea in all conditions. It follows that it is important to have a full understanding of a vessel while calculating the ship motion. This becomes difficult when the calculations are for a replenishment ship with large bodies of fluid accommodated. Through several experiments discussed in this paper it was shown in all cases tested that the inertia of a fluid is less than that of a solid, consistent with the earlier analytical work of Lee [1]. This adds further complications in calculating ship motion characteristics. A common practice during these calculations has been to treat all bodies within the vessel as a solid. The results from the experimentation and CFD methods conducted show that this would be an inaccurate assumption to make as the inertia for a fluid was measured to range between 3% and 77% for an un-baffled square cross section container. It has been observed that the fraction between a fluid's inertia and a solid's inertia is very dependent upon container configuration and position. The results of the experiments are broadly consistent with the analytical results of Lee [1] and therefore, use of the approach outlined by him to account for fluid inertia is recommended.

ACKNOWLEDGEMENT

This work was conducted principally as an undergraduate honours thesis project by the lead author. The thesis task was posed by NTB, Department of Defence. More details of the thesis can be obtained by contacting the authors.

A more detailed literature review is contained in the thesis and the additional references drawn upon [8-25] have been included here for completeness.

References

[1] G. J. Lee, "Moment of Inertia of Liquid in a Tank," KRISO, Daejeon, 2014.

[2] E. C.Tupper, Introduction to Naval Architecture, Burlington: Elsevier, 2004, pp. 218-232.

[3] J. Journee, "Fluid Tanks and Ship Motions," Kyushu University, Mekelweg, 2000.

[4] M. Grimm, W. Smith and D. Fortescue, "The Influence of Roll radius of Gyration Including the Effect of Inertia of Fluids on Motion Predictions," *Pacific 2017 International Maritime Conference,* 2017.

[5] J. Fu, "Effects of Biodiesel Contamination on Oxidation and Storage Stability of Neat and Blended Hydroprocessed Renewable Diesel," *Energy & Fuels,* p. 5179, July 2015.

[6] Australian Defence Standard, DEF(AUST)5240D Aviation Turbine Fuel Specification, Melbourne: DPS, 2007.

[7] The Shell Company of Australia Limited, "Material Safety Data Sheet F-44/AVCA/FSII," Hawthorn, 2013.

[8] Masa Yards, Significant Ships of 1990, T. Knaggs, Ed., London: RINA Ltd, 1990.

[9] Burmeister & Wain Skibsvaerft A/S, Significant Ships of 1991, London: RINA Ltd, 1991.

[10] Finnyards, Significant Ships of 1992, London: RINA Ltd, 1992.

[11] F. T. Dodge, "The New "Dynamic Behavior of Liquids in Moving Containers"," SWRI, Texas.

[12] M. Levadou, "Operational Guidence to Avoid Parametric Roll".

[13] K. McTaggart, "SHIPMO7: An Update Strip Theory Program for Predicting Ship Motions and Sea Loads in Waves," Defence Research Establishment Atlantic, 1997.

[14] N. Salvesen, "Ship Motions and Sea Loads," The Society of Naval Architects and Marine Engineers, New York, 1970.

[15] R. T. Schmitke, "Ship Sway, Roll and Yaw Motions in Oblique Seas," SNAME Transactions, 1978.

[16] I. A. Sheikh, "Parametric Roll Instability of Ships," Faculty of Mathematics and Natural Scienes University of Oslo, Oslo, 2008.

[17] A. Vredeveldt and J. Journee, "Roll Motions of Ships due to Sudden Water Ingress, Calculations and Experiments," 2001.

[18] H. N. Abramson, "The Dynamic Behavior of Liquids in Moving Containers," National Technical Information Service, Washington, 1966.

[19] R. A. Ibrahim and I. M. Grace, "Modeling of Ship Roll Dynamics and Its Coupling with Heave and Pitch," Hindawi Publishing Corporation, 2009.

[20] M. V. Ricketts and P. A. Gale, "On Motions, Wetness and Such: The USS Midway Blister Story," in *Transactions*, vol. 97, New Jersey, The Society of Baval Architects and Marine Engineers, 1990, pp. 429-471.

[21] S. Kruger and F. Kluwe, "A Simplified Method for Esitmation of the Natural Roll Frequency of Ships in Heavy Weather," TU Hamburg, Harburg.

[22] Y. S. Shin, V. L. Belenky, J. R. Paulling, K. M. Weems and W. M. Lin, "Criteria for Parametric Roll of Large Containerships in Longitudinal Seas," The Society of Naval Architects & Marine Engineers, Washington, 2004.

[23] Y. A. Cengel, J. M. Cimbala and R. H. Turner, "Thermal-Fluid Sciences," New York, McGraw-Hill, 2012, pp. 605-607.

[24] R. E. Howe, "Free Running Seakeeping Experiments," University of New South Wales Canberra, Canberra, 1999.

[25] P. Hayes, "Stability Requirements of Naval Landing Craft," University of New South Wales Canberra, Canberra, 2014.

The Influence of Roll Radius of Gyration Including the Effect of Inertia of Fluids on Motion Predictions

Martin Grimm[1], Warren Smith[2], and Darris Fortescue[3]

[1] NTB, Department of Defence, Australia. martin.grimm@defence.gov.au
[2] UNSW Canberra at ADFA, Australia. w.smith@adfa.edu.au
[3] CASG, Department of Defence, Australia. darris fortescue@defence.gov.au

Computer based ship motion predictions using strip theory and similar methodologies have become routine for many naval architects over the last three decades. Aside from definition of the wave and operating conditions of interest, other inputs to such codes are typically straightforward definitions of hull and appendage geometries, load conditions and control parameters for fins, rudders or anti-roll tanks. While displacement and centre of gravity data can be obtained relatively easily from stability analysis, either at the design stage or from a stability book once a vessel is in service, obtaining accurate roll, pitch and yaw radius of gyration data is invariably more difficult.

Discussed in this paper are approaches to determining radius of gyration information for ships, with a focus on roll radius of gyration. In association with a companion paper [1], accounting for the influence of fluids in tanks when estimating radius of gyration is also addressed. Finally, the sensitivity of motion predictions to variations in assumed roll radius of gyration is examined using a strip theory method.

NOMENCLATURE

b	=	width of tank [m]
B	=	beam of hull of vessel [m]
C	=	constant
f	=	liquid to solid inertia ratio
g	=	gravity [m/s^2]
GM	=	metacentric height [m]
h	=	height of tank [m]
H	=	mass item height [m]
i	=	subscript for weight element
I_{44}	=	roll moment of inertia [kg.m^2]
I_{55}	=	pitch moment of inertia [kg.m^2]
I_{66}	=	yaw moment of inertia [kg.m^2]
k_4	=	roll radius of gyration [m]
k_5	=	pitch radius of gyration [m]
k_6	=	yaw radius of gyration [m]
L	=	mass item length [m]
L_{WL}	=	length on waterline [m]
LCG	=	Longitudinal Centre of Gravity [m]
m	=	mass [kg]
T	=	draft [m]
TCG	=	Transverse Centre of Gravity [m]
T_{n4}	=	roll natural period [s]
VCG	=	Vertical Centre of Gravity [m]
W	=	mass item width [m]

INTRODUCTION

While most input data for ship motion predictions can be obtained with relative ease, the pitch, yaw and particularly roll radius of gyration of ships and other marine craft are rarely readily available. At the design stage, ship weight estimates at best typically include the lightship and variable load mass estimates and their corresponding VCG, LCG and occasionally also TCG. Any seakeeping calculations at that stage are therefore invariably based on 'rule of thumb' equations that estimate the pitch, yaw and roll radius of gyration as a fraction of either the ship length or its beam.

While such rules of thumb may be reasonable for ships of a common type on which those estimates were based, such guidance becomes invalid outside of its bounds. Guidance for monohulls would not be valid for catamarans, SES, SWATHs or hydrofoils. Similarly, data relevant to powered craft would not be valid for sailing craft, even if only considering monohulls. A further complication is that tankers have a fluid cargo that makes up a large percentage of their mass yet this does not behave in the same manner as solid cargo [1].

It is not immediately apparent how much error is introduced in ship motion predictions by using rule of thumb assumptions for inertia properties of ships.

Existing rule of thumb guidance on inertia properties is reviewed in this paper. The application of a simple first principles based approach for the more accurate estimation of inertia properties, including in the case of ships with significant fluid loads, is also outlined.

RULE OF THUMB GUIDANCE FOR INERTIA

Guidance on ship inertia estimation is often provided in ship motion code manuals. For example, the SHIPMO7 manual [2] suggests pitch radius of gyration (k_5) as 25% of ship length and assumes yaw radius of gyration (k_6) is equal to that of k_5. Guidance for 'dry' roll radius of gyration (k_4) of a small warship is that this lies between 32% and 36% of ship beam. For the later ShipMo3D code [3], guidance for k_5 and k_6 remains as 25% of length while the default dry k_4 is suggested as 40% of maximum beam. Similarly, the FREDYN manual [4] proposes approximate input values of 24% of L_{WL} for k_5 and k_6 and an equation for estimating k_4:

$$k_4 = \left((0.4\,(B + H)^2/12 + 0.6\,(B^2 + H^2)) - (T - 0.5H - VCG)^2 \right)^{0.5} \qquad (1)$$

where "B" is the beam of the ship, "H" it depth amidships and "T" the draft.

Through their guidance for natural roll period of ships ($T_{n4} = 2.27.B/(g.GM)^{0.5}$), SNAME [5] imply that for ships of typical form the wet k_4 may be 36.1% of beam. However, they note this can vary by as much as 20% for ships of unusual form. Aircraft carriers, where the flight deck is much wider than the hull beam, are given as an example where the wet k_4 may be as large as 41.3% of hull beam. A further example is provided for submarines where wet k_4 may be as low as 31.8% of beam, corresponding to the hull diameter. The Society of Allied Weight Engineers [6] provide radius of gyration estimates for a range of USN surface ships and submarines, some of which are summarized within Table 1. Another common reference [7] indicates k_5 is usually in the range from 20% to 25% of L_{WL} with the later edition [8] providing estimates of 30% of waterline beam for k_4 and 22.5% of L_{WL} for k_5 and k_6 in the case of conventional ships. Considering sailing craft, Marchaj [9] suggests that while the rig may account for only about 5% of the displacement of a modern yacht, it may contribute 60% to 70% of the total roll moment of inertia and in turn this results in a relatively large wet k_4 relative to beam with examples ranging from 62% to 78% of maximum beam. The US Navy [10] has $T_{n4} = C.B/(GM)^{0.5}$ where "B" is maximum beam at or below the waterline and "C" is usually 0.69 to 0.89 s/m$^{0.5}$. However examples are also provided using C = 0.82 for a destroyer and C = 0.725 for an aircraft carrier. When these coefficients are expressed in relation to the wet k_4, this corresponds to a range from 34.4% to 44.4% of beam with the destroyer at 40.9% and carrier at 36.1% of beam. The latter is seemingly inconsistent with the advice in [5].

A more elaborate means of estimating the wet roll radius of gyration of ships has been proposed by IMO [11]. In this case the wet k_4 is dependent not only on beam, but also on draft and length:

$$T_{n4} = 2.C.B/(GM)^{0.5} \tag{2}$$

where:

$$C = 0.373 + 0.023(B/T) - 0.043(L/100) \tag{3}$$

and "T" is draft in metres, "B" is breadth of hull and "L" is the ship length.

Re-expressing the equation in terms of k_4, we have:

$$k_{4\,wet} = [0.373 + 0.023(B/T) - 0.043(L/100)].(g^{0.5}/\pi).B \tag{4}$$

This equation suggests the wet k_4 increases for wide shallow draft ships. However this may be due to the increase in roll added mass inertia rather than any inherent characteristic of the mass distribution of ships as a function of B/T. The equation also suggests that as ships become longer (larger) their roll inertia as a proportion of beam decreases. This might be attributed to the relative proportions of superstructures and hull depths as ships become larger. The origins and range of validity of this equation are unknown to the authors.

DIRECT CALCULATION APPROACH FOR INERTIA

The lead author was involved in the early design phase of the NQEA-Danyard developed Seajet high-speed passenger-vehicle catamaran ferries (Figure 1). These catamarans were optimized for seakeeping performance and as a consequence were the subject of a seakeeping model test program. This necessitated the reliable estimation of the roll, pitch and yaw moments of inertia to ensure the test model was realistically ballasted. As the rule of thumb guidance for monohulls was unsuitable for this catamaran, the weight estimate for the vessel was extended to incorporate the calculation of these moments.

In subsequent years, this approach was also applied to estimate the moments of inertia for hydrofoil model testing (Figure 2), hydrographic ships (Figure 3) and replenishment ships (Figure 4). Some of these predictions, as well as a selection provided by the Society of Allied Weight Engineers (SAWE) are provided in Table 1 to compare against the 'rule of thumb' guidance discussed in the previous section.

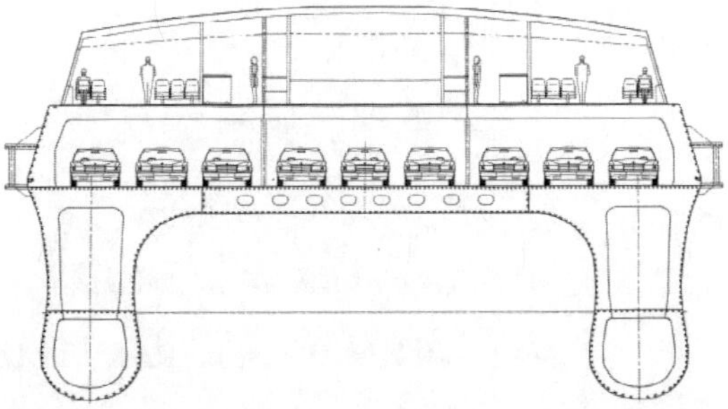

Figure 1. NQEA-Danyard Seajet passenger-vehicle catamaran with midship section

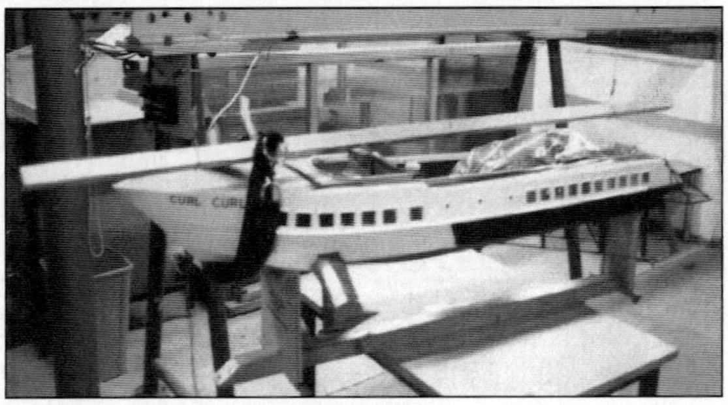

Figure 2. RHS140 hydrofoil and corresponding model for seakeeping tests at AMC

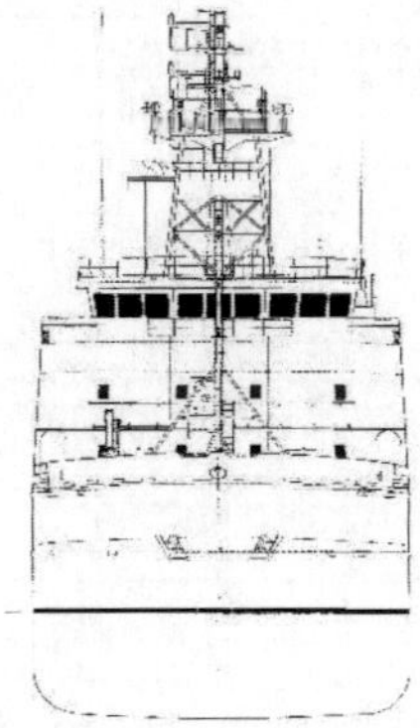

Figure 3. RAN Leeuwin Class Hydrographic Ships and corresponding frontal profile

Figure 4. HMAS Sirius (RAN photo) and corresponding section profile

Table 1: Radius of Gyration estimates for various Vessels

Vessel	k_4 (% B)	k_5 (% L_{WL})	k_6 (% L_{WL})	Notes
Arleigh Burke class Destroyer	38.9	25.2	25.1	via SAWE
Oliver Hazard Perry class Frigate	36.2	24.4	24.3	via SAWE
Ticonderoga class Cruiser	40.4	25.3	25.2	via SAWE
Wasp class LHD	42.0	25.6	25.6	via SAWE
San Antonio class LPD	40.5	23.8	23.8	via SAWE
Avenger class MCM	38.1	24.3	24.4	via SAWE
Mean of TAGOS19 & 23 SWATH	43.5	29.1	31.1	via SAWE
Nuclear submarines surfaced	36.6	24.1	24.1	via SAWE
Nuclear submarines submerged	34.8	26.0	26.0	via SAWE
Leeuwin Class Hydrographic Ship	38.8	24.3	21.7	Mean of Full Load Departure & Intermediate
RHS140 Surface Piercing Hydrofoil	40.1	28.9	29.1	Full Load
HMAS Sirius - Replenishment Ship	29.7	-	-	Full Load Departure

ESTIMATING RADIUS OF GYRATION OF SHIPS

For monohull ships, roll, pitch and yaw radius of gyration can be estimated, where appropriate, from the rules of thumb as already noted.

In the case of pitch and yaw radius of gyration, more accurate ship specific values can also be derived easily from the longitudinal weight distribution curves that are also required for assessing the longitudinal strength of larger ships. In this case, the pitch moment of inertia (I_{55}) can be calculated by summing the contributions of all the strips of weight over the length of the ship:

$$I_{55} = \Sigma\,(I_{55\ OWN\ i}) + \Sigma\,(\,m_i\,.\,(LCG_i^2 + VCG_i^2)) - m.(LCG^2 + VCG^2\,) \qquad (5)$$

In this equation, the first term represents the pitch moment of inertia of each mass item about its own centre of gravity, the second term represents the pitch inertia of the item about the origin for LCG and VCG definition while the final term corrects the total inertia to be relative to the overall CG location of the ship. This can easily be implemented in a weight estimating spreadsheet. In turn the 'own inertia' of each strip can be approximated by the following equation assuming the strip weight is spread uniformly across a prism of a given length and height. The former corresponding to the length of each strip in the weight distribution and the latter can be taken as the depth of the hull at that section:

$$I_{55\ OWN\ i} = m_i\,.\,(L_i^2 + H_i^2)/12 \qquad (6)$$

As the VCG of the individual strips will not be calculated as part of a longitudinal weight estimate, a reasonable approximation of the inertia will be obtained from the following simplified formula if the length of the ship is large compared to its depth:

$$I_{55} = \Sigma\,(I_{55\ OWN\ i}) + \Sigma\,(m_i\,.\,LCG_i^2) - m.LCG^2 \qquad (7)$$

The Pitch Radius of Gyration is then simply:

$$k_5 = (I_{55}/m)^{0.5} \qquad (8)$$

In cases of patrol boats and frigates where pitch radius of gyration has been calculated in this manner based on the longitudinal weight distribution, our experience is that this invariably remains close to the 25% L_{WL} guidance.

As a first approximation, yaw moment of inertia (I_{66}) and hence the yaw radius of gyration for long slender ships can be considered to be equal to that of pitch.

The Roll Moment of Inertia (I_{44}) and corresponding Radius of Gyration (k_4) cannot be so readily estimated from ship weight estimates. In this case, it becomes necessary to expand the weight estimating spreadsheet to explicitly define weight elements in their correct relative transverse locations. At this stage, it becomes worthwhile to directly estimate not only the roll, but also the pitch and yaw moments of inertia as part of the same process. Assuming all weight items can be idealized as rectangular prisms of uniform density, the weight estimating spreadsheet needs to be expanded to include the definition of not only the mass, VCG, LCG and TCG of items, but also their effective length, width and height. Items such as engines, if not on the centreline, must be individually defined, as must, for example, port and starboard side shell plating. The weight estimate therefore becomes significantly longer, though not substantially more effort to prepare. As well as the pitch moment of inertia listed above, the roll and yaw moments of inertia become:

$$I_{44} = \Sigma\,(I_{44\ OWN\ i}) + \Sigma\,(m_i.(VCG_i^2 + TCG_i^2)) - m.(VCG^2 + TCG^2) \qquad (9)$$

$$I_{66} = \Sigma\,(I_{66\ OWN\ i}) + \Sigma\,(m_i.(LCG_i^2 + TCG_i^2)) - m.(LCG^2 + TCG^2) \quad (10)$$

where:

$$I_{44\ OWN\ i} = m_i.(W_i^2 + H_i^2)/12 \quad (11)$$

$$I_{66\ OWN\ i} = m_i.(L_i^2 + W_i^2)/12 \quad (12)$$

and Roll and Yaw Radius of Gyration are:

$$k_4 = (I_{44}/m)^{0.5} \quad (13)$$

$$k_6 = (I_{66}/m)^{0.5} \quad (14)$$

As the weight estimating spreadsheet is simultaneously used to calculate the centre of gravity location, care needs to be taken to ensure the item TCGs maintain an appropriate sign convention for items to port or starboard such that the total TCG is correctly calculated. For a laterally symmetric ship, TCG should be zero. Discrete items such as seating, passengers or vehicles on deck do not need to be separately itemized. They can be treated as a group provided their distribution reasonably represents a prism of uniform density. So, for example, seating in any saloon on either side of an aisle can be treated as one uniformly distributed mass.

DETERMINING ROLL RADIUS OF GYRATION FROM SALLY TEST

Once a ship is built, the roll radius of gyration can also be derived from measurements of the roll period of the ship in still water when subject to an initial roll disturbance such as through sallying tests in parallel with inclining experiments. However, care needs to be taken to avoid the condition of such a test from influencing the result such as due to blockage effects in a dry dock and the radius of gyration derived in this manner needs to be recognized as being a 'wet' rather than 'dry' radius of gyration [12]. This approach is not further reviewed in this paper.

FLUIDS IN TANKS

The preceding discussion considered the calculation of radius of gyration when all items on the ship are considered as rigid prismatic bodies. However many ships, such as tankers and replenishment ships have fluid cargoes that make up a significant portion of their total displacement. As noted in [1], fluids in tanks do not behave as a rigid body, even when tanks are pressed full. To account for the reduced inertia of the fluid in large cargo or fuel tanks, the calculations presented in [13] as generally confirmed experimentally in [1] can be applied to computations of moments of inertia and radius of gyration. The approach proposed is to correct the moment of inertia of the fluid in the tank about its own CG by the ratio of fluid to solid inertia as calculated analytically for a range of tank cross section shaped [13]. The equations implemented in a spreadsheet calculation therefore become (for the roll inertia case):

$$I_{44} = \Sigma\,(f.\,I_{44\ OWN\ SOLID\ i}) + \Sigma\,(m_i.(VCG_i^2 + TCG_i^2)) - m.(VCG^2 + TCG^2) \quad (15)$$

where the factor f is the ratio of liquid to solid inertia, and can be estimated from polynomial trend lines derived from [13].

For rectangular tanks:

$$f = -2.299.(h/b)^4 + 6.2738.(h/b)^3 - 4.8795.(h/b)^2 + 0.0604.(h/b) + 1 \quad (16)$$

For octagonal tanks:

$$f = -2.6857.(h/b)^4 + 7.4307.(h/b)^3 - 5.8909.(h/b)^2 + 0.1621.(h/b) + 1 \quad (17)$$

and the solid inertia of octagonal form tanks is:

$$I_{44\ OWN\ i} = 0.0661.m_i.(b_i^2 + h_i^2) \quad \text{(approx.)} \quad (18)$$

For elliptical tanks:

$$f = -2.7588.(h/b)^4 + 7.5947.(h/b)^3 - 5.9947.(h/b)^2 + 0.1586.(h/b) + 1 \quad (19)$$

and the solid inertia of elliptical tanks is:

$$I_{44\ OWN\ i} = m_i.(b_i^2 + h_i^2)/16 \quad (20)$$

where in all cases h/b represents the ratio of height to breadth or breadth to height of a tank, whichever is the lowest.

These values of f, for varying h/b or b/h values are illustrated graphically in Figure 5:

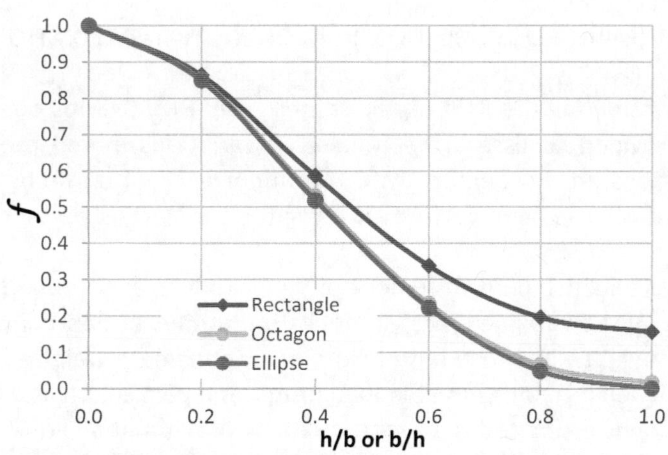

Figure 3. Ratio of Liquid to Solid Inertia as function of h/b from [13].

In cases where tank cross section shapes fall between these geometry cases, interpolation based on judgement is suggested.

It is not anticipated that it will be necessary to account for liquid effects when estimating the pitch and yaw moments of inertia and radius of gyration as the lever arms between the ship centre of gravity and the CG of the individual weight items will dominate. However, the same principles as for roll would also apply.

INFLUENCE OF VARIATION IN ROLL RADIUS OF GYRATION ON MOTION PREDICTIONS

Given the additional effort required to estimate the roll radius of gyration, a natural question is how much does this estimate, and more specifically the assumption made regarding tank fluid inertia (solid vs fluid) have on the transverse motions of a ship, especially where fluid cargo makes up a large proportion of the total displacement?

One published example is the case of the modification of the USS Midway [12]. When blisters were proposed to be added to the ship to improve its survivability as part of a modification program, seakeeping analysis was undertaken to assess the motion characteristics post modification. However, using the normal rules of thumb, the input to the analysis had assumed the roll moment of inertia of the ship would increase in proportion to the increased beam of the hull due to the addition of the blisters. In practice, the radius of gyration increased only due to the mass distribution of the blisters themselves. A compounding error was that the 'wet' roll radius of gyration was input where the program was expecting a 'dry' radius of gyration. The seakeeping assessment didn't identify any issue of concern for the modification. However, the roll inertia of the modified ship was in fact significantly lower than input. The result was that the roll stiffness of the ship was greater than anticipated and this compromised flying operations as described in [12]. This example highlights the need to more accurately determine the roll radius of gyration of ships when undertaking seakeeping analysis.

CASE STUDY OF IMPACT OF VARIATION IN k_4 ON MOTION PREDICTIONS

A case study considering the RAN replenishment ship HMAS Sirius operating in a full load condition is provided to assess the impact of variations in roll radius of gyration on motion predictions and to demonstrate the importance of correctly estimating k_4. The motion predictions have been performed using the SHIPMO7 strip theory code [2].

HMAS Sirius has a lightship displacement of 9,464 tonnes and a full load displacement of 44,969 tonnes. Of the variable loads, some 30,360 tonnes or approximately 67% is fuel in cargo tanks. The lightship k_4 estimated from an approximate weight distribution is 38.4% of ship beam. When including variable load items, and particularly the cargo fuel load, the radius of gyration is estimated to vary from 28.7% of beam if all liquid loads are assumed to have no roll inertia about their own CG through to 32.9% of beam if tank liquids are all treated as being solid. A summary of cases examined is provided in Table 3. The largest k_4 value assessed, 40% of beam, corresponds with the upper bound of rule of thumb guidance.

Table 3: Radius of Gyration Estimates for HMAS Sirius case

Load case	Radius of Gyration as % of Beam	Notes
Lightship	38.4	From approximate mass distribution
Full Load with no liquid inertia about own CG included	28.7	Combining lightship with variable load distribution
Full Load with some allowance for liquid inertia about own CG included	29.1	As above, but implementing method in this paper assuming cargo tanks as octagonal cross section
Full Load with realistic liquid inertia about own CG included	29.7	As above, but implementing method in this paper assuming cargo tanks as rectangular cross section
Full Load with liquid loads treated as solids	32.9	This will over-estimate the roll inertia of the ship and result in a predicted roll natural period that is too large
SHIPMO7 dry RoG ShipMo3D dry RoG	32 – 36 40	Guidance in manuals for 'small warships'

The cross section geometry of the cargo fuel tanks of HMAS Sirius is shown in Figure 5. The centerline bulkhead is a corrugated arrangement to avoid the need for internal stiffeners within the tanks. It is apparent that these tanks are not entirely rectangular in cross section. However, they also fall short of being octagonal in section (Figure 5). The fluid inertia characteristics of the tanks could therefore be expected to lie between those predicted for a rectangular and an octagonal cross section tank.

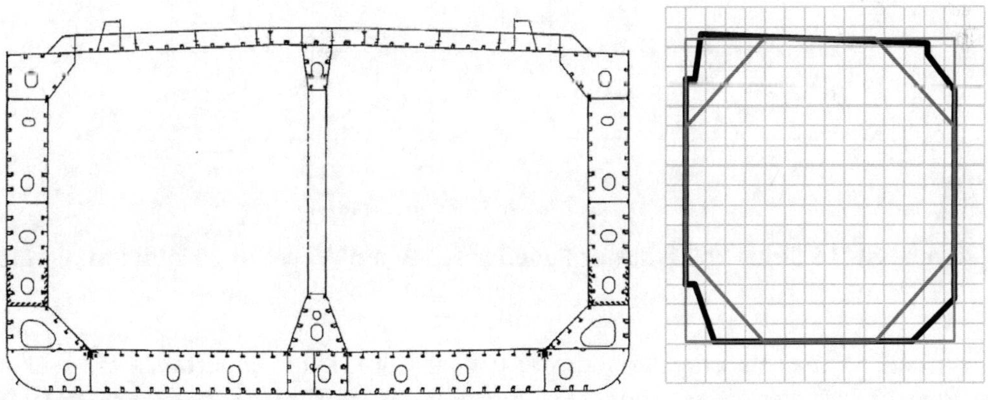

Figure 5. HMAS Sirius Midship Section (left) and Cross Section of HMAS Sirius cargo tank in comparison with rectangular (red) and octagonal (green) tank configurations (right).

Motion predictions are only presented for a narrow selection of conditions in regular and irregular seas. While HMAS Sirius as a modified commercial product tanker, has a relatively full form that could be expected to have reasonable wave damping for larger wave frequencies, it has only relatively small span bilge keels of approximately 0.26m width. As SHIPMO7 may over-predict roll response at wave frequencies near resonance for cases with little roll damping, the width of the bilge keels was doubled for the purposes of presenting the following comparisons.

Roll transfer functions for 10 knots in beam seas are shown in Figure 6. The significant variation in roll natural period due to only a variation in k_4 is clear. The lowest roll natural period of approximately 12.4s corresponds to the assumption of no inertia of the tank fluid about its own centre of gravity and the largest period of approximately 16.0s equates to k_4 assumed to be 0.4B. The predicted peak at resonance also increases with an increase in k_4. It is anticipated that this in part is due to the relative reduction in roll damping for the slower roll motions associated with larger roll natural periods. Notwithstanding that the width of the bilge keels has been doubled for the SHIPMO7 computations when compared to the actual ship, the predicted large resonant roll peak when k_4 is assumed to be 0.4B warrants validation against model tests.

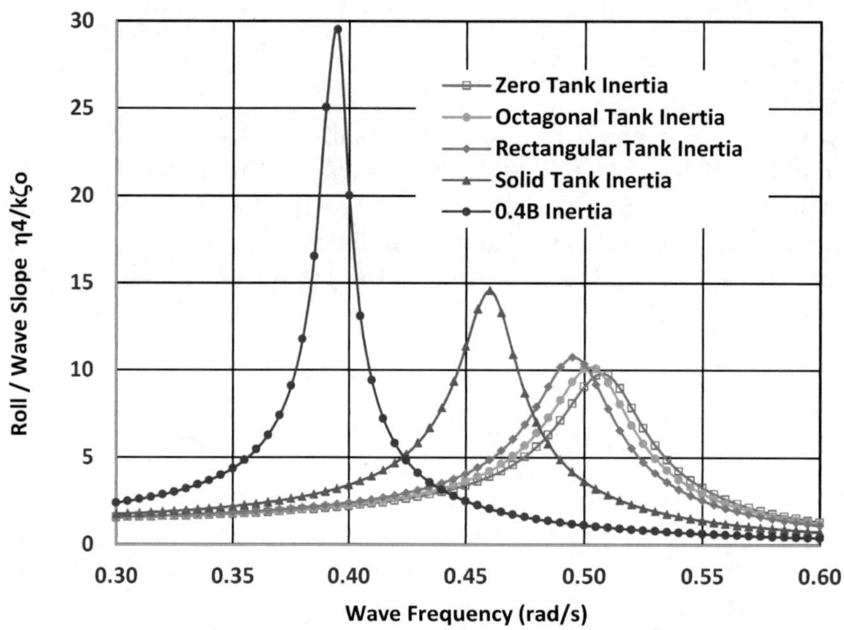

Figure 6. HMAS Sirius Roll Transfer Functions in Beam Waves at 10 knots for various k_4

As motions in realistic irregular seas are typically of greater importance for seakeeping assessments, a single seaway with a Significant Wave Height (H_S) of 4m and Modal Period (T_0) of 12.3s was examined. This combination is representative of average conditions that could exist at the top of Sea State 5 in ocean areas around Australia. A unidirectional Bretschneider sea spectrum has been assumed. While the peak of the wave amplitude spectrum of this sea corresponds approximately with the peaks of the roll transfer functions for the cases that account for the fluid in the tanks, the peak of the wave slope spectrum for this sea is at a much shorter period.

First, considering the SSA roll (Figure 7), this remains relatively consistent in beam seas regardless of the assumption for k_4, with the main exception being the largest k_4 case of 0.4B. However in stern quartering seas (30-60 degrees relative heading), the assumption of treating fluid cargo as solid is also shown to introduce significant errors in the predictions. The larger roll response in following seas for the cases with larger k_4 may related to the wave encounter frequency more closely matching the roll natural frequency of the ship in those conditions.

The SSA Lateral Force Estimator (LFE) at the bridge for the same seas are comparable in character to the SSA roll predictions and as such are not presented separately. However since lateral accelerations are governed by both the roll amplitude and the roll frequency then as is apparent from Figure 6, on the one hand larger values of k_4 will reduce the roll natural frequency and hence also reduce LFE. This is counterbalanced by the increase in roll response amplitude near resonance for larger k_4 which increases LFE.

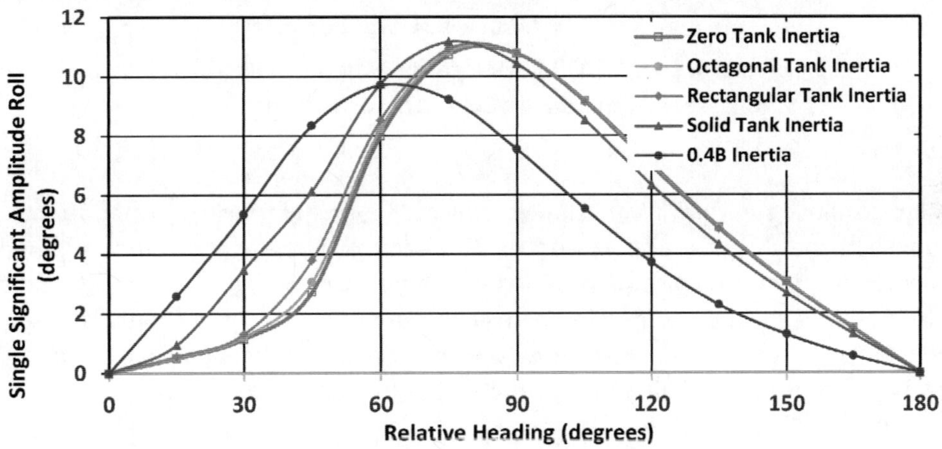

Figure 7. SSA Roll in H_S = 4.0m & T_0 = 12.3s Beam Seas at 10 knots for various k_4

A more direct seakeeping measure than LFE is the rate of lateral Motion Induced Interruptions (MII) of a crew member standing facing forward (or aft) on the bridge for the same seas (Figure 8). While the assumptions for k_4 based on either ignoring the inertia of the cargo tank fluid, treating the tanks as octagonal or rectangular all result in similar MII predictions across the range of headings, the predictions begin to deviate significantly once all the fluid cargo is considered to be solid. The rule of thumb k_4 value of 0.4B is seen to vary MII predictions even more substantially compared to the more accurate k_4 predictions. As noted earlier, with this assumed large roll radius of gyration, the roll natural period is increased and so the roll motions are 'softer' and somewhat slower, even for a fixed wave modal period, thus leading to reduced MII levels.

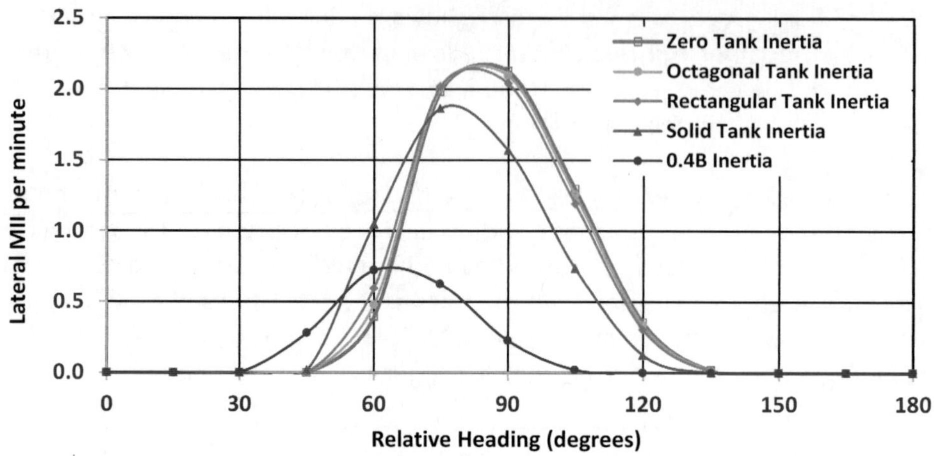

Figure 8. Lateral MII in Bridge for H_S = 4.0m and T_0 = 12.3s Beam Seas at 10 knots for various k_4

Finally, to illustrate that a variation in roll radius of gyration not only affects the roll response of the ship, but also influences other coupled degrees of freedom of motions, a yaw transfer function is also provide for the same cases in Figure 9. For consistency with Figure 6, this is again presented in beam seas, even though yaw response of a ship is only modest in such seas. While ships have no yaw stiffness and hence no yaw natural periods, the yaw response of the ship tends to mimic the roll response.

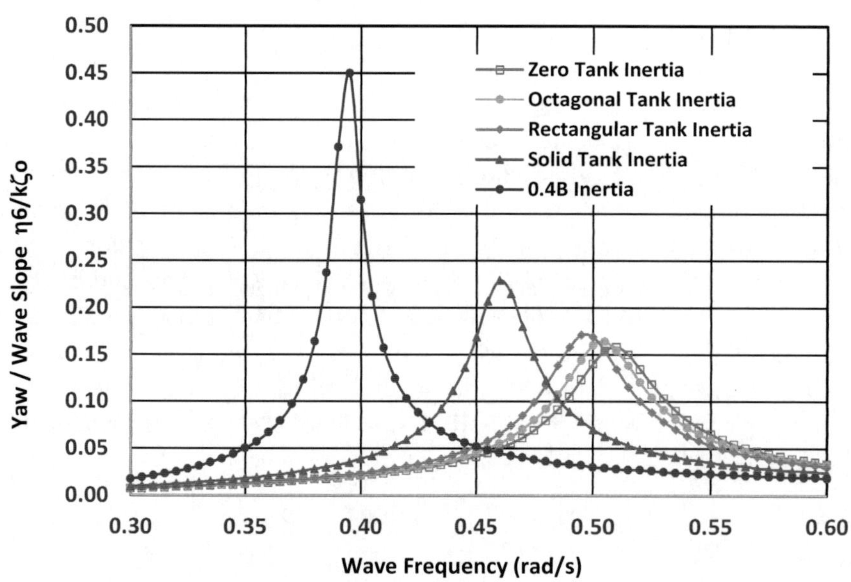

Figure 9. Yaw Transfer Functions in Beam Waves at 10 knots for various k_4

CONCLUSION

Aside from reviewing rule of thumb equations, a method has been presented for the calculation of radius of gyration values for ships based on adaptation of ship weight estimates. A methodology to also account for the inertia of fluids in tanks is also provided.

A case study demonstrating how variations in estimates of roll radius of gyration affect the lateral plane motions of a ship is provided and this shows the importance of correctly estimating the roll radius of gyration in order to generate reliable ship motion predictions.

ACKNOWLEDGEMENT

This paper is an extension of work originally undertaken in the early 1990's by the lead author while on secondment to NQEA Australia Pty Ltd. It also consolidates discussions with colleagues within Navy Technical Bureau, Seastate Pty Ltd and with UNSW Sydney and Canberra staff on derivation of radius of gyration. The ongoing assistance of Kevin McTaggart in providing advice concerning the use of SHIPMO7 and the support of colleagues at the Defence Science and Technology Group in providing validation data is also acknowledged. The work by Darris Fortescue on one aspect of this subject served as the motivation to prepare this overarching discussion.

REFERENCES

[1] Fortescue, Darris; Grimm, Martin; Smith, Warren; "Experiments investigating the Mass Moment of Inertia of Fluids in Roll", Pacific 2017 International Maritime Conference, Sydney, October 2017.

[2] McTaggart, Kevin; "SHIPMO7: An Updated Strip Theory Program for Predicting Ship Motions and Sea Loads in Waves", DREA Technical Memorandum 96/243, March 1997.

[3] McTaggart, Kevin; "ShipMo3D Version 1.0 User Manual for Simulating Time Domain Motions of a Freely Maneuvering Ship in a Seaway", DRDC Atlantic Technical Memorandum 2007-172, October 2007.

[4] De Kat, J.O; Oomen, A.C.W.J; van Daalen, E.F.G; van't Veer, A.P; "FREDYN: A Computer Program for the Simulation of a Steered Ship in Extreme Seas and Wind - Version 9.0 / 9.1 - Theory Manual", Marin, February 2003.

[5] Lewis, Edward V; Principles of Naval Architecture, 2nd Revision, Volume 3, Motions in Waves and Controllability, SNAME, 1989.

[6] Society of Allied Weight Engineers, Inc, "Weight Estimating and Margin Manual for Marine Vehicles", Recommended Practice Number 14, 22 May 2001.

[7] Lloyd, A.R.J.M; "Seakeeping - Ship Behaviour in Rough Weather", Ellis Horwood Ltd, 1989.

[8] Lloyd, A.R.J.M; “Seakeeping – Ship Behaviour in Rough Weather”, A.R.J.M Lloyd, 1998.

[9] Marchaj, C.A; “Seaworthiness – The Forgotten Factor”, Adlard Coles Ltd, 1986.

[10] NAVSEA; “Interface Standard for Shipboard Systems – Ship Motion and Attitude (Metric)”, DOD-STD-1399(NAVY) Section 301A, 21 July 1986.

[11] IMO MSC/Circ.707, “Guidance to the Master for Avoiding Dangerous Situations in Following and Quartering Seas”, 19 October 1995.

[12] Ricketts, Myron V; Gale, Peter A; “On Motions, Wetness, and Such: The USS Midway Blister Story”, SNAME Transactions, Vol. 97, 1989, pp. 429-472.

[13] Lee, Gyeong Joong; "Moment of Inertia of Liquid in a Tank", International Journal of Naval Architecture and Ocean Engineering, 2014 (http://dx.doi.org/10.2478/IJNAOE-2013-0168).

Ship Roll Reduction Using Water-flow Induced Coriolis Effect

Mario P. Walker and Masaaki Okuma
Tokyo Institute of Technology
walker.m.aa@m.titech.ac.jp

ABSTRACT

Ships are subjected to motions which can disrupt on-board operations and damage equipment. Roll motion in particular is of great importance due to low damping conditions which may lead to capsizing. Several techniques have been investigated to reduce rolling. These include anti-roll tanks, fin stabilisers and bilge keels. However these systems are not without their challenges. For example water-flow in anti-roll tanks creates complications and for fin stabilisers and bilge keels, an extremely large size is required to produce any significant damping creating operational challenges. This paper proposes and investigates a method to reduce rolling by inducing Coriolis effect using water-flow in the radial direction. The system is modelled with two degrees of freedom, one degree for parametric roll of the ship and one degree for the water-flow. Parameters from a Ro-Ro vessel is used as these are prone to capsizing and extensive research have been conducted on these over the years. The vessel is investigated under free roll decay and excited conditions by beam waves. The analysis is conducted under various mass ratios and velocity of water-flow. The proposed system is found to produce substantial damping and roll reduction which increases with increase in any of the parameters. The proposed system provides a simple approach to reduce ship rolling. Water-flow control is fairly simple as the water flows in only one direction with a constant velocity. This is a starting point for designing an effective and practical system. For this to be a viable approach further investigations are needed to address challenges that present themselves.

1. INTRODUCTION

Ships and other ocean-going vessels are subjected to motions which can disrupt on-board operations and equipment, damage machinery and can also reduce the efficiency of the propulsion system. Important among these motions is the roll motion, as it is a major contributing factor to capsizing. Additionally, low damping conditions during rolling makes it important to address for ship safety and operation and this motion can become severe even in relatively moderate weather conditions without any apparent pre-warning for the crew onboard [1]. In [2] it was highlighted that transverse accelerations from roll motion interrupts the on-board crew performance and in severe cases will completely prevent performance. Roll motion can also cause sea sickness, cargo damage and limit the capability to handle equipment on board which can be detrimental to naval vessels performing weapon operations, launching or recovering systems, landing airborne systems and sonar operations [3, 4].

Researchers have developed several methods over the years to minimise ship motions. These systems are generally classified into two categories: internal and external systems, and can be of a passive or active nature [5]. Furthermore these systems operate in one of two ways, either the system manipulates the centre of gravity of the ship in order to provide some restoring moment or the system generates the restoring force itself and apply directly to the unstable vessel. Common internal systems include anti-roll tanks and gyroscopic systems whereas common external systems include bilge keels, fin stabilisers and rudder control. These systems will be described in more details in the next section.

However, these systems are not without their challenges. For example water flow in anti-roll tanks creates complications. Additionally for most internal systems there are the issues of weight and volume [5]. For external systems like bilge keels and fin stabilisers, an extremely large size is required to produce any significant damping creating operational challenges. They are also ineffective at zero forward speed [6, 7].

This paper proposes and investigates a method to reduce rolling by inducing Coriolis effect using water-flow in the radial direction of the vessel. When structures undergo rotational motion, any coupled motion in the radial direction induces Coriolis effect which may be used to reduce the rotation of the structure.

The system is modelled with two degrees of freedom. The parametric roll of the vessel is modelled as a single-degree-of-freedom (SDOF) system while the water-flow provides the second degree of freedom of the system. The equations of motion for the coupled effect of the rolling of the ship and the flow of the water is derived. Numerical simulations are carried out with the equations of the coupled motion to study the behaviour of the ship roll motion and damping effect and for comparison the approximated analytical results for a linear system are presented. The problem as presented should be considered as a preliminary into developing a strategy for active control of ship roll motion using flow induced Coriolis effect. This concept provides a simple approach to reduce ship rolling. Water-flow control is very simple as the water flows in only one direction with constant velocity.

2. ROLL REDUCTION SYSTEMS

Since the 1800s researchers have been investigating roll-damping devices. In 1861, William Froude studied the effect of a water-filled tank on the roll motion of a ship which led to the proposal of anti-roll tank systems. Since then researchers have proposed and implemented several devices and system to reduce the rolling of vessels. This section presents common approaches to date to reduce the ship rolling and highlights the advantages and disadvantages of such systems.

2.1. Internal Systems

Internal stabilisaton systems are usually weight-based systems and work on the principle of generating moments to counteract the undesired (roll) motion by carefully manipulating weight within the vessel. Common approaches are anti-roll tanks and gyroscopic stabilisers.

2.1.1. Anti-roll Tanks

The most common type of anti-roll tank in use is the U-tube tank which consists of two reservoirs connected by a water duct. Anti-roll tanks reduce ship rolling via motion of fluid within the tank which has the same period of the ship motion but a quarter of period behind the rolling motion, as a result the fluid motion produces a counteractive moment [2]. These systems can either be passive or active through valves and pumps. The principal advantages of this system are there are no hydrodynamic drag effect and they are effective at zero forward speed [5]. However the disadvantages are they occupy large spaces and produce weight issues, as well as stability issues due to sloshing effects [2, 5].

2.1.2. Gyroscopic Stabilisers

Gyroscopic stabilisers consist of a large rotating flywheel which generates moments via its inertial property to eliminate rolling of the vessel. This method is used to a lesser extent in comparison to other methods. The principal advantages of this system type are like anti-roll tanks they provide no hydrodynamic drag effect and operate at zero forward speed [5]. However the drawbacks are they increase the weight of the vessels, produce large stress on the hull and has limited stabilisation capability [2].

2.2. External Systems

External stabilisation systems generally work on the principle of hydrodynamic interactions to generate forces and moments outside the hull of the ship to reduce the undesired ship roll motion [5]. Common approaches are bilge keels, fins and rudder control.

2.2.1. Bilge Keels

Bilge keels are simply just long narrow ridges mounted to the bilge which generate drag forces that act perpendicular to them to counteract ship roll motion. These devices are passive in nature. The principal advantages are they can still have some effect at low speeds (generally in the range of 10-20% roll reduction) [8], they are lightweight and provide no space occupancy [5]. However, they create hydrodynamic drag, are exposed to elements in the water so vulnerable to damage, not appropriate for all ships and have dependence on forward speed, hence ineffective at zero forward speed [2].

2.2.2. Fin Stabilisers

Fin stabilisers consist of a pair of hydrofoils mounted on the side of the ship's hull which can be either passive or actively controlled, to generate hydrodynamic lift to counteract the ship roll motion. The principal operational advantages are they have high performance roll reduction and easy control system design. However the drawbacks are they create hydrodynamic drag and are ineffective at low and zero forward speeds. Additionally, for effective performance they require control system with sensors and powerful hydraulic actuators [2]. They are also exposed to elements in the water, including other vessels passing by so vulnerable to damage, creates the possibility of underwater noise pollution which can affect sonar systems and in severe conditions may cause dynamic stall [2].

2.2.3. Rudder Control

The ship's rudder can be used to mitigate roll motions by creating short rudder impulses [1]. This is an active control technique and is usually a feature of autopilot. The advantages of this type of system are they have medium-high performance, they do not occupy large spaces, require no physical modifications to vessels and can be combined with other stabilisers to increase performance. However, they are Ineffective at low and zero forward speeds, create hydrodynamic drag and for high performance requires sophisticated control systems as well as rudder machinery up-grade [2].

2.3. The Proposed Concept

The system investigated in this work is an active control system where water flows in the radial direction through some channel of the ship to generate restoring forces to mitigate ship roll motion.

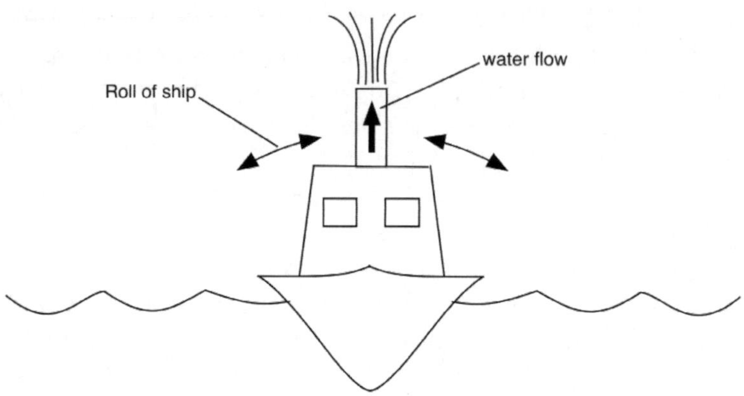

Fig.1 Simple illustration of proposed stabilising system using water flow

Since the vessel will be in water, sea openings can be placed on areas immersed in water (for example the base of the hull) which will allow water in to be pumped into the stabilising system. The advantages of such type system are no complicated control is necessary as the water is simply pumped in one direction with a constant velocity, they create no hydrodynamic drag effect, operate at zero forward speed, provide no weight issues, no stability issues as the water flow will always generate positive damping effect and have no dependency on any external parameters other than the volume of water flow and the degree of roll.

3. PROBLEM FORMULATION

In this section the equations of the coupled motions of rolling of a ship and fluid flow through the proposed stabilising system are highlighted.

3.1. Parametric Roll of Ship

The parametric roll of a ship can be modelled as a single-degree-of-freedom (SDOF) system as given by the popular equations:

Free Roll:

$$(I_{44} + A_{44})\ddot{\phi} + B_{44}\dot{\phi} + C_{44}\sin\phi = 0 \tag{1}$$

Excited Roll:

$$(I_{44} + A_{44})\ddot{\phi} + B_{44}\dot{\phi} + C_{44}\sin\phi = M_w \tag{2}$$

where I_{44} is ship inertia, A_{44} is added mass, B_{44} is the damping in the rolling direction and C_{44} is the restoring moment. For simplicity it is very common to assume these parameters to be constant [9]. M_w is the external excitation by the waves. Also $(\dot{\ })$ represents a derivative with respect to time.

For the case of this study a Ro-Ro ship is considered to show the application of the proposed concept as these vessels are quite prone to capsizing. The parameters for this is given in Table 1.

Table 1. Principal of a Ro-Ro ship [10].

Principal particulars	**Symbol**	**Quantity**
Length overall	L_{OA}	192.60 m
Length between perpendiculars	L_{BP}	177.60 m
Beam	B	28.00 m
Depth	D	18.00 m
Draught	D	7.50 m
Displacement	Δ	22012 t
Transverse meta-centric height	GM	2.66 m
Vertical centre of gravity	KG	13.98 m
Vertical centre of buoyancy	KB	4.40 m
Block coefficient	C_b	0.60

In this paper the ship will be excited by beam waves modelled as harmonic excitation. Also the nonlinearity in the restoring moment and the damping are neglected.

3.2. Coupled Effect of Fluid Flow

We introduce the coupled effect due to water-flow induced Coriolis effect.

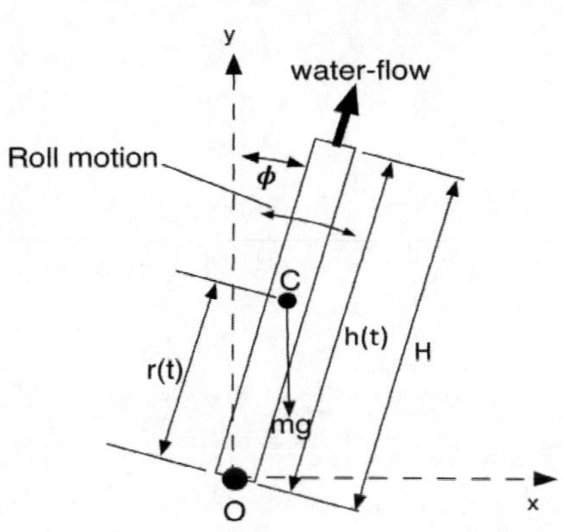

Fig. 2. Illustration of the water-flow coupled effect.

Figure 2 shows the water flowing through the channels while the ship experiences roll motion. We derive the equations of motion for the effects of the water-flow using Lagrangian mechanics.

Consider the water-flow with a mass m with pivot at O through the channel as shown in Figure 2. The centre of mass C is located at a distance $\mathrm{r(t)}$ from the pivot, where $\mathrm{r(t)}$ is half the height of the water-flow $\mathrm{h(t)}$. The set of generalised coordinates are the angular displacement ϕ of $\overline{OC}$ from the y-axis, and the radial height of the centre of mass C of the fluid from the pivot O.

The kinetic energy (T)

$$\mathrm{T} = \frac{1}{2}m\dot{r}^2 + \frac{1}{2}mr^2\dot{\phi}^2 \tag{3}$$

The potential energy (V)

$$\mathrm{V} = \mathrm{mgr}\cos\phi \tag{4}$$

Using Lagrangian and the Euler-Lagrange equation and considering only the ϕ coordinate since we are only concerned with the roll motion,

$$\mathrm{L} = \mathrm{T} - \mathrm{V} \tag{5}$$

$$\frac{d}{dt}\left(\frac{\partial L}{\partial \dot{\phi}}\right) - \frac{\partial L}{\partial \phi} = 0 \tag{6}$$

This gives

$$\mathrm{m}r^2\ddot{\phi} + 2mr\dot{\phi}\phi - mgr\sin\phi \tag{7}$$

Therefore Equation (1) for free roll becomes

$$(I_{44} + A_{44} + J)\ddot{\phi} + (B_{44} + 2mr\dot{r})\dot{\phi} + (C_{44} - mgr)\sin\phi = 0 \tag{8}$$

and Equation (2) for excited roll becomes

$$(I_{44} + A_{44} + J)\ddot{\phi} + (B_{44} + 2mr\dot{r})\dot{\phi} + (C_{44} - mgr)\sin\phi = M_w \tag{9}$$

where $\mathrm{J} = \mathrm{m}r^2$ is the inertia of the water-flow.

3.3. Non-dimensionalised Equations of Motion

We now present the equations of motion in a generalised non-dimensional form. We describe the variables as

$$\tau = \mathrm{t}\sqrt{\frac{C_{44}}{I_{44}+A_{44}+J}} \tag{10}$$

$$\dot{t} = \sqrt{\frac{C_{44}}{I_{44}+A_{44}+J}} \tag{11}$$

$$\bar{r} = \frac{r}{KG} \tag{12}$$

Therefore we get

$$\dot{r} = \left(KG\sqrt{\frac{C_{44}}{I_{44}+A_{44}+J}}\right)\frac{d\bar{r}}{d\tau} \tag{13}$$

$$\dot{\phi} = \left(\sqrt{\frac{C_{44}}{I_{44}+A_{44}+J}}\right)\frac{d\phi}{d\tau} \tag{14}$$

$$\ddot{\phi} = \left(\frac{C_{44}}{I_{44}+A_{44}+J}\right)\frac{d^2\phi}{d\tau^2} \tag{15}$$

Substituting Equations (13)-(15) into Equations (8) and (9), rearranging and letting

$$\alpha = \frac{1}{C_{44}} \quad (16)$$

we get

Free Roll:

$$\frac{d^2\phi}{d\tau^2} + \left(B_{44}\alpha\dot{t} + 2m\bar{r}\alpha\dot{t}KG^2\frac{d\bar{r}}{d\tau}\right)\frac{d\phi}{d\tau} + (1 - mg\bar{r}\alpha KG)\sin\phi = 0 \quad (17)$$

Excited Roll:

$$\frac{d^2\phi}{d\tau^2} + \left(B_{44}\alpha\dot{t} + 2m\bar{r}\alpha\dot{t}KG^2\frac{d\bar{r}}{d\tau}\right)\frac{d\phi}{d\tau} + (1 - mg\bar{r}\alpha KG)\sin\phi = M_w\alpha \quad (18)$$

Taylan [11] found the external excitation M_w as

$$M_w = \omega_e^2\alpha_m(I_{44} + A_{44})\cos(\omega_e t) \quad (19)$$

where ω_e is the frequency of encounter and α_m is the maximum wave slope.

To find ship inertia I_{44} we use Weiss formula

$$\mathrm{i} = \sqrt{\frac{I_{44}}{\Delta}} \approx 0.4B \quad (20)$$

Similar to [10] we assumed the added mass A_{44} to be 20% of ship inertia. B_{44} in this case is such that the ship has a typical hydrodynamic roll damping of 10% and $C_{44} = \Delta gGM$. It should be noted that for fluid flow we assumed laminar flow.

4. RESULTS AND DISCUSSION

In this section the results are presented. For the first assessment the vessel is investigated under free roll decay with and without the proposed stabilising system. The numerical analysis is carried out using the 4th Order Runge-Kutta Method and for comparison the approximated analytical results for the linearised model of the system is also presented. The system is analysed for the cases of mass ratios (μ) 0.01, 0.03 and 0.05 where

$$\mu = \frac{m}{\Delta} \quad (21)$$

and velocities of water-flow 20 ms^{-1}, 40 ms^{-1} and 60 ms^{-1}.

4.1. Free Roll Decay

The system is observed under free roll decay with and without the system. The vessel is initially inclined at an angle $\phi_0 = 60^o$. For comparison the nonlinear model is linearised and an approximate analytical solution is also presented using equation

$$\phi(\tau) = \phi_0 e^{-\zeta\omega\tau\sqrt{\frac{I_{44}+A_{44}+J}{C_{44}}}} \cos\left(\sqrt{1-\zeta^2}\omega\tau\sqrt{\frac{I_{44}+A_{44}+J}{C_{44}}} + \varphi\right) \quad (22)$$

where ζ and ω are the damping ratio and natural frequency respectively of the system, assuming constant water-flow height ($h(t) = h(max)$) and φ is the phase angle. Figures 3-5 show these results for the different cases of mass ratio and velocity.

From the results we see that the proposed stabilising system provides substantial damping effect which increases with the velocity and mass ratio of the water-flow. The system increased the damping ratio in the range of 0.140 - 0.720 for the smallest mass ratio and velocity to the largest mass ratio and velocity. Additionally we see that the numerical simulation is in good agreement with the linear analytical solution.

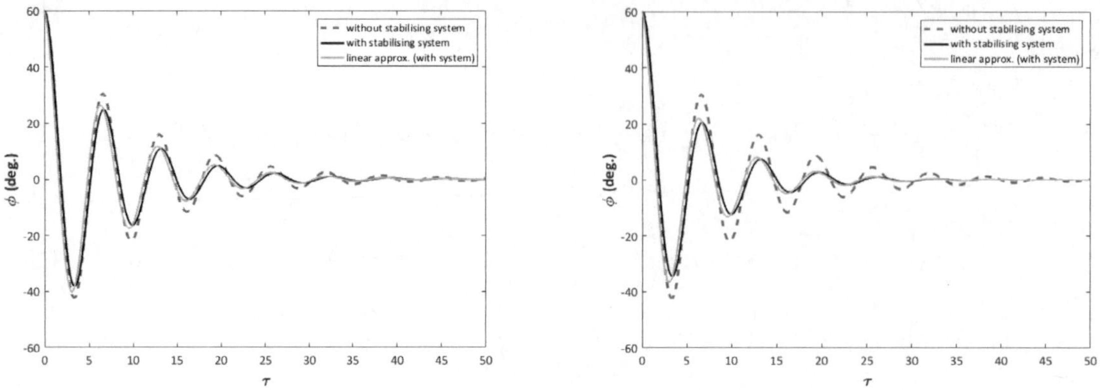

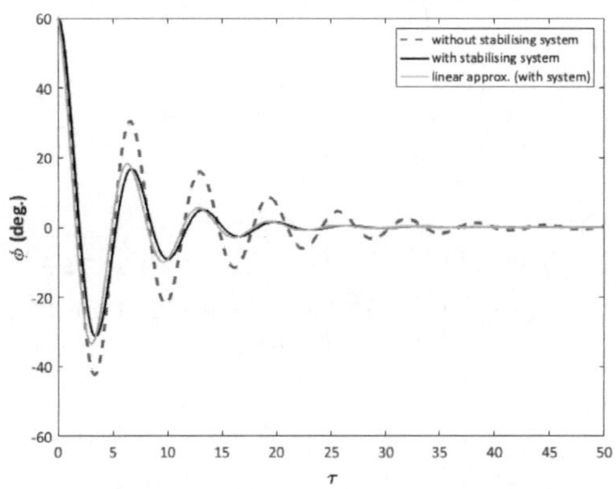

Fig. 3. Free roll decay with and without the stabilising system and comparison with linear approximation of the model with the stabilising system on for μ=0.01. Top left: velocity =20 ms^{-1}, top right: velocity = 40 ms^{-1} and bottom left: velocity = 60 ms^{-1}.

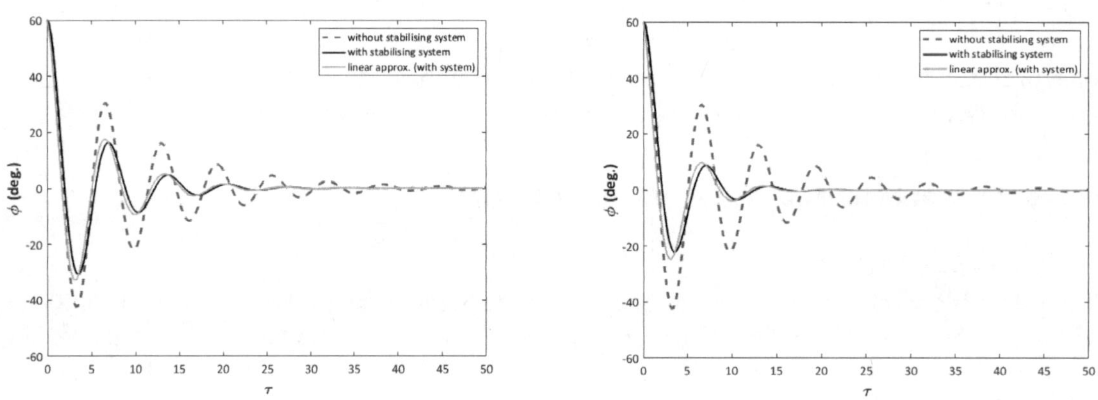

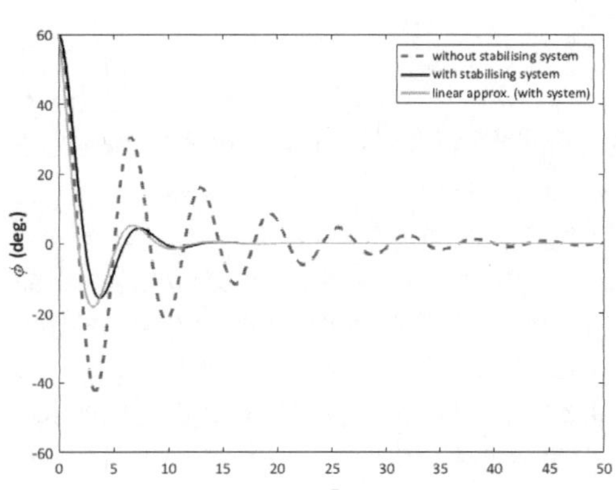

Fig. 4. Free roll decay with and without the stabilising system and comparison with linear approximation of the model with the stabilising system on for μ=0.03. Top left: velocity = 20 ms^{-1}, top right: velocity = 40 ms^{-1} and bottom left: velocity = 60 ms^{-1}.

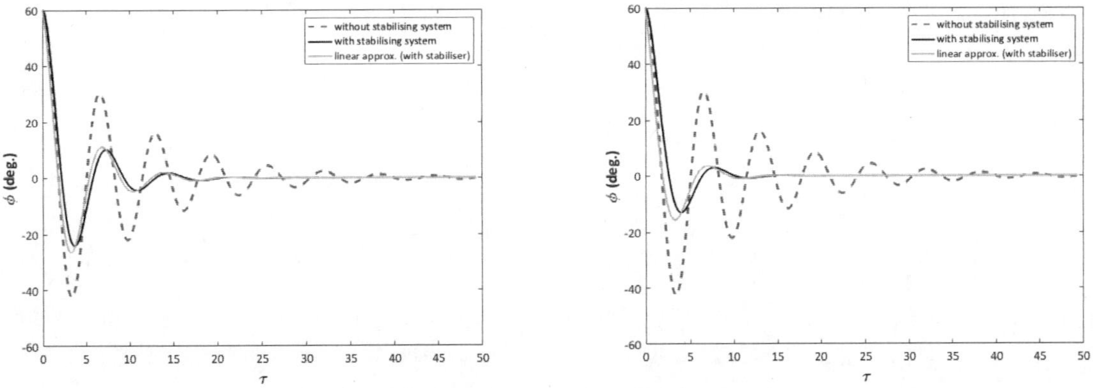

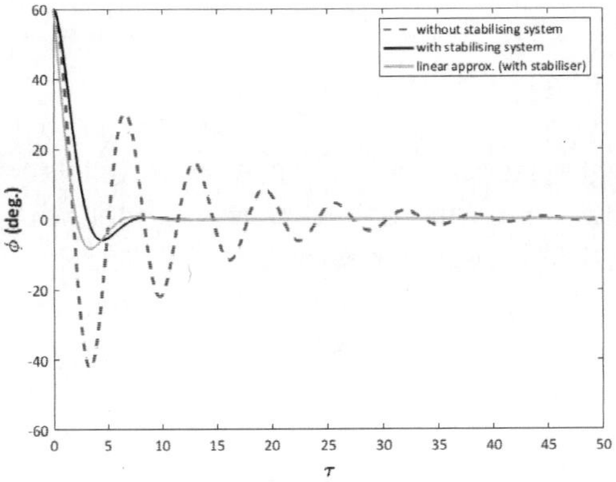

Fig. 5. Free roll decay with and without the stabilising system and comparison with linear approximation of the model with the stabilising system on for μ=0.05. Top left: velocity = 20 ms^{-1}, top right: velocity = 40 ms^{-1} and bottom left: velocity = 60 ms^{-1}.

4.2. Excited Roll Motion

The system is additionally observed under excited motion from beam waves modelled using Equation (19). For the entirety of the analysis we use maximum wave slope $\alpha_m = 1/30$ and wave encounter frequency (ω_e)

$$\omega_e = \sqrt{\frac{C_{44}-mgr}{I_{44}+A_{44}+J}} \tag{23}$$

The parameters of the mass ratio and velocity are similar to those of the free roll decay analysis and for comparison the approximated linear steady-state results are presented using

$$\phi(\tau) = \frac{M_w\alpha}{\sqrt{\left((1-mg\bar{r}\alpha KG)-\omega_e^2\right)^2+\left(B_{44}\alpha\dot{t}+2m\bar{r}KG^2\alpha\dot{t}^2\frac{d\bar{r}}{d\tau}\right)\omega_e^2}}\sin\left(\omega_e\tau\sqrt{\frac{I_{44}+A_{44}+J}{C_{44}}}-\varphi\right) \tag{24}$$

The results show a substantial reduction in roll motion with the stabilising system, in particular for the larger mass ratios and velocities (between 15.7% - 71.4% reduction).

Figures 6-8 show these results for the different cases of mass ratio and velocity.

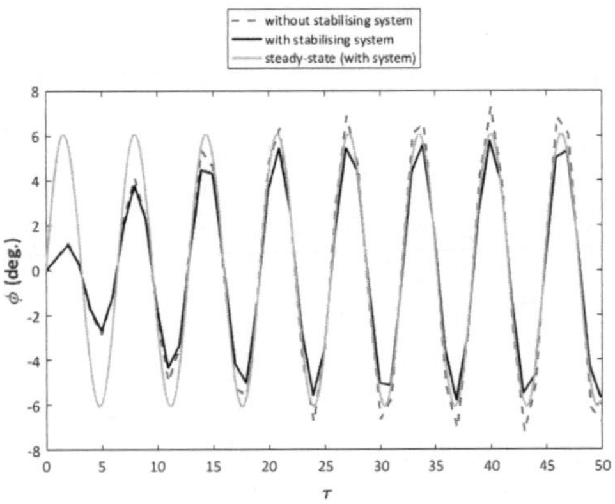

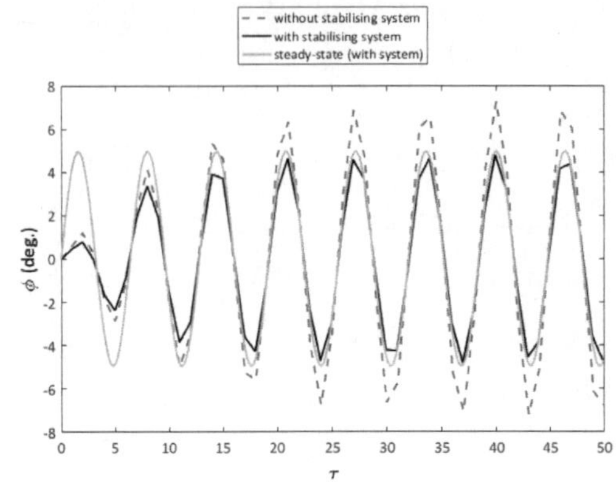

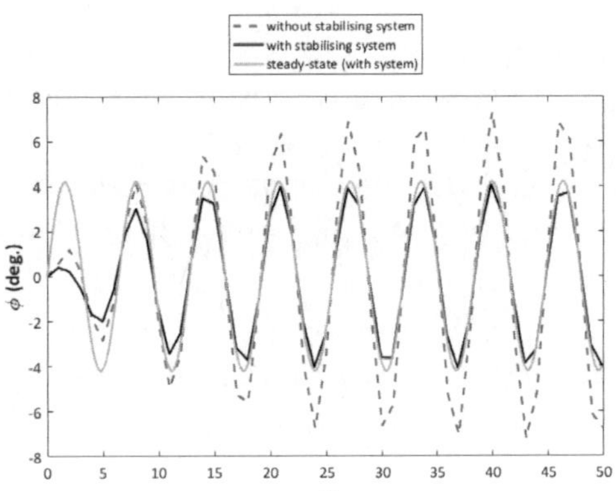

Fig. 6. Excited roll motion with and without the stabilising system and comparison with linear steady-state approximation of the model with the stabilising system on for μ=0.01. Top left: velocity = 20 ms^{-1}, top right: velocity = 40 ms^{-1} and bottom left: velocity = 60 ms^{-1}.

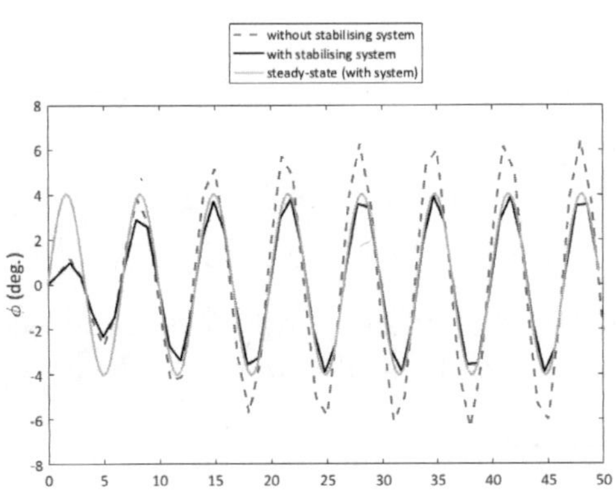

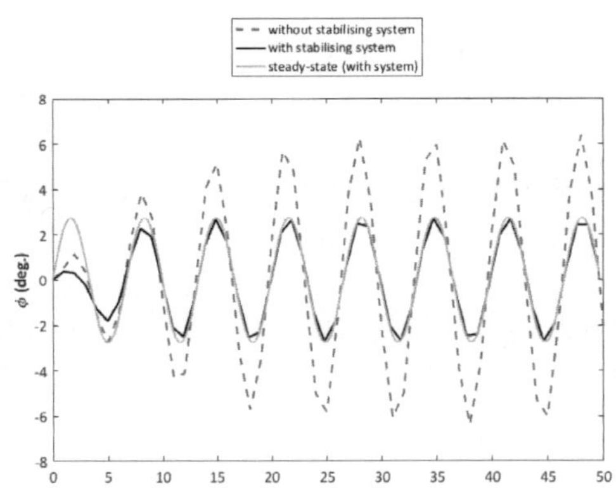

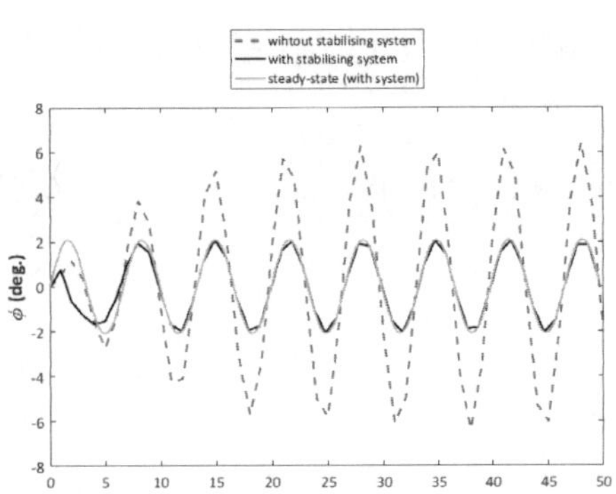

Fig. 7. Excited roll motion with and without the stabilising system and comparison with linear steady-state approximation of the model with the stabilising system on for μ=0.03. Top left: velocity = 20 ms^{-1}, top right: velocity = 40 ms^{-1} and bottom left: velocity = 60 ms^{-1}

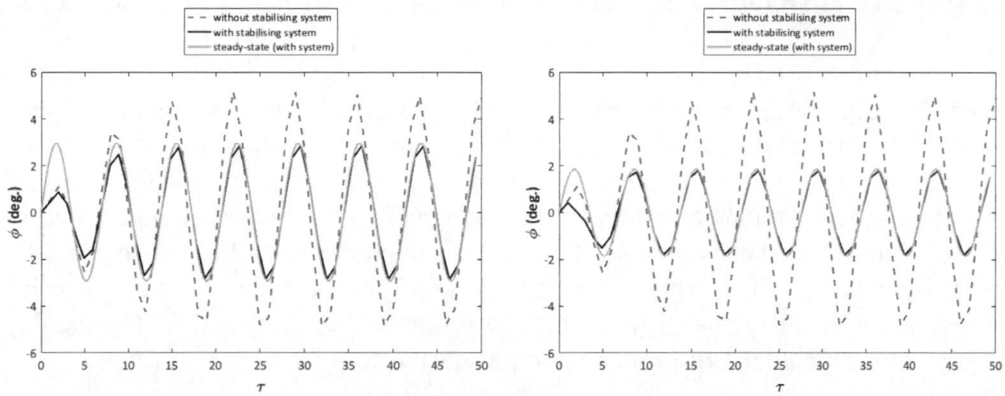

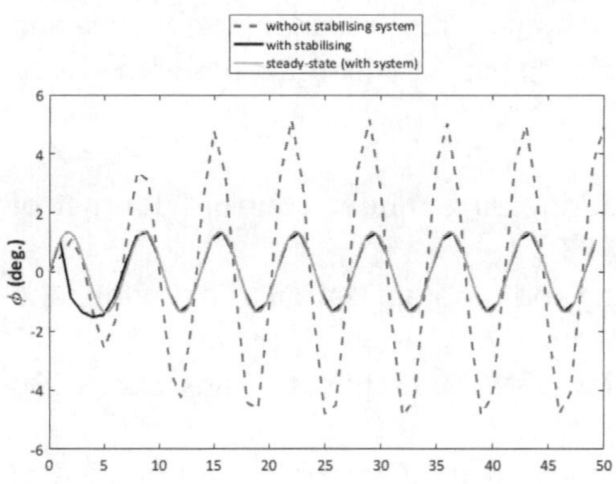

Fig. 8. Excited roll motion with and without the stabilising system and comparison with linear steady-state approximation of the model with the stabilising system on for μ=0.05. Top left: velocity = 20 ms^{-1}, top right: velocity = 40 ms^{-1} and bottom left: velocity = 60 ms^{-1}.

4.3. Comment On The Limitations of The Proposed Stabilising System

As Coriolis force is dependent on the velocity of flow and the degree of angular motion, the efficiency of roll reduction using the proposed system is expected to increase with the velocity of fluid-flow as seen in the results. Additionally, the effect also increases with increased mass ratios. As such the performance of the system is limited by the volume and velocity of fluid-flow which is dependent on the limitations of the pumping system. At the present moment with the available technology, this may pose a problem for larger vessels, however this should be quite a feasible concept for smaller vessels.

5. CONCLUSION

A new concept for roll reduction of ships has been presented and discussed in this paper. The system uses water-flow in the radial direction to induce Coriolis effect to create a restoring moment. The system is modelled using two degrees of freedom, one degree for parametric roll of the ship and one degree for the water-flow. A Ro-Ro ship with a typical hydrodynamic roll damping of 10% is used as an example for the purpose of the analysis. The system is analysed under free roll decay and excited roll motion in beam waves for mass ratio 0.01 - 0.05 and velocities 20 ms^{-1} - 60 ms^{-1}. For comparison the approximated linear analytical results are also presented.

The results showed that the stabilising system can substantially increase the damping ratio (damping up to 72%) if the volume and velocity of water-flow are sufficiently large. And the results also showed a substantial reduction of roll angle (more than 71% reduction in roll angle).

The proposed stabilising system provides a fairly simple approach to reducing roll motion and requires no complicated control system, provides no hydrodynamic drag, operates in zero forward speed, no weight issues and should have no stability issues as the water flow will always generate positive damping effect and has no dependency on any external parameters other than the volume and velocity of water flow and the degree of roll. The system provides substantial roll reduction providing the pump capacity is available for the volume and velocity of water-flow. This is an attractive concept for roll reduction, in particular at present for smaller vessels as they would require smaller pump capacities.

The work presented here should be considered as a preliminary for designing an effective and practical system using fluid-flow to induce Coriolis effect to provide restoring moment.

REFERENCES

1. Söder, Carl-Johan, et al. "Parametric roll mitigation using rudder control." Journal of marine science and technology 18.3 (2013): 395-403.
2. Perez, Tristan. "Ship motion control: course keeping and roll stabilisation using rudder and fins." Springer Science & Business Media (2006).
3. Monk, K. "A warship roll criterion." Royal Institution of Naval Architects Transactions 130 (1988).
4. Perez, Tristan, and Mogens Blanke. "Ship roll motion control." IFAC Conference on Control Applications in Marine Systems (2010).
5. Townsend, N. C., A. J. Murphy, and R. A. Shenoi. "A new active gyrostabiliser system for ride control of marine vehicles." Ocean engineering 34.11 (2007): 1607-1617.
6. Lewis, Edward V. "Principles of naval architecture second revision." Jersey: SNAME (1988).
7. Lloyd, A. R. J. M. Seakeeping: Ship behaviour in rough weather. Ellis Horwood Limited (1989).
8. Sellars, Frank H., and John P. Martin. "Selection and evaluation of ship roll stabilization systems." (1991).
9. Himeno, Yoji. "Prediction of ship roll damping-a state of the art." University of Michigan (1981).
10. Surendran, S., et al. "Non-linear roll dynamics of a Ro-Ro ship in waves." Ocean Engineering 32.14 (2005): 1818-1828.
11. Taylan, M. "The effect of nonlinear damping and restoring in ship rolling." Ocean Engineering 27.9 (2000): 921-932.

Estimation of added mass moment of inertia in roll motion through numerical simulation

S.S. Kianejad, Australian Maritime College, seyed.kianejadtejenaki@utas.edu.au
Hossein Enshaei, Australian Maritime College, hossein.enshaei@utas.edu.au
D. Ranmuthugala, Australian Maritime College, d.ranmuthugala@amc.edu.au

ABSTRACT

Dynamic ship stability is an important subject in the field of Naval Architecture due to lack of understanding may result in extreme roll motion. Direct investigation of dynamic stability modes such as parametric rolling is possible through the CFD analysis but is costly due to the long simulation time, while other numerical methods like panel methods can speed up the process. Success in using these methods depend on the solution of motion's equation and accurate estimation of each term including mass moment and added mass moment of inertia, damping and restoring moments. At the present time, there is a lack of precise estimation of added mass moment of inertia, which is usually accounted for 10-25% of mass moment of inertia. This paper presents an innovative method to calculate added mass moment of inertia by exciting a ship model through series of harmonic moments using CFD simulations. At maximum roll angle, where the acceleration is maximised and angular velocity is zero (damping moment is negligible), the total moments are measured. By ascertaining the restoring and harmonic excited moments momentarily, the added mass moment of inertia could be determined with greater accuracy. In this investigation and for the purpose of analysis, the influence of forward speeds, coupled heave, pitch and roll motions, and various encounter frequencies are taken into account.
Keywords: Added mass moment of inertia, CFD, harmonic excited roll motion, coupled motion.

INTRODUCTION

When a ship has acceleration in water, the hydrodynamic added mass acts on the ship. The hydrodynamic added mass is significant both for structural and hydrodynamic investigations. Considering accurate magnitude of the added mass improves the simulation of ship behaviours. There are different methods for calculating the added mass of a ship in roll, heave and pitch motions [1]. McGoldrick [2] conducted an analytical simulation to estimate the added mass of an oscillating ship. The proposed method has two stages. In the first stage, the added mass of 2D cross-section at different ship stations is computed. While in the second stage, the added mass is corrected by a factor (J) in considering 3D flow effects to estimate the 3D added mass. However, wetted surface area and oscillation mode changes the magnitude of J factor. This method provides an acceptable accuracy in the case of displacement/conventional ships. However, the method is not suitable for passenger or RO-RO ships, where the natural frequency (due to change the number of passenger and weight of cargo) and especially the J factor is changing in different loading conditions (the J factor for any loading condition and wetted surface is unique and determining this factor is time consuming). The same limitation exists for ships with special hull form. In another approach,

Kaydıhan et al. [3] performed 3D acoustic finite elements method to simulate the flow. In order to improve the computation Wilken et al. [4] and Tong [5] combined the FEM and BEM methods to reduce the number of elements. This method has an advantage of not being limited to the conventional ship and determines the 3D effects of flow on the ship. To date, there are several methods to estimate the added mass including projection method [6], multi-pole method [4] and panel method [7], which are based on BEM. Due to several advantages, numerical approaches are preferred to analytical approaches, but they should be validated before being practically applied. To improve the simulation of flow around the ship Discusser et al. [8] utilised Reynolds-averaged Navier-Stokes (RANS) to estimate the added mass.

Salvesen et al. [9]used the strip theory to compute two-dimensional sectional added mass in roll motion and compared the results with experimental values. The two-dimensional sectional added mass was integrated over length of the ship to extract the 3D added mass moment of inertia. Das et al. [10] utilised the proposed method by Salvesen et al. [9]and in order to integrate the two sectional added mass, applied the frank close-fit method and experimental results. The strip theory which is based on potential flow does not capture the impact of viscosity and turbulent flow around the ship especially in roll motion. Subramanian [11] incorporated an empirical correction to consider the effect of viscosity in the strip theory. The magnitude of the virtual mass moment of inertia for displacement ships has estimated as 10-25 percent of mass moment of inertia by potential flow and strip theory approaches [12-16].

In order to consider the viscous effect, Jaouen et al. [17] conducted CFD simulation to estimate the added mass moment of inertia of a ship hull section. They extracted the added mass using Fourier series by submerging the hull section and vanishing the restoring moment. Judge [18] carried out model test in different roll forced angle at different forward speeds and frequencies. They observed that added mass is independent to the frequency at high range of frequencies, while it was dependent on both forward speeds and frequency at low range of frequencies. Judge et al. [19] conducted experiments using forced roll oscillation to estimate hydrodynamic coefficients including virtual mass moment of inertia. They investigated the impact of forward speed, roll amplitude and frequency on virtual mass moment of inertia.

In this study, the added mass moment of inertia of a container ship was calculated by CFD simulations. The model was excited by external sinusoidal moments over a range of frequencies using different exciting moment magnitudes. Despite this, the effects of forward speed, single (restricting the model just roll motion in experiment is not easy) and coupled motions on added moment of inertia were investigated. More information about harmonic excited roll motion can be found in [20].

Ship geometry and model scale

A model of Post-Panamax containership was employed for the numerical simulation, with and without appendages. In the case of adding appendages, the model is equipped with a propeller, full shape of rudder and bilge keels in parallel sides of the model in middle section. The bilge keels are set in five parts having 0.042Lwl length and a width of 0.008 B_{wl}. The model is equipped with a large bow flare and transom stern. This model was tested in Hamburg ship model basin [21] for estimating the damping coefficient and resistance in full load and ballast conditions. The 3D perspective and body plan of the model are depicted in Fig. 1, and main characteristics of the ship and model are presented in Table 1. To make a model of Post-Panamax containership, scale factor of 59.467 was opted.

Table 1. Main characteristics of model and full-scale ship

Main dimension	Full scale	Model
L_{pp} [m]	355	5.9697
L_{wl} [m]	360.91	6.0691
B_{wl} [m]	51	0.8576
D [m]	14	0.2354
CB	0.6544	0.6544
V [m3]	165868.5	0.7887
KM [m]	25.05	0.4213
GM [m]	1.37	0.023
KG [m]	23.68	0.3983
T_0 [s]	38.17	4.95
I_{xx} [m]	20.25	0.34
I_{yy} [m]	88.19	1.483
I_{zz} [m]	88.49	1.488

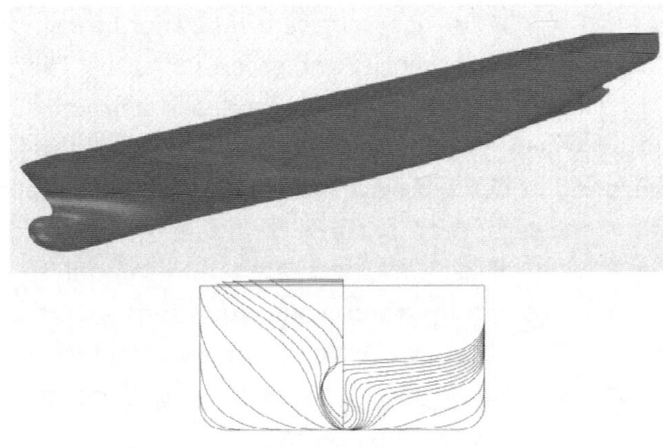

Fig. 1. Bare hull geometry of Post-Panamax containership

NUMERICAL MODELLING

The following sections discuss details of numerical simulation methods selected for this study.

Governing equations and physics modelling

The solver utilises the averaged continuity and momentum equations. The aforementioned equations for incompressible flow in terms of tensor form and Cartesian coordinates are as follows [22]:

$$\frac{\partial(\rho \bar{u}_i)}{\partial x_i} = 0 \tag{1}$$

$$\frac{\partial(\rho \bar{u}_i)}{\partial t} + \frac{\partial}{\partial x_j}(\rho \bar{u}_i \bar{u}_j + \rho \overline{u_i' u_j'}) = -\frac{\partial \bar{p}}{\partial x_i} + \frac{\partial \bar{\tau}_{ij}}{\partial x_j} \tag{2}$$

$$\bar{\tau}_{ij} = \mu\left(\frac{\partial \bar{u}_i}{\partial x_j} + \frac{\partial \bar{u}_j}{\partial x_i}\right) \tag{3}$$

Where:

$\rho \overline{u_i' u_j'}$: Reynolds stress tensor	$\bar{u}_i$: time-averaged velocity	$\bar{P}$: time-averaged pressure
ρ: density	i and j: direction of flow	μ: viscosity
$\bar{\tau}_{ij}$: mean viscous stress tensor		

The selected solver is based on finite volume approach and a predictor-corrector method is employed to maintain a correlation between continuity and momentum equations. Turbulence model is required to address the uncertainty of the stress tensor. The standard k–ε turbulence model was selected for this study. This turbulence model reduces the simulation running time in comparison to other turbulence types such as SST and K-ω [23].

The "volume of fluid" (VOF) approach has been used for modelling the free surface, which is a simple multiphase approach. The quality of mesh around the free surface has the capability of solving interfaces between two phases. There is no need for extra modelling, while VOF method is selected. This is because VOF can be utilised for simulating flows with different phases, in which each phase has a large structure with almost a low contact area with other phases. The VOF considers the same equations for single phase or multiphase conditions. They also have the same velocity and pressure values. The boundary condition characteristics such as velocity, pressure and volume fraction are functions of the flat waves (calm water condition without wave), which are set in the free surface defining section. The free surface in a flat wave or other types of waves varies during the simulation. To capture a sharp interface among phases, the second-order convection scheme was selected.

The solver utilises the segregated flow model to solve the governing equations in an uncoupled condition, where convection terms were discretised by the second order upwind scheme, where the SIMPLE algorithm was selected throughout the solution. To simulate a real ship behaviour in the sea, dynamic fluid body interaction (DFBI) was used.

Time step of solver

Courant number was utilised to specify the time step. The courant number (CFL) is a proportion of physical time step (Δt) to the mesh cell dimension (Δx) per mesh flow velocity (U). For having numerical stability the courant number should be less than one in each cell.

$$CFL = \frac{U\Delta t}{\Delta x} \quad (4)$$

For implicit unsteady simulations like resistance and ship motions calculations, the time step can be selected based on the flow specifications [23] in order to improve the stability and reduce the simulation time.

Mesh generation and structure

To capture the large motions, the overset mesh approach was employed. This approach consists of two regions including overset and background. Overset region moves with the body while surrounding and holding the body inside the motionless background mesh [24]. The size of mesh inside the overset region is small enough to capture boundary layer, flow separation during body motion, wave making and vortices around the body.

To generate the mesh four types of meshers were used including trimmed, prism layer, surface and automatic surface repair meshers. The trimmed mesher creates a high-quality mesh for complex geometries, while the prism layer mesh was utilised to generate orthogonal prismatic mesh next to the body for capturing the velocity gradient and boundary layer. To enhance the quality of a surface to create the high quality of volume mesh, the surface remesher was used. Finally, the automatic surface repair was used to remove the geometry problems left after surface remeshing. The volumetric control zones were also created to increase the number and quality of the mesh around the body and in the free surface to capture the complex flow characteristics. To prevent divergence of the solution, the cell size of the overset region and the background were matched using an overlap volumetric block. Fig. 2 shows a cross section of computation mesh.

Boundary and initial conditions

To diminish the running time and boost the accuracy of simulation results, suitable initial and boundary conditions are imperative. The velocity inlet boundary condition was set in upstream of the model (positive x direction) while the pressure outlet boundary was set in downstream (opposite x direction) to prevent any backflow. The other boundaries including sides, top and bottom were set as velocity inlet to prevent generating the velocity gradient resulted from wall and flow interaction. The initial flow velocity at each inlet boundary condition is set according to the velocity in flat wave condition. Therefore, flow in all lateral boundaries are directed to the outlet boundary so the rate of flow reflection at those boundaries are minimised. The initial hydrostatic pressure of flat wave was selected for outlet boundary.

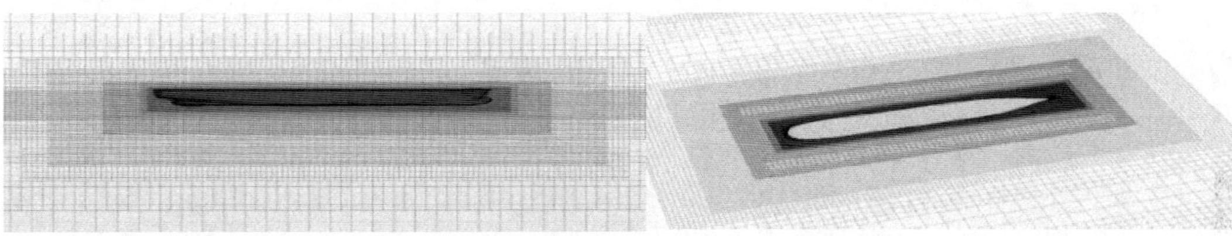

Fig. 2. Cross section of computation mesh

Coordinate system

Earth-fixed and body local coordinate systems were adopted to measure the model's motions. Primarily, forces and moments exerted on the model are calculated by solving the flow around the model. In the next step, the forces and moments are transferred into the body local coordinate system, which is set at the centre of gravity. The velocity and acceleration of the model are calculated solving the motion's equations and transferred back to the earth-fixed coordinate system to detect the new location of the model and its mesh system. During this process the overset mesh was adjusted at each time step as suggested by Simonsen et al. [25].

RESULTS AND DISCUSSION

Numerical uncertainty analysis

To improve the reliability of the simulation output, it is necessary to specify the level of uncertainty. In verification method advised by Stern et al. [26], numerical uncertainty U_{SN}, consists of iterative convergence uncertainty U_I, grid-spacing uncertainty U_G and time-step uncertainty U_T, which is formulated by the following equation:

$$U^2_{SN} = U^2_I + U^2_G + U^2_T \tag{5}$$

Since the simulation for this study was set in calm water condition, the uncertainty of U_I and U_T was negligible and the time step is selected based on Courant number, which is small enough to minimise the time-step uncertainty. However, grid-spacing as a major source of uncertainty was investigated in the case of applying 20 Nm external exciting moment at frequency of 1.3 rad/s. Three different mesh configurations with a refinement ratio of $r_G = \sqrt{2}$ were considered as shown in Table 2. The grid uncertainty of different configurations was performed based on Richardson extrapolation [27]. Simulation results for cases of coarse, medium and fine configurations are shown by S_3, S_2 and S_1 respectively and variation of simulation results are calculated as:

$$\varepsilon_{G32} = S_3 - S_2 \tag{6}$$

$$\varepsilon_{G32} = S_2 - S_1 \tag{7}$$

$$R_G = \varepsilon_{G21} / \varepsilon_{G32} \tag{8}$$

Table 2. The number of mesh in different configurations.

	Background	Overset	Total
Fine (1)	3473981	4323846	7797827
Medium (2)	2476403	3164113	5640516
Coarse (3)	1404137	2005080	3409217

The numerical convergence ratio was calculated using equation 8. Four typical conditions can be predicted for convergence ratio: (i) monotonic convergence ($0<R_G<1$), (ii) oscillatory convergence ($R_G<0$; $|R_G|<1$), (iii) monotonic divergence ($R_G>1$), and (iv) oscillatory divergence ($R_G<0$; $|R_G|>1$). For the cases (iii) and (iv) numerical uncertainty cannot be computed. For the case (ii) uncertainty can be computed based on bounding error with upper limit S_U and lower limit S_L using the equation below:

$$U_G = |\frac{1}{2}(S_U - S_L)| \tag{9}$$

In the case (i), the generalised Richardson extrapolation should be adopted to compute the numerical error $\delta^*_{RE_{G1}}$ and the order of accuracy P_G as shown in below equation:

$$\delta^*_{RE_{G1}} = \frac{\varepsilon_{G21}}{r_G^{p_G} - 1} \tag{10}$$

$$p_G = \frac{\ln(\varepsilon_{G32} / \varepsilon_{G21})}{\ln(r_G)} \tag{11}$$

Correction factor C_G, which is defined by the equation below, determines the method of uncertainty calculation.

$$C_G = \frac{r_G^{p_G} - 1}{r_G^{p_{Gest}} - 1} \tag{12}$$

If the C_G is close to 1, then the solutions are close to the asymptotic range. The numerical error δ^*_{SN}, benchmark result S_C and uncertainty U_{GC} can be computed from the following set of equations:

$$\delta^*_{SN} = C_G \times \delta^*_{RE_{G1}} \tag{9}$$

$$S_C = S - \delta^*_{SN} \tag{10}$$

$$U_{GC} = \begin{Bmatrix} (2.4(1-C_G)^2 + 0.1)|\delta^*_{RE_{G1}}| \leftrightarrow |1-C_G| < 0.125 \\ |1-C_G||\delta^*_{RE_{G1}}| \leftrightarrow |1-C_G| \geq 0.125 \end{Bmatrix} \tag{11}$$

If the value of C_G is much greater than 1, then solutions are away from the asymptotic range and numerical uncertainty should be computed by:

$$U_{GC} = \begin{Bmatrix} (9.6(1-C_G)^2 + 1.1) | \delta^*_{RE_{G1}} | \leftrightarrow | 1-C_G | < 0.125 \\ (2|1-C_G|+1) | \delta^*_{RE_{G1}} | \leftrightarrow | 1-C_G | \geq 0.125 \end{Bmatrix} \quad (12)$$

The results of numerical simulation for each mesh configuration and the amplitude of uncertainty for grid-spacing are presented in Table 3. It is apparent that the uncertainty in roll angle and roll moment are more than uncertainty in angular acceleration. Further simulations were performed based on the fine mesh configuration but changes were not significant.

Table 3. Mesh convergence study for roll angle, total roll moment and angular acceleration.

	S1	S2	S3	RG	$\delta^*_{RE_{G1}}$	UG (%S1)
Roll (Deg)	25.1343	25.3739	25.7486	0.639445	0.578446	5.784456
Mtot. (N.m)	100.731	100.426	104.101	-0.08299	-	-7.36335
Acc. (Rad/S^{-2})	0.90948	0.92497	0.94499	0.773726	0.037396	0.373962

The physical time of simulations was considered as 20 seconds. It is worth mentioning that roll motion characteristics including roll angle, angular acceleration and roll moment have almost same amplitude after a couple of cycles and the computation of virtual mass moment was performed at the following cycles and not at starting cycles. The simulations were carried out in unsteady condition and the magnitude of motion characteristics was changed at different cycles but the variation after the starting cycles was not significant. The variation is related to the interaction between the model and wave generated by the model during harmonic excited roll motion as it exists in experiments. Before conducting the simulation, a 30 seconds simulation was carried out to study the variation of motion characteristics. The variation of motion characteristics after a couple of cycles was not remarkable, therefore the simulations were performed just for 20 seconds to reduce time of simulation. The simulations were run for 37 cases and each case took 80 hours. That is 2960 hours simulations by 32 CPUs and 32G memory. Reducing the physical time to 20 seconds, reduced the overall simulation time.

Different Froude numbers and frequencies

The variation of roll angle, angular acceleration and roll moment of bare-hull model tested under 30 Nm exciting moment using frequencies less than resonance frequency are presented in Fig. 3 to Fig. 5. It was observed that increasing the frequency and reducing the forward speed, increases the roll angle, acceleration and roll moment, particularly after the second cycle. It was also obvious that effects of frequency in changing the roll motion characteristics are significant compared to the forward speed when it approaches model natural frequency. The effect of forward speed is noticeable in higher frequency. The magnitude of roll angle increases twice when changing from frequency 0.9 to 1.3 rad/s. At constant frequency of 1.3 rad/s the roll angle increases by 5 degrees when Froude number changes from 0.19 to zero.

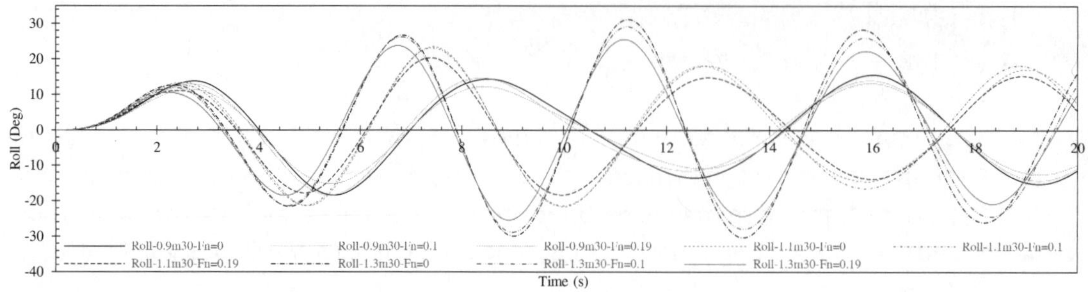

Fig. 3. Roll angle trajectory of the model under 30 Nm exciting moment with frequency of 0.9, 1.1 and 1.3 rad/s in Fn=0, Fn=0.1 and Fn=0.19.

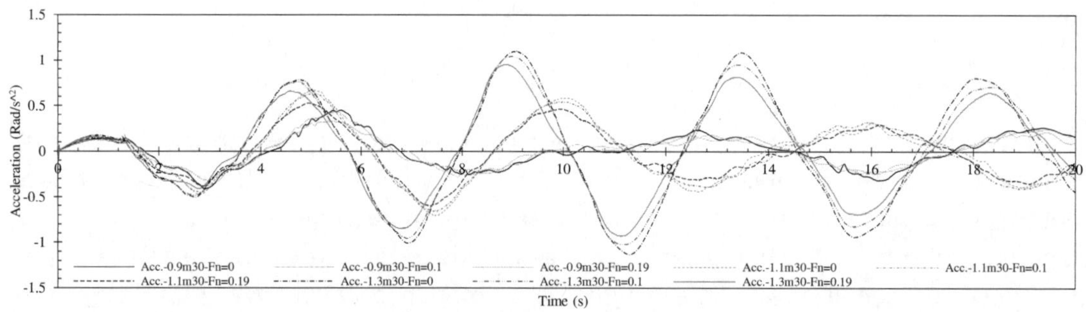

Fig. 4. Angular acceleration trajectory of the model under 30 Nm exciting moment with frequency of 0.9, 1.1 and 1.3 rad/s in Fn=0, Fn=0.1 and Fn=0.19.

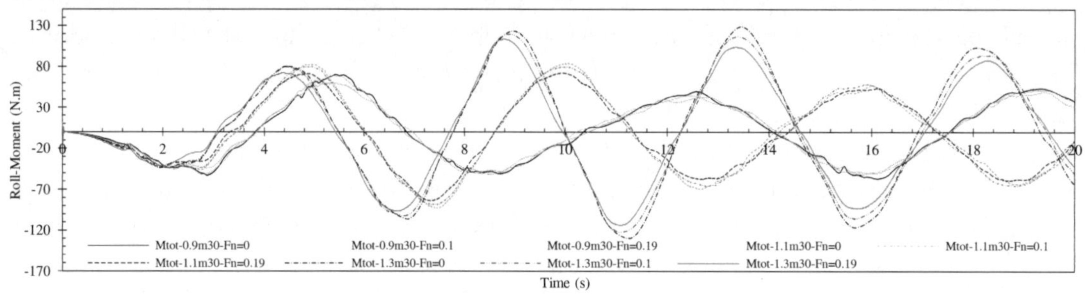

Fig. 5. Roll moment trajectory of the model under 30 Nm exciting moment with frequency of 0.9, 1.1 and 1.3 rad/s in Fn=0, Fn=0.1 and Fn=0.19.

The variation of roll angle, acceleration and roll moment of the bare-hull model tested under 20 Nm exciting moment over a range of frequencies from 1.3 to 1.7 rad/s are illustrated in Fig. 6 to Fig. 8. Although motion characteristics recorded at the first cycle is not noticeable s, but in the next cycles the changes are significant. It is apparent that at frequency closer to the resonance frequency the model experiences larger roll angle, acceleration and roll moment. The response for frequency of 1.7 rad/s, which is far from resonance frequency declines in the following cycles. The natural frequency of bare-hull model is slightly higher full hull model which is about 1.4 rad/s.

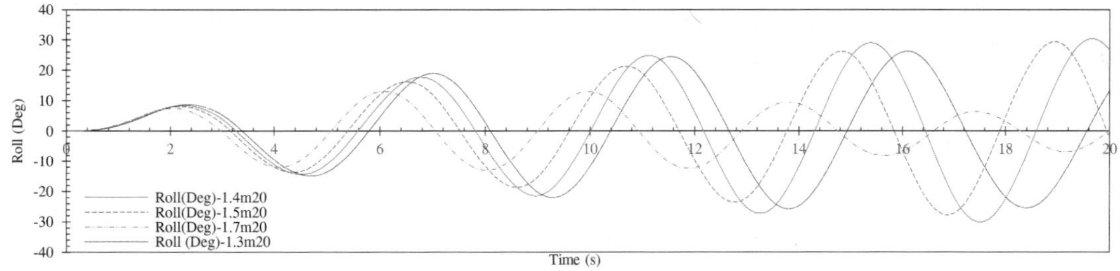

Fig. 6. Roll angle trajectory of the model under 20 Nm exciting moment with frequencies ranging from 1.3 to 1.7 rad/s.

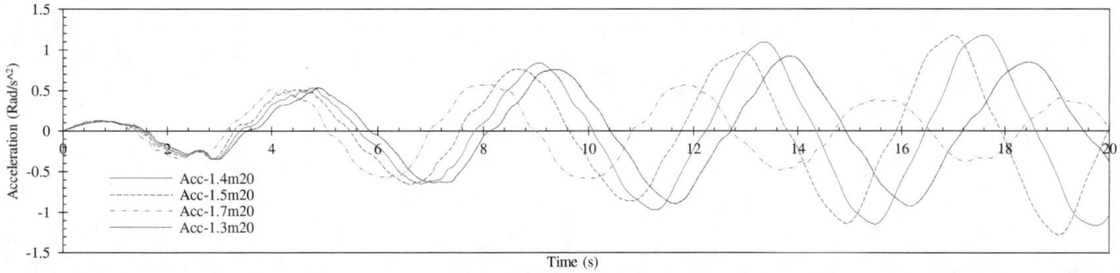

Fig. 7. Angular acceleration trajectory of the model under 20 Nm exciting moment with frequencies ranging from 1.3 to 1.7 rad/s.

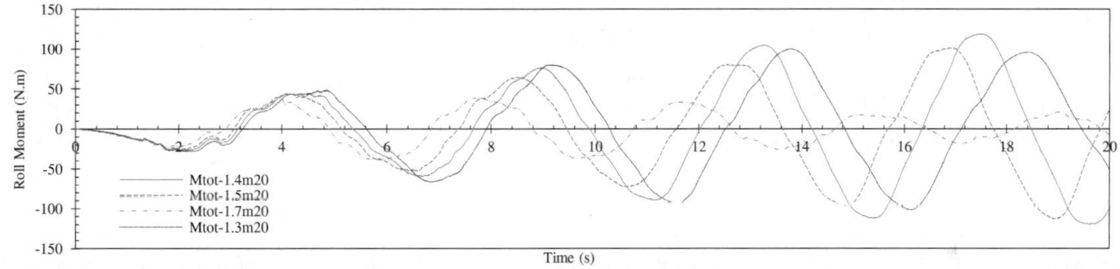

Fig. 8. Roll moment trajectory of the model under 20 Nm exciting moment with frequencies ranging from 1.4 to 1.7 rad/s.

Effect of Froude number on single and coupled motions

To investigate the effects of forward speed on the roll angle, angular acceleration and roll moment, the bare-hull model was tested under 30 Nm exciting moment at frequency of 1.3 rad/s as shown in Fig. 9 to Fig. 11. From the roll angle trajectories, it is apparent that the model has greater roll angle without a forward speed. Increasing the Froude number to 0.19 reduces the maximum roll angle. Coupling the heave motion to the roll motion increases the maximum roll angle, though, the increase is not significant. However, the maximum roll angle while considering coupled roll and pitch motions is more significant and is close to a condition when all three roll, heave and pitch motions are active. It is also observed that the maximum roll angle measured in coupled conditions compared with the single roll motion is more in the case of no forward speed, than the case with forward speed. The same pattern is expected for angular acceleration and roll moment (see Fig. 10 and Fig. 11). With forward speeds or single degree of freedom, magnitude of both angular acceleration and roll moment reduces. In case of no forward speed, the acceleration and roll moment of coupled pitch and roll condition are similar to the coupled roll, heave and pitch condition. It shows that although pitch motion is influential in case of no forward speed, the effect of heave motion is not significant. However, at the higher forward speeds heave and pitch motions have almost same contribution on increasing the acceleration and roll moment.

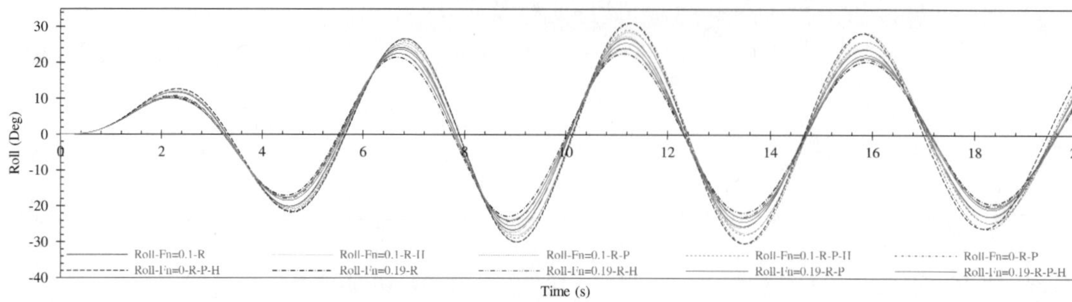

Fig. 9. Roll angle trajectory of the model under 30 Nm exciting moment at 1.3 rad/s frequency in Fn=0, Fn=0.1 and Fn=0.19, considering both coupled and uncoupled conditions. Roll, heave and pitch motions are shown by R, H and P respectively.

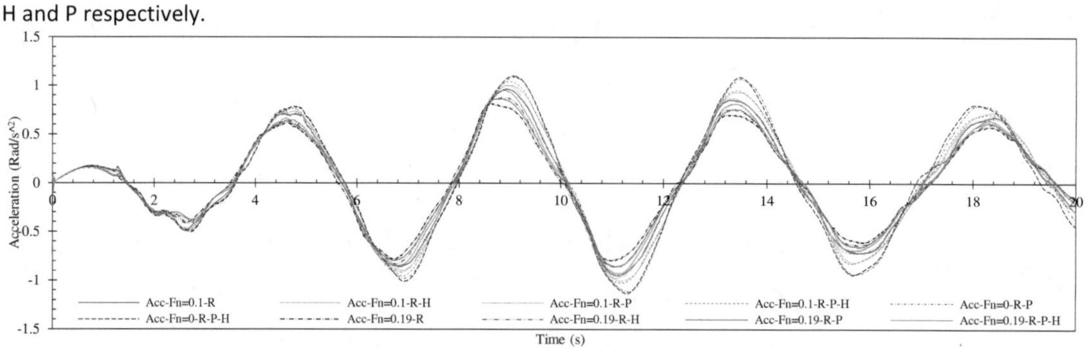

Fig. 10. Angular acceleration (Acc.) trajectory of the model under 30 Nm exciting moment at 1.3 rad/s frequency in Fn=0, Fn=0.1 and Fn=0.19, considering both coupled and uncoupled conditions. Roll, heave and pitch motions are shown by R, H and P respectively.

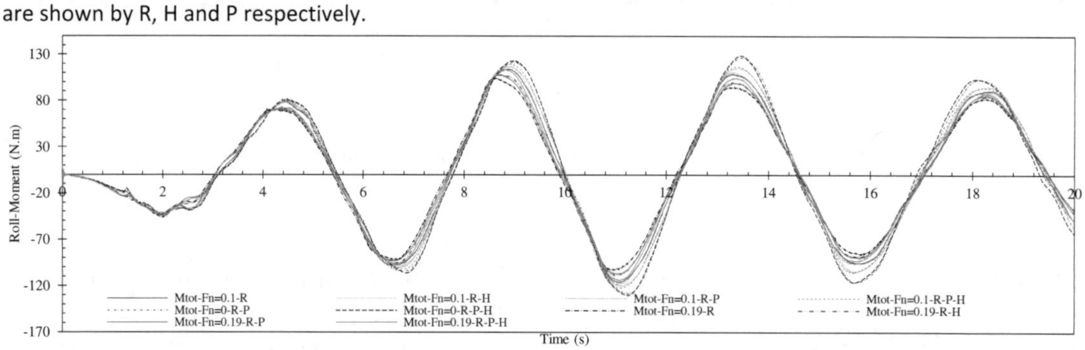

Fig. 11. Roll moment (Mtot) trajectory of the model under 30 Nm exciting moment at 1.3 rad/s frequency in Fn=0, Fn=0.1 and Fn=0.19, considering both coupled and uncoupled conditions. Roll, heave and pitch motions are shown by R, H and P respectively.

Effect of appendages

The records of the roll angle, acceleration and roll moment of the model in both the bare hull and full hull conditions can be seen in Fig. 12 to Fig. 14 in diverse forward speeds. In these simulations the different magnitude of exciting moments were considered at a constant frequency of 1.3 rad/s. At the beginning the roll angle difference between the bare and full hull is small, whereas, in the following cycles are noticeable. The same pattern are seen for acceleration and roll moment. The appendages prevent model from experiencing large roll angles, because they create moment opposing the direction of motion. It is obvious that increasing the exciting moment increases the roll angle, but in the cases of 15 and 20 Nm, the maximum roll angle per exciting moment is more than the cases of 10 and 30Nm. The relationship is quite nonlinear since the exciting moment of 10 Nm imposes 11.5 degrees roll, while with the 15 Nm experiences 21 degrees. Increasing forward speed, adding appendages and reducing the exciting moment all reduce the acceleration and roll moment. Fluctuation of the acceleration and roll moment are higher with low exciting moment and becomes smoother at higher exciting moment.

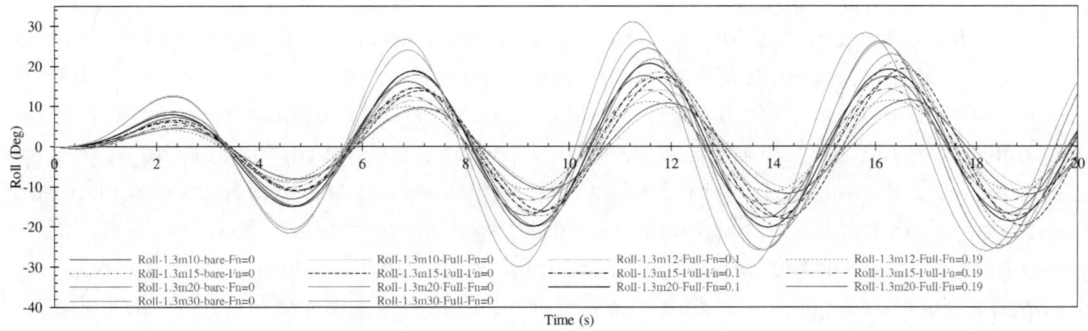

Fig. 12. Roll angle trajectory of the bare-hull and full model under different exciting moments at 1.3 rad/s frequency and in Fn=0, Fn=0.1 and Fn=0.19.

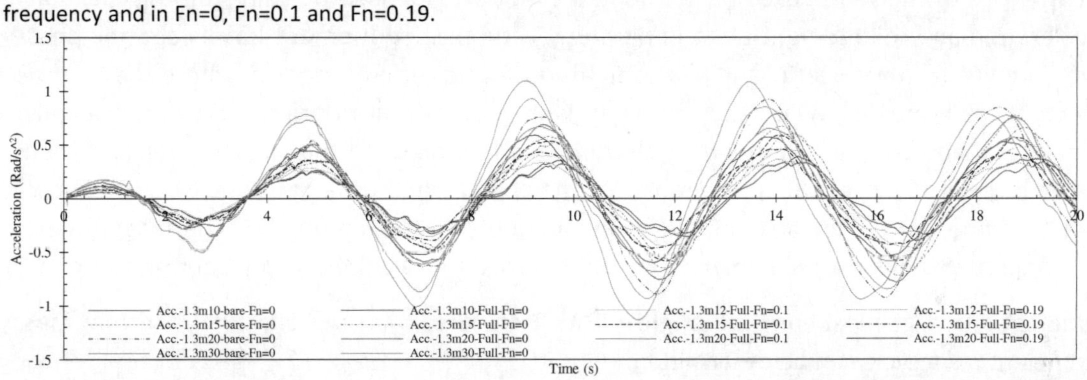

Fig. 13. Angular acceleration trajectory of the bare-hull and full model under different exciting moments at 1.3 rad/s frequency and in Fn=0, Fn=0.1 and Fn=0.19.

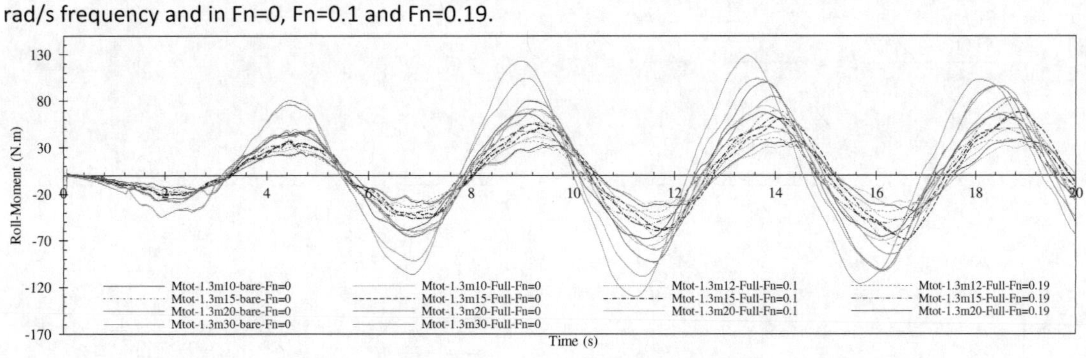

Fig. 14. Roll moment trajectory of the bare-hull and full model under different exciting moments at 1.3 rad/s frequency and in Fn=0, Fn=0.1 and Fn=0.19.

Added mass moment of inertia calculation method

The roll motion of a ship is governed by the equation of motion developed by Newton's second law:

$$(I_{44}+\delta I_{44})\frac{\partial^2\varphi}{\partial t^2}+N_{44}\frac{\partial\varphi}{\partial t}+S_{44}\varphi=M_{E44}(t) \tag{17}$$

Where the mass moment and added mass moment of inertia coefficients are I_{44} and δI_{44} respectively. N_{44} is the roll damping coefficient, S_{44} illustrates the restoring coefficient and M_{E44} is external exciting moment.

For the range of frequencies lower than the resonance frequency, the external exciting moment and the total moment of inertia are the two contributing terms to generate the roll motion, while damping and restoring moments oppose the roll motion. However, for range of frequencies higher than resonance frequency and at the large roll angles the actions of these terms are different. Exciting moment at higher frequency generates higher acceleration in shorter time and causes the exciting moment act against roll motion. That is mainly due to a large phase shift generated between roll motion and exciting moment once the model has developed a large roll angle. Therefore, the restoring, damping and exciting moments oppose the moment of inertia to reduce the roll angle.

According to the harmonic exciting moment, the model responses including roll angle, angular acceleration and roll moment were in harmony with exciting moment having specific phase shift. The virtual mass coefficient was computed for one cycle of motion, where the start-up effects were removed. When the roll angle is at its maximum during a cycle, the angular velocity is zero (damping moment is negligible at that point), so the virtual mass coefficient can be calculated by Eq. 18. In this equation, the ω and magnitude of I_{44} and M_{E44} are known over the time, since they are defined as initial inputs to the simulation. The magnitude of restoring coefficient (S_{44}) is known from the hydrostatic calculation, considering the quasi-static condition at maximum roll angle (φ_a). Therefore, by using Eq. 18 the virtual mass coefficient for a cycle can be determined.

$$\delta I_{44}=\frac{M_{E44}(t)-S_{44}\varphi}{\frac{\partial^2\varphi}{\partial t^2}}-I_{44}=\frac{S_{44}-\frac{M_{E44}(t)}{\varphi_a}}{\omega^2}-I_{44} \tag{18}$$

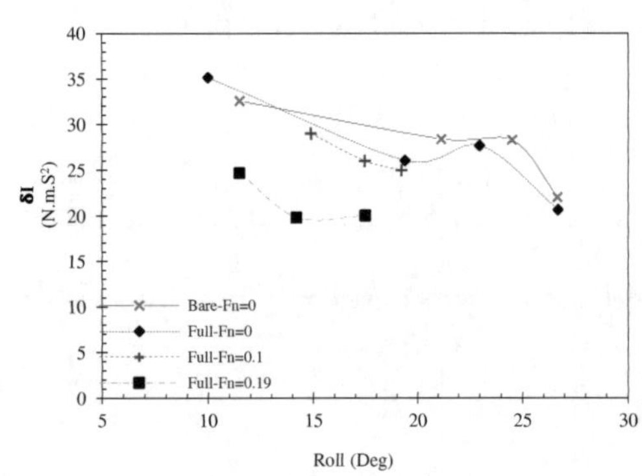

Fig. 15. The added mass moment of inertia coefficients of bare-hull and full model in different forward speeds at frequency of 1.3 rad/s.

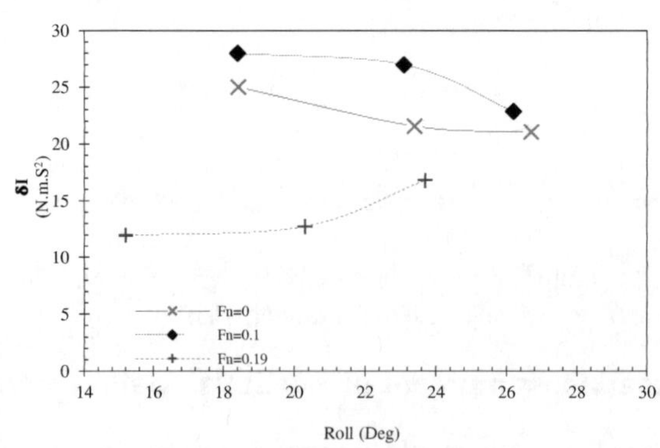

Fig. 16. The added mass moment of inertia coefficients of bare-hull model. For the first point on each curve frequency is 0.9 rad/s, second point 1.1 rad/s and last point 1.3 rad/s.

The added mass moment of inertia coefficients in both bare hull and full hull conditions over different forward speeds are shown in Fig. 15. It was observed that increased roll angle reduces the added mass moment of inertia coefficient. For the same angle, induced exciting moment for the bare hull is larger than the full hull due to the phase shift between the exciting moment and the roll angle. The phase shift is larger in full hull condition as shown in Fig. 12, while other parameters remained almost the same for both conditions. Therefore, equipping the model with appendages slightly reduces the added mass moment of inertia coefficient in the case of no forward speed. That is due to the generation of separation and wakes by appendages, so accelerated volume of flow during roll motion was reduced.
As depicted in Fig. 15, forward speed reduces the added mass moment of inertia coefficient and reduction is greater for the higher forward speed. The effects of different forward speeds at frequencies less than resonance frequency on the added mass moment of inertia coefficient are presented in Fig. 16. The frequency is changing from 0.9 to 1.3 rad/s, while the exciting moment is constant at 30 Nm. The added mass moment of inertia coefficient in case of Fn=0.1 is larger than Fn=0 and slightly reduces while roll angle increases. The behaviour is quite opposite in the case of Fn=0.19, where its magnitude is smaller initially and follows the roll angle increment.

Table 4. The effect of different frequencies on added mass moment of inertia coefficients under 20 Nm exciting moment.

Fn	**φ** (Deg)	**ω** (Rad/S)	**δI** (N.m.S^{-2})
0.00	26.2	1.3	24.97
0.00	28.95	1.4	24.73
0.00	26.25	1.5	29.22
0.00	13.11	1.7	35.84

Table 5. The effect of different forward speeds, single and coupled motions on added mass moment of inertia coefficients under 30 Nm exciting moment at 1.3 rad/s frequency. (Roll, heave and pitch motions are shown by R, H and P respectively).

Fn	**DOF**	**φ** (Deg)	**δI** (N.m.S^{-2})
0.00	R-P-H	26.68	21.05
0.00	R-P	26.66	23.13
0.10	R	24.22	19.2
0.10	R-H	24.92	19.8
0.10	R-P	25.47	23.45
0.10	R-P-H	26.19	22.86
0.19	R	21.5	6.89
0.19	R-H	22.58	10.26
0.19	R-P	22.56	14.71
0.19	R-P-H	23.7	16.81

The results for investigation of the effects of various frequencies on added mass moment of inertia coefficient of the bare-hull model are presented in Table 4. It was observed that in frequencies higher than resonance frequency, the added mass moment of inertia coefficient is larger. Frequency 1.5 rad/s is in the region dominated by inertia where the magnitude of moment of inertia and acceleration are larger than frequency of 1.3 rad/s which is in a region dominated by hydrostatics. It is apparent that virtual mass moment coefficient increases away from resonance frequency. The results for the virtual mass moment coefficients of the bare-hull model, with and without forward speeds, considering single and coupled motions under

30 Nm exciting moment at frequency of 1.3 rad/s is shown in Table 5. The results indicate that restricting the model to just roll motion in all conditions of forward speeds reduces both the maximum roll angle and virtual mass moment coefficients. At low Froude number, coupled heave and roll motions slightly increases the roll angle and virtual mass moment coefficient, but in high Froude number the increase is remarkable. At low Froude number, coupled roll and pitch motions has larger virtual mass moment of inertia coefficient, whereas in higher Froude number coupled roll, heave and pitch motion has larger virtual mass moment of inertia coefficient.

CONCLUSION

CFD simulations were carried out to estimate the added mass moment of inertia coefficient of a container ship model. The model was tested in different situations of bare-hull and full hull, different forward speeds, external exciting moments over a range of frequency and considering up to three degrees of freedom. The study investigated the influential factors in estimating added mass moment of inertia coefficient for the roll motion. The results indicate that bare hull model has larger added mass moment of inertia coefficient compared to full hull. Moreover, increasing the forward speed and development of roll angle reduces the added mass moment of inertia coefficient at frequencies close to the resonance frequency. For frequencies greater than the resonance frequency, due to the model develops larger acceleration, the added mass moment of inertia coefficient is larger in comparison to low frequencies at the similar roll angle. The results indicate that restricting the model to just the roll motion decreases the roll angle and the added mass moment of inertia coefficient under a same exciting moment. At low Froude number, the coupled roll and pitch motion has larger added mass moment of inertia coefficient, while at higher Froude number the combination of roll, heave and pitch condition generates larger added mass moment of inertia coefficient.

The findings of this study is extremely useful for naval architecture in designing a hull form to be less vulnerable in extreme sea conditions. These findings are equally important for modifying the operational guidance used by shipmasters in avoiding dangerous situations at sea.

More research is needed to better understand the impact of increasing the Reynolds number on the magnitude of virtual mass and flow patterns around the body. The established method can be used to estimate the virtual mass moment of inertia of full-scale and multiple-hull ships.

REFERENCES

1. Korotkin, A.I., *Added masses of ship structures*. Vol. 88. 2008: Springer Science & Business Media.
2. McGoldrick, R.T., *Ship vibration*. 1960, DTIC Document.
3. Kaydıhan, L., B. Uğurlu, and A. Ergin, *A hydroelastic investigation into the dynamic response characteristics of bulk carriers.* Advances in Marine Structures, 2011: p. 33.
4. Wilken, M., et al., *Efficient calculation of the effect of water on ship vibration.* Analysis and Design of Marine Structures. Proceedings of MARSTRUCT, 2009: p. 93-101.
5. Tong, Z., et al., *Dynamic behavior and sound transmission analysis of a fluid–structure coupled system using the direct-BEM/FEM.* Journal of Sound and Vibration, 2007. **299**(3): p. 645-655.
6. Wilken, M., et al., *Efficient calculation of fluid structure interaction in ship vibration.* Advances in Marine Structures, London, 2011: p. 75-85.
7. Vernon, T.A., B. Bara, and D. Hally, *A surface panel method for the calculation of added mass matrices for finite element models*. 1988, DTIC Document.

8. Discusser, O., et al. *DYNAMIC RESPONSE*. in *18TH INTERNATIONAL SHIP AND OFFSHORE STRUCTURES CONGRESS*. 2014.
9. Salvesen, N., E. Tuck, and O. Faltinsen, *Ship motions and sea loads.* Trans. SNAME, 1970. **78**(8): p. 250-287.
10. Das, S., P. Sahoo, and S. Das, *Determination of roll motion for a floating body in regular waves.* Proceedings of the Institution of Mechanical Engineers, Part M: Journal of Engineering for the Maritime Environment, 2006. **220**(1): p. 41-48.
11. Subramanian, R., *A time domain strip theory approach to predict maneuvering in a seaway*. 2012, University of Michigan.
12. Bikdash, M., B. Balachandran, and A. Navfeh, *Melnikov analysis for a ship with a general roll-damping model.* Nonlinear Dynamics, 1994. **6**(1): p. 101-124.
13. Grinnaert, F., J.-Y. Billard, and J.-M. Laurens, *KGmax curves associated with second generation intact stability criteria for different types of ships.* Journal of Marine Science and Application, 2016. **15**(3): p. 223-235.
14. Schumacher, A., S.R. e Silva, and C.G. Soares, *Experimental and numerical study of a containership under parametric rolling conditions in waves.* Ocean Engineering, 2016. **124**: p. 385-403.
15. Taylan, M., *Static and dynamic aspects of a capsize phenomenon.* Ocean engineering, 2003. **30**(3): p. 331-350.
16. Wright, J. and W. Marshfield, *Ship roll response and capsize behaviour in beam seas*. 1979.
17. Jaouen, F., A. Koop, and G. Vaz. *Predicting roll added mass and damping of a ship hull section using CFD.* in *OMAE*. 2011.
18. Judge, C. *Frequency dependence of hydrodynamic coefficients in roll*. in *Proceedings of the 29th American Towing Tank Conference, Annapolis, MD*. 2010.
19. Judge, C.Q. and J.A. Judge, *Measurement of hydrodynamic coefficients on a planing hull using forced roll oscillations.* Journal of Ship Research, 2013. **57**(2): p. 112-124.
20. Handschel, S. and M. Abdel-Maksoud, *Improvement of the Harmonic Excited Roll Motion Technique for Estimating Roll Damping.* Ship Technology Research, 2014. **61**(3): p. 116-130.
21. Moctar, O.e., V. Shigunov, and T. Zorn, *Duisburg Test Case: Post-panamax container ship for benchmarking.* Ship Technology Research, 2012. **59**(3): p. 50-64.
22. Ferziger, J.H., M. Peric, and A. Leonard, *Computational methods for fluid dynamics*. 1997, AIP.
23. Tezdogan, T., et al., *Full-scale unsteady RANS CFD simulations of ship behaviour and performance in head seas due to slow steaming.* Ocean Engineering, 2015. **97**: p. 186-206.
24. Field, P.L., *Comparison of RANS and Potential Flow Force Computations for the ONR Tumblehome Hullfrom in Vertical Plane Radiation and Diffraction Problems*. 2013, Virginia Tech.
25. Simonsen, C.D., et al., *EFD and CFD for KCS heaving and pitching in regular head waves.* Journal of Marine Science and Technology, 2013. **18**(4): p. 435-459.
26. Stern, F., et al., *Comprehensive approach to verification and validation of CFD simulations-Part 1: methodology and procedures.* Transactions-American Society of Mechanical Engineers Journal of Fluids Engineering, 2001. **123**(4): p. 793-802.
27. Jin, Y., et al., *RANS prediction of FLNG-LNG hydrodynamic interactions in steady current.* Applied Ocean Research, 2016. **60**: p. 141-154.

Innovative Experiments for Investigating Maritime Hydrodynamic Problems

Gregor Macfarlane and Alan Fleming
Australian Maritime College, University of Tasmania.
gregorm@amc.edu.au

ABSTRACT

The past two decades has seen a marked shift in the primary purpose of performing physical scale model experiments in hydrodynamic facilities to help solve a range of maritime hydrodynamic problems. It is now commonplace for experimental programs to utilise complex measurement techniques to investigate highly non-linear phenomena, often to provide high-quality data from a controlled environment to aid the validation of numerical techniques, such as potential flow and computational fluid dynamic (CFD) models. This has resulted in a significant increase in the usefulness of relatively small- to medium-sized experimental facilities due to their versatility and cost effectiveness in (a) developing and verifying innovative measurement techniques, and (b) applying these techniques to perform complex experiments to solve hydrodynamic problems for the maritime industry. This paper presents examples of several state-of-the-art measurement techniques that can now readily be performed within the National facilities at the Australian Maritime College, including the 100 m long towing tank and 35 m long x 12 m wide shallow water wave basin. Also described are examples of how these techniques have recently been applied to a range of industry-driven research projects.

INTRODUCTION

Gone are the days when the most common experiments performed in the Australian Maritime College (AMC) towing tank are simple calm water resistance tests to predict the powering required for a new design or modification to an existing ship hull. These experiments, which were once the primary reason for having such a facility, accounted for almost 50% of the projects undertaken during the first decade of the AMC facility's operation (1985 to 1995). The past two decades has seen a gradual but clear shift towards much more complex experiments, to such an extent that the once "bread and butter" calm water resistance experiment occupies only about 5% of the available time in AMC's heavily utilised towing tank and shallow water wave basin (technical details related to these hydrodynamic facilities can be found at [3]).

There has been an observed decline in demand for basic hull resistance testing as established naval architects acquire reliable model and full scale data for their 'stock' hull forms, but this is a relatively small part of the move towards much more complex and challenging experiments. Measurement technologies and analysis techniques for testing ship models and other marine structures have rapidly developed in recent years. This has been partly driven by the introduction of new digital measurement technologies and the increasing demands of the maritime industry in some emerging areas of application. A significant proportion of this effort has been to provide vital support in validating numerical techniques, such as potential flow codes and computational fluid dynamic (CFD) models.

This paper briefly describes some of these measurement and analysis techniques and how they have been used to solve a wide range of applications for the maritime industry.

MEASUREMENT TECHNOLOGIES

A description of several new measurement techniques, and the hydrodynamic problems they have been applied to solve, are described in this section. They have been arranged in three broad groups; the first two are based on the primary type of measurement methodology adopted: stereo videogrammetry and particle imaging velocimetry. The third group covers several examples where the primary focus is on interaction effects when multiple vessels operate in close proximity to one another. All experiments follow International Towing Tank Conference (ITTC) recommended procedures [13].

Stereo Videogrammetry

Surface flow and wave field analysis is an emerging area of interest in the fields of coastal hydrodynamics and marine renewable energy. AMC has developed a bespoke stereo videogrammetry system to quantify wave patterns and elevations over a wide surface area. This non-contact system uses two strategically located digital cameras and a precise calibration technique to monitor the motions of a large quantity of custom-made positively-buoyant ultraviolet fluorescent flake particles (of approximate size 5 x 5 x 0.8 mm). Videogrammetry data is processed using the 'SurfaceFlow' module of the software DaVis (version 8.3.1) [31].The accuracy of the technique has been validated against measurements from many conventional resistive type wave probes.

This technique has recently been used to capture radiation and diffraction effects from the presence of multiple submerged wave energy converter (WEC) models [25]. The purpose of the radiation experiments was to quantify the effect of the energy radiated from one WEC model in the array on the others. The experiment consisted of a single 'active' model with up to five other 'static' models in an array configuration. The active model was driven for ten sinusoidal cycles, in both pure heave and surge independently, with the effect of the radiated waves being measured on the passive models in the array. A monopolar pattern is characterised by a perfectly axisymmetric wave emitting from the device. This is the characteristic pattern for a heaving device, or from a practical point of view, the effect of the dynamic pressure in an oscillating water column (OWC) chamber. On the other hand, the dipole pattern has two poles generating waves of same amplitude but opposite phase. This is the characteristic pattern for a surging or pitching device. A photograph from these experiments in shown in Figure 1 and an example of the resultant dipole radiated wave fields acquired from stereo videogrammetry can be seen in Figure 2.

In what is believed to be a world-first, AMC has pioneered the adaptation of digital videogrammetry for underwater operation to quantify the deformation of a flexible membrane. The membrane was an air/water interface for a novel sea-floor mounted WEC, the concept of which has been patented by the Australian company Bombora Wave Power [28, 1] (refer Figure 3 for a photograph of the scale model used in the experiments). This short range underwater trinocular videogrammetry system consists of three synchronised cameras mounted in separate underwater housings acquiring images of the same location at a frame rate of 50Hz. Equi-spaced markers of diameter ~4mm on a 15mm rectilinear grid on the membrane surface were then solved to give the time series motion of the membrane. This data has validated a CFD model of the WEC system which has allowed non-linear effects to be explored and to estimate the stresses in the membrane under the action of idealised hydrodynamic and system loads [14].

Particle Imaging Velocimetry

Particle Imaging Velocimetry (PIV) is a flow measuring technique which allows full field measurement in a plane of velocity fields without interfering the flow itself. The fluid is seeded with very small neutrally buoyant fluorescing particles and a laser generated light sheet is used to illuminate the particles in the area of investigation. Two images are captured at a small time offset which are compared using cross correlation to determine the displacement of the fluid and therefore the fluid's velocity. This technology has been adopted to investigate a wide variety of maritime applications, particularly in the field of marine renewable energy.

Possibly the most promising of the various marine renewable energy technologies currently being developed is the oscillating water column (OWC). Traditionally, the majority of physical experiments conducted within Australia and around the world have treated various OWC geometries simply as a 'black box' where the input energy (ocean waves) is known and the available output energy is predicted, usually through the simple measurement of quantities such as air pressure, air velocity and water level within the chamber. One of the advances that AMC has made is the application of PIV to quantify the flow fields inside the OWC 'black box' [9, 10]. When applied to experimental analysis of OWCs, standard 2-dimensional PIV can reveal phenomena associated with the transformation of the oscillatory flow of the wave to the linear (or near linear) oscillation of the free surface within the OWC. This has led to the optimization of the geometry of the device for improved performance [9] and a new understanding of the energy balance (the flow of energy from incoming waves into the device and extracted at the power take-off)[11]. In doing so, viscous energy losses (energy converted into other forms, such as heat, which cannot be extracted) due to the generation of vortices were identified and we found that these could be reduced by modifying the geometry of the column.

PIV experiments have also been performed to compare the performance characteristics between differing OWC geometries with the aim to investigate conversion losses and device performance by modifying the underwater geometry [8]. Four models were tested in total; a base model (used in much of the abovementioned studies) and three variations that had additional segments to afford different chamber length and lower/upper lip angles (10, 20 and 30 degrees). An example of the output is shown in Figure 4 where the velocity fields for the base geometry and one of the variations are presented. This study concluded that replacing sharp edges with a softer transition is effective at reducing energy loss via the production of vorticity and angling the upper-lip upwards improves energy capture. Both of these improvements have been implemented within the revised geometry adopted by the first full-scale commercial device currently under development by Wave Swell Energy [32].

Similar PIV data has also been acquired to investigate the effect that wave type (regular, polychromatic and irregular wave spectra) and 3-dimensional flow effects have on the power-generating performance of OWC WECs [21, 22, 23]. It is also being used to validate

2D and 3D CFD models for undertaking parametric studies on OWC design. These numerical models have successfully been used to assess many design and performance parameters for stationary, near-shore and floating offshore OWCs [6]. Recent work has used the validated 3D CFD model to assess the survivability of an offshore floating-moored OWC which experiences mooring line failure in extreme seas [7].

Experiments using PIV have also been applied to assess a very unusual maritime hydrodynamic scenario presented by the new and novel transhipment concept known as the floating harbour transhipper (FHT). The FHT is a concept of transferring bulk materials from shore to an offshore floating transfer facility by use of a low draught transfer vessel, which has the potential to revolutionise the export of bulk cargo from remote areas where no deep-water port currently exists. The FHT concept is described in detail elsewhere [17, 18, 19] but a key aspect of the FHT is the open stern well dock on the transhipper which provides shelter for a shallow draught feeder vessel that shuttles cargo between the shore-based facilities and the FHT, and onto an export vessel moored alongside the FHT.

The FHT well dock represents a tightly confined space in which the feeder vessel must manoeuver. The cross-sectional area of the feeder vessel is relatively large compared to that of the well dock so a large body of water is displaced when the feeder vessel enters the well dock, all of which must exit via the limited space between the feeder vessel and the well dock floor and sides. The resultant flow velocities around the feeder and its propulsors makes for very unusual operational conditions that the Master of the vessel must understand in order to safely manoeuvre the feeder in and out of the well dock. Acquiring PIV measurements inside the well dock as a feeder vessel enters (and exits) has increased our understanding of the complex flows involved and led to some innovative design features being implemented within the well dock that have greatly mitigated the operational challenges (Figure 5) (findings will soon be published).

In a very different project, PIV has been used to measure velocity fields around a series of tidal turbines to assess the performance and effects of parameters such as blade shape, number of blades and the interaction between the turbine hub and blades. The commercial nature of this work has prevented this from being published at this stage.

The International Towing Tank Conference [13] have developed a standard series of 2D and 3D experiments from which to benchmark PIV measurements and analysis between facilities. Results from tests performed in the AMC tank to quantify the wake flow around the tip of a surface piercing flat plate at an angle of incidence was studied using 2D PIV and found to have excellent agreement with results from the much larger facilities of MARIN and INSEAN used by the European Hydro Testing Alliance [2].

A key factor in the successful use of either stereo videogrammetry or PIV for all of the above-mentioned projects are the characteristics and quality of the seeding particles. AMC have perfected the in-house production of fluorescing particles that are customised to the precise needs of the experiment. For example, the fluorescent particles used in the OWC tests (Figure 4) were all within the range of 38 – 75 microns in diameter, doped with Rhodamine dye and neutrally buoyant. These particles fluoresce at a different wavelength to the excitation wavelength from the PIV laser; enabling blocking of reflected laser light from the acquired images through the use of optical high pass filters. Filtering the reflected light enables capture of the flow field in close proximity to reflection from bubbles, vortex filaments with entrained air and the free surface, all of which may occur during some of the experiments described here. The method of seeding the water prior to any experiment must also be considered: this is relatively simple for a case like the stationary OWC models, but is a challenging task to obtain the desired and consistent seeding density when capturing unsteady phenomena from a high-speed hydrofoil that may require running nearly the full length of the 100 m long towing tank.

The relatively large and positively-buoyant particles (flakes) used in the stereo videogrammetry experiments seen in Figure 1 have also been used for other projects where there has been a need to capture the free surface waves. For example, to investigate the generation of waves for a surfing wave pool [29] and extremely rare ship-generated soliton waves in a restricted channel - the latter has resulted in an online educational tool [12].

Multi-Vessel Interaction: Motions & Loads

In this section, four examples of multi-vessel interaction are described. In each case the forces and moments experienced by the ship models are measured using load cells and all 6 degree of freedom motions using a non-contact optical infrared video motion capture system. We have acquired and developed a large suite of load cells and force balances to suit every application, from small single-axis force transducers, to small 6-DOF force balances (where small forces are expected), to a large 6-DOF force balance that can restrain a ship model that displaces 800 kg and is exposed to large incident waves. For motion measurements, our wave basin has a fixed 8-camera Qualisys system with inbuilt calibration markers to allow rapid set-up and calibration [27]. We also have an 8-camera portable system that can be used in either the tank, basin or off-site.

A scenario that we are regularly asked to study is the interaction between large cargo ships when manoeuvring within very restricted waterways. For example, when one ship passes a berthed ship in a shipping channel it causes hydrodynamic interactions which can result in the berthed ship ranging on its moorings. This is often a critical factor in the design and/or operation of ports where entering or departing ships are required to pass those already moored. Not only is it important to accurately model the channel width and depth, but recent work has highlighted the importance of the local bathymetry surrounding the berthed ship as the type and location of banks and wharves can have a significant effect on the berthed ship motions and loads [5]. It is not unusual for this berthed ship - passing ship interaction to restrict loading/ unloading operations, and in extreme cases it has resulted in damage to mooring infrastructure, vessels and dangerous situations for crew and passengers [30].

For the past 15 years, we have generally addressed these problems by performing physical scale model experiments to accurately measure the interaction forces and moments and use this as input to a numerical simulation software package to predict the motions and mooring loads experienced by the berthed ship due to a passing ship. It is common to consider various different mooring arrangements (each of which may typically involve 16 mooring lines, of specific type, quality and pretension) and to assess this against PIANC and OCIMF safety criteria [26, 24]. In recent years we have used our experimental data to successfully validate CFD predictions using an inviscid double body model, thus providing a cheaper and quicker alternative to experiments for acquiring reliable and very site-specific interaction forces and moments [4].

Another ship interaction scenario investigated, usually for defence applications, are replenishment at sea (RAS) operations. These typically involve the transfer of fuel, supplies and/or personnel between vessels operating in close proximity at moderate forward speeds (approximately 12 to 16 knots). Due to the close proximity of the vessels, hydrodynamic interaction forces develop between the hull forms which impact both the ship motions and the manoeuvring capability of the vessels (Figure 6). Recent developments have seen these experiments also include incident waves at various headings to assess the effect of the sea state [20].

The interaction between surface ships during a RAS operation, such as that described above, is not dissimilar to the scenario when dealing with submarine applications. Fully submerged missions often necessitate an autonomous underwater vehicle (AUV) and submarine to operate in close proximity in order to launch, recover, and recharge the AUV, in addition to the ability to communicate data between them during operations. The interaction effects acting on an AUV operating close to a moving submarine have been investigated, predominantly using CFD but suitably validated and supplemented through physical model experiments [15, 16]. The resulting simulation model is intended to be coupled with a control system in a dynamic manoeuvring simulation to evaluate the motion behaviour of the AUV and develop the necessary algorithms to maintain the desired trajectory of the vehicle when in operation near a moving submarine.

The physical experiments in the towing tank, with known constant forward speed, use a horizontal planar motion mechanism (HPMM) capable of generating horizontal motion (pure sway and/or yaw) on the AUV model. The AUV model was mounted to the HPMM using a 'sting' arrangement that connects to the model through the aft end, with the forces and moments acting on the AUV model recorded via two 6-DOF load cells located inside the model. The submarine model was mounted directly onto the carriage by means of rigid supports as shown in Figures 7 (CFD model) and 8 (experimental set up).

The fourth (and final) example of multi-vessel interaction covered in this paper involves the FHT concept previously outlined in Section 2.2, which considers the interaction effects between three ship models all in close proximity: the FHT (mothership), the feeder vessel that will be docked inside the FHT well dock, and an ocean-going export vessel that moors alongside the FHT (Figure 9). One aspect of the current investigation is to quantify the relative motions between the docked feeder vessel and the FHT, as this must be understood to determine the operational window in terms of maximum sea state. Related to this is the operation of materials handling equipment onboard the FHT to extract the bulk cargo from the docked feeder vessel. It has already been determined that the sheltered location inside the well dock significantly reduces the motions of the feeder vessel when compared to a more traditional side-by-side transhipment operation, thus decreasing downtime due to inclement weather [17]. A similar study is underway to assess the relative motions between the FHT and the export vessel moored alongside.

SUMMARY

This paper has discussed the growing demand for more advanced measurement techniques and how they are being used to solve complex maritime hydrodynamic problems. Examples provided cover a wide range of specific applications that have adopted stereo videogrammetry and PIV to investigate hydrodynamic phenomena previously beyond conventional measurement technology. Also discussed are several novel experimental programs to study interaction effects between multiple vessels operating in close proximity.

Complex experiments such as these, especially those that acquire vast quantities of digital images over a prescribed area, are often best suited to relatively small to medium-sized facilities due to the high cost and time involved to develop and fine tune the measurement techniques. The facility staff and researchers at AMC will continue to utilise their versatile and cost effective facilities to develop innovative measurement techniques to solve hydrodynamic problems for the maritime industry.

ACKNOWLEDGEMENTS

Many of the experimental techniques described in this paper have been developed through projects supported by the Australian Government through the Australian Research Council (Grant numbers ARC LP0990307, LP110200129, LP130100962, LP150100502) and the Australian Renewable Energy Agency (ARENA).

We acknowledge our colleagues and collaborators who contributed to the work outlined in this paper. Where appropriate, the relevant publications have been referenced.

When preparing for any physical experiment, but especially when they are highly novel and as complex as all those outlined here, good technicians are worth their weight in gold. Many personnel assisted with this work, but the talents and efforts of AMC technicians, Liam Honeychurch and Kirk Meyer, deserve special mention.

REFERENCES

[1] Algie, C., Fleming, A.N. and Ryan, S., Experimental and numerical modelling of the Bombora wave energy converter, *Proc. 3rd Asian Wave & Tidal Energy Conf.*, Singapore, pp. 246-253, 24-28 October 2016.

[2] Ashworth Briggs, A., Fleming, A., Ojeda, R. and Binns, JR,, Tracking the vortex core from a surface-piercing foil by particle image velocimetry (PIV) using fluorescing particles, *Proc. 19th Aust. Fluid Mechanics Conf.* 8-11 December, Melbourne, pp. 1-4. 2014.

[3] Australian Maritime College (AMC), http://www.amc. edu.au/facilities/towing-tank, http://www.amc.edu.au/ facilities/model-test-basin, [accessed 22 July 2017]

[4] Denehy., S. Duffy, J.T, Ranmuthugala, D. and Renilson, M, Squat in berthed ship – passing ship interaction for restricted water cases, *Proc. 4th Int. Conf. on Ship Manoeuvring in Shallow and Confined Water*, Hamburg, Germany, 23-25 May 2016.

[5] Denehy., S. Duffy, J.T, Renilson, M. and Ranmuthugala, D., An experimental and numerical investigation into berthed ship – passing ship interaction in a channel: methods to increase safe passing ship speeds, *PIANC Year Book 2016*.

[6] Elhanafi, A., Fleming, A., Macfarlane, G.J. and Zhi Leong, Underwater geometrical impact on the hydrodynamic performance of an offshore oscillating water column–wave energy converter, *Renewable Energy*, 105, pp. 209-231, 2017.

[7] Elhanafi, A., Macfarlane, G.J., Fleming, A. and Zhi Leong, Intact and damaged survivability of an offshore floating–moored OWC, *Proc. 27th Int. Ocean and Polar Engineering Conference*, San Francisco, California, USA, 25-30 June 2017.

[8] Fleming, A. & Macfarlane, G.J., Experimental flow field comparison for a series of scale model oscillating water column wave energy converters, *Marine Structures*, 52, 2017.

[9] Fleming, A., Macfarlane, G.J., Penesis, I., Bose, N. and Hunter, S., Experiment based underwater geometry optimisation programme for an oscillating water column ocean wave energy converter, *Proc. 18th Aust. Fluid Mech. Conf.*, Launceston, 3-7 Dec. 2012a.

[10] Fleming, A., Penesis, I., Goldsworthy, L., Macfarlane, G.J., Bose, N. and Denniss, T., Phase averaged flow analysis in an oscillating water column wave energy converter, *Journal of Offshore Mechanics and Arctic Engineering*, Vol. 135, No.2, May 2013.

[11] Fleming, A., Penesis, I., Macfarlane, G.J., Bose, N. and Denniss, T., Energy balance analysis for an oscillating water column wave energy converter, *Ocean Engineering*, Vol. 54, pp 26-33, July 2012b.

[12] Forrest, S., Soliton Waves, BEng Thesis, UTAS, 2016. https://www.youtube.com/watch?v=D14QuUL8x60 [accessed 22 July 2017]

[13] ITTC, International Towing Tank Conference, Quality Systems Manual, http://ittc.info/downloads/quality-systems-manual/ [accessed 22 July 2017].

[14] King, A., Algie C, and Ryan, S. (2015), Modelling of large deformations in membranes for analysis of wave energy converters, *3rd Symposium on Fluid-Structure-Sound Interactions and Control*, July 2015.

[15] Leong, Z.Q., Ranmuthugala, D., Penesis, I. and Nguyen, H., Quasi-static analysis of the hydrodynamic interaction effects on an autonomous underwater vehicle operating in proximity to a moving submarine, *Ocean Engineering*, 106, 175-188, 2015.

[16] Leong, Z.Q., Ranmuthugala, D., Penesis, I. and Nguyen, H., Transient analysis of hydrodynamic interaction effects on an autonomous underwater vehicle in proximity of a moving submarine, Trans. RINA, Vol. 157, Part A4, *Int. J. Maritime Eng.*, Oct-Dec 2015.

[17] Macfarlane, G.J., Ballantyne, R.J., Ballantyne, S.E. and Lilienthal, T., An experimental study on the relative motions between a floating harbour transhipper and a feeder vessel in regular waves, Trans RINA, Vol. 154, Part A2, *Int. J. of Maritime Eng.*, Apr-Jun 2012.

[18] Macfarlane, G.J., Johnson, N.T.M., Clarke, L.J., Ballantyne, R.J. and McTaggart, K.A., The floating harbour transhipper: new-generation transhipment of bulk ore products, Proc. ASME 2015 *34th Int. Conf. on Ocean, Offshore and Arctic Engineering*, OMAE2015, St. John's, Newfoundland, Canada, May 31-June 5, 2015.

[19] Macfarlane, G.J., Matsubara, S., Clarke, L.J., Johnson, N.T.M. and Ballantyne, R.J., Transhipment of bulk ore products using a floating harbour transhipper, *Proc. Australasian Coasts & Ports Conference*, Auckland, New Zealand, 15 - 18 September 2015.

[20] Mathew, J.L., Sgarioto, D.E., Duffy, J.T., Macfarlane, G.J., Denehy, S.P. and van Walree, F., An experimental study of ship motions during replenishment at sea operations between a supply vessel and a LHD, [under review] Trans. RINA, *Int. J. of Maritime Eng.*

[21] Mitchell Ferguson, T., Macfarlane, G.J., Penesis, I., Fleming, A.N., PIV investigation of 3-dimensional flows within an oscillating water column, *Int. J. of Marine Energy*, 2015.

[22] Mitchell Ferguson, T., Penesis, I., Macfarlane, G.J. and Fleming, A.N., A PIV investigation of OWC operation in regular, polychromatic and irregular waves, *Renewable Energy*, 103, pp. 143-155, 2016.

[23] Mitchell Ferguson, T., Penesis, I., Macfarlane, G.J. and Fleming, A.N., Improving OWC performance prediction using polychromatic waves, *Energy*, 93 Part 2, 2015

[24] OCIMF, Mooring equipment guidelines", 3rd Edition, Livingstone, United Kingdom: *Witherby Seamanship International*, ISBN: 9781905331321, 2008.

[25] Penesis, I., Manasseh, R., Nader, J-R., De Chowdhury, S., Fleming, A., Macfarlane, G.J. and Hasan, MK., Performance of ocean wave-energy arrays in Australia", *Proc. 3rd Asian Wave & Tidal Energy Conf.*, Singapore, pp. 246-253, 24-28 October 2016.

[26] PIANC, Criteria for movements of moored ships in harbours, a practical guide, Working Group 24 Report, Permanent Tech. Committee II, Suppl. Bulletin no 88, 1995.

[27] Qualisys, http://www.qualisys.com/ [accessed 22 July 2017].

[28] Ryan, S., Algie, C., Macfarlane, G.J., Fleming, A.N., Penesis, I. and King, A., The Bombora wave energy converter: a novel multi-purpose device for electricity, coastal protection and surf breaks, *Proc. Aust. Coasts & Ports Conf.*, Auckland, 15-18 Sept. 2015.

[29] Schmied, S.A., Binns, J.R., Renilson, M.R., Thomas, G., Macfarlane, G.J. and Huijsmans, R. A novel method for generating continuously surfable waves – comparison of predictions with experimental results, *Journal of Offshore Mechanics and Arctic Engineering*, Vol. 135, Issue 3, August 2013.

[30] Seelig, W. N., Passing ship effects on moored ships, Tech. Report, TR-6027-OCN, Naval Facilities Eng. Service Center, Port Hueneme, California, USA, 2001.

[31] SurfaceFlow, Product Manual 1003019 for DaVis 8.3, LaVision GmbH, 2015

[32] Wave Swell Energy, http://waveswellenergy.com.au/ [accessed 22 July 2017].

Figure 1 A photograph of stereo videogrammetry experimental set-up. The cameras and UV lights are mounted on the raised trusses and the surging active WEC model lies submerged beneath the pink fluorescing particles floating on the water surface.

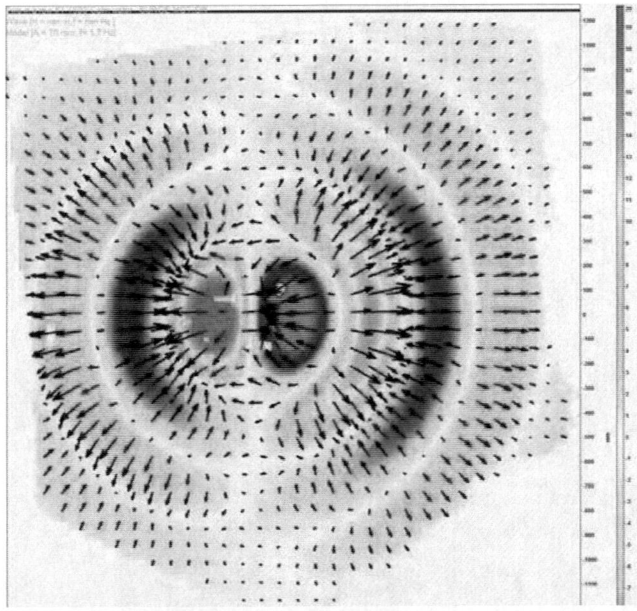

Figure 2 Example of experimental measurement of a dipolar radiated surface wave field from the submerged surging WEC model using stereo videogrammetry [25].

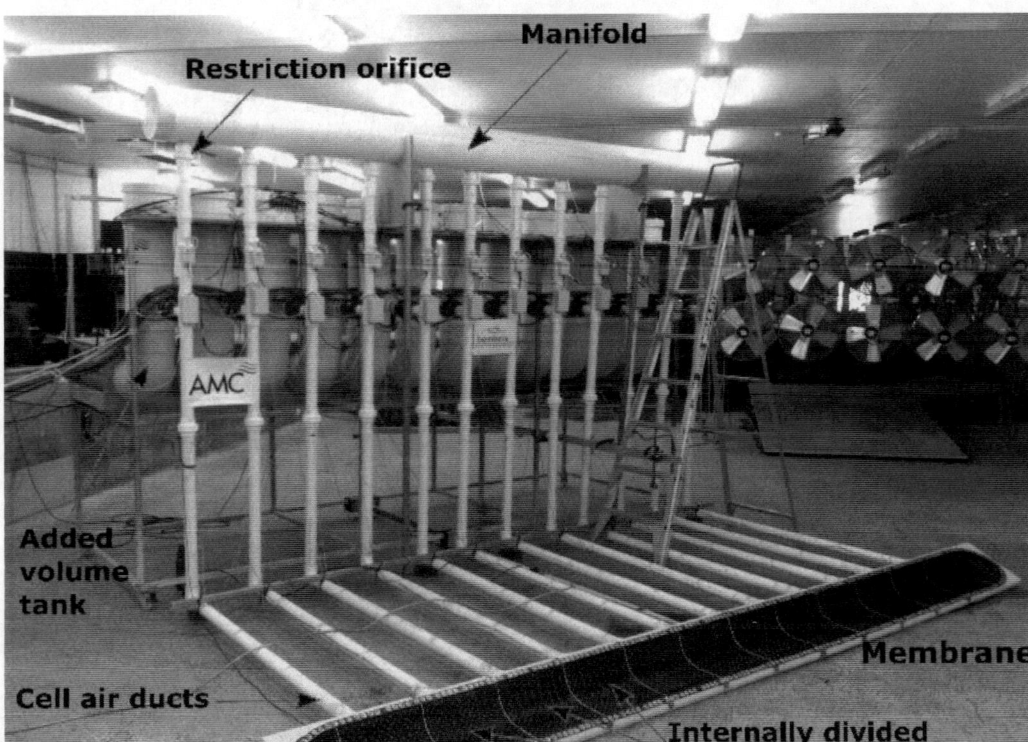

Figure 3 A photograph of the experimental set-up for the Bombora WEC model (while the basin was drained). The flexible membrane is the black material in the foreground. The pipework behind the WEC model provided a means of measuring air flows and pressures in the system (and to correctly scale the required air volumes) [1].

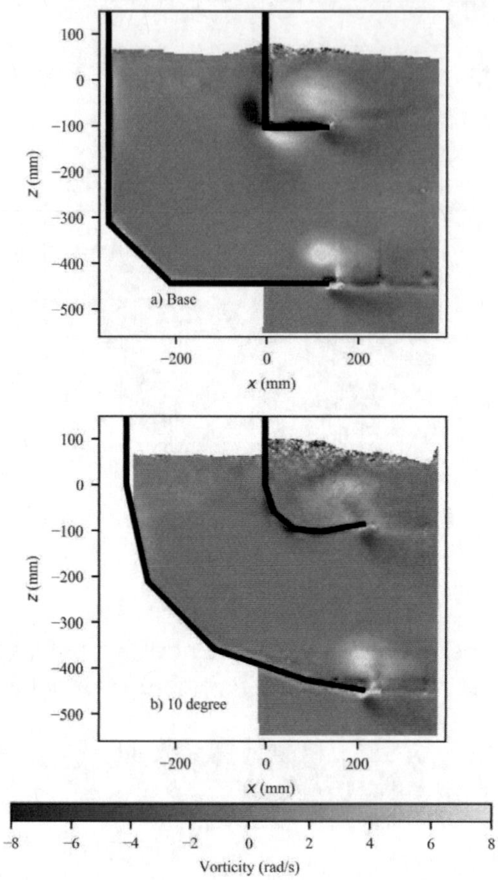

Figure 4 PIV results showing the average vorticity (energy losses) within two different OWC geometries. The significant reduction in losses around the upper lip is evident between the base geometry (top) and the smoother transition of the 10 degree lip case (bottom). The PIV data can also be used to examine the average kinetic energy of the velocity fields which can identify inefficiencies as an uneven distribution is likely to result in further energy loss [8].

Figure 5 Photograph showing the scale model of the feeder vessel (right) about to travel astern to enter the stern well dock of the FHT (left). In addition to acquiring PIV data of the complex flow inside the well dock as the feeder performs this manoeuver, the dynamic effects on the hull resistance (powering) and motions were also quantified.

Figure 6 Photograph of RAS experiments to investigate the influence that both lateral and longitudinal separations between vessels have on the interactions. It was found that this can lead to large supply vessel (left) roll motions in head seas and that variations in lateral and longitudinal separation between vessels had a large influence on the peak roll response of the supply vessel [20].

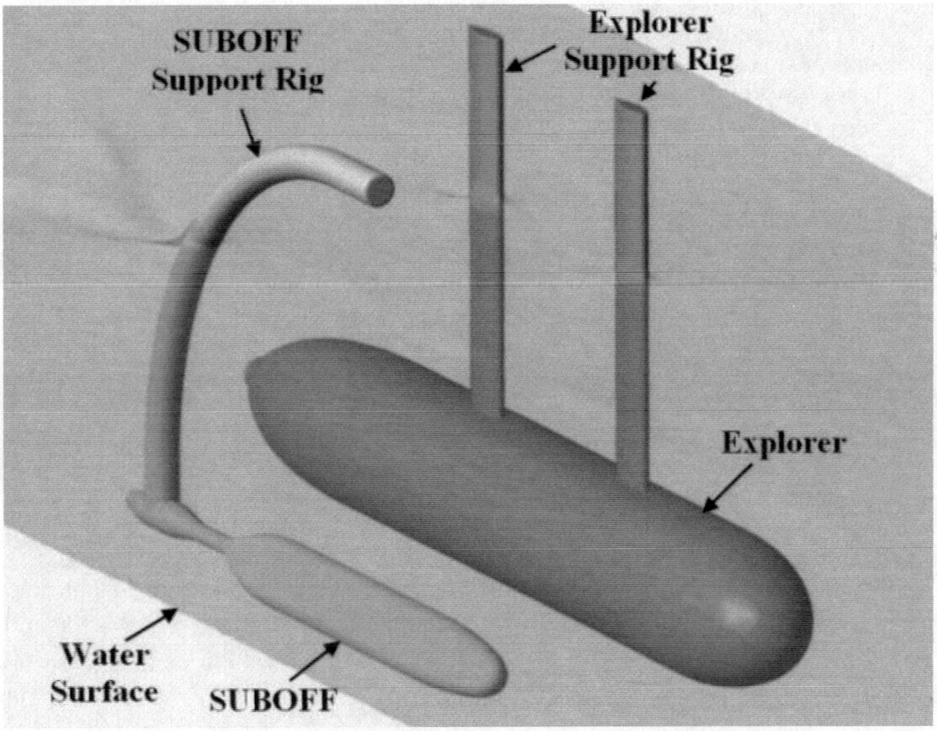

Figure 7 CFD model of the experimental setup for an AUV (SUBOFF) operating in close proximity to a submarine (Explorer). The support rigs used during the physical experiments are included in the CFD simulation (as shown) to quantify their effect on the measurements and to assist in describing the physical test rig. The experiments considered various lateral separations and longitudinal positions for the two models. Further scenarios were investigated using CFD, once validated against the experimental data [15].

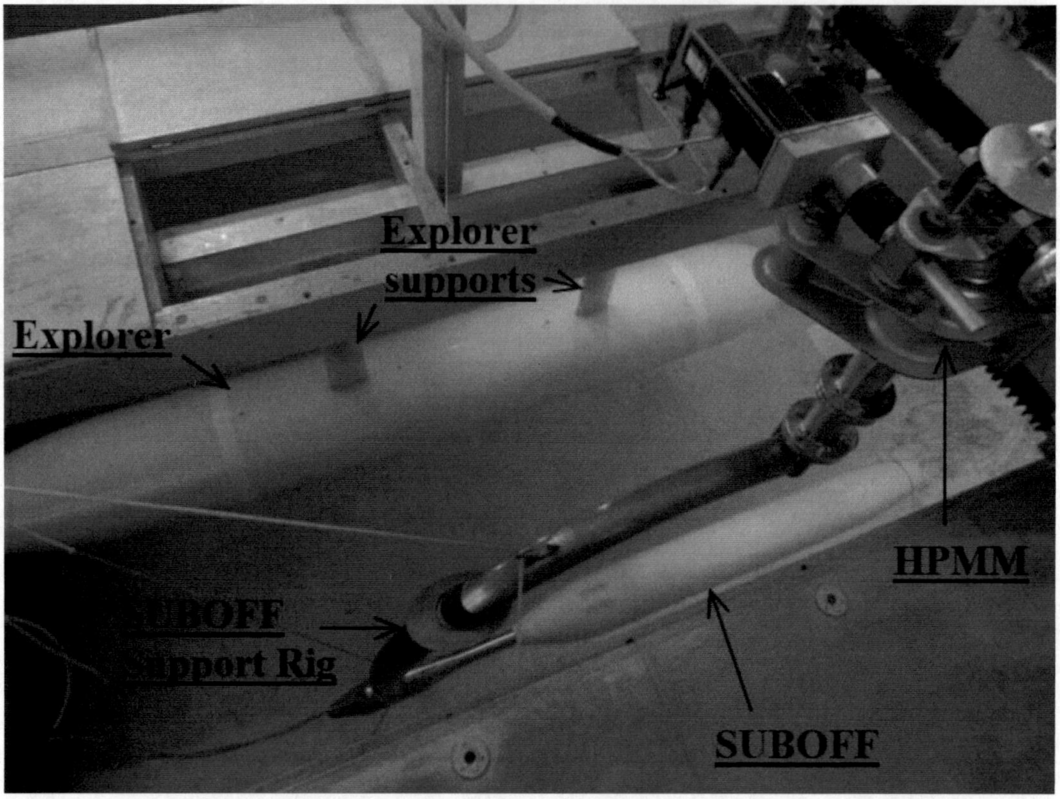

Figure 8 Photograph of the experimental setup for an AUV when operating in close proximity to a submarine [15].

Figure 9 A photograph from recent experiments involving the 5m long scale model of the FHT, feeder vessel and ocean going vessel in the AMC shallow water wave basin. The bow of the feeder vessel can be seen protruding from the FHT well dock (left). Both Panamax and Capesize ocean going vessels are under consideration: the smaller Panamax size is shown here. A minimum of four markers (silver spheres) are fixed to each of the three models to allow the Qualisys optical motion tracking system to measure all six degrees of motion for each ship model. The incident waves seen here are regular head seas, but the study is investigating a number of irregular wave spectra and wave headings.

Slow Speed Depth Control of a Submarine Under Waves

Paul Crossland, QinetiQ Ltd, pcrossland@qinetiq.com

ABSTRACT

It is important to understand the behaviour of a submarine under the influence of surface waves at the early stages of design because of the impact on whole boat design including any depthkeeping requirements associated with control surfaces, compensation tanks and pumps and their appropriate control algorithms. The main concern to the designer is how to accommodate the effects of these wave forces which could mean that if the planes and trim systems are poorly designed they may be unable to control the submarine in certain wave conditions. This paper describes an experimental and numerical approach to quantifying the first and second order wave effects on a submarine operating at slow speed under the influence of waves and how this approach can provide guidance on both hydrodynamic and hydrostatic control mechanisms.

INTRODUCTION

Today's military submarine can typically only operate in a very restricted portion of the world's ocean depths; the typical collapse depths of submarines from World War II were around 280m,, (representing a depth around 4 times the length) far less than the depths of The World's Oceans. The only way of ensuring that a submarine design is capable of operating within these rather tight boundaries is by understanding the manoeuvring characteristics early on in the design process; this is crucial in reducing the risks of producing designs that are unsuitable for the environment in which they are expected to operate. As far as manoeuvring and control is concerned, most of the work has been targeted towards understanding the performance of a submarine operating in deep water (away from the effects of the free surface) [see Ray *et al*, 2008]; many methods, reflecting the state-of-the-art at the time, have been applied and many have been used throughout the design process. However, submarines do operate at shallow depths, under waves, in which they experience complex oscillatory motions due to the sea load effects. These effects, particularly for depthkeeping, manifest themselves as a time varying suction force that can cause the submarine to be drawn to the surface risking broaching; or the submarine, when in a heavy trim state due to compensating for the suction effects, coupled with the effects of hull compressibility from the depth excursion may be significant unless rapid de-ballasting, through pumping, can be achieved to regain a safe trim state. Therefore, it is important to understand the behaviour of a submarine under the influence of surface waves at the early stages of design because of the impact on whole boat design including any depthkeeping requirements associated with control surfaces, compensation tanks and pumps and their appropriate control algorithms. By way of qualifying the motion characteristics of a submarine under waves, [Crossland *et al*, 2012], tested a free running model of an X-plane concept design in head seas. Figure 1 shows typical motion responses, in this case for a sea state 3. The motion of the submarine clearly contains frequencies that are typical of the waves themselves (so called 1st order responses) but also very low frequencies beyond those contained in the wave spectrum (called 2nd order responses), [Musker *et al*, 1988].

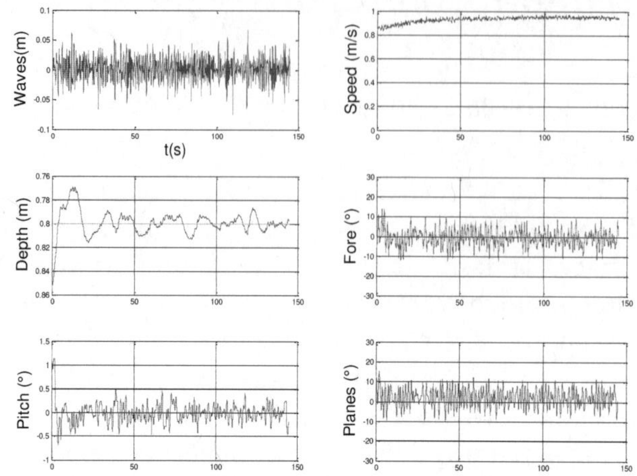

Figure 1: Motion responses in a sea state 3 [Crossland et al, 2012]

To provide insight into the extent of this suction force the reported discussion following the paper by Veillon et al, [1996] suggested that for a 10,000 tonne submarine at a depth of 50m, these non-linear forces can require around 20 to 30 tonnes of compensation to balance the submarine. Crossland [2013] presented a method for generating profiles of excess mass (or mass required to balance the suction force) varying with speed, heading, sea states and spectral formulations that can be used to create an understanding of the surface suction phenomena for a submarine design for potential parametric modelling and inclusion in submarine simulation codes. However, these pre-generated tables cannot account for all possible environmental scenarios that the submarine could encounter and, whilst, the non-zero mean suction force may be a dominate term, the tables do not account for any low frequency variation in this suction force.

MODELLING SUBMARINES UNDER WAVES

Most of the work undertaken in modelling the response of a submarine under the free surface relates to autopilot design and understanding submarine autopilot performance under waves. Byström [1988] developed a simulation of a deeply submerged submarine and a submarine at a snorting condition under the influence of regular and irregular waves. The author looked to develop a control system for the deeply submerged submarine, applying a Kalman filter to remove the measurement noise and environment noise (waves assumed to be white noise) near the free surface. The incremental force (due to the waves) on an element of the body was given as the sum of the drag force due to the water velocity flowing passed the body and the inertia force rather like the formulation of the Morison equation, modified by Byström [1998] for a moving body. This formulation (at least the zero-speed equation) is traditionally used in problems relating to offshore structures, [Wilson, 2003]. This is an empirical result whereby the coefficient terms can be determined experimentally or in the case of the inertia coefficient can be obtained from potential flow theory but does not include diffraction terms. Veillon *et al* [1996] developed a method based on the use of a physical model in a towing tank; in this case there was a need to develop a dynamic control system that could compensate for the reduced effectiveness of the hydroplanes at low to zero speeds. The main aim of this work was to assist in sizing the stabilisation system and deriving specific control algorithms. In order to help size the actuators approximations to the first and second order forces where made based on coefficients derived from numerical techniques, in the first instance, then physical model tests. The more detailed analysis of the verification of the stability of the autopilot and the identification of the performance of the stabilisation system was done using physical model tests. The clear drawback of this approach is the cost associated with undertaking physical model tests.

Mandzuka [1998] considered the behaviour of a submarine under waves as consisting of a low and high frequency response; the low frequency response consisting of a reduced set of hydrodynamic derivatives from a standard coefficient based model such as Gertler and Hagen [1967] and modelled the "high frequency response" including the first and second order vertical forces on a submarine, as a set of four equations with empirical coefficients. Mandzuka updated the surface suction equations to account for a phase difference between the heave force and pitch moment to define a model that was suitable for depthkeeping autopilot control design (the approach was limited to the vertical plane). The instantaneous heave force and pitch moment equations consisted of first and second order contributions, with constant coefficients.

$$Z_{wave}(t) = Z_1(t) + Z_2(t)\,, M_{wave}(t) = M_1(t) + M_2(t)$$

$$Z_1(t) = \sum_{i=1}^{N_w} C_{Z1} \nabla\rho(1.5sin^2\mu + 1)\,(1 - 0.02Ucos\mu)F_{1i}sin\omega_{ei}t$$

$$Z_2(t) = -\sum_{i=1}^{N_w}\sum_{j=1}^{N_w} F_{1i}sin\omega_{ei}t.F_{1j}sin\omega_{ej}t\frac{\nabla\rho}{g}\frac{(sin^2\mu + 3)}{10^{|sin\mu|}}C_{Z2}(1 - 0.04Ucos\mu)$$

$$M_1(t) = -\sum_{i=1}^{N_w} C_{M1}\mathrm{L}\nabla\rho\,(1 - 0.02Ucos\mu)Sgn(cos\mu)F_{1i}sin\omega_e t$$

$$M_2(t) = -C_{M2}L\theta(t)Z_2(t)$$

The term F_{1i} is the force due to the wave components, attenuated to account for the depth of the submarine.

$$F_{1i} = a_i\omega_i^2 e^{-\frac{\omega_i^2 H}{g}}$$
$$a_i = \sqrt{2S(\omega_i)\delta\omega}$$
$$\omega_{ei} = \omega_i - \frac{\omega_i^2 U}{g}cos\mu$$

Hirom [1977] developed a similar set of equations for estimating the effects of the waves on a submarine but considered first order forces and moments for the full six degrees of freedom; only heave force and pitch moment had second order terms.

As part of European collaborative project, a series of experiments were reported by Martinussen [2006] to measure the first and second order forces and moments on a fully captive submarine model under regular waves. The six degrees of freedom first order forces and moments response amplitude operators (RAOs) were non-dimensionalised as:

$$F'_{RAO} = \frac{F_{RAO}}{\rho g L \zeta_w} \quad \text{and} \quad M'_{RAO} = \frac{M_{RAO}}{\rho g L D \zeta_w}$$

The means of the second order heave force and pitch moment were non-dimensionalised as:

$$Z'_{MEAN} = \frac{Z_{MEAN}}{\rho g L \zeta_w^2} \quad \text{and} \quad M'_{MEAN} = \frac{M_{MEAN}}{\rho g L D \zeta_w^2}$$

These data can be used to derive the coefficients in the force and moment equations for this particular EUCLID design. Figure 2 shows the case of zero speed and beam regular waves; those forces and moments that are predicted to be identically zero have been omitted from the figure. By choosing suitable values for the coefficients the first and second order forces and moments can be replicated quite well. However, since the captive model had the pitch angle fixed at zero the second order pitch moment is identically equal to zero. Therefore, also shown in the figure is a case where there is a small pitch angle error on the model; then a coefficient can be also chosen to replicate the subsequent second order pitch moment shown in red.

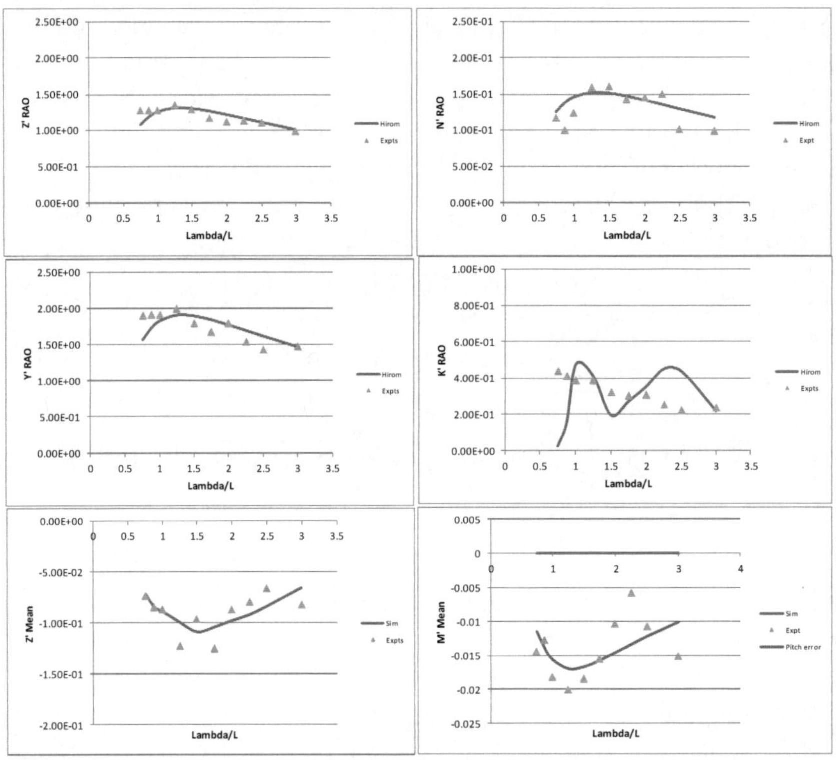

Figure 2: Forces and moments from captive model tests.

SUBMARINE CONTROL

A submerged submarine is required to manoeuvre in the vertical and horizontal plane, thus its control surfaces (appendages) are designed to allow a level of controllability in this 3D space. Typically, in the horizontal plane, agility is usually required to achieve higher rates of turn (defined, in this context, as the ratio of steady yaw velocity to the rudder angle achieved during the turn) and lower tactical diameters. However, for reasons of safety, in the vertical plane a reduced level of agility is typically required, mainly due to the dangers of large pitch angles and depth excursions associated with large angles on the planes due to mechanical failure or damage. So, a submarine will usually have a set of appendages or control surfaces that provide a suitable horizontal plane turning performance yet ensuring safe manoeuvring in the vertical plane. These control surfaces can be controlled manually (helm) or by a suitable autopilot (for both the vertical and horizontal planes). An autopilot is simply a control algorithm that is designed to regulate the output by keeping it at a desired value, (In the context of a submarine this will be maintaining heading or depth; if the input changes, a heading or depth change is required, then the algorithm responds by demanding a rudder angle or hydroplane angle respectively). A system with no regulation is referred to as an open loop system, which is the case when a submarine is undertaking a turn with no depth control; the submarine would experience a depth change during the heading change, [Kimber and Crossland, 2008]. A closed loop system is essentially an open loop system with some feedback, in the case of a heading change and the output from the heading gyro would be providing feedback into the autopilot.

The submarine dynamics could be represented by the quasi-steady state mathematical model such as Gertler and Hagen [1967]; the external disturbances that include the forces and moments due to the waves would be represented by the model described earlier. The hydrodynamic characteristics of control surfaces (hydroplanes and rudders) and their hydraulic actuators form a key component of the entire system model in that they represent the main source of constraints that limit the performance of control surfaces. In the case of a submarine the constraints will be the size of the control surfaces, the maximum angle and the maximum rate. The schematic suggests that the forces and moments due to the control surfaces are external to the submarine dynamics model, this is only for illustration since the terms involving the force and moment contributions to the submarine body are a part of the overall quasi-steady state mathematical model and so form part of the submarine dynamics model. The sensors provide the mechanism for the feedback, in the horizontal plane this would be a heading gyro and the vertical plane would be pressure sensors to measure depth.

The design of the controller, typically, requires an understanding of the performance limitations and any design objectives for the control system, [Perez, 2003]. In the context of an autopilot controlling a deeply submerged submarine, the performance limitations of such a system may be characterised by the limitations of the servo mechanism controlling the appendages and any design objectives such as limitations on the amount of control surface movement for example. For the work presented here, a depth autopilot was developed based on a control design technique known as H_∞ loop-shaping. There are two modes of operation; depth-keeping control provides independent control of depth and pitch, while the depth-changing control provides independent control of pitch angle and trajectory angle. The degree of independence in both cases is limited by the maximum travel of hydroplanes. The design uses the method developed by Glover-McFarlane [1992] in the MATLAB® Robust Control Toolbox. The H_∞ controller design approach is suitably robust to be able to account for some changes in the dynamics of the submarine.

For a submarine operating close to the free surface, a depth control autopilot would use onboard pressure sensors to determine the depth error, [Burcher and Rydill, 1995]. However, the pressure variations caused by a wave travelling over the submarine and the first order response of the submarine would be picked up by these onboard sensors and will be interpreted as a depth error that requires correcting unless the system is designed appropriately. If a control algorithm has been poorly designed, the depth autopilot would cause the control surfaces to respond to this depth error by demanding significant plane angles that can lead to unnecessary cycling of the planes. In this case, control surface saturation can readily occur, impacting on the performance of the planes if the control algorithms have been designed without the consideration of the 1st order wave effects. Specifically for this work, a notch filter was included in the control system design to reduce the excessive duty cycle on the control surfaces by preventing the autopilot from responding to the first order wave frequencies. This notch filter algorithm used the pitch signal as its input, and identified the dominant frequency within a specified band. Thus, it requires a means of tracking the frequency, the block diagram of the frequency tracker is shown in Figure 3.

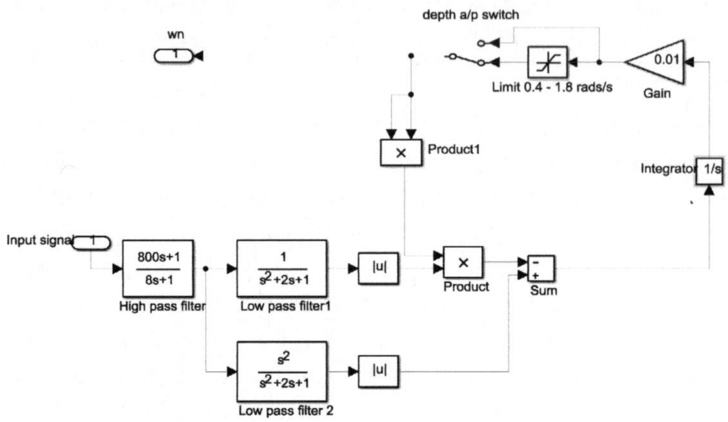

Figure 3: Frequency tracker block diagram

Once the peak frequency has been identified by the frequency tracker, a notch filter is generated, specifically to remove the wave frequency components from the depth and pitch signals that are then fed into the autopilot. The autopilot is tuned to keep depth error to a minimum, using mainly the bow planes for depth control thus allowing bow plane deflection to inform on the trim condition of the submarine for the near surface operation.

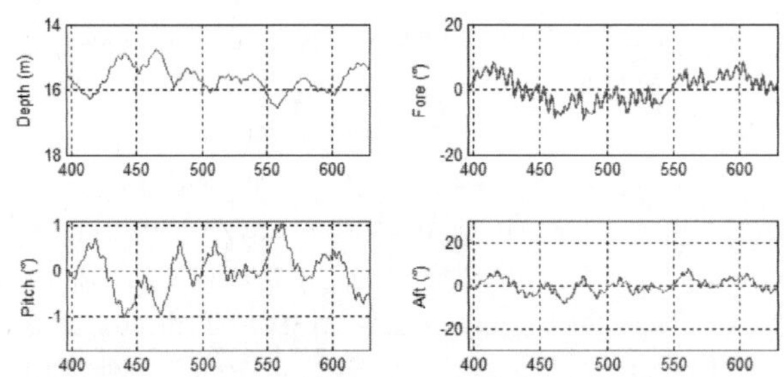

(a) depth-keeping response without wave filter

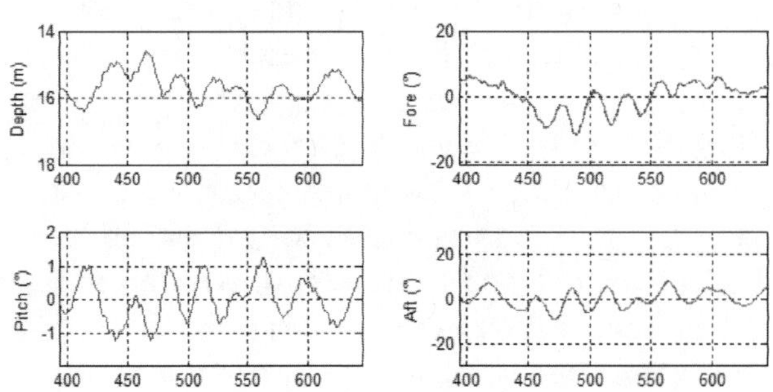

(b) depth-keeping response with wave filter

Figure 4: Depth-keeping response in head seas with $H_{1/3} = 0.88m, T_0 = 7.5s$

Figure 4 shows an example of the EUCLID submarine at a depth of 15.7m in an ITTC sea state 3, head seas, at 10 knots both with and without the notch filter being employed. A static 10 tonne compensation has been included which is why the non-zero offsets on the planes are small. The effects of not including the notch filter are apparent in the top figure where there are higher frequency components leading to potential issues with wear and tear on the hydraulic system as a consequence of this increase in duty cycle.

Another issue is that as the speed of the submarine reduces the effectiveness of the control surfaces also reduce; it has been reported that typical submarine hydroplanes can generate about 2 kN/(knot)^2 which severely limits the ability of the hydroplanes at slow speeds. Figure 5 shows an example of the same design at a depth of 20m in the same sea state 3 this time at 5 knots.

The hydroplanes are beginning to struggle to maintain depth in this case. Therefore, alternative methods of depth control are required where hydrodynamic control becomes ineffective. One possible solution is to use hydrostatic control using compensation tanks. Figure 6 shows such a system, in this case a single tank located at the centre of buoyancy of the submarine. The valve on the compensation tank is controlled by a PID controller using depth error. In this example the flow rate through any pumps is limited to 0.5m^3/s and the compensation tank has a capacity of 15 tonnes (i.e., +/- 7.5 tonnes).

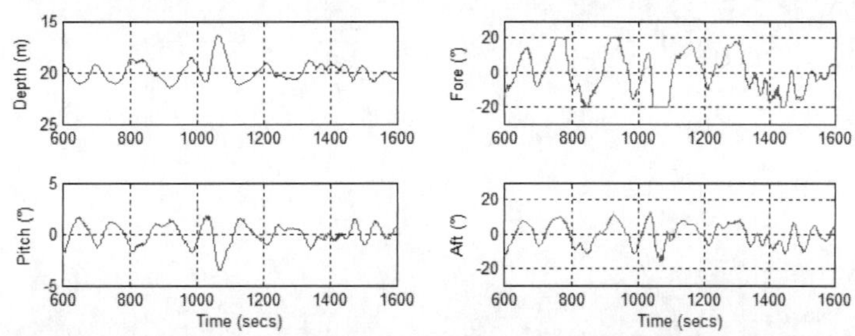

Figure 5. Depth-keeping response in sea state 3 at 5 knots.

Figure 7 shows an example of the EUCLID submarine at 0.5 knots at depth of 30m in a sea state 3. The example shows a case where there is no active compensation and depth control is by planes only; 1 tonne of additional ballast has been included to assist the planes. Without this ballast the submarine would have broached almost immediately. The figure then shows an example of active control using compensation with no planes to assist.

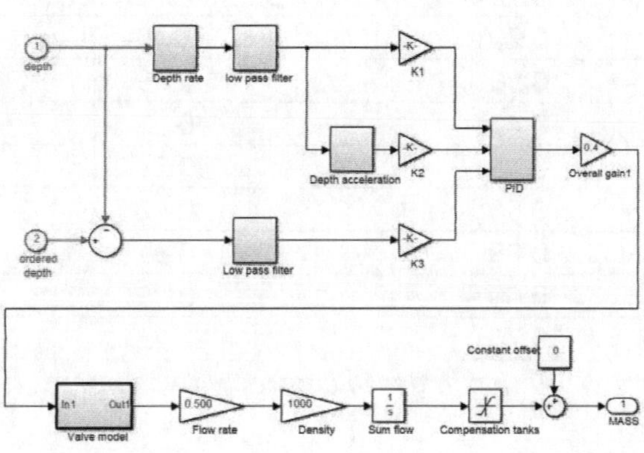

Figure 6: PID controller for an active compensation system.

The figure shows that the planes are succeeding, but struggling, to hold depth up to around 1000 seconds, then, the planes fail to maintain depth. However, rather than broaching the submarine sinks; this is due to the slowly varying second order force reducing to an out of trim (the 1 tonne of additional ballast) that is too significant to be carried on the planes. Whilst depth control is clearly lost in the case of non-active compensation, the depth rate probably is sufficiently low to enable manual adjustment of the ballast to recover from this unplanned depth excursion. For this same sea state in the example in Figure 7 a number of calculations were undertaken varying the compensation rate. The results are in Table 1.

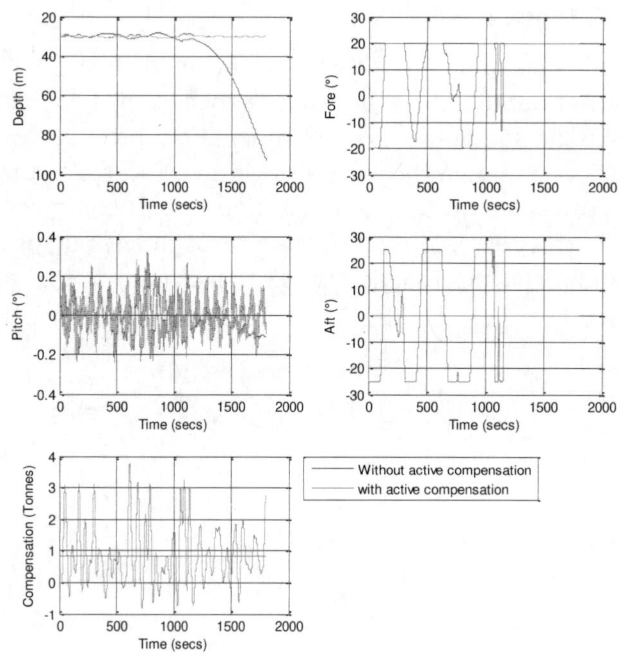

Figure 7: Slow speed depth control in sea state 3.

Compensation rate (m^3/s)	Depth standard deviation (m)
0.10	Submarine broached
0.20	1.40
0.30	0.79
0.40	0.62
0.45	0.59
0.50	0.60
0.55	0.67
0.60	0.79
0.70	1.23

Table 1: Variation in depth control with compensation rate

As the flow rate reduces the depth control decreases as expected, however, this is also the case for the higher compensation rates suggesting that the controller is becoming unstable.

MODEL IMPROVEMENTS

The equations developed by Hirom [1977] cannot fully define the wave induced forces on a submarine for the wide range of depths of submersion, speeds and headings. For example, the shape of the curve for the first order heave force varying with frequency is defined by the function F_{1i} the other terms in the equation are constant for a particular speed and heading (i.e., independent of wave frequency).

The problem of the vertical suction force on a submarine under waves is somewhat similar to that of the horizontal drift forces that are such a challenge on some offshore structures, Pinkster [1979]; both are large volume structures with very little in the way of restoring force. The horizontal slow drift oscillations can create large forces on the anchoring systems of offshore structures which have been illustrated by Faltinsen and Løken [1979] as associating a drift force in each "regular part" of the irregular wave train. The magnitude of the forces may not be large but if the mean period is close to the natural frequency of surge, sway or yaw of the structure, then resonance may occur, [Faltinsen and Løken, 1979].

By the principal of linear superposition, an irregular wave (to first order) can be described as N discrete sine waves at equally spaced frequencies ω_i with amplitudes a_i and phase components ε_i, Pinkster [1979]:

$$\varsigma(t) = \sum_{i=1}^{N_w} a_i \cos(\omega_i t + \varepsilon_i)$$

The square of the wave elevation is therefore:

$$\varsigma^2(t) = \sum_{i=1}^{N_w} \sum_{j=1}^{N_w} a_i a_j \cos(\omega_i t + \varepsilon_i) \cos(\omega_j t + \varepsilon_j)$$

Consider the specific case of bi-chromatic waves (where $N_w = 2$), $\varsigma^2(t)$ will have frequency components at $2\omega_1, 2\omega_2, \omega_1 + \omega_2, \omega_1 - \omega_2$ and a non-zero mean. Since, the second order forces and moments are proportional to the square of the wave amplitude, the second order forces will therefore consist of number of components at each of these different frequencies; the equations used by Hirom implicitly assumes that all these components contribute to the second order force to the same extent. Pinkster [1979] presented the low frequency component of the square of the wave elevation as a linear superposition of a number of bi-chromatic waves:

$$\varsigma_L^2(t) = \frac{1}{2} \sum_{i=1}^{N_w} \sum_{j=1}^{N_w} a_i a_j \cos\left((\omega_i - \omega_j) + (\varepsilon_i - \varepsilon_j)\right)$$

Each component has amplitude $\frac{1}{2}(a_i a_j)^{1/2}$, frequency $(\omega_i - \omega_j)$ and phase $(\varepsilon_i - \varepsilon_j)$

The slowly varying drift force was represented as:

$$F^{(2)}(t) = \sum_{i=1}^{N_w}\sum_{j=1}^{N_w} a_i a_j P_{ij} \cos\left((\omega_i - \omega_j) + (\varepsilon_i - \varepsilon_j)\right) + \sum_{i=1}^{N_w}\sum_{j=1}^{N_w} a_i a_j Q_{ij} \sin\left((\omega_i - \omega_j) + (\varepsilon_i - \varepsilon_j)\right)$$

Where P_{ij} and Q_{ij} are second order transfer functions also known as Quadratic Transfer Functions (QTFs). The special case of $\omega_i = \omega_j$ accounts for the mean drift force as:

$$\overline{F}_H = \sum_{i=1}^{N_w} a_i^2 P_{ii}$$

Where P_{ii} is the mean drift force in regular sinusoidal waves with unit amplitude and frequency ω_i.

Newman's approximation [1974] has been adopted for modelling the slow-drift problem of moored floating vessels; the QTF values can be obtained using the following:

$$F_{ij} = \frac{1}{2}\sqrt{F_{ii} + F_{jj}}$$

where $F_{ij} = \sqrt{P_{ij}^2 + Q_{ij}^2}$

This approximation means that slowly varying second order drift forces on offshore structures can be obtained from the solution to the problem in regular waves. This methodology avoids the requirement to undertake a significant number of computations in bi-chromatic waves. However, Matos et al [2011] suggested that for semi-submersible motions in the vertical plane the use of Newman's approximation may not be advisable, largely due to the considerably lower natural periods of heave, roll and pitch (typically 20 – 80s) compared to the natural periods of drift (generally above 100s), which might be the case with a submerged submarine. These QTFs can be obtained from numerical predictions or model tests which can then be used in the simulation model to provide a prediction of the slowly varying second order forces and moments.

Numerical Methods

Matos et al [2011] presented a series of computations of the low frequency forces on a PETROBRAS semi-submersible using the commercial boundary element method code, WAMIT, Lee [1995]. With this code, different hydrodynamic approximations were tested and compare the predicted responses with model tests.

The formulation of a time domain solution of the wave induced forces on a submarine was developed by Musker et al [1988] and implemented into a QinetiQ in-house code called PERISIM (outlined by Crossland [2013]). The definition of the problem can be formulated from the equations associated with mass continuity, body kinematics, free surface kinematics and free surface dynamics in terms of the velocity potential. The fluid is assumed to be inviscid, irrotational, incompressible and infinitely deep. That is the total velocity potential can be regarded as consisting of components due to the incident wave potential ϕ_I and the disturbance potential associated with the position and motion of the submarine, ϕ_B.

Musker *et al* [1988] computed a total potential to satisfy the continuity and body kinematic equations and to compute ϕ_B to satisfy only the calm water free surface condition (with the assumption that the free surface is rigid. Musker *et al* [1988] represented the submarine hull by 330 panels with point sources. To represent the body boundary better, the source singularity, itself, was then positioned inside the body along the surface normal. The incident wave potential in deep water was defined by Stokes [1847] for irregular waves as:

$$\phi_I = \sum_{i=1}^{N_w} \left[\frac{a_i g}{\omega_i} \sin(k_i x' - \omega_i t + \beta_i) \exp(-k_i x') \right]$$

In [Musker *et al,* 1988], it was noted that the non-lifting potential flow model related only to the naked body and pointed out that any effects due to the appendages were beyond the scope of the work in the paper – except in so far as the appendages effects has been modelled in terms of their components in the set of hydrodynamic derivatives. Subsequent updates led to the appendages been modelled using an empirical database containing lift and drag coefficients of various NACA sections [Abbot and Von Doenhoff, 1949]. A series of improvements to PERISIM has been undertaken including the implementation of the method developed by Hess and Smith [1962]; this is an alternative method for calculating the velocity potential around the surface of an arbitrary body S with the steady free stream flow. The method of determining the velocity distribution on a lifting surface is based on a classical approach of using the knowledge of the spanwise circulation, Γ. Glauert [1947] approximated Γ by a trigonometric expansion that Lloyd [1989] applied to modelling submarine appendages. In determining the contribution of the appendages to the total force on the submarine, the above method is used to extract the unsteady lift and drag forces on each appendage as a consequence of the unsteady flow velocities due to the waves; the forces imparted by the appendages due to a steady velocity, as a consequence of forward speed for example, are already accounted for in the appendage terms in the quasi-steady state equations.

Model Tests

Crossland [2013] presented a comparison between measured data from captive model test in regular waves and predictions. Now, only the non-zero mean second order force can be derived directly from regular wave tests with the QTFs being estimated from the Newman approximation shown earlier. However, the captive model tests don't fully represent the second order forces on a submarine under waves – largely due to the first order motion being restrained. To correctly capture second order forces, the model needs to be allowed to move in response to the first order wave excitation but is restrained against the second order effects. Therefore, a series of captive model tests were undertaken in the Ship Tank at QinetiQ Haslar using a model of the EUCLID submarine. In order to correctly capture the first order motions and second order forces, the model needed to be lightly restrained using two towing swords (or guide swords) that are attached to the model and the carriage. The towing assembly restrained the model in the horizontal plane and surge but allowed it to freely move in the vertical direction whilst being lightly restrained using a suitable spring system. Figure 8 shows the arrangement of the aft tow post and the general set-up with the model attached; the forward tow post, which would bear most of the towing load, is at a forward position to ensure sufficient directional stability. The tow posts were attached to the model by a suitable hinge joint arrangement to allow the tow post to pivot at the attachment point. The tow rig needs to be carefully designed to ensure that the resonant frequency of this spring-mass system is sufficiently far from the wave frequencies that are being tested; in this way, the springs supply stiffness (but negligible damping) in the vertical plane. The towing arrangement was rigged with a set of compression springs providing the stiffness in two directions; each tow post had two springs where only one end of each spring was rigidly fixed to a load cell. In this configuration only one spring on each tow post provided the restoring force (when that spring was under compression) – when one spring was under compression the other one would provide no force at all see Figure 8.

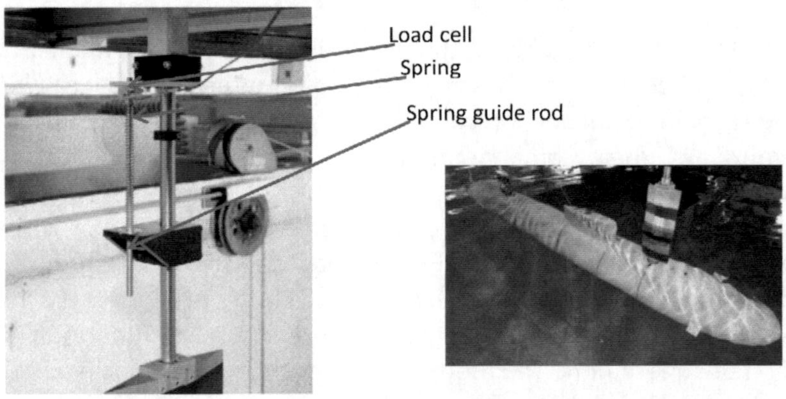

Figure 8: Aft tow post spring assembly and model set up

The arrangement means that as the stern of the model rises (as a consequence of the suction force) the lower spring will compress providing the restoring force. Furthermore, the compressed spring will impart a force on the load cell from which this component of the suction force can be measured. The choice of spring stiffness is really a balance between ensuring the natural frequency of the system is far away from the wave encounter frequency and that the spring is stiff enough to offset the second order suction force without significant compression. The damped natural period was found to be around 12 seconds (at model scale).

Tests were undertaken in calm water over a range of speeds, head seas regular waves, bi-chromatic waves and irregular waves. The method of analysis for the regular wave and bi-chromatic tests is that from the 23rd ITTC Ocean Engineering Committee, [2002] whereby the means of extracting the amplitude and phase of all the harmonics is to use a non-linear least squares analysis.

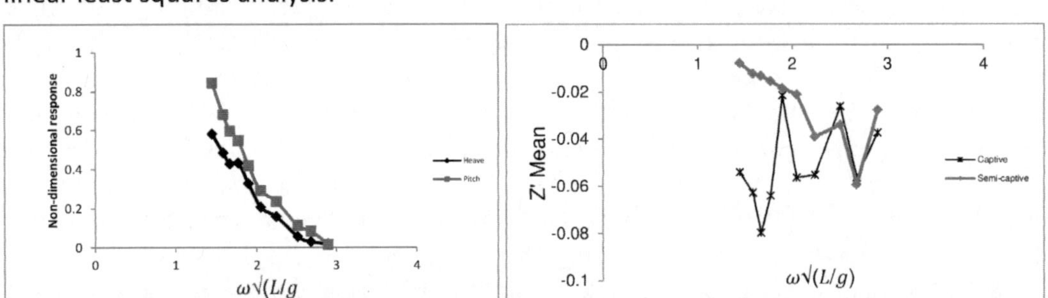

(a) first order motion (b) mean heave force

Figure 10: Regular waves at zero speed

Figure 10 shows the first order motion responses (heave and pitch) and mean heave force of the model in regular waves; also shown is the mean heave force measured during the fully captive model tests of Martinussen [2006]. The comparison of the mean heave force shows that for all but the waves with the highest frequency the results from the semi-captive tests are different to those from fully captive tests. Figure 11 shows the heave QTF derived from bi-Chromatic wave tests; it is this function that the second order slowly varying heave force can be determined.

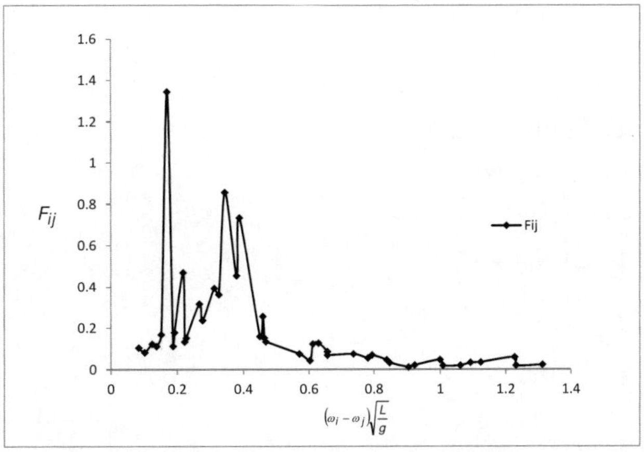

Figure 11: QTF for heave force in bi-chromatic waves at zero speed

SUMMARY

Described here is a potential approach that supports the design of systems for depth-keeping control of a submarine under the influence of waves. The methodology is reliant on a credible mathematical model for the wave induced forces and moments on the submarine. The simplified approach to modelling surface suction described here is currently being updated to include a higher fidelity mathematical model and to include the use of Quadratic Transfer Functions to support the design of any control system.

NOTATION

a_i	-	amplitude of the ith wave component in the irregular wave
A_j	-	amplitude of the jth harmonic
C_{Z1}	-	non-dimensional coefficient for first order heave force
C_{M1}	-	non-dimensional coefficient for first order pitch moment
C_{Z2}	-	non-dimensional coefficient for second order heave force
C_{M2}	-	non-dimensional coefficient for second order pitch moment
D	-	representative diameter of the submarine
F_{RAO}	-	Force amplitude
$k_i = \frac{2\pi}{\lambda_i}$		wave number.
H	-	depth of immersion
$H_{1/3}$	-	Significant wave height
L	-	submarine length
M	-	number of harmonic responses (m=2 for up to second order)
M_p	-	maximum overshoot
M_{RAO}	-	Moment amplitude
M_{MEAN}		mean pitch moment

N	-	number of points in the time history
N_w	-	number of wave frequencies
T_0	-	Modal period
t_s	-	settling time defined as the time taken for the system to settle to a specified level of $\pm x\%$
t_r	-	time taken for the output to rise from 10% to 90% of the step
t_p	-	time taken to reach the peak value
$x(t)$	-	measured time history
$\overline{x}$	-	mean of the measure time history
Z_{MEAN}		mean heave force
β_i	-	phase angle in waves
ρ	-	density,
ζ_w	-	wave amplitude.
∇	-	volume of displacement
μ	-	heading of the submarine relative to the waves (0 degrees is following seas and 180 degrees is head seas)
ω_i	-	i^{th} wave frequency
ω_e	-	fundamental frequency
ϕ_j	-	phase of the jth harmonic
ϕ_I	-	incident wave potential
ϕ_B	-	disturbance potential
λ_i	-	wavelength

REFRERENCES

1. 23rd International Towing Tank Conference Ocean Engineering Committee. Analysis procedure for model tests in regular waves. No, 7.5 – 02 07 - 03.2, 2002.
2. Burcher, RK., Rydill, LT. Concepts in submarine design, Cambridge University Press, 1994. ISBN 0 521 41681 7.
3. Byström, L., Adaptive control of a submarine in a snorting condition in waves. RINA Warship'88: Conventional Naval Submarines, London, 1988.
4. Crepel JL., Bovis AG., New trends in the design of steering and diving control systems. RINA Warship'91: Naval Submarines 3, London, May 1991.
5. Crossland P., Kimber NI., Thompson N., Understanding the Manoeuvring Performance of an X- plane Submarine in Deep Water and Near the Free Surface. Pacific 2012: International Maritime Conference, Sydney, January 2012.
6. Crossland P., Profiles of Excess Mass for a Generic Submarine Operating Under Waves. Pacific 2013: International Maritime Conference, Sydney, November 2013.
7. Faltinsen OM., Løken AE., Slow drift oscillations of a ship in irregular waves, Applied Ocean Research, Vol 15, No.1, 1979
8. Gertler, M., Hagen GR., Standard equations of motion for submarine simulation. Naval Ship Research and Development Center Report No. 2510, June 1967, David Taylor Model Basin, Carderock, MD, USA.

9. Glauert, H., The elements of aerofoil and airscrew theory, Cambridge University Press, 1947.
10. Hess JL., Smith AMO., Calculation of non-lifting potential flow about arbitrary three-dimensional bodies. Douglas Aircraft Division Report No. E.S. 40622, March 1962.
11. Hirom, CPJ., The mathematical analysis of submarine motion near the free surface. Admiralty Experiment Works Internal publication, AEW TR 77015, Haslar, March 1977.
12. Kimber, NI., Crossland, P., Advances in experimental techniques for understanding the manoeuvring performance of submarines. RINA Warship 2008: Naval Submarines 9, Glasgow, June 2008.
13. Kimber, NI., Thompson, N., Submarine motions in confined waters, EUCLID RTP 10.17. Analysis of captive model experiments and manoeuvring simulations, QinetiQ/07/01163, April 2007, QinetiQ Proprietary.
14. Lee, C.H., WAMIT Theory Manual. Report No.95-2.Massachusetts Institute of Technology. 1995
15. Lloyd, ARJM., A theory for submarine manoeuvring SUBSIM (Issue 3). Admiralty Research Establishment Internal Publication, Gosport, 1989.
16. Mandzuka S., Mathematical model of a submarine dynamics at the periscope depth. Brodogradnja 46, (1998), 2, pp129-138.
17. Martinussen K., Submarine motions in confined waters – tests in the Ocean Basin Labatory close to the surface in oblique waves. EUCLID/RTP 10.17/MARINTEK/WE-2.5/TR/1/1, Marintek Report 810009.00.01, March 2006.
18. McFarlane, DC., Glover K. A Loop Shaping Design Procedure using Synthesis, IEEE Transactions on Automatic Control, vol. 37, no. 6, pp. 759–769, June 1992.
19. Musker, AJ., Loader, PR., Butcher, MC., Simulation of a submarine under waves. International Shipbuilding Progress, 35, No. 404, 1988, pp 389-410.
20. Perez T., Performance Analysis and Constrained Control of Fin and Rudder-Based Roll Stabilizers for Ship. PhD Thesis School of Electrical Engineering and Computer Science, University of Newcastle, Australia, June 2003.
21. Pinkster, JA., Mean and low frequency wave drifting forces on floating structures. Ocean Engineering, Vol 6, pp593-615, 1979.
22. Ray, A., Seshadri V., Singh SN., Sen D., Manoeuvring studies of underwater vehicles – a review. International Journal of Maritime Engineering, 2008.
23. Stokes, GG., On the theory of oscillatory waves, Transactions, Cambridge Philosophical Society 8, 1847.
24. Veillon A., Aillard JM., Brunet, P. Submarine depth control under waves: an experimental approach. RINA Warship '96: naval Submarines 5 – the total weapon system, London, 1996.
25. Wilson, JF., Dynamics of Offshore Structures (Second Edition). ISBN 0481264679, Published by John Wiley & Sons, 2003.

© Copyright QinetiQ Limited 2017

The Whipping Response of a Submerged Free-Free Cylinder Due to Underwater Explosions

Steven De Candia[1]; Roberto Ojeda[1]; Warren Reid[2]; Max Ratcliffe[3]; Jonathan Binns[1]

[1]Australian Maritime College, University of Tasmania; [2]Defence Science and Technology Group; [3]Babcock Australasia

steven.decandia@utas.edu.au

ABSTRACT

Underwater explosions (UNDEX) generated by torpedos and mines present one of the most significant threats to the survivability of naval vessels. A non-contact UNDEX event produces two distinct pressure loading phenomena: an initial short duration, high frequency shock wave followed by a relatively low frequency pulsating bubble containing the gaseous products of detonation. The pulsating bubble may excite global, beam-like, structural response modes of the vessel, known as UNDEX induced whipping. Depending on the size and position of the explosive charge relative to the vessel, the maximum strain levels during whipping may exceed those during the initial shock phase. Due to the large costs of full-scale explosive trials, development of validated numerical modelling procedures for UNDEX whipping and shock response are required to assess the survivability of naval vessels. An experiment has been conducted to investigate the UNDEX whipping response of a fully submerged cylindrical pipe structure considering the effects of different charge sizes, stand-off distances and longitudinal positions along the hull. It was found that the longitudinal stand-off location has a significant influence on the whipping response and that the peak whipping response may not always occur at the stand-off point.

Keywords: Underwater explosion, structural response, near-field, whipping

NOMENCLATURE

UNDEX	Underwater explosion	DST	Defence Science & Technology Group
W	Charge mass	ONR	Office of Naval Research
R	Stand-off distance	DNA	Defence Nuclear Agency
A	Maximum bubble radius	NAVSEA	Naval Sea Systems Command
R/A	Stand-off to bubble ratio	USA	Underwater Shock Analysis
BM	Bending Mode		

INTRODUCTION

Underwater explosion (UNDEX) induced whipping is a global vessel bending response, induced by a pulsating bubble of the explosive gas by-product. A whipping response has the potential to compromise the overall integrity of the hull, e.g. breaking the keel or buckling the hull plate on a surface vessel (Figure 1), or initiating interframe shell yielding or overall collapse failure of the pressure hull on a submarine [1].

Figure 1 The whipping response of HMAS Torrens from a heavy weight torpedo sinking exercise, and the buckled hull plate on USS Princeton due to whipping from a mine strike.

The explosive charge mass (*W*) and stand-off distance (*R*) are the two most significant variables in UNDEX [2]. The combination of these are the measurement of severity [3], which for a spherically spreading shock wave may be calculated by the ratio of $W^{1/3}/R$. For bubble effects, event regimes are considered as being near-field or far-field, defined by the ratio of R to the maximum bubble radius (*A*) [4], where an event is considered near-field when the ratio of $R/A < 2$. Whipping can be induced by the periodic loading from the bubble pulse in far-field events, however the whipping response can be more severe from near-field UNDEX as the fluid flow from the bubble expansion and pulse loadings could be coupled with the fluid flow from the vessel's whipping response [5]. If present, the bubble-structure coupling can potentially amplify the whipping response, particularly if the pulsating frequency of the bubble is close to the relevant natural response frequencies of the vessel.

Ideally, naval vessels are assessed against the UNDEX threat using analytical and numerical models. These models are in turn validated by experimental studies, often on a scaled down simplified structure representing a surface vessel or submarine. To have confidence in the analytical results, the analysis method employed needs to be validated against experimental observations, which often have some limitations due to their scaled down and/or simplified representation, that will be discussed in the following section.

In this paper, a number of experiments are discussed. These have been conducted on cylindrical, air-tight structures. The paper provides the details of the setup and an overview of the results of an UNDEX induced whipping experiment, recently conducted on a generic submerged slender cylindrical hull.

OVERVIEW OF PREVIOUS CYLINDRICAL HULL UNDEX EXPERIMENTS

An analytical response of a vessel to UNDEX induced whipping was first presented by Chertock [6] for both floating and submerged structures. Chertock made assumptions to simplify this solution: He ignored the effects of bubble migration, and assumed a uniform mass distribution. The theoretical analysis was based around a set of experimental observations from UNDEX tests conducted against a floating barge and submerged cylinder [7]. The 2.51 m long, 0.206 m diameter cylinder was stiffened, and sealed by end plates. The cylinder was subjected to UNDEX events ranging from 1.2 to 8.2 grams of TNT, at depths of 3.66 to 275 m, and stand-off distances of 0.3 to 1.52 m; with all events considered far-field. This experiment also explored the effect of longitudinal stand-off point on the whipping response, but only for the case of the surface barge. The cylinder experiments showed how the variation of depth and charge size could initiate 2-node and 3-node bending responses, however the analytical theory over-predicted the 3-node bending response by almost 30%. A review by Bannister [8] noted that there was no indication of how well these experiments and the analytical theory would practically scale when extrapolated to full-scale. Furthermore, Bannister questioned as to how well this theory would apply to near-field bubble-structure interaction, particularly when whipping response is coupled with the pulsating bubble.

Experimental work since then has primarily focused on the response of internal equipment and the extent of localised hull damage. A combined effort of the Defence Nuclear Agency (DNA), the Office of Naval Research (ONR), and the Naval Sea Systems Command (NAVSEA) produced the DNA/ONR 33.6 test model which consisted of a submerged 6.43 m long, 0.845 m diameter, ring stiffened cylinder that was subjected to a series of UNDEX events for numerical analysis benchmarking purposes [9]. This vessel was tested under three configurations: no internal equipment, symmetrical mass distribution, and asymmetrical mass distribution of internal equipment. The measurements and analysis conducted on this experiment only considered the early time shock response, resulting in an insufficient time history to determine whether whipping occurred.

The shock response on small submerged cylinder sections has been explored, specifically looking at far-field response and modelling techniques [10], close-in bubble collapse contact responses [11, 12], and the response of internal decks and their modelling [13, 14]. However, none of the above experiments were able, or designed to consider any significant whipping effects on the structural response. van Aanhold et al. [15] conducted experiments on a large floating stiffened cylindrical hull that briefly displayed a whipping response, however they did not discuss the whipping in any significant detail and the reported data ranges were sanitised, making it unsuitable for validation or benchmarking studies by a third-party.

A limited number of full-scale experiments have been conducted, but were not as well documented as generic tests discussed above. The shock response of full-scale submarines and onboard equipment has been documented from WWII tests in [16], but the measurement and data acquisition equipment used at the time was in its infancy, and the data was case specific to the submarine vessels tested. The 'Steel Mosquito' from Hellqvist [17] was an 11 m long, 5.7 m diameter scale submarine section that investigated the localised panel damage, and the response to UNDEX events of structure and equipment such as ballast tanks, batteries, hull penetrations, and resiliently mounted decks. However, this low length to diameter ratio section did not exhibit any notable whipping response. Very recent work from Liu et al. [18] presented two experiments on complex water and double hull structure interfaces. The more complex of these experiments consisted of a liquid oxygen tank, resiliently mounted within a large stiffened pressure hull. This hull was surrounded by an elliptical sectioned light casing that was free-flooded. Two smaller independent pressure hulls were positioned port and starboard side of the large pressure hull, separated by the free-flooded area of the light casing. While no whipping response was predicted or observed by the outer hull, it was predicted that the resiliently mounted tank would exhibit an oscillatory response that was amplified at the bubble pulse. However, the experimental results showed much heavier damping of this cylindrical tank response with lower amplification resulting from the bubble pulse load.

Of all experiments discussed here, none have been suitable to determine the nature of near-field UNDEX induced whipping, or the effect on the whipping response of longitudinal stand-off points other than amidships. In the following section, we give an overview of a submerged whipping experiment, designed to explore the whipping response of a cylindrical hull to near-field UNDEX loads.

OVERVIEW OF A NEAR-FIELD WHIPPING EXPERIMENT

As previously noted, there is no validation of the applicability of older analysis approaches when scaled-up, or for near-field UNDEX scenarios where there is a coupled response between the bubble and vessel structure. The effects of the longitudinal placement of the charge have also been ignored for submerged cylinders. To investigate this, an experiment was conducted on a generic submerged unstiffened cylinder to determine the elastic whipping response from near-field UNDEX at amidships and for various longitudinal stand-off positions. The cylindrical whipping vessel in Figure 2 was 12 m long, 0.4 m nominal diameter with a wall thickness of 6.35 mm. Weighted endcaps were used to tune the vessel response to the range of the pulsating bubble frequency. The vessel was constructed from cold rolled steel with a yield stress of 350 MPa. The total vessel mass including the onboard measurement equipment was 1548 kg.

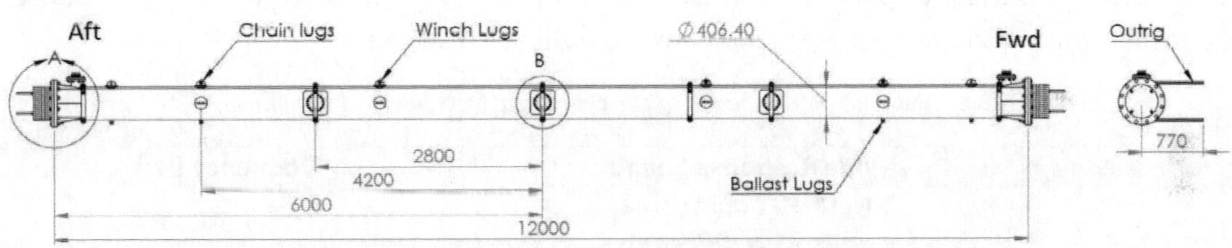

Figure 2 Cylindrical whipping vessel.

A total of eight UNDEX events were conducted at the Defence Science & Technology Group (DST) former underwater explosion test facility, which was a flooded quarry with approximate dimensions of 100 m x 40 m and a depth of 16 m [19]. Discussed here are six event pairs that were conducted at different longitudinal positions along the hull, shown in Figure 3. Each event pair consists of one 0.25 kg or 0.043 kg Pentolite charge, at 1.3 m or 0.8 m stand-off respectively. All charges were detonated from outriggers mounted at the vessel's portside, in line with the vessel's centreline. The vessel and charges were submerged to 5 m depth, with the quarry floor approximately 11 m below the vessel's centreline. The maximum bubble diameters were calculated using similitude equations from [20] as 0.9 m and 0.5 m, for 0.25 kg and 0.043 kg charge sizes respectively, shown to scale in blue in Figure 3. The R/A ratio for all events was less than 2, therefore all events were considered near-field.

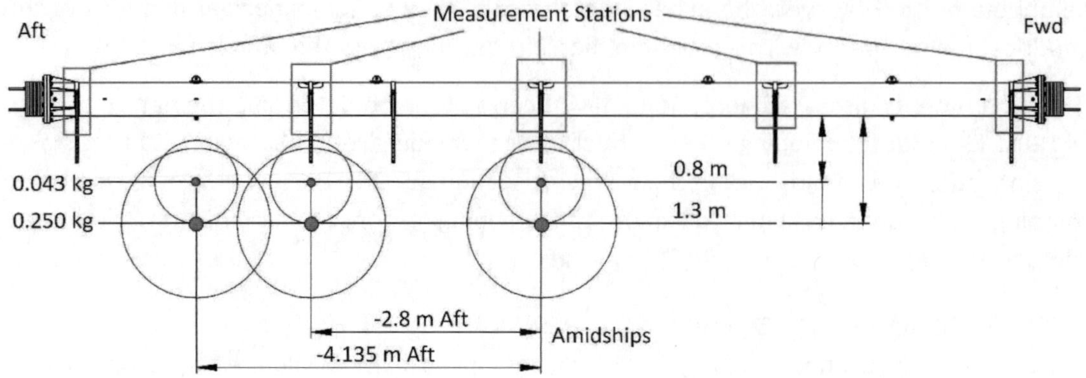

Figure 3 Topside view of measurement locations and charge placements, with one charge used for each event.

A modal analysis was undertaken using the LS-Dyna Underwater Shock Analysis (USA) code [21] to determine the theoretical submerged natural frequencies and mode shapes of the structure. The first three bending mode shapes of the cylindrical free-free beam structure are shown in Table 1. The first two mode shapes were used to determine the locations of the charges and the measurement stations.

The bubble pulsation period was calculated using similitude equations from [20], which resulted in a frequency of 7.11 Hz and 12.78 Hz for 0.25 kg and 0.043 kg charge sizes respectively. The similarity between the calculated bubble loading, and structural response frequencies created the potential for the coupled bubble-structure response.

Table 1 Bending mode shapes of the cylindrical free-free beam hull.

Mode Response Shape	Frequency (Hz)
2-Node Bending (BM1) 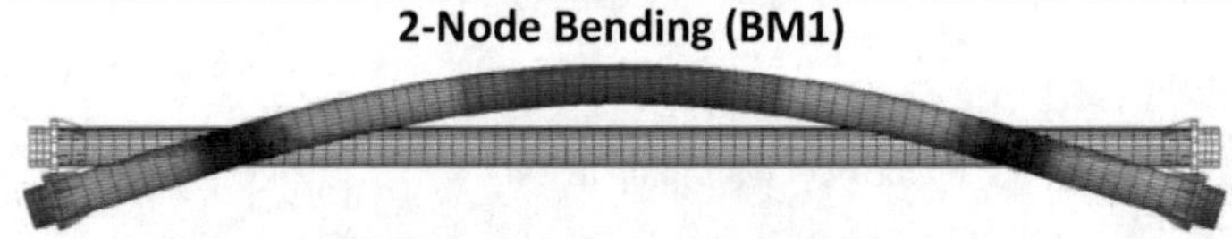	6.85
3-Node Bending (BM2) 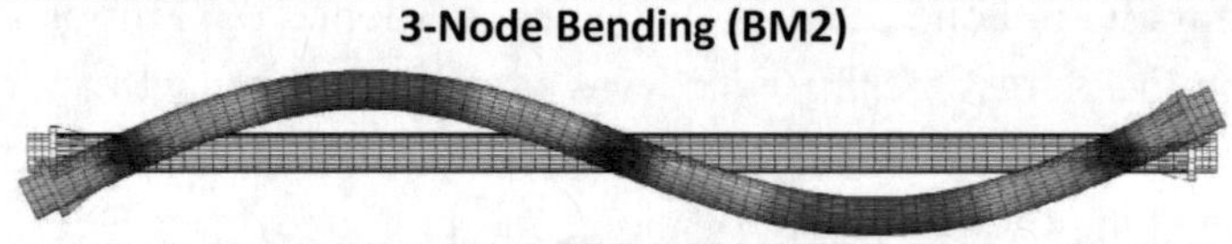	20.32
4-Node Bending (BM3) 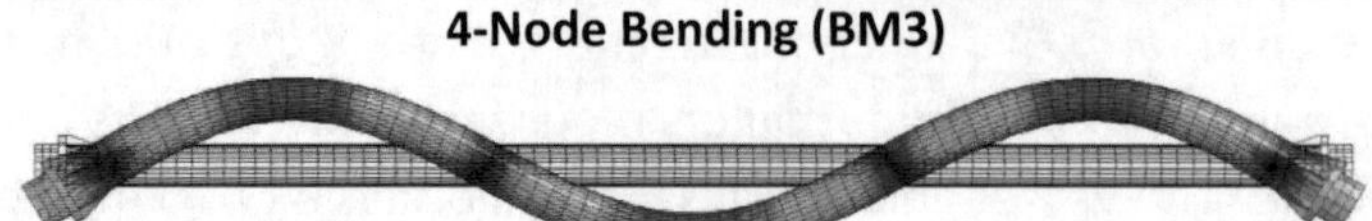	41.56

EXPERIMENTAL RESULTS

Axial strain measurements were taken on the portside at five measurement stations shown in Figure 3. The maximum recorded whipping strains for 0.25 kg and 0.043 kg charge events have been presented in Figures 5 and 6 respectively, for the three longitudinal stand-off charge locations: amidships; -2.8 m aft; and -4.135 m aft.

From the 0.25 kg results presented in Figure 4, it is shown that the amidships charge stand-off location produced the highest response, which approached the theoretical yield strain limit of the hull material. Notably the -2.8 m aft charge stand-off location showed a similar trend in the strain response levels at all stations, except amidships. This event also saw amplification of the whipping response at the 2.8 m forward station, located away from the charge. This amplification is attributed to positive interference between the first and second bending mode responses at this location. As the second bending mode is antisymmetric about amidships, the interference of modes was destructive at the 2.8 m aft location during the peak whipping phase.

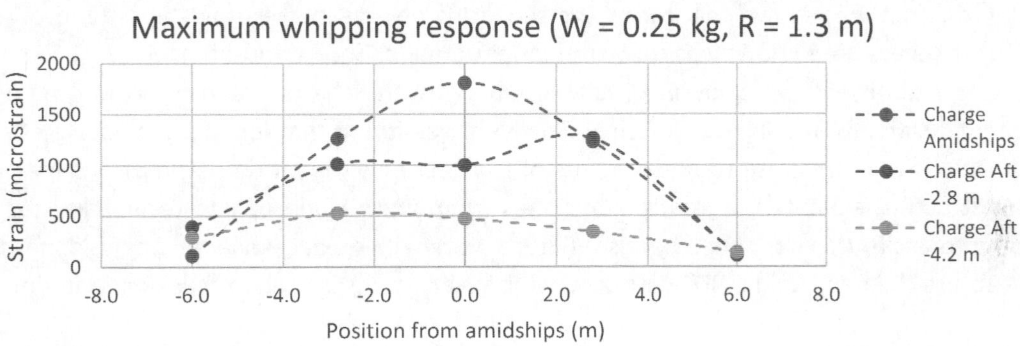

Figure 4 Maximum measured strain from whipping response for 0.25 kg charge at 1.3 m stand-off.

The results for the smaller 0.043 kg charge events, presented in Figure 5, also saw the highest whipping response from the amidships charge stand-off location. However, the smaller charge size and spherically spreading load excited the third bending mode of the structure, resulting in deconstructive interference at amidships, resulting in the peak responses occurring at the 2.8 m forward and aft locations. The -2.8 m aft charge stand-off saw a similar response level across the amidships and 2.8 m forward and aft positions, with an overall reduced response compared to the amidships stand-off. The amplification of whipping response seen in the 0.25 kg event was not seen in the 0.043 kg event.

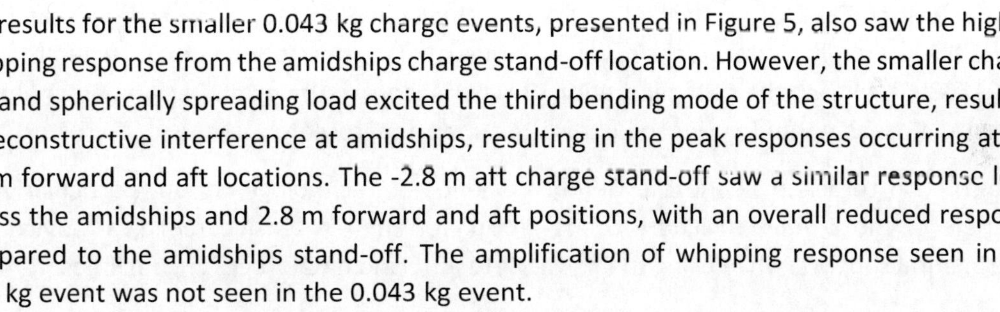

Figure 5 Maximum measured strain from whipping response for 0.043 kg charge at 0.8 m stand-off.

Both charge sizes induced a greatly diminished whipping response at the -4.135 m longitudinal charge stand-off location. This is attributed to the close proximity of the charge stand-off point to the nodal point of the first bending mode response, which precluded this mode from contributing to the overall whipping response of the structure.

CONCLUSION

A review of literature has identified that analysis of structural whipping responses may not be sufficiently well predicted at larger scales and for near-field UNDEX events. The effect of longitudinal stand-off location has also not been explored in any significant detail on a generic structural model. An experiment was conducted that demonstrated an elastic whipping response of a generic submerged cylinder structure to near-field UNDEX. Here the effects of different longitudinal stand-off locations on the vessel's whipping response have been explored and it was shown that for all the scenarios discussed, the amidships stand-off was the most severe case. However, for some longitudinal charge stand-off scenarios, the peak whipping response may not occur directly in-line with the charge stand-off location, due to the amplification of modal interference. Finally, it was found that longitudinal charge stand-off locations that are close to nodal points of the first bending mode shape of this free-free beam structure exhibited a greatly diminished whipping response, due to inability for the loading to excite this bending response. Future work will look at how well analysis methods can capture the near-field UNDEX induced whipping responses seen in these experiments.

ACKNOWLEDGEMENTS

This work was conducted with support from an Australian Government Research Training Program scholarship and the ARC Research Training Centre for Naval Design and Manufacturing (RTC-NDM). The RTC-NDM is a University-Industry partnership established under the Australian Research Council Industry Transformation grant scheme (ARC IC140100003), to train the next generation of Naval Manufacturing and Design personnel and provide the naval manufacturing industry with supporting research.

We also wish to acknowledge the Defence Science & Technology Group, particularly the members of the Dynamic Military Loads group, for their extensive technical advice and experimental support, while using their UNDEX testing and data processing facilities.

REFERENCES

1. Burcher, R. and L.J. Rydill, *Concepts in submarine design*. Vol. 2. 1995: Cambridge University Press.
2. Cole, R.H., *Underwater Explosions*. 1948, Princeton, New Jersey: Princeton University Press.
3. Reid, W.D., *The Response of Surface Ships to Underwater Explosions*. 1996, Defence Science and Technology Organisation: Melbourne, Victoria.
4. Hunter, K.S. and T.L. Geers, *Pressure and velocity fields produced by an underwater explosion.* The Journal of the Acoustical Society of America, 2004. **115**(4): p. 1483-1496.
5. Best, J.P., *The dynamics of underwater explosions*, in *Department of Mathematics*. 1991, University of Wollongong: University of Wollongong. p. 184.
6. Chertock, G., *The Flexural Response of a Submerged Solid to a Pulsating Gas Bubble.* Journal of Applied Physics, 1953. **24**(2): p. 192.
7. Chertock, G., *Transient Flexural Vibrations of Ship-Like Structures Exposed to Underwater Explosions.* The Journal of the Acoustical Society of America, 1970. **48**(1B): p. 170-180.
8. Bannister, K.A., *Whipping analysis techniques for ships and submarines.* The Shock and Vibration Bulletin, 1980. **50**: p. 83.

9. Ranlet, D. and R.P. Daddazio. *A submerged shock response problem suitable for use as a benchmark*. in *Proceedings of the 68th Shock and Vibration Symposium*. 1997.
10. Kwon, Y. and P. Fox, *Underwater shock response of a cylinder subjected to a side-on explosion.* Computers & structures, 1993. **48**(4): p. 637-646.
11. Brett, J.M. and G. Yiannakopoulos, *A study of explosive effects in close proximity to a submerged cylinder.* International Journal of Impact Engineering, 2008. **35**(4): p. 206-225.
12. Brett, J.M., G. Yiannakopoulos, and P.J. van der Schaaf, *Time-resolved measurement of the deformation of submerged cylinders subjected to loading from a nearby explosion.* International Journal of Impact Engineering, 2000. **24**(9): p. 875-890.
13. Yiannakopoulos, G., *Small Scale Elastic Deformation Studies from an Underwater Explosion on a Submerged Thin Walled Platform Cylinder*, in *Proceedings of the 77th Shock & Vibration Symposium*. 2006, Shock and Vibration Information Center: Monterey, CA.
14. Yiannakopoulos, G., *Fluid and Structure Interaction Studies on a Submerged Platformed Cylinder near an Underwater Explosion*, in *Proceedings of the 78th Shock & Vibration Symposium*. 2007, Shock and Vibration Information Center: Philadelphia, PA.
15. van Aanhold, J., W. Trouwborst, and J. Vaders, *Underwater Shock Response Analysis of a Floating Cylinder*, in *Proceedings of the 79th Shock & Vibration Symposium*. 2008, Shock and Vibration Information Center: Orlando, FL.
16. Welch, W.P., *Mechanical shock on naval vessels*. 1946: US Government Printing Office.
17. Hellqvist, K., *A large-scale submarine shock test carried out as part of the Swedish Shock Design Development Program*, in *The Shock and Vibration Bulletin*. 1981, Shock and Vibration Information Center: Naval Research Laboratory, Washington D.C. p. 129-136.
18. Liu, G.-z., et al., *A numerical method for double-plated structure completely filled with liquid subjected to underwater explosion.* Marine Structures, 2017. **53**: p. 164-180.
19. Brett, J.M., et al., *An experimental facility for imaging of medium scale underwater explosions*. 2003, DSTO Platforms Sciences Laboratory: Fishermans Bend, Victoria, Australia.
20. Swisdak, M.M., Jr, *Explosion Effects and Properties. Part II. Explosion Effects in Water*. 1978, Naval Surface Weapons Center: Silver Spring, MD, USA.
21. DeRuntz, J.A. *The underwater shock analysis code and its applications*. in *Proceedings of the 60th Shock and Vibration Symposium*. 1989. Virginia Beach, Virginia: Shock and Vibration Information Center.

The Ungoverned Space: Call for Improved Test & Maintenance to Comply with Regulations & Go Beyond

Dr Carl Stephen Patrick Hunter
Coltraco Ultrasonics

The maintenance and testing of watertight hatches, doors and multiple cable transits on commercial and military vessels is essential.

A lack of proper servicing of these seals can lead to deterioration which can endanger the ship, cargo and lives of the crew through flooding and the potential of capsizing. Leaking hatch-covers are generally accepted as a principal cause of cargo damage. A 2001 report from the Institute of Marine Engineers suggests that "hatch cover deterioration or mal-operation could jeopardise the lives of seafarers and may contribute to the loss of vessels. 33% of large claims by value in general cargo and bulk carrier vessels are caused by leaking hatch covers according to the Swedish P&I Club. They also noticed that there is an increased number of costly wet damage claims. These claims are most commonly seen on bulk carriers and container vessels. Wet damage claims make up 19.56% of claims in total and the poor condition of cargo hatch covers is the greatest contributor to this. Even on new ships, hatch covers may only be truly weather-tight in the static "alongside" condition and cannot be guaranteed in dynamic sea state conditions.

Hatches leak for a variety of reasons, but mainly because of poor maintenance or failure to seal them properly after loading.

A watertight hatch cover is designed to prevent the passage of water in either direction under a head of water for which the surrounding structure is designed. Many mariners may think hatches are indestructible, robust, monolithic structures, thereby failing to appreciate the reality of the small tolerances on panel alignment and gasket compression. It is better to think of hatches as complex, finely-made structures, to be handled with care. All types of seals, experience dynamic stresses as part of their operational lifetime. For example, 4mm wear on the steel-to-steel contact is sufficient to damage rubber sealing gaskets beyond repair; 5mm sag along the cross-joint can cause a large gap between the compression bar and gasket.

The risk is worsened by the ageing nature of many ships, particularly bulk carriers. There is also a degree of bending/deformation that naturally occurs in ships during travel, which puts pressure on hatch covers and can damage sealing. A recent wave of inexperienced crew members has swept across the shipping industry as a cost-saving mechanism, leading to a decrease in quality of vessel maintenance and hatch cover testing.

Emma Maersk - 2013

The case study of the Emma Maersk exemplifies the danger of improper servicing. A severe leakage occurred on the container ship in February 2013 when it was loaded with 14,000 containers. The leakage was caused by a mechanical break-down of a stern thruster, creating the shaft tunnel to flood, as well as leading to severe ingress of water in the aft part. Eventually there was flooding of the main engine room. This was because of non-effective cable penetration sealings. In a sudden blast, four cable penetration sealings in the watertight bulkhead gave way to the water pressure followed by a massive ingress of seawater. Shortly after this, the other three cable penetration sealings also failed, resulting in an even larger ingress of water into the engine room. This led to approximately USD 45 million worth of damages and towage cost.

Figure 19: Seawater ingress from propeller shaft tunnel into the main engine room via burst cable penetration sealings in the watertight bulkhead, viewed from the starboard side of the engine room
Source: Maersk Line

Figure 17: Seawater ingress from propeller shaft tunnel into the main engine room via burst cable penetration sealings in the watertight bulkhead, viewed from the port side of the engine room

Source: DMAIB, Feb 2013, URL <http://www.dmaib.com/Ulykkesrapporter/Flooding%20of%20engine%20room%20-%20EMMA%20M%C3%86RSK%20on%201%20February%202013.pdf>

The risk of misunderstanding the significance of testing watertight integrity

The cost of improper watertight seals to life, vessel and cargo is detrimental. The need for servicing of the hatch covers, doors, multiple cable transits (MCTs) and all watertight seals onboard is clear.

High pressure watertight doors are designed for pressures in excess of 100' (30.5m) (this is equivalent to a pressure of 299 kPa/ or 3 bar). Typical applications include protection for ship power plants, high-pressure chambers, hydro-electric plants and diving/decompression chambers which are often found on oil platforms and support vessels. In these sensitive areas, a small 1m^2 watertight door is often expected to be able to resist the weight of 30 tonnes of water, without buckling or allowing the protected compartment to flood. Therefore, the requirement for frequent and accurate preventative maintenance on these seals cannot be overstated. Watertight doors are critical to preserving the separation of watertight compartments so flooding can be contained to a local area. Manufacturers of watertight doors often have their own tolerances which allow a certain amount of water to flow through a door per minute at a certain pressure.

Regulations

The regulations fully state the need to ensure watertight integrity, with periodic inspections required by SOLAS such as two included for ease of reference below:

SOLAS Reg II-1/13-1.1 – Openings in Watertight Bulkheads and Internal Decks in Cargo Ships

- "Where penetrations of watertight bulkheads and internal decks are necessary for access, piping, ventilation, electrical cables, etc., arrangements are to be made to maintain the watertight integrity."

SOLAS Reg II-1/21.3 – Periodical Operation and Inspection of Watertight Doors, etc., in Passenger Ships

- The watertight doors and all mechanisms and indicators connected therewith[sic] shall be periodically inspected at sea at least once a week.

Traditional Testing Methods have Limitations

The traditional testing methods are insufficient to ensure full watertight integrity. Owing to being well known methods this paper will just briefly outline how they are conducted and their limitations before suggesting an alternative method.

During chalk testing, chalk is applied to the compression seals and the hatches are closed, sealed and reopened. On examination, any irregularity in the chalk pattern implies that improper sealing is occurring in that section of the seal. Chalk testing is ineffective, because even if seals are touching there is no way of indicating if there is sufficient pressure between them to create a complete seal. Chalk testing also cannot be performed on MCTs or air vent seals which cannot be easily open and closed manually or at all.

During high pressure hose testing a surveyor is present inside the hold. A jet of pressurized water is aimed all the way round the hatch-cover. Any water ingress which occurs will be seen by the surveyor. This test is time consuming and inefficient, requiring two crew members for testing, as well as significant clean up. Hose testing provides almost no indication of the severity and/or location of leaks, and provides no recordable, verifiable readings. Furthermore, ecological concerns have been raised concerning pollute run off.

Gas pressure test are most common for airtight testing. Simply put, the airtight compartment is sealed and then pressurised above atmospheric pressure. Depending on the standard, the compartment is then left pressurised for a degree of time while the internal pressure is monitored. An airtight compartment is expected to hold and maintain the internal pressure for a fixed amount of time. Common issues include rendering the compartment unusable whilst the gas pressure test is underway. Furthermore, no information on leak site locations is inferred, obligating the operator to conduct a thorough and tedious manual inspection of all seals should a leak be identified. Some operators combine this test with 'Bubble testing' where soap mixture is sprayed on seals in the hope that the escaping gas will produce bubbles at the leak site. This is simply impractical due to the size of many compartments and its inherent unreliability.

Innovative Alternative Method: Ultrasonics

The technology is highly sensitive and accurate capable of testing watertight integrity of seals to a very high standard. Ultrasonic Testing (UT) is being rapidly adopted as a safe, easy and accurate method of identifying seal leak sites and has been quickly adopted by the Royal Navy, among others, to quickly identify any incorrectly installed MCTs.

The sounds we hear consists of physical vibrations that move through air. Ultrasound is no different, except its frequency is beyond the range of human hearing. Whilst the highest frequency a human can hear is around 20 kHz, ultrasound generators produce signals above this frequency around 40kHz. It is used by bats and dolphins in nature, submarines in defence to locate themselves and others by decoding the emitted pulse being bounced back, and here for seal integrity testing.

This technology is well suited for seals, including NBCW ones, that are meant for high pressure watertight protection. Ultrasonic watertight integrity testing is a proven, pin-point accurate method. UT for the purpose of leak detection is utilised in the form of portable ultrasonic watertight integrity test indicators for inspecting hatch covers in the shipping industry, doors, multiple cable transits, scuttles, shell doors, flanges, bulkheads and other watertight compartments on multiple vessel types, offshore and in defence.

This is understood in the regulations:

SOLAS Reg II-1/11.1 – Initial Testing of Watertight Bulkheads states "where a hose test is not practicable [sic] it may be replaced by [sic] an ultrasonic leak test or an equivalent test. In any case, a thorough inspection of the watertight bulkheads shall be carried out."

The Value of Ultrasonic Testing: Mathematically Proven Accuracy

Different models of equipment vary in the market, but in general UT equipment consists of two main units; a generator and a receiver. The generator produces an ultrasonic modulated tone which is positioned on one side of the seal. The receiver is then used by the operator from the far side of the closed seal. If at any point the seal is imperfect, the ultrasonic signal will be able to pass though the seal through the leak, which can then be detected by the receiver. One model of ultrasonic watertight integrity testers is capable of detecting leak apertures as small as 0.06±0.02mm in size (the Portascanner WATERTIGHT, see end section). UT seals can provide the sensitivity to detect the smallest leaks, and are often displayed in two different types of readout scale: linear and decibel. The linear scale provides an intuitive measurement of the leak size, whereas the decibel scale allows for comparison of standards set by international classification societies.

Research into UT for leak detection has been extensively carried out by Coltraco Ltd. who found that the theory relating the area of a leak site to volume flow rate through said leak is well understood. The volume flow rate, Q, through a hole is given by:

$$Q = A\sqrt{2gh}$$

where A is the area of the hole, g is the acceleration due to gravity on Earth and h is the height of the head of water above the leak. Therefore, it can be expected that a linear relationship between flow rate and hole area exists. Furthermore, their research found that a linear relationship exists between signal leakage out of a certain sized hole (in a watertight box) and the quantity of water able to enter that hole which further proves the accuracy of the flow rate – hole area relationship.

The use of UT is far more efficient than the traditional methods described in the previous section, taking less time and requiring no clean up while being in a portable, light-weight model for ease of use. Due to its convenience, UT tests can be conducted more frequently and can contribute to the safety management system (SMS) and preventative maintenance procedures on board. Moreover, the accuracy of this method is unrivalled, and leak sites can be identified and specifically located quickly by the operator. Furthermore, ultrasonic indicators use safe and green technology; it does not violate any environmental codes, which also means there are no IATA transportation restrictions. Ultrasonic technology is being rapidly adopted as a safe, easy and reputable method of identifying seal leak sites in a highly accurate and convenient way.

Ultrasonic Watertight Seal Integrity Testing: Customer Evidence and Experience

One company wishing to improve their seal testing to prevent water ingress undertook extensive global market research until they chose the ultrasonic method. They researched the market and chose their preferred model partially due to its unique numerical linear reading. They discovered that its maximum open hatch value of 31775 can be satisfactorily broken down to provide exact results: 0 – 5 linear numerical reading equated to "watertight" and 6 – 100 linear numerical reading equated to "weathertight" and 101+ meant "strong likelihood of there being a weak or damaged seal or a leak". This company (who wished to remain anonymous) is a leading offshore oil and gas exploration and production company with 100+ rigs. They have chosen to embed this ultrasonic watertight integrity test indicator into their Safety Management Regime to improve the safety of their rigs, platforms and vessels. They are using UT to test their watertight and weathertight doors.

A second example of how UT can provide precise readings is from the shipping industry: a marine surveyor in the Caribbean wrote a feedback report in 2014 supporting the use of UT. Here are a few extracts: "Customers recognized that the instruments are more reliable and provide results that may be repeated easily and often, rather than using hose testing." "In our extensive experience using (the UT model chosen) in all these different types of environments and on different types of vessels, we found the surveyor's time to carry out the surveys has been significantly reduced and carrying out the surveys is far safer. We have also carried out tests of holds on behalf of charterers to demonstrate the cargo worthiness of the vessel, and this has resulted in more pre-cargo condition surveys being carried out using ultrasonics". "The lesser time taken to carry out these tests is valuable to owners and ship managers in that they can demonstrate without lengthy visual inspections that the vessel is ready in all respects to carry the cargo, meaning cargo loading can commence earlier. The results determined by electronic instrumentation are also extremely valuable to vessel owners, charterers, and cargo interests because they can all be assured the cargo worthiness has been determined to a high degree thereby assuring all that the chance of cargo damage by water ingress is a very low probability".

Further examples of how UT can provide a data recordable audit trail are illustrated below. Figure 2 is the sheet used for a bulk carrier vessel by its ship manager. Figure 3 is the sheet used by BMT Surveys for their servicing to record the data and save the results of the inspections for future requirements.

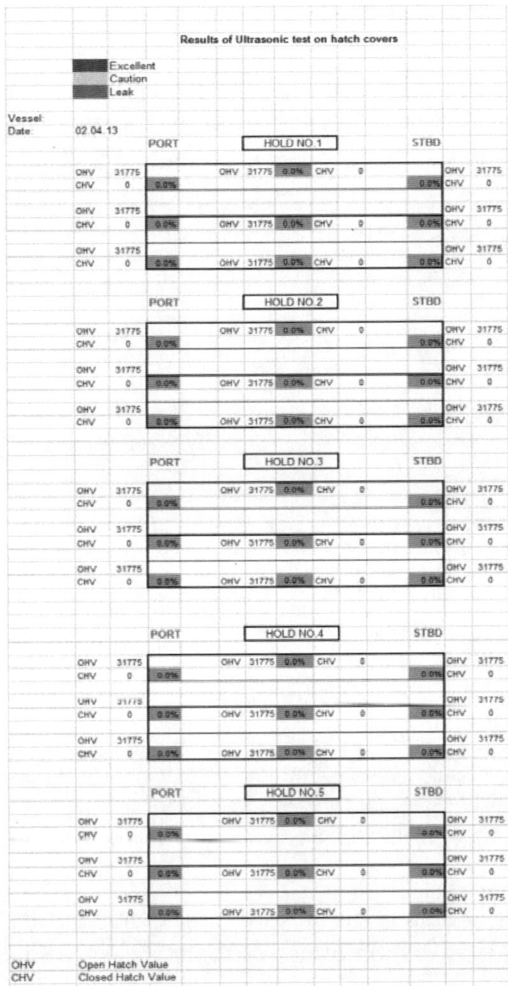

Results of Ultrasonic test on hatch covers

Excellent
Caution
Leak

Vessel:
Date: 02.04.13

PORT HOLD NO.1 STBD
OHV 31775 OHV 31775 0.0% CHV 0 OHV 31775
CHV 0 0.0% 0.0% CHV 0
OHV 31775 OHV 31775
CHV 0 0.0% OHV 31775 0.0% CHV 0 0.0% CHV 0
OHV 31775 OHV 31775
CHV 0 0.0% OHV 31775 0.0% CHV 0 0.0% CHV 0

PORT HOLD NO.2 STBD
OHV 31775 OHV 31775 0.0% CHV 0 OHV 31775
CHV 0 0.0% 0.0% CHV 0
OHV 31775 OHV 31775
CHV 0 0.0% OHV 31775 0.0% CHV 0 0.0% CHV 0
OHV 31775 OHV 31775
CHV 0 0.0% OHV 31775 0.0% CHV 0 0.0% CHV 0

PORT HOLD NO.3 STBD
OHV 31775 OHV 31775 0.0% CHV 0 OHV 31775
CHV 0 0.0% 0.0% CHV 0
OHV 31775 OHV 31775
CHV 0 0.0% OHV 31775 0.0% CHV 0 0.0% CHV 0
OHV 31775 OHV 31775
CHV 0 0.0% OHV 31775 0.0% CHV 0 0.0% CHV 0

PORT HOLD NO.4 STBD
OHV 31775 OHV 31775 0.0% CHV 0 OHV 31775
CHV 0 0.0% 0.0% CHV 0
OHV 31775 OHV 31775
CHV 0 0.0% OHV 31775 0.0% CHV 0 0.0% CHV 0
OHV 31775 OHV 31775
CHV 0 0.0% OHV 31775 0.0% CHV 0 0.0% CHV 0

PORT HOLD NO.5 STBD
OHV 31775 OHV 31775 0.0% CHV 0 OHV 31775
CHV 0 0.0% 0.0% CHV 0
OHV 31775 OHV 31775
CHV 0 0.0% OHV 31775 0.0% CHV 0 0.0% CHV 0
OHV 31775 OHV 31775
CHV 0 0.0% OHV 31775 0.0% CHV 0 0.0% CHV 0

OHV Open Hatch Value
CHV Closed Hatch Value

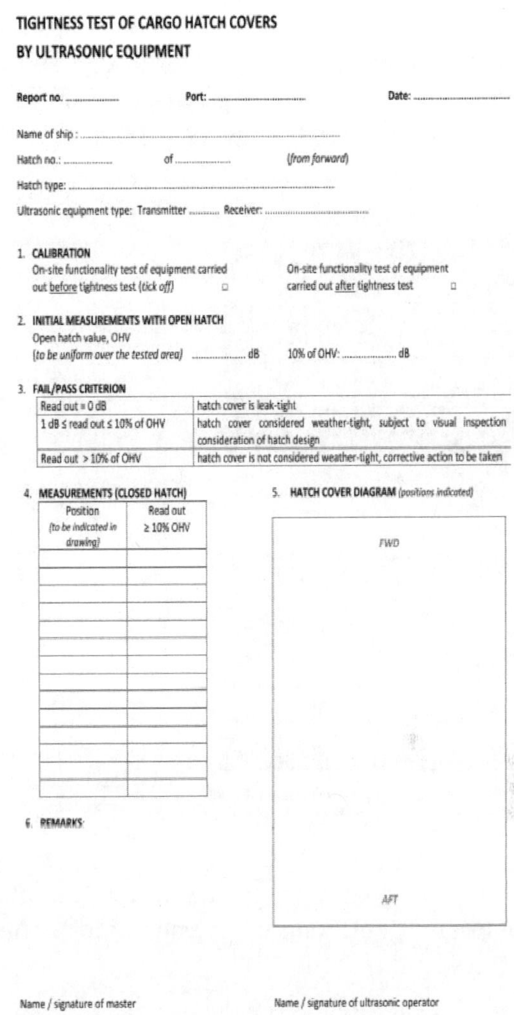

TIGHTNESS TEST OF CARGO HATCH COVERS
BY ULTRASONIC EQUIPMENT

Report no. .... Port: .... Date: ....

Name of ship : ....

Hatch no.: .... of .... (from forward)

Hatch type: ....

Ultrasonic equipment type: Transmitter .... Receiver: ....

1. CALIBRATION
On-site functionality test of equipment carried out before tightness test (tick off) □
On-site functionality test of equipment carried out after tightness test □

2. INITIAL MEASUREMENTS WITH OPEN HATCH
Open hatch value, OHV
(to be uniform over the tested area) .... dB 10% of OHV: .... dB

3. FAIL/PASS CRITERION

Read out = 0 dB	hatch cover is leak-tight
1 dB ≤ read out ≤ 10% of OHV	hatch cover considered weather-tight, subject to visual inspection consideration of hatch design
Read out > 10% of OHV	hatch cover is not considered weather-tight, corrective action to be taken

4. MEASUREMENTS (CLOSED HATCH)

Position (to be indicated in drawing)	Read out ≥ 10% OHV

5. HATCH COVER DIAGRAM (positions indicated)

FWD

AFT

6. REMARKS:

Name / signature of master

Name / signature of ultrasonic operator

Limitation of Portable Ultrasonic Inspections

The large issue here is that ships are only tested before and after one or perhaps several journeys and yet a leak could occur at any point in between testing and continue unnoticed until the next inspection.

For example, carrying perishable cargo can be particularly problematic simply because watertight integrity amongst several other aspects needs to be maintained to ensure the cargo is not ruined. Fresh fruits and vegetables can over ripen if not cared for properly and maize can sprout if moisture enters the cargo hold. Other cargos such as coal or grain can become liquid-like when mixed with enough water, leading to extreme instability of the ship as the cargo sloshes around. This can lead to severe ship damage and even capsize.

The limitations of UT extend to the fact that UT is conducted only while the vessel is in a static position in dock/port and makes no recognition of how the vessel is affected by its dynamic positions at sea. A vessel generates its leak sites due to load states, sea states, wind states, and dynamic movement. The severity is amplified within a vessel structure constantly changing by varying sea, wind, load states, cargo types and dynamic stresses. There is a great deal of bending and deformation that naturally occurs in ships during travel. The dynamic movement of a vessel is illustrated through analysis of stills from a video showing the stress and effect on a vessel during passage from Suez Canal to Singapore in severe weather conditions as shown by the analysis in the inset 1, 2 and 3 in Figure 1 below (source: Coltraco Ltd).

Inset 1: This still (1m:20s) shows wire frames highlighting rectangular sections along the vessel which will be used for comparison with Inset 2 (below)

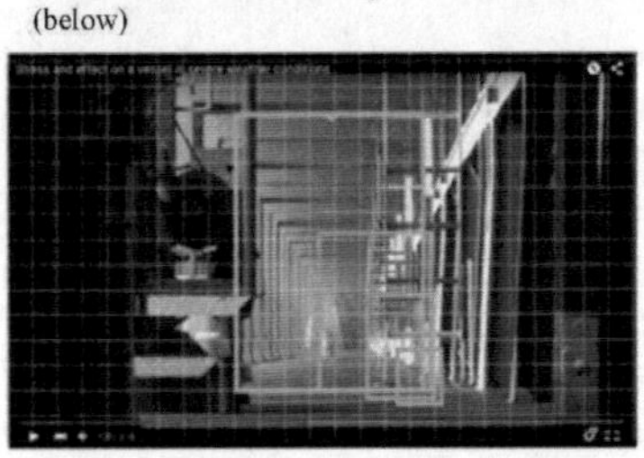

Inset 2: Still from (1m:30s) with rectangular boxes, combined with grids to highlight the movement

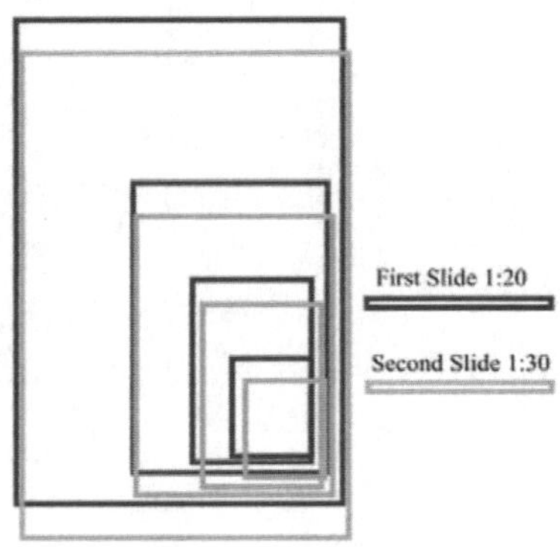

Inset 3: This image overlays the rectangular sections from Inset 1 and 2 to show the difference between 1m:20s and 1m:30s.

Given the limitations of traditional watertight integrity testing methods that also only occurs while the vessel is docking, it becomes a further necessity to monitor leak sites as they develop whilst the vessel is at sea.

The future: 24/7 constant monitoring

The main issue is that without continuous monitoring, a leak could still occur at any point in between tests during a journey, and a crew member is still needed for testing. Research is being carried out by enterprises such as Coltraco Ltd. to create the most comprehensive and safe solution using ultrasonic technology.

It was found that a comprehensive, autonomous continuous monitoring system for the watertight integrity of a ship's cargo hatches, weathertight doors and other seals can be developed. One that is capable of automatically detecting emerging leak sites, alerting officers and crew to the location and severity of the leak site and logging all data for future review. Coltraco is looking to see this technology fully utilised by 2020, as part of a pioneering wider vessel system integration and shore-based data management system.

It is becoming clear that UT is developing into an integral component of watertight integrity testing for ship hatch covers most notably and is best positioned to secure the safety of vessels in the shipping Industry. Ultrasonic readings from hatches / doors / MCTs can also be logged and trended over time which helps increase safety and can also be used for insurance companies. There is more work to be done, but the future must be condition-based monitoring of hatch covers, using UT to provide continuous monitoring. Such a system has the potential to centralise a vessel's watertight and airtight integrity into a single alarm system. When used on a warship, such a system would also allow the status of the ship to be transmitted for an analysis and response by supporting vessels.

The future of UT is increasingly moving towards constant and even automatic monitoring of the performance of all types of seals, especially critical ones. Furthermore, in the near future, low power ultrasonic generators will allow submarine crews to easily conduct watertight integrity tests without the risk of alerting enemy vessels to their presence by picking up the ultrasonic noise through the hull.

An Overview into Products Available in the Market

For portable ultrasonic watertight integrity testers, Coltraco has developed the Portascanner® WATERTIGHT which is the industry standard in ultrasonic watertight integrity testing. It is accurate to identify leak site apertures to 0.06mm that is a world-leading accuracy and used to ensure watertightness of vessels to protect crew, vessel, cargo and cargo claims. The unit is capable of conducting watertight tests with the cargo in-situ and produces no water run-off. The unit is light and intuitive to use, with magnetic base to ease the use of its miniature ultrasound generator, and removing the need for the operator to go to the tank top – saving further time to test.

An extension to this portable unit is the Permascanner® Hi-Life which is an ultrasound generator magnetically attached inside the cargo hold to provide continuous ultrasound signal output. Its battery is sufficient for up to 10 days' voyage time (e.g. Liverpool to New York, 3000 NM). During this period, the Generator remains in the hold. In fair-calm weather, the crew or engineer can go out on deck with the Portascanner® Receiver to test the hatches to see how they are reacting in the dynamic conditions of the changing sea, weather and load states. This is beneficial to identify any potential weak seal compressions, ahead of becoming leak sites. Following a period of rough weather, the personnel can also conduct similar tests to determine if water has entered the cargo hold without the need to open the hatch cover. This is the first capability of its type in the world: an innovative solution for constantly monitoring hatch-covers.

For a fixed monitoring system, the Permascanner® DYNAMIC is Coltraco's fixed ultrasonic watertight integrity test indicating system for the dynamic testing of hatch covers, holds, bulkheads, cable transits and watertight doors and compartments at sea. Derived from its sister semi fixed system above, Permascanner® DYNAMIC is designed to quickly identify leak sites in their extent and precise location, whilst at sea. It allows a pattern of readings to be recorded against varying sea states, enabling a true picture of the seal integrity to be recorded over time. It fulfils the requirements of a continuous monitoring system as mentioned in the previous section and is the industry's first ever solution to be introduced into the market.

Conclusion

In conclusion, greatly reducing the chance of water ingress occurring by implementing safer and improved methods of watertight integrity testing, it may be possible to save lives, asset and vessel integrity at sea. In the particularly risky conditions of the Naval world, surface and sub-surface vessels require all doors, hatches, MCTs and NBCW seals to be protected at all times. The introduction of a continuous watertight integrity monitoring system is the only reliable solution to obtain certainty that the seals are behaving the way they should whether the vessel is at dock or on operations at sea.

Effect of Pitting Corrosion on the Ultimate Strength and Buckling Behaviour of Stiffened Aluminium Panels.

Lenore M. Pedrina, Defence Science and Technology Group,
Lenore.Pedrina@dst.defence.gov.au
Daniel Dehtyriov, Monash University, ddeh1@student.monash.edu
Seref Aksu, Defence Science and Technology Group, Seref.Aksu@dst.defence.gov.au

ABSTRACT

Assessment of the ultimate strength of stiffened panels plays an important role in determining the load characteristics, structural capacity, reliability and life of a structure. Degradation of the hull material due to corrosion can greatly affect the life of type assessment. Historically, degradation due to corrosion has been modelled by thinning of the material or approximated by removal of material from areas of the modelled structure. Modelling corrosion as thinning of material may be considered as appropriate for steel structures that are mostly affected by general corrosion. However, pitting corrosion is typically observed in internal structures of high speed aluminium naval craft. Therefore for aluminium structures, a more precise method of modelling the pitting corrosion is required.

This paper examines a numerical assessment methodology of the impact of pitting corrosion on the ultimate strength of aluminium stiffened panels under axial compressive loading. The pitting corrosion damage was modelled as elliptical voids and its effects were analysed through the parameters of the pitting profile (pit diameter), pit depths and pitting intensity that was characterised by degree of pitting. Finite element modelling and analysis of stiffened panels, representative of critical ship sections, was carried out utilising COMSOL Multiphysics 5.2a software. The analysis showed that; the individual pits acted as significant stress raisers; ultimate strength decreased linearly with increases in the degree of pitting (number of pits) and decreased nonlinearly with an increase in the pitting profile (pitting diameter) and the pitting depth.

NOMENCLATURE

A_p	Cross Sectional Area of the pit	E	Modulus of Elasticity
a or a	Panel length between beams	γ	ratio between corroded and entire stiffener web heights
α	ratio between corroded and entire plate widths	h_w	Stiffener web height
B	Total plate width	λ	Column slenderness ratio
b	Panel bay width	L	Total Panel length
b_f	Stiffener flange width	n	Number of pits
b_p'	Heat Affected Zone extent on plate	r	Radius of gyration
b_s'	Heat Affected Zone extent on web	ρ	Density
β	Plate slenderness ratio	$\sigma_{Y,eq}$	Equivalent yield stress
		σ_y	Yield strength
		σ_u	Ultimate strength
		t	Plate thickness

t_w	Stiffener web thickness	w_0	Deflection
t_f	Stiffener flange thickness	z_{fillet}	Distance from fillet weld
υ	Poisson's ratio		

INTRODUCTION

The Royal Australian Navy (RAN) currently operates a fleet of 13 Armidale Class Patrol Boats (ACPBs). The ACPBs were designed and constructed by Austal Ships to meet Det Norske Veritas (DNV) High Speed, Light Craft and Naval Surface Craft (HSLC) rules [1]. Marine-grade aluminium alloys were used as the construction material of the ACPBs. The use of aluminium alloys in marine applications is increasing, on account of their high strength-to-weight ratio, and excellent formability [2].

Corrosion is one of the most aggressive forms of material degradation. Aluminium alloys have been found to be susceptible to corrosion and when exposed to a corrosive environment such as a chloride solution, aluminium alloys are vulnerable to highly localised or pitting corrosion. This occurs when the applied protective or natural protective oxide coating is damaged or as the chloride ions in salt water break down the protective oxide layer, allowing the environment to contact the metallic surface. Under cyclic loads a corrosion pit acts as a stress raiser and can become a crack initiation site, resulting in the reduced endurance of the metal [3]. The geometry and distribution of the pitting corrosion can influence the local stress field of the pits, and can degrade the fatigue life of the structure [4, 5]. Further, pitting corrosion can have a significant impact on the ultimate strength.

The ACPB fleet has experienced significant corrosion problems, shortly after commissioning, with through hull corrosion in the vicinity of the keel in the bilge areas along the length of the ships and corrosion penetrations at higher levels of the hull at sealed voids. As a result, significant repairs to the hull structure due to pitting corrosion have been carried out across the ACPB fleet [6], and extended maintenance activities have reduced the availability of the boats for operations.

Moreover, increased pressure to have structures last longer and do so in a safe and efficient manner, has led to a drive for more accurate life prediction calculations. To this end the effects of material degradation must be incorporated into critical failure modes such as fatigue and ultimate strength affecting life of a structure; and to better inform through life maintenance practices.

This paper presents findings of an investigation into the effects of pitting corrosion on the ultimate strength of a welded structure through extensive use of Finite Element (FE) modelling and analysis. The FE modelling and analysis explicitly models pits in a stiffened panel representative of a typical bottom structure of a high speed naval vessel.

Modelling of Corrosion

A ship is a complex structure, designed to withstand the forces and moments resulting from extreme operational loads throughout its lifetime. The global strength and stiffness of a ship's hull-girder are governed by the strength of the individual plate and stiffener elements assembled to comprise the overall hull girder structure [7, 8]. Age-related forms of degradation, such as corrosion is a parameter of influence in ultimate strength analysis.

The influence of corrosion in structural integrity assessment has been investigated predominantly in steel structures. A detailed literature review of corrosion modelling in FEA analysis is being undertaken in a DST Group internal report [9]. The review has identified a number of methodologies of incorporating corrosion into the FEA analysis of naval structures. A summary of the review findings follows- Paik et al. [10, 11] investigated the ultimate strength characteristics of steel plates degraded by pitting corrosion. Paik et al. utilised regression analysis of computed results to derive closed-form design formulae for the ultimate strength of pitted plates under axial tension, axial compression and edge shear. Paik and Thayamballi [12] compared the effect of modelling pitting corrosion idealised as general corrosion and as individual pits. Kim et al. [13] considered two corrosion wastage models to compare the ultimate strength behaviour of the midship section of double hull oil tankers; time-dependent corrosion wastage and uniform corrosion. Teixeira et al. [14] conducted a parametric study of probabilistic modelling of ship plates with non-uniform corrosion. Using non-linear FEM analysis, the load-shortening behaviour of plates with non-uniform diminution of thickness due to corrosion was calculated.

These studies considered varying degrees of simplifications in the modelling of corrosion such as removal of areas of material, thinning the 'affected' area. This is considered adequate for areas of general corrosion, however for marine grade aluminium alloys corrosion occurs much more frequently in the form of pitting in structural members and/or plating. The modelling approach for pitting corrosion and the impact on ultimate strength and buckling behaviour is discussed in the next section.

MODELLING APPROACH

The numerical modelling and simulation approach in this paper is based on extensive use of finite element modelling and analysis. All ultimate strength and buckling assessment was carried out utilising commercially available COMSOL Multiphysics software with the Structural Mechanics; Non-linear structural materials; and Fatigue modules. Solidworks CAD software was used to construct panel models when the complex pit geometry was being incorporated.

Stiffened Panel model

A stiffened panel was considered to determine the ultimate strength capacity of a structure. The stiffened panel was based on a generic panel that can be found in most ships which comprises longitudinal stiffeners/frames and transverse beams similar to that shown in Figure 1a. A cross-section view of a longitudinal stiffener is shown in Figure 1b.

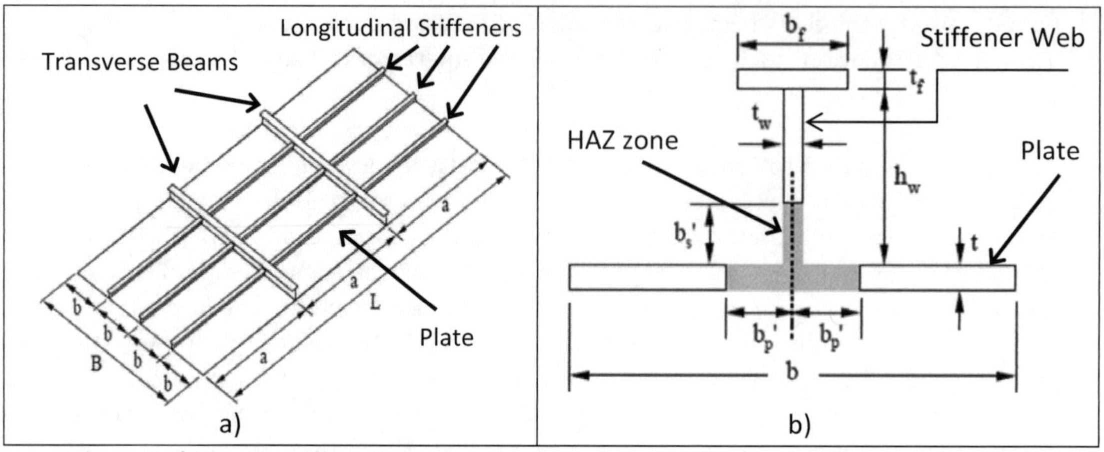

Figure 1 a) Generic stiffened panel structure and nomenclature and b) Cross-section of longitudinal 'T' Stiffener with idealised HAZ.

Indicated on this figure is the Heat Affected Zone (HAZ). To accurately model the buckling of this stiffened panel, the change in material properties in the heat affected zone (HAZ) was incorporated in the model. The extent of the HAZ was approximated using the equations proposed by Wang et al. [15]

$$z_{fillet} = \min\left(3t_A, 20 + \frac{t_A}{3}\right), \mathbf{mm} \text{ where } t_A = \min\left(1.5t_w, \frac{t+t_w}{2}\right), \mathbf{mm} \quad \text{Equation 1}$$

Employing Equation 1, it was estimated that the HAZ extended 18 mm from the fillet weld in both the plate and the stiffener. The reduction in the yield strength through this region was estimated to be on average 35.5 %. No other material properties are assumed to change within the HAZ.

Explicit modelling of the corrosion pits requires a high mesh density with a large number of degrees of freedom. As such the base panel needed to be simplified and refined to allow the maximum computational resources to be available for the modelling and analysis of pitting corrosion. To simplify the panel model, it was assumed that the transverse beam did not need to be modelled accurately for linear buckling analysis – similar effect was achieved by setting the out of plane deflection to zero where the panel and beam met. This assumption led to the base panel being a single span-double bay model with the transverse beam out of plane deflection restraints and symmetric periodic conditions. (see Figure 2). The physical dimensions of the model are given in Table 1.

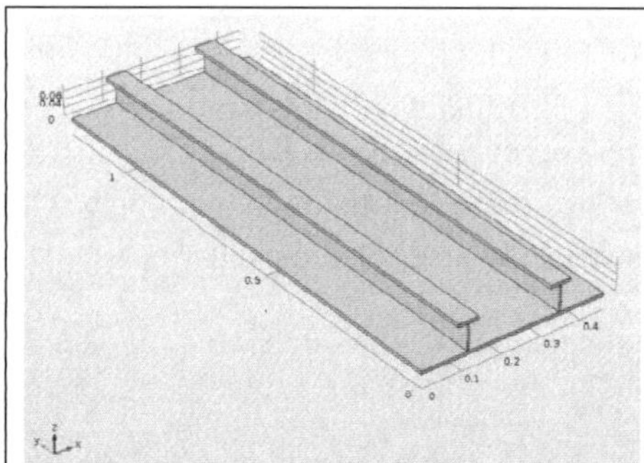
Figure 2 Single span-double bay model

Table 1 Dimensions of the panel model

Parameter	Nomen-clature	Value
Plate Bay Width	b	0.23 m
Plate Length	L	1.2 m
Plate Thickness	t	0.008 m
Stiffener Webb Thickness	t_w	0.004 m
Stiffener Webb Height	h_w	0.062 m
Stiffener Flange Thickness	t_f	0.008 m
Stiffener Flange Width	b_f	0.050 m

Material Properties

Aluminium alloy 5083 in the H116 temper (AA5083-H116) and aluminium alloy 6082 in the T6 temper (AA6082 T6) were selected as materials for the plate and the longitudinal stiffeners, respectively. These aluminium alloys are widely used in the construction of fast planing ships and catamarans. The material properties used in the models are outlined in Table 2 and the calculated specific geometric-material properties for the panel are given in Table 3.

Table 2 Material Properties of the Aluminium Alloys used in the model.

Property	AA5083-H116	AA6082-T6
Modulus of Elasticity, E [GPa]	71	70
Density, ρ [kg/m^3]	2660	2700
Yield Strength, σ_y [MPa]	215	250
Poisson's Ratio, υ	0.33	0.33

Table 3 Geometric-Material properties of the selected stiffened panel

Geometric-Material Property	Value
Equivalent Yield Stress [MPa] $\sigma_{Y,eq}$	224.0
Radius of Gyration [m] r	0.574
Plate slenderness ratio β	1.582
Column slenderness ratio λ	0.374

Loading and Boundary Conditions

The boundary conditions need to be derived and applied so that they do not influence the solution but do model collapse behaviour as it would happen in reality. Thus a variety of boundary conditions were tested and compared to results of experimental and numerical work published by Paik et al. [16]. This resulted in the following boundary conditions:

- Both the stiffener and plate are simply supported along their edges and the out of plane deflection is zero.
- Symmetry conditions can be imposed at the longitudinal face boundaries due to the use of symmetric stiffeners.

Corrosion Pit Modelling

Pit geometry is complex, particularly for aluminium alloys where pitting corrosion can be the result of different mechanisms, each with their own unique morphology [17]. Pits therefore are unlikely to be geometrically simple, with rough boundaries and severe undercutting and thus potentially having deleterious effects on strength. Furthermore, pit diameters, depth and the extent of undercutting are not uniform as seen in Figure 3a which shows illustrations of some common pit geometries.

The DST Group literature review [9] has also identified various approaches detailing the specific modelling of pits for fatigue assessment studies in the open literature. A summary from the review follows-

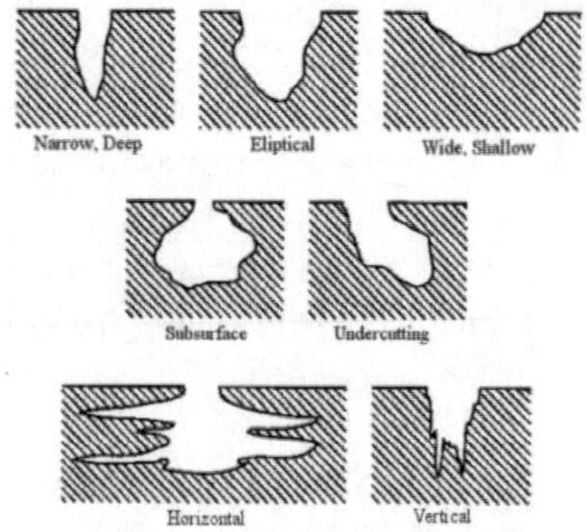

a) Various common Pitting Geometries [18]

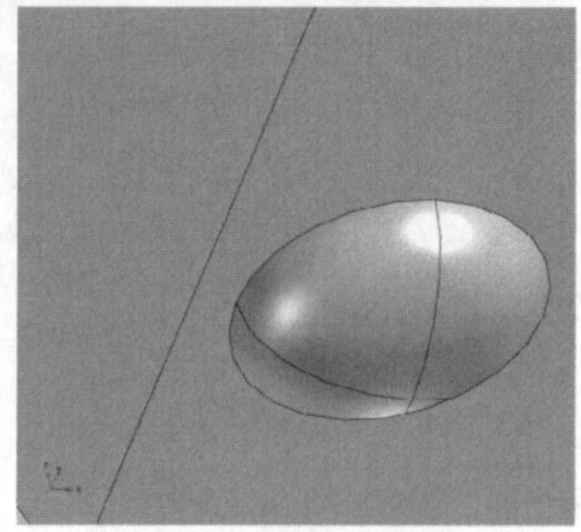

b) Example of a modelled pit with elliptical geometry

Figure 3 Common pit geometries and the modelled pit geometry.

Rezig et al. [3] calculated the stress concentration factor (SCF) of crack-initiating corrosion pits using FEM with each pit modelled as a semi-ellipse. Similarly, Almaraz and Mora [4] related the SCF to the proximity of two and three neighbouring corrosion pits that initiated cracks under uni-axial loading. The SCFs of the pits, modelled as hemispherical shapes, were calculated using a very fine mesh of solid elements. Silva et al. [19] determined the ultimate strength of plates subjected to random corrosion wastage using FEM by representing corrosion pits as having a prismatic cross-section. Nakai et al. [20] considered the shape of the pits to be circular cones in their FEA. Lv et al. [21] constructed a 3D solid FEM model, with pitting corrosion geometry created in a Computer Based Design (CAD) package based on experimental data. Ok et al. [22] found that although the most common shape of corrosion pits is semi-spherical, owing to modelling this shape being complex and computationally intensive, the authors simplified the corrosion pits as having a rectangular shape. However, the validity of this simplification applied to aluminium has not been established and is lacking. Indeed, the characterisation and efficient assessment of the effects of localised corrosion is an on-going research area.

In order to approximate the varying geometries some assumptions were made:

- Individual pits act as stress raisers in the stiffened panels
- Sharp discontinuities resulting in numerical singularities are to be avoided in FEA
- Undercut regions and regions of material with thin cross sections are non-load bearing.

An elliptical geometry was therefore chosen as the approximation to pitting corrosion. Ellipses can model narrow deep pits, wide shallow pits and with the assumption that undercut regions were non-load bearing and equivalent to removed material, undercut pits were also well approximated by an ellipse. An example of a modelled elliptical pit is shown in Figure 3b. The model allows the effects of varying pit diameter, depth, intensity (number of pits) and profile to be investigated. It is noted that the pits were not only modelled in the plate but also modelled in the stiffener web.

For this model a mesh was required that was fine enough to predict long wavelength buckling, recognise the initial imperfections to ensure buckling occurs, accommodate the complex geometry of the pitting corrosion while avoiding stiff behaviour. It was determined that on the un-corroded panel a structured mesh of 6720 elements was sufficient to converge to an accurate solution. When corrosion was incorporated in the model the mesh was modified to an unstructured tetrahedral mesh consisting of 130692 elements. The unstructured mesh is required to accurately model the pit geometries and profiles.

FEA APPROACH

COMSOL 5.2a Multiphysics software was utilised for the linear and non-linear Finite Element (FE) analyses. In buckling and plastic collapse analysis employing FEA, it is necessary to provide the initial imperfections. These initial imperfections are necessary to provide the initial 'disturbances' allowing the buckling modes to be excited. Although numerical instability in an unstructured mesh can be used to allow buckling to occur, introducing initial deflections is far more likely to excite realistic buckling modes in stiffened panels.

These initial imperfections are traditionally introduced by explicitly modelling the imperfections due to welding to give initial disturbances which is complex and takes the form of a double sinusoidal series [23]. A CAD model incorporating these initial imperfections and adjusted geometry is then utilised for the FEA. Due to the computational expense of the corrosion pit modelling to explicitly model the initial imperfections, an alternate method to introducing these initial imperfections was proposed and is outlined below.

A linear buckling analysis is performed on the model to obtain the first eigenmode of the panel. It is assumed that this is the true buckling shape of the panel and the eigen model solution is then scaled and used as the initial imperfections in the non-linear analysis. The scaling factor for the linear solutions was determined by using an equation proposed by Smith et. al. [18]. Smith proposed an equation to calculate the maximum magnitude of the welding imperfections, see Equation 2

$$w_{0,max} = 0.1\beta^2 t \qquad \text{Equation 2}$$

Where the plate slenderness value, β is calculated as

$$\beta = \frac{b}{t}\sqrt{\left(\frac{\sigma_y}{E}\right)} \qquad \text{Equation 3}$$

The eigen modal solution is also a double sinusoidal curve and so yields similar results to the traditional method of modelling the welding imperfections. Figure 4 shows a schematic of the two methods.

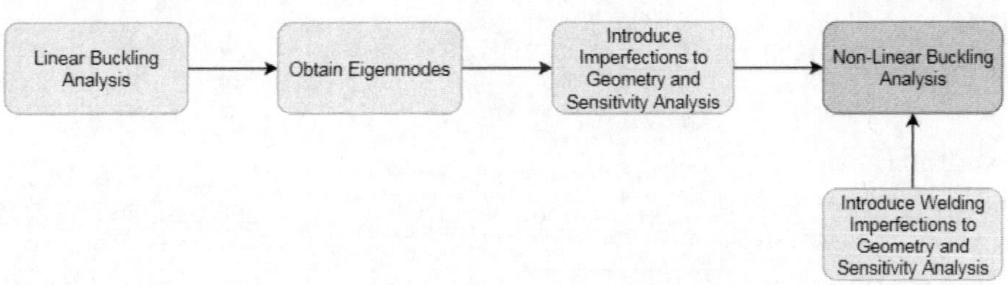

Figure 4 Schematic of the two approaches of introducing geometric imperfections to a non-linear buckling analysis problem.

Applying Equation 2 to the geometry of the stiffened panel used in this study, the maximum deflection was estimated to be 2 mm. Figure 5a shows an eigenmodal solution for one of the corroded panels that was used as the initial imperfections in the non-linear analysis, the deflection is not scaled in the figure. Note also that this is a double bay model incorporating a transverse stiffener. This model was used to validate the assumption that the out of plane and axial displacement of the transverse stiffener is indeed negligible. The first buckling mode is shown.

Once the eigenmodal solution is found this was then scaled to use as the initial imperfections in the non-linear solution. The non-linear finite element analysis considers the following criteria in the analysis:

- Material Non-Linearities with Elastic-Perfectly Plastic Model and Von Mises Failure Criterion (Large Plastic Strain Theory);
- Softening in the Heat Affected Zone;
- Geometric imperfections through consideration of Initial distortion of plating between stiffeners, column type initial distortion of the stiffener and angular initial distortion of the stiffener;
- Large deformation theory.

The ultimate strength was determined by incrementing the load applied to the panel until the solver failed indicating a negative stiffness matrix. All results were normalised against the ultimate strength of an un-corroded panel. Figure 5b shows the resulting buckling displacement for the non-linear solution on an un-corroded panel at the ultimate buckling strength. The post buckling Stress-Strain curve for an un-corroded panel is shown in Figure 5c. Both the stable and unstable regions are plotted, all panels modelled in this study undergo unstable post-buckling behaviour. The ultimate strength for the un-corroded panel was found to be σ_u=160 MPa and all results were normalised against this value.

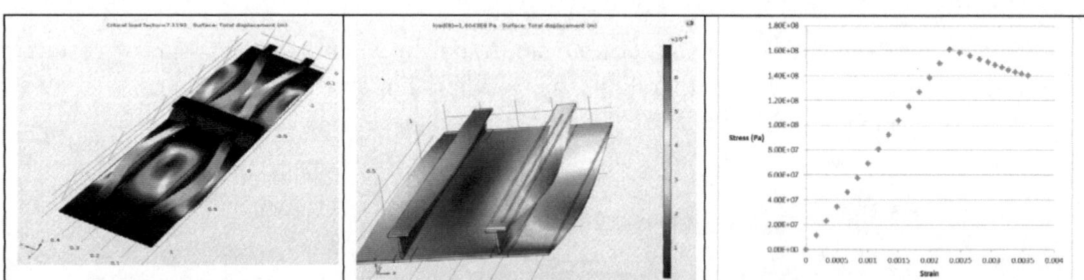

a) Fundamental mode shape of eigenmodal solution

b) Non-linear solution for an un-corroded panel

c) Post buckling Stress-Strain curve for an un-corroded panel

Figure 5 a) Eigenmodal solutions for an un-corroded panel b) Non-linear solution and c) post buckling Stress-Strain curve for an un-corroded panel.

The un-corroded and a large number of corroded panels were numerically modelled and analysed to determine the loss in ultimate strength due to pitting corrosion through the use of non-linear analysis. All models experienced unstable post-buckling behaviour. Large deformations of the structure are undesirable in application so the point of interest of the analyses was the ultimate strength, the strength prior to collapse. The effects of corrosion parameters (pitting profile, pit depth and degree of pitting) were investigated.

The effect of pitting profile

In an actual corroded stiffened panel, the size of the pits would vary randomly. As such, the size (diameter) of pits in both plate and stiffener was varied in the FE models by adopting a simple modelling approach where the largest diameter pits were located closest to the stiffener and plate intersection on both the plate and stiffener. Four different pitting profiles with different pit diameters were considered in the analysis, simulating different stages of corrosion (Figure 6). Table 4 provides variation in pit diameters (in the order starting from plate/stiffener interface and moving away) for the selected 4 pitting profiles.

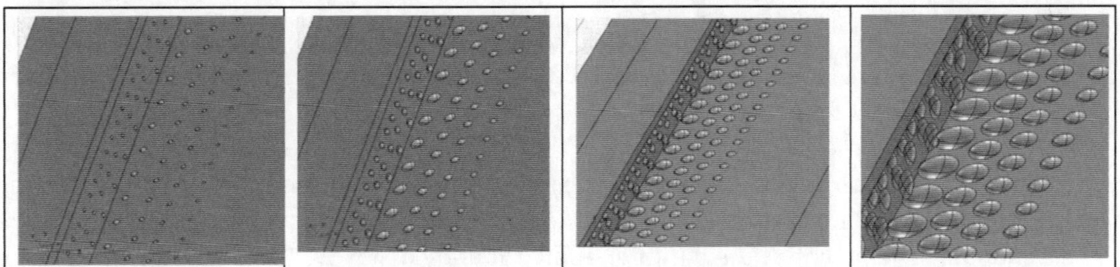

a) Profile 1 b) Profile 2 c) Profile 3 d) Profile 4

Figure 6 Varying the Pit Diameter in the FE model

Table 4 Variation of Pit diameter in the Plate and Stiffener

Pitting Profile	Plate		Stiffener	
	Pit Diameter (mm)	Average Pit Diameter (mm)	Pit Diameter (mm)	Average Pit Diameter (mm)
1	2.5, 2.0, 1.7, 1.4	1.9	2.0, 1.7, 1.4	1.7
2	5.0, 4.0, 3.3, 2.9	3.8	4.0, 3.3, 2.9	3.4
3	7.5, 6.0, 5.0, 4.3	5.7	6.0, 5.0, 4.3	5.1
4	10.0, 8.0, 6.7, 5.7	7.6	8.0, 6.7, 5.7	6.8

The effect of pit depth

Similar to pit diameter, the depth of the pits would vary in the plate and stiffener. A similar approach was employed where the maximum pit depth was considered closest to the plate/stiffener intersection with gradual reduction of depth away from the plate/stiffener intersection. Maximum pit depths modelled on the plate were 4 mm, 6 mm, 8 mm and 10 mm while the corresponding maximum depths on the stiffener were 2 mm, 3 mm, 4 mm and 5 mm. It is worthwhile to note that the pits of 8 mm and 10 mm deep in the plate and 4 mm and 5 mm in the web resulted in perforations on the plate and the stiffener web as the thickness of the stiffener web was 4 mm while the plate was 8 mm thick.

The effect of Degree of Pitting (DoP)

The third parameter investigated with respect to pitting was the degree of pitting (DoP). This is a measure of the intensity of pits per unit area. The use of the definition introduced by Paik et al [10, 11] (the ratio of corroded surface area to plate area) was made. However, Paik's formulation was modified to include corrosion taking place also on the stiffener web.

Thus,

$$DoP(\%) = \frac{100}{a(b_p+h_w)}\sum_{i=1}^{n} A_{pi} \qquad \text{Equation 4}$$

Pitting corrosion is not uniform across the entire surface of the stiffened panel. Based on in-service experience of corrosion affected naval ship panels, the pitting corrosion was most frequently found around the intersection of the stiffener and plate on one side only where bilge water can pool and allow corrosion to take place. This is reflected in the selected pitting profiles shown in Figure 6. Thus, Equation 4 is modified to take account of localized degree of pitting:

$$L-DoP(\%) = \frac{100}{a(\alpha b_p+\gamma h_w)}\sum_{i=1}^{n} A_{pi} \qquad \text{Equation 5}$$

Where α and γ are fractions of the plate width and stiffener height that are affected by the pitting, respectively.

In this study, the pitting was primarily located at the plate-stiffener intersection and only extended across approximately a third of the plate width from the stiffener, and extended up the height of the stiffener approximately half way. Thus, α and γ in Equation 3 are taken as 0.33 and 0.5, respectively. The effect of degree of pitting was investigated by considering 7 pitting intensity levels covering a range of number of pits from 102 to 306. Both the DoP and L-DoP values for the selected pitting profiles and the selected pitting intensity levels are given in Table 5. Examples of pitting intensity for a given pitting profile are shown in Figure 7 (a-g). L-DoP provides a better representation of the intensity of the pitting corrosion in the localized area. DoP however is a better parameter to use in assessing the effects of corrosion on ultimate strength of ship panels as it related directly to the fraction of material removed.

Table 5 DoP (%) and L-DoP (%) values for the selected pitting intensity and pitting profiles.

Pitting Intensity	Number of Pits	Profile 1		Profile 2		Profile 3		Profile 4	
		DoP	L-DoP	DoP	L-DoP	DoP	L-DoP	DoP	L-DoP
1	102	0.32	0.8	1.36	3.1	3.13	7.1	5.61	12.6
2	136	0.42	1.0	1.81	4.1	4.15	9.4	7.44	16.8
3	170	0.52	1.3	2.27	5.1	5.18	11.8	9.20	21.0
4	204	0.63	1.6	2.72	6.2	6.21	14.2	10.98	25.2
5	238	0.73	1.8	3.18	7.1	7.23	16.5	12.77	29.4
6	272	0.84	2.1	3.63	8.2	8.26	18.9	14.55	33.6
7	306	0.94	2.4	4.09	9.3	9.29	21.2	16.33	37.8

Figure 7 Examples of varying the pitting intensity (number of pits) on the stiffened panel for a given pitting profile.

ANALYSIS AND RESULTS

To investigate the effects of corrosion parameters (pitting profile, pit depth and degree of pitting), a total of 112 corroded stiffened panel models and one un-corroded model were created and analysed. The effect of degree of pitting on the ultimate strength of the stiffened panel for the selected pit depths and the pitting profiles are presented in Figure 8. From Figure 8 we can observe the relationship between the degree of pitting and the ultimate strength, where an increase in the number of pits on the panel for a given pitting profile and pit depth results in a decrease in the ultimate strength. The relationship between DoP and ultimate strength was found to be approximately linear for all pitting depths and diameters. As the pitting depth is increased, the gradient of the linear fit decreases (to a larger negative value), Implying that for deeper pits, an increase in pitting intensity (number of pits) has a significantly larger impact on ultimate strength.

The variation in ultimate strength of the stiffened panel with pitting depth for the selected pitting intensity levels characterised by the degree of pitting and the pitting profiles are presented in Figure 9. For a constant degree of pitting and pitting profile, it was found that the relationship between the ultimate strength and the pitting depth can be approximated by a 2nd order polynomial. The goodness of the fits for all the curves was very good with the R2 values of 0.98 and above. As the pit depths are increased, a significant decrease in ultimate strength was observed (up to 12% reduction compared to that of un-corroded panel). This finding is in line with Paik and Thayamballi's [12] conclusion that the post-collapse behaviour of a corroded steel plate diverged significantly from the un-corroded result.

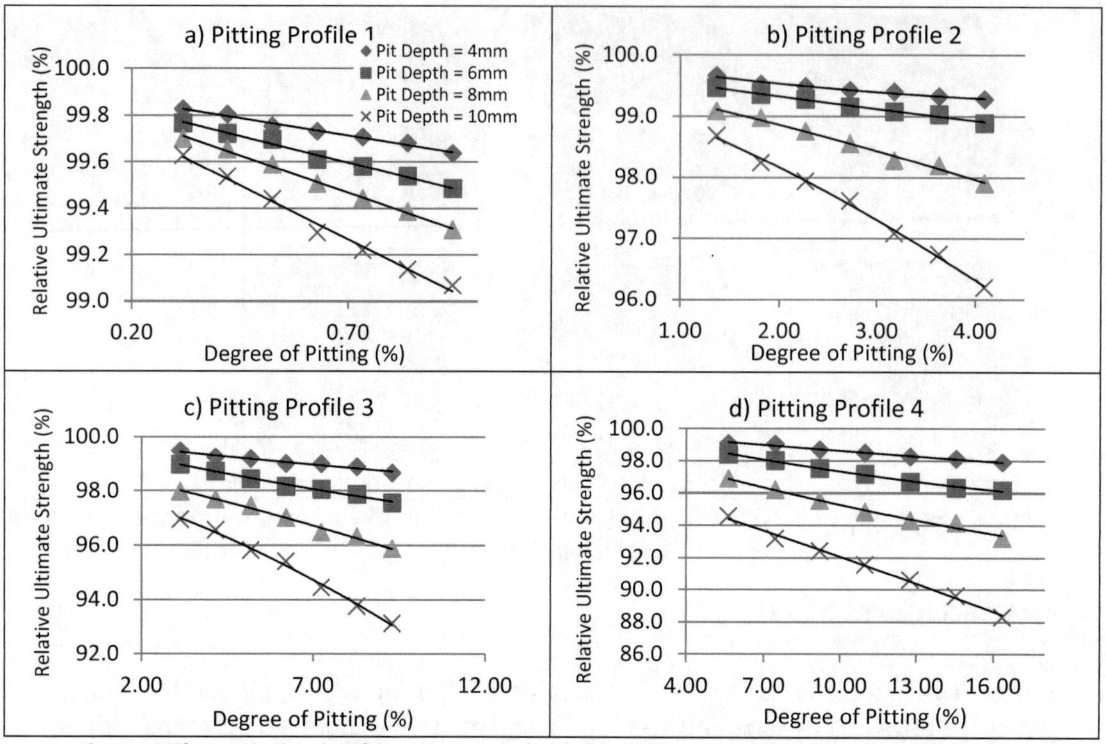

Figure 8 The variation in ultimate strength of the stiffened panel with pitting intensity charecterised by Degree of Pitting (DoP) for the selected pit depths and pitting profiles; a) Pitting Profile 1, b) Pitting Profile 2, c) Pitting Profile 3 and d) Pitting Profile 4.

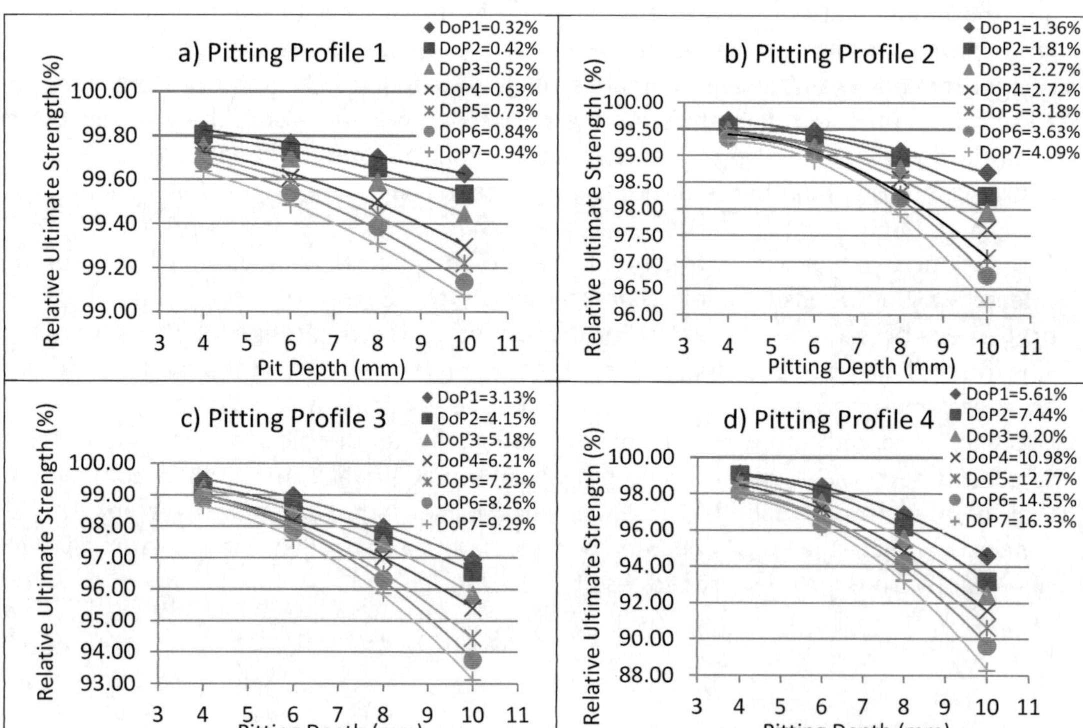

Figure 9 The variation in ultimate strength of the stiffened panel with pitting depth for the selected pitting intensity levels characterised by DoP and pitting profiles; a) Pitting Profile 1, b) Pitting Profile 2, c) Pitting Profile 3 and d) Pitting Profile 4.

Similarly, Figure 10 shows the variation in ultimate strength of the stiffened panel with pitting profile for the selected pitting intensity levels characterised by degree of pitting and the selected pit depths. As can be seen from Figure 10, for a given degree of pitting (number of pits) and a given pit depth, the relationship between pitting profile (pit diameter) and the reduction in ultimate strength was found to be well approximated by a 2nd order polynomial. Again, goodness of the fits for all the curves was very good with R^2 values well above 0.98.

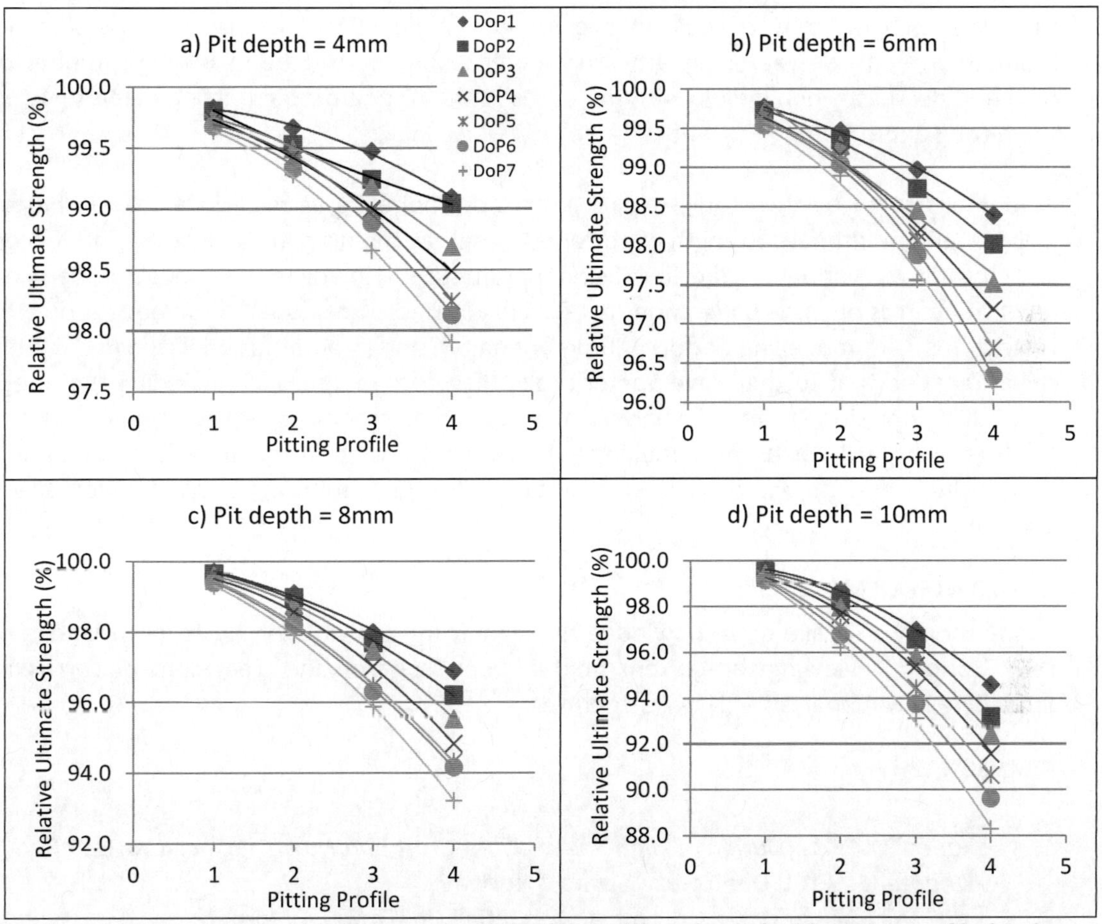

Figure 10 The variation in ultimate strength of the stiffened panel with pitting profile for the selected pitting intensity levels characterised by DoP and pit depths; a) Pit depth = 4 mm, b) Pit depth = 6 mm, c) Pit depth = 8 mm and d) Pit depth = 10 mm.

For all the corrosion cases considered, the degree of pitting varied between 0.32 and 16.33. It is expected that the relationships developed will hold true when extrapolated for small increases in DoP beyond 16.33. However, for large DoP values, a further investigation is required as the relationships displayed strong dependency with degree of pitting.

CONCLUSIONS AND FUTURE WORK

The assessment of ultimate strength of stiffened panels plays an important role in determining the load characteristics, structural capacity, reliability and life of a structure.

Degradation of the hull material due to corrosion can greatly affect the life of type assessment.

A numerical assessment methodology applied to the effect of pitting corrosion on the ultimate strength and buckling strength of aluminium stiffened panels under axial compressive loading was discussed. The pitting corrosion damage was modelled as elliptical voids and its effects were analysed through the parameters of pitting profile (pit diameter), pit depths and pitting intensity characterised by degree of pitting. COMSOL Multiphysics 5.2a software was utilised for non-linear FEA. The analysis showed that; ultimate strength decreases linearly with increases in the degree of pitting (number of pits) and decreases nonlinearly with increases in the pitting profile (pitting diameter) and the pitting depth up to perforation.

As demonstrated by this study, explicit corrosion modelling is necessary to gain better insight into the ultimate strength of stiffened panels as the ultimate strength is influenced significantly by various pitting corrosion parameters. However, this level of detailed (explicit) corrosion modelling cannot be incorporated into global ship models due to restrictions in modelling effort, time penalty and computational power. It is recommended that further development take place on an approach including the effect of pitting corrosion in global analysis of naval ship structures relying heavily on the findings of explicit corrosion modelling such as is presented in this paper. Development may include assessment of corrosion pitting modelling approaches in the aerospace domain.

ACKNOWLEDGEMENT

The authors would like to acknowledge the significant contribution by Teresa Magoga of DST Group in reviewing the current field of knowledge on the treatment of corrosion modelling in FEA analysis of Naval Structures.

BIBLIOGRAPHY

1. *Rules for Classification of High Speed, Light Craft and Naval Surface Craft*, in *Part 3 Chapter 1*, 2011, Det Norske Veritas: Norway.
2. du Toit, M. and K. Mutombo, *Reduction in Fatigue Strength of Arc Welded Aluminium 5083-H111 on Immersion in NaCl.* Advanced Materials Research, 2014. **891-892**: p. 1469-1475.
3. Rezig, E., P.E. Irving, and M.J. Robinson, *Development and early growth of fatigue cracks from corrosion damage in high strength stainless steel.* Procedia Engineering 2010. **2**(1): p. 387-396.
4. Almaraz, G.M.D. and R.P. Mora, *Ultrasonic fatigue testing on high strength steel: Effect of stress concentration factors associated with corrosion pitting holes.* International Journal of Damage Mechanics, 2012. **22**(6): p. 860-877.
5. Kolios, A., S. Srikanth, and K. Salonitis, *Numerical Simulation of Material Strength Deterioration due to Pitting Corrosion.* Procedia CIRP, 2014. **13**: p. 230-236.
6. *Navy patrol boat repairs blow out by $45m*, The Australian, January 6, 2016.
7. Hughes, O. and J. Paik, *Ship Structural Analysis and Design*. 2010: The Society of Naval Architects and Marine Engineers

8. Benson, S., J. Downes, and R. Dow, *Ultimate strength characteristics of aluminium plates for high-speed vessels.* Ships and Offshore Structures, 2011. **6**(1-2): p. 67-80.
9. Magoga, T. and L.M. Pedrina, *Incorporation of corrosion in Fatigue assessment of Aluminium Ship Structures-Review.* (In Preparation), Defence Science and Technology Group, Internal Report.
10. Paik, J.K., J.M. Lee, and M.J. Ko, *Ultimate shear strength of plate elements with pit corrosion wastage.* Thin-Walled Structures, 2004. **42**(8): p. 1161-1176.
11. Paik, J.K., J.M. Lee, and M.J. Ko, *Ultimate compressive strength of plate elements with pit corrosion wastage.* Proceedings of the Institution of Mechanical Engineers, Part M: Journal of Engineering for the Maritime Environment, 2003. **217**(4): p. 185-200.
12. Paik, J.K. and A.K. Thayamballi, *Ultimate Limit State Design of Steel-Plated Structures*. 2003, Chichester, UK: J. Wiley.
13. Kim, D.K., et al., *Effect of corrosion on the ultimate strength of double hull oil tankers - Part II: hull girders.* Structural Engineering Mechanics, 2012. **42**(4): p. 531-549.
14. Teixeira, A.P., C. Guedes Soares, and G. Wang, *Probabilistic modelling of the ultimate strength of ship plates with non-uniform corrosion.* Journal of Marine Science and Technology, 2013. **18**(1): p. 115-132.
15. Wang, X., et al., *Buckling and Ultimate Strength of Aluminum Plates and Stiffened Panels in Marine Structures*, in *Fifth International Forum on Aluminum Ships*. 2005: Tokyo, Japan.
16. Paik, J.K., C. Andrieu, and C. H., *Mechanical Collapse Testing on Aluminium Stiffened Plate Structures for Marine Applications.* Marine Technology, 2008. **45**(4): p. 228-240.
17. Szklarska-Smialowska, Z., *Pitting Corrosion of Aluminum.* Corrosion Science, 1999. **41**: p. 1743-1767.
18. Smith, C.S., P.C. Davidson, and J.C. Chapman, *Strength and Stiffness of Ships' Plating Under In-Plane Compression and Tension.* Royal Institution of Naval Archetects Transactions, 1988. **130**: p. 277-296.
19. Silva, J.E., Y. Garbatov, and C. Guedes Soares, *Ultimate strength assessment of rectangular steel plates subjected to a random localised corrosion degradation.* Engineering Structures, 2013. **52**: p. 295-305.
20. Nakai, T., H. Matsushita, and N. Yamamoto, *Effect of pitting corrosion on the ultimate strength of steel plates subjected to in-plane compression and bending.* JOURNAL OF MARINE SCIENCE AND TECHNOLOGY, 2006. **11**(1): p. 52-64.
21. Lv, S.-l., et al., *Influence of morphology of corrosion on fracture initiation in an aluminum alloy.* Materials & Design, 2013. **45**: p. 96-102.
22. Ok, D., Y. Pu, and A. Incecik, *Artificial neural networks and their application to assessment of ultimate strength of plates with pitting corrosion.* Ocean Engineering, 2007. **34**(17–18): p. 2222-2230.
23. Rahmdel, S., et al., *A Novel Stepwise Method to Predict Ultimate Strength Reduction in Offshore Structures with Pitting Corrosion* Advances in Mechanical Engineering, 2015. **7**(8): p. 1-10.

Prediction of Ultimate Strength of Locally Corroded Plates Using ANFIS Model

Nagi Abdussamie, Mohamed Daboos, Roberto Ojeda
National Centre for Maritime Engineering and Hydrodynamics, Australian Maritime College, University of Tasmania, nagia@utas.edu.au

ABSTRACT

In recent years, increasing attention has been paid to the effects of pitting corrosion in ships' hull structures. Unlike general corrosion, where the thickness loss is normally assumed to be uniform, pitting corrosion is extremely localised which poses difficulties in the ultimate strength assessment of structures suffering from it. The use of an Adaptive Neuro-Fuzzy Inference System (ANFIS) method is proposed in this paper to predict the ultimate strength reduction of locally corroded plates. Published ultimate strength data sets for unstiffened plates with pitting corrosion subjected to uniaxial in-plane compressive loads were used to create and test ANFIS models. A number of input variables were used including the ratios of the plate slenderness, pit breadth to plate width, pit length to plate length and pit depth to plate thickness. Rule-based fuzzy sets were used for mapping the inputs to the output which is the reduction in the ultimate strength. Different types of membership functions were tested to find the best accurate model. The two-sided Gaussian-type function was found to be more effective and less sensitive to the sample size than other functions tested in this study. It was concluded that the developed model can predict the ultimate strength of structures locally corroded.

Keywords: Marine structures; pitting corrosion; adaptive neuro-fuzzy inference system.

INTRODUCTION

For any vessel and/or offshore structure, pitting is one of the most hazardous forms of corrosion and often occurs in locations where it cannot be readily observable (Bhandari et al., 2015). Furthermore, due to the nature of the service of offshore structures, and since it is not common to inspect these structures during an out-of-water dry-docking, the assessment of pitting corrosion requires a conservative approach. However, the characteristic of the pitting corrosion is that it is extremely localised and its penetration is relatively deep in comparison with the thickness of the area affected. Pitting corrosion can in turn reduce the strength of marine structures of both steel and aluminium. Compared to steel, aluminium structures have very good resistance to corrosion even if they are unpainted and operated in seawater for several years. Nevertheless, there are conditions in which corrosion does occur and sometimes lead to a rapid deterioration of the strength of aluminium structures. It is, therefore, important to predict the residual strength of structures affected by pitting corrosion.

Pitting corrosion has been experimentally and numerically investigated for many decades and extensive understanding about this issue has been gained (Frankel and Sridhar, 2008). However, there is a lack of in-depth knowledge of pitting measurement and pitting modelling. Unlike general corrosion, the thickness loss is not uniform throughout the structure which makes difficult to assess the ultimate strength of a locally corroded plate. The buckling/ultimate strength of structures, in the case of uniform/ general corrosion, can be predicted by reducing the original thickness of the structure. Several formulae were empirically derived to estimate the ultimate strength of plates with general corrosion (Bhandari et al., 2015). However, the calculation of strength degradation caused by pitting corrosion is a more complex endeavour for which relatively few guidelines and research information is currently available.

For structural analysis, Finite Element Analysis (FEA) has become the most powerful tool. Paik et al. (2003, 2004) conducted a series of FEA studies of pitted plates and found that when pit area coverage is larger than 30%, the ultimate strength can be reduced by up to 50%. Ok (2006) has investigated the ultimate strength of square plates with pitting corrosion using a series of nonlinear FEA. The effects of several parameters such as pit width, depth, length and its transverse location on ultimate strength was systematically studied using 256 numerical simulations. It should be noted that running FEA simulations is computational and cost intensive activity, as a reason to develop an alternative technique in order to assess the structural integrity of locally corroded structures.

Artificial Neural networks (ANNs) have recently been developed for solving structural engineering problems. Ok et al. (2007) used an ANN method to predict the ultimate strength reduction of unstiffened plates subjected to uniaxial in-plane compression with pitting corrosion. Ok et al. (2007) found that their proposed technique is able to predict the ultimate strength of the plates tested. Even though the ANN technique can easily learn from datasets, it is hard to interpret the information acquired using such technique; the meaning of each neuron and its weight is quite difficult to comprehend (Mathur et al., 2016).

An alternative to the ANN is the Adaptive Neuro-Fuzzy Inference System (ANFIS) technique. The ANFIS method combines the properties of both neural networks and fuzzy logic which makes it advantageous over the aforementioned techniques. Unlike the ANN method, fuzzy logic technique cannot learn from the datasets, but it is easy to interpret the outcomes of fuzzy logic-based models, as they utilise linguistic terms and *if-then* rules rather than numeric. Fuzzy logic is a mathematical tool for modelling inaccuracy of the real world applications in which variables have degrees of falsehood or truthfulness represented by a range of values between 0 (false) and 1 (true). For instance, the outcome of an operation may mean “probably true”, “possibly true” or “probably false” (Abdussamie et al., 2016).

The objectives of this paper are to present and discuss results obtained by the ANFIS method in predicting the ultimate strength reduction of locally corroded plates. Published ultimate strength datasets (Ok, 2006) for unstiffened plates subjected uniaxial in-plane compressive loads were used to construct and test ANFIS models. The input variables included the ratios of the plate slenderness, pit breadth to plate width, pit length to plate length and pit depth to plate thickness. Proper rule-based fuzzy sets were used for mapping the inputs to the output which is the reduction in the ultimate strength of the plate. ANFIS outputs were then compared with FEM and ANN results.

METHODOLOGY

For a crisp set, the membership degree of an element x in set A can be expressed by a characteristic function $\mu_A(x)$ as (Mathur et al., 2016):

$$\mu_A(x) = \begin{cases} 1 \text{ if } x \in A \text{ (full-membership)} \\ 0 \text{ if } x \notin A \text{ (non-membership)} \end{cases} \quad (1)$$

In fuzzy logic, an element may partially belong to one set or more, for instance, the membership function $\mu_A(x)$ of a fuzzy set A can take values in the interval [0, 1]. As shown in Figure 1, the ANFIS models developed for the prediction of ultimate strength reduction of a locally corroded plate composed of four parts including fuzzification, rule base, inference engine and the de-fuzzification interfaces (see e.g., Sivanandam et al., 2007; Mathur et al., 2016).

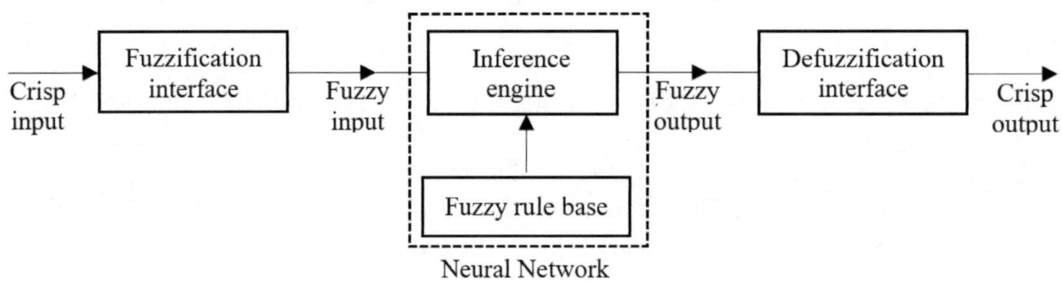

Figure 1. Block diagram of ANFIS model.

In ANFIS models, the crisp input value is converted to fuzzy inputs using membership functions which are then fed into a neural network. Different types of membership functions exist including, e.g., triangular, trapezoidal and Gaussian functions. Once the network is trained, appropriate rules can be generated and fired from the neural network. As the configuration of ANFIS models allows only for one output value to be produced at a time, the output from all rules are combined to yield a single crisp output value (Mathur et al., 2016). The Takagi–Sugeno fuzzy inference method implemented in ANFIS MATLAB was used to calculate the final output value which is the weighted average of all rule outputs (Sivanandam et al., 2007):

$$Z = \frac{\sum_{i=1}^{N} w_i f_i}{\sum_{i=1}^{N} w_i} \quad (2)$$

where N is the number of rules, w_i is the firing strength and f_i is the estimated outcome for ith rule.

IMPLEMENTATION OF ANFIS TECHNIQUE

Ok (2006) conducted a series of nonlinear FE models using ANSYS software in order to investigate the effect of pitting size and location on the ultimate strenght. An unstiffened square plate of 1 m x 1 m x 20 mm with a yield strenght of 235 MPa was used for these FE analyses. Ok (2006) have considered different patterns of pitting corrosion by creating the nearest individual pits into one group (Figure 2). In this study, pits on single edge (denoted as *S*) was selected to create and test ANFIS models.

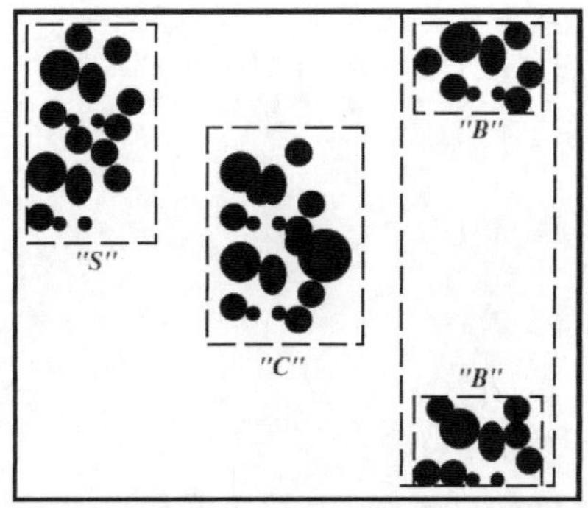

Figure 2. Patterns of pitting corrosion (Ok, 2006): S = single edge, C = centre and B = both edges.

ANFIS models developed in this study consist of four inputs and one output as illustrated in Figure 3. The input variables include the ratios of the plate slenderness (x_1), pit breadth to plate width (x_2), pit length to plate length (x_3) and pit depth to plate thickness (x_4), see Table 1. The ultimate compressive strength reduction caused by pitting corrosion, which is the output of ANFIS model, can be estimated from:

$$R = \frac{\sigma_c}{\sigma_0} \quad (3)$$

where σ_c is the ultimate strenght of corroded plate obtained by FEA and σ_0 is the ultimate strenght of uncorroded plate (σ_0 range: 231.67 – 280.70 MPa).

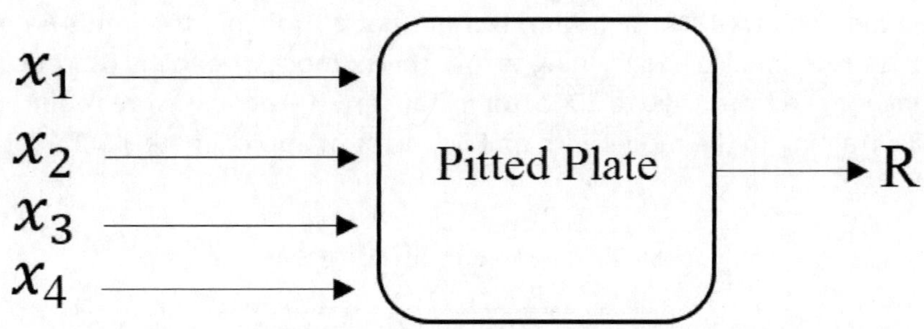

Figure 3. Block diagram showing the inputs and output for locally corroded plate.

Table 1. Definition of input variables of locally corroded plate model.

Input	Definition	Tested range	Unit
x_1	Plate slenderness parameter	1.65 – 2.58	[-]
x_2	Pit breadth to plate width ratio	0.00 – 0.40	[-]
x_3	Pit length to plate length ratio	0.00 – 1.00	[-]
x_4	Pit depth to plate thickness ratio	0.00 – 0.50	[-]

In order to test the performance of ANFIS models and provide a reliable statistical test between the actual values and ANFIS predictions, both the Mean Absolute Error (MAE) and Root Mean Squared Error (RMSE) are introduced as given by Eqns. 4 and 5, respectively:

$$MAE = \frac{1}{n}\sum_{i=1}^{n}|Z_i - R_i| = \frac{1}{n}\sum_{i=1}^{n}|e_i| \quad (4)$$

$$RMSE = \sqrt{\frac{1}{n}\sum_{i=1}^{n} e_i^2} \quad (5)$$

where the absolute error is given by $|e_i| = |Z_i - R_i|$, in which Z_i is the predicted value, R_i is the true value and *n* is the number of data points.

In order to train the data of ultimate strength, the original data of 136 FEA cases were randomly split into training and testing datasets. Three conditions were tested in order to examine the effect of sample size on the performance of ANFIS models, see Table 2.The ANFIS models were designed using MATLAB's Fuzzy Logic Toolbox. In order to create fuzzy sets, different types of membership functions (MF) were tested in this study including triangular MF, trapezoidal MF and the two-sided Gaussian MF. The two-sided Gaussian MF was found to be more effective and less sensitive to the sample size than other types tested in this study, and hence was selected for the analyses. Figure 4 shows the membership functions of the model inputs for condition 1 (85% training and 15% testing) using the two-sided Gaussian functions where 3 MFs were generated for each input. A number of rules (81 rules = 3^4) were created using the grid partition method, and back propagation algorithm was employed to train data. During model training, the input data is mapped several times to minimise the training error. The number of epochs (iterations) required for mapping ranged from 1000 to 2000, for instance, 1000 epochs were required to achieve an effective training to the model with an RMS error of approximately 1.2% for condition 1 (Figure 5).

Table 2. Testing conditions in ANFIS models.

Condition	Training (%)	Testing (%)
1	85	15
2	74	26
3	60	40

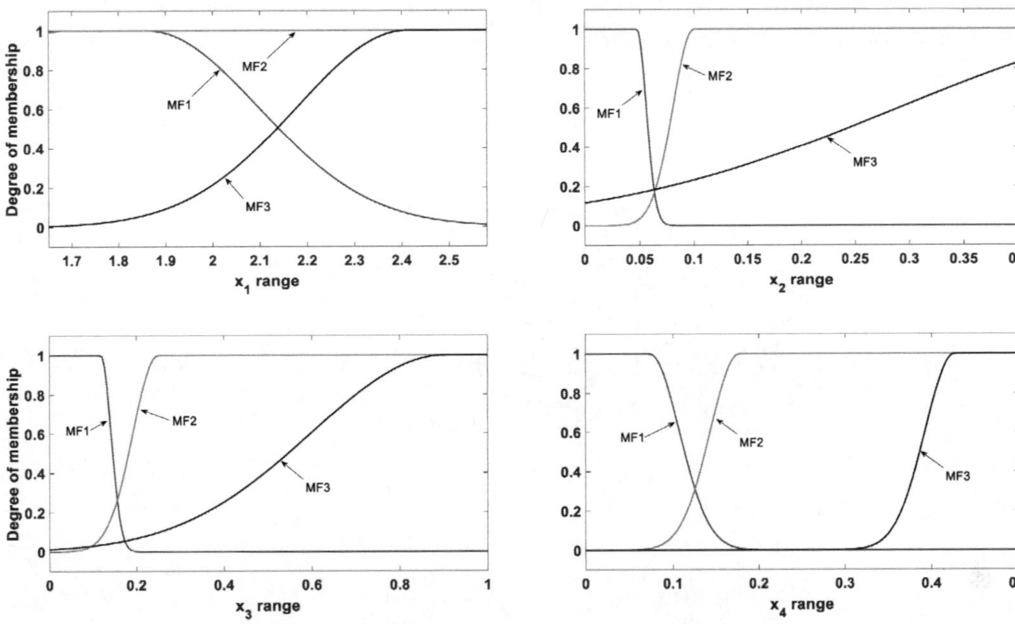

Figure 4. Membership functions (MF) of inputs for condition 1.

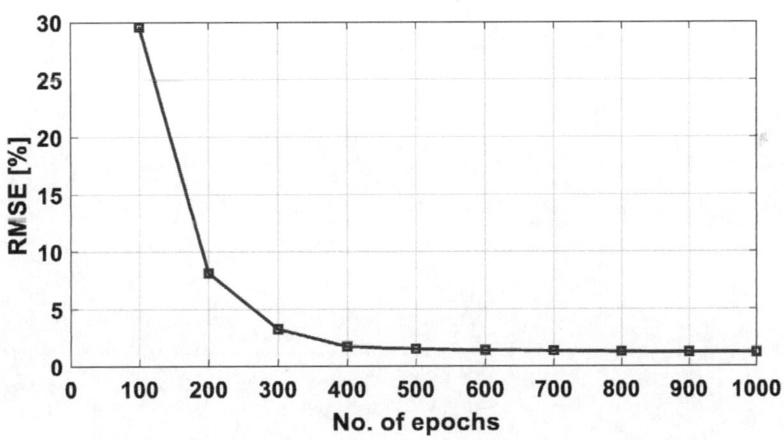

Figure 5. Training error versus number of epochs for condition 1.

RESULTS AND DISCUSSION

The results for conditions 1 – 3 obtained by the ANFIS method are discussed in this section. Figure 6 shows ANFIS predictions of training data against observed data for condition 1 (85% training data). An RMSE of approximately 1.2% for training was achieved which demonstrates a high level of accuracy using the two-sided Gaussian MF. Figure 7 shows a comparison between ANFIS outputs (FIS output) and the tested data of FEM results (Testing data) where 20 data points (15% testing data) were used to validate ANFIS results. Good estimates were obtained by the ANFIS model in comparison with FEM results with a mean absolute error of 0.0153. Table 3 presents the obtained RMSE and MAE values for all conditions. It should be noted that condition 3 required 2000 iterations to achieve the same amount of training error obtained for conditions 1 and 2. ANFIS outputs were also compared with ANN results obtained by Ok (2006), as shown in Figure 8, where good match can be seen between the present results predicted by ANFIS model and ANN results.

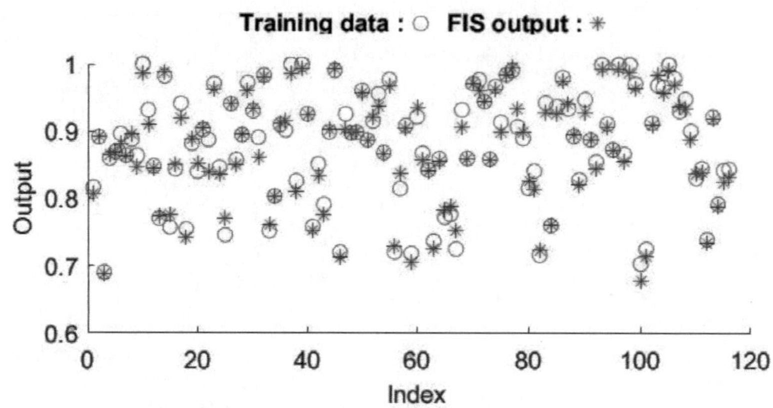

Figure 6. Comparison between training data and ANFIS predictions for condition 1.

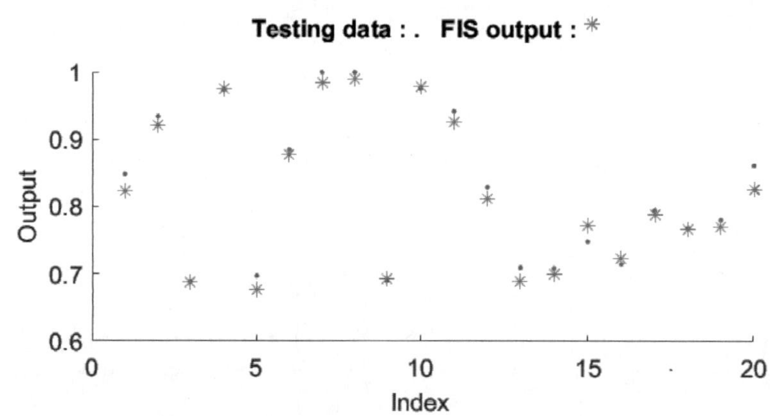

Figure 7. Comparison between testing data and ANFIS predictions for condition 1.

Table 3. Training and testing errors for all conditions.

Condition	Training RMSE (%)	Testing MAE (-)
1	1.23	0.0153
2	1.21	0.0123
3	1.11	0.0154

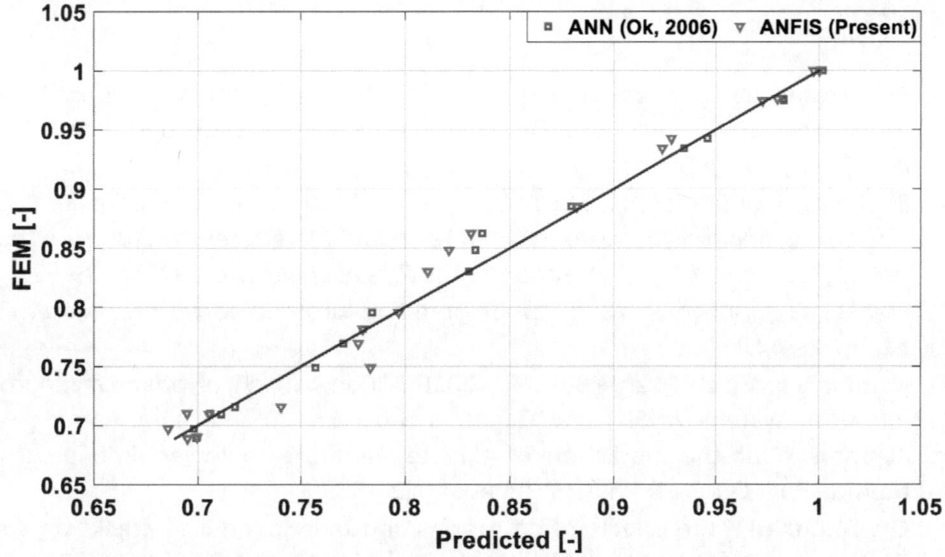

Figure 8. Comparison of ANFIS output with FEM and ANN results for condition 1.

CONCLUSIONS

The use of an Adaptive Neuro-Fuzzy Inference System (ANFIS) method is proposed in this paper to predict the ultimate strength reduction of locally corroded plates. ANFIS models were created and tested using published ultimate strength datasets for unstiffened plates with pitting corrosion under uniaxial in-plane compressive loads. The input variables included the ratios of the plate slenderness, pit breadth to plate width, pit length to plate length and pit depth to plate thickness. Different types of membership functions were tested to find the best accurate model. Based on the findings reported in this paper, the following conclusions were drawn:

- The two-sided Gaussian curve membership function was found to be more effective and less sensitive to the sample size than other types tested in this study with a maximum training error of 1.23%.

- ANFIS outputs were found to agree well with both FEM and ANN results; a maximum mean absolute error was found to be 0.0154 which is evident for the validity of the ANFIS method in predicting the ultimate strength reduction of locally corroded structures.

- Overall, the developed model can predict the reduction in the ultimate strength of plates suffering from pitting corrosion. Using the ANFIS method to validate additional ultimate strength results for structures with pitting corrosion is recommended for further investigations.

REFERENCES

- Abdussamie, N., Daboos, M. and Chai, S. (2016). "Assessment of operational risks of LNG and FLNG vessels during manoeuvring in bad weather conditions," *Proc 12th ISOPE Pacific/Asia Offshore Mechanics Symposium*, ISOPE.
- Bhandari, J. and Khan, F. and Abbassi, R. and Garaniya, V. and Ojeda, R (2015). "Modelling of pitting corrosion in marine and offshore steel structures - A technical review," *Journal of Loss Prevention in The Process Industries*, **37**, pp. 39-62.
- Frankel, G., Sridhar, N. 2008. "Understanding localised corrosion," Mater. Today, **11**, pp. 28-44.
- Mathur, N., Glesk, I. and Buis, A. (2016). "Comparison of adaptive neuro-fuzzy inference system (ANFIS) and Gaussian processes for machine learning (GPML) algorithms for the prediction of skin temperature in lower limb prostheses", Medical Engineering & Physics, **38**, pp. 1083-1089.
- Ok, D. (2006). "The effects of corrosions and fatigue induced cracks on strength degradation in aging ships," PhD thesis, School of Marine Science and Technology, University of Newcastle upon Tyne, UK.
- Ok, D., Pu, Y. and Incecik, A. (2007). "Artificial neural networks and their application to assessment of ultimate strength of plates with pitting corrosion," Ocean Engineering, **34,** pp. 2222-2230.
- Paik, J. K., Lee, J. M. and Ko, M. J. (2003). "Ultimate compressive strength of plate elements with pit corrosion wastage," Journal of Engineering for the Maritime Engineering, **217**, pp. 185-200.
- Paik, J. K., Lee, J. M. and Ko, M. J. (2004). "Ultimate shear strength of plate elements with pit corrosion wastage," Thin-Walled Structures, **42**, pp. 1161-1176.
- Sivanandam, S., S. Sumathi and S. Deepa (2007). "Introduction to fuzzy logic using MATLAB," Springer.

A paradigm shift in AUV capabilities – Boeing's Echo Voyager XLUUV

Dan Tubbs, Deputy Director Advanced Technology Programs, Boeing
<Dan.Tubbs@boeing.com>
David Allen, International Programs Manager, Boeing
<david.allen18@boeing.com>
Shane Rogers, Deputy Director Experimentation, Boeing
<Shane.Rogers@boeing.com>

ABSTRACT

The need to conduct meaningful missions in a wide range of environments in the maritime domain is driving demand for increased Unmanned Undersea Vehicle (UUV) range, persistence and payload capability. Further, operations in contested environments require UUVs that can operate at long range for extended periods with the ability to self-deploy and conduct host independent operations.

This paper describes the Boeing Echo Voyager Extra Large UUV (XLUUV) development history and high-level concept of operations. Echo Voyager provides the necessary range, payload and persistence to successfully operate in the Asia Pacific region. As a complementary capability to manned submarines, surface, and maritime patrol assets, Echo Voyager has the flexibility to conduct Manned-Unmanned Teaming (MUM-T) / Unmanned-Unmanned Teaming (UMUM-T) operations and this paper will outline some of the potential benefits.

FUTURE MARITIME OPERATING CONTEXT

The maritime operating environment is becoming increasingly complex, particularly in subsurface operations within the Asia Pacific region. The significant build-up of submarine forces in the region, coupled with ongoing regional tensions, are contributing factors.

Navies are embracing the concept of bringing Unmanned Undersea Vehicles (UUVs) to bear as assets that enable optimization of force structure, and provide the flexibility to execute the multitude of tasks set forth by their respective governments. The US Navy (USN) Unmanned Undersea Vehicle Master Plan (UUVMP) states that "the long-term UUV vision is to have the capability to:

1. deploy or retrieve devices,
2. gather, transmit, or act on all types of information, and
3. engage bottom, volume, surface, air or land targets"[i]

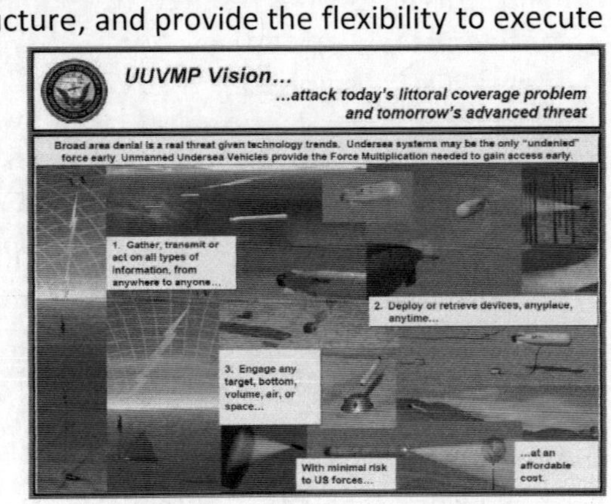

Figure 1- USN UUV Master Plan

While there have been classified updates to the Master Plan, the vision still holds true. The UUVMP further highlights that:

> *UUVs should be used in applications where they increase performance, lower cost, enable missions that cannot be performed by manned systems, or reduce the risk to manned systems. The characteristics of UUVs that may facilitate meeting these principles include their ability to put sensors in an optimal position in both the vertical and horizontal dimensions, autonomy, endurance, low-observability, expendability, and standoff or reach from the launch platform.*[ii]

This view is shared by NATO Sea Centre of Excellence. They define specific missions that can be undertaken by UUVs as being:[iii]

- Intelligence, Surveillance and Reconnaissance (ISR);
- Mine Countermeasures (MCM);
- Anti-Submarine Warfare (ASW);
- Inspection/Identification (ID);
- Oceanography/Hydrography;
- Communication/Navigation Network Nodes (CN3);
- Payload Delivery;
- Influence Activities (IA);
- Time Critical Strike (TCS).

As UUV technology and familiarity of operations increases, so will their importance. The potential to realize game changing capability drove Boeing to build UUVs with greater capability, increased autonomy, and larger payloads. This paper focuses on what Boeing has been developing over the past decade to deliver a paradigm shift in UUV capability.

BOEING HISTORY IN UNMANNED UNDERWATER VEHICLES

Boeing has a 50-year history in the design, development, and manufacture of undersea systems including high-end UUV and Unmanned Surface Vehicle (USV) programs. These were completed for a variety of government and commercial customers with a wide-range of missions and purposes in mind. This history has resulted in considerable expertise in integrating complex UUV systems and specialist knowledge in the areas of guidance and control systems, fault tolerant architectures, launch/recovery and deployment systems, and acoustics processing.

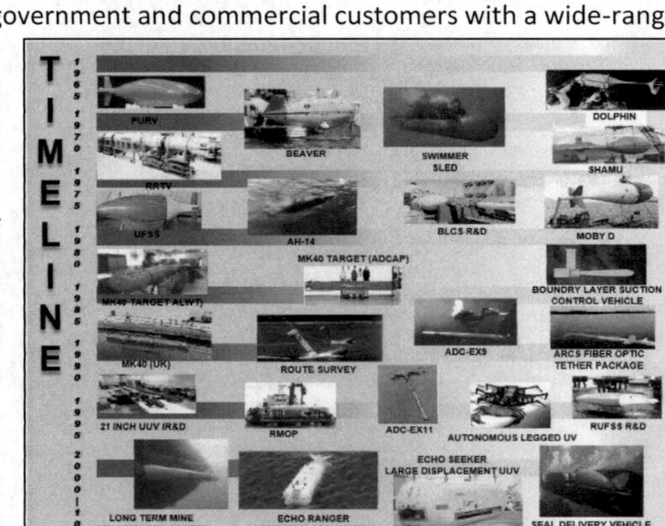

Figure 2- Boeing UUV History

An example of one of the complex and innovative systems produced by Boeing in the UUV domain for the US Navy is the Long-term Mine Reconnaissance System (LMRS). Designated as the AN/BLQ-11, it was developed to be launched from a submerged SSN's torpedo tube, search for tethered and/or bottom mines while also surveying underwater objects for up to 60 hours[iv], and then recovered back into the torpedo tube.

During 2006/2007 the AN/BLQ-11 successfully completed underway launch and recovery from various USN 688 Los Angeles class submarines, along with completing a number of other UUV technical firsts. The challenges for covert launch and recovery of an asset such as the AN/BLQ-11 are non-trivial, and the ability to successfully complete this program (under the direction of PMS406) is a prime example of the competence of Boeing's UUV division.

Figure 3- Boeing AN/BLQ-11 LMRS

Echo Series of UUVs

By the early 2000s, Boeing had identified the need for large displacement UUVs capable of carrying greater payloads with sufficient power onboard to fulfil more demanding missions. Commencing with the Echo Ranger Large Displacement Unmanned Undersea Vehicle (LDUUV), Boeing began development of UUVs that provide the ability to test varied payloads required by end-users in a variety of operating environments. Echo Ranger's major specifications are highlighted in Figure 4. Launch and recovery is achieved from shore facilities (towed to an operating area) or from a suitably equipped surface ship. Submerged communication is via an acoustic modem, with longer range command and control made possible by various radio frequency (RF) links.

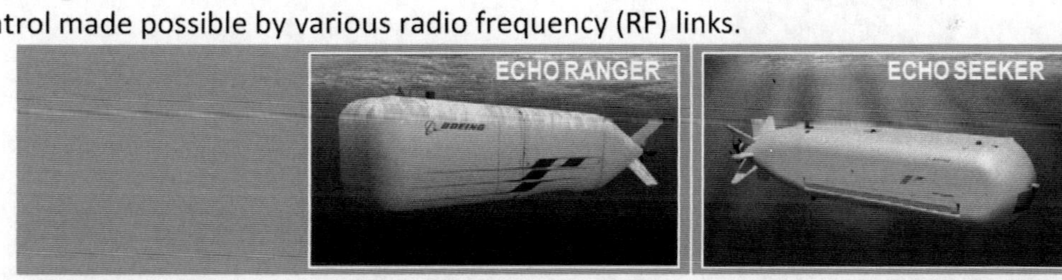

VEHICLE SPECIFICATIONS:		
Weight in Air:	11,690 lb (5,300 kg)	30,000 lbs (13,608 kg)
Envelope:	18.5 ft x 4.2 ft x 4.2 ft (5.6 m x 1.3 m x 1.3 m)	30 ft x 6 ft x 6 ft (9.1 m x 1.8 m x 1.8 m)
Maximum Depth:	10,000 ft (3,000 m)	20,000 ft (6,000 m)
Maximum Speed:	6 knots (14.5 km/hr)	6 kts (14.5 km/hr)
Minimum Speed:	1.5 knots (2.8 km/hr)	2.5 kts (4.63 km/hr)
Optimal Speed:	2.5 - 3 knots (4.6 – 5.6 km/hr)	2.5 -3 kts (4.6–5.6 km/hr)
Continuous Run:	28 - 150 hours dependent on battery	50-75 hours
PAYLOAD SPECIFICATIONS:		
Size:	60 in x 50 in x 20 in (1.5 m x 1.25 m x 0.5 m)	15 ft x 6 ft x 2.2 ft (4.6 m x 1.8 m x 0.7 m)
Volume:	Approx. 21 ft^3 (0.6 m^3)	Approx. 170 ft^3 (4.8 m^3)
Capacity:	Approx. 975 lbs dry (440 kg)	Approx. 6,000 lbs dry (2,722 kg)

Figure 4- Echo Ranger and Echo Seeker Specifications

Echo Ranger has been tested and observed by a number of agencies. However, its most extensive operations were undertaken in the Gulf of Mexico conducting bathymetric surveys for the oil and gas industry. As of 2017, Echo Ranger has completed approximately 1,000 hours of submerged operations. With a 975lb (440 Kg) dry weight payload capacity, in excess of 3000m diving depth and more than 40 hours of endurance, Echo Ranger proved to be a highly capable test bed as UUV technologies progressed.

Next in line was the Echo Seeker LDUUV. Echo Seeker's specifications are shown in Figure 4. Most notable are the increases in size (30 ft), payload (6,000 lbs or 2,722 kg), persistence (75 hours) and diving depth (6000 m). With Echo Seeker, Boeing refined UUV fault tolerance and redundancy architectures to enable long endurance operations, and incorporated obstacle avoidance capabilities using a forward looking sonar and complex autonomous obstacle avoidance algorithms. This included implementation and verification of underwater terrain following that allows Echo Seeker to maneuver at slow speeds with altitude separation of approximately 10 feet (3.3m) from the sea floor.

Echo Seeker began trials in 2013 and has been at sea as a test and operational asset since early 2015. Meanwhile, Boeing had completed early designs for a game-changing XLUUV. The Echo Voyager XLUUV is the third UUV in the Echo series and represents a paradigm shift in UUV capability for both military and commercial end-users. The phased approach Boeing has used while developing the Echo series of UUVs is shown in Figure 5. Each new asset has achieved significant technical advances in persistence, payload, power, diving depth, and autonomy to realize the vision desired by sophisticated operational users.

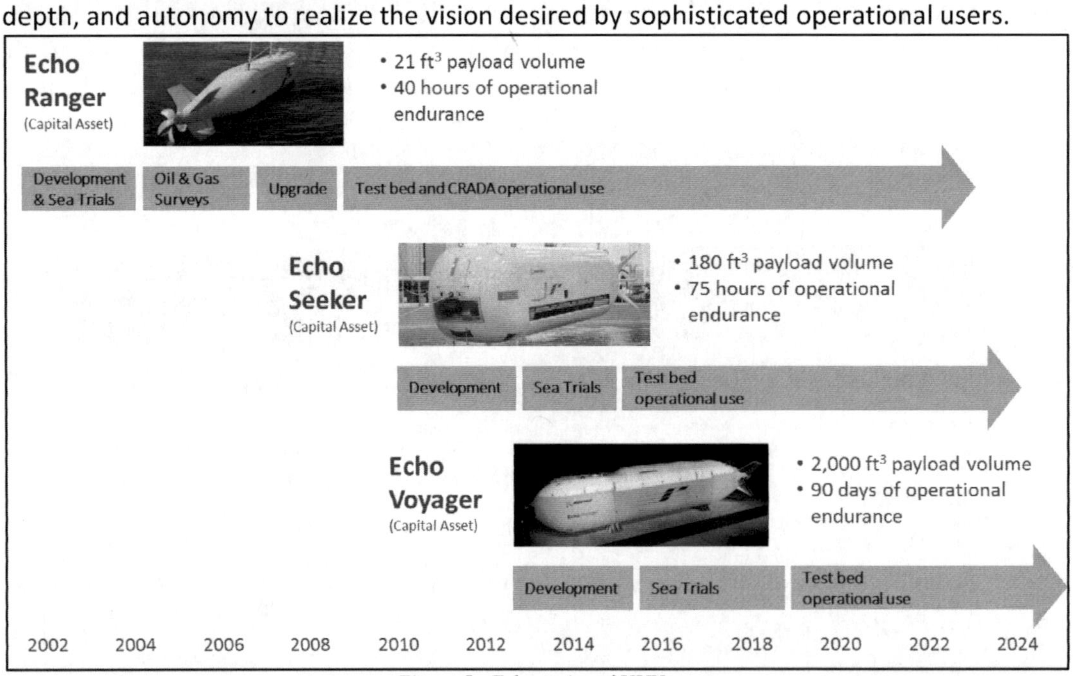

Figure 5 - Echo series of UUVs

Echo Voyager (XLUUV)

Echo Voyager is a multi-mission, long endurance UUV designed to act as a complementary capability to maritime forces above, on, and below the ocean surface. It represents a quantum leap in Autonomous Underwater Vehicle (AUV) operations, and possesses the necessary characteristics to cover an array of mission sets spanning military and commercial domains. It can also be employed as a disruptive capability in the area of Manned-Unmanned Teaming (MUM-T) and Unmanned-Unmanned Teaming (UMUM-T).

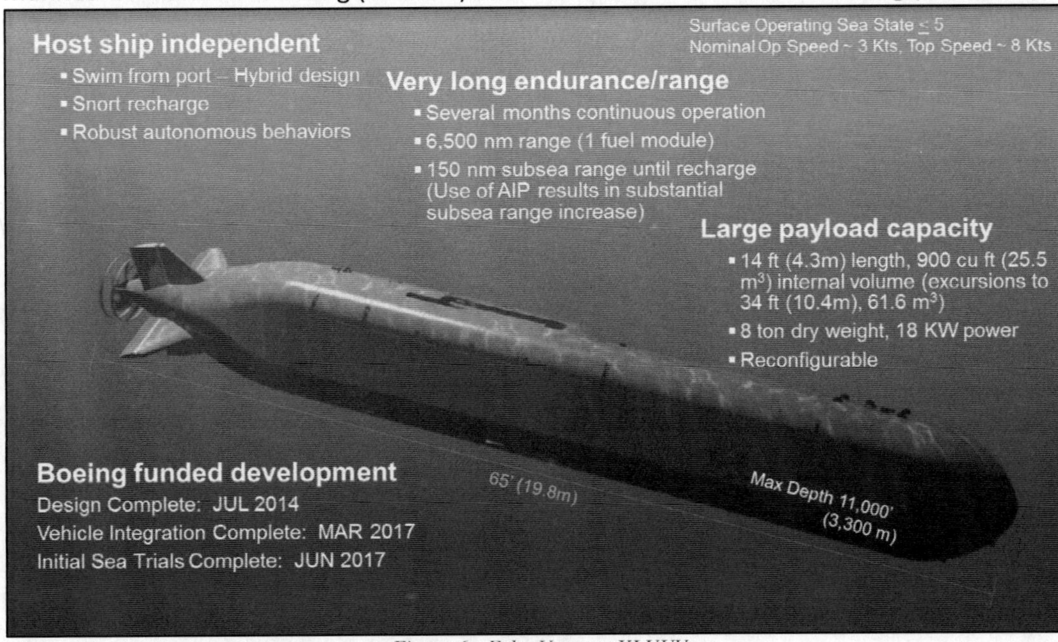

Figure 6 - Echo Voyager XLUUV

Echo Voyager is essentially an autonomous, diesel-electric submarine with broad specifications detailed in Figure 6. It is designed to self-deploy from port, but is also sea, road, rail, and air transportable. This provides flexibility from a deployment standpoint, and allows for rapid transportation when needed. Realizing that launch, recovery, and control of many current UUVs requires the support of tender vessels, Boeing has designed Echo Voyager with host-independent operations in mind from the beginning. This has many advantages from a safety, resource, and operational flexibility standpoint – including the covert launch and recovery of XLUUVs.

Echo Voyager has unsurpassed range and endurance when compared to other currently fielded UUVs. On a single fuel module Echo Voyager has a range of approximately 6500nm, and can operate for several months without the need to return to port for servicing and maintenance. Mission profile dependent, and operating at a nominal speed of approximately three knots with the currently employed battery technology, Echo Voyager possesses a subsea range of 150nm (the range at which the diesel engines must then be run to recharge the batteries). Alternative batteries, propulsion technologies (such as Air Independent Propulsion [AIP]), and fuel cell technologies are under investigation to substantially increase Echo Voyager subsea range and dramatically lower the indiscretion ratio.

Perhaps the most important capability of Echo Voyager is its large payload capacity. Payload modules have been designed up to 34ft in length, which provides space for an unprecedented suite of potential payloads and or additional fuel modules - potentially doubling range and patrol time. The payload space is further complemented by the power generation capability provided as a facet of the XLUUV's diesel-electric design. Currently, up to 18 KW of power is available to the payload section, which far exceeds the power available to current operational UUV payloads. Echo Voyager is also capable of mounting external payloads to the outer hull and on either the main or auxiliary masts to provide sensors external to the casing.

Due to depth and precision navigation capabilities that allow for terrain following, Echo Voyager can develop and refine seabed maps to help other maritime assets with underwater navigation. Echo Voyager has benefitted from Boeing's long history in this area, as these capabilities have been rigorously tested over time and now form a critical part of Echo Voyager's mission set.

A high-level of redundancy has been designed into Echo Voyager, leveraging Boeing's significant experience in UUV development and operation. In addition, all equipment in Echo Voyager is housed in pressure vessels, designed to be accessible via removable fairings to enable pier-side maintenance. This allows Echo Voyager to operate and be sustained by a small footprint of trained professionals.

Current Status

Echo Voyager completed in-pool testing early in Q1 2017 and quickly progressed begin initial sea trials. The initial trials program has been completed off the coast of Southern California, and Echo Voyager has already amassed more than 670 hours underway and has achieved depths in excess of 1000 feet (300m).

Echo Voyager's trials program provided the opportunity to test and certify key subsystems. This included the platform's discharge and recharge cycle, surface and subsurface navigation, sensor operation, and command and control functionality via a number of communications bearers.

Figure 7- Echo Voyager Trials

Boeing has also responded to a number of Request for Proposals (RFPs) following the launch of Echo Voyager. As XLUUV forms part of the USN's UUV Master Plan, it is expected that further Request for Proposals will be forthcoming as UUV policy, capability needs, and confidence in their operation grows. As RADM William Merz (OPNAV N97) stated in Oct 2016[v]:

> *"I'm tired of waiting...Give me what you have, I will take it to sea, I will give you feedback and you can continue to develop and evolve, and then I'll take that to sea and give you more feedback."*

This rapid prototype and development cycle should see a rapid increase in UUV capability and allow assets to be fielded to complement the fleet in being.

XLUUV CONCEPT OF OPERATIONS

The size and endurance of Echo Voyager opens up a vast array of operational possibilities. As mentioned earlier in this paper, the focus for the Echo Voyager XLUUV is to act as a complementary capability for other maritime assets. Echo Voyager is a cost-effective solution that, with the right payloads, can take on the dull, dirty, dangerous and deep missions that are likely to be a significant part of future maritime operations. By doing so, it frees up the highly specialized, scarce and expensive assets (such as manned submarines) to undertake the missions where they are needed most and provide the greatest benefit for the fleet.

A small sample of some of the basic missions that could be undertaken by Echo Voyager are described in the following sections.

Surface ISR

Echo Voyager possesses the capability to act as a collection asset operating both inshore and in open water environments. With payloads capable of assisting in the completion of Visual Intelligence (VISINT), Communications Intelligence (COMINT) and Electronic Intelligence (ELINT) missions Echo Voyager can be a powerful forward deployed asset to help determine pattern of life and provide Indications and Warnings (I&W) back to headquarters and/or other forward deployed assets.

Figure 8 –Surface ISR

Forward Package Delivery

Payloads under development for Echo Voyager allow conduct of missions for forward precision delivery of packages. The ability to achieve this in a covert manner is potentially a powerful asset to Special Forces (SF). Forward deployed packages can be placed, marked and accessed by SF teams on demand. Allowing for additional equipment to be accessed when needed most. This allows for the covert insertion of more prepared and better equipped SF teams into demanding littoral environments.

Figure 9 –Forward Package Delivery

Critical Infrastructure Inspection

With significant diving depth, extreme endurance, and large payload capacity Echo Voyager is ideally suited to conduct inspections of critical infrastructure. This includes inspection and repair of:

- subsea communications cables;
- oil and gas pipelines;
- oil and gas well heads and other equipment;
- maritime and environmental sensors.

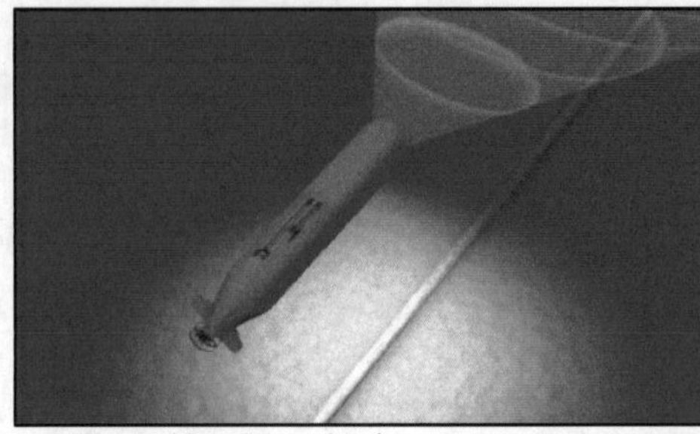

Figure 10 – Critical Infrastructure Inspection

In addition, Echo Voyager has the size and power available to act as a deep diving MOSHIP (Mother Ship) for Remotely Operated Vehicles (ROVs), although in this highly specialized case the remote operation would be controlled from Echo Voyager utilizing Unmanned-Unmanned Teaming (UMUM-T) capabilities.

An extension of this capability could be to assist in the conduct of deep sea Search and Rescue operations. This could include assisting in case of a DISSUB (Disabled Submarine) by surveying the area, providing valuable information for the DISSUB's location, the presence of debris, obstructions and how it is lying on the seafloor. This information is critical to the success of rescue efforts. A further option would be to use Echo Voyager to search for downed aircraft following ocean crashes. Post cessation of the black box signal, Echo Voyager can continue the search at depths up to 3000m for weeks on end. Hugging the sea floor and taking regular images which are processed onboard. Echo Voyager can be programmed to return to periscope depth (PD) and provide information back that meets certain search criteria (e.g. aircraft debris, localization of a black box signal etc.).

Maritime Sensor Deployment

Precision placement of maritime sensors at depth is another Echo Voyager mission. This allows for the deployment of sensor networks, including the ability to provide a limited trials range capability at relatively short notice and in the specific environment desired by test and regulatory authorities.

Figure 11 – Maritime Sensor Deployment

Intelligence Preparation of the Operational Environment (IPOE)

Missions that contribute to IPOE are vital for providing Advance Forces with operational awareness. This includes beach survey missions prior to Amphibious or Special Forces lodgment, identification of suitable anchorages and bottom composition, tidal movements, and the presence of debris and/or obstructions. While military focused in nature, the same capabilities can be utilized in Humanitarian Assistance and Disaster Relief (HADR) operations when responding to natural disasters. An example is conducting beach and harbour surveys in the immediate aftermath of a tsunami striking land.

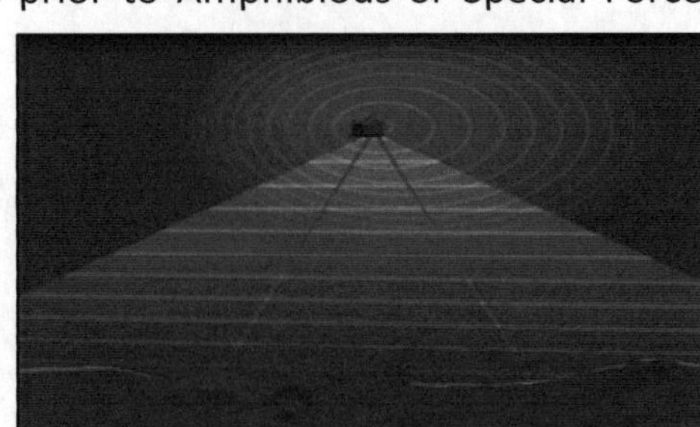

Figure 12 – IPOE

UAV Operations

Figure 13 – UAV Operations

The use of small UAVs as an asset for ISR activities is recognised as a potential force multiplier for advance forces. The ability to covertly deploy UAVs from Echo Voyager allows the end-user to operate a UAV in close proximity to contested areas. Control of the UAV can be exercised by multiple methods, but importantly the UAV is able to launch from close range. This minimizes the warning time for threat forces and increases the UAV's time over the target area. This concept is similar to the Defense Advanced Research Projects Agency (DARPA) Upward Falling Payload[vi] concept, with the major difference being that Echo Voyager has the ability to deploy UAVs at will, which allows for a more dynamic response to ISR needs.

Mobile Acoustic Intelligence (ACINT)

Figure 14 – Mobile ACINT

An Echo Voyager fitted with hull-mounted and towed array sonars make it a potential mobile acoustic intelligence (ACINT) platform. While its relatively slow speed would not be suited to trailing high speed advance noise sources, the capability to perform ACINT in contested areas is desirable. In addition, the large payload bay presents an opportunity for other sensors to be added to complement the sonar capability and increase the probability of detection for submerged tracks. Acting in this role, Echo Voyager provides valuable cueing for ASW operations.

ASW Training

Modern navies devote considerable time to ASW training in order to develop and maintain required high-end warfighting skills. Typically, this places a strain on the submarine fleet as they are typically few in number, are heavily tasked and consequently have to deal with competing priorities. While there is a definite need for manned submarines as training assets, there is an opportunity to complement the ASW training program for both maritime and air assets with CASEX (Combined Anti-Submarine Exercise) configured Echo Voyagers.

Figure 15 - ASW Training

Such a capability could emulate signatures, respond to stimuli, and provide a persistent training asset that could effectively hold in exercise areas for very long periods of time. This would free up manned submarines from having to undertake basic CASEXs, allowing those expensive and more capable assets to be tasked with the higher value and more demanding missions.

Cable/Array Deployment

Echo Voyager's ability to conduct precision underwater navigation, close to the seabed and at great depths allow it to lay cables and sensor arrays on the seafloor. There are numerous potential benefits associated with this capability. First and foremost is the potential for Echo Voyager to lay acoustic arrays and infrastructure on the seafloor that contribute to the development of an improved subsurface picture.

Figure 16 - Cable/Array Deployment

COLLABORATION OPPORTUNITIES

As the application of Unmanned Vehicles (UxVs) is further considered by military end-users, there are multiple opportunities for collaborative development on systems and sub-systems that will further enable the development of required UUV technologies. Underwater communications, propulsion, battery technology, and development of advanced payload/sensor packages are all potential collaboration areas. This is not an exhaustive list, as the mission sets Echo Voyager can execute both now and in the future are extremely diverse, however one of the benefits of having already fielded Echo Voyager is that Boeing can quickly test and incorporate new technologies as they mature. As a consequence, potential game-changing technologies can be made accessible to navies in an accelerated timeframe.

CONCLUSION

The maritime operating environment is increasingly complex. The prevalence of contested environment capabilities in the Asia Pacific region in particular dictates that a real emphasis must be placed on the ability for assets to be able to operate in both contested and congested areas for extended periods of time. Submarine capabilities continue to play a critical role in the future, but manned submarines will require significant investment (and quantities) in order to be survivable and capable of achieving all required missions.

Complementary UxV capabilities have the potential to be a powerful force multiplier and enabler for maritime assets, particularly as MUM-T concepts evolve. UUVs combined with manned submarines will enhance force effectiveness, however XLUUVs capable of operating alone, at extended range, for long periods provide an additional effect. They provide an ability to take on some tasking traditionally achieved by manned submarines, thereby providing navies with the ability to do more in a cost effective manner. Critically, this allows optimal use of manned submarine assets.

The Echo Voyager XLUUV breaks the capability curve when it comes to UUV performance. It provides a cost-effective, low-risk platform that enables navies to make best use of their assets, undertake specialized missions, and complicate the battlespace for potential adversaries. With the introduction of Echo Voyager to operational service a paradigm shift in AUV capability has truly arrived.

[i] The Navy Unmanned Undersea Vehicle (UUV) Master Plan, 9 Nov 2004, pp xv, xvii

ii The Navy Unmanned Undersea Vehicle (UUV) Master Plan, 9 Nov 2004, p.5

iii NATO Combined Joint Operations from the Sea Centre of Excellence Guidance for developing Maritime Unmanned Systems (MUS) capability, 9 Jul 2012, p.8

iv http://auvac.org/configurations/view/14. AUVAC Systems Spec Sheet for LMRS)website viewed 17 Jun 17)

v https://news.usni.org/2016/10/31/navy-seeking-uuv-advances-to-field-today-to-inform-ssnx-design-in-2020s. Statement by RADM Merz to Navy Submarine League 27 Oct 16 (website viewed 17 Jun 17)

vi https://www.darpa.mil/program/upward-falling-payloads DARPA Upward Falling Payload program (website viewed 7 Jul 2017)

Northrop Grumman Demonstrates High Area Clearance Rate with Unmanned Mine Hunting System at Unmanned Warrior

Gene Cumm
Northrop Grumman Corporation Undersea Systems
gene.cumm@ngc.com

THE ADVANTAGE OF UNMANNED SYSTEMS

Remote offboard systems bring multiple advantages with regards to their autonomy, persistence, and extended endurance in all weathers and environments. The ability of unmanned systems to operate at higher tempo inside the mine danger area increases search coverage rate, improves environmental characterization and assessment, and allows the capture of very high resolution seabed data. Furthermore, unmanned vehicles can be collaboratively tasked so as to perform multiple missions in series – or in parallel where multiple systems are available – using different sensor / effector payloads.

Increased safety of personnel is another key benefit. The current political environment, sensitive to attrition, demands that ricks to high-value ships and their crews are reduced as much as possible. The ability to deploy an unmanned system into the mine danger area, so extending the reach of intervention, removes the mine detection and classification process to a safe distance – and in doing so effectively removes the man from the minefield.

Unmanned systems have a lower logistical requirement for deployment, operation, and recovery, and require relatively simple ship integration with minimal platform impact. This makes them ideally suited to operations from non-specialist craft of opportunity, and supports rapid theatre entry using modular and highly portable containerized mission packages.

Fielding unmanned MCM systems also yields substantial acquisition and through-life cost savings when compared to conventional platform-based MCM assets, thereby dramatically reducing the mine's asymmetrical advantage. Manning requirements are reduced, and the pre-deployment overhead associated with work-up and signature ranging is massively reduced.

Specifically, the use of Unmanned Surface Vessels (USVs) with towed sensors is the fastest way to conduct mine warfare missions.

UTILITY OF UNMANNED SURFACE VESSELS FOR MINE CLEARANCE MISSIONS

USVs make rapid mine clearance missions feasible and cost-effective – today. Northrop Grumman Undersea Systems demonstrated this fact at the Royal Navy Unmanned Warrior exercises by deploying the AQS-24B Mine Hunting Sonar system integrated with an ATLAS ARCIMS USV. This system performed detection and classification and finished the mission much faster than any other surface vessel. Furthermore, USVs with towed sensors are suitable for augmentation with hydrographic capabilities. Overall, using USVs with towed mine hunting systems allows sailors to get extremely close to dangerous mine targets from as far away as possible.

USVS WITH TOWED SENSORS MAKE HIGH-SPEED MINE CLEARANCE ACHIEVABLE AND AFFORDABLE

Technology has developed to the point where the incorporation of USVs into a force structure is feasible. In fact, incorporating USVs is advantageous because USVs can perform a host of much needed missions including force protection, ASW, hydrography, and MCM in an amphibious construct. Furthermore, USVs with towed sensors can significantly speed mine clearance operations. For an opposed landing, combining the use of Unmanned Underwater Vessels (UUVs) and USVs is ideal. Using UUVs for the early covert activity in the shallow water regions close to shore, and USVs for the overt, deeper water search provides the maximum effectiveness for the mine hunting operation.

The U.S. Navy has adopted the concept of USVs with towed sensors through their Mine Hunting Unmanned Surface Vessels (MHUs). Also, such systems have been demonstrated to be effective in coalition exercises including the Belgian North Sea Unmanned MCM trial and the Royal Navy Unmanned Warrior exercises. These deployed mine hunting USVs feature the AQS-24B, which has an operating speed range 6 to 18 knots, High Speed Synthetic Aperture Sonar (HSSAS), and Laser-Line-Scan for optical mine ID at 10 knots. The AQS-24B system provides long range detection, localization, classification, and identification capabilities. Plus, the use of Sonar and Laser simultaneously provides maximum swath coverage with no gaps.

AQS-24B ON ATLAS ARCIMS USV: FULL INTEGRATION FOR UNMANNED WARRIOR EXERCISE 2016

At the Royal Navy Unmanned Warrior exercise in Scotland, Northrop Grumman demonstrated the benefits of using a USV with the AQS-24B to rapidly search a mine field. The AQS-24B system completed a 3 square nautical mile search in a complex, rocky terrain bottom environment in less than 4 hours, whereas surface ships required 5 times as long for the same search using hull mounted VDS. The only other USV-based competitor completed the same search in 7 hours. Northrop Grumman's performance at the Unmanned Warrior exercise proved the ability to remotely launch, operate and recover the AQS-24B in a challenging sea state 3 (SS3) condition from a USV. It also demonstrated the modularity of the system design: our team completed system installation onto ARCIMS in 7 days and started towing on day 8. This is the first time the system was installed on the ARCIMS platform.

POST – UNMANNED WARRIOR DEVELOPMENT

After demonstrating the capabilities of the AQS-24B system at the Unmanned Warrior trials, the AQS-24B team pivoted towards integrating the AQS-24B Mine Hunting Sonar technology into a larger system capable of detecting, localizing, classifying, identifying, and neutralizing mines.

USE OF AQS-24B CAPABILITY IN A SINGLE SORTIE DETECT TO ENGAGE SCENARIO

Northrop Grumman demonstrated the phases of Single Sortie Detect to Engage at the Belgian trials in June 2017. The system operated in a very coarse sand, high turbidity, and high current environment and completed a 4 square nautical mile search in under 4 hours. The AQS-24B with its high resolution SAS provided detection, localization, classification, and identification in real time using SAS and laser data forwarding. This enabled the engagement of targets with the Atlas SeaFox neutralizer, from the same platform, during the same Sortie, without having to return to the host platform or shore to conduct mission analysis and re-roll of mission package. This demonstrated the Single Sortie Detect to Engage mission capability.

COTS HYDROGRAPHIC SURVEY CAPABILITY WITH AQS-24B

Having proven that a cohesive Single Sortie Detect to Engage system is feasible with a USV and the AQS-24B, the AQS-24B team is exploring options for diversification and enhancement. Notably, Hydrographic software can be used to augment the AQS-24B HSSAS.

Dual-Mission Capability at a Fraction of the Cost

The integration of HSSAS and hydrographic technology provides a cost-effective augmentation of a military-grade MCM towed sonar system with a bathymetry system and minimum modification to the host platform. This increases the effectiveness of both MCM/Hydro missions. The unmanned platform can be operated as a hydrographic-only system, it can be operated as a hydrography system that can switch into sonar and laser imaging modes in response to conditions encountered, or it can be operated as an MCM sonar system with single-pass bathymetry, volume, and bottom imaging. This multimode adapts readily to a range of operations without retrieval and reconfiguration of the USV, maximizing the effectiveness of the single USV platform with minimal personnel support requirements. If deployed correctly from a larger multi-mission platform, the costs to procure and operate the system will be a fraction of existing legacy systems.

Rapid Environmental Assessment and MCM in Challenging Conditions

Using a new data forwarding communications link architecture, MCM Mission Commanders will receive mine threat contact logs before the MCM/Hydro platform, and the data stored on its hard drive, returns to base. The AQS-24B's capabilities will be vital in situations where high search rates are required, providing significantly decreased MCM detect to engage timelines. A USV based MCM/Hydro system can also be more effective than UUV-based MCM systems in regions where high currents preclude the use of slow-moving UUVs.

The addition of hydrographic sensor technology can meaningfully compound the utility to USVs with towed sonar sensors.

THE NEXT STEP IN MARITIME INTEROPERABILITY: COLLABORATIVE C3

Northrop Grumman has engineered a vision in which undersea dominance is achieved by drawing the contributions of many together in a unified and greater whole. The result is a common, interoperable command, control, and communications (C3) framework that serves as an essential enabler for undersea dominance.

In the next several years, a new generation of unmanned and increasingly autonomous offboard systems will replace many current MCM platforms, offering significant advances with regard to operational tempo, safety, and cost. Even so, it is apart that no single nation will possess the resources necessary to fully address the MCM challenge; a collaborative effort is essential.

Northrop Grumman believes that current circumstances demand a novel and transformative approach to that enables coalition navies to establish a common, interoperable MCM command, control, and communications (C3) framework at the same time that they recapitalize their individual MCM capabilities. This collaborative architecture will allow each navy to play to its component strengths, while ensuring high levels of interoperability at the operational and tactical level.

SUMMARY

Unmanned operations reduce risk and logistical overhead while enhancing endurance and autonomy. With regard to mine hunting missions, USVs compliment UUVs. UUV performance is limited in high speed and high current conditions and UUVs require lengthy PMA timelines. However, they provide excellent data collection capabilities for covert and deep water scenarios.

USVs can provide multi-mission support to amphibious operations and use of the AQS-24B installed on a USV enables rapid mine clearance operations. An AQS-24B-based system enables mine hunting at up to 18 knots and mine ID at up to 10 knots. Furthermore, integration of the AQS-24B and hydrographic capabilities enables simultaneous mine hunting and IPOE.

These operations are feasible and cost-effective today due to the ability to rapidly integrate existing technologies. USV operations drive the ability to get extremely close to targets from as far away as possible. Moving forward, interoperability and collaboration between coalition navies must drive efforts to achieve undersea dominance.

Achieving Integrated Command and Control of Unmanned Systems – The Unmanned Warrior Experience

Lisa Hunter-Yeats, BSc,
Working for QinetiQ Ltd UK

SYNOPSIS

Unmanned vehicles (UxVs) are doing things that have never been done before, creating new possibilities, challenging the way navies operate. Individually, these systems are capable. But can they be made to work together, integrating with, and into, manned platforms, and operating across air, surface and subsurface? A preceding paper introduced the Autonomy Exploitation and Realisation (ACER) programme and its architecture, outlining the development to date and setting out the overall programme objectives. This paper reports back on the work undertaken as part of Unmanned Warrior in October 2016, namely the rapid design, build, integration and fielding of 3 ACER command and control demonstration systems and their active employment at the centre of the largest unmanned systems demonstration of its type ever undertaken. Key technical features, including the use of open standards, the use of heterogeneous systems from different manufacturers, the deployment of systems at sea and ashore and the inclusion of an in service Combat Management System, are described, and the principal achievements and challenges are outlined. More detail is given on the operational feedback from RN operators and also the nature of the scenarios supported. The paper continues with a look ahead, outlining how the ACER programme is proceeding, fully benefitting from the learning that took place at UW16, drawing in developments from across the Dstl research programme and targeting the key challenge of managing the flow of robust, useable information to the Command without driving up crew sizes or overloading key operators. The final part of the paper is given over to an outline of how ACER is now part of a wider multinational Autonomy Challenge, set to take place in Australia in 2018.

INTRODUCTION

Unmanned Warrior (UW16), within Joint Warrior (2), 2016 presented the Research and Development community with a unique and unexpected opportunity to showcase an early implementation of the RN approach for integrating off board assets and their payloads into the Combat System. As part of the Dstl Maritime Autonomous PLatform Exploitation (MAPLE) demonstrator programme, the response to this opportunity was project ACER (Autonomous Control Exploitation and Realisation), designed specifically to support UW16 with the aim of delivering end to end Command and Control.

The wider context to UW16 and the Royal Navy's push to adopt maritime autonomous systems[1, 2] (MAS) is of rapidly developing technology and ever increasing cost of building and

[1]MAS are defined as off-board vehicles or equipments that operate in the maritime and littoral environment

manning warships. The latter, alongside constant pressure on defence budgets, has resulted in an RN Surface Flotilla one third of the size it was 50 years ago. In stark contrast the numbers (and tonnage) of commercial vessels continues to grow, and the UK economy is increasingly dependent on overseas trade. In order to combat these disparate trends unmanned vehicles have been identified as a route by which the Royal Navy can "buy back mass", so that a single ship can have the same impact as multiple ships which are not operating unmanned vehicles. "Buying back mass" in this manner is not going to be achieved by operating a single UAV from a single warship; it will only be achieved by operating a squad or swarm of 10 or more unmanned systems, so that a single warship can truly dominate an area of sea thousands of square miles in size. Currently the deployment and operation of such a collection of offboard systems would require multiple operators per vehicle. After accounting for watch rotations, a deployment of 10 such systems might require 75 – 100 staff under current concepts of operation. This is not only prohibitively expensive, it is also not physically possible from current warships.

Currently the operation of unmanned systems, particularly unmanned air vehicles (UAVs), is highly reliant on employing narrow field of view, full motion video. This naturally has a high dependence on a human operator to detect, identify and track objects of interest in a highly interactive manner with the unmanned vehicle. Realising the future vision of a MAS enabled Royal Navy will require increased levels of autonomy, taking the operator out of direct control of the unmanned vehicles, so that a single operator can command and control multiple vehicles. When considering the Ops Room of a modern warship it is apparent that the Royal Navy already has the concept of using a tactical picture based command and control to provide a single view of a large number of targets operating in a defined area. In addition, there are increasing developments in RADAR that are suitable for deployment from a UAV, plus automated target recognition and tracking in Electro-Optical and Infra-Red systems. It is therefore not a significant step to realise that the route to reducing the required number of MAS operators is for unmanned systems to provide track information on the contacts that they hold within their vicinity, with tracks from multiple vehicles being integrated into a single tactical picture. This is analogous to the way that Link 22 and Link 16 enable contacts from multiple sources to be integrated into a single tactical picture. The automated integration of unmanned vehicle sourced track information into a tactical picture enables the operator to focus on the bigger picture. It is this integration, alongside the wider planning, tasking and management of unmanned systems that has formed the focus for ACER.

without the physical presence of human operators, although this does not preclude operators being necessarily engaged with the remote operation of the system, and the associated C2, handling and maintenance facilities. This is distinct from the autonomy increasingly being introduced for the control of onboard systems such as propulsion machinery or for remote compartment monitoring.

[2]An autonomous system is capable of understanding higher level intent and direction. From this understanding and perception of its environment, such a system is able to take appropriate action to bring about a desired state. It is capable of deciding a course of action, from a number of alternatives, without depending on human oversight and control, although these may still be present. Although the overall activity of an autonomous system will be predictable, individual actions may not be.

MAPLE PHASE 2 – PERSISTENT ARCHITECTURE

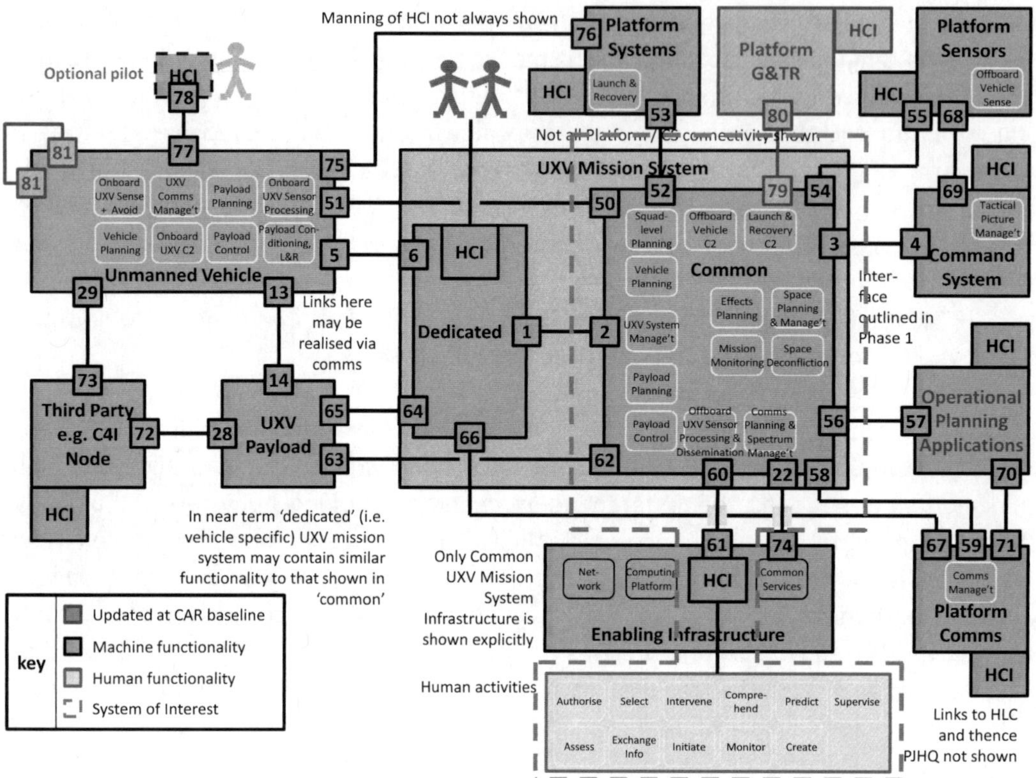

MAPLE Phase 2 Persistent Architecture

The Persistent Architecture (PA) partitions the system into specific blocks that separates concerns for:

- the unmanned vehicle and its payload;
- the dedicated aspects of the unmanned vehicle such as its ground control station (GCS) that, for example, has been certified for flight safety;
- linkage to the maritime combat system (i.e. sensors, G&TR, communications systems, and the command system) as provided by a RN surface ship;
- the enabling infrastructure as provided by the Open Architecture Combat System (OACS).

MAPLE PHASE 3 – ACER

The MAPLE Phase 1 & 2 team of QinetiQ, BAE Systems, Thales and SeeByte continued to work together on Phase 3, ACER demonstration delivery in UW16. The team leveraged the knowledge gained during the earlier study phases, capitalised on leadership of NATO UxV C2 working groups and was able to pull through key elements of the ongoing Dstl Open Architecture Combat System (OACS) programme. The team additionally had the advantage of being actively involved in a number of UxV development and integration programmes sponsored across UK MoD as well as within the international environment. Through making use of the breadth of the various organisations expertise, the ACER team was able to bring UxV experience from across the air, surface and subsurface domains as well as across military tasks.

The ACER objective as captured in the statement of need is:

'The user shall be able to integrate\ unmanned vehicle (UxV) mission planning and C2 into an Open Architecture Combat System, exploiting off-board UxV sourced data working in at least two different domains, in at least two warfare environments for at least two different types of UxV.'

In practical terms, this end to end capability was illustrated through the implementation of a number of specified information flows across equipment boundaries, enabling demonstration of Command Team tasking of UxVs and integration of UxV sourced data into the compiled tactical picture. The implementation in UW16 represented a significant advance on the limited integration previously achieved in fielded unmanned systems and allowed meaningful user engagement on future concepts of employment and on the functionality required in the ACER application set. The system solution implemented for ACER enabled an early assessment of the architectural development work conducted during the previous Phases of MAPLE and the realisation of an effective command and control node to support the Unmanned Warrior demonstrations.

This breadth of knowledge in the development of new and novel techniques, combined with the team's deep experience of integration & maintenance of in-service equipment and the way the trials were conducted, were a key factors in the success of this Phase of the MAPLE Programme.

ACER Component Parts

The scope of the system is summarised in the diagram below:

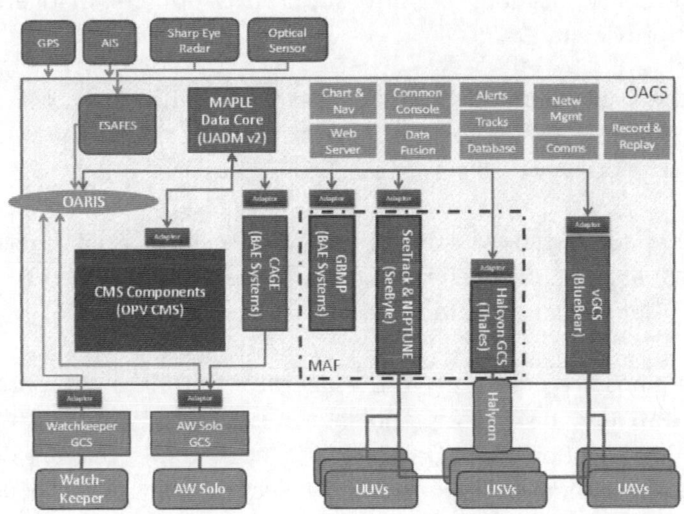

Fig 1: ACER C2 Demonstration Architecture

The yellow box indicates the boundary of Open Architecture Combat System (OACS). This is the enabling infrastructure that provides the computing resource for hosting ACER applications (the dark blue boxes). In addition OACS provides a set of open interfaces and standards to enable the rapid integration of capability to include:

- Open Architecture Radar Interface Standard (OARIS): includes use cases to allow sensor reports to be input into the combat management system (CMS); adapters are used to convert sensor data from its raw format into OARIS (the brown ellipse).

- Alert Management Service (ALMAS): allows third party applications to send alerts to the command system (e.g. DNA(2) / CMS-1) for the attention of the command.

- Naval Navigation System Interface (NNSI): allows third party applications to read navigational position and attitude information.

- Open Geospatial Standards: provides services defined by Open Geospatial Consortium (OGC) standards to third party applications, to include images and vector datasets.

- Data Distribution Service (DDS) middleware: provides the integration technology for new tactical message-based interfaces within the combat system, based on a publish / subscribe technology where a common information model is shared amongst a federation of components.

The services provided by OACS are depicted by the brown boxes.

Goal Based Mission Planner (GBMP)
GBMP developed by BAE Systems provides a set of algorithms to support an operator planning missions for multiple vehicles and coordinating their activities. The GBMP HCI allows the operator to input mission objectives (e.g. mine search goal, area patrol goal, convoy protection goal), and constraints such as area exclusion zones and expected deployment and recovery locations. It also supports a number of overlays such as the shared chart server provide by OACS.

GBMP supports the multi-tier planning approach as defined by Maritime Autonomy Framework. This manages planning complexity by separating the overall 'who does what and when' (Tier 1) from the task and vehicle specific 'how' (Tier 2).

The planning process supported via the HCI follows the steps of goal decomposition, task assignment & scheduling and plan rehearsal. During the rehearsal step GBMP shares plans with the connected Tier 2 planners associated with each of the assets.

Generated mission plans are presented to the operator both in a geospatial view, showing computed routes between task areas, and in a mission timeline view representing the scheduling of task completion for all assets. The operator is responsible for reviewing generated plans, editing them as required and selecting plans for execution. GBMP also supports monitoring of mission progress when feedback is provided from assets during execution or from the CMS tactical picture.

Enhanced Situational Awareness From Existing Sensors (ESAFES)

ACER also utilised components from the Dstl situational awareness integration test facility (SA-ITF), including its radar, IR and EO sensors. The enhanced situational awareness from existing sensors (ESAFES) fusion engine (the green box) was used to fuse radar and IR information in order to provide a single sensor feed into the Combat Management System (CMS).

ESAFES is a sensor fusion engine developed by QinetiQ that fuses plot sensor data from 2D and 3D radar systems, 1D sensors such as infra-red search and track (IRST) and radar electronic support measures (RESM) sensors. ESAFES was adapted to use the OARIS and NNSI interfaces and virtualised in order to run on OACS. ESAFES provides the primary organic sensor data into the CMS to aid Command situational awareness.

The specific UxVs, their ground control stations (GCS) and planners where not hosted by OACS and are shown outside of the yellow box.

The ACER software additionally included a Data Model (termed the ACER Data Core) for all UxV information including mission plans, assets and contacts. The OACS services were used to provide track input, operational alerts, charts and additional military layers (AML). A set of open standards and their associated Interface Control Documents (ICDs) were generated, defining generic information sets appropriate to vehicle types across the various warfighting domains. Adaptors for all vehicle mission management or ground control stations were written to ensure the open interfaces including the Data Core were used enabling this implementation to be a genuine step towards implementation of the persistent architecture vision from MAPLE.

Physically, the initial ACER build was an adaption of an existing ISO containerised system - the Situational Assessment - Integration Test Facility (SA-ITF) with an additional ISO Container being procured to provide a more comfortable “Command Team” working environment, allowing easier user interaction.

Fig 2: SA ITF prior to ACER fit out

Fig 3: Layout of ACER afloat

Functionally, the ACER implementation was targeted at a number of specific tasks within the overall UW16 scenarios. These tasks were chosen based upon the availability of known vehicles, sensors and the compatibility of associated command software.

Development Approach

All development activity was performed at the supplier's base sites, including initial testing. Each application supplier was responsible for developing an adaptor interface (against the appropriate published ICD) to their application. This enabled the application to communicate with all other ACER elements using the open architecture protocols forming a key part of the OACS and MAPLE design approach.

An incremental development and delivery approach was adopted whereby multiple "drops" of the Data Core were distributed to all. This approach enabled early proof of the adaptor interfaces and the timely identification of misinterpretation of specifications.

Early testing included communication with the Data Core and running the OACS virtualised environment and appropriate utilisation of the OACS services. Each supplier was responsible for the collection of evidence from integration and factory acceptance tests appropriate to the support of contract acceptance. Functional acceptance was completed at Portsdown Technology Park (PTP) through a combination of presentation of the evidence collected during integration testing and factory acceptance testing and a real-time demonstration. Physical acceptance was conducted through inspection (and demonstration) of the ISO containers at PTP and final acceptance followed the delivery of a final ACER report (post UW16).

ACER & UNMANNED WARRIOR 16

The participation of ACER in UW16 provided a unique opportunity to demonstrate the C2 of unmanned vehicles across numerous operational scenarios and vignettes. In total, 3 separate ACER instantiations participated, the initial build was deployed afloat on Serco Denholm's NORTHERN RIVER (SDNR), participating in the ISTAR theme, whilst ashore in Kyle of Lochalsh a second variant, procured by the DE&S mine countermeasures (MCM) and hydrographic capability team, participated in the MCM theme. The third variant, a BAE Systems installation was deployed at UW16 HQ in Benbecula, again supporting the ISTAR theme.

Each of the variants was deployed with specific objectives but all 3 allowed meaningful user engagement on future concepts of employment and on the functionality required to support future exploitation of unmanned vehicles and their payloads. For the afloat deployment, the key focus was on assessing the command and control (C2) of UxVs from a naval platform as enabled by ACER and on assessing the level of integration enabled by ACER of UxV sourced data into the combat system tactical picture generated by the SA-ITF sensor suite.

A joint industry and ministry of defence (MOD) trials team deployed to SDNR over the period 3rd to 20th October 2016, under the direction of the Dstl sponsor, to participate in UW16. During the first week of the deployment ACER Afloat took part in a WATCHKEEPER demonstration in Cardigan Bay, transiting North on completion to the Minches to conduct the main two week phase of UW16. This duration allowed a much longer trials programme than would normally be achieved and it also permitted human factors assessments of the ACER solution at all 3 installations.

The ACER system on SD Northern River was manned by Military Advisors from Dstl supported by Industry personnel as required. The system was exercised 05-06 October 2016 in Cardigan Bay for the Watchkeeper UAV demonstration, and in the Minches 10–20 October 2016 as part of UW16. The military operations included:

- Persistent Wide Area Surveillance (PWAS)
- Maritime Interdiction Operations (MIOPS)
- Fleet Protection
- Support to Landing
- Choke Point Transit

The UXVs that interoperated with ACER included:

- Thales Watchkeeper UAV (location, radar tracks, video, EO video);
- Boeing / Insitu ScanEagle UAV (EO video);
- Leonardo SOLO Rotary UAV (location, optical/ radar/ ESM/ AIS tracks, EO video);
- Lockheed Martin Indago UAV (EO video);
- Boeing / Schiebel Camcopter Rotary UAV (EO video);
- BAE Systems PAC950 USV (location, radar/ AIS tracks, EO/IR video);
- Dstl Maritime Autonomy Surface Testbed (MAST) Bladerunner USV (location, radar/ AIS tracks, EO/IR video).

From the analysis of the observations on SDNR and captured data, a number of conclusions were drawn with the following main findings:

- Enabling the UxV location to be displayed in the CMS tactical picture allows the Principal Warfare Officer (PWO) and the rest of the command team to employ all the embedded CMS functionality that is usually employed to monitor missions, de conflict battlespace, identify collision risks and direct and task organic and non-organic UxV assets;

- ACER was able to successfully exploit the majority of sensor information provided by UxV controlled by or operating in support of SDNR during UW16;

- Command situational awareness was significantly enhanced through the integration of UxV location and sensor tracks including automatic identification system (AIS), electromagnetic support measure (ESM) and radar tracks. Importantly, these tracks could be correlated, both manually and automatically with SA-ITF organic sensor and AIS tracks, subject to correlation rules being met;

- UxV video from electro optic (EO)/infra-red (IR) sensors made a highly significant contribution to the identification and recognition of the ACER Afloat CMS tactical picture.

MCM Theme

The ACER MCM C2 system was manned during UW16 (01-21 October 2016) by RN serving personnel from the Maritime Autonomous Systems Trials Team (MASTT) supported by Industry personnel as required (see photo over page). The military exercises at Kyle of

Lochalsh focussed on the MCM Theme with a major element being the Hellbay 4 trials (01-15 October 2016). These were the concluding part of a multi-year TTCP (The Technical Cooperation Program) initiative to develop and demonstrate autonomous behaviours within a MCM mission scenario, and comprised a set of 8 technical objectives shared between the 4 nations participating (UK, US, Canada, Australia). Specifically Objective 2 that utilised the ACER MCM C2 demonstrator had the following aims:

- demonstrate the ability for a heterogeneous squad of unmanned vehicles to co-operatively deliver an MCM mission, based upon an overarching goal;
- demonstrate the ability to use a UAV as a communications relay between a shore based C2 station and a deployed squad of vehicles enabling over the horizon operations.

In all, 32 recommendations were made on the preparation and conduct of similar trials in the future and for follow on research activity to support the development of UxV C2.

More widely, the level of collaboration between industry partners and with customers was a key factor in the achievements of MAPLE Phase 3. An agile, flexible and focussed approach to programme delivery enabled the team to respond to short term requirement changes. The overall ACER programme was delivered in 7 months to 2 different customers. Across the 3 instantiations, the following was achieved:

- 15 different types of UxVs from 12 organisations and 6 different nations were integrated with ACER and data integration was achieved with (at least) 25 UxVs; a quick look analysis confirmed successful integration and correlation of multiple data sources into a single tactical picture.
- In support of ISTAR operations up to 7 vehicles were simultaneously communicating with ACER.
- In support of MCM operations/trials a total of 10 vehicles were simultaneously communicating, over 3 domains (Air, Surface and Sub-Surface) and from 3 nations (UK, USA and Canada).
- ACER was able to successfully exploit the majority of sensor information provided by UxVs controlled from or operating alongside SD Norther River during UW16.
- A wide range of USVs and UAVs were integrated with ACER. The levels of integration varied; at the lowest level of integration a live video feed was integrated into the

system; further levels of integration enabled "Own UxV Track Position" and "Tracks" of objects detected from the vehicle to be passed into ACER.

In human factors terms users / observers rated the ACER displayed information as safe and easy to understand, and considered that, where information was provided, it was generally essential to UxV management. Whilst users / observers described a number of specific useful features associated with UxV C2 they also made a number of recommendations for their improvement.

The following high level key conclusions have been drawn:

- ACER demonstrated a viable route to integrate information from unmanned vehicles into the CMS of RN ship's and how those UxVs (from multiple suppliers) could be controlled from a workstation hosted on OACS.
- The use of open interfaces and standards coupled with the approach to making information available to all hosted applications enabled the rapid integration of UxV capability.
- Integration with ACER at UW16 and the ability to display UxV information on a tactical display allowed the RN to more easily appreciate and visualise the capability offered by unmanned and autonomous vehicles.
- The integration of sensor information from unmanned vehicles can provide significant enhancement to situation awareness in an RN Ops Room.
- UW16 demonstrated UxV technology is significantly more robust than perceived by many senior officers and is capable of delivering credible military capability today.
- Unmanned vehicle technology and autonomy presents a range of Human Factor issues, which will pose significant challenges for the transition into service of UxVs.
- Developing a set of interface standards enables interoperability with the unmanned vehicles of coalition nations.
- At UW16 ACER demonstrated the potential for industry to deliver military capability as an UxV based service, utilising a deployable C2 facility.
- ACER and the success of UW16 demonstrated collaborative effort of participants delivers an output significantly greater than the sum of individual contributions.

The following high level recommendations are offered to inform future MAPLE activity:

- The next phase of development should build on the integration of sensor data demonstrated during UW16 and aim to demonstrate / investigate how the command team can best task unmanned vehicles from the operations room.
- Use of a Synthetic Environment should be considered for assessing the appropriateness of the various implementation options for the command of vehicles to minimise overheads and maximise experimentation options.
- Investigations should include assessment of the level of automation or autonomy that the operator finds helpful and how far that autonomy can be extended without the operator feeling he is "out of control".
- Significant advantage was provided to the MCM UW16 Theme through taking advantage of the ACER build programme – the next phase of MAPLE should take every opportunity to identify other research programmes from which it can adopt, adapt or integrate output to gain a similar advantage in the next phase.

ONWARDS TO MAPLE PHASE 4

Phase 1 saw feasibility studies and some preliminary architecting ahead of a more comprehensive architecting effort in Phase 2 – this gave rise to what is referred to as the Persistent Architecture[3]. Phase 3 has been outlined above, with its focus on practical demonstration at UW16. In line with Dstl's original intention, Phase 4 will refine and develop the architecture in parallel with further demonstration and user engagement. Throughout there has been and will remain a close relationship with developments in the NATO industry community, with respective programmes feeding into each other as new developments are made. The main phases of the programme are illustrated schematically in Figure 4 below.

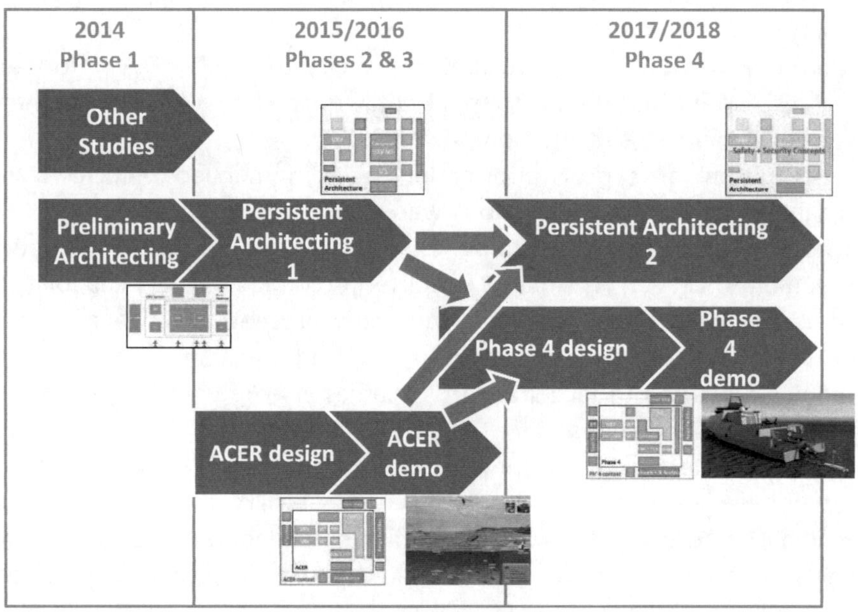

Fig 4: MAPLE Roadmap

As a direct result of the successful integrated demonstrations during Unmanned Warrior 16 (UW16), confidence in the technological maturity and the operational utility of unmanned vehicles has significantly increased throughout the stakeholder community. Despite this, there remain a number of technical problems and operational concepts still requiring attention prior to entry into service. Phase 4 seeks to address these issues through a mixture of study, design, experimentation and demonstration. The ambition is putting the Maritime Autonomous PLatform Exploitation (MAPLE) solution onto a Royal Navy warship within this decade.

[3] A Persistent Architecture is an enduring architecture description which enables and informs the development and consideration of different architectural forms and epochs within a domain. These may include current solutions, future developments, demonstrator rigs and prototypes, and envisioned concepts. Typically it is used to capture and control the evolution of current solutions towards an envisioned architecture-based form.

In order to enable operational command of UxVs without introducing manning overhead to the Operations Room of existing warships, we have identified a number of key areas (consistent with the Persistent Architecture) where application development or enhancement over that which exists in bespoke UxV management systems is most likely to yield results. Building on the outputs of MAPLE Phase 2, this project phase will also seek to mature the non-functional architectural elements which are not yet sufficiently well understood for an implementation to be proposed. The key areas include security, safety and communications.

Phase 4 will adopt an iterative and spiral development approach whereby small increments of "autonomy" can be envisaged, tested with a Stakeholder community and then assessed in terms of operational utility and feasibility of delivery. The plan is for a three staged programme:

- Stage 4A: Alongside the maturation of the PA this stage will focus on the establishment of the Synthetic Environment (SE) and the enhancement needed to the ACER system to support a live event in Cardigan Bay in September 2017 (CB17);
- Stage 4B: This stage will focus on the continued maturation and validation of the PA and enhancements to ACER. This stage will culminate in a SE Demonstration trial to help prepare for The Technical Cooperation Programme (TTCP) event in 2018, Autonomous Warrior 2018 (AW18). The TTCP event scheduled for Jervis Bay in Australia in 2018 presents a unique opportunity to demonstrate the MAPLE capability;
- Stage 4C: This stage will focus on the final preparations for the TTCP event and completion and reporting of the work strands.

On completion of MAPLE 4 the overall technology readiness level (TRL) of MAPLE will be TRL 5/6, realised by demonstrations in both a relevant Synthetic Event and a representative Live Event. To achieve this it is recognised that the system configuration tested will have to match closely the intended implementation on an RN warship. At phase end, Dstl will also have a roadmap describing a recommended approach to bring into service the integration of UxVs into the Command Chain of an RN platform. As the ACER system is hosted on the OACS Shared Infrastructure a natural progression would be to be put subsequent implementations of MAPLE (an enhanced ACER) on the OACS sandpit proposed for the T23. The sandpit would allow MAPLE to be demonstrated in the same operating environment as the platform and facilitate integration with operational UxV feeds, such as from Watchkeeper.

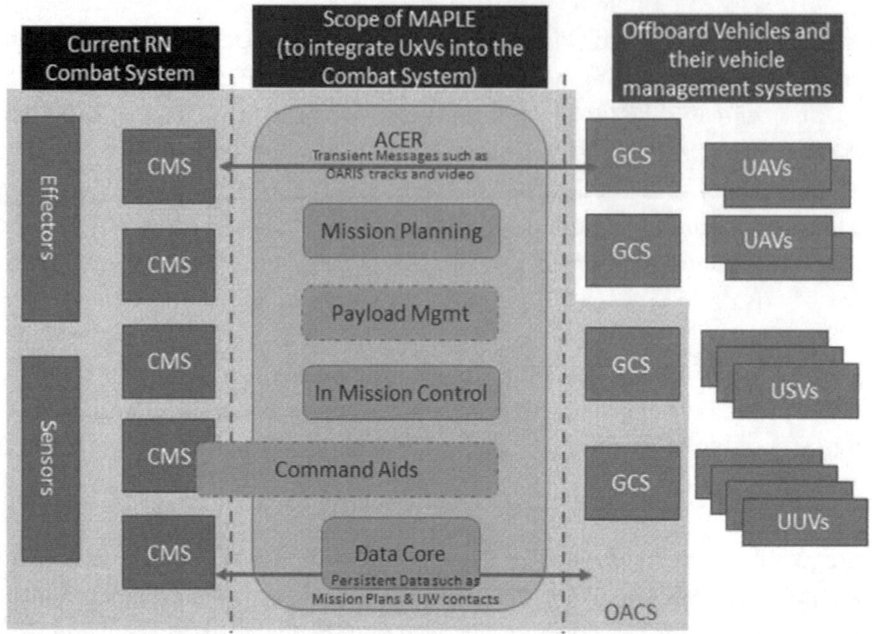

Figure 5: Scope of MAPLE as an Integration Path for UxVs into a Royal Navy Combat System

Looking more broadly, Dstl are determined to reduce time and costs of integration through the use of recognised open standards or NATO STANAGs, eliminating the proliferation of bespoke solutions and ensuring interoperability with coalition partners. There are already a number of STANAGs employed to define the connectivity interface between UAVs and their ground control stations, such as STANAG 4586 – Control of UAS motion imagery payloads, or STANAG 4609 – Digital motion imagery standard. However, neither of these STANAGs includes the concept of a track table owned by the UxV, or the ability to share the track table between the UxV and a central tactical picture. It is for this reason that the MAPLE work has adopted the OMG endorsed OARIS (Open Architecture RADAR Interface Standard) and will be expanding this interface standard in Phase 4 to support the levels of autonomy required to drive manpower out of the system.

CONCLUSIONS

Unmanned, increasingly autonomous systems under the banner of MAS continue to be seen as central to the future sustainment of maritime operational advantage. Regaining mass without increasing complements requires intelligent adoption of autonomy and an integrated approach to command and control (C2), bringing together organic platform sensors and effectors and traditional C2 with a network of intelligent offboard systems. For the UK, the MAPLE programme is progressively exploring the problem space, with the ACER demonstrator representing a genuine advance on current integration practice for unmanned systems and its deployment as part of UW16 was a notable achievement. More specifically:

- ACER demonstrated a viable route to integrate information from unmanned vehicles into the CMS of RN ships (or a deployable C2 node) and how those UxVs (from multiple suppliers) could be controlled from a workstation hosted on OACS.
- The use of open interfaces and standards coupled with the approach to making information available to all hosted applications enabled the rapid integration of UxV capability.
- Integration with ACER at UW16 and the ability to display UxV information on a tactical display allowed the RN to more easily appreciate and visualise the capability offered by unmanned and autonomous vehicles.
- The integration of sensor information from unmanned vehicles can provide significant enhancement to situation awareness in an RN Ops Room.

With the clear goal of MAPLE integration on an RN warship, Phase 4 of the MAPLE programme is further developing its architecture to address key non-functional considerations and exploring and developing new functionality in the ACER demonstrator. This new functionality will be tested both synthetically and in live events, most notably in a multi-national demonstration in Australia in 2018 – Autonomous Warrior. A key focus moving forwards is the development of the set of interface standards enables that interoperability with the unmanned vehicles of coalition nations.

REFERENCES

1. P Smith et al, 'Architectural developments to enable the integration of unmanned vehicles into the maritime system of systems', INEC 2016 Bristol
2. W Biggs et al, 'Systems engineering enablers for maritime intelligent systems', FAAW 2015 Bath
3. 'Global Marine Technology Trends', Lloyds Register/QinetiQ/University of Southampton 2015.
4. https://en.wikipedia.org/wiki/Offset_strategy
5. Royal Navy vision for Maritime Autonomous Systems D/Navy//MARCAP/SCCS/10-14 dated 5 Aug 14
6. R E J Westgarth et al, 'Human interaction and integration with future warships', Warship 2015: Future Surface Vessels, 10-11 June 2015, Bath
7. DCDC, Joint Doctrine Publication (JDP) 0-01.1 United Kingdom Supplement to the NATO Terminology Database 8th Edition dated September 2011
8. 2012 NRAC Study – How Autonomy Can Transform Naval Operations
9. T Rabbets et al, 'Developing a new Combat System architecture for Royal Navy surface combatants', MAST 2007 Genoa

A Practical Demonstration of ASW Using Unmanned and Autonomous Vehicles.

Robin Campbell
QinetiQ
rdcampbell@qinetiq.com

INTRODUCTION

Articles in the open press indicate that many navies are considering using Maritime Autonomous Systems (MAS) for a range of military tasks. The strategic drive to embrace MAS for military tasks in the maritime domain generally focuses on their potential to distance or remove human operators from dull, dirty, dangerous tasks in a number of warfare scenarios; typically Anti-Submarine Warfare (ASW), Rapid Environmental Assessment (REA), Intelligence, Surveillance, Target Acquisition & Reconnaissance (ISTAR) and Mine Countermeasures (MCM). However, within the UK, the adoption of MAS in the Maritime Sector has been slow and has principally been driven by MCM with the application of commercial off the shelf (COTS) equipment originally designed for off-shore industries.

This paper details 2 examples of the use of MAS in ASW scenarios. The first is taken from the ASW theme of Unmanned Warrior 16 and is focused on the use of Unmanned Surface Vehicles (USV). The second is from a series of proof of concept trials conducted by QinetiQ at the British Underwater Test and Evaluation Centre (BUTEC) using a COTS UUV and a COTS thin line towed array to detect and track dived targets. The paper concludes with an assessment of the future development of MAS for ASW scenarios based on observations and experience from the 2 events described in this paper.

UW 16 - AN OVERVIEW

Background to UW16.

In October 2017 the First Sea Lord (1SL) challenged industry to demonstrate the use of MAS to deliver military capability within the construct of a large scale UK exercise, JOINT WARRIOR (JW162). 40 organisations responded to the challenge and planning started for what would later become Unmanned Warrior 16 (UW16). A broad 'top down' structure was established by the MoD around 5 themes; namely, Command & Control (C2), ASW, Hydrographic & Oceanographic Intelligence Collection (GEOINT), ISTAR and MCM, with participants in UW16 coalescing into these five communities of interest dependant on their interests and capabilities i.e. participants volunteered equipment demonstrations rather than responding to a technical challenge/requirement from the Ministry of Defence (MoD). Each theme nominated an industry lead to work in partnership with a MOD lead to develop a coherent programme of demonstrations within their theme. Fig 1 below gives an indication of the range of organisations which showed early interest in UW16.

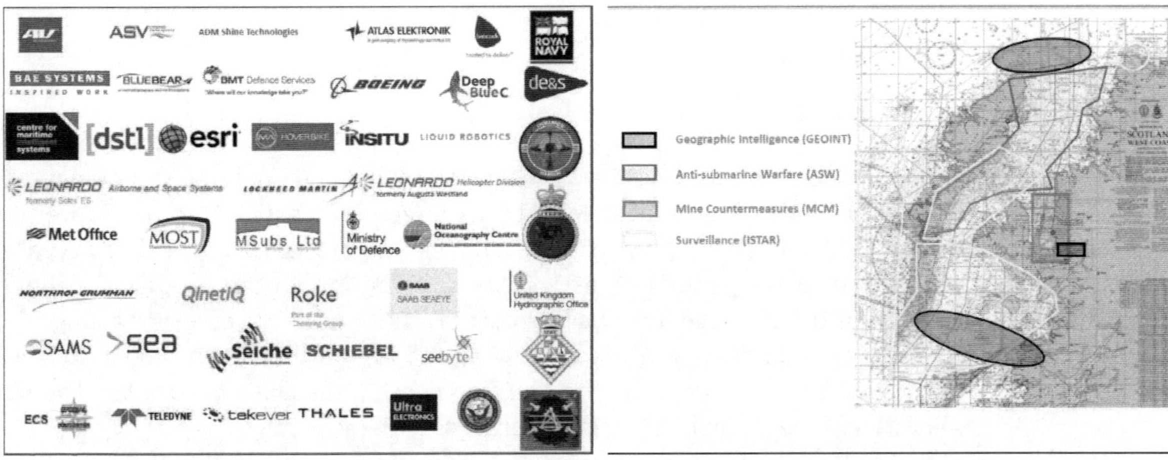

Figure 1 UW16 Participants (left) and Themes and Area of Operation (Right).

Following nearly 2 years of planning, UW16 took place off the north west coast of Scotland between 9 Sep 16 and 20 Oct 16 with the bulk of the activity occurring in the last 2 weeks of that period to coincide with JW162. More than 40 organisations showcased their systems at UW16, including teams from industry, academia, allied Navies and the RN and in total over 50 unmanned systems took part in demonstration events across the 4 capability themes (MCM, ASW, GEOINT and ISTAR). Fig 1 provides an overview of areas of activity for UW16.

Interaction with JW16.

As previously indicated, UW16 was delivered within the wider construct of JW162, a large scale military exercise held over a 2 week period predominantly off the Northwest Coast of Scotland with over 32 Ships and 68 aircraft participating from 20 nations. In addition to the programme of trials and demonstrations planned for UW16, UW16 platforms also received operational tasking from exercise controllers for JW162 to conduct a range of military tasks and the information generated by UW16 platforms during these tasks was used in near real time by the JW162 exercise controllers to inform the exercise scenario for JW162.

UW16 - ASW THEME

Introduction.

The ASW theme was conducted in the Northern Minch using Stornoway as a base port for activity (see Fig 1). The aim was to explore whether MAS could provide an enduring unmanned ASW barrier across a 'choke point', which would be capable of detecting and reporting underwater contacts attempting to cross the barrier. In keeping with the wider context of JW162, the barrier was set across the approaches to Stornoway Harbour, which for exercise purposes was assessed to be the base for a hostile submarine or Unmanned Underwater Vehicle (UUV) fleet. The aim of the barrier was to detect if a hostile submarine/UUV attempted to break out of Stornoway and to report this event quickly enough to enable manned assets to be tasked to engage the hostile submarine.

Event Participants.

The main participants in the ASW theme were:

- Four Liquid Robotics SHARC wave gliders owned and operated by Boeing Defence UK, each fitted with an SEA Ltd manufactured passive Thin Line Towed Array (TLTA). The SHARCs also had the capability to process acoustic data in real time and detect targets of interest, before transmitting contact reports. For the duration of UW16, target reports from the SHARCs were relayed by long range Wave Relay Radio from SHARC to SHARC and to SD Warden which simulated an on task MPA. Target reports were then relayed back up the chain of command and/or were used to vector in local assets to prosecute the target.

- A single Autonaut long endurance USV fitted with a Seiche passive TLTA and a radar ESM threat warner, was supplied and operated by Autonaut Ltd. Similarly to the SHARC, information from the Autonaut was transmitted by long range Wi-Fi to SD Warden and acted upon.

- A further Liquid Robotic SHARC, fitted with multi-parameter water chemistry and a 360° video/stills camera (ISTAR sensors), was also supplied and operated by Boeing Defence UK. The system was capable of transferring the information generated by the GEOINT SHARC to update in real time environmental data on the ASW SHARCs and Autonaut to optimise search tactics but this level integration/automation was not demonstrated during UW16.

- For the majority of the exercise a SAAB AUV 62 target was used to simulate the hostile submarine/UUV. The AUV 62 was supplied and operated by SAAB. The AUV 62 was successfully deployed 7 times and on each occasion it completed run plans in the order of 6 hours duration at varying speeds and depths. On one occasion the AUV 62 was launched and recovered from Stornoway Harbour, successfully navigating to the start point of the run plan for the exercise, completing the exercise and returning to Stornoway Harbour. On all other occasion the AUV 62 was launched and recovered by SD Warden, however these operations were subject to sea state limitations imposed by SD Warden.

- SD Warden, is a commercial vessel chartered to support UW16. Originally designed as a torpedo recovery vessel with a stern ramp and large working deck, this vessel proved to be a versatile and essential part of the UW16 ASW team.

Figure 2 Participants in UW16 ASW Theme

Clockwise from Top left, Liquid Robotics SHARC, Autonaut, SD Warden and AUV 62

Command and Control.

The Autonaut and SHARC long endurance USVs (henceforth referred to collectively as USVs), were deployed and recovered by commercial charter vessels. Once deployed, the SHARCs maintained a “swarm” at sea for 13 days of continuous operation in conditions up to sea state 5. The USVs deployed in UW16 were all autonomous platforms capable of station keeping and/or following a planned navigation track without operator intervention. In line with current best practice, the USVs participating in UW16 were monitored remotely 24/7 from locations around the UK and in the US via commercial Iridium satellite communications network, which was used to telemeter vehicle status to those monitoring the vehicles and if necessary to allow for updates to the mission plan to be sent to the vehicles. Most, if not all, of the vehicles participating were fitted with AIS receivers and have collision avoidance algorithms to avoid oncoming traffic. As previously mentioned, tactical data was transmitted by Wave Relay Radio from the SHARCs and by Wi-Fi from the Autonaut to SD Warden, which was fitted with a commercial satellite communications fit that provided high bandwidth connectivity to the other themes in UW16 and the exercise controllers of JW16.

As part of the exercise planning process a grid was devised to delineate areas which USVs could then be allocated to as required, this planning grid formed the building block for the following scenarios:

- A single line barrier drawn up across a choke point.

- A single ‘receding line’ barrier. This was a barrier drawn up diagonally across a choke point, the idea being that if contact is made by the leading elements of the barriers, those further down the line can reposition to maintain contact on the target as it passes through the barrier.

- Double barrier operations, effectively 2 parallel single line barriers to provide increased depth of coverage and the ability to localise contacts and track them for longer periods of time. This approach could also be used to coordinate additional assets either working between the lines of USVs or on the flanks to prosecute targets detected and localised by the USVs.

Due to operating constraints on the manned units participating in the trial, the trial was paused during the hours of darkness. This daily pause in trial activity provided time for the USVs (which stayed on station 24/7 during the trial) to re-position for the next day's evolution if required. Whilst this was an inconvenience for trial planning, the limitations imposed on manned vessels served to highlight one of the unique qualities of unmanned platforms; namely their persistence on task.

Results Seen.

Over the course of UW16, there were 10 days where targets were available for assets in the ASW theme to track (7 days with AUV 62, 2 days with EMATT and one day with an SSK from JW16). Of the 9 objectives set for the ASW theme of UW16, 6 were achieved. The remaining 3 objectives could not be achieved due to either lack of manned assets to interact with or the USV not being fitted with the required capability. In all 3 instances the outstanding objectives could have been met had suitable assets/equipment been available. It is perhaps also worth noting that UW16 was conducted in the shallow waters of the Minch and saw a range of weather conditions, with sea states varying from 1 – 5 and given the exercise was conducted off the West Coast of Scotland, periods of persistent heavy rain. In addition there were high levels of shipping activity during UW16 (normal levels of merchant and fishing vessel activity, augmented by above average numbers of military vessels participating in JW16); in short a challenging ASW environment with high ambient noise levels. Notwithstanding the challenging scenario, the ASW Theme in UW16 clearly demonstrated the ability of USVs with passive towed arrays to detect and track a dived platform, a high level summary of the result achieved is detailed below:

- Maintenance of a persistent sonar barrier in busy coastal waters for 13 days in sea states up to and including sea state 5.

- Detection ranges out to 3.5 nautical miles of the AUV 62 and the transfer in near real time of target reports from the USVs to the chain of command. On at least one occasion a contact report generated by the USV swarm, was used by the exercise controllers for JW16 to generate intelligence reports to inform the exercise scenario.

- The use of contact reports from the USV Swarm to direct 3rd party platforms and sensors to localise and track underwater targets.

- Detection of a dived SSK. As part of JW162 one of the participating SSKs was routed through the UW16 ASW theme area at periscope depth. The SSK was successfully detected by the USVs, and the target reports that were generated enabled a simulated Unmanned Air vehicle (UAV), in this case a light aircraft fitted with a LIDAR payload, to be cued to detect and localise the target. The successful detection of a live submarine using MAS at the first attempt was, it is argued, a potent demonstration of the future capability of unmanned platforms.

- The ability for Industry to provide an ASW barrier capability, using commercial vessels and a 'pop up' shore side facility demonstrated the relative ease with which MAS can deliver enduring ASW tasks around static facilities such as ports and harbours. This, however, is a double edged sword and illustrates the very low barriers to entry for those wishing to deploy this type of capability.

- The inclusion of an ESM surveillance capability on the Autonaut and a GEOINT fitted SHARC, demonstrated the flexibility of these platforms and the potential for multi-role USV in the future and/or swarms of single role platforms acting in collaboration to deliver a military effect.

- The exercise illustrated how the addition of intelligent on board analysis functions to USVs can significantly ease the workload for ASW qualified personnel required to monitor the output of the USVs.

- The slow moving nature of the USVs makes them vulnerable to being removed from the water by an adversary or even a curious fisherman; there was one instance of this during UW16. This highlights the need to consider the protection of information and or software deployed on USV given this apparent vulnerability.

ASW EXPERIMENTATION AT BUTEC

Introduction.

Anti-Submarine Warfare (ASW) is often cited as one of the potential use cases for Unmanned Underwater Vehicles (UUV). For a UUV to realistically conduct this role it must be able to sense the environment around it, make sense of what it is seeing and respond to its vision of the real world. QinetiQ have undertaken a series of proof of concept trials at the British Underwater Test and Evaluation Centre (BUTEC) to demonstrate that a COTS UUV towing a COTS TLTA can detect and track an underwater target.

Approach to trial.

The overall proof of concept was conducted as a series of 4 trials over a period of a year, with each trial demonstrating incremental advances in functionality leading to a full system demonstration in the final trial. An IVER 3 UUV manufactured by OCEANSERVER was selected as the vehicle for the trial principally because it has an open architecture and there are multiple examples in the open press of 3rd party software and sensors (including towed bodies) being integrated into an IVER. The choice of array was therefore limited to something that the IVER could tow comfortably, which initial trials with dummy loads indicated was in the region of 30 meters length. With this limitation in mind an 8 element digital array manufactured by Seiche Ltd was selected to support the trial.

- Trial 1 sought to characterise the performance of the array in terms of acoustic performance and self noise (both flow and electrical noise). The conduct of this trial also demonstrated an early version of the data recording system to be used in later trials.

- Trial 2 demonstrated the ability of the UUV to tow the Seiche Array, and the capability of that combination of sensor and platform to capture and record the acoustic signature of a dived target; in this case a mobile target radiating representative broad band and narrow band signatures. Data from this trial was processed post event.

- Trial 3 demonstrated the integration of beamforming and signal processing to the IVER to enable the creation of a sonar picture in real time on the vehicle.

- Trial 4, the final trial in the sequence has yet to be conducted at the time of writing this paper but the aim is to integrate simple search detect and track algorithms to exploit sonar processing functionality demonstrated in Trial 3 to enable the IVER to respond autonomously to underwater targets.

System Integration.

Completion of the programme of trials required the following modifications to the IVER 3:

- Removal of the as fitted Side Scan Sonar processing to create the space budget for the Towed Array Power Controller, Towed Array Data Interface / Controller and a Sonar Processor.

- Provision of a data recording systems to cope with the 25 – 40 Gb of data produced during each day on range, which required the existing hardware architecture of the IVER 3 to route the Ethernet output of the Seiche array to the 456 Gb Hard drive which is fitted to the vehicle.

- Implementation of range tracking functionality and abort controls to satisfy range safety requirements. This functionality was delivered by exploiting features in the Woods Hole Oceanographic Institute (WHOI) Micro Modem, supplied with the vehicle and under the control of the Sonar Processor.

- Trim and Balance of vehicle to account for the additional weight of processing and the shift in centre of gravity caused by the towed load.

- Integration of sonar processing functionality, delivered by a Xlinix Zynq XC7Z00 / Red Pitiya System on Chip (SoC) processor, which was installed on a mounting panel to provide the necessary connectors, interfacing circuits and voltage references. The signal processing within the IVER provides a frequency domain beam former and Fast Fourier Transforms (FFT) to deliver 8 channels of green screen data over the band of interest.

- Development of, search, detect and track algorithms, which will be subsequently hosted on the sonar processor and integrated with existing sonar processing functionality.

Figure 3 IVER modified in workshop (Top Left) Seiche Array (Top Right) IVER 3 (Bottom Right) and IVER and Array in RIB prior to launch

A schematic of the integration of the towed array and additional processing to the IVER3 is shown at below.

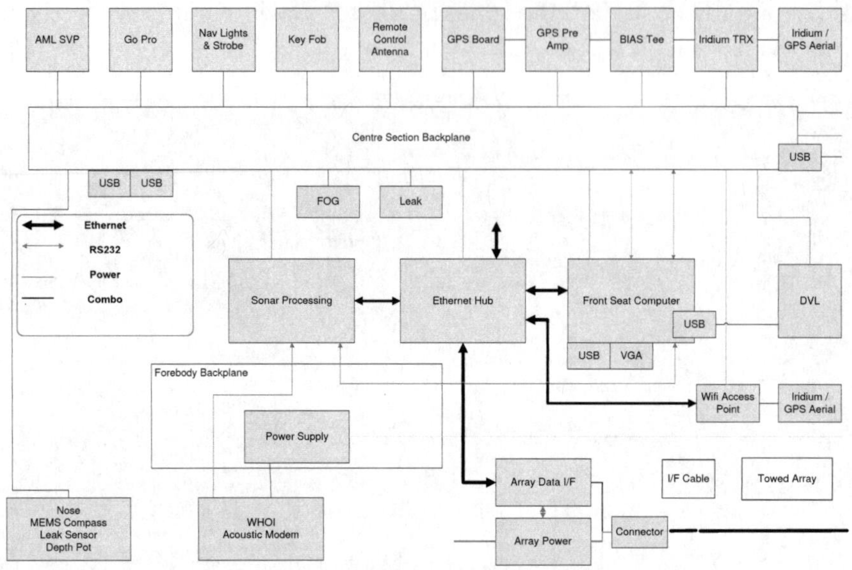

Figure 4 Block Diagram of IVER 3 and Additional Processing

Results.

The results of the trials conducted to date have demonstrated that the combination of vehicle and sensor can:

- Detect a dived platform out to a range of 4000 meters.
- Process the acoustic data in real time and with sufficient detail to identify features in the acoustic signature that human operators would exploit to conduct target motion analysis (TMA) on a manned platform.
- Perform simple search detect and track behaviours of an underwater target autonomously.

Notwithstanding the success of this trial, in terms of a proof of concept, the combination of vehicle and sensor used in this trial are far from being an operational capability and it is anticipated that any operational system will require larger arrays and consequently larger vehicles with longer endurance to pull such arrays. In addition this trial has focused on the collection and exploitation of data by a single vehicle and did not include work to off board the data i.e. unlike the previous example from UW16, data captured by the IVER 3 was not used to cue 3rd party sensors or platforms to the target.

WHAT NEXT?

The concept of technological change having a significant impact on Society and the way mankind lives is not new e.g. Stone Age, Iron Age or the steam turbine and the industrial revolution. All of the changes were deeply disruptive, and, as is widely commented on elsewhere, the pace of change has accelerated with each successive 'revolution.' Autonomous systems are part of the next 'revolution'[1]. Hence competitive advantage is there to be gained for the nation(s) that embrace the new opportunities fastest. However, it is not the technology itself which disrupts; rather it is the innovative applications enabled by technology which are disruptive[2]. Hence if military forces are to enjoy the potential benefits on offer from autonomy, they will need to adapt their doctrine and concepts of operation to embrace and exploit autonomous technology. However, the military are seen as a conservative body and the perceived resistance to embracing new technology is driven by a lack of understanding and trust of new technology and what it can deliver[3].

So historically how have society and the military dealt with disruptive technology? In short: experimentation, an iterative process of trial and error to improve understanding and develop new ways of exploiting technology. An integral part of which is establishing trust in new technology. So whilst UW16 has contributed to building the necessary trust in the technology and has contributed to an appetite to do more(both in the UK and abroad[4]) the impetus to embrace MAS will falter without an ongoing programme of experimentation to develop the technology and systems further to achieve their full potential.

So, if there is to be more experimentation to increase the necessary understanding and trust in Autonomous vehicles in the ASW domain, what should it be? Before answering that, it is worth highlighting that government spending on R&D significantly lags behind industry spending (Google spent in the order of $12B on R&D in 2016[5], in the same period the UK MoD spent £1.7B[6]). One could therefore draw the conclusion that government funded R&D should focus on those niche areas that industry will not deliver. With that in mind, the lessons learnt from the ASW experimentation described in this paper would indicate that the following need to be advanced before MAS can be deployed as part of an ASW capability:

- Doctrine and Concepts of Operation. Industry is clearly not going to develop military doctrine and this is something that the military will need to do for itself. What is clear is that MAS, in most cases, cannot be considered as small cheaper unmanned versions of current manned systems. Furthermore, early indications are that MAS will not lead to a reduction in manpower rather they will force a re-distribution and reskilling of the existing manpower. What is also clear is that MAS will need to operate with manned vehicles for at least the next 50 years[7]. So if MAS do not replace manned platforms and don't reduce the overall manpower budget, what can they do? Evidence to date would suggest that their key

[1] Lloyds' Register, University of Southampton & QinetiQ – Global Maritime Technology Trends 2030.

[2] Dr Heiko Borchert et al, Waiting for disruption?! Undersea autonomy and the challenging nature of Naval innovation, dated 10 Feb 2017.

[3] Naval Research Advisory Committee Report – How Autonomy can Transform Naval Operations – dated Oct 2012.

[4] Report to Congress, Autonomous Undersea Vehicle Requirement for 2025 dated Feb 2016.

[5] https://www.strategyand.pwc.com/innovation1000, site accessed 15 May 17.

[6] UK Office of National Statistics, Finance and Economics Statistical Bulletin, dated 5 Jan 2017

[7] Procurement decisions made in the UK e.g. Queen Elizabeth Carrier and/or Successor, would suggest that manned platforms will be a part of the order of battle for the RN for the next 30 – 50 years.

attributes are persistence (the ability to keep performing the same task 24/7 for periods measured in weeks) and flexibility (as evidenced by UW 16, the vehicles are just a truck that can carry different sensors/capabilities) which potentially enables them to extend the sensor range of manned platforms. In addition they are relatively cheap; therefore could be deployed in large numbers to restore "mass" to military forces that have been hollowed out by repeated cuts in defence spending. Experience from UW16 would suggest that the development of ASW Doctrine and CONOPs for MAS should focus on exploiting their unique selling points (persistence, flexibility and mass) to deliver tasks that manned platforms cannot deliver or might not be able to deliver in the near future; rather than a strategy that uses unmanned platforms to duplicate tasks already delivered by manned platforms.

- Manpower and training. A review of employment websites would indicate that the basic educational requirement for entry level jobs working with autonomous systems is a Bachelor of Engineering in subjects such as Computer Science, Robotics or Electronic Engineering. However, most Navies have a workforce, the majority of which do not have a degree standard education, hence there is disparity between the skill levels industry is designing the current set of MAS for and those that the Navies have at their disposal. It is therefore argued that there is work to be done to either re-skill military MAS operators and/or work to simplify the operation of MAS to reduce the academic qualification required to operate them. An interim solution which some navies have employed is the creation of dedicated military trials teams to develop the initial cadre of qualified personnel from which to grow a core competency in the operation of Autonomous Vehicles and to support the wider development of Doctrine.

- Equipment. There are a number of areas where the current commercial off the shelf MAS potentially do not have the level of performance desired to deliver an ASW capability within current doctrine e.g. Speed and endurance, sensor technology, information assurance and autonomous behaviour/artificial Intelligence. Of these weaknesses arguably Industry is better placed to lead technology that will increase the speed and endurance of vehicles (e.g. there is a pressure to improve battery technology for electrical cars to get the $/KW figure below a threshold value that makes them economically valid and/or the KW/Kg figure to a point where the endurance of electrical cars is an attractive alternative to petrol vehicles). Similarly work by the likes of Google and Tesla to make autonomous vehicles mainstream will, it is argued, provide spin offs in autonomous behaviours for military applications. Where industry will be less inclined to invest in R&D (obviously with the exception of the Defence Industry) will be the sensing technology required by ASW units so it is perhaps here that the defence community should focus. Similarly cyber security or aspects of this subject, are an area where defence standards might be more exacting than commercial standards. But again development in equipment will (should?) be driven by development in Doctrine and CONOPS; what does the military want to achieve with MAS?

- Procurement. Advances in MAS have benefited greatly from advances in computing and telecommunications, markets which are driven by consumer demand and can be characterised as evolving rapidly e.g. Moor's Law. So it is reasonable to assume that MAS will evolve quickly and the recent appearance of a number of commercial micro UUVs .e.g. Riptide or Ecosub, would appear to support this assertion. It is therefore argued that the traditional platform centric procurement process which is focused on delivering fully operational and proven systems, will need to adapt to enable the low rate fielding of technology in parallel with experimentation. Such an approach would allow for the early pull through of innovation to the war fighter whilst providing operational feedback to inform the development process.

CONCLUSION

The events described in this paper have achieved their aim of raising awareness of autonomous technology and demonstrating their potential utility in a number of military scenarios (one of which was Anti-Submarine Warfare). They have contributed to improving the level of trust and understanding of autonomous systems for the military, however, there remains work to be done before a truly autonomous ASW capability could be deployed by the military. Key to this will be an iterative programme of trials and demonstrations (do a little, learn a lot and move forward) where industry, academia and the military work together to develop new concepts of operation for autonomous technology and /or develop autonomous systems to support new and innovative concepts of operation.

Lessons Learned from HMNZS CANTERBURY's Service in the Royal New Zealand Navy

Commander Euan Henderson, Royal New Zealand Navy

Euan.henderson@nzdf.mil.nz

ABSTRACT

Ten years of operational service in the Royal New Zealand Navy has more than justified the acquisition of the multi role vessel HMNZS CANTERBURY. The flexibility and capability that the ship has shown in response to cyclone events in the Pacific Islands and to earthquakes within New Zealand repeatedly demonstrates that the ship has satisfied a range of the requirements identified for it under the Project Protector acquisition. But the operations to date have not been without difficulties and the use of a commercial design for a vessel in a military context has realised a number of challenges. This paper will reflect on HMNZS CANTERBURY's project and service history and will highlight a number of successes including the significant support to Humanitarian and Disaster Relief (HADR) operations that the ship and crew have responded to since commissioning. However, also using the HADR operations and other exercises as a back drop, some of the limitations of CANTERBURY will be examined. The paper will briefly look at the future of the ship and the ongoing expansion of CANTERBURY's capability with a particular emphasis on how the ship is a significant contributor to the New Zealand Defence Force's Joint Task Force.

Disclaimer: This paper is based upon the author's opinions, drawing on publicily available material, and is not the official view of either the New Zealand Defence Force (NZDF) or New Zealand Ministry of Defence.

THE RNZN'S DEVELOPMENT OF STRATEGIC MARITIME SEALIFT

CANTERBURY's acquisition initially began as early as the 1970s when the purchase of a sealift ship was mooted in order to support New Zealand's defence and foreign policy in the South Pacific, as well as supporting United Nation's operations.[i] The 1987 Defence White Paper highlighted this with the need for a Logistic Support Ship being identified in order to transport and sustain the New Zealand Army's Ready Reaction Force around the Pacific and South East Asian regions.[ii] In 1988 studies identified the need for a ship of approximately 12,000 tonnes displacement, capable of transporting up to 200 troops and their heavy equipment (including helicopters), and importantly, being able to deliver this load onto a beach or conventional wharves.[iii] By 1991 the expected benign security scenarios identified in the Defence White Paper of that year had further identified a gap in transport and force projection capabilities, but the approximate cost of NZ$240 million identified in the 1987 White Paper probably slowed acquisition.[iv] The Air and Sea Transport Review of 1991 discounted chartering due to the poor availability of suitable ships along with the time to lease and position them and concluded that there was no doubt that military sealift was a core capability element if the New Zealand Defence Force (NZDF) was to meet the defence tasks specified in the 1991 White Paper.[v] The Review also reduced the capability of the proposed transport vessel by suggesting that the less ambitious task of operating into ports which had wharf facilities might be sufficient[vi] which enabled existing commercial vessels to be considered alongside a specialist vessel.

With these reviews and requirements as background, suitable new build and used commercial vessels were examined as options. A 24,000 tonne trans Tasman vessel, the UNION ROTORUA, was a lead contender based on its ability to fully transport the Army's requirements and pedigree of open ocean voyaging. An alternative was a Mercandian 1500 vessel which could transport 50 per cent of the Army's load as the ship's displacement was 10,500 tonnes, but would have smaller through life costs than the UNION ROTORUA.[vii] Crucially though, the Mercandian 1500 class were designed for trade in the North Sea and it was reported that 'unless fully laden, is very uncomfortable at sea in the South Pacific'.[viii] With hindsight this observation was to prove to be the undoing of the ship ultimately selected and commissioned to fill the requirement, HMNZS CHARLES UPHAM, in 1995. It is interesting to note here that the lesson of purchasing a ship that was not designed for the role and environment it was to operate in, relatively lightly loaded in the Pacific and Southern Ocean, was missed when it came to approving the design of HMNZS CANTERBURY some 10 years later.

Prior to discussing HMNZS CHARLES UPHAM in RNZN service, it is worth examining a further option for the sealift capability, the leasing of HMAS TOBRUK. Acquired by Australia to also fulfill a permanent sealift requirement, she was commissioned into RAN service in April 1981. In late 1993 the Australian Government offered to lease TOBRUK to the NZDF. While senior Defence personnel identified the political benefits of agreeing to the proposal, TOBRUK was described at the time as 'a 25 year old ship in a 10 year old body' with support for the orphan being seen as difficult over the remaining life of the ship.[ix] TOBRUK, at just under 6,000 tonnes displacement, also did not have the load capacity of the Mercandian Class ships. However, she did provide the capacity to land equipment across the shore using her ability to beach herself and, when operating with landing craft, undertaking ramp to ramp marriages at the stern of the vessel. Despite growing political pressure via increasingly advantageous offers for lease, sale and even for free,[x] TOBRUK's relatively high staffing levels and concerns over the maintenance burden eventually halted the deal.[xi] Even though it was noted that '..TOBRUK can do many things that even a converted [Mercandian Class Ship] will not be able to do' the NZDF acquired the ex MERCANDIAN QUEEN II on 16 December 1994 for NZ$14.15 million.[xii]

Commissioned into the RNZN on 18 October 1995 as HMNZS CHARLES UPHAM, one of the crucial aspects of the acquisition, and one that was fully understood prior to purchase, was that an additional NZ$46 million was needed (and nominally approved by Government) for modifications to the ship to increase its modest military capability.[xiii] Prior to the modification project starting, the 1997 Review of CHARLES UPHAM reiterated that a number of fundamental capabilities of the ship were lacking, particularly in the need for offloading cargo where unimproved wharves were located.[xiv] To be flexible the ship needed organic cranes and deck strengthening in order to use heavy forklifts internally and the inability to get cargo across a beach was also a significant constraint to operations.[xv]

In parallel with the modification project being developed, sea experience was being gained with the ship, showing a number of basic deficiencies. As noted by the Air and Sea Transport Review team, when lightly loaded CHARLES UPHAM was expected to experience significant motion issues (particularly roll), and this quickly became apparent with even the experienced personnel onboard finding conditions uncomfortable.[xvi] Coupled with main engine reliability concerns, the ship was withdrawn from service in August 1996 in order to undertake interim projects to reduce the motion of the ship and increase engine reliability.[xvii]

By April 1998 the funding for the modification package had not been secured due to the 1997 Asian Financial Crisis and a stringent Defence budget. Consequently the NZDF examined options for leasing the ship back to a commercial operator until such times as the funding could be secured.[xviii] On 12 May 1998 the ship sailed on a two year lease amid multiple reviews of the acquisition, all with the same conclusion, that the ship would be suitable given the modification package being implemented.[xix] Finally, in July 1999, Cabinet gave approval in principle for the conversion process to begin. However, a change to Government in 1999 led to a number of Defence reviews and these culminated in The Government's Defence Policy Framework of June 2000. The two options regarding CHARLES UPHAM were; undertake the modification package in order to achieve the most cost effective sealift solution; or to review the generic requirements for sealift, including the ability to undertake operations across the beach and identify available alternatives.[xx]

By this time the lessons of the NZDF's involvement in the East Timor operation had also been identified and these were rolled into the subsequent 2000 Sealift Review. What was apparent from this review was that the NZDF required strategic sealift (delivery of heavy equipment to a fixed port, generally outside the operational area) as well as, and just as importantly, tactical sealift (delivery of heavy equipment to the area of operations where generally port facilities are not available, for example beaches). A modified CHARLES UPHAM was still deemed able to undertake both types of sealift, albeit with constraints. However, on 2 April 2001, the Sustainable Defence Plan, developed by the Treasury and the Department of the Prime Minister and Cabinet, was approved by the Government which endorsed the proposal to sell the ex HMNZS CHARLES UPHAM when it came off its charter. Subsequently it was sold in July 2001 for NZ$8.6 million.[xxi]

The events surrounding the acquisition of HMNZS CHARLES UPHAM have been examined in detail because a number of the lessons that can be learned from the ship's time in RNZN service feed into the decisions around HMNZS CANTERBURY's acquisition and current Naval projects. The requirement for a ship capable of both strategic and tactical sealift had again become apparent during the East Timor conflict in particular (as well as Bosnia and Bougainville[xxii]), reinforcing the studies dating from 1988 that highlighted this requirement, rather than just a strategic sealift ship alone. The design of the ship's hull and an understanding of the operations that the ship would be expected to undertake is, and continues to be, critical in the selection process. Acquiring and modifying an existing ship comes with a small initial cost, but requires significant investment to meet the needs of the military user due to the modifications needed. And finally, significant decisions concerning military equipment and capability will always have a political dimension which must be understood and appreciated, as seen with the discussions around HMAS TOBRUK. While it is disappointing that the full vision for CHARLES UPHAM was never achieved, many of the characteristics that were desired for the ship were folded into the design of HMNZS CANTERBURY. In many ways HMNZS CHARLES UPHAM was a steep learning curve for Ministry of Defence (MoD) and NZDF staff in all aspects of the acquisition process and especially in fulfilling the NZDF's Logistic Support Ship requirement.

ACQUISITION OF A MULTI ROLE VESSEL

The change in Government in 1999 led to a series of reviews of defence policy. Not only was the Sealift Review completed and released, but the multi government agency focused Maritime Patrol Review was also underway, with the results released in February 2001.[xxiii] Late in 2001 the Maritime Forces Review was also presented to Government which included '...the need for an appropriate sealift capability, including the use of such a capability for disaster relief and other tasks in the South Pacific...' The recommendations of the Maritime Forces Review set the scene for the commencement of Project Protector, which included the acquisition of a multi-role vessel and patrol vessels with a funding cap of NZ$500 million.[xxiv]

By May 2003 requests for proposals were issued to six previously short listed companies and were received in October of that year. In February 2004 the evaluations had been completed by the NZDF and MoD teams and a preferred bidder identified who could provide the mix of ships required for under NZ$500 million.[xxv] By July 2004 a fixed price contract had been agreed with Tenix Defence Pty Ltd for a multi role vessel, two offshore patrol vessels and four inshore patrol vessels.[xxvi] What was interesting for the tender process was that requirements for the ships were based on functional and performance specifications (FPS) and key user requirements rather than a long list of specific requirements. According to a review by John Coles into the Safety and Functionality of HMNZS CANTERBURY, this was designed to allow industry to provide innovative solutions to meet the NZ$500 million budget. However, as noted by Coles, the contract negotiated did not incorporate the FPS, included different verification and acceptance criteria and consequently lacked the mechanisms to enable compliance with the FPS.[xxvii]

Construction on the Multi Role Vessel (MRV) began with steel being cut on 15 April 2005 by Merwede in the Netherlands, as a subcontract to Tenix, and was based on a commercial Roll on/Roll off (Ro-Ro) ferry, Ben my Chree, based in the Irish Sea. Fundamentally, the subsequent shortfalls in performance that affected the ship can be attributed to an 'insufficient appreciation of the constraints to the ship's operations imposed by the selection of a commercial [Ro-Ro] design'.[xxviii] 'The Ben my Chree is a 'short/fat' ship that operates across the Irish Sea where conditions are akin to coastal waters, where the seas are generally short crested as compared with deep oceans like those around New Zealand.

Attempts to de-risk the design were undertaken by the Project Team prior to contract signature. A report by BMT Defence Services Ltd completed in April 2004 identified high level risks that included sea keeping, stability, aviation operations and aircraft handling, powering and electrical generation, structural integrity and aspects of design that impact on sea keeping issues and operational effectiveness.[xxix] Only some of these issues were put to Tenix and due to the limited responses provided, Tenix was confirmed as the preferred bidder.[xxx]

Concern over the MRV's sea keeping dominated the acquisition process after contract signature. In February 2005 serious concerns about the dynamic behavior of the MRV in high sea states emerged. The contract with Tenix was suspended on 24 March 2005 (prior to construction work starting) and then recommenced on 17 May following acceptance of Tenix's remedial plans, which included simulation modelling.[xxxi] This was formally provided in October 2005, six months into the ship's construction and was followed by tank testing of a scaled model. Results from the Maritime Research Institute Netherlands (MARIN) were published in February 2006, 10 months into the ship's build.[xxxii] Tank testing revealed similar results to the simulations, as well as the propensity for the boat alcoves of the MRV to be effected by water ingress in high sea states, with subsequent damage to the ship's boats. Debate on the results took place over the subsequent months. However, without enough staff, and of the right technical capability (no embedded naval architect for instance), any early intervention opportunities were lost.[xxxiii] By the time the real impact of the MARIN results concerning the ship's performance in higher sea states was fully appreciated, the ship was effectively completed and the sea keeping performance dictated by the original design was already pre-determined.[xxxiv]

It is the author's view that in retrospect the incorrect sequencing of major project steps was the key issue that led to a number of the problems that have affected CANTERBURY. In essence, construction got ahead of design and remained ahead of it for the rest of the build process. No overall detailed design review was able to be completed during the project. Sections of design were reviewed, but by the time these were done some construction was already completed.

In addition, the review by Coles suggested that the '..acceptance of the ship off contract illustrates vividly the drive to deliver HMNZS CANTERBURY and get her into service at the earliest opportunity'.[xxxv] In the RNZN this was the 'Ready Day One' mantra with a delivery date designed to reduce the training and operational loss of the Leander Class Frigate CANTERBURY rather than following an extensive period of acceptance trials and testing, which should have occurred. 'The incompleteness of the verification and acceptance process immediately before planned acceptance' was highlighted in the Coles report into the Safety and Functionality of HMNZS CANTERBURY.[xxxvi]

Along with issues relating to the design and acceptance into service, project governance was also highlighted by the Coles report. As well as inadequate project performance reporting, there were a number of shortcomings in the governance of the MRV acquisition, exacerbated by some strained relationships between the MoD and NZDF.[xxxvii] Linked to this was the size of the team being used on the project, which was under sized and lacking specific skill sets.[xxxviii] In total, the entire Protector team in Williamstown, Tenix Defence's main office, never reached more than 12, with a maximum of 2 project personnel in the Netherlands covering the MRV build.

Lastly, the governance and project structure that was in place at that time was built around focusing on acquiring the steel platform as quickly and as well priced as possible. As such, the full capability elements required to ensure the acquisition was able to be fully exploited on delivery were missing. Reflecting on the completeness of the full PRICIE[1] elements on the day of delivery, it is the author's view that a number of these were lacking given that it has taken 10 years of service in order to release an agreed Operational Capability Statement showing that introduction into service has been completed.[xxxix]

DELIVERY, PROBLEMS AND REMEDIATION

Commissioned as Her Majesty's New Zealand Ship CANTERBURY on 12 June 2007, the vessel was to encounter a number of problems over the next two years. A hint to these was seen on the delivery voyage to Australia with the ship's civilian engineer noting bow slamming and swamping of the starboard boat alcove. The high profile loss of boats from their alcoves and the tragic loss of a sailor led to the John Coles directed report into the Safety and Functionality of HMNZS CANTERBURY being commissioned on 20 December 2007.[xl] While finding that the ship was fundamentally safe, Coles recommended remedial work be undertaken in order for the ship to meet the expectations required of the vessel and that some operating limitations would have to be accepted.[xli] In order to provide further safety assurance, a retrospective safety case was commissioned that aided in the RNZN's understanding of the ship's inherent capabilities and shortfalls and this flowed into a number of CANTERBURY related publications including the Operational Capability Statement.

In May 2008 a warranty claim was put to BAE Systems Defence Australia Pty Ltd (BAE), who had acquired Tenix Defence Ltd in June 2008. In early 2010 BAE and the New Zealand Government agreed on compensation for the deficiencies that had been identified across the Protector Project ships.[xlii] By mid 2009 plans were begun for remediating CANTERBURY, in order to bring the ship closer to the capability envisioned, using the anticipated compensation funding. For CANTERBURY, the remediation requirements were numerous, ranging from moving the boat alcoves to additional tie down points for helicopters, from installing extra ballast to reduce the ship's roll speed, to installing new mission systems including a new weapon mount and control system. As of late 2016 the remediation project was completed and CANTERBURY was finally declared as having her Full Operational Capability, officially completing the acquisition process begun in 2003.[xliii]

[1] PRICIE: **P**ersonnel, **R**esearch and Development, **I**nfrastructure/organisation, **C**oncepts of operations/doctrine, **I**nformation technology, **E**quipment and Logistics.

OPERATIONS AND EXERCISES

Despite the capability limitations on delivery, CANTERBURY has proven its worth in a number of exercises including the Rim of The Pacific (RIMPAC), Talisman Sabre and the Southern Katipo series of graduated amphibious based exercises designed to prove, and improve, the sealift capability of the ship and the NZDF. Vitally, CANTERBURY has provided exceptional support to a number of Humanitarian and Disaster Relief (HADR) operations, easily fulfilling the vision for a sealift vessel first discussed in the 1970s. Working within the capability envelope that has been slowly expanded as remediation work has progressed, CANTERBURY's capabilities and limitations have had to be learned by an organization that has not had organic sealift previously, either of the strategic or tactical variety. Consequently, in parallel with physical changes to the ship, cultural changes have taken place in the NZDF concerning the ship. The learning opportunities and cultural changes started early in the ship's life in service with the 2009 Samoan tsunami and have accelerated due to the Christchurch and Kaikoura earthquakes and cyclone responses in Vanuatu and Fiji in particular.

CANTERBURY has become the centre piece of the NZDF's HADR response capability around the South Pacific as well as performing well in the NZDF's Joint Amphibious Task Force, while accepting that the ship is not designed for tactical amphibious operations. Indeed, in the 2010 and 2016 Defence White Papers, CANTERBURY is placed as the corner stone of the NZDF's Joint Capability[xliv] and is seen to be until the ship is replaced.[xlv] The integration and understanding of CANTERBURY's capabilities continues to be a work in progress across the NZDF but has been an exceptional catalyst for the three New Zealand single services to perform as One Force. While originally designed for only the carriage of Royal New Zealand Air Force (RNZAF) NH90 helicopters, it is intended for NH90s to operate from CANTERBURY as a moving platform, providing a significant increase in operational capability.[xlvi] Additional work is underway on improving the command and control capability of the ship, with the Wideband Global SATCOM (WGS) system being installed soon, providing a quantum leap in data flow.

GOVERNANCE IMPROVEMENTS

Along with the cultural and material changes within the NZDF in order to fully exploit the capability that is available in CANTERBURY, the NZDF and MoD have also learned from Project Protector concerning Governance and the need for improvement.[xlvii] The transformation from what existed in the early 2000s to the current governance arrangement is no less impressive than the physical changes that have occurred on CANTERBURY and addresses many of the issues highlighted in the Coles Report concerning this area.

Along with the Coles Report, the 2009 Review of the NZDF's and MoD's Procurement Process in particular has been a catalyst for change that continues to this day. While not directly attributable to the problems encountered with CANTERBURY, the issues identified with the Protector and NH90 helicopter procurements spawned the review.[xlviii] It was comprehensive, honest in its recommendations and has laid the foundations for the transition to Integrated Project Teams (IPTs) and many other governance changes that exist today, including establishment of the Capability Management Board chaired by the Secretary of Defence and the Chief of Defence Force.

Of the recommendations of the 2009 review and a number of others since, possibly the most important is greater co-operation between the MoD and NZDF throughout the entire capability management life cycle. In particular this outlines how the MoD is responsible for policy and acquisition and the NZDF for introduction into service activities, while acknowledging the cross over between the two in delivering Defence capability.[xlix] As a result of the direction in the 2010 Defence White Paper, and other lessons learned, the Capability Management Framework has been updated to take account of current best practice in industry. The MoD acquisition division has been renamed capability delivery, has increased in the numbers of personnel and, as mentioned above, Integrated Project Teams have been established for all major acquisitions starting at approval to initiate and finishing at an agreed point in the operational release of a capability.[l] Key to this is that projects are now led by long serving civilians with skills in project management and procurement who are linked with policy analysists and uniformed subject matter experts.[li] While still small in number, the issues of subject matter expertise identified by Coles have been addressed with, for example, naval architects in each maritime project.

Capability Branch has fostered greater interaction with MoD staff and the three services. The author has seen the excellent co-operation this has engendered between the three services, creating lasting relationships in the process. In 2008, Coles was critical of the oversight provided to the NZDF and the MoD in the Protector Project. Almost 10 years on, the MoD and the NZDF has, and continues to, improve governance and leadership, project management and project process and execution.[lii] While the journey toward excellence continues, it is clear from a number of recent projects undertaken in the new environment, that success is being attained.

LESSONS FROM OPERATIONS AND EXERCISES

After 10 years of service it is timely to consider what key lessons the NZDF and MoD has learned from CANTERBURY, especially if the ship is seen as a further rung on the amphibious capability learning ladder.

Operational Concept and Design. It is critical to get the operational concept and design right for the environment and the operational role the equipment will be undertaking. It is the author's view that as a multi role vessel, CANTERBURY is a "jack of all trades but master of none" which may be a reflection of budget constraints and the desire to get as much capability as possible into one vessel. If possible, a single, or multiple vessels, should have a dedicated role, as demonstrated by the AOTEAROA decision to purchase a tanker with a small amount of logistics support capability, rather than another CANTERBURY with fuel tanks. CANTERBURY will always be constrained by heavy seas as the ship's design places inherent limits on what the ship can safely deal with. In service this is mitigated by careful planning and intense meteorological support with the acceptance that shelter in harbor will be required at times. If the ship is going to be lightly loaded compared to the parent design, then ballasting and crew comfort compensation needs to be addressed. This lesson is now well understood within the NZDF and the MoD as the author witnesses first hand the criticality of getting the tanker AOTEAROA's design right before cutting steel using a number of embedded naval architects to provide subject matter expertise in the arena.

Additional Vessels. Given the service that CANTERBURY has already provided, it almost appears that there is more than one in NZDF service. While the ship has indeed been present in numerous exercises and operations over the last 10 years, the ship is an orphan. With orphan vessels come the issues of when to undertake maintenance so as to minimize the loss of the vessel to contingency events. An example of this was when CANTERBURY was being fitted with 200 tonnes of top weight, as a remediation activity, and was unavailable for sea when the Samoan Tsunami occurred. The personnel involved in the project managed through sheer hard work and dedication to get the ship back to a seaworthy condition to allow CANTERBURY to deploy to aid New Zealand's neighbours in their time of need. It is the author's view that the answer to this issue is to have more than one vessel. Certainly a second vessel would provide a constant response and support ability for all Government directed activities and numerous training, amphibious experience and support synergies would also be a by-product.

Military Capability. While CANTERBURY performs a number of amphibious type operations safely, these often come with constraints, particularly when supporting advanced force opertions. The commercial nature of the vessel means that it is not hardened against a military threat, was not designed as a command and control platform and has limited self defense capabilities. A recent example of the risk such vessels might face was the unarmed United Arab Emirate's Navy ship HSV-2 SWIFT, a commercially designed and built sealift vessel, which was attacked in October 2016 while on operations, resulting in significant damage.[liii]

THE REPLACEMENT FOR CANTERBURY

CANTERBURY will continue to be the cornerstone of the NZDF's Joint Amphibious Task Force for many years to come and culturally the organisation still has much to learn about the Joint effect, how best to operate the ship and importantly, what does a replacement to CANTERBURY need in the future to remain the centre piece of the Joint effect. At 10 years old the ship is nearing its mid life upgrade point and discussions around replacement options may need to be commenced soon if the ship's life is closer to 20 than 30 years in service. In the author's opinion the key capabilities for the replacement are as follows. Firstly, the discussion around a replacement for CANTERBURY must commence with whether the proposed budget covers the cost for at least two vessels ensuring that HADR and military operations are continuously covered in New Zealand's neighbourhood. A docking ability for landing craft will be vital, as will a large flat deck for multiple simultaneous helicopter operations. The ships will need to be bigger than CANTERBURY in order to accommodate the dock, helicopter hangar, more vehicle lane metres and the numbers of personnel needed to simultaneously operate the embarked flight, landing craft, command and control systems and the ships themselves. In addition, the ships will need to have some resistance to militarily induced damage through improved survivability and have an effective defensive suite for self protection. The future ships will possibly have 500 to 1000 New Zealander's onboard. Loss of a ship, its equipment and the personnel onboard would have a massive impact on New Zealand's society and therefore the requirement for military build standards and defensive weapons need to be considered in an uncertain future.

CONCLUSION

After ten years of service HMNZS CANTERBURY has more than paid back the NZ$130 million price tag with the humanitarian and disaster relief support the ship and her crew have provided to thousands of people affected by cyclones and earthquakes. 10 years of operations with an organic sealift ship has also allowed the NZDF to develop further along the path of amphibious operations in the South Pacific Ocean with considerable cultural change within the three services being seen in order to create a Force for New Zealand based around the ship. The difficult gestation of CANTERBURY has also provided a number of lessons to be learned at the corporate governance and management level that are, and will continue to, assist in making expensive acquisitions in the MoD and NZDF more efficient and effective. Using a commercial design for a military vessel may involve compromises around capability versus cost which must be understood and accepted and this conundrum will be an ongoing consideration for large maritime acquisitions. From a purely maritime project management and engineering perspective lessons around sea keeping, design, programmatics and ensuring the people with the right skill sets are included in the project have been learned, which are now paying dividends in the current Integrated Project Team construct. In addition, the lessons learned from the operations and exercises that CANTERBURY has been a part of are helping to shape any replacement vessel, or more likely vessels, in the future. After a difficult beginning and a target for criticism, HMNZS CANTERBURY has weathered the storm and now embodies the NZDF's desire to be three services acting as a Force for New Zealand and will do so for many years to come.

BIBLIOGRAPHY

Aurecon, (2009). *Review 09 Procurement Process Review.* Retrieved from http://www.defence.govt.nz/assets/Uploads/defence-review-2009-released-aurecon-procurement-process-review.pdf.

Coles, J. (2008). *Independent Review of the Safety and Functionality of HMNZS CANTERBURY*. Retrieved from http://www.defence.govt.nz/publications/?start=63.

Government Printer, (1987). *Defence of New Zealand: Review of Defence Policy*. Wellington, New Zealand: Government Press.

Greener, P. (2009). Timing is Everything: The Politics and Processes of New Zealand Defence Acquisition Decision Making. *Canberra Papers on Strategy and Defence, No. 173,* 53 – 72.

Jane's International Defence Review, (2016). *Canterbury tales re-told: RNZN multirole vessel delivers*. Retrieved from http://www.defence.govt.nz/what-we-do/delivering-defence -capability/defence-capability-projects/protector-remediation/

Ministry of Defence. (2004). *Project Protector: Review of a Major Multi Agency Project.* Wellington, New Zealand: Government Press.

Ministry of Defence and New Zealand Defence Force. (2016). Defence Capability Change Action Programme. *DCCAP Generic Presentation Dec 2016.*

Ministry of Defence, (2014). *Defence Capability Plan*. Retrieved from http://www.defence.govt.nz/publications/publication/defence-capbility-plan-june-2014.

Ministry of Defence, (2010). *Defence White Paper 2010.* Retrieved from http://www.defence.govt.nz/publications/publication/defence-white-paper-2010-november.

Ministry of Defence, (2016). *Defence White Paper 2016.* Retrieved from http://www.defence.govt.nz/publications/publication/defence-white-paper-2016.

Ministry of Defence, (2015). *Major Projects Report 2015*. Retrieved from http://www.defence.govt.nz/publications/publication/major-projects-report-2015-volume-1.

Royal New Zealand Navy. (2016). *NZBR 63 (Volume 3) – Operational Capability Statement – Multi Role Vessel (MRV)*. Retrieved from http://org/m-navy/Publications/NZBR%20063%20(Vol%203)%20-%20Operational%20Capability%20Statement%20-%20Multi%20Role%20Vessel%20(MRV).pdf.

Yemen: Houthis claim attack on UAE military vessel. (2016, October 2). Al Jazeera. Retrieved from http://www.aljazeeera.com/news/2016/10/yemen-houthis-claim-attack-uae-militaery-vessel-161001210036896.html

REFERENCES

[i] Greener, P. (2009). Timing is Everything: The Politics and Processes of New Zealand Defence Acquisition Decision Making. *Canberra Papers on Strategy and Defence. No. 173.* Canberra, ACT: ANU E Press, p. 53.
[ii] Government Printer, (1987). *Defence of New Zealand: Review of Defence Policy*. Wellington, New Zealand: Government Press. p. 35.
[iii] Greener, P. (2009). Timing is Everything: The Politics and Processes of New Zealand Defence Acquisition Decision Making. *Canberra Papers on Strategy and Defence. No. 173.* Canberra, ACT: ANU E Press, p. 53.
[iv] Greener, P. (2009). Timing is Everything: The Politics and Processes of New Zealand Defence Acquisition Decision Making. *Canberra Papers on Strategy and Defence. No. 173.* Canberra, ACT: ANU E Press, p. 53.
[v] Greener, P. (2009). Timing is Everything: The Politics and Processes of New Zealand Defence Acquisition Decision Making. *Canberra Papers on Strategy and Defence. No. 173.* Canberra, ACT: ANU E Press, p. 54.
[vi] Greener, P. (2009). Timing is Everything: The Politics and Processes of New Zealand Defence Acquisition Decision Making. *Canberra Papers on Strategy and Defence. No. 173.* Canberra, ACT: ANU E Press, p. 53.
[vii] Greener, P. (2009). Timing is Everything: The Politics and Processes of New Zealand Defence Acquisition Decision Making. *Canberra Papers on Strategy and Defence. No. 173.* Canberra, ACT: ANU E Press, p. 55.
[viii] Greener, P. (2009). Timing is Everything: The Politics and Processes of New Zealand Defence Acquisition Decision Making. *Canberra Papers on Strategy and Defence. No. 173.* Canberra, ACT: ANU E Press, p. 55.
[ix] Greener, P. (2009). Timing is Everything: The Politics and Processes of New Zealand Defence Acquisition Decision Making. *Canberra Papers on Strategy and Defence. No. 173.* Canberra, ACT: ANU E Press, p. 57.
[x] Greener, P. (2009). Timing is Everything: The Politics and Processes of New Zealand Defence Acquisition Decision Making. *Canberra Papers on Strategy and Defence. No. 173.* Canberra, ACT: ANU E Press, p. 58
[xi] Greener, P. (2009). Timing is Everything: The Politics and Processes of New Zealand Defence Acquisition Decision Making. *Canberra Papers on Strategy and Defence. No. 173.* Canberra, ACT: ANU E Press, p. 59.
[xii] Greener, P. (2009). Timing is Everything: The Politics and Processes of New Zealand Defence Acquisition Decision Making. *Canberra Papers on Strategy and Defence. No. 173.* Canberra, ACT: ANU E Press, p. 61.
[xiii] Greener, P. (2009). Timing is Everything: The Politics and Processes of New Zealand Defence Acquisition Decision Making. *Canberra Papers on Strategy and Defence. No. 173.* Canberra, ACT: ANU E Press, p. 61.
[xiv] Greener, P. (2009). Timing is Everything: The Politics and Processes of New Zealand Defence Acquisition Decision Making. *Canberra Papers on Strategy and Defence. No. 173.* Canberra, ACT: ANU E Press, p. 62.
[xv] Greener, P. (2009). Timing is Everything: The Politics and Processes of New Zealand Defence Acquisition Decision Making. *Canberra Papers on Strategy and Defence. No. 173.* Canberra, ACT: ANU E Press, p. 62.
[xvi] Greener, P. (2009). Timing is Everything: The Politics and Processes of New Zealand Defence Acquisition Decision Making. *Canberra Papers on Strategy and Defence. No. 173.* Canberra, ACT: ANU E Press, p. 63.
[xvii] Greener, P. (2009). Timing is Everything: The Politics and Processes of New Zealand Defence Acquisition Decision Making. *Canberra Papers on Strategy and Defence. No. 173.* Canberra, ACT: ANU E Press, p. 63.
[xviii] Greener, P. (2009). Timing is Everything: The Politics and Processes of New Zealand Defence Acquisition Decision Making. *Canberra Papers on Strategy and Defence. No. 173.* Canberra, ACT: ANU E Press, p. 65.
[xix] Greener, P. (2009). Timing is Everything: The Politics and Processes of New Zealand Defence Acquisition Decision Making. *Canberra Papers on Strategy and Defence. No. 173.* Canberra, ACT: ANU E Press, p. 65.
[xx] Greener, P. (2009). Timing is Everything: The Politics and Processes of New Zealand Defence Acquisition Decision Making. *Canberra Papers on Strategy and Defence. No. 173.* Canberra, ACT: ANU E Press, p. 66.
[xxi] Greener, P. (2009). Timing is Everything: The Politics and Processes of New Zealand Defence Acquisition Decision Making. *Canberra Papers on Strategy and Defence. No. 173.* Canberra, ACT: ANU E Press, p. 68.
[xxii] Greener, P. (2009). Timing is Everything: The Politics and Processes of New Zealand Defence Acquisition Decision Making. *Canberra Papers on Strategy and Defence. No. 173.* Canberra, ACT: ANU E Press, p. 71
[xxiii] Ministry of Defence. (2004). *Project Protector: Review of a Major Multi Agency Project.* Wellington, New Zealand: Government Press. p. 1.
[xxiv] Ministry of Defence. (2004). *Project Protector: Review of a Major Multi Agency Project.* Wellington, New Zealand: Government Press. p. 2.
[xxv] Ministry of Defence. (2004). *Project Protector: Review of a Major Multi Agency Project.* Wellington, New Zealand: Government Press. p. 4.
[xxvi] Coles, J. (2008). *Independent Review of the Safety and Functionality of HMNZS CANTERBURY.* Retrieved from http://www.defence.govt.nz/publications/?start=63. p. 1.
[xxvii] Coles, J. (2008). *Independent Review of the Safety and Functionality of HMNZS CANTERBURY.* Retrieved from http://www.defence.govt.nz/publications/?start=63. p. 27.
[xxviii] Coles, J. (2008). *Independent Review of the Safety and Functionality of HMNZS CANTERBURY.* Retrieved from http://www.defence.govt.nz/publications/?start=63. p. 5.

[xxix] Coles, J. (2008). *Independent Review of the Safety and Functionality of HMNZS CANTERBURY*. Retrieved from http://www.defence.govt.nz/publications/?start=63. p. 5.
[xxx] Coles, J. (2008). *Independent Review of the Safety and Functionality of HMNZS CANTERBURY*. Retrieved from http://www.defence.govt.nz/publications/?start=63. p. 5.
[xxxi] Coles, J. (2008). *Independent Review of the Safety and Functionality of HMNZS CANTERBURY*. Retrieved from http://www.defence.govt.nz/publications/?start=63. p. 6.
[xxxii] Coles, J. (2008). *Independent Review of the Safety and Functionality of HMNZS CANTERBURY*. Retrieved from http://www.defence.govt.nz/publications/?start=63. p. 6.
[xxxiii] Coles, J. (2008). *Independent Review of the Safety and Functionality of HMNZS CANTERBURY*. Retrieved from http://www.defence.govt.nz/publications/?start=63. p. 7.
[xxxiv] Coles, J. (2008). *Independent Review of the Safety and Functionality of HMNZS CANTERBURY*. Retrieved from http://www.defence.govt.nz/publications/?start=63. p. 6 and 8.
[xxxv] Coles, J. (2008). *Independent Review of the Safety and Functionality of HMNZS CANTERBURY*. Retrieved from http://www.defence.govt.nz/publications/?start=63. p. 8.
[xxxvi] Coles, J. (2008). *Independent Review of the Safety and Functionality of HMNZS CANTERBURY*. Retrieved from http://www.defence.govt.nz/publications/?start=63. p. 9.
[xxxvii] Coles, J. (2008). *Independent Review of the Safety and Functionality of HMNZS CANTERBURY*. Retrieved from http://www.defence.govt.nz/publications/?start=63. p. 14.
[xxxviii] Coles, J. (2008). *Independent Review of the Safety and Functionality of HMNZS CANTERBURY*. Retrieved from http://www.defence.govt.nz/publications/?start=63. p. 11.
[xxxix] Royal New Zealand Navy. (2016). *NZBR 63 (Volume 3) – Operational Capability Statement – Multi Role Vessel (MRV)*. Retrieved from http://org/m-navy/Publications/NZBR%20063%20(Vol%203)%20-%20Operational%20Capability%20Statement%20-%20Multi%20Role%20Vessel%20(MRV).pdf
[xl] Coles, J. (2008). *Independent Review of the Safety and Functionality of HMNZS CANTERBURY*. Retrieved from http://www.defence.govt.nz/publications/?start=63. p. 1.
[xli] Coles, J. (2008). *Independent Review of the Safety and Functionality of HMNZS CANTERBURY*. Retrieved from http://www.defence.govt.nz/publications/?start=63. p. 2.
[xlii] Jane's International Defence Review, (2016). *Canterbury tales re-told: RNZN multirole vessel delivers*. Retrieved from http://www.defence.govt.nz/what-we-do/delivering-defence -capability/defence-capability-projects/protector-remediation/. p. 4

[xliii] Royal New Zealand Navy. (2016). *NZBR 63 (Volume 3) – Operational Capability Statement – Multi Role Vessel (MRV)*. Retrieved from http://org/m-navy/Publications/NZBR%20063%20(Vol%203)%20-%20Operational%20Capability%20Statement%20-%20Multi%20Role%20Vessel%20(MRV).pdf
[xliv] Ministry of Defence, (2014). *Defence Capability Plan*. Retrieved from http://www.defence.govt.nz/publications/publication/defence-capbility-plan-june-2014. p. 14.

[xlv] Ministry of Defence, (2016). *Defence White Paper 2016*. Retrieved from http://www.defence.govt.nz/publications/publication/defence-white-paper-2016. p. 47.
[xlvi] Ministry of Defence, (2014). *Defence Capability Plan*. Retrieved from http://www.defence.govt.nz/publications/publication/defence-capbility-plan-june-2014. p. 29.
[xlvii] Aurecon, (2009). *Review 09 Procurement Process Review.* Retrieved from http://www.defence.govt.nz/assets/Uploads/defence-review-2009-released-aurecon-procurement-process-review.pdf. p. 5.
[xlviii] Aurecon, (2009). *Review 09 Procurement Process Review.* Retrieved from http://www.defence.govt.nz/assets/Uploads/defence-review-2009-released-aurecon-procurement-process-review.pdf. p. 5.
[xlix] Ministry of Defence, (2010). *Defence White Paper 2010*. Retrieved from http://www.defence.govt.nz/publications/publication/defence-white-paper-2010-november. p. 87
[l] Ministry of Defence and New Zealand Defence Force. (2016). Defence Capability Change Action Programme. *DCCAP Generic Presentation Dec 2016.* p. 9.
[li] Ministry of Defence and New Zealand Defence Force. (2016). Defence Capability Change Action Programme. *DCCAP Generic Presentation Dec 2016.* p. 8.
[lii] Ministry of Defence. (2004). *Project Protector: Review of a Major Multi Agency Project.* Wellington, New Zealand: Government Press. p. 12.
[liii] Yemen: Houthis claim attack on UAE military vessel. (2016, October 2). Al Jazeera. Retrieved from http://www.aljazeeera.com/news/2016/10/yemen-houthis-claim-attack-uae-militaery-vessel-161001210036896.html

Large Displacement Frigate Total Ownership Cost Model

Malcolm Waugh
WAMA Alliance Management Office
malcom.waugh4@defence.gov.au

ABSTRACT

Arguing the case for large frigates is a study into the benefits and costs for frigate size surface combatants using a parametric study of hull forms of 4000 tonnes and 6000 tonnes given steel is cheap and air is free.

Immediately of benefit is the improved seakeeping of a larger combatant however this is weighed against the fuel cost, from additional resistance due to wetted surface area. Whilst damage stability can be optimised in compartment sub division through zoning the required additional fire pumps increase requisite maintenance. Habitability is improved within the larger combatant however air conditioning and lighting demands increase. The larger combatant allows for additional weapon and sensor systems plus survivability against modern anti-ship missiles can be significantly improved however this is where costs escalate. Production engineering becomes easier due to greater space for pipework runs but these pipe runs grow in length. Margin management becomes less of an issue if the starting point is greater margins. For every benefit there is a cost and vice versa.

A Total Ownership Cost Model, across the lifecycle from design to disposal, is presented that explores the design simplicity, production engineering efficiencies, modular construction ingenuity, fuel savings resident within a slender hullform and maintenance agility available within the larger combatant. For a true comparison, care is taken to not present a militarily weak 4000 tonne combatant against a strong 6000 tonne combatant. The paper will demonstrate the case, or otherwise, for large frigates is sound whereby benefits outweigh costs.

INTRODUCTION

Chief of Navy, Admiral Barrett spoke of the five tenants of national naval capability at his Navy and the Nation presentation [1]. He stated "the centrality of **deterrence** as an outcome of decisive **lethality**; the importance of **availability** and **sustainment**, and of course, **affordability**". To maintain deterrence and lethality in the mission to fight and win at sea requires a never ending cycle of upgrading with modern weapons and sensors plus defeating modern weapons. Capability drives acquisition and at some point in the spend profile it is prudent and efficient to plan for ship replacement.

Total Ownership Cost (TOC) is akin to Asset Management and the iceberg model that depicts the impost of poor early design decisions being realised in-service. Driving down the Unit Production Cost in the shipbuilding phase (CAPEX) is desirable but not if it results in greater TOC due to high in-service costs (OPEX). A 30 year view is prudent.

TOTAL OWNERSHIP COST

Affordability is the performance balance variable that delivers maximally capable warships yet remains affordable through acquisition, sustainment and operations phases of the lifecycle. In pursuit of that affordability, without a TOC perspective, warfighters might view capability as the only cost decision variable. From a systems engineering perspective TOC is a decision or weighting factor in alternate systems review. Principal decision criterion, backed by strong metrics and analysis, set out to improve the life-cycle affordability. The aim is to not purchase fewer ships but to improve or increase the value gain in lethality through plans to mitigate each cost driver and for the Mission System to be balanced to the Support System. TOC breaks the acquisition paradigm exclusive focus.

The total or true ownership cost (TOC) contributors are the Unit Production Cost (UPC) and the Life Cycle Cost (LCC) contributors of crew wages and fuel use for operations (O) in addition to maintenance (M) cost drivers with complete disposal or demilitarisation viewed as the final maintenance cost. The equation for TOC is developed as:

$$TOC_n = UPC_n + LCC_n$$

For Option 1 $\quad TOC_1 = UPC_1 + O_1 + M_1$

Where UPC is the ship Unit Production Cost + amortised non-recurring engineering
O comprises crew costs (CC) and fuel usage (FU)
M comprises planned maintenance (PM) and defect repairs (DR)

For Option 1 $\quad TOC_1 = UPC_1 + CC_1 + FU_1 + PM_1 + DR_1$

Assuming Crew costs are constrained to 150 personnel for all options therefore constant
Disposal costs are less than +/- 1% of TOC for all options therefore constant

Then $\quad TOC_1 = UPC_1 + FU_1 + PM_1 + DR_1$

Comparisons can then be made for a number of designs based on TOC. If the planned maintenance can be achieved considerably cheaper through life with refresh cycle costs minimised then the UPC may be higher and still achieve neutrality. If the fuel usage is higher in one option then savings offsets need to be found in maintenance and/or achieving a lower UPC. When viewed as a 30 year project it is within the realm of the possible that a *big empty ship* can have a lower TOC to a small frigate that presents as *shrink wrap around a combat system*. The TOC model explores the art of the possible.

Modelling or predictions start with comparisons and these are examined. The tool presents high level thresholds to be achieved across design and construction, fuel usage, planned maintenance and defect repair cost contributors. The literature search identifies the common relationship of 25% to each cost contributors of acquisition, maintenance, fuel usage and crew [2, 3, and 4]. Opportunities via continuous shipbuilding, form following function, modularisation and flexibility taken to the extreme are reviewed given TOC based decisions at these early project points present the optimised LCC opportunity.

REQUIREMENTS DRIVE COST

In the simplest expression – customer requirements drive warship costs. Transitioning quickly from requirements to the design solution can lock in cost drivers for life. The financial impost of early UPC decisions on LCC are exhibited in acquisition courses as an iceberg, as presented in Figure 1, to deliberately display the consequences of decisions.

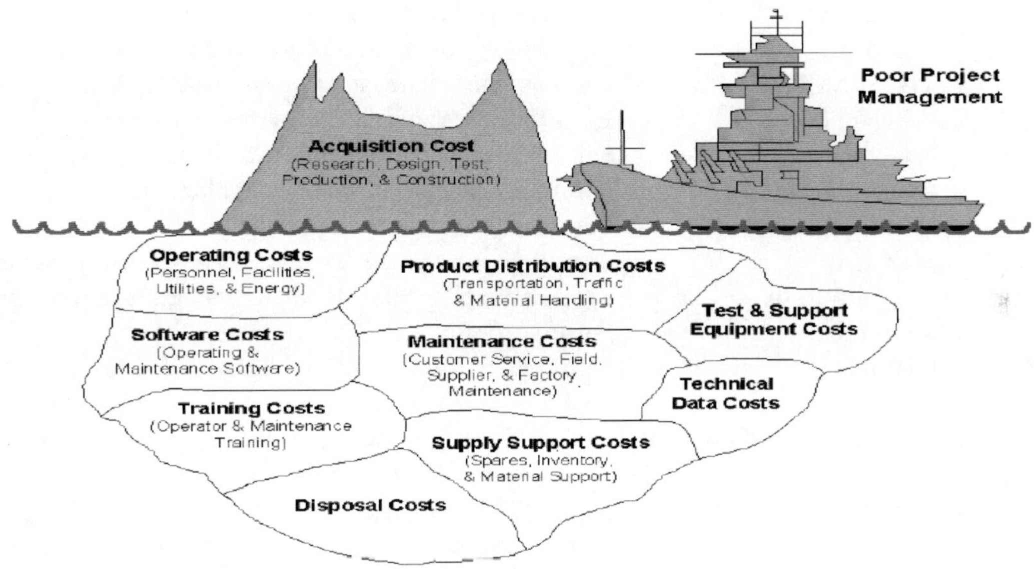

Figure 1 The danger of early UPC decisions on LCC (TLSMAN, 2001)

Sir Rowland Baker was a great innovator of the Royal Corps of Naval Constructors [5] and he often maintained that any new design should be 25 % innovation and 75 % well-tried practice as "more novelty would make it unreliable and less would make it obsolete in concept". D K Brown [6] expressed this philosophy more bluntly in coining the phrase "Innovate or Die". Therefore a mix of evolution and revolution is required.

Haydon [7] indicates in 2005 costs "a helicopter equates to about 400 tonnes of displacement and adds as much as $20 million to the sail away cost of the ship .... a large hull mounted sonar, such as SQS 53, equates to about 500 tonnes and over $60 million in total cost". The relationship between frigate size and effectiveness is both about the laws of physics and the laws of lethality.

Immediately of benefit is the improved seakeeping ability of a larger vessel however this is weighed against the fuel cost, from the additional resistance. Whilst damage stability can be optimised in compartment sub division the required additional fire pumps increase requisite maintenance. Habitability can be improved within the larger vessel however air conditioning plants need to be upsized. Whilst production engineering becomes easier due to space for pipework runs these pipe runs grow in length. Margin management becomes less of an issue to manage if the design includes greater through life margins. The ship in-service costs range from additional hull preservation through to wharf space.

Designing to requirements functional and performance specification that meets current threats but may not meet the long term need over a 25 to 30 year in – service life can be achieved through modularity and/or flexibility being resident within the ship design. RAND [8] define the difference where modularity has a construction focus whilst flexibility allows modernisation for new missions. However by its nature a modular platform does have a degree of flexibility. Examples are:

Module construction	Hull and structure steel blocks
Electrical power	Rafts for all power generation and switchboards
Machinery rafting	Platform auxiliary systems zone based
Hotel functions	Rafts for galley, medical facilities, laundry etc
Crew accommodation	Self-contained cabins similar to cruise ship practice
Open architecture	Future proofing for combat and platform systems
Mission modules	Combat system command and control centres
Main gun	Optional modules for 3" or 5" naval gun with magazine
Missile systems	Drop in modules such as Mk 41 Vertical Launch System
Multi-function mast	Future proofing with spare space
Mission bay	Spare space for dedicated mission equipment

The lesson to learn is that modularity and/or flexibility that influences LCC can be a lost opportunity quite early in the design solution if focused on UPC. The flexibility features identified above do not lock in LCC at UPC. Modularity benefits do begin at design and construction and are carried through into reduced time under major maintenance where remove and replace is more expedient that overhaul in-situ. Technology refresh, via update maintenance, and capability increase, via upgrade installation are flexibility in design payback benefits. Flexibility is a valuable attribute where there is uncertainty and there is uncertainty in future threats considering the 25 to 30 year ship in-service life.

The six earliest considerations that pull together the design of a warship can be judged to be stability, seakeeping, structural strength, speed, stealth and survivability however maximising the positives of one can come at negative design consequences of the others. The first three design elements, stability, seakeeping and structural strength, could be considered as lever arms within a triangle and the balanced design residing in the middle as a compromise. This is allowing form to follow function and not presenting a small frigate as *shrink wrap around a combat system* wherein design ratios are sub optimal.

FORM SHOULD FOLLOW FUNCTION

Arguing that form should follow function starts with the concept of systems engineering whereby requirements and needs analysis, or capability, are identified. The argument can become somewhat circular if form, such as overall length, has inherent flexibility via modularisation to address the pace of evolution / revolution. Function is the generic capabilities required, such as the anti-submarine role, whereas form is the open architecture style combat management system combined with a large multi-purpose mission bay that allows for plug-and-play infrastructure. Futureproofing functions can be resident in the *big empty ship*. The authors view is that form should follow function.

STEEL IS CHEAP – AIR IS FREE

In the design phase there are trade-offs with regard to performance that still meet the customer requirements. Recalling the float – move – fight construct major cost drivers are typically not realised in the float or move contributors but restricted to fight where warfighting includes the combat system, offensive and defensive weapons and a range of sensors. The literature search identifies and reinforces *steel is cheap* in simple terms.

- Keane and Sandberg [9] identify "limits on acquisition cost have been translated into a fixed displacement, a constrained hull length, an allowable enclosed hull volume or some other criterion which usually constrains hull size and principle dimensions, although the acquisition cost of the hull is less than 10 percent of the total ship cost".

- Brown [10] in comparing a 107 metre frigate of 2700 tonnes to a 125 metre frigate of 2835 tonnes concluded "the longer ship will burn some 550 tonnes of fuel per year less than the shorter ship .... the additional cost of structure will be paid off in less than three years".

- Material costs, as a proportion of the total construction costs, were analysed by Dorey [11]. He found total materials cost 63 % with the balance, of 37 %, accounted for by labour overheads and profit. From the materials cost of 63 % the weapon systems comprised 43 % and a further 15 % on machinery. Structural costs for materials are the 5 % remaining.

- Wade and Pattison [12] identify costs for a frigate to consist of total platform 46 %, weapon equipment 30 % with 24 % allocated to shipyard services and costs. The hull contribution, within the total platform of 46 %, was presented as 11 %.

- Structural cost optimisation and estimation was explored by Heather [13] highlighting "typically for a warship the structure represents about 45 % of the weight but only about 10 – 12 % of the Unit Production Cost".

LENGTH – DISPLACEMENT RATIO

Frigates or surface combatants in general, are often classified as having a high speed round bilge (HSRB) hullform. The most important hull form parameter for resistance optimisation of a HSRB hullform is the Displacement / Length Ratio. Design lanes for Prismatic Coefficient (Cp) and the Displacement / Length Ratio, to assist in resistance optimisation, are presented in the classic work by Saunders [14].

Saunders work is useful in plotting the optimum value for Cp and the Displacement / Length Ratio. For a Froude Number of 0.4 the resistance optimised frigate would have a Displacement / Length Ratio of 52 and a Cp of 0.58 as in Figure 2. Evolution of the ANZAC Frigate, from design to contemporary displacements, is overlayed.

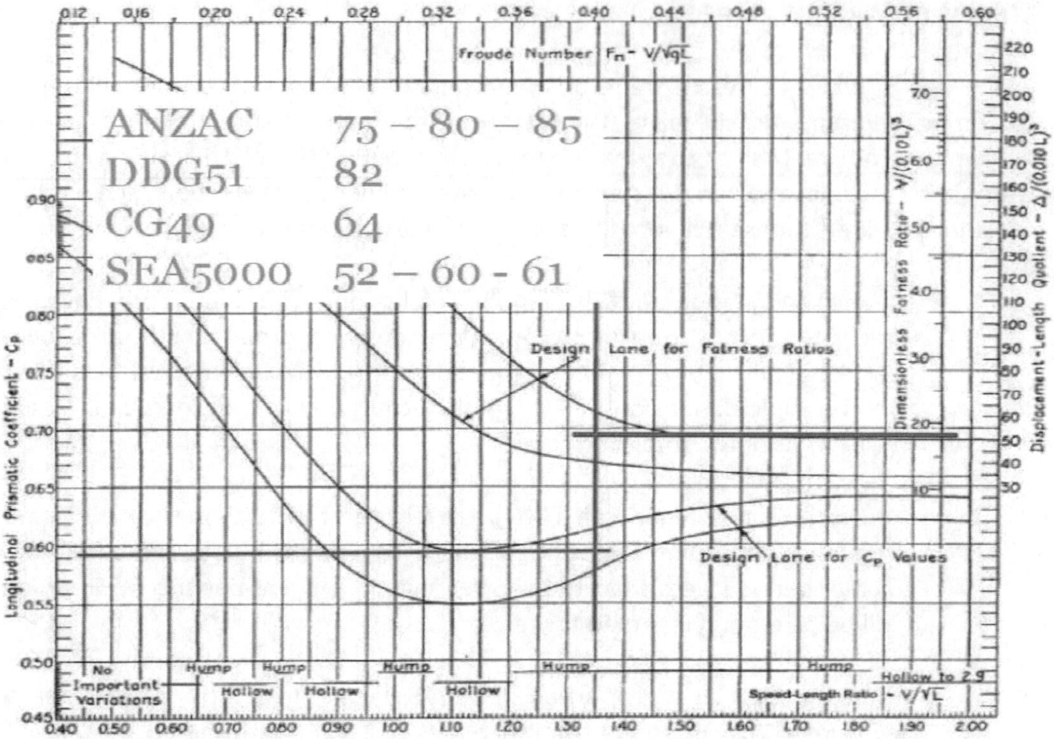

Figure 2 Design lanes for high speed round bilge warships (Saunders)

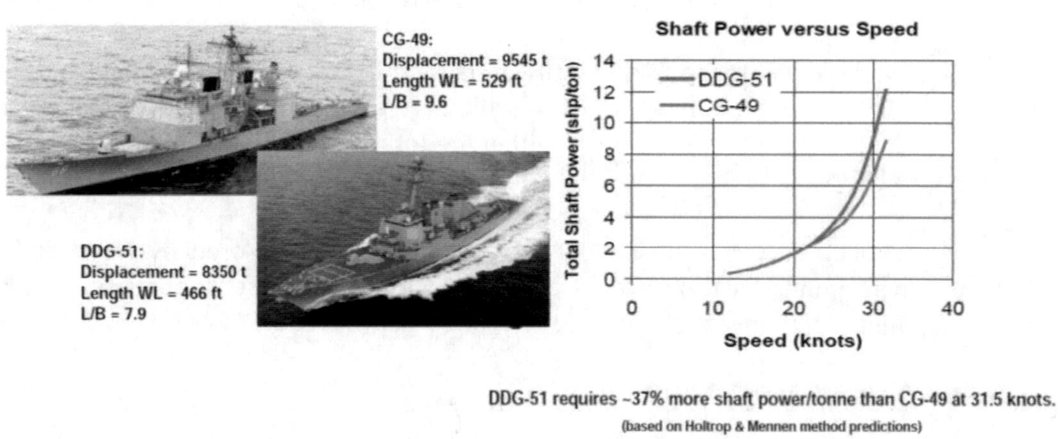

Figure 3 DDG-51 and CG-49 speed – power curve comparison (Grimm)

Grimm [15] identifies the penalty that the USN DDG 51 Class incurs in required shaft horsepower, to that of the heavier displacement CG 47 Class, due to the inferior Displacement / Length Ratio and low length-beam ratio as presented in Figure 3. The design lanes developed by Saunders in 1957 remain valid to this day and Watson [16] identifies savings in fuel costs, through "improved lines and proportions of ship main dimensions reducing resistance can be 8 %". Savings in required power and/or fuel consumption can be found where form is not arbitrarily fixed early in the design.

STABILITY

The stability of a ship, or its resistance to heeling forces, is the measure of its righting lever (GZ). Importantly, increasing the ships length to improve seakeeping, at the expense of decreasing beam to remain interior volume neutral, will have a negative effect on GZ and the recovery from roll will be slower. Hull form parameters contribute to GZ, in either a positive or negative effect. These hull form parameters are recognisable in the ages old *short – fat versus long – thin surface combatant argument for a given displacement*. Given the presence of topweight from superstructure design and the emerging preference for large Phased Array Radar panels there may be a requirement for the smaller 4000 tonne hull form frigate to have a sea water compensated fuel system and/or solid ballast installed.

SEAKEEPING

There is a loss of operational time when the frigate cannot perform its mission in higher sea states. Increasing length, in simple terms, gives the best improvement on seakeeping. Roll stabilisation systems can be listed, very generally, in decreasing order of effectiveness of active fin stabilisers, passive tanks, rudder roll stabilisation and bilge keels. Effective roll reduction devices, individually optimum for a particular speed domain but collectively capture the total speed range, can all easily be incorporated into the larger 6000 tonne hull form to increase seakeeping ability.

STRENGTH

Structural strength is withstanding sea loads and fatigue over the service life. In the 2000 tonne delta between a smaller 4000 tonne hull form and a larger 6000 tonne hull form a range of structural strength design options can be selected that would collectively significantly increase the global structural strength of a frigate. The superstructure can be all high strength steel as opposed to aluminium for an additional 150 tonnes and continue to the side shell to design out structural discontinuities and reduce radar reflection for a further 50 tonnes additional weight. Frame spacing can be closer thereby increasing local strength for instances when tugs push a little too hard.

SPEED

The Displacement / Length ratio has a *sweet spot* for frigate sized warships in the resulting speed – power profile. The advantages of form following function were detailed above.

STEALTH (RADAR CROSS SECTION REDUCTION)

It is vital to reduce the radar cross section in the new design. As vessel height increases the radar horizon, the first minimisation iteration is to opt for a long and low profile. The second minimisation iteration requires miro scale design to reduce the sources of radar reflection. Increasing the internal volume allows for upper deck clutter reduction.

SURVIVABILITY

Survivability is linked to resilience and duplication / separation of vital capability. It is the *ability to fight hurt*. A size increase can be used to advantage to improve fighting hurt.

CONSTRUCTION EFFICIENCIES

A continuous ship building enterprise has all of the economies of scale attributes that contribute to driving down the UPC. The commercial advantages include a consistent acquisition strategy environment and long term sub-contractor / supplier relationships. The non-recurring engineering costs can be amortised over a longer period over a longer production run. The supply chain advantages come from economies of scale. The learning curve benefits from a secure skilled workforce, that trains apprentices in both blue collar and white collar disciplines, can also be realised. Industry capacity can be smoothed and optimised rather than boom-bust cycles. There appears to be no disadvantage associated with this strategic commitment.

The *learning curve* can be exploited in a continuous shipbuilding approach whereby manhours decrease as the ship number increases. However these ships have to be sister ships with very few changes in configuration between each other. Too many design changes decreases the learning curve gained efficiencies. Furthermore, if the same workforce is utilised the repetition effect can further drive down manhours. A continuous build strategic approach of *one class* with modularisation and flexibility has easy potential to offer significant UUC savings. Further construction efficiencies include:

> Shipbuilders rule of 1-3-5-10 [17] to be left shifted as much as practicably possible where for every 1 hour work at assembly level is 3 hours at the module level and 5 hours in the block consolidation level and 10 hours of rework from a failed builders trial. Increase the degree of fit out at the earliest stage.
>
> Embrace modularity and flexibility for the reasons outlined earlier for design simplicity, ease of construction, raft installation and futureproofing all of which **drive down UPC and LCC achieving capital efficiency**.
>
> Density is a cost driver with the most expensive frigates being low in displacement and the hull structure referred to as a shrink wrapped combat system. It costs more in design effort hours and production squeeze to work in tight spaces. Upsizing is easier to work in therefore quicker resulting in a 10% UPC saving [18].
>
> Doerry [19], in referring to the *tyranny of the shaft line* identifies there are many disincentives to continue designing frigates where the traditional propulsion power is delivered from the prime mover, through a gearbox, down a shaft line, into a thrust block to drive a propeller. Mattick [20] indicates “ship displacement can often be reduced by at least 8 % ... when prime movers can be mounted anywhere in the ship”. The *tyranny of the shaft line* can be broken by matching *requirements pull* with *technology push* within an all-electric ship.
>
> Defence SA [21] conclude “a rolling build program of one class of warship could substantially reduce costs to the conceivable point that it is no more expensive to build one class of more capable warship than it is to build a mix of capable and less capable warships”.

COST MODELS

Comparisons are often the only starting point available for UPC and/or LCC estimating. Price and Coolahan [22] identify the difficulty in estimating in the early design phases when parametric studies or analogy are the only methods available. The transition from assumptions through to actuals requires real data. The degree of evolution and revolution within the new design influences the cost estimate data validity through risk and uncertainty. Cost evaluation needs to be based on requirements (functional – performance – construction infrastructure - operations – maintenance) hence the transition from coarse estimate to detailed price based on the proportion of analogy – parametric – build-up and extrapolation.

Requirements creep is identified by Peer [23] as "the tendency over time of navies to replace older ships with more capable and more complicated ships. This often means that simple traditional tasks are now undertaken by extremely capable and expensive ships". The RAND study [24] noted that the naval ship cost growth "has been upwards by as much as 10 percent per year". This study was focused on the difference between military and commercial shipbuilding with an excellent comparison in labour hours. However, comparison modelling via cost per tonne simple metrics can lead to wrong conclusions. Warship complexity requires more design effort, construction hours and procurement of expensive technology. The RAND study [24] acknowledges the complexity driver difference between military and commercial vessel construction costs whereby the naval cost per unit volume is "100 times the commercial cost per unit of volume ... only cruise ships come close in price per tonne to the least expensive warship".

The genesis of the Littoral Combat Ship (LCS) was to design and deliver a cheap replacement to the FFG 7 Class of frigate in large numbers that would free up DDG 51 Class from routine patrol missions. The US Government Accountability Office review [25] revealed the result to be sub optimal. The GAO review found that "per ship life cycle costs per year for the LCS program are nearing or may exceed those for other surface ships including guided missile frigates and multi mission destroyers". What was expected to be realised from the LCS concept was a cheap replacement to the FFG 7 Class. The divide in actual LCC versus design concept LCC remains to be reconciled.

A common cost model allows for easier comparisons of tender responses. The Technical Subject Code (TSC 100 Series) is most useful in cost control, weight control and a range of other performance reporting metrics as well as a serving as a configuration management tool through a parent-child construct of system-equipment-component identification structure. The TSC100 Series [26] is used in the TOC model depicted.

CALCULATING TOC

The developed model has the ANZAC Frigate as the baseline where TOC = 1 with weighting pre set within the acquisition, fuel and maintenance contributors. Factors can be changed by the user to see what efficiencies must be achieved within the contributors to better the TOC = 1 for the baseline. Figure 4 shows the 6000 tonne variant result.

THE 6000 TONNE VARIANT IS "THE BIG EMPTY SHIP" EMBRACING MODULARITY & FLEXIBILITY

$TOC_1 = UPC_1 + FU_1 + PM_1 + DR_1 + CC_1$

PLEASE LEAVE THESE CELLS UNCHANGED
THESE CELLS MAY HAVE THE FACTOR CHANGED

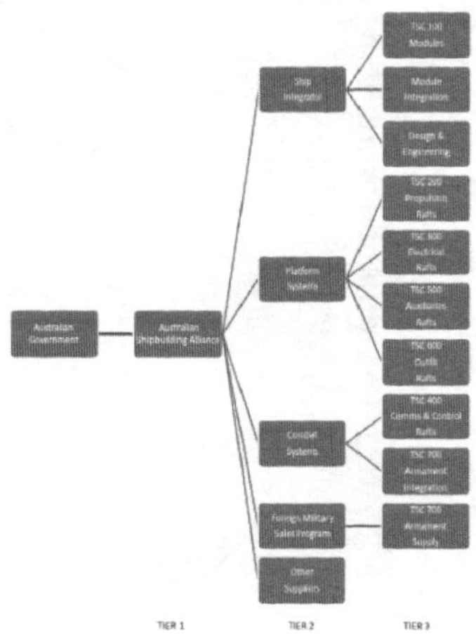

	100 SERIES	UNIT PRODUCTION COST: DESIGN FACTOR	DESIGN WEIGHT %	CONSTRUCTION FACTOR	CONSTRUCTION WEIGHT %	ACCEPTANCE FACTOR	ACCEPTANCE WEIGHT %	FUEL USAGE FACTOR	FUEL USAGE WEIGHT %	MAINTENANCE: PLANNED FACTOR	PLANNED WEIGHT %	REPAIR FACTOR	REPAIR WEIGHT %
00	GENERAL												
100	STRUCTURE	0.8	20	1.2	5	1	10			0.8	5	0.8	5
200	PROPULSION	0.8	20	1.1	20	1	10	1.2	70	0.8	20	0.8	20
300	ELECTRICAL	0.5	10	0.8	15	0.5	10	1.1	30	0.8	15	0.8	15
400	C2	0.5	10	0.8	20	0.5	20			0.8	20	0.8	20
500	AUXILIARIES	0.5	20	0.8	5	0.8	10			0.8	5	0.8	5
600	OUTFIT	0.5	10	0.8	5	0.5	10			0.8	5	0.8	5
700	ARMAMENT	1	10	1	30	1	30			1	30	1	30
800	VARIABLE LOADS												
	(CHECK SUM)		100		100		100		100		100		100
	SUB TOTALS	0.67		0.94		0.78		1.17		0.86		0.86	
	WEIGHTING	15%		80%		5%				70%		30%	
	TOTAL	0.8915						1.17		0.86			

RECALL TOC CONTRIBUTORS ARE 25% ACQUISITION + 25% FUEL + 25% MAINTENANCE + 25% CREW COSTS

	ACQUISITION	FUEL	MAINTENANCE	CREW
BASELINE %	22.2875	29.25	21.5	25

NEW TOC	0.980375

Simple maintenance
Quick maintenance
Easy removal routes

Increased fuel consumption for the larger hullform
Increased fuel consumption for additional lighting etc

The design is easier / quicker due to greater space
Ships structure 20% additional cost but steel is cheap
Larger propulsion plant required at 10% additional cost
Time savings across elec and auxiliary due to quick raft instal
Cost savings due to large number of small simple equipment
Easier to test raft mount equipment pre installation

SUMMARY

There is a 2.0% TOC reduction (saving over the lifecycle)

CONCLUSION

It is possible to achieve a Total Ownership Cost saving within the larger hullform due to design ease and construction simplicity despite the additional fuel requirement as maintenance is also quick and easy.

ASSUMPTIONS

1. Armament is constant across options (factor = 1 and weighting unchanged) so as to present equivalent weapon systems thereby avoiding capability creep
2. Command and communications is containerised on rafts for offsite assembly and testing before "plug and play" integration onboard. Refresh is via container removal and replacement
3. All auxiliary equipment on platform system rafts is in the order of 100kg max with soft patches above to facilitate easy removal and replacement
4. The "tyrany of the shaftline" is understood where freeform with propulsion pods can save design and integration but not included in this present study
5. The 100 series components for maintenance need further assessment to determine the contribution of each system on costs and time

ABSOLUTE COST

If a frigate costs $800M to acquire and this represents 25% of TOC then a 2% saving is $64M over 30 years in absolute cost and $576M for a class of 9 ships.
Another way to consider the TOC saving is for $576M to be visualised as 2/3 of the 9th frigate being nil cost.

FURTHER WORK

The Basic Equation – TOC Neutrality

- What are the cost drivers
- Where are the savings to be had
- What do the cost savings weigh
- What is the optimum weight spend for neutrality

Figure 4 TOC Model Output – 6000 Tonne Variant with zonal architecture

In simple terms should fuel usage for propulsion increase by 20%, due to increased resistance from the larger hullform and heavier displacement, then a comparable 20% saving needs to be found in the planned maintenance and defect repair contributor. This 20% saving can be found in having room to work due to less density and complexity, wide passageways, simple 2 Deck access, trouble-free removal routes due to zone based machinery rafts comprising smaller plant rather than cutting open the hull. However this 20% maintenance saving can only be harvested if this is resident within the design in the first instance. The design must consider the LCC to achieve the savings.

Embracing modularity and flexibility for the reasons outlined earlier for design simplicity, ease of construction, raft installation and futureproofing all drive down UPC and LCC achieving capital efficiency. Rafting and zonal architecture, studied by Rains [27] was predicted to *achieve half a ship* more affordability in a 10 ship class should electrical and auxiliary machinery component rafting be embraced. By extension, fully embracing modularity and flexibility through standard interfaces across the full range of examples above will significantly reduce the cost of a 25 to 30 year in – service life.

In the example of Figure 4 the acquisition contribution has been reduced by 2% from the typical 25% TOC. Whilst 2% may seem small the value of the saving is $16M if the ship cost is $800M.

CONCLUSION

The financial impost of early Unit Production Cost decisions on Life Cycle Costs can be modelled and displayed clearly for Total Ownership Cost visualisation. Form should be allowed to follow function provided the design is futureproofed to remain abreast of technology. The examples of being penalised for sub optimal design lanes for Prismatic Coefficient and the Displacement / Length Ratio can be recovered in a larger hullform.

The big empty ship has merit in design simplicity and less density. The simplicity presents saving prospects in design, production engineering, planned maintenance and defect repair to offset additional fuel usage. The big empty ship can be better than TOC neutral.

ACKNOWLEDGEMENTS

The views expressed herein are those of the author. The research article does not represent a return to in house design within the Australian Government Department of Defence. This academic study on frigate design and requirements selection, allowing form to follow function, may not accord with Government direction for the SEA5000 Future Frigate Project. I acknowledge that Defence serves Government and this study does not usurp Government policy or direction in the SEA5000 Future Frigate Project. It is an academic study.

The assessment of the performance of the DDG 51 Class comes from lessons learnt published in freely available journal articles and is fully referenced. No criticism is made of this class of warship as it continues to be the workhorse of the US Navy.

REFERENCES

1. Barrett, Admiral, 2016, *The Navy and The Nation*, CN Address to Australian Defence Magazine Conference, 09 February 2016, at: http://www.navy.gov.au/sites/default/files/documents/20160209_ADM_Conference_Navy_and_the_Nation.pdf, accessed 27 March 2017

2. West, J., 2016, *Asset Management – An Overview Course*, BAE Systems, presented July 2016

3. TLSMAN, Introduction to Lifecycle Costing Analysis, Volume 6 Part 1, September 2001

4. NATO Code Of Practice for Life Cycle Costing SAS-069 Booklet, at: https://www.sto.nato.int/publications/STO%20Technical%20Reports/Forms/Technical%20Report%20Document%20Set/docsethomepage.aspx?ID=2242&FolderCTID=0x0120D5200078F9E87043356C409A0D30823AFA16F6010066D541ED10A62C40B2AB0FEBE9841A61&List=92d5819c-e6ec-4241-aa4e-57bf918681b1&RootFolder=https://www.sto.nato.int/publications/STO%20Technical%20Reports/RTO-TR-SAS-069, accessed August 20, 2016

5. Baker R., contribution to, *Some problems in the construction of warships today*, Transactions of the Institute of Naval Architects, Volume 101, 1959

6. Brown D.K., 1991, *Innovate or Die*, Journal of Naval Engineering, Volume 33, Number 1, 1991

7. Haydon, P., 2005, *Choosing the right fleet mix: Lessons from the Canadian Patrol Frigate selection process*, Maritime Issues, Canadian Military Journal, Volume 9, Number 1, pages 66 - 75

8. RAND RR696, 2016, *Designing Adaptable Ships – Modularity and Flexibility in Future Ship* Designs, Rand Corporation, Santa Monica, California

9. Keane, Robert and Sandberg, William, 1984, *Naval Architecture For Combatants, A Technology Survey*, Naval Engineers Journal, September 1984, pages 47 - 64

10. Brown, David, 1985, *The value in reducing ship motions*, Naval Engineers Journal, March 1985, pages 41 - 46

11. Dorey, A.L., 1992, *Affordable new warships*, Paper Number 3, International symposium on affordable warships, 11 – 12 May 1992

12. Wade, S.J. and Pattison, N.P., 1990, *Towards affordable solutions for warship ownership*, International symposium on the future for surface warships, 4 – 6 June 1990

13. Heather, R.G., 1990, *Preliminary design considerations for fast warships*, International symposium on the future for surface warships, Paper 16, 04 June 1990, London

14. Saunders, H.E., 1957, *Hydrodynamics in ship design*, New York, 1957

15. Grimm, M., 2010, *Destroyer hullform optimisation for minimum resistance*, May 07, 2010

16. Watson, D.G.M., 1998, *Practical Ship Design*, Elseiver Ocean Engineering Book Series, Elseiver Science Ltd, Oxford UK

17. RAND RR767, 2014, *Keeping Major Naval Ship Acquisitions On Course – Key Considerations For Managing Australia's SEA5000 Future Frigate Program*, Rand Corporation, Santa Monica, California, page 128

18. Depetro, A. and Hoey, R., 2012, *Rapid Generation and Optimisation of Ship Compartment Configuration based on Life Cycle Cost*, BMT Design and Technology, Australia, at: http://www.bmtdesigntechnology.com.au/downloads/bmt-papers/rapid-generation-and-optimisation-of-ship-compartment-configuration-based-on-life-cycle-cost-and-operational-effectiveness/, accessed July 21, 2016

19. Doerry, N., 2006, *Institutionalising the electric warship*, Naval Engineers Journal, 2006 Number 4, pages 57 - 64

20. Mattick, D.J., 1995, *More concepts, concepts, concepts*, Journal of Naval Engineering, Number 36 Volume 1, 1995, pages 1 - 12

21. Defence SA, 2009, *Naval Shipbuilding: Australia's $250 Billion Nation Building Opportunity*, Defence SA Advisory Board, December 2009

22. Price, C.R. and Coolahan, J.E., 2011, *Modelling of life cycle, operations and support costs*, Prepared by the life cycle cost modelling subcommittee for the National Defense Industrial Association, February 2011

23. Peer, D., 2012, *Estimating the costs of naval ships*, Canadian Naval Review, Volume 8 Number 2 (Summer 2012), pages 4 - 8

24. RAND MG236, 2005, *Differences between military and commercial shipbuilding – Implications for the United Kingdom's Ministry of Defence*, Rand Corporation, Santa Monica, California

25. GAO, 2014, *Littoral Combat Ship, Report to Congressional Committees*, GAO-14-447, July 2014

26. SECNAV, 2012, *Ship Design and Material*, 9000 Series, commencing at page 72, at: https://www.uscg.mil/records/docs/SECNAV5210_2.pdf, accessed 26 May2015

27. Rains, D.A., 1996, *Naval Ship Affordability*, Naval Engineers Journal, July, 1996

Waste Heat Recovery of Marine Diesel Engines

Noam Olshina[1], Chris Manzie[1], Michael Brear[1], Peter Hield[2], Owen Tregenza[2]
1 The University of Melbourne, Vic, Australia
2 Defence Science and Technology, Vic, Australia
nolshina@student.unimelb.edu.au

ABSTRACT

The overall energy efficiency of maritime platforms is increasingly important, for both commercial and military applications. The energy available on board a maritime platform is limited by the size of its fuel tanks. Using the available energy efficiently provides increased platform range and endurance, reduced fuel tank volume, reduced fuel use and reduced exhaust emissions. Typically, 30% of the energy in the fuel is lost through the engine exhaust. This research project focuses on recovering some of this energy using a turbocompounding system consisting of a radial flow turbine coupled to an electrical generator. This system is retrofittable to existing diesel engines and only marginally increases the engine weight, volume and complexity. Additionally, the reduced exhaust gas temperature reduces the vessel's thermal signature. A previously developed model based control technique is extended to include control of the waste heat recovery system. This enables the engine and turbocompounding system to operate optimally to minimise fuel use, or alternatively to maximise power output, while ensuring that engine operational limits are not exceeded.

INTRODUCTION

Improving the energy efficiency of maritime vessels results in environmental, economic and for the military, strategic benefits. From an environmental perspective, the amount of engine emissions is directly correlated to fuel consumption. These emissions are regulated by the International Maritime Organisation [1]. Since the introduction of these regulations, the emission limits have become, and continue to be, progressively tightened. Achieving the future emission limits will require significant technological advances.

From an economic perspective, a commercial operator is motivated to reduce fuel consumption in order to reduce costs. The military considers the 'fully burdened cost of fuel', which also includes the transport, storage, protection of the fuel, which can significantly exceed the commodity price alone [2].

Increasing the energy efficiency of naval platforms presents several strategic opportunities. Maintaining a given fuel tank capacity while improving energy efficiency results in increased vessel range and endurance. Alternatively, there is potential for reduced fuel tank size, which can either lead to smaller vessels, or repurposing the space for other uses such as additional combat systems. In submarines, the battery charge time may be reduced, minimising the submarine's indiscretion ratio. Waste heat recovery (WHR) systems have the added advantage of reducing the exhaust temperature, thereby reducing the thermal signature of the vessel [3].

WHR is a class of technologies for improving the energy efficiency of engines by targeting the energy contained in the exhaust which would otherwise be wasted. Of the total energy contained in the fuel, approximately 40% is converted into useful work, 30% lost through the hot exhaust gasses and the remainder lost through other systems. WHR technologies include thermodynamic cycles, thermoelectrics and turbocompounding. Each offers distinct advantages and disadvantages [4]. The research presented in this paper focuses on turbocompounding, which involves the addition of a secondary power turbine in the exhaust stream of an engine. Turbocompounding provides competitive fuel efficiency gains compared to other WHR technologies with minimal increase in engine weight and complexity [4].

Diesel-electric submarines are required to operate at constant speed to minimise acoustic emissions, as well as maximise engine reliability and longevity, however, this has been difficult to achieve in practice. Air is drawn through a snorkel mast and exhausted underwater through the hull. This results in a low intake air pressure and a high and varying exhaust back pressure as ocean waves pass over the submarine (Figure 1). Consequently, the engine experiences high and varying exhaust gas temperatures and engine speed fluctuations, increasing engine wear and reduced reliability, resulting in a demanding maintenance schedule [5].

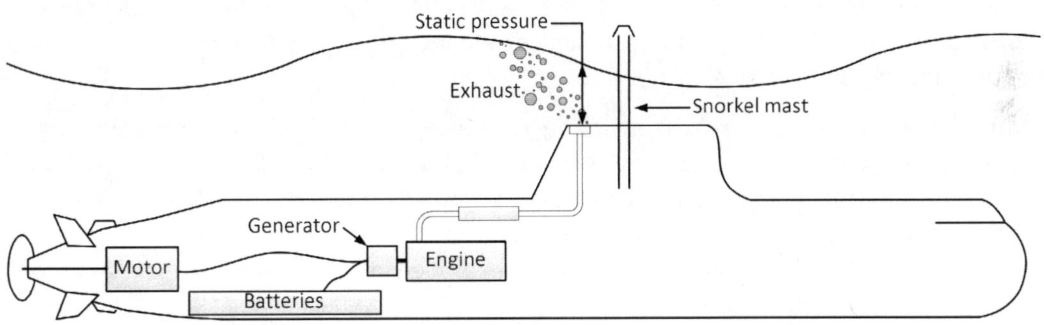

Figure 1: Schematic diagram illustrating how ocean waves cause engine back-pressure fluctuations.

Naval surface vessel engines are required to provide a variable power output consisting of the hotel load and propulsive power demand, and must respond to changes in demand quickly to meet mission objectives. In doing so, the engine operates over a large range of set points with a significant amount of time transitioning between these.

Unlike naval applications, commercial vessel powertrains are required to operate at steady state for the vast majority of their lives, and without the challenges of the unique operating environment of submarine engines. Therefore, the expected performance and idiosyncrasies of commercially available marine turbocompounding systems, such as that developed by MAN Diesel & Turbo [6], operating in challenging naval environments are unknown. However, this commercial experience provides additional motivation for and guidance in exploring this form of WHR for naval applications, and ultimately the development of a suitable naval turbocompounding system.

The use of advanced control methodologies such as model predictive control (MPC) offers attractive options for operating the engine, allowing for the fulfilment of multiple objectives simultaneously, while satisfying system constraints. These objectives may include maintaining engine speed, maximising power or minimising fuel use, all while ensuring specified engine operating limits are not exceeded. Constraints may also include emission constraints, which, as previously outlined for commercial shipping, have regulated limits. MPC is a novel control strategy for maritime engine control.

Previous work undertaken by Defence Science and Technology and the University of Melbourne (UoM) explored the use of MPC to control submarine engines subjected to varying exhaust back pressure [7] as well as for fixed diesel engines [8]. This study demonstrated the potential of MPC as a control solution for these applications, with distinct advantages over traditional PID control.

In this paper, the current and future research by DST and UoM is presented which explores the use of turbocompounding WHR in a marine diesel engine to maximise vessel range and endurance. The advantages of WHR will be maximised using MPC to ensure optimal performance is obtained while guaranteeing adherence to engine operation limits, even under the arduous submarine operating conditions.

TURBOCOMPOUNDING

Early turbocompounding technologies were first introduced into aircraft engines in the 1940's. Since the 1990's, turbocompounding systems have been adapted to heavy commercial vehicle engines [9]. These turbocompounding systems rely on a direct mechanical or hydraulic coupling between the turbine and the engine drive shaft to deliver the recovered energy to the driveshaft, and are known as mechanical turbocompounding (MTC). These have demonstrated fuel efficiency improvements of approximately 3% to 6% [9].

More recently, efforts have been made to develop electrical turbocompounding (ETC) systems for commercial markets and have generated interest within the academic community, often with industry collaboration, for example in [10]–[12]. These systems couple the turbine directly to a generator. Various engine and electrical turbo-compounding configurations have been developed or proposed. A selection of ETC configurations is shown in

Figure 2 and include:

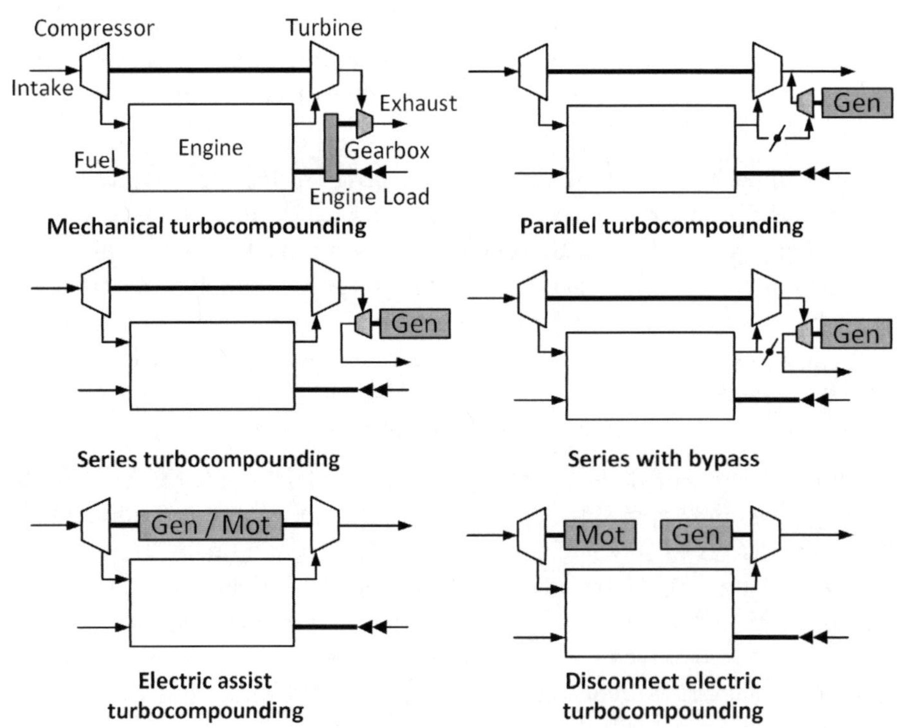

Figure 2: Schematic diagram of various turbocompounding arrangements

- Series electric turbocompounding (STC), with the turbine installed in series with the turbocharger turbine, with or without a wastegate before the STC.
- Parallel electric turbocompounding (PTC), with the turbine installed in parallel with the turbocharger turbine.
- Electric assist turbocompounding (EATC), with the generator incorporated around the turbocharger shaft between the turbine and compressor.
- Disconnected electric turbocompounding (DETC), with the turbocharger shaft replaced by a generator powered by the turbine and a motor driving the compressor.

One drawback of MTC is the parasitic effect of the additional power turbine at low engine speed and load conditions, which limit the applicability of such systems to engines that operate at nominal conditions for the majority of their operation. Nevertheless, ETC systems offer better efficiency gains compared to MTC of up to 10%.

STC suffer from similar parasitic losses as MTC since the power turbine is also positioned within the main exhaust stream, however this may be offset with the addition of a waste-gate valve. PTC avoids this effect by placing the power turbine in parallel with the main exhaust stream. However, PTC yields a smaller efficiency gain than STC since only a portion of the exhaust gasses pass through the power turbine.

Since EATC was first described in [13], there has been much interest in the performance and control for this type of system. When the energy available for extraction in the exhaust is in excess of that required to drive the compressor, the generator extracts this excess energy. This system has the added advantage that at low engine speed, the generator can be used as a motor to power the compressor, increasing boost pressure and improving engine performance. This capability can also be used to reduce turbocharger lag, thereby improving engine transient performance too.

In a recent paper, Pasini *et. al.* describe a DETC system [12], which provides for additional degrees of freedom with which to control the engine. The performance of DETC was compared to EATC at steady-state conditions and found brake specific fuel consumption improvements of 8% and 6% respectively.

CONTROL ORIENTED ENGINE MODELLING

In order to use model based control techniques, as well as simulate the engine, a suitable mathematical model of the turbocompounded engine must be developed. Such a model must, with a reasonable degree of accuracy, predict the engine dynamics and include all relevant system inputs, internal states and outputs. For control applications, the predictions made using the model must be computed within a time frame significantly shorter than the corresponding system dynamics. This class of models are known as control oriented models. The development of a control oriented model begins with establishing a high order model followed by a model reduction process to systematically simplify the model while retaining sufficient accuracy.

There are many engine modelling techniques and each has its own accuracy vs. simplicity trade off. Mean value engine models (MVEMs) are widely used as a basis for control oriented models. These models describe the average engine behaviour over one or more engine cycles. A MVEM is composed as a network of basic submodels with each submodel representing a distinct engine component such as a manifold, turbocharger compressor or turbine, etc. For any type of component, for example an intake and exhaust manifold or length of pipe, we can use essentially the same submodel, tuned to be representative of that component's individual characteristics. This modelling approach enables the development of a set of submodels that can be arranged and calibrated to suit any engine.

The literature available describing various automotive MVEMs is extensive, including both spark and compression ignition engines. For example, refer [14]–[16]. MVEMs developed for marine engines are less common than for automotive applications. A fourth order MVEM was derived in [17] to aid in the calibration of the speed governor gains of a 2-stroke turbocharged marine engine. A comparison of two low order MVEMs for a 2-stroke marine diesel engine is presented in [18], and the same author uses these models to investigate the engine's transient dynamics in [19]. A 4-stroke marine engine model using a hybrid MVEM/zero-dimensional model in order to improve model accuracy from a MVEM alone is presented in [20].

With regards to turbocompounded engines, the majority of research found has been conducted with 1-D simulations using commercial software, and are not suitable as control oriented models due to their high computational burden. A MVEM of a heavy-duty vehicle diesel engine with hydraulically coupled MTC is presented in [21]. A MVEM for an EATC system for heavy duty engines has only recently been developed in [22].

In this research project, the prediction model used in the MPC controller will be derived from the developed high fidelity MVEM. This may be achieved using a systematic model reduction process which rigorously determines how to sufficiently simplify the high fidelity MVEM so that the MPC optimisation problem (described below) can be solved within a single sampling timestep while retaining an acceptable degree of model accuracy. This process has been used successfully for automotive engines [14].

MODEL PREDICTIVE CONTROL

Model predictive control (MPC) techniques encompass a family of advanced control schemes which share defining characteristics. A mathematical model of a system to be controlled is used to determine the optimal sequence of inputs (and corresponding system states and outputs) along a fixed prediction horizon, while ensuring that system constraints are satisfied. The control input is determined using constrained optimisation techniques to find the minimum of a predefined cost function, which encapsulates the control objective.

Although the conceptual basis for MPC had been recognised and developed throughout the mid 20th century, MPC was only first applied industrially in the late 1970's in the petrochemical refining industry. Since then, the popularity of MPC increased throughout the 1980's with a growing number of applications and companies providing MPC software and support. The computational power available during this period was sufficient for these applications due to their slow dynamics, however, remained a limiting factor in the uptake of MPC in other industries. Academic interest followed this initial uptake and first began to provide guarantees of stability for practical MPC systems since the late 1980's. Various MPC algorithms were in use during this time, and generally make use of quadratic cost functions penalising deviations of state, input and/or outputs from predetermined set points, similar to the tracking MPC formulation presented below [23].

During the start of the 21st century, and with significant and ongoing advancement in computational power, research focus has been directed towards applicability of MPC to systems with "fast" dynamics and fast sampling rates in the order of tens of milliseconds to seconds, for example, within automotive and aerospace applications.

In tracking MPC, the objective is to guide the system to operate at a pre-determined set point corresponding to input $u^*$and state x^*. The optimisation problem determines the optimal sequence of inputs along the prediction horizon, $\{u_0, u_1, \dots, u_N\}$, by finding the minimum of the cost function

$$\min \sum_{i=0}^{N} (x_i - x^*)^T Q(x_i - x^*) + (u_i - u^*)^T R(u_i - u^*), \tag{1}$$

given the constraints

$$x_0 = x(k), \tag{2}$$

$$x_{i+1} = f(x_i, u_i), \tag{3}$$

$$y_i = g(x_i, u_i), \tag{4}$$

$$y_{i,min} \leq y_i \leq y_{i,max}, \tag{5}$$

$$x_N = x^*. \tag{6}$$

u, x and y are the vectors describing the system inputs, states and outputs respectively. The subscript i denotes the i'th prediction along the prediction horizon of length N. The functional in the summation of (1) describes a multi-dimensional paraboloid whose shape is defined by coefficients in matrices Q and R, which are used to 'tune' the behaviour of the controller to achieve the desired system behaviour. This function has its unique minimum at the set point (u^*, x^*).

Constraint (2) is the current state at time instant k, (3) and (4) are discrete time prediction model for the system's states and outputs respectively, (5) includes minimum and maximum constraints on the systems outputs and (6) is known as a 'terminal constraint', which is required to guarantee the stability of the controlled system.

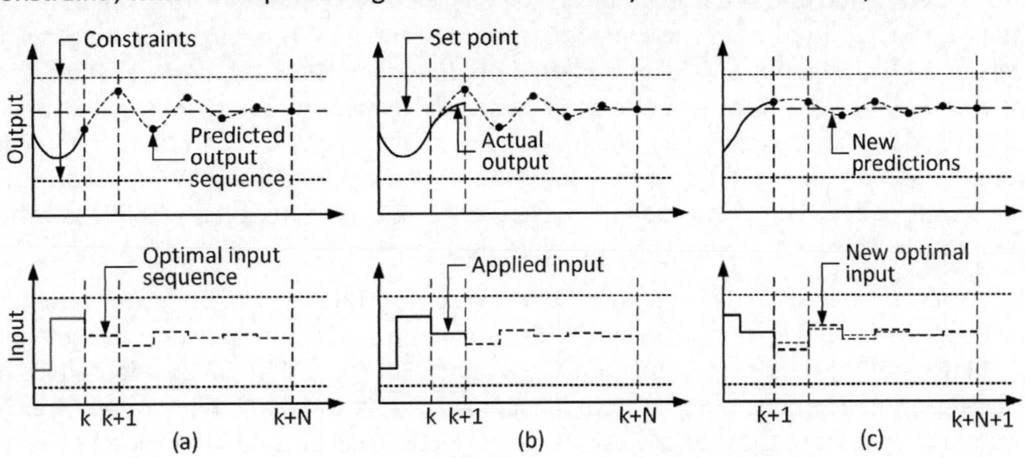

Figure 3: Graphical representation of MPC algorithm.

Once this optimisation problem is solved (Figure 3a), the first input in the sequence of inputs, u_0, is applied to the system (Figure 3b). Sensors are used to measure the states resulting from this input at the next sampling time, which is then used as the initial condition (2) in the optimisation problem for the next time step (Figure 3c).

The primary advantage of MPC over other control strategies is its framework is inherently suited to control multiple input, multiple output systems and ensure constraint satisfaction, a feature of MPC that is not shared by traditional PI control methodologies. Ensuring constraint satisfaction is of particular importance as many systems yield their optimal performance when operated at or near system constraints [24].

Given the advantages of MPC over traditional control methodologies described above, MPC is an ideal control methodology for engines with integrated WHR systems. This type of system is a multi-input, multi-output system, which can be easily incorporated into the MPC formulation using a state-space prediction model in (3) and (4). Constraints on, for example, engine speed and exhaust temperature, or minimum or maximum actuator positions, are also explicitly incorporated into the MPC formulation. Constraints may also include limits on engine emissions [8].

Furthermore, robust MPC formulations may be used to ensure that the control is robust to unmodelled disturbances, such as sensor noise and inaccuracies in the prediction model. Without robust MPC formulations, these disturbances may result in unacceptable constraint violations. One approach is to use 'constraint tightening'. First proposed in [25], constraint tightening reserves some control capability to be used to counteract the effect of disturbances by progressively tightening the constraints along the prediction horizon.

In the context of turbocompounded engines, the manipulated inputs include the volume of fuel injected into each cylinder for each cycle, the load on the engine's driveshaft, the commanded position of the wastegate valve and the turbogenerator speed set point. The outputs may include engine power output, fuel consumption and other variables which are required to be constrained, such as actuator limits or maximum engine speed.

UPCOMING & CURRENT RESEARCH

Experimental test bed

The experimental test bed used in this research includes a Hedemora V6 submarine diesel engine test facility, located at HMAS Cerberus (Figure 4). This engine is a six-cylinder version of the Hedemora V18 engines used on the Collins Class submarines. The engine's driveshaft is coupled to a variable fill hydraulic dynamometer. The engine speed is controlled by a speed governor, which maintains engine speed by controlling the fuel flow rate. The engine is fitted out with a full suite of sensors, and LabVIEW is used for high level control and data acquisition. A buttery valve has been installed within the exhaust pipe to simulate wave induced backpressure fluctuations [26].

In order to improve the engine's fuel efficiency, DST is exploring the use of a turbocompounding system installed on the engine in a PTC configuration as a means of WHR (Figure 5). This system is provided by Bluebox Energy and has been adapted from an electrical turbocompounding system developed by Bowman Power Group. This ETC system is comprised of a radial flow turbine directly coupled to a high speed generator, along with suitable power electronics to convert the generated power to an output of 400VAC, 3-phase, 50 Hz / 60Hz and associated cooling and safety systems. The cooling system consists of a coolant oil loop servicing the generator, oil/water plate heat exchanger and a water loop with an air cooled radiator. The turbogenerator speed is controlled by a proprietary PI control system which regulates the load applied to the generator. The wastegate valve is an 80NB Somas butterfly valve complete with pneumatic actuator.

Figure 4: Hedemora V6 test facility

Figure 5: Bluebox Energy turbogenerator

Experimental plan

A preliminary experimental phase where the turbogenerator will be commissioned and tested will be conducted prior to integration with the Hedemora V6 engine (Figure 6Figure 4). This work will be conducted in the new DST test facility located at DST's Fisherman's Bend site. An adjacent air processing plant will be used to deliver high pressure, high temperature air representative of engine exhaust to the test facility. A portion of the air will be directed either through the turbogenerator via an actuated wastegate valve and the remainder through an actuated bypass valve, which will be used for hardware-in-the-loop simulation of the engine. A third actuated valve located in the exhaust will be used for hardware-in-the-loop simulation of wave induced back pressure fluctuations.

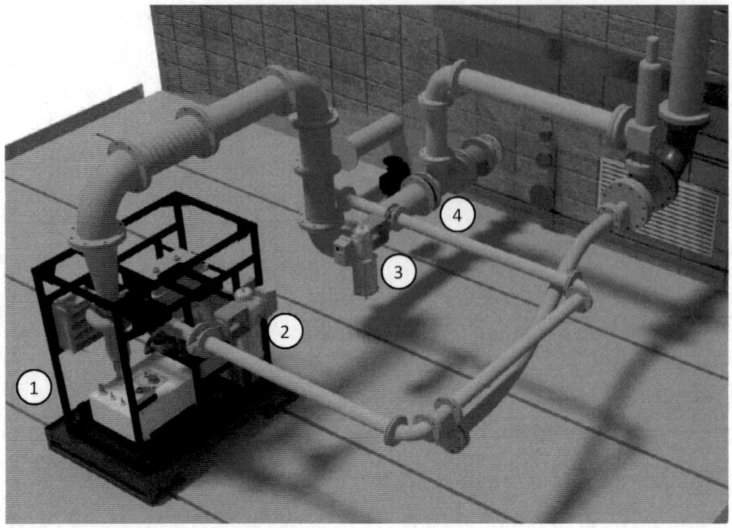

Figure 6: Pipework and turbogenerator arrangement in the DST test cell. (1) Turbogenerator, (2) wastegate valve, (3) bypass valve and (4) back pressure valve

A secondary experimental phase entails integrating the turbogenerator onto the existing engine test bed, followed by subsequent experiments quantifying the benefit of the turbocompounding system, as well as the efficacy of the developed MPC controller.

Development of MPC formulation

Broomhead *et. al.* [27] developed an advanced robust MPC formulation for the control of systems which are best operated in a periodic fashion. A submarine engine is one such example since the periodic back pressure fluctuations results in periodic actuation of the engine's inputs. This MPC formulation requires knowledge of the wave period. The formulation may also be used for a wide variety of nonperiodic systems, such as a surface vessel engine, by adopting a period equal to a single controller time step.

The formulation in [27] made use of linear time invariant prediction models of the form

$$x_{i+1} = Ax_i + Bu_i + B_d d_i + w \tag{7}$$

where d_i is the known modelled periodic disturbance (such as the wave induced back pressure), w includes the unmodelled disturbance and A, B and B_d are the system matrices. Such a model provides sufficient predictive accuracy for an engine operated close to a fixed set point, such as a submarine engine. However, it is unlikely to be suitable for surface vessel applications, which as previously described, requires the engine to operate over a large range of set points. For these applications, linear time varying (LTV) or nonlinear prediction models may be required.

One aim of this research project is to further develop the controller presented in [27] to include LTV or nonlinear prediction models to extend the application to surface vessel engines, as well as make it robust to changes in wave period, which is required for successful operation in real-world sea conditions.

CONCLUSION

The objective of this project is to improve the energy efficiency of maritime platforms using turbocompounding for waste heat recovery, and in doing so, provide advantageous environmental, economic and strategic outcomes. Strategic benefits include maximizing vessel range and endurance, the possibility to reduce fuel tank and engine size, reduced thermal signature, and for submarines, improved indiscretion ratio. Using MPC as a control solution for turbocompounded marine engines maximises the energy efficiency gains, while ensuring the turbocompounded engine is operated within its specified limits to improve its reliability and longevity. The application of these technologies to military applications is novel and challenging given the unique operational requirements of both surface and submarine vessels.

REFERENCES

[1] International Maritime Organization, *Revised MARPOL Annex VI: Regulations for the Prevention of Air Pollution from Ships*, no. October. 2008.

[2] M. Schwartz, K. Blakeley, and O. R. Rourke, "Department of Defense Energy initiatives : Background and issues for congress," 2012.

[3] H. M. A. Schleijpen and F. P. Neele, "Ship exhaust gas plume cooling," in *Proc. SPIE, Targets and Backgrounds X: Characterization and Representation*, 2004, vol. 5431, pp. 66–76.

[4] D. V. Singh and E. Pedersen, "A review of waste heat recovery technologies for maritime applications," *Energy Convers. Manag.*, vol. 111, pp. 315–328, 2016.

[5] P. Hield, "The effect of back pressure on the operation of a diesel engine," Australian Government Department of Defence - Defence Science and Technology Organisation, Technical Report DSTO-TR-2531, 2011.

[6] MAN Diesel & Turbo, "Waste Heat Recovery System (WHRS) for Reduction of Fuel Consumption, Emissions and EEDI"

[7] T. J. Broomhead, C. Manzie, P. Hield, and M. J. Brear, "An experimental investigation of additional actuators on a submarine diesel generator," *Control Eng. Pract.*, vol. 55, pp. 26–37, Oct. 2016.

[8] T. J. Broomhead, C. Manzie, P. Hield, R. C. Shekhar, and M. J. Brear, "Economic model predictive control and applications for diesel generators," *IEEE Trans. Control Syst. Technol.*, vol. 25, no. 2, pp. 388–400, 2016.

[9] H. Aghaali and H.-E. Ångström, "A review of turbocompounding as a waste heat recovery system for internal combustion engines," *Renew. Sustain. Energy Rev.*, vol. 49, pp. 813–824, 2015.

[10] A. T. C. Patterson, R. J. Tett, and J. McGuire, "Exhaust heat recovery using electro-turbogenerators," *SAE Int.*, vol. 2009-01–16, 2009.

[11] D. T. Hountalas, C. Katsanos, and V. T. Lamaris, "Recovering energy from the diesel engine exhaust using mechanical and electrical turbocompounding," *SAE Tech. Pap. Ser.*, vol. 2007-01–15, 2007.

[12] G. Pasini, S. Frigo, and M. Antonelli, "Electrical turbo compounding applied to a CI engine: A numerical evaluation of different layouts," in *Proceedings of the ASME 2016 Internal Combustion Engine Fall Technical Conference*, 2016.

[13] U. Hopmann and M. C. Algrain, "Diesel engine electric turbo compound technology," *SAE Int.*, vol. 2003-01–22, 2003.

[14] T. J. Broomhead, C. Manzie, M. J. Brear, and P. Hield, "Model reduction of diesel mean value engine models," *SAE Int.*, vol. 2015-01–12, 2015.

[15] L. Eriksson and L. Nielsen, *Modeling and Control of Engine and Drivelines*, 1st ed. Wiley, 2014.

[16] A. Gambarotta, G. Lucchetti, and I. Vaja, "Real-time modelling of transient operation of turbocharged diesel engines," *Proc. Inst. Mech. Eng. Part D J.*

Automob. Eng., vol. 225, no. 9, pp. 1186–1203, 2011.
[17] D. N. Malkhede, B. Seth, and H. C. Dhariwal, "Mean value model and control of a marine turbocharged diesel engine," *SAE Tech. Pap.*, vol. 2005-01–38, 2005.
[18] G. P. Theotokatos, "A comparative study on the cycle mean value modelling of a large two-stroke marine diesel engine," in *Proceedings of the 2nd International Conference on Maritime and Naval Science and Engineering*, 2009, pp. 107–112.
[19] G. P. Theotokatos, "Ship Propulsion Plant Transient Response Investigation using a Mean Value Engine Model," *Int. J. Energy*, vol. 2, no. 4, pp. 66–74, 2008.
[20] F. Baldi, G. Theotokatos, and K. Andersson, "Development of a combined mean value-zero dimensional model and application for a large marine four-stroke diesel engine simulation," *Appl. Energy*, vol. 154, pp. 402–415, 2015.
[21] O. Flärdh and M. Gustafson, "Mean Value Modelling of a Diesel Engine with Turbo Compound," Linköping University, Linköping, Sweden, 2004.
[22] D. Zhao, E. Winward, Z. Yang, J. Rutledge, and R. Stobart, "Control-oriented dynamics analysis for electrified turbocharged diesel engines," *SAE Tech. Pap.*, vol. 2016-01–06, 2016.
[23] S. J. Qin and T. A. Badgwell, "A survey of industrial model predictive control technology," *Control Eng. Pract.*, vol. 11, no. 7, pp. 733–764, 2003.
[24] M. L. Darby, M. Harmse, and M. Nikolaou, "MPC: Current Practice and Challenges," *IFAC Proc. Vol.*, vol. 42, no. 11, pp. 86–98, 2009.
[25] J. R. Gossner, B. Kouvaritakis, and J. A. Rossiter, "Stable generalized predictive control with constraints and bounded disturbances," *Automatica*, vol. 33, no. 4, pp. 551–568, 1997.
[26] O. Tregenza and P. Hield, "Diesel Engine Back Pressure Controller," in *Pacific 2012, International Maritime Conference*, 2012.
[27] T. J. Broomhead, C. Manzie, R. C. Shekhar, and P. Hield, "Robust periodic economic MPC for linear systems," *Automatica*, vol. 60, pp. 30–37, 2015.

Advanced Coating Solutions for the Marine Industry

Andrew S.M. ANG[1,5], Hugo HOWSE[2], Matthew LEIGH[3], Richard PIOLA[4], Wayne NEIL[4], Christopher C. BERNDT[1,5], and Scott A WADE[1,5]

[1] Swinburne University of Technology, Melbourne, Australia.
[2] United Surface Technologies Pty Ltd, Melbourne, Australia
[3]. MacTaggart Scott Australia Pty Ltd, South Australia, Australia.
[4] Defence Science and Technology Group, Melbourne, Australia.
[5] Defence Materials Technology Centre (DMTC), Melbourne, Australia
Email: aang@swin.edu.au

ABSTRACT

Thermal spray technology is an additive surface engineering solution that offers benefits for the marine industry, such as improving corrosion performance, and also addresses the environmental protection agencies regulations of lowering VOC emissions. The potential for anti-corrosive solutions using the thermal spraying process are widely recognised in the shipping and marine industry. Steel super structures and vessels, for example, have been coated with zinc or aluminium based metal composites to help minimise corrosion. In the first part of this paper we will provide an overview of thermal spray coatings that have been used in the marine industry and discuss other potential applications. The authors have undertaken investigations into thermal spray technology to provide functional coatings for marine hydraulic applications. This has included studies of the corrosion and biofouling performance of the latest generation High Velocity Oxygen Fuel (HVOF) high quality carbide based coatings that can be applied to hydraulic components subjected in immersed marine environment. The results of this work, including field trials at sites around Australia, will also be presented.

INTRODUCTION

Thermal spray technology is a generic term for a group of coating processes that involve the acceleration, impact and deposition of molten, semi-molten or solid particles onto a substrate. These particles produce morphological features that are commonly known as "splats" [1]. A coating can be generated if the accelerated droplets or particles can (i) plastically deform or rapidly cool into thin lamellae on impact, (ii) adhere to the surface, and (iii) overlap and interlock into a consolidated coating during the solidification process (see Figure 1). The coating is usually generated in multiple passes and can range in thickness from 50 µm to a few millimetres as required. The coating material may be metallic; ceramic oxide or carbide; polymeric, or composite, in the form of powder, wire or rod [1, 2].

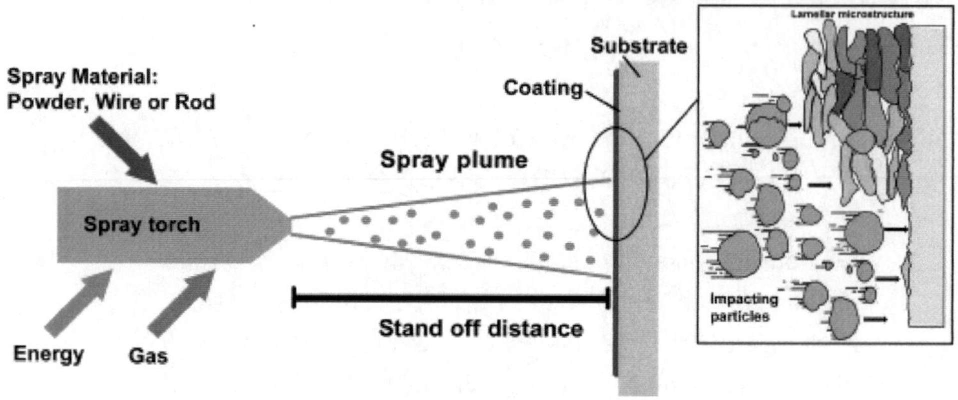

Figure 1: Schematic of the thermal spray coating process

COMPARISION OF THERMAL SPRAY WITH OTHER COATING PROCESSES

A distinguishing feature of thermal spray methods is that coating materials can be applied onto substrates that are unsuited for welding because of their composition or tendency to distort; as well as an ability to deposit thin layers that are not possible for welding processes. Hence, repair procedures can be applied to restore heat sensitive components to their original dimensions. These coatings can be turned or ground to finished size and often outlast a brand new part.

Thermal spray processes also generally do not face restrictive standards and laws on environmental pollution and health hazards; in contrast the electro-deposition or plating process has attracted increasing concerns over the use of hazardous substances [3]. For instance, the acid plating bath used in hard chrome plating contains Cr^{6+} ions (from chromic acid), which is believed to be primarily responsible for the increased lung cancer rates observed in workers who were exposed to high levels of chromium in workroom air. A broad restrictions of compliance with increasingly stringent regulations must be addressed. Thus, spent plating solutions must be removed of all toxic materials by precipitation and evaporation before discharging the water to the sewer system. This is an expensive and technically onerous process..

The versatility of the thermal spray process is also a key advantage over other additive surface engineering solutions. For example, thermal spray does not require a vacuum chamber infrastructure that is mandatory for the chemical vapour deposition (CVD) or plasma vapour deposition (PVD) techniques. Hence, thermal spray technology can be applied in the field in a cost effective manner. A wide range of metal alloys, ceramics and composites can be applied to impart functional properties such as corrosion and wear performance to the component. Table 1~~Table 1~~ provides a summary comparison of the various coating processes available for an engineer to apply additive surface coatings.

Table 1: Comparison of characteristics of coating process technologies

Characteristics	Thermal spray	Plating	CVD	PVD	Hardfacing/ weld overlay
Operating cost	Low to moderate	Low	Moderate	Moderate to high	Moderate
Equipment cost	Low to moderate	Low	Low to moderate	Low to moderate	Low
Process environment	Atmospheric to soft vacuum	Aqueous solution	Atmospheric to medium vacuum	Hard vacuum	Atmospheric
Coating geometry	Line-of-sight	Omni-directional	Omni-directional	Line-of-sight	Line-of-sight
Coating thickness	50 μm to cm	10 μm to mm	0.1 μm to mm	0.1 μm to 20 μm	100 μm to cm
Substrate temperature	Low to moderate	Low	Moderate to high	Low	Moderate to high
Adherence	Good mechanical	Moderate mechanical, very good chemical	Very good chemical, excellent diffusional	Moderate mechanical, good chemical	Very good diffusional
Surface finish	Coarse (R_z up to 200 μm) but can be superfinished	Coarse to glossy (R_z up to 10 μm)	Smooth to glossy (R_z up to 5 μm)	Smooth to high gloss (R_z up to 5 μm)	Course to rough but can be superfinished
Coating materials	Polymers, metals, ceramics and composites	Metals	Metals, ceramics, polymers	Metals, ceramics, polymers	Metals

STANDARDS FOR APPLICATION OF THERMAL SPRAY COATINGS

Thermal spray coatings for the marine industry are a widely accepted solution for corrosion protection, particularly the galvanic (sacrificial) cathodic protection of metal structures as well as for the dimensional reclamation of engineering components. Common coating materials applied are aluminium, zinc, and their alloys via the thermal spray process variants such as flame wire and twin wire arc spraying (which will be described in detail in a later section). In fact, these processes of thermal spraying can be readily adopted since they are accepted within the various standards and codes found in Table 2~~Table 2~~.

Table 2: Various international and national standards for application of thermal spray coatings

Standard	Title	Details
BS ISO 14921	Procedures for the application of thermally sprayed coatings for engineering components	Specifies the general procedure when a thermally sprayed coating is applied to enhance the surface properties of a component or to reclaim worn and non-conforming parts.
BS ISO 2063	Thermal spraying - Metallic and other inorganic coatings - Zinc, aluminium and their alloys	Describes the characteristic properties and gives test methods for coatings obtained by the spraying of zinc and aluminium and their alloys for the general purpose of corrosion protection of iron and steel.
MOD UK – DEF STAN 02-828	Requirements for thermal spray deposition of metals and ceramics for engineering purposes	Defines the requirements for thermal spray deposition of coatings using automatic or semi-automatic methods onto engineering components in order to enhance the surface properties or reclaim worn, damaged and non-conforming parts.
NACE No. 12/AWS C2.23M/SSPC CS 23	Specification for the application of thermal spray coatings (metallizing) of aluminium, zinc, and their alloys and composites for the corrosion protection of steel	Requirements for surface preparation, coating application, repairing coating defects, measurement of coating thickness, adhesion testing of the coating, and application of sealers and topcoats over the thermally sprayed metal coating for the preservation and maintenance of steel structures and components.
AWS C2.18	Guide for the protection of steel with thermal sprayed coatings of aluminium and zinc and their alloys and composites	Covers job description, safety, consumable materials, surface-preparation and thermal spray equipment, quality control equipment, operator training and qualification, a step-by-step surface preparation and thermal spraying application method with quality control checkpoints, maintenance and repair of thermal spray coatings, and a job control record.
MIL-STD-1687A (SH)	Thermal spray processes for naval ship machinery applications.	Covers thermal-spray processes for machinery element repair. Includes requirements for the qualification of thermal spray procedures and operators, requirements and guidance for use of thermal spray material and equipment, quality assurance requirements, and descriptions of applicable qualification tests.
MIL-STD-2138A (SH), (withdrawn with no replacement)	Metal sprayed coatings for corrosion protection aboard naval ships.	Covers the requirements for the use of metal-sprayed coatings for corrosion control applications on board naval ships. Includes certification of facilities, certification of operators, application procedures, metal-spray procedures, and production quality assurance.

BACKGROUND OF TWIN WIRE ARC AND HIGH VELOCITY OXYGEN FUEL PROCESSES

The wire arc spray (TWA) process originated from a patent of M.U. Schoop in 1910. It is a reliable, economical and commonly used method for depositing metal coatings for corrosion protection applications. The principal technical details are simple: a direct current electric arc is formed between two electrically charged consumable electrode wires that converge at a point and a high velocity gas jet propels the molten material toward the substrate (see Figure 2). The temperature within the arc jet (~7,000 K) is sufficient to melt the consumable electrodes, which are atomised into molten particles and accelerated onto the substrate. Thus, the thermal efficiency of the wire arc spray is greater than other thermal spray processes because the particles are in a fully molten state when they enter the spray jet. The molten particles start cooling down immediately after leaving the arc zone; therefore, coating properties such as porosity and adhesion can be affected [4]. Deposition efficiencies of up to 99.5% have been reported but more typical values lie in the range of 58% due to practical application limitations. Wire arc spraying is not limited to metallic wires; for instance, cored wires consisting of hard powdered materials such as carbides or amorphous alloys that are contained within a metal based sheath can be used.

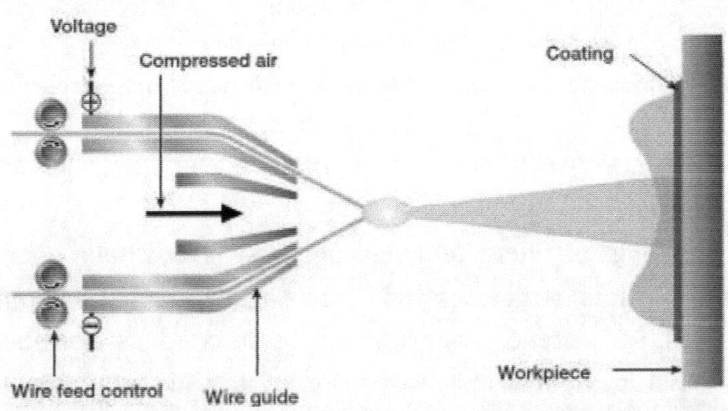

Figure 2: Operating principal of TWA torch (courtesy of Oerlikon Metco).

The High Velocity Oxygen Fuel (HVOF) process uses high kinetic energy and controlled thermal output to produce dense, low porosity coatings that exhibit high bond strengths, low oxides and relatively low roughness as-sprayed finishes [5, 6]. The HVOF method represents the current state-of-the-art in thermal spray, and is capable of producing very high quality coatings. The operating characteristics are based on a continuous combustion of the fuel gas mixture supplied at high volumetric flow to the burning chamber of the torch. Process gas options include propylene, propane, hydrogen or natural gas used in gas-fuelled systems and kerosene in liquid-fuelled systems [7], (see Figure 3).

Depending on the fuel gas used and torch design, the combustion jet temperatures reach ~4,000 K and the deposition energies are in the order of 288 to 1,000 MJ/ hour (80 to 278 kW). After passing the nozzle, the gas velocities are in excess of 1,800 m/s, resulting in the formation of a 'shock' diamond pattern within the exiting flame jet. The feedstock, in powdered form, is usually axially fed through the torch using an inert carrier gas such as nitrogen. The ignited gases surround and heat the powdered spray material as it exits the torch and particle velocities may reach 800 m/s [8, 9]. HVOF coatings often exhibit densities of >95% and adhesion values greater than 10,000 psi (69 MPa) [1, 10].

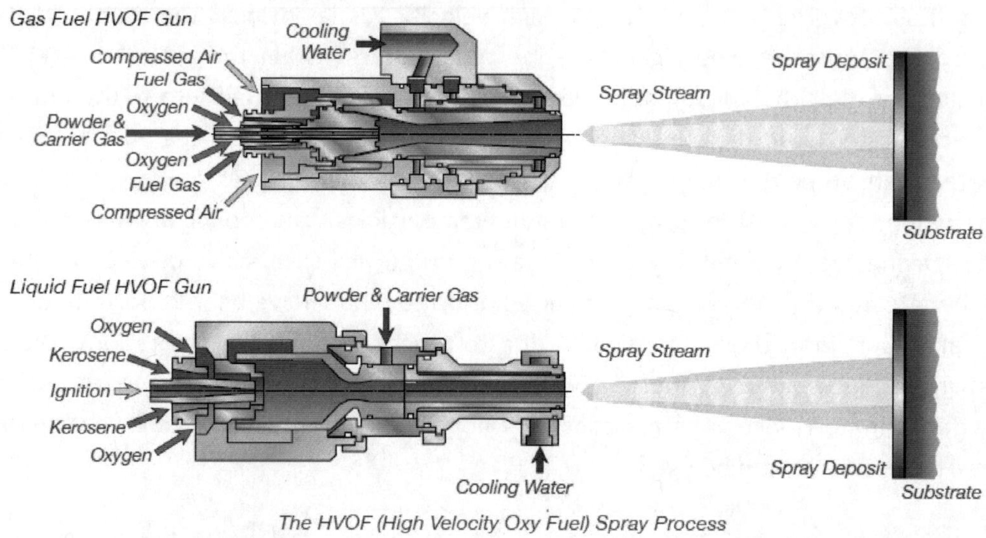

Figure 3: Operating principal of gas fuel and liquid fuel HVOF spray torches (courtesy of Oerlikon Metco).

THERMAL SPRAY COATINGS FOR CATHODIC PROTECTION OF METAL STRUCTURE AGAINST CORROSION

Thermal spray coatings of zinc, aluminium and their alloys have been used to protect a diverse range of surfaces, structures and vessels against corrosion applications extending back to the 1930s. For instance, sacrificial corrosion coatings were been deposited onto the St. Denis canal lock gates in France (zinc coated in the 1930s) and the locks and navigation gates of the Panama Canal (both zinc and aluminium coated) [11]. At Japan's Haneda airport in Tokyo Bay, steel pipe piles as well as the bridge like structure for the guiding lights are protected by thermally sprayed aluminium coatings [12]. In New Zealand, the Auckland Harbour Bridge had a thermally sprayed zinc coating with three top coats of phenolic paint [13] when it was constructed in 1956. Tiong et al [14] has claimed that oil and gas operators have reduced the corrosion allowance by 3 mm on a number of its offshore structures by adopting thermal spray aluminium. These coatings when properly applied and with the use of specific sealer systems, will provide a service life in excess of 30 years with zero maintenance required.

Coatings of zinc, aluminium and their alloys are typically deposited by two thermal spray process variants: Flame wire spraying or twin wire arc spraying. The latter is often the preferred coating method due to advantages including superior coating adhesion, greater deposition efficiency, higher spray rate and reduced application costs.

The typical TWA system has been designed for high-throughput coatings. The arc wires are typically driven by a gearbox arrangement with an air or electric motor. Application rates of up to 52 kg/hour of zinc or 12 kg/hour of aluminium can be achieved, which makes the process ideal for coating large areas such as boat hulls, offshore oil platforms and other large steel structures. The coating finish is typically very fine, about 10-15 µm RM and is often used in conjunction with a paint top coating or sealant system. Similar to TWA, the flame spray wire system also consists of a drive unit with a motor and drive rollers for feeding the wire. The oxygen-acetylene gas valves and air cap control the combustion temperature and atomization, respectively. Compared with TWA, wire flame spraying is generally slower (e.g. application rate 10-20 kg/hour of zinc) and costly due to the higher cost of the oxygen-fuel gas mixture compared to the cost of electricity. One advantage, however, is that these systems are field portable and may be used to apply quality metal coatings in difficult to reach places.

In the marine environment where high chloride levels are present, aluminium and its alloys (i.e. Al-5Mg), are used for coatings, Table 3. Application areas include seawater immersion and splash zones and when applied properly, have the ability to replace paint systems because of their predictable service life, increased effectiveness, and lower life-cycle costs. Seal coats and paint topcoats may be used to add a further degree of protection to the thermal spray coating systems used in seawater immersion. Marine corrosion tests of thermal sprayed zinc, aluminium and Zn-Al coatings have been performed by the Japan Association of Corrosion Control (JACC) since 1985. It was found after a period of 18 years that a TWA aluminium coating of 175 µm with a sealant system as well as the aluminium flame sprayed, primer treated and 3 mm thick urethane resin system were effective under most marine exposure conditions; including the aggressive tidal, splash and sea air zones [12].

Table 3: Minimum coating thickness recommended for TWA aluminium and Al-5Mg in accordance to ISO 2063

Environment	Thickness loss of low-carbon steel after first year of exposure	Thickness (µm)			
		Aluminium		Al-5Mg	
		Unsealed	Sealed	Unsealed	Sealed
Saltwater immersion	> 150 µm	200	150	250	200
Marine atmosphere	80-200 µm	200	100	250	200

THERMAL SPRAY NON-SKID COATINGS FOR SHIP DECKS

Aluminium TWA coatings have been used successfully on metal substrates to prevent corrosion and impart non-skid properties. Historically, these maritime assets that require high friction surfaces used polymer based coating technology. Polymer coatings, however, offer limited service life/readiness under conditions of high traffic, wear and impact. The high friction polymer coating surfaces are created with an aggregate protruding from a base coat of resin to prevent slips by pedestrians and the increase of traction for machinery and vehicular traffic. However, these coating systems do not offer any galvanic protection to the substrate and if the coating spalls, the area is susceptible to corrosion. Moreover, with the potential high temperature imposed by modern aircraft at take-off areas, the low thermal stability of resin systems hinders the mission durability. The high friction surfaces are required in areas such as aircraft hangar bays, flight decks, frigate decks and ship ramps.

The superior hardness and roughness of aluminium-based TWA coatings confers their selection for non-skid applications. The non-skid coating system is achieved by first applying aluminium with the thermal spray system. An additional spray pass of aluminium is then applied using reduced atomization air pressure. The lower air pressure allows for the deposition of larger spray particles that produce a rougher surface. Non-skid coatings should be sealed with thin film epoxies. Non-skid ceramic-metal coatings have been successfully applied on to U.S. navy USS Wasp surface ship [15]. The coating which is based on TWA spraying of ceramic and aluminium produces a surface more resistant to extreme heat and has improved endurance to the highly aggressive wear and tear of flight operations.

HIGH VELOCITY OXYGEN FUEL COATINGS FOR MARINE HYDRAULIC COMPONENT RESTORATION

Maritime hydraulic machine components, especially exposed sections of shafts, splines and valves, see regular wetting and drying or full immersion of seawater. This produces problems with biofouling and corrosion on these critical components and often degrades operational performance. A review of the shipbuilding and ship repair industry found that an engineered coating layer that offers high wear resistance as well as biofouling and corrosion protection is the preferred engineering solution for the aforementioned components. Many surface engineering techniques exist, but the thermal spray process variant of high velocity oxygen fuel (HVOF) has significant advantages over other coating solutions such as: (1) electroplated hard chrome since the inherent microcracks can cause corrosion of the base material and (2) plasma spray coatings are usually composed of a ceramic material that exhibit high porosity and requires a sealant. These two coating options (i.e. electroplated hard chrome and plasma spray coatings) are also generally brittle and have lower bond strength compared to HVOF coatings. A third coating option is: (3) plasma transfer arc or laser cladding; however the formation of large brittle, diffusion zones, which presents the risk of part distortion due to high heat flux cause concern.

HVOF coatings are employed by the United States Naval Shipyards, United Kingdom Naval shipyards as well as applied commercially by marine hydraulic companies such as Rexroth

Bosch, Hunger Hydraulic Group, Nurmi Cylinders Oy and the STI Group. HVOF coatings are also offered as standard coatings on many high-performance applications include coatings of metallic materials applied on pump shaft bearing journals, severe service hydraulic cylinders, turbine rotor shaft labyrinth seal areas, impeller fit areas and oil fit areas.

Recent developments of HVOF spray systems [2] have aimed predominantly at optimising the combustion, minimising fuel consumption, and increasing the particle acceleration. This has led to the availability of liquid fuelled HVOF systems that impart more kinetic energy to the injected particles compared to gas fuelled HVOF systems [16]. Consequently, the coatings have reduced interconnected porosity and metal binder oxidation. Kerosene fuelled HVOF systems also offer high spray rates, system robustness and lower running cost, all of which are of commercial advantage to a manufacturing applicator, see Figure 4~~Figure 4~~. As such, HVOF has become the method of choice to replace electrolytic hard chrome (EHC) plating on components for commercial and military aircraft [17].

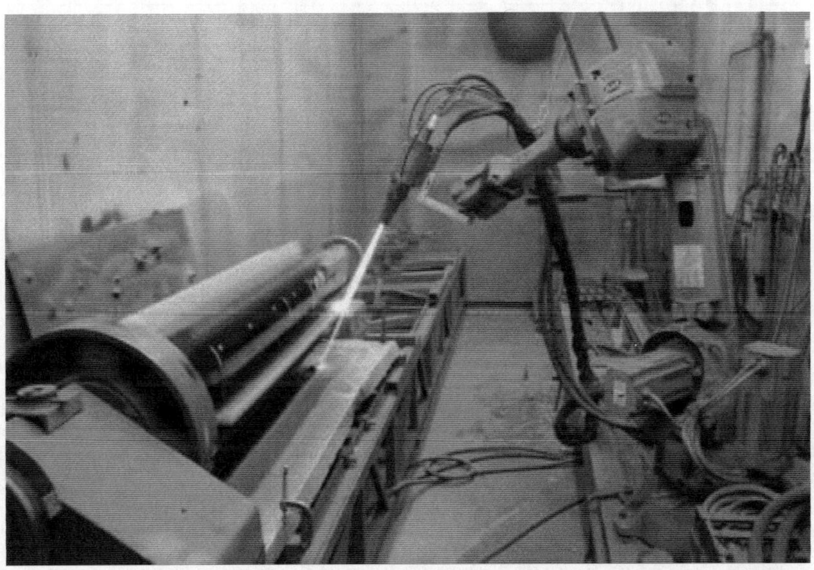

Figure 4: Kerosene fuelled HVOF system used for refurbishment of hydraulic components.

A project using HVOF coatings for the refurbishment of damaged marine hydraulic components is currently being undertaken by the authors. The potential productivity improvement and/ or time savings achieved by refurbishment over replacement with a brand new part is a key factor in this work as can be seen in Table 4~~Table 4~~. Data in Table 4 has been compiled from relevant Australian companies that operate within the marine hydraulics industry. Refurbishment can save up to 13 weeks in turnaround and generally is a fraction of the cost of manufacturing a brand new component. In our work, we also reported single layer HVOF carbide based coatings with corrosion resistant binders that have dense, low porosity microstructures (< 5%), high bond strengths and low oxide content_[18, 19]. The novelty of adopting a single coating layer, instead of duplex layers offers further potential for a reduction in repair time and cost.

Table 4: Comparison of typical time required to manufacture a hydraulic shaft versus refurbish the shaft. Note: Data is surveyed from relevant Australian companies that operate within the marine hydraulics industry.

	Newly Manufactured Cylinder	**Hydraulic Cylinder Shaft 40 mm OD x 2 metres long**
No	**Process**	**Estimated time required**
1	Strip and survey	5 days (can be 2-5 days)
2	Identify scope of work	1 week (quotes, POs to all new suppliers takes additional time)
3	Procurement of material	10 weeks (can be 4-16 weeks, depends on what is purchased)
4	Machining of material	12 weeks (can be 12-20 weeks)
5	Assembly of equipment	1 week
6	Testing and inspection	1-2 weeks
	TOTAL:	Minimum 20 weeks

	Refurbished Cylinder	**Hydraulic Cylinder Shaft 40 mm OD x 2 metres long**
No	**Process**	**Estimated time required**
1	Strip and Survey	5 days (can be 2-5 days)
2	Identify scope of work	2-3 days (less as sourcing new material/suppliers takes more time)
3	Send out items for refurbishment	3 weeks (can be 2-4 weeks, depends on workload)
4	Assembly of equipment	1 week
5	Testing and inspection	1-2 weeks
	TOTAL:	Minimum 7 weeks

Current work by the authors on HVOF coatings is focused on field testing of the corrosion and biofouling properties; whilst also looking at mechanical properties and evaluating the coatings' performance when used in hydraulic applications after an extended immersion period (see Figure 5~~Figure 5~~). More than 100 coated samples have been distributed among~~Test samples have been deployed at~~ various marine sites around the Australian coastline to measure rates of biofouling accumulation and corrosion of the HVOF coatings. These test sites were Williamstown in Victoria, Cairns in Queensland and Port Adelaide in South Australia to represent a wide variety of environmental conditions that the coatings could be exposed to. Samples were deployed in late 2015, with regular inspections being undertaken until the final removal of samples in late 2016 (see Figure 6).

~~The results of twelve month field tests validated two potential HVOF carbide-based coatings that could be used for marine hydraulic applications, particularly when the shaft material is Monel® K500. For these coatings (WC-NiCr or WC-Hastealloy® HVOF) there~~

~~were no delamination failures. The biofouling rate was also approximately 2-3 times less than those of the control coatings (plasma spray alumina-titania) exposed to the same test conditions. Furthermore, a modified ASTM C633 coating adhesion test also indicated no detrimental loss of coating adhesion strength following 12 month immersion. Surface metrology results also reveal that the surface roughness of these HVOF coatings did not vary significantly over the field testing trials and is within the specification requirements of a functional hydraulic shaft.~~

Figure 5: HVOF coated samples undergoing field immersion testing (left), cleaned HVOF samples after field testing (right)

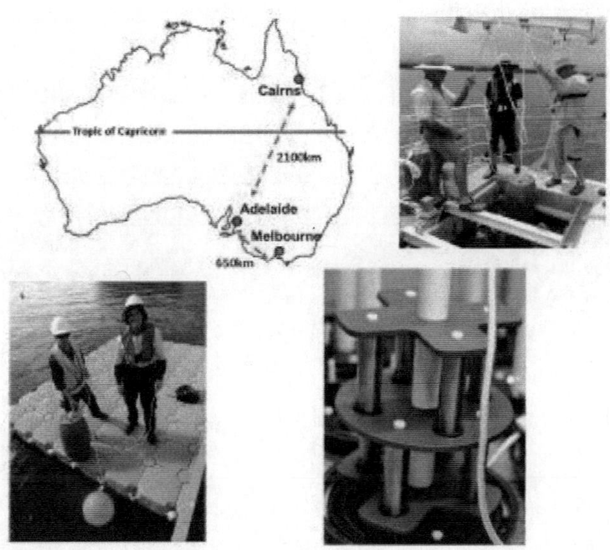

Figure 6: Details of field testing locations and deployment photographs

The results of 12 month field tests validated two potential HVOF carbide-based coatings that could be used for marine hydraulic applications, particularly when the shaft material is Monel® K500. For these HVOF coatings (WC-NiCr or WC-Hastealloy®), there were no coating delamination failures and biofouling products could be removed from the coatings. From both coating's cross sectioned microstructure, the substrate-coating interface, as well as the free edge of the coating remained unaffected by the seawater immersion or the biofouling attachment (see Figure 7).

Figure 7: (A) Substrate-coating interface for WC-Hastealloy® and (B) Free-edge of WC-NiCr coatings after 12 months immersion in seawater field testing.

The biofouling rate of WC-NiCr or WC-Hastealloy® coatings, were approximately 2-3 times less than those of the control coatings (plasma spray alumina-titania) exposed to the same test conditions over a 4 month period in Melbourne site and more than 3 times less than the biofouling rate in Cairns and Adelaide, see Figure 8. The mean total fouling coverage (soft and hard fouling) on samples across all sites ranged from 1.8±0.9% (WC-Hastealloy®, Adelaide) to 86.8±5.8% (alumina-titania, Cairns). Overall, the control alumina-titania samples recorded significantly greater levels of fouling at all locations relative to HVOF coated samples (ANOVA, $F(3,84) = 9.744$, $p = 0.010$; Tukey's $p < 0.001$).The fouling at Cairns was greater than for Melbourne and Adelaide (ANOVA, $F(2,84) = 10.045$, $p = 0.012$; Figure 8). This implies that the nickel based HVOF coated components studied in the trial will require less frequent cleaning in operation situations compared to components using the current coating materials.

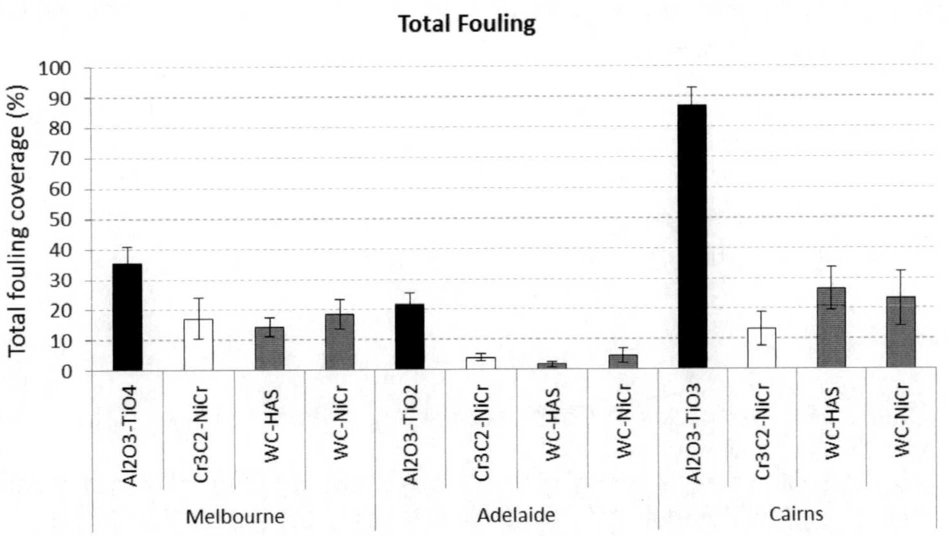

Figure 8: Total biofouling coverage across 3 field test sites for 4 months.

Tensile adhesion testing to measure the coating adhesion strength found that the HVOF coatings consistently had bond strengths at least 2.5 times greater that of the control plasma sprayed coating, which fails at approximately 25 MPa. The coating adhesion also did not deteriorate due to the field immersion. A modified ASTM C633 coating adhesion test indicated no detrimental loss of coating adhesion strength of HVOF test samples following 12 months immersion. This is complimented with the high surface hardness of greater than 1000HV for the HVOF coatings, which allows a high scraping force to be applied to clean the HVOF coating surfaces without concerns of surface damage. Surface metrology results also reveal that the surface roughness of these HVOF coatings did not vary significantly over the duration of the field trials and is within the specification requirements of a functional hydraulic shaft. Overall, these HVOF coatings have great potential to provide improved performance compared to the current coating technology used for marine hydraulics.

CONCLUSION

This paper has discussed selected thermal spray processes that have a proven track record of providing surface engineering solutions for the shipping and marine industry. These include the thermal spray coating of steel super structures and vessels with aluminium or zinc based metal composites to help minimise corrosion. Thermal sprayed alumina-aluminium coatings have also been used to provide heat resistant and durable non-skid surfaces on ship decks. Finally, the authors have conducted investigations into the use of HVOF coatings for improved corrosion and biofouling performance of marine hydraulic components. These studies, including field exposure trials around Australia, have found that single layer carbide based coatings can be successfully applied to hydraulic components subjected to an immersed marine environment. These HVOF coatings provide the necessary corrosion and biofouling performance required by industry and can be used as a productive and cost effective refurbishment option for the marine platforms and infrastructure.

ACKNOWLEDGMENTS:

The authors would like to acknowledge the support of the Defence Materials Technology Centre (DMTC). The DMTC was established and is supported under the Australian Government's Defence Future Capability Technology Centres Program.

REFERENCES

1. Davis, J.R., *Handbook of Thermal Spray Technology*. 2004, Materials Park, OH, USA.: ASM International.
2. Ang, A.S.M., et al., *Thermal Spray Maps: Material Genomics of Processing Technologies*. Journal of Thermal Spray Technology, 2013. **22**(7): p. 1170-1183.
3. Gibb, H.J., et al., *Clinical findings of irritation among chromium chemical production workers*. American Journal of Industrial Medicine, 2000. **38**(2): p. 127-131.
4. Jandin, G., et al., *Correlations between operating conditions, microstructure and mechanical properties of twin wire arc sprayed steel coatings*. Materials Science and Engineering A, 2003. **349**(1-2): p. 298-305.
5. Schmid, R.K., et al., *New HVOF Developments at Sulzer Metco*. Sulzer Technical Review, 2004. **86**(1): p. 4-7.
6. Thorpe, M.L. and H.J. Richter, *A pragmatic analysis and comparison of HVOF processes*. Journal of Thermal Spray Technology, 1992. **1**(2): p. 161-170.
7. Kamnis, S. and S. Gu, *3-D modelling of kerosene-fuelled HVOF thermal spray gun*. Chemical Engineering Science, 2006. **61**(16): p. 5427-5439.
8. Dongmo, E., M. Wenzelburger, and R. Gadow, *Analysis and optimization of the HVOF process by combined experimental and numerical approaches*. Surface and Coatings Technology, 2008. **202**(18): p. 4470-4478.
9. Li, M. and P.D. Christofides, *Computational study of particle in-flight behavior in the HVOF thermal spray process*. Chemical Engineering Science, 2006. **61**(19): p. 6540-6552.
10. Li, C.J. and Y.Y. Wang, *Effect of particle state on the adhesive strength of HVOF sprayed metallic coating*. Journal of Thermal Spray Technology, 2002. **11**(4): p. 523-529.
11. Paul, S., *Corrosion Control for Marine-and Land-Based Infrastructure Applications*. ASM Handbook—Thermal Spray Technology, 2nd ed.; Robert, CT, Jr., Ed: p. 248-252.
12. Kuroda, S., J. Kawakita, and M. Takemoto, *An 18-year exposure test of thermal-sprayed Zn, Al, and Zn-Al coatings in marine environment*. Corrosion, 2006. **62**(7): p. 635-647.
13. Mandeno, W. *Thermal metal spray for bridges: a new zealand perspective*. in *Steel Innovations Conference, Christchurch, New Zealand*. 2013.
14. Tiong, D.K.-K. and H. Pit. *Experiences On Thermal Spray Aluminum (TSA) Coating On Offshore Structures*. in *CORROSION 2004*. 2004. NACE International.

15. Sanchez, J.M. *CoatingsPro's Greatest Hits: Anti-Slip Deck Coatings*. 2017; Available from: http://www.coatingspromag.com/articles/steel-coatings/2017/02/greatest-hits-anti-slip-deck-coatings.

16. Zhang, D., S.J. Harris, and D.G. McCartney, *Microstructure formation and corrosion behaviour in HVOF-sprayed Inconel 625 coatings.* Materials Science and Engineering A, 2003. **344**(1-2): p. 45-56.

17. Legg, K.O. and J.P. Sauer, *Use of Thermal Spray as an Aerospace Chrome Plating Alternative. Final Report, Grant No. N00173-98-D-2006, D.O. 0002*, 2000.

18. Ang, A.S.M., et al., *Development of Processing Windows for HVOF Carbide-Based Coatings.* Journal of Thermal Spray Technology, 2015. **25**(1-2): p. 28-35.

19. Ang, A.S.M., et al., *Manufacturing of nickel based cermet coatings by the HVOF process.* Surface Engineering, 2016. **32**(10): p. 713-724.

Underwater Glider Performance at Model-Scale and Full-Scale Reynolds Numbers

Max Haase[1] , max.haase@utas.edu.au
Gregory Seil[2,3], gregory.seil@dst.defence.gov.au
Ron Allum [4], ron@ronallum.com
David Battle [5] david.battle@missionsystems.com.au

[1]Australian Maritime College, University of Tasmania
[2] Advanced VTOL Technologies Pty Ltd
[3] Maritime Division, Defence Science and Technology Organisation
[4] Ron Allum Deep Sea Services Pty Ltd
[5] Mission Systems Pty Ltd

ABSTRACT
Underwater gliders are used for autonomous collection of oceanographic data. These vehicles glide through the water column in a vertical saw-tooth trajectory by alternating their net buoyancy. It is important to achieve high Lift to Drag if long range and endurance are sought. However, such gliders are likely to have significant areas of laminar flow over their hull. This may be critical when predicting the performance of the full-scale glider from the results of physical model-scale testing.

In this study, computational fluid dynamics (CFD) and model-scale testing was used for predicting the Lift to Drag ratio and assessing the static stability of a blended wing-body glider in pitch and yaw. Both fully turbulent and transitional flows were studied. The glider was simulated for a range of Angle-of-Attack at Reynolds numbers ranging from 1.0 x 10^5 – 2.4 x 10^6, based on the length of the glider.

As the Reynolds number was increased there was a significant increase in Lift to Drag ratio and improved pitch stability. The glider was unconditionally stable in yaw for all cases modelled. Reynolds number and physical scale therefore has a significant effect on the performance of underwater gliders. This implies that the performance of the full-scale glider cannot be simply correlated with model-scale results, but requires model testing or numerical prediction at full-scale Reynolds numbers.

INTRODUCTION

Glider Hydrodynamics

Underwater gliders are used for autonomous collection of oceanographic data. They glide through the water column following a vertical saw-tooth-like trajectory by alternating their *net* buoyancy.

For descent, the glider must become *negatively* buoyant. This is achieved by filling a buoyancy tank. As the glider descends, it develops lift which allows it to achieve a force equilibrium (Figure 1) and a constant velocity (U) at a glide angle (γ) relative to the horizontal. In this force equilibrium, the lift (L) balances the component of the buoyancy force (B) perpendicular to the glide path. Similarly, the buoyancy force in the direction of the glide path is balanced by the drag force (D). This can be expressed as,

$$L = B\cos\gamma, \quad D = B\sin\gamma \tag{1}$$

The glide angle is related to the Lift to Drag (L/D) ratio by,

$$\gamma = tan^{-1}(1/(L/D)) = tan^{-1}(1/(C_L/C_D)) \tag{2}$$

In Equation 2, the lift coefficient (C_L) and drag coefficient (C_D) are defined by,

$$C_L = \frac{L}{\frac{1}{2}\rho U^2 S_w}, \quad C_D = \frac{D}{\frac{1}{2}\rho U^2 S_w} \tag{3}$$

where ρ is the density of the water and S_w is the planform wing area.

For ascent, the glider must become *positively* buoyant. This is achieved by expelling water from the buoyancy tank. As with the case of decent, the glider again develops lift, but in the opposite sense to that shown in Figure 1, since the buoyancy force now acts upward rather than downward.

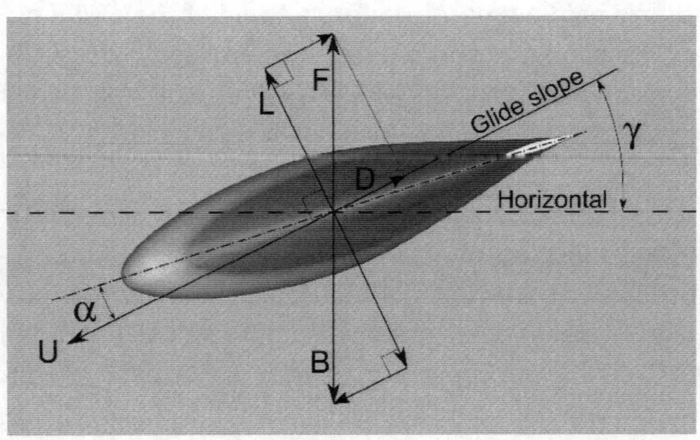

Figure 1: Glider force equilibrium diagram

In the past, a variety of gliders have been used for oceanographic applications, such as the Slocum (Stommel [1]), Sea glider (Eriksen et al. [2]), and Spray (Sherman et al. [3]). These gliders all comprise elongated bodies of revolution with short wings and are known as "profiling gliders", since their primary task is to measure vertically changing profiles of water column properties such as sound speed.

A further recent development of underwater glider technology has been the blended wing-body concept for long range operations, such as the Liberdade/Z-Ray glider first developed by Jenkins and D'Spain [4], or the BWBUG (blended wing-body underwater glider) proposed by Sun et al. [5]. In both cases, the objective has been to minimise the total wetted area with respect to the lifting surface in order to maximise the Lift to Drag (L/D) ratio, and hence the range and endurance of the glider. From Equation 2, an increase in L/D reduces the glide angle and hence increases the range of the glider for a given number of descent/ascent cycles to depth. The blended wing-body concept also allows for the carriage of larger payloads when compared to winged bodies of revolution – see Jenkins & D'Spain [4].

For predicting the hydrodynamic performance of a full-scale underwater glider, free-flying model tests can deliver valuable insights. However, it is well known that Reynolds numbers, defined here as

$$Re = \frac{\rho U c}{\mu} \tag{4}$$

for a characteristic length (centreplane wing chord) c and dynamic viscosity μ, are a major determinant of the flow regime, and hence hydrodynamic parameters such as lift and drag. Not surprisingly, L/D ratios are observed to decrease substantially with decreasing Re (McMaster [6]). Thus, alternative methods of scaling are required in predicting the performance and stability of full-scale gliders from model tests.

DST Group Study

With the growing interest in the potential applications of underwater gliders for acoustic surveillance, DST Group undertook a low-cost hydrodynamic study of a blended wing glider shape in order to gain a greater understanding of their achievable Lift to Drag ratios and glide stability. The configuration chosen was inspired by the Liberdade/Z-Ray vehicle developed by Jenkins and D'Spain [4], which was, in turn, heavily influenced by the flying wing aircraft designs of the Horten brothers in the 1940's [7].

The design consists of tapered NACA 650XX laminar flow wing sections [8] swept back to maximise the rear-ward displacement of the Centre-of-Pressure (CoP). Since stability is a function of the position of the CoP in relation to the Centre-of-Gravity (CoG), it was decided that a scale model was an essential adjunct to CFD modelling, due to the ease with which the CoG could be adjusted to assess its impact on pitch and yaw stability. Hence, the 1.0 - metre wing span scale model shown in Figure 2 was constructed by Ron Allum Deepsea Services (RADS) for pool testing.

Computational Fluid Dynamics (CFD) was conducted to assess steady-state L/D performance, as well as pitch and yaw stability, in both fully turbulent and transitional flow. CFD modelling was first performed for a Reynolds number of 2.4×10^6, representative of the full-scale glider. This provided the L/D ratio and CoP over a range of Angles-of-Attack (AoA, α). To take the effect of Reynolds number into account, a laminar/turbulent transition model was used for the numerical simulations, which was validated against tabulated data for specific NACA wing sections. Additional CFD was performed at Reynolds numbers relevant to the pool tests in order to gain better physical insight into the model's hydrodynamics and assess the relevance of model-scale results to full-scale glider performance.

Figure 2 Scale model of the "Minray" glider

Overview

In this paper, we consider the variation of L/D ratio with Reynolds number and the underlying hydrodynamic causes of this variation, as well the implications in terms of the design and operation of underwater gliders.

In the following section, the CFD modelling methodology is presented. The L/D ratio over a range of Angles-of-Attack at a full-scale Re = 2.4×10^6 is calculated to identify the AoA providing the highest Lift to Drag ratio. The variation of L/D ratio, lift coefficient, drag coefficient and CoP over a range of Reynolds numbers at α = 6° is then considered. This is then related to the flow over the glider by examining the distribution of static pressure and skin friction in conjunction with the surface flow topology. Finally the implications in terms of the design and operation of the glider are considered, together with the limitations of CFD modelling.

COMPUTATIONAL FLUID DYNAMICS MODELLING

CFD modelling was performed using the SIEMENS PLM CFD software STAR-CCM+. The mesh was generated utilising Pointwise V16. Steady-state Reynolds-Averaged Navier-Stokes (RANS) equations in STAR-CCM+ were used.

Modelling of Turbulence and Laminar-to-Turbulent Transition

Turbulence was modelled using the SST-k-ω model in conjunction with the empirically-based γ-Re_θ laminar-turbulent transition model. In STAR-CCM+ the γ-Re_θ model solves additional transport equations for turbulent intermittency (γ) and momentum thickness Reynolds number (Re_θ). At the Australian Maritime College, the γ-Re_θ model has been successfully validated for flows over flat plates (Lee [9]) and for a bow of a wave-piercing catamaran scale model (Pearson [10], Schnabel [11]). It was also used for the nose cone improvement of an AUV for extending its operating range (Webster [12]). For the simulations, zero turbulence intensity was considered for the incoming flow, as this was deemed applicable for an undisturbed deep-ocean environment. For the model-scale testing, no precise information on the background turbulence was known, thus zero intensity was assumed. The glider surface was considered hydraulically smooth.

Computational Grid

The glider was modelled in a spherical domain 80 wing-chord lengths in diameter. A fully-structured grid of approximately 5.5 million cells was created for the half-model. The sub-layer-scaled distance from the wall (y^+), defined as,

$$y^+ = \frac{\rho u_\tau y}{\mu}, \qquad u_\tau = \sqrt{\frac{\tau_w}{\rho}} \tag{5}$$

was less than unity over the surface of the glider, thereby placing it deep in the viscous sublayer of the glider's boundary layer. This is of particular importance for the accurate resolution of velocity gradients and hence wall shear stress adjacent to wall surfaces. In Equation 5, u_τ is the friction velocity, τ_w the wall shear stress and y the normal distance from the wall.

The CFD grid of the glider must be sufficiently fine to model the laminar-to-turbulent transition using the γ-Re_θmodel. Of particular importance is the resolution of the grid in the chordwise and wall-normal direction. The grid resolution in the spanwise direction is less critical.

In order to select an appropriate number of grid cells in the chordwise and wall-normal directions, a grid dependency study was performed for a NACA 65035 foil section for α = 4° using a two-dimensional C-type mesh. The predicted C_L, C_D, L/D and transition location on the pressure and suction side of the section were plotted with increasing grid refinement. When these quantities ceased to change with further grid refinement, the results for C_L and C_D were compared with data from Abbott and Doenhoff [8].

The minimum number of chordwise and wall-normal gridlines, giving grid insensitive results was then applied to the glider. The same mesh was used for different inflow conditions, where the inlet conditions were adjusted to achieve the desired Angles-of-Attack or side-slip angles.

RESULTS

Variation of Performance with Angle-of-Attack at Full-Scale Reynolds Number

Calculations were initially performed for the glider over a range of AoA under the assumption that the flow is fully turbulent from the leading edge. The variation of predicted L/D with AoA is shown in Figure 3 and appears to peak around α = 6°, which was the value chosen for further study. At this AoA, the calculation was repeated with the transition model enabled. This yielded a 6.9% increase in C_L and a reduction of 18.1% in C_D, resulting in a 31.4% increase in L/D. This suggests that maintaining laminar flow over part of the chord can result in significant increase in L/D and hence range. Increasing the Reynolds number by 50% to 3.2×10^6 under the assumption of fully turbulent flow results in an increase in L/D of 5.1% and is therefore a much smaller increase than that predicted using the transition model.

Variation of Performance with Reynolds Number

The glider performance at α = 6° was calculated using the transition model for four "model-scale" Reynolds numbers: 1.0×10^5, 2.0×10^5, 3.4×10^5 and 5.0×10^5. These being representative of the operational Reynolds numbers of the pool model shown in Figure 2.

Figure 4 shows the difference between the values of L/D, C_L , C_D and the location of the CoP, relative to their values at full-scale (Re = 2.4×10^6). The corresponding values are presented in Table 1. Note that in Figure 4, the above model-scale Reynolds numbers, when plotted on a logarithmic scale ($log_{10}(Re)$), have values of 5.00, 5.30, 5.53 and 5.70, respectively. On this scale, the logarithm of the full-scale Reynolds number is 6.38. In Figure 4, the variation in the location of the CoP is plotted as a proportion of chord length relative to the location at full-scale.

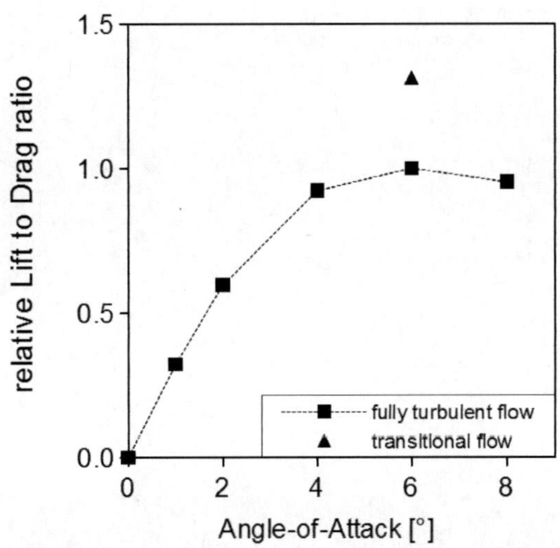

Figure 3 Variation of Lift to Drag (L/D) ratio for a fully turbulent and a transitional solution, relative to the value at α=6° at Re = 2.4 x 10^6 at varied Angle-of-Attack.

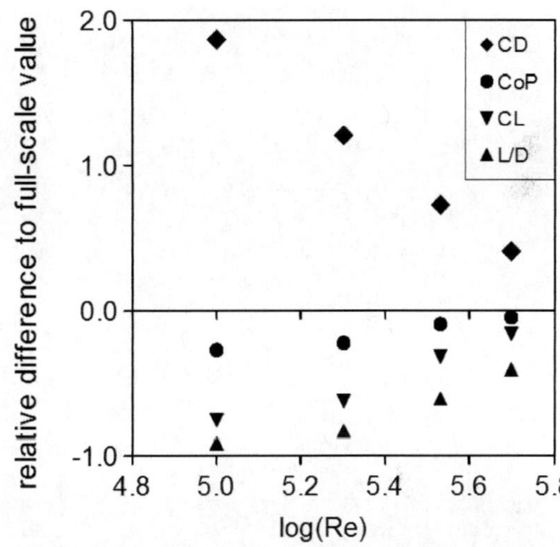

Figure 4 Variation of Lift to Drag (L/D) ratio, lift coefficient (C_L), drag coefficient (C_D) and location of Centre-of-Pressure (CoP) with Reynolds number, relative to values at full-scale Reynolds number at α=6°.

Table 1 Relative difference of lift and drag coefficient (C_L , C_D), location of Centre-of-Pressure (CoP) and Lift to Drag ratio (L/D) at model-scale Reynolds numbers with respect to the full-scale.

log(Re)	CoP	C_L	C_D	L/D
5.00	-0.27	-0.75	1.87	-0.91
5.30	-0.22	-0.62	1.21	-0.83
5.53	-0.09	-0.32	0.73	-0.60
5.70	-0.05	-0.16	0.41	-0.40

Figure 4 shows that as the Reynolds number is decreased from its full-scale value, C_L decreases and C_D increases, which results in the observed overall L/D decrease. The proportional change is greater for C_D than for C_L; hence C_D exhibits greater sensitivity to Reynolds number. The CoP can be seen to move forward, which would act to make the glider less stable at low Reynolds numbers. At the lowest Reynolds number, L/D drops by 91%! The CoP was then predicted to be 0.27 chord lengths forward of the full-scale CoP.

Estimated glide angles calculated by observations of pool testing (Figure 5) suggested an operational Reynolds number of approximately 3.4×10^5 and a reduction in L/D of 60% relative to full-scale values, which is in agreement with the numerical predictions. The scale model was found to be stable in both pitch and yaw, however, when appropriately ballasted.

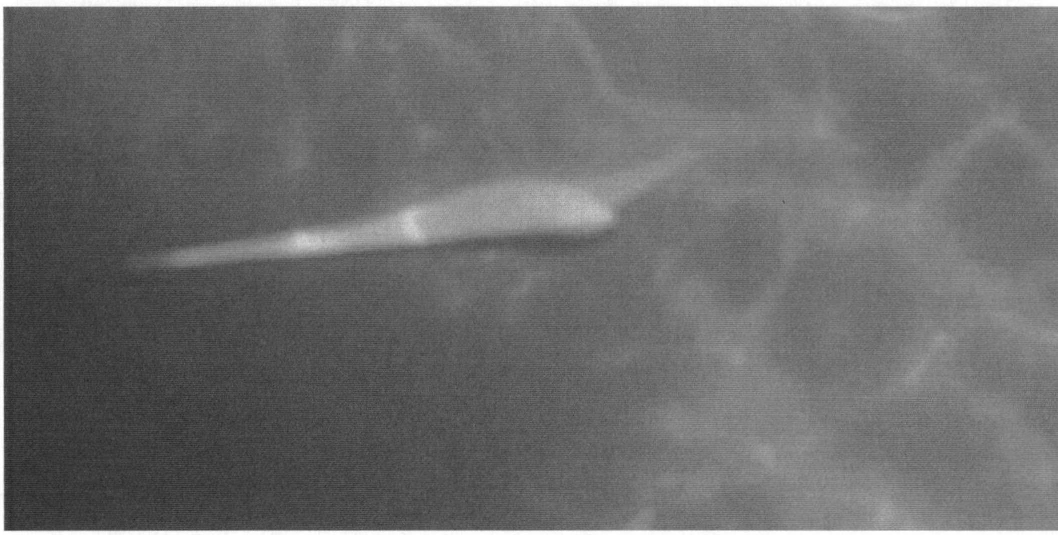

Figure 5 Pool testing of the model glider.

Relationship between Performance and Flow over the Glider

The above-noted changes in L/D, C_L and C_D with Reynolds number, as well as the location of the CoP, are the result of changes in flow topology, which will now be considered.

Figure 6 shows the distribution of static pressure and skin friction coefficient over the glider at a model-scale Reynolds number of Re = 3.4×10^5 at α = 6°. The skin coefficient (C_f) and the static pressure coefficient (C_p) are defined as,

$$C_p = \frac{p - p_o}{\frac{1}{2}\rho U^2}, \qquad C_f = \frac{\tau_w}{\frac{1}{2}\rho U^2} \tag{6}$$

where p is the static pressure and p_o the reference static pressure. Skin friction lines (shown in white) have also been plotted. These are based on the shear stress vector on the gliders surface and provide information about the flow direction. Lines of laminar

flow separation (negative bifurcation line) and fully turbulent re-attachment (positive bifurcation line) are indicated in the figure. Figure 6 is discussed in detail to facilitate the interpretation of Figures 7 and 8.
In Figure 6, the left-hand side of the glider represents the suction side and the right-hand side the pressure side. The flow over the glider from its leading edge is initially laminar, with the pressure dropping with distance along the chord. As the pressure begins to increase beyond the location of minimum pressure, the laminar flow separates as indicated. On the suction-side, this occurs further upstream than on the pressure-side. On the pressure side, the flow remains laminar to the wing tip, but this is not the case for the suction side.

Downstream of the separation line, C_f becomes negative, indicating a local reversal of the flow direction. On the pressure side, the flow re-attaches at the locations indicated, and at the trailing edge being in a fully turbulent state. Overall, poor flow is exhibited on the suction-side of the glider to a greater extent than on the pressure side.

The distribution of skin friction coefficient and static pressure distribution over the glider at $\alpha = 6°$ for varying Reynolds numbers is shown in Figures 7 and 8, respectively. In both figures, the skin friction lines illustrate the flow topology. Again, the left-hand and right-hand sides of the graphic represent the suction and pressure sides respectively.

Figure 7 shows that as the Reynolds number is increased, the region of separated flow on both the pressure and suction sides of the glider decrease. There appears to be relatively little change in the chordwise location of the flow separation (negative bifurcation) line over most of the wing span. Differences in the extent of flow separation are, therefore, driven primarily by changes in the chordwise location of the reattachment line.

Near the wing tip, the separation line moves toward the leading edge with increasing Reynolds number. On the pressure side, the flow near the wing tip is laminar from the leading edge to the trailing edge at the two lowest Reynolds numbers. As the Reynolds number is increased, the flow separates close to the trailing edge and the location of this separation line moves upstream with further increases in Reynolds number. For a fully turbulent solution, no flow separation was observed and consequently no flow separation or re-attachment lines are present.

Figure 8 shows relatively little variation (relative to changes on the suction side) in the pressure distribution on the pressure side upstream of the separation line, with changes in Reynolds number. Note that this is a region of laminar flow. Downstream of the separation line, the pressure increases with increasing Reynolds number. The most significant change with Reynolds number is in the development of the low pressure area on the suction side over the forward half of the chord. This is the primary cause for the increase in lift, although the increasing average pressure over the pressure side of the glider also contributes, but to a much lesser extent.

Figure 7 shows that upstream of the separation line, C_f decreases with increasing Reynolds number to well below values for fully turbulent flow, as would be expected for a laminar flow. An increased Reynolds number is associated with the development of a thinner boundary layer, relative to the glider's length. This results in greater acceleration and higher velocities over the glider and hence lower static pressures. Consequently, lower pressures occur on the suction side of the glider with increasing Reynolds number. As the Reynolds number increases, the size of the low pressure region on the suction side increases. This results in the downstream movement of the CoP and acts to increase the static pitch stability of the glider.

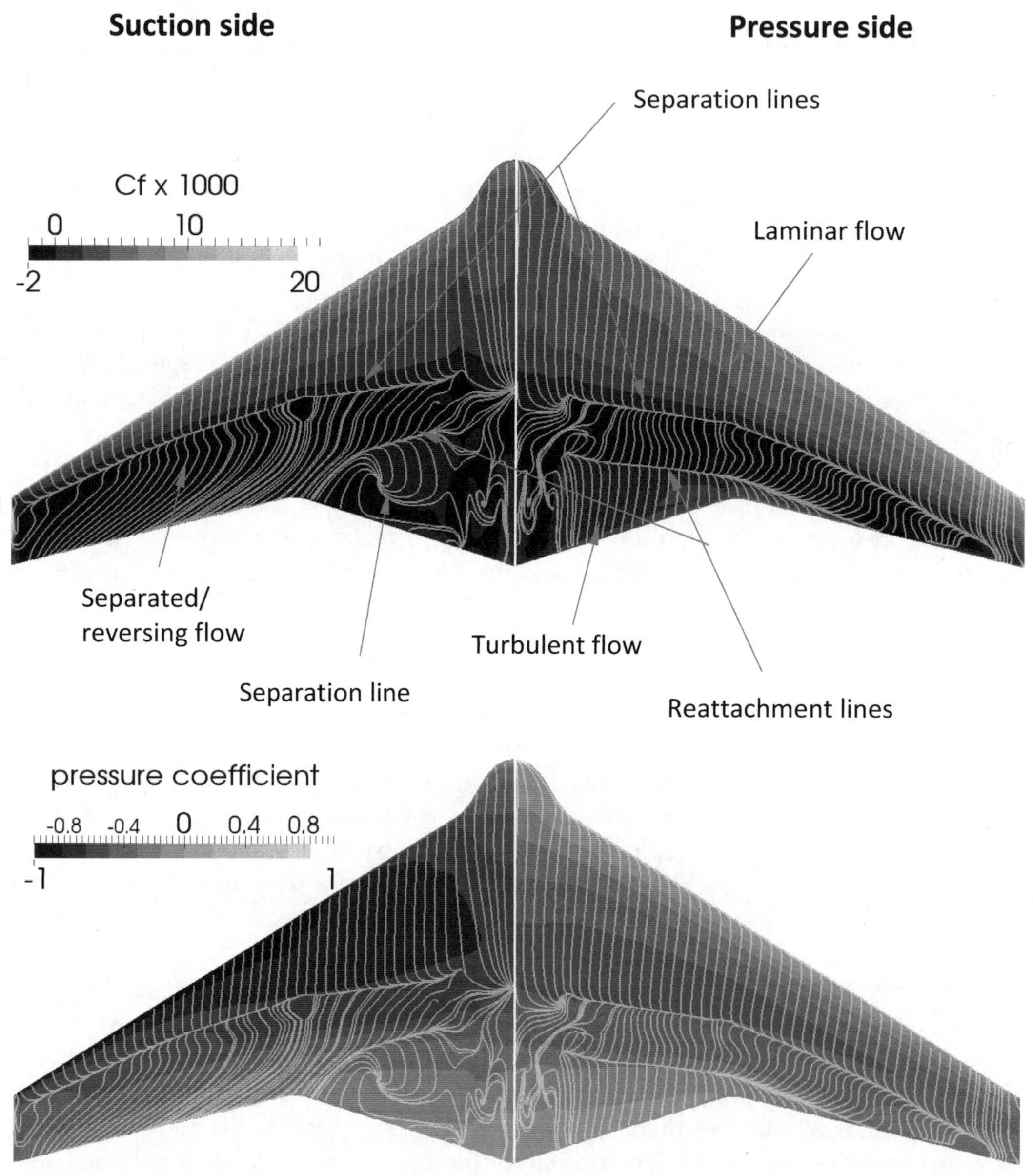

Figure 6 Interpretation of flow topology adjacent to the surface of the glider, Re = 3.4×10^5, α = 6°. Surface skin friction lines shown in white. Contours coloured by skin friction coefficient (top) and pressure coefficient (bottom).

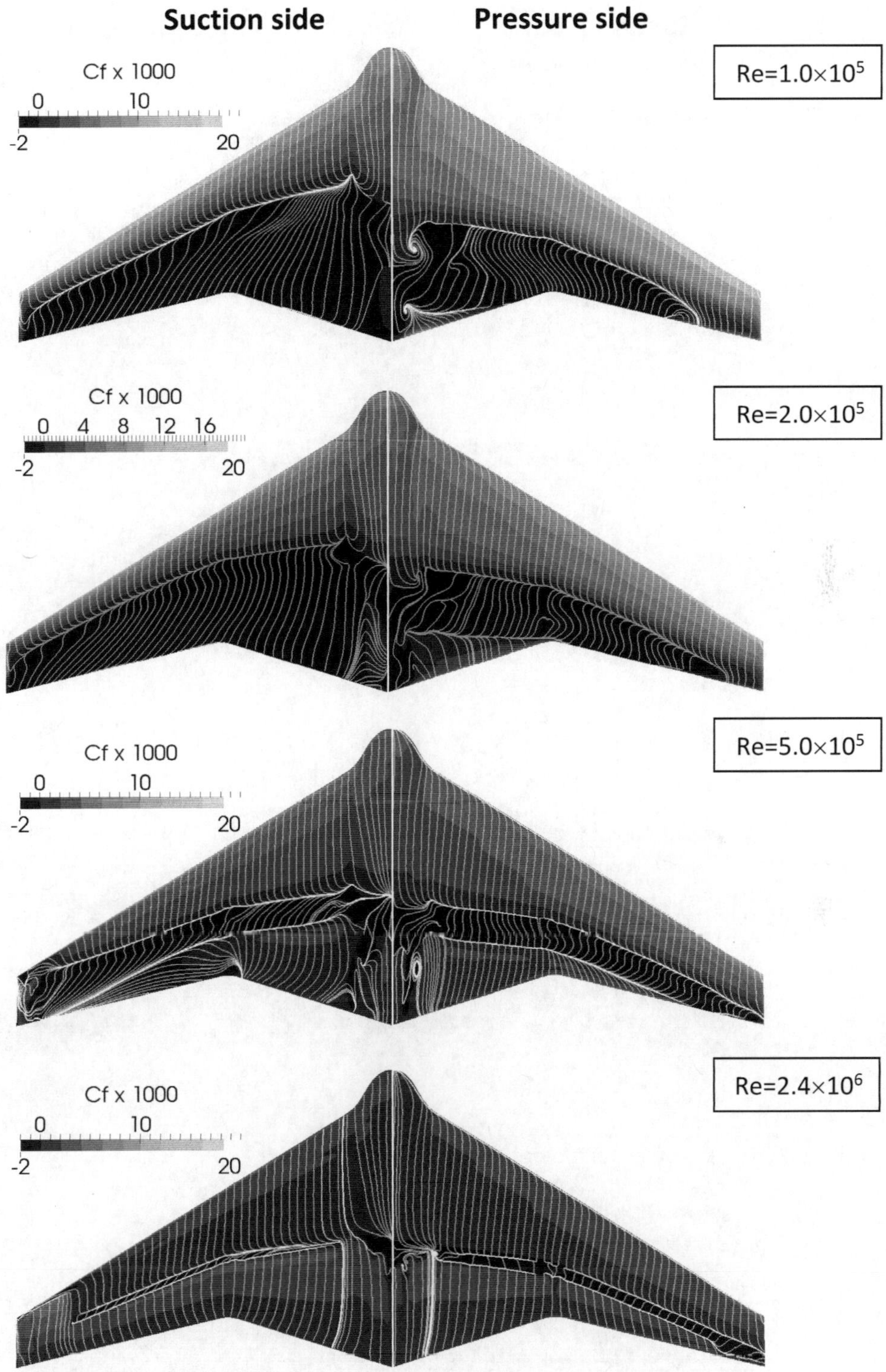

Figure 7 Surface skin friction lines (in white) showing the variation in flow topology adjacent to the surface of the glider with Reynolds number, α=6°. Contours coloured by skin friction coefficient.

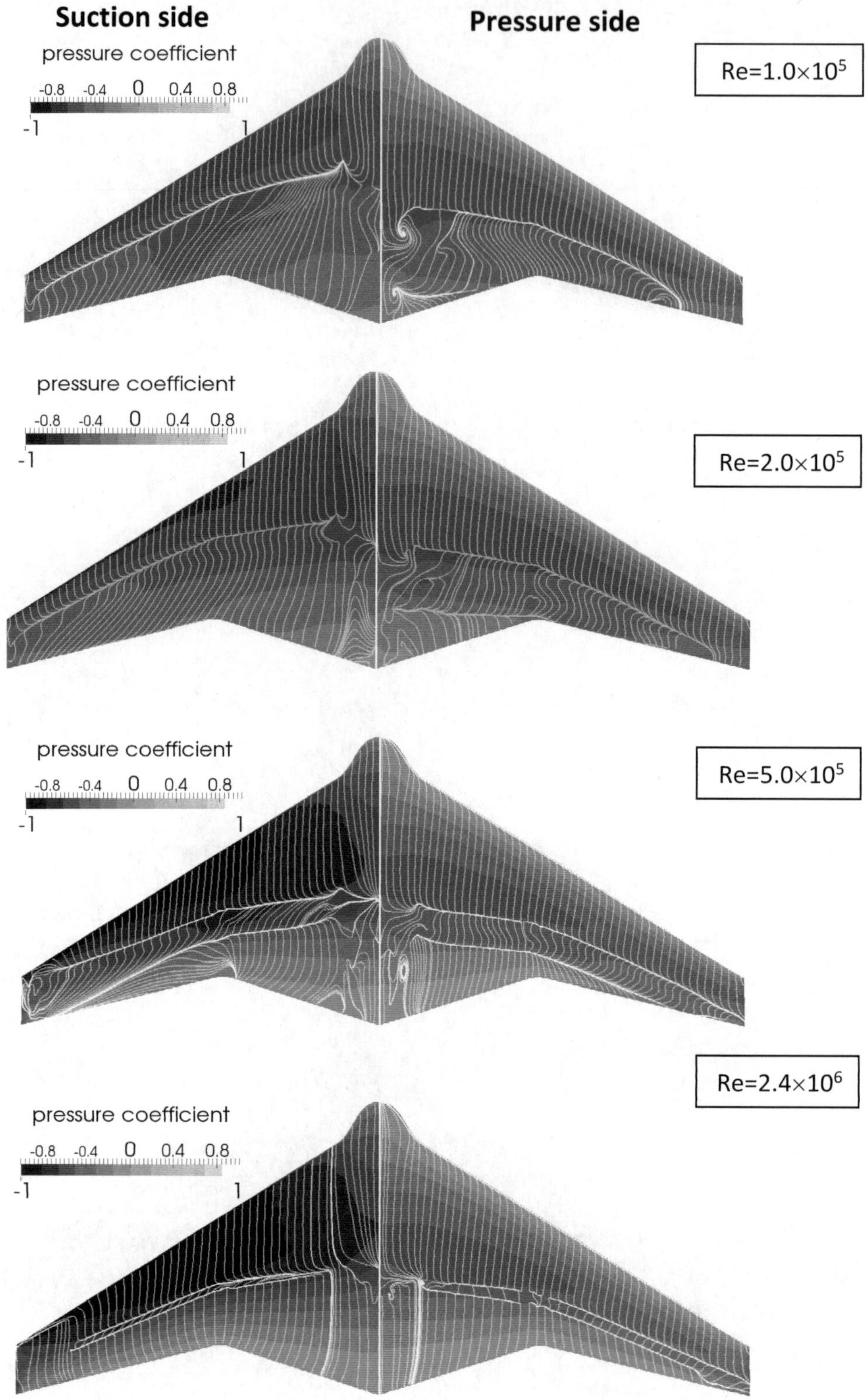

Figure 8 Surface skin friction lines (in white) showing variation in flow topology adjacent to the surface of the glider with Reynolds number, α=6°. Contours coloured by static pressure coefficient.

DISCUSSION OF RESULTS

The application of correlation-based transition models is potentially problematic at critical Reynolds numbers and for complex geometric shapes due to the more complex three-dimensional flow structures. Other mechanisms of transition such as attachment line instability at the leading edge and cross-flow instability, due to the sweep of the wing, may be present (see Saric et al. [13]). These additional mechanisms are usually not accounted for by correlation-based transition models and can introduce uncertainty in the results.

However, since the numerical prediction of reduced L/D at model scale concurred with observations during pool testing, it is assumed that the transition modelling was sufficiently accurate. Also, the variation of performance with varied Reynolds numbers appears physically reasonable. For the full-scale glider, the fully turbulent solution provides a worst-case scenario if the flow is not as laminar as predicted: which could be due to higher free stream turbulence in the fluid (promoting by-pass transition) or roughness on the glider's surface.

Flow separation occurs where the pressure starts to rapidly increase in the stream-wise direction. Since this change in pressure gradient is known to cause flow separation and laminar-to-turbulent transition it would be unphysical to have larger areas of laminar flow for this given pressure distribution. As a consequence at the full-scale Reynolds number the transitional cases can be considered the best-case scenario, while the fully turbulent solution would be the worst-case scenario.

DESIGN IMPLICATIONS

The presence of large regions of laminar flow over the glider presents obvious challenges when attempting to scale results from model-scale to full-scale, and suggests that physical model testing needs to be performed at realistic (i.e. full-scale) Reynolds numbers for the outcomes to be anything more than qualitative. It is also necessary that the free stream turbulence in the wind tunnel or flow tank be representative of that encountered by the glider in the ocean, which would be minimal at depth.

It should also be noted that the prime motivation for model testing (and, in fact, CFD modelling) in this study was the assessment of the *stability* of the selected glider shape, rather than its detailed performance. Though qualitative in nature, free-flying pool tests represent a worst-case in terms of stability and gave sufficient confidence to progress the construction of a full-scale test vehicle. In a complementary sense, CFD simulations have delivered valuable insights into the flow dynamics and helped to explain *why* performance and stability are so strongly affected by Reynolds number. Thus, in cases where pool or wind-tunnel testing at full-scale Reynolds numbers is not possible, CFD provides a way predicting the potential performance gap between model-scale and full-scale, once its accuracy has been validated against the former (Haase et al. [14]).

CONCLUSIONS

The results of this numerical study clearly show the performance benefits, in terms of L/D and hence range, of maintaining laminar flow over a significant region of the surface of a full-scale blended wing-body underwater glider with a smooth surface. At an AoA of 6°, the increase in L/D was found to exceed 30% relative to the value obtained for fully turbulent flow.

As Reynolds numbers are reduced, for example at reduced forward speed or in model-scale testing, it was found that there are significant changes in the flow over the glider which result in reduced L/D. These flow phenomena include reduced lift due to a thicker boundary layer, and larger areas of separated flow. Reducing the Reynolds number from a full-scale value of 2.4×10^6 to 3.4×10^5 led to a reduction of 60% in L/D, which concurred with observations during model-scale tests. Further reduction in Reynolds number to 1.0×10^5 resulted in a 91% decrease in L/D. A reduction in static pitch stability due to the forward movement of the CoP of up to 0.27 chord lengths relative to its full scale location occurred at a Reynolds number as low as 1.0×10^5.

Finally, pool testing of an underwater glider at low Reynolds numbers can be viewed as a conservative approach to performance and stability assessments, representing a worst case scenario. CFD simulations can help by providing a more realistic estimate to the full-scale performance when validated against model-scale experiments.

ACKNOWLEDGEMENTS

This research was conducted by AMC Search, the commercial arm of the Australian Maritime College at the University of Tasmania and was fully funded by the Maritime Division of the Defence Science & Technology Group (DSTG). The authors would like to thank Dr David Pook for his advice on transition modelling.

REFERENCES

1 Stommel, H. (1989), "The SLOCUM mission", *Oceanography* 2 (1), 22-25.

2 Eriksen, C. C., Osse, T. J., Light, R. D. , Wen, T. , Lehman, T. W. , Sabin, P. L. , Ballard, J. W., Chiodi, A. M., (2001), "Seaglider: A long-range autonomous underwater vehicle for oceanographic research", IEEE J. Oceanic Engin. 26 (4), 424-436.

3 Sherman, J., Davis, R. E. , Owens, W. B. , Valdes, J., (2001), "The autonomous underwater glider Spray", IEEE J. Oceanic Engin. 26 (4), 437-446.

4 Jenkins, S. and D'Spain, G. (2016), "Autonomous Underwater Gliders" in Springer Handbook of Ocean Engineering, Eds Manhar R. Dhanak and Nikolaos I. Xiros. 301-322.

5 Sun, C., Song, B., Wang, P., Zhang, B. (2016), "Energy Consumption Optimization of Steady-State Gliding for a Blended-Wing-Body Underwater Glider" OCEANS 2016, MTS/IEEE, 19-23 Sept 2016, Monterey, USA.

6 McMasters, J. H. (1974), "An analytic survey of low speed flying devices: Natural and man-made", Tech. Soar. 3 (4), 17-42.

7 Myhra, D. (2004), "The Horten Brothers and their All-Wing Aircraft", Schiffer Publishing Ltd., ISBN: 9780764304415, 320 pp.

8 Abbott, I. H., Von Doenhoff, A. E., (1959) "Theory of Wing Sections" Dover, NY. 705 pp.

9 Lee, K. S., (2014), "Tackling the laminar-turbulent transition at medium-speed catamarans using computational fluid dynamics"; BEng Thesis, AMC, University of Tasmania.

10 Pearson, M. (2015), "Modelling laminar-turbulent transition on wave-piercing catamarans for enhancing experimental techniques", BEng Thesis, AMC, University of Tasmania.

11 Schnabel, P. (2017), "Impact of laminar flow on the viscous drag of wave-piercing catamarans at model scale"; intermediate MSc thesis, AMC, University of Tasmania / University of Rostock.

12 Webster, L. (2016), "Investigation into the effect of nose- and tail-cone shape on drag of AUV considering laminar-turbulent transition"; BEng Thesis, AMC, University of Tasmania.

13 Saric, W. S., Reed, H. L., White, E. B. (2003), "Stability and Transition of Three-Dimensional Boundary Layers" Annual Review Fluid Mechanics, 35, 413-440.

14 Haase, M., Zurcher, K., Davidson, G., Binns, J. R., Thomas, G., Bose, N. (2016), "Novel CFD-based full-scale resistance prediction for large medium-speed catamarans", Ocean Engineering, 111, 198–208

Recent Advances Utilizing Autonomous Underwater, Surface and Aerial Vehicles for Operational Field Logistics.

Neil Trenaman, Ocean Aero Inc., ntrenaman@oceanaero.inc
Eric Patten, Ocean Aero Inc., eapatten@oceanaero.us

SYNOPSIS:

In August 2016, Ocean Aero Inc. participated in the U.S. Navy's second Advanced Naval Technology Exercise (ANTX), which featured more than 30 participants operating under the theme of "Cross Domain Communications and Command and Control," fostering collaboration with industry, academia, and other warfare centres. The participants' complementary technologies effectively demonstrated the benefits of managing and controlling operations across the air, sea, and sub-sea domains to optimize the U.S. Navy's collective effectiveness throughout the maritime environment.

The Ocean Aero's Submaran™ S10 Autonomous Unmanned Surface and Subsurface Vehicle (AUSSV) teamed with Lockheed Martin's "Marlin" Autonomous Underwater Vehicle (AUV) and a Lockheed Martin "Vector Hawk" Unmanned Aerial Vehicle (UAV) to showcase the combined capabilities of utilizing autonomous underwater, autonomous surface and autonomous aerial vehicles for communications, and command and control via underwater acoustic and aerial radio links.

A description of the exercises will be presented to illustrate successful deployments in a variety of locations during the ANTX-2016.

In April 2017, Ocean Aero Inc., and the U.S. Navy's Space and Naval Warfare Systems Center Pacific (SPAWAR SSC PAC), Unmanned Maritime Vehicles Lab collaborated in the Advanced Naval Technology Exercise (ANTX) -2017 "Ship to Shore Maneuver Exploration and Experimentation Exercises (S2ME2).

SPAWAR SSC PAC proposed an exercise be conducted during ANTX 2017 to demonstrate how cooperative AUSV/UAV operations can be coordinated to provide bathymetric survey data for amphibious landings.

A description of this exercise will be presented to further illustrate the advances of utilizing autonomous underwater, surface and aerial technologies for operational surveillance and near real-time environmental updates to support field command logistics.

ANTX – 2016: "CROSS-DOMAIN COMMUNICATIONS, COMMAND AND CONTROL"

PARTICIPATING TECHNOLOGIES

Cross-domain communications between unmanned maritime systems is the next advancement in two-way communications capabilities at sea.

Ocean Aero Inc. teamed with Lockheed Martin Technologies to demonstrate the versatility and adaptability that unmanned systems bring to complex missions. Lockheed Martin (NYSE: LMT) successfully launched Vector Hawk, a small, unmanned aerial vehicle (UAV), on command from the Marlin MK2 autonomous underwater vehicle (AUV) during a cross-domain command and control event hosted by the U.S. Navy. In addition to Marlin and Vector Hawk, the Ocean Aero Submaran™ S10, a hybrid autonomous underwater and surface vehicle (AUSV) provided surface reconnaissance and surveillance.

During the Annual Navy Technology Exercise (ANTX) activities in August 2016, the Ocean Aero Submaran™ S10 relayed instructions to the Lockheed Martin Marlin from a ground control station via underwater acoustic communications. Following these instructions, the Marlin launched the Vector Hawk using a specially-designed canister from the surface of the Narragansett Bay.

Following launch, Vector Hawk successfully assumed a mission flight track. All three autonomous vehicles—Marlin, the Submaran™ S10 and Vector Hawk—communicated operational status to the ground control station to maintain situational awareness and provide a means to command and control all assets.

The Ocean Aero Submaran™ S10 is the first wind and solar-powered hybrid autonomous unmanned surface and subsurface vehicle designed for extended autonomous ocean observation and data collection.

Ocean Aero's Submaran™ S10

Built for speed, efficiency and maneuverability, the Submaran™ S10 is powered by a unique composite wing sail. The combination of wind and solar power enables the Submaran™ S10 to travel further and faster – with the navigating agility for lengthy station keeping and prolonged monitoring.

With a hull length of 4.2m (13.8'), a payload capacity of 23kg (50lbs) and a top speed of 5.5kts and the ability to transition from surface sailing to fully submerged in 2-3 minutes, the Submaran™ S10 is ideally suited for autonomous missions that require speed, stealth, long endurance and persistence under demanding ocean conditions.

Lockheed Martin's Vector Hawk is designed for canister or hand-launch in all-weather, maritime environments to provide customers with an organic, tailored intelligence, surveillance, and reconnaissance (ISR) capability at the moment they need it.

Lockheed Martin "Vector Hawk"

In addition to its configuration versatility, Vector Hawk is capable of fully autonomous flight and landing, which enables operators to shift their focus from flying the aircraft to managing the mission.

The four-pound Vector Hawk can fly for 70-plus minutes, at line-of-sight ranges up to 15 kilometres. Operators can recover and re-launch the Vector Hawk in a matter of minutes (including changing the system's battery). Vector Hawk is built on an open architecture to enable rapid technology insertion and payload integration.

The Marlin MK2 is a battery powered, fully autonomous underwater vehicle that is 10 feet long with a 250-pound payload capacity, 18 to 24 hours' endurance, depth rating of 1000 feet and weighs approximately 2,000 pounds. Its open architecture design and modularity allow new mission packages to be quickly integrated into Marlin to meet emerging customer needs.

Lockheed Martin "Marlin MK2"

Also during the three-day event, the Marlin surveyed a sunken barge using the Coda Octopus Echoscope© 3D imaging sonar using our unique signal processing to create accurate 3D models. Teams on the ground used that data to create a 3D printed model of the barge. The Marlin can quickly generate accurate, hi-resolution, 3D, geo-referenced models, giving users a clear view of subsea structures.

Scope and Outcome of the Exercise.

During the 2016 Annual Navy Technology Exercise (ANTX) in Rhode Island's Narragansett Bay, a three-day Naval exercise conducted by Lockheed Martin and Ocean Aero Inc. demonstrated an alternative to the traditional U.S. Navy chain of command, deferring administrative mission control from Naval and Air Force commanders to a series of coordinating autonomous vehicles. During the display, an unmanned surface vehicle coordinated the launch of an unmanned aerial vehicle from an unmanned submersible — the world's first multi-domain, autonomous unmanned vehicle chain.

The exercise began with instructions from a ground control station to Ocean Aero's Submaran™ S10 a wind and solar-powered autonomous unmanned surface and subsurface vehicle (AUSSV) that can be outfitted with a variety of sensors for different surveillance applications.

The Submaran™ S10 (AUSSV), in turn, relayed command instructions to Lockheed Martin's submerged Marlin MK2 autonomous underwater vehicle (UAV) via underwater acoustic communications. The 10 ft. battery-powered UAV is equipped with sophisticated sensors and robust autonomy and is able to cruise for 24 hours, descend up to 1,000 feet and manage a maximum payload of up to 250 lbs. The Marlin cruised below the surface of the bay to a designated launch site as instructed.

Once surfaced, the Marlin followed supplementary instructions to ready its payload, a canister containing a Lockheed Martin Vector Hawk a lightweight, multipurpose autonomous drone (UAV) capable of surveillance and delivering a scalable payload up to 9 miles (line of sight). The Vector Hawk features a field-reconfigurable open architecture to enable rapid technology and payload integration.

The UAV can be field-reconfigured to fit multiple mission environments, including fixed-wing, vertical take-off and landing (VTOL), and tilt-rotor enabling VTOL with transition to fixed wing flight. Once on its predetermined flight path, the autonomous Vector Hawk maintains communication and data linkage using adaptable high-bandwidth software-defined radio, mesh networking (including 3G, 2G, and LTE cellular), and employs over-the-air reconfiguration. The system also incorporates fail-safes to ensure that it can safely return to the user or auto-land when communication or power failures occur.

Meanwhile, the Marlin and UAV transmitted electro-optical (EO) and infrared (IR) video to the Submaran™ S10, which relayed images to the ground station. The Submaran™ S10 provided additional surface reconnaissance and surveillance for the mission.

This is the first time that three autonomous vehicles in three different domains (air, surface, and underwater) have worked together to execute a mission."

The operational status of each vehicle was communicated to and monitored by the ground station, allowing operators full situational awareness and control over each asset.

The collaborative demonstration is another step forward to realizing a future where different unmanned systems work in cooperative operations to support first responders, military operations, and commercial users.

Future civilian and military applications for the coordinating of autonomous vehicles is endless as it helps shift operational focus from commanding and coordinating vehicles to conducting the mission at hand.

ANTX – 2017: "Ship to Shore Maneuver Exploration and Experimentation (S2ME2)"

In April, 2017 Marines at Camp Pendleton field-tested more than 50 different new technologies ranging from palmtop mini-drones to self-driving amtracs, from wireless networks to precision-guided mortar shells.

Technologies that do well could graduate to a more formal Operational Evaluation (OPEVAL) or to further testing.

The project had high-level support from the acting Navy Secretary Sean Stackley, who used to head Navy Department Research, Development and Acquisition (RDA).

The exercise provided a unique opportunity for war fighters to assess emerging technologies and innovative engineering in support of amphibious assault operations.

Six Missions

The Marines organized the experiment not by technical categories but by missions, which required experts in different fields from different agencies and companies to integrate disparate technologies towards a single purpose.

Six mission areas were defined and given codenames:

1. **Shield:** early intelligence (and) reconnaissance, using, for example wide-ranging swarms of robotic scouts in the air, sea, and land, which would allow Marines to identify far more landing sites and potentially bypass defenders by coming ashore in unexpected places.
2. **Spear:** threat identification, e.g. covert drones coming in for a closer look with high-powered sensors and sending detailed data back using hard-to-intercept transmissions.
3. **Dagger:** follow-on reconnaissance and threat elimination, e.g. more drones and manned platforms marking obstacles and mines.
4. **Cutlass:** maneuver ashore, e.g. unmanned boats carrying Marines ashore at high speed or unmanned amtracs swimming in on their own power, with expendable decoy drones.
5. **Broadsword:** combat power ashore, e.g. battlefield 3D printing of spare parts and unmanned ground vehicles providing fire support or carrying supplies.
6. **Battle Axe:** amphibious C4ISR (Command, Control, Communications, Computers, Intelligence, Surveillance, & Reconnaissance), e.g. high bandwidth networks, resisting to jamming and hacking, that can tie the whole operation together.

Participating Technologies

On April 19th – 28th 2017, Ocean Aero Inc. teamed with the San Diego based Space and Naval Warfare Systems Center Pacific (SPAWAR Systems Center Pacific or SSC Pacific) to demonstrate Cooperative Unmanned Systems Hydrographic Survey Mission (CUSHM) capabilities.

The backbone of the mission was supported by two unmanned autonomous technologies housing a variety of specific sensors and communication systems designed to seamlessly integrate data collection and data delivery in field conditions, simulating early reconnaissance in advance of an autonomous amphibious landing exercise. The two unmanned autonomous technologies are:

1) The Ocean Aero Inc., Submaran™ S10 an Autonomous Unmanned Surface & Subsurface Vehicle (AUSSV).

Ocean Aero's Submaran™ S10

2) The Pulse Aerospace Vapor 55, a battery powered, single rotor helicopter capable of autonomous flight and waypoint navigation, provided by SPAWAR SSC Pacific. Payloads can vary from integrated electro-optical infrared cameras, to communications relay packages, to a host of other sensors. For the S2ME2 exercise, the Vapor 55 carried an in-house designed communications package capable of carrying data from the in-situ Submaran™ S10 UASSV to a shore based or ship-based Command & Control (C2) node, and relay mission commands.

Pulse Aerospace Vapor 55

Scope and Outcome Aim of the Exercise

The Scope of the Cooperative Unmanned Systems Hydrographic Survey Mission (CUSHM) addressed aspects in three of the six primary mission area (Shield, Spear & Dagger), described above.

By way of demonstration, the missions' task was to simulate a real-world operation whereby an over the horizon force preparing for an amphibious landing, gathered intelligence and threat identification information by means of stealthy deployment of autonomous, unmanned technologies.

The Ocean Aero Submaran™ S10 AUSSV, equipped with an acoustic multi-beam sonar, was deployed from a remote location to collect hi-definition imagery of the sea bed in a designated area of interest in the littoral zone offshore of a potential amphibious landing site.

The multi-beam data provided near-real time bathymetric images detailing possible hazards such as mine like objects and subsurface structures designed to thwart amphibious landing craft gaining access to the beach.

The high resolution data was then transmitted to the Vapor 55 UAV, which had flown autonomously from a simulated over horizon asset, hovering above the Submaran™ S10.

On completion of the data transfer from the Submaran™ S10, the Vapor 55 UAV returned to its point of departure, the data set recovered and then processed to provide acoustic image maps of the sea floor and identify hazards and threats.

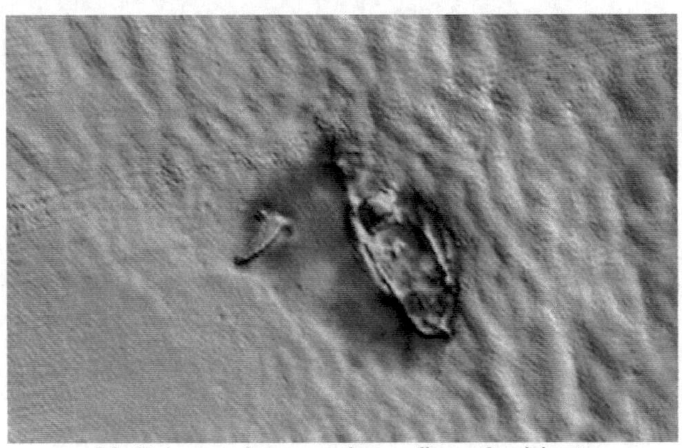

Multi-beam image of sea floor & objects

Summary:

Rapid advances in the development of sensors that are specifically designed to be integrated and hosted by autonomous unmanned aerial and waterborne vehicles for intelligence, reconnaissance and surveillance missions will help drive the acceptance and use of these technologies for Operational Field Logistics.

The utilization of autonomous technologies in air, sea and subsea environments to support cross-domain communications, command and control was successfully demonstrated in the 2016 and 2017 ANTX events.

ACKNOWLEDGEMENTS:

The authors would like to thank Lockheed Martin and SPAWAR for their participation and contribution with Ocean Aero Inc. for the two ANTX exercises.

Persistent Autonomous Unmanned Surface Vessels powered by the ocean

Author(s) Robert Dane, Lloyd Breckenridge, OCIUS robert@ocius.com.au

Ocius sun, wind and wave powered persistent Unmanned Surface Vessel (USV), are autonomous drones that can navigate around the seas for long periods and cover vast areas, conducting long duration autonomous maritime surveillance missions. They are connected and intelligent platforms that can carry a customer's "blackbox" equipment and sensors for varying missions including ASW, monitoring maritime boundaries, and inspecting the seabed.

In 2015, Ocius was selected by the Australian Defence Department and Thales as having the most capable USV to use in a Capability Technology Demonstrator trial for towing a 'thin line' array. This trial will be complete by September 2017 and all OCIUS milestones and KPIs are being achieved or exceeded ahead of schedule.

Ocius was selected because its Bluebottle USV, named after the Australian marine animal that uses its body as a sail, has a meaningful speed of advance in all sea conditions, can carry a 200kg customer at 5-6knots and power that array with a continuous 50W average and peak loads of kW.

Competing existing persistent autonomous unmanned surface vessels USVs are suited to collecting oceanographic data, a less demanding application where speed, power, payload, and advancing in all conditions are not vital to the mission.

The design features that give Ocius's "Bluebottle" USV its advantages in power, payload, performance and 'advance in all conditions' design are the result of years of solar sailing and boat building experience, and are protected by 5 patents.

The Design & Safety Challenges of a Lithium-ion Main Storage Battery for Conventional Submarines

Aidan Depetro
BMT Design & Technology
adepetro@bmtdt.com.au

INTRODUCTION

The current geo-political environment dictates that conventional submarines remain a popular choice for many nations, however current underwater range and endurance is a key disadvantage of the diesel-electric design. The application of emerging battery technology seeks to address this weakness and provide significant tactical and operational advantages to conventional submarine operators.

At the forefront of emerging battery technology is lithium-ion batteries which can offer far superior performance compared to alternative technologies. The challenge facing submarine designers is the integration of this disruptive technology into conventional submarines without compromising reliability or safety. Of particular concern are the fire safety implications of lithium-ion batteries, a technology with limited application in the maritime domain. This creates unknown risks in an already challenging and unforgiving underwater environment.

This paper considers the issues associated with the integration of two lithium-ion cell size classes (small 30-50Ah and large 400+Ah), and two different chemistries (Nickel Cobalt Aluminium-oxide and Iron Phosphate), focused on battery compartment arrangement and fire risk.

A concept design is proposed for a large diesel-electric submarine with a lithium-ion main storage battery, including battery compartment arrangement, and used as the basis for computational fire modelling of thermal runaway scenarios for each of the different battery types. The results provide an understanding of the operational risks introduced by use of lithium-ion batteries and of the most suitable prevention, detection and suppression systems. This is a key first step in enabling the critical advantages offered by this technology to be exploited without compromising system integrity and safety.

QUANTITATIVE RISK ANALYSIS – LITHIUM-ION BATTERIES

Overview

Fire risk analysis of lithium-ion batteries in use as the main storage battery of a conventional submarine has been undertaken. The analysis utilised Fire Dynamics Simulator (FDS) to model the consequences associated with thermal runaway of lithium-ion cells within a test concept submarine design. The analysis was conducted in line with the International Fire Engineering Guideline (IFEG).

Operational Scenario

The scenario considered for analysis is operation of the submarine beyond periscope depth at a nominal underwater patrol speed. Batteries are undergoing discharge in order to power the submarine's hotel load and main propulsion system.

Area of Analysis

The analysis and simulation will be restricted to one of the main storage battery compartments. The conditions inside this compartment will be measured and the consequence will be assessed in terms of the impact on the submarine as a whole.

Analysis Objectives

Lithium-ion batteries are a unique fire hazard, as inherent in their design is fuel (organic electrolyte), oxygen (metallic oxide cathode material) and the necessary means to generate heat (faults, failures and defects causing short-circuits and exothermic reactions). Therefore, once thermal runaway is initiated within a lithium-ion cell, it is likely to continue until all fuel or oxygen is expended regardless of the cooling and oxygen depriving effects applied by a fire suppression system.

As detailed below, the battery compartment is usually an unmanned space and it is expected that a fixed fire suppression system will be activated automatically in the event of a battery fire, however there are several unknowns:

a. What is the likelihood that thermal runaway will spread to adjacent cells given different cell sizes and chemistries, and what is the magnitude of the consequence?
b. In the event of cascading thermal runaway, what degree of cooling is required to ensure adjacent compartments are not compromised and that structural integrity is maintained?
c. To what degree could the battery compartment be heated in the event of cascading thermal runaway and what implications might this have for the submarine?
d. What effect does cell chemistry and size have on the severity of thermal runaway fires?

Before attempting to design a suitable fire safety system, it is important to understand the nature of the fire risk associated with the use of a large number of lithium-ion batteries in a submarine. Therefore, answering the above questions forms the key objectives of the analysis.

CONCEPT SUBMARINE

General

The submarine is a conventional, diesel-electric submarine and typical of the large, long-range designs that have been operated by the RAN over the last 20 years.

Principal Characteristics

The principal characteristics as calculated by the Submarine Design Exploration Application[1] are as follows:

a. Length – 78.8m;
b. Diameter – 8.8m;
c. Displacement – 4,500 tonnes;
d. Maximum propulsion power – 5500 kW;
e. Hotel load (patrol) – 220kW;
f. Propulsion power (patrol) – 130 kW;

[1] The Submarine Design Exploration Application was developed by BMT Design & Technology. It utilizes a combination of empirical and parametric calculations to determine key submarine characteristics based on user input design requirements.

g. Indiscretion ratio – 15%.

The hotel load and propulsion power requirements refer to the power required to operate the submarine's systems and maintain speed whilst on patrol. Note that 'indiscretion ratio' refers to the percentage of time the submarine must spend at periscope depth snorting to recharge batteries whilst on patrol. These key parameters determine the amount of energy required to be stored on the submarine to satisfy operational requirements, and therefore the number of batteries/cells that the submarine must operate and carry.

A sketch of the concept design is show in Figure 1.

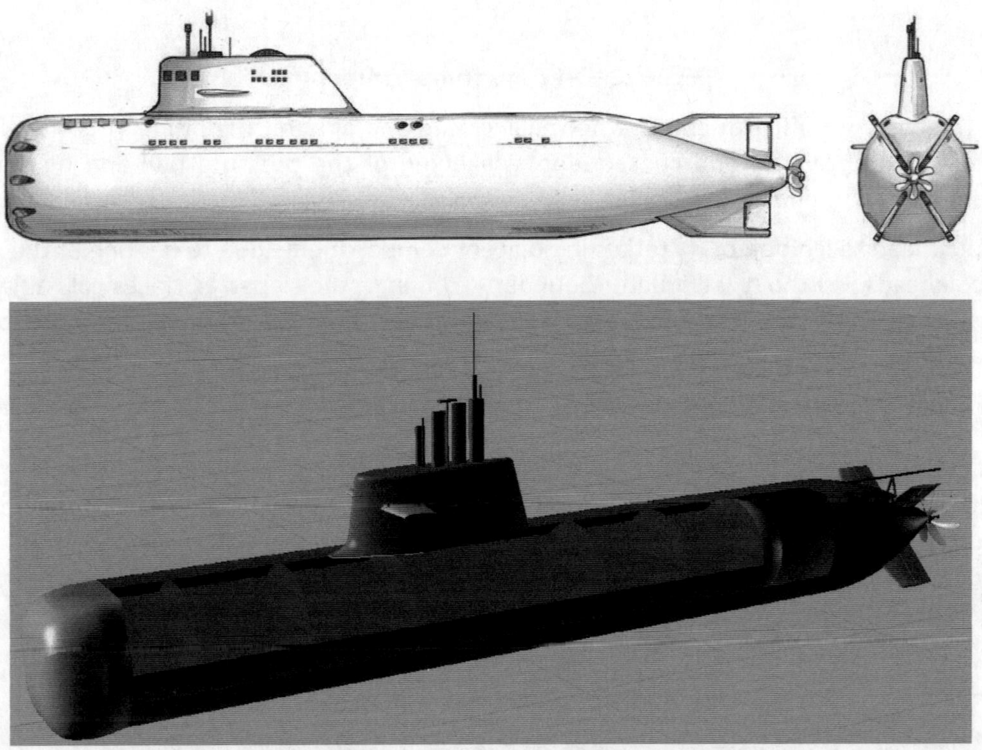

Figure 1 – Future Submarine Concept

Battery Compartment Location

Assuming a traditional diesel-electric submarine design, the submarine has two battery compartments, one forward and one aft. Adjacent compartments on the same deck include void and tank spaces aft and typically unmanned machinery spaces forward. Compartments on the deck above include normally manned machinery, accommodation and munition spaces.

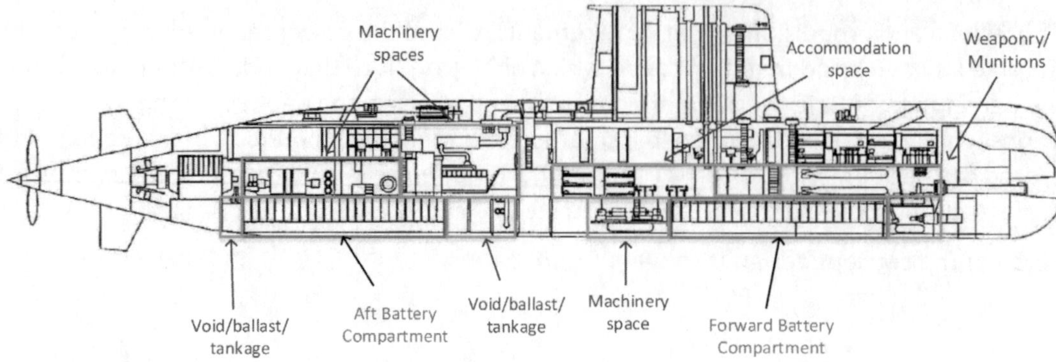

Figure 2 – Battery compartment location [1]

Given the battery compartments are normally closed and airtight, the primary risk is related to heating, overpressure and consequent weakening of the pressure hull and heating of adjacent compartments.

This could lead to ignition of materials in adjacent compartments due to heating of the deck and bulkheads. Without adequate boundary cooling, the inner surfaces of adjacent compartments may become hot enough to ignite secondary fires or compromise other submarine systems.

Battery State, Design and Configuration

The battery types to be modelled are summarised in Table 1.

It is important to note that thermal abuse testing data of the specific batteries listed above is not publicly available. Critical temperature and other design fire parameters were derived from published data based on Sandia National Laboratories' testing of very similar batteries [2].

Table 1 – Battery Specification

Chemistry	Geometry	Energy Capacity	Product Number	# Cells Required[1]	Len. (mm)	Width / Diam. (mm)	Height (mm)	Cell Mass (kg)	Critical Temp.[2] (°C)
Nickel Cobalt Aluminium oxide (NCA)	Cylindrical	45 Ah	HP 602030 NCA [3]	130,000	60	60	232	1.5	150
		485 Ah	HE 1682100 NCA [4]	12,050	168	168	270	13.0	150
Iron Phosphate (LFP)	Cylindrical	38 Ah	HP 602030 LFP [3]	173,000	60	60	232	1.45	240
	Prismatic	400 Ah	SP-LFP400AHA [5]	16,450	450	71	285	14.3	240

1. Based on energy storage requirements of the submarine, as calculated by the Submarine Design Exploration Application. Number of cells provides equivalent storage capacity for each battery type.
2. The temperature at which thermal runaway is initiated [2].

General Objectives

The key submarine mission objectives in the event of a fire are assumed to be:

a. Prevent the loss of the submarine;
b. Prevent loss of life and minimise injury where possible;
c. Maintain stealth and continue mission if possible.

In general, the main objective would be to control the fire, recover maximum functionality of the submarine and either continue the mission or return to base without compromising stealth.

Hazards

Hazards include:

a. Loss of propulsion power, leading to depth excursion, resulting in loss of submarine;
b. Spread of fire and smoke to other compartments, leading to multiple system failure, injury, fatality and loss of submarine;
c. Weakening of pressure hull due to heat, leading to loss of structural integrity, resulting in loss of submarine.

Compartment Layout and Arrangement

The arrangement of batteries proposed here is loosely based on the configuration adopted for most conventional submarines, mainly in terms of physical location in the submarine, as shown in Figure 3. Note that this is traditionally for lead-acid batteries and the arrangement of lithium-ion cells will differ dependent on their number, size and design.

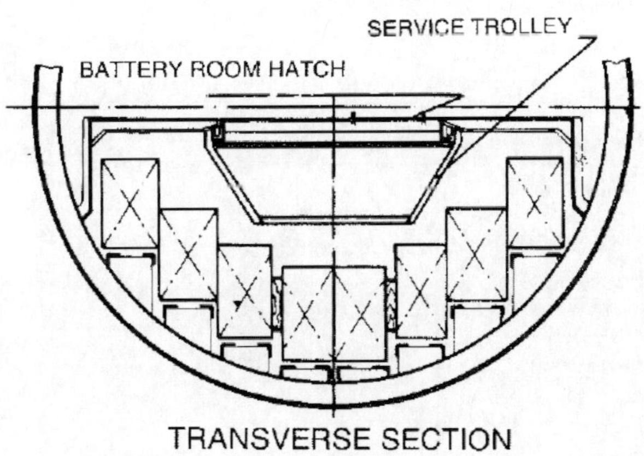

Figure 3 – Conventional submarine battery compartment arrangement [6]

With respect to battery layout and integration into the submarine, the key differences between lead-acid batteries and the lithium-ion cells under consideration include:

a. Significant difference in size - Small lithium-ion cells are 400 times smaller than lead-acid cells, and large lithium-ion cells are around 50 times smaller.
b. Significant increase in the number of cells required (700 - 800 lead-acid cells compared to 100,000+ small lithium-ion cells and 10,000+ large lithium-ion cells);
c. Small lithium-ion cells are required to be grouped into modules to form battery packs;
d. Lithium-ion cells have different electrical configurations and wiring requirements;
e. Lithium-ion cells have different cooling requirements.

These key differences necessitate a significant departure from traditional conventional submarine battery layout. As there is no prescribed method or standard for the layout and integration of lithium-ion batteries on conventional submarines, general guidance for arrangement of large lithium-ion battery arrays in other marine applications has been used. Section 5 of the DNV GL Guideline for Large Maritime Battery Systems [7] has informed the trial design as follows:

a. Battery compartment dedicated room - as per standard conventional submarine design, the main storage battery compartments are dedicated, isolated spaces;
b. Battery system - the battery system architecture has been devised in-line with the Guideline, as illustrated in Figure 4;
c. Minimisation of cell thermal runaway propagation - cell modules or sub-packs have been grouped to minimise the fuel load and maximum HRR in order to decrease the likelihood of module to module propagation;
d. Electrical isolation - cell module groupings allow for isolation at a suitably granular level;
e. Cell cooling for the prevention and mitigation of thermal runaway - the layout of modules and battery pack arrays maximises the cell casing surface area able to be cooled;
f. Inspection and maintenance - cell module groupings, module layout and battery pack array arrangement (including pathways in between arrays) allows for easy access and incremental isolation of the battery system to facilitate regular inspection and maintenance, see Figure 6.

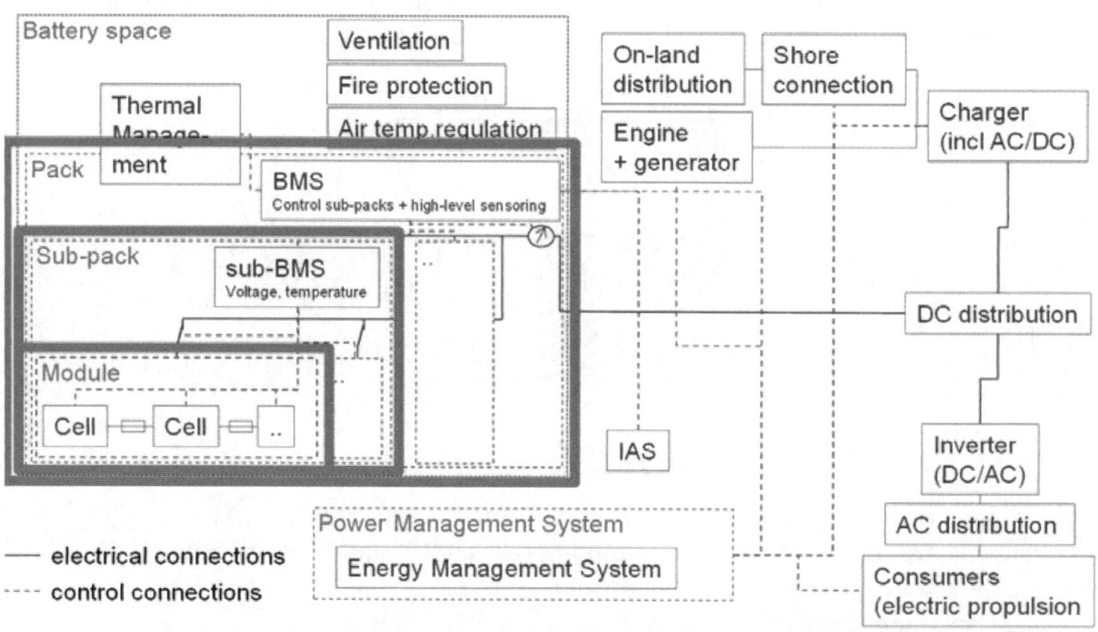

Figure 4 - Battery System Architecture [7]

Due to the extremely high number of cells required and the requirements noted above, cells have been arranged in their modules as a series of vertically stacked arrays to achieve a feasible arrangement within the limited space of the submarine.

For the purposes of this fire risk analysis, a concept layout has been devised. The physical configuration of the small (38/45Ah) cells within the designated battery compartments is shown in Figure 6.

Figure 5 – 3D Concept Model showing Aft Battery Compartment

[Note: Outer casing and pressure hull is hidden]

Note the walkways in between arrays and at the ends of the compartment. This provides quick and easy access to any module in the compartment, whilst providing several egress pathways in the event of an emergency.

The small cells are grouped into modules or 'cell module sub-pack strings' [7] of eight or nine with the cylindrical cells lying side-by-side, flat on the shelves within the arrays as shown in Figure 8. As stated above, this minimises the total fire load per module and provides adequate access for maintenance, monitoring and cooling.

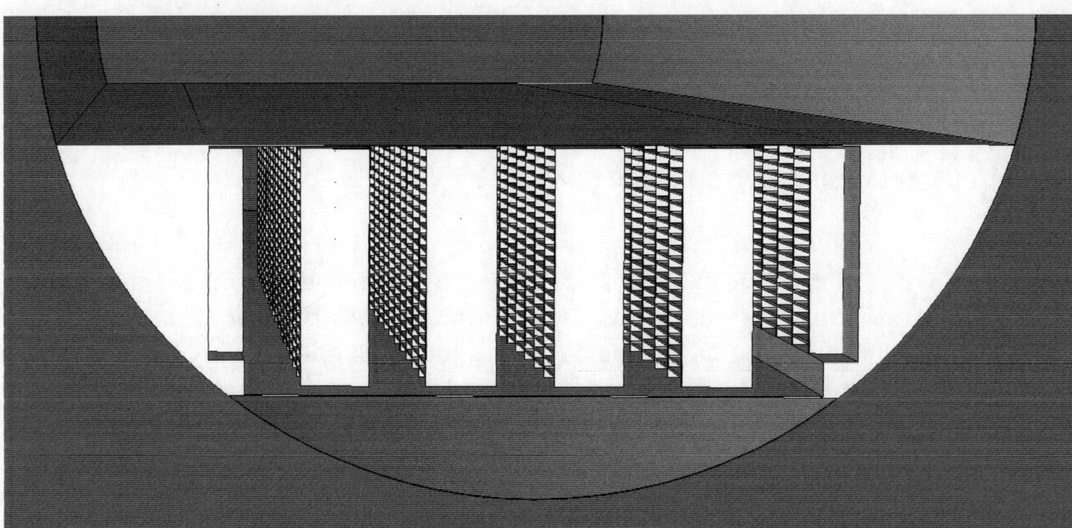

Figure 6 – Aft Battery Compartment showing cell arrays (small cells)

[Note: Bulkheads and outer hull are hidden]

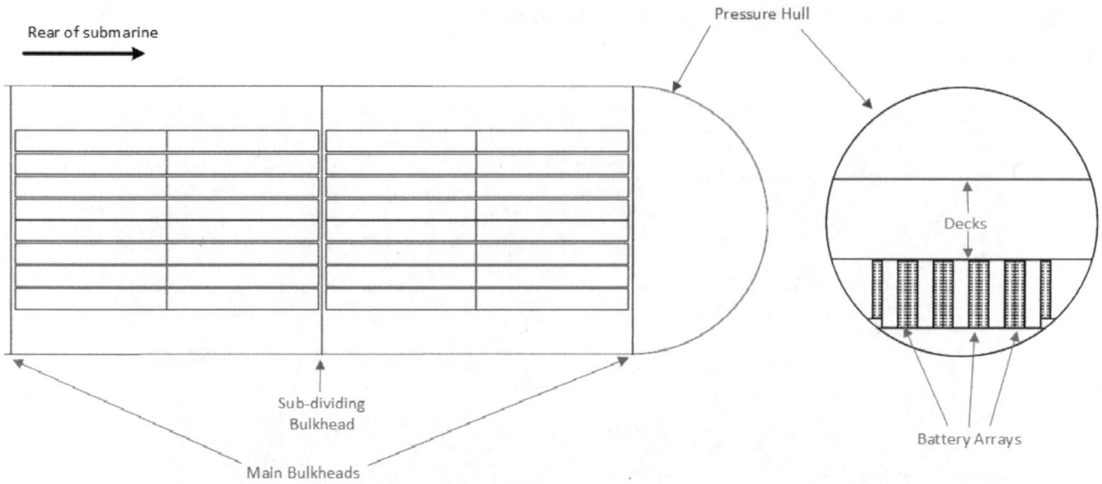

Figure 7 – Aft Battery Compartment Plan and Front Views

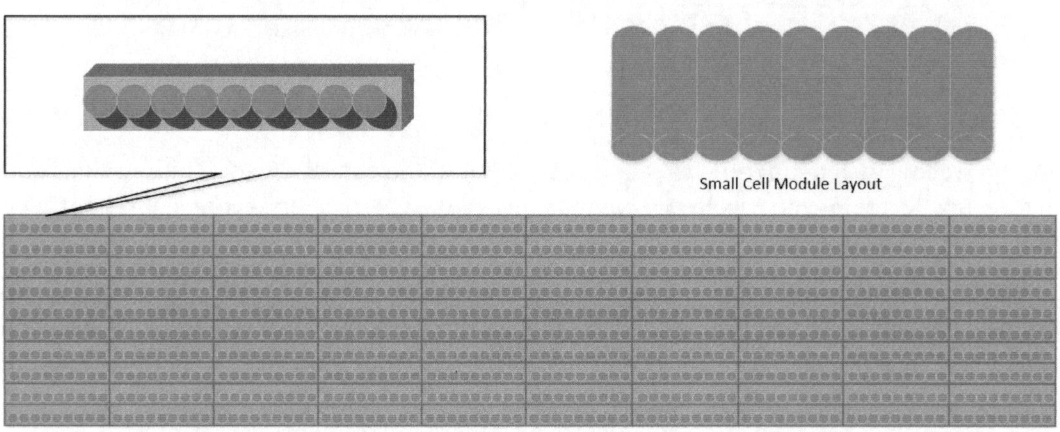

Figure 8 - Small Cell Module Layout

The arrangement for large (400/485Ah) cells is the same, except the modules consist of a single cell and the battery array shelving is sized to suit.

THERMAL RUNAWAY SIMULATION RESULTS

Overview

The results of the four CFD simulations are detailed in this section. Row and column numbering, and example module numbering is shown in Figure 9. Note that the computational domain features only one third of the compartment due to computational constraints. The inner boundary of the domain, near the origin, was set to be "open".

The initiating cell undergoing thermal runaway from simulation time 0.0s is situated in the middle of array SB_3. The module number varies, depending on the size of the array which is differs depending on cell or cell module geometry. All references to time within the results refer to the time elapsed since thermal runaway commenced in the initiating cell.

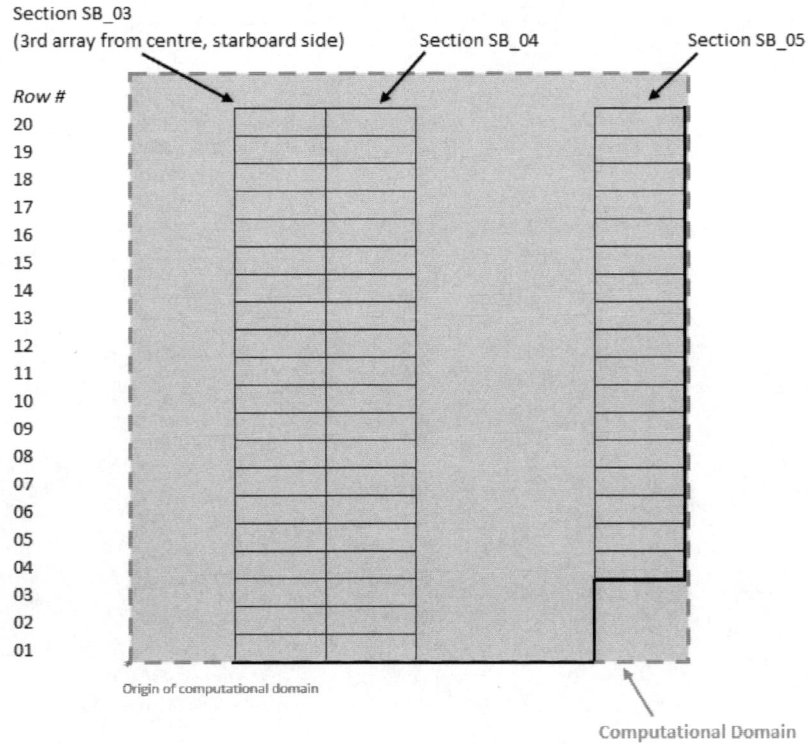

Figure 9 - Front elevation showing array and row numbers (small cells shown)

Example of Results - Small NCA Cells (45Ah)

The rate of thermal runaway propagation is shown in Figure 10. The temperature isosurface and temperature contours at different time intervals are shown in Figure 11 and Figure 12. The pink isosurface denotes a temperature of 50°C and the grey isosurface 150°C. Module surfaces marked red are those undergoing thermal runaway.

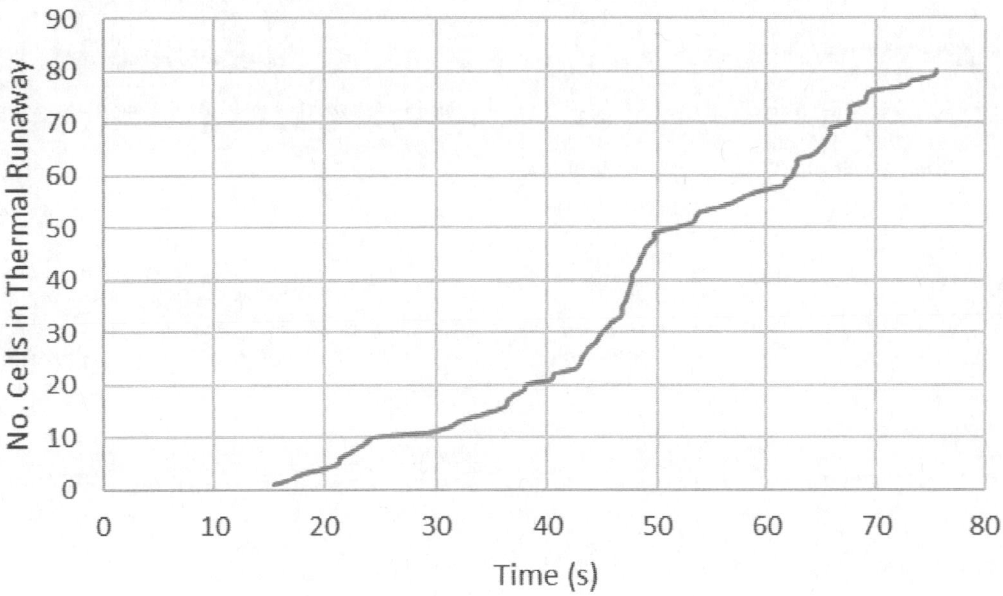

Figure 10 - Small NCA rate of propagation

a) View of SB_3 at 15s (smoke layer forming)

b) View of SB_3 at 30s (smoke layer heating to critical temperature)

c) Side view looking forward at 90s (smoke layer descending)

d) Side view looking forward at 150s (fire burning out, smoke layer retracting)

Figure 11 - Small NCA temperature isosurface

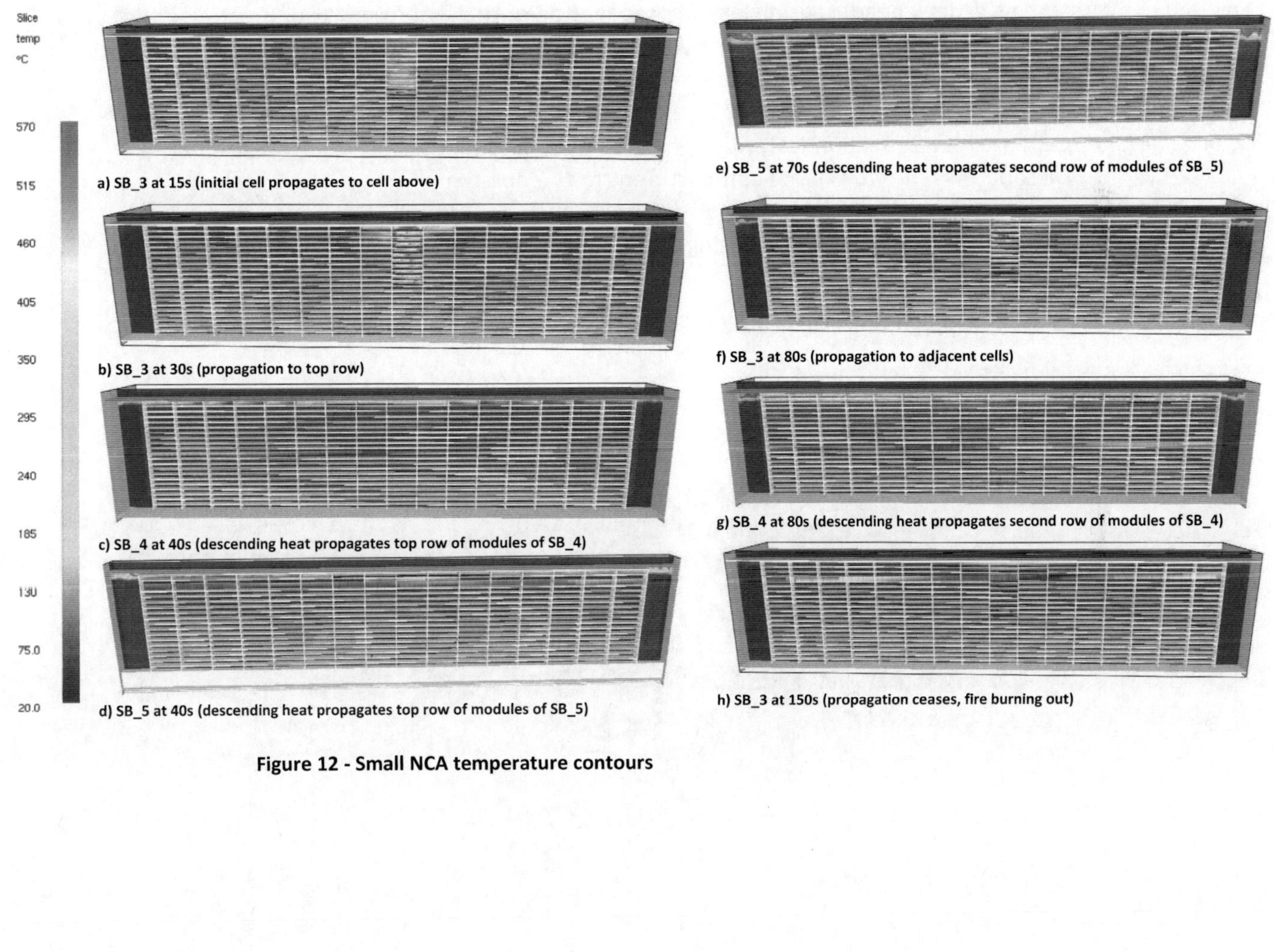

a) SB_3 at 15s (initial cell propagates to cell above)

b) SB_3 at 30s (propagation to top row)

c) SB_4 at 40s (descending heat propagates top row of modules of SB_4)

d) SB_5 at 40s (descending heat propagates top row of modules of SB_5)

e) SB_5 at 70s (descending heat propagates second row of modules of SB_5)

f) SB_3 at 80s (propagation to adjacent cells)

g) SB_4 at 80s (descending heat propagates second row of modules of SB_4)

h) SB_3 at 150s (propagation ceases, fire burning out)

Figure 12 - Small NCA temperature contours

SUMMARY RESULTS AND DISCUSSION

Propagation of Thermal Runaway

The results in all simulations show a similar pattern of thermal runaway propagation:

a. A cell fails, leading to thermal runaway;
b. Heat released by the failing cell rises to the top of the compartment, heating the cells directly above;
c. Upon reaching critical temperature, thermal runaway is propagated to the cells above which in turn release more heat;
d. A smoke layer begins to form at the top of the compartment which spreads evenly along the deck head, causing lateral propagation of thermal runaway along the length of the top row of the battery array;
e. Smoke begins to fill the compartment, causing the heat to descend toward the bottom of the compartment;
f. Cells are heated by the descending smoke layer, causing thermal runaway to cascade from the top of the compartment to the bottom;
g. Each additional cell that undergoes thermal runaway increases the temperature and volume of the smoke layer, causing further propagation of thermal runaway to the cells below.

The order of propagation is illustrated in Figure 13.

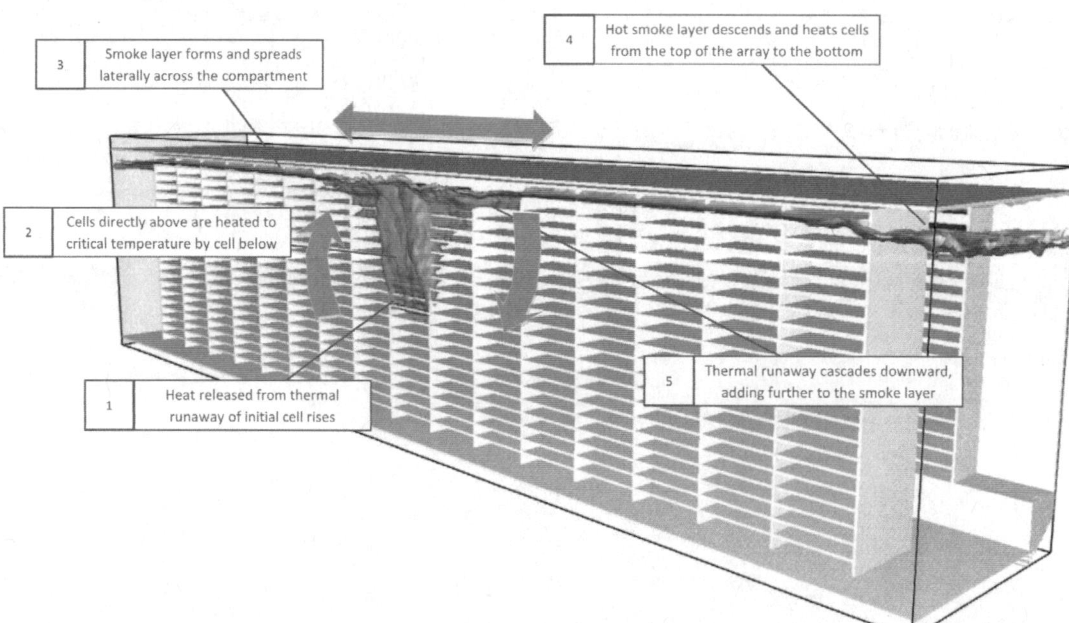

Figure 13 - Order of thermal runaway propagation within the battery compartment

In summary, the buoyancy of the heat and exhaust gases from the failing cell results in an initial upward spread of thermal runaway. The natural spread of smoke as it fills the compartment then causes a lateral spread of thermal runaway across the battery array. Finally, the descending smoke layer causes a downward cascade of thermal runaway from the uppermost cells to the bottom.

If the initial upward spread of thermal runaway creates a deep enough smoke layer, lateral spread of thermal runaway to the uppermost cells in the array could theoretically start a chain reaction ending in thermal runaway of every cell in the compartment. The likelihood of reaching this critical tipping point is dependent on the location of the failing cell. If the failing cell is located higher in the array, the initial upward propagation of thermal runaway will affect fewer cells and the resultant smoke layer may not descend low enough to cause lateral spread to the top row of cells.

Rate of Propagation

The rate of propagation can be gauged by the total number of cells undergoing thermal runaway over time. A comparison of propagation rates across each of the cell types is shown in Figure 14.

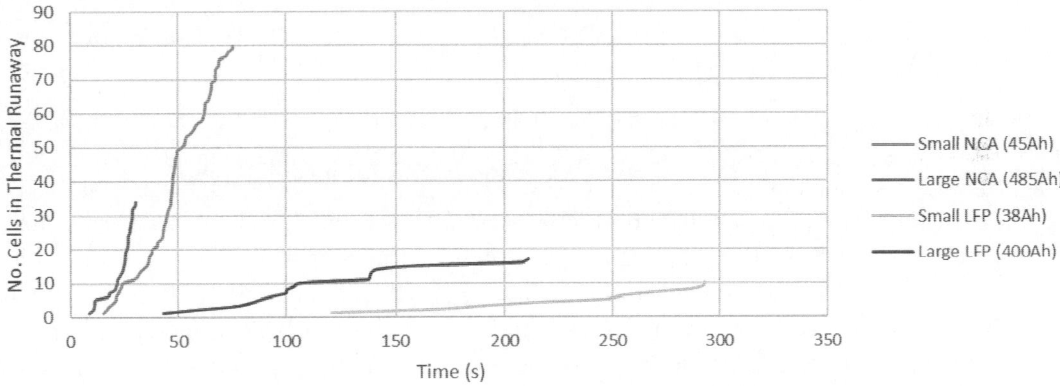

Figure 14 - Rate of thermal runaway propagation across all cell types

The average propagation rate for each cell type is shown in Table 2.

Table 2 - Average thermal runaway propagation rate

Cell Type	Average Propagation Rate (cells/second)
Small NCA (45Ah)	1.06
Large NCA (485Ah)	1.12
Small LFP (38Ah)	0.03
Large LFP (400Ah)	0.08

The NCA cells have a significantly greater propagation rate than the LFP cells. This is mainly attributed to their higher energy release and lower critical temperature. The relatively low energy release and higher thermal runaway threshold of the LFP cells results in a much slower spread of thermal runaway and a more manageable fire.

Large cells have higher propagation rates than their corresponding small cells due to their increased energy output. Prevention of cell to cell propagation is therefore more difficult for large cells, making small cells a comparatively more manageable option.

It should be noted that the total number of propagated cells was limited by use of an open boundary on the in-board side of the computation domain. This is particularly evident in the case of the Large NCA cells, as discussed above. Despite this limitation, the relative difference in propagation rates between cell types is clear.

Heat Release Rate

The global HRR measures the total heat released by all cells at a given time. The global HRR for all cells is shown in Figure 15.

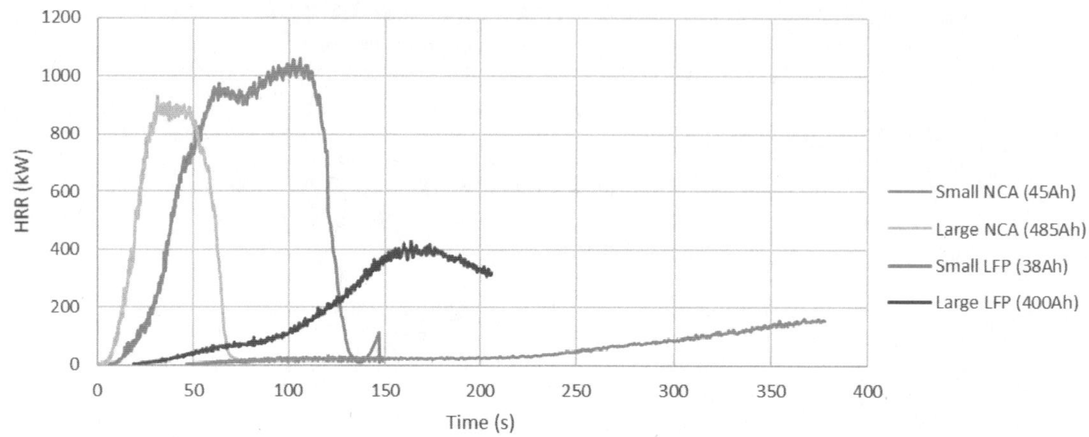

Figure 15 - Global HRR for all cell types

The NCA cells have a significantly higher global HRR than the LFP cells. This is due to a combination of the NCA chemistry having higher individual cell HRR in thermal runaway and a lower thermal runaway threshold. A lower thermal runaway threshold results in higher propagation rates and more cells burning at once which combined with increased heat release per cell results in maximum HRR of up to 1MW. By comparison, the Large LFP cells' peak HRR is more than 50% lower at 400kW.

In terms of cell size, larger cells were expected to have a higher global HRR than smaller cells of the same chemistry due higher energy storage capacity and consequently higher individual cell HRR. This was true for the LFP cells, however NCA cells showed slightly higher HRR for the small cells. This is mainly due to cell geometry and the open boundary used in the simulation which limited smoke layer depth. The Large NCA cells are four times taller than the Small NCA cell modules which situated the top row of NCA cells in arrays SB_3 and SB_4 below the smoke layer. This prevented further propagation of the cells and limited the overall HRR. In a full compartment simulation, it is expected that more of the Large NCA cells would undergo thermal runaway resulting in a higher global HRR.

SUBMARINE OPERATIONAL CONTEXT

Thermal Runaway Consequences

In the operational scenario being considered for this study, the submarine is operating at depth on battery power when the initial cell failure occurs. If unable to be mitigated, the simulation results show that, for all cell types, the failure of a single cell results in cascading thermal runaway which inevitably escalates to an uncontrollable fire and likely loss of the submarine.

In both the case of the Small and Large NCA cells, downward propagation and loss of the compartment happens within a minute. This makes it incredibly difficult to mount an effective response and bring the situation under control, regardless of the fire safety measures employed on the submarine.

In comparison, LFP cells take in excess of five minutes to commence downward propagation. With adequate detection, this grants the crew and any automatic systems a reasonable amount of time in which to affect a response to bring the fire under control.

In any case, as the DNV-GL Guideline [7] recommends, preventing cell to cell or module to module propagation is critical to maintaining control of the fire, minimising damage, recovering the submarine and resuming the mission. As a minimum, lateral and downward propagation must be avoided as this is the tipping point after which preventing further thermal runaway becomes insurmountable.

Potential Mitigation Measures

Fire safety measures were not part of the simulation scope, however based on the simulation results, potential mitigation measures can be postulated and are discussed in this section.

Battery Cooling System

Depending on the ambient environment within the battery compartment, a cooling system may be necessary to regulate the temperature of cells during normal operation, however may not be sufficient to handle a thermal runaway event.

Once thermal runaway is initiated, the self-sustaining chemical reactions taking place in the cell are difficult to stop. An active thermal management system which provides cooling direct to each cell or cell module could be designed to provide adequate extraction of heat from a cell undergoing thermal runaway in order to stop propagation to neighbouring cells and prevent cell rupture or explosion.

This could be one of the more effective mitigation measures, as it prevents cell to cell propagation and is applied directly to the cell which eliminates issues related to obstructions, such as the module housing and battery array structure, which limit the effectiveness of other mitigation measures.

Battery Management System

A BMS which provides persistent monitoring of battery health and performance is critical to early identification of malfunctioning cells which could go into thermal runaway. This could be a highly effective mitigation measure as it provides advanced warning, giving the crew time to isolate or remove the effected cell, provide additional cooling or take other steps to stop cell to cell propagation or prevent thermal runaway altogether.

Compartment Design

The simulation results show that propagation from array to array occurs via the smoke layer which travels along the deck head of the compartment. Structural features such as baffles or recesses could be designed into the compartment to stop or slow the spread of smoke to other arrays.

This may help to slow the propagation of thermal runaway and isolate the affected battery array. However, this does not prevent cell to cell propagation and would not serve as a standalone mitigation strategy.

Insulation

Insulation of cells or cell modules can help prevent or slow the propagation of thermal runaway in two ways:

a. Slowing or limiting the heat released by the cell undergoing thermal runaway;
b. Slowing or limiting the heat transferred to neighbouring cells.

This may not be sufficient as a standalone measure however, but could be combined with other mitigation measures such as an active cooling system.

Hypoxic Atmosphere Control System

Maintaining a hypoxic atmosphere in the battery compartment could help to reduce fire intensity. This would be most effective in limiting the impact of hot, flammable gasses vented by cells which ignite upon contact with oxygen.

This may help to slow the escalation of a battery fire, however would need to be part of a strategy which employs other measures to prevent cell to cell propagation of thermal runaway.

Fixed Fire Suppression System

Due to the self-sustaining nature of thermal runaway, a fixed suppression system will have limited impact on the propagation of thermal runaway. With the exception of a water mist system, this is also due to the fact that the battery array structure and module housing obstructs the application of suppressant directly onto the cells.

A water mist system may prove to be more effective than other suppression system types for fighting thermal runaway fires. Assuming that nozzles are appropriately placed and configured, the fog created by a water mist system could provide cooling to cells undergoing thermal runaway as well as adjacent cells, and effectively slow or prevent the propagation of thermal runaway. More work is required in this area to evaluate the effectiveness of a water mist system in this application and investigate suitable droplet sizing, dispersion rates and other system parameters

Regardless of suppression type, a fixed fire suppression system will help to minimise a battery compartment fire, and will be necessary to suppress the burning of materials other than the battery cells such as cables, electronic equipment and other combustible materials.

Fire Rated Construction

Fire rated construction may help to preserve the structural integrity of the battery compartment and the submarine as a whole for longer. It is also something that will mitigate the spread of fire to adjacent compartments through heating of compartment boundaries. This is a measure that only becomes effective when the fire becomes reasonably large, and will not have any impact on the propagation of thermal runaway.

CONCLUSIONS

Due to the relative immaturity of lithium-ion battery technology and increased probability of failure, lithium-ion batteries are the highest risk proposition for next generation diesel-electric submarines with respect to fire safety.

Preliminary design calculations and development of the trial design showed that a relatively high number of lithium-ion cells will be required to power a large conventional submarine, ranging between 15 and 200 times the number of lead-acid cells required for equivalent energy capacity. This creates a challenging engineering and fire safety design problem.

Thermal runaway propagation within the battery compartment was shown to occur with all cell types, with each simulation exhibiting similar dynamics. Initially, thermal runaway propagates to the cells above the failing cell due to the rising heat. Formation of a smoke layer causes lateral propagation across the top row of a battery array, and finally a downward cascade of thermal runaway is experienced as the heated smoke layer descends.

Providing small cells (30Ah - 50Ah) are packaged in suitably sized sub-packs or modules, they could provide a significantly safer option from a fire safety perspective compared to larger (400Ah+) cells. Whilst larger cells provide reduced complexity and packaging requirements, it is expected that they will pose a higher fire risk with an increased probability and rate of thermal runaway propagation from cell to cell and array to array. Due to lack of large cell testing to date, this is based on extrapolation of small cell fire test data and needs to be qualified with fire testing of large lithium-ion cells.

Despite the increased number of cells required to achieve equivalent storage capacity, LFP cells appear to be the safest and most logical choice for a conventional submarine main storage battery. This is due to the increased thermal stability and relatively benign thermal runaway characteristics compared to NCA cells and other chemistries.

CFD simulation results indicated that all cell types were capable of cascading thermal runaway. This shows that, irrespective of cell size or chemistry, thermal runaway of a single cell or module within a lithium-ion main storage battery can lead to an uncontrollable fire and probable loss of the submarine, if left unmitigated. The Small LFP cells were found to have the lowest propagation rate and HRR, suggesting this cell type could be the most suitable and manageable from a fire safety perspective.

In terms of maximising fire safety outcomes for lithium-ion batteries, the most effective mitigation measures are those able to prevent cell to cell or module to module propagation of thermal runaway. Of those mitigation measures discussed in the study and based on simulation results, an active thermal management or cell cooling system and Battery Management System will be integral in a fire safety strategy for lithium-ion batteries.

Research into lithium-ion battery fire test data showed there is a lack of consistent data, particularly for the large format and high capacity cells under consideration for future submarines. Further battery testing is required to provide a more consistent and comprehensive set of data to facilitate modelling and design development.

REFERENCES

[1] Australian Department of Defence, An Introduction to Collins, Canberra, Australia: Australian Department of Defence, 1991.

[2] C. J. Orendorff, J. Lamb, L. A. M. Steele, S. W. Spangler and J. Langendorf, "SANDIA REPORT SAND2016-0486 - Quantification of Lithium-ion Cell Thermal Runaway Energetics," Sandia National Laboratories, Albuquerque, New Mexico, 2016.

[3] EAS - Germany, *HP 602030 NCA - 45 Ah/ 162 Wh Lithium Ion Ce ll,* Nordhausen, 2013.

[4] EAS - Germany, *HE 1682100 NCA - 485 Ah/ 1746 Wh - Prototype Lithium Ion Ce ll,* Nordhausen, 2013.

[5] Sinopoly, *Sinopoly Lithium Ion Battery Specifications - Model: SP-LFP400AHA,* Hong Kong, 2015.

[6] U. Gabler, Submarine Design, Austria: Bernard & Graefe Verlag, 2000.

[7] DNV-GL, "DNV GL Guideline for Large Maritime Battery Systems," DNV-GL, Oslo, Norway, 2014.

[8] Naval Ships' Technical Manual Chapter 555 - Volume 2 Submarine Firefighting, Fifth Revision, USA: Naval Sea Systems Command, United States Navy, 1998.

[9] Telegraph Media Group Limited, "Nuclear submarines hit by more than 200 fires in the past 25 years," 8 July 2012. [Online]. Available: http://www.telegraph.co.uk/news/uknews/9384766/Nuclear-submarines-hit-by-more-than-200-fires-in-the-past-25-years.html.

[10] Telegraph Media Group Limited, "Sailors killed in Indian submarine explosion," 14 August 2013. [Online]. Available: http://www.telegraph.co.uk/news/worldnews/asia/india/10241917/Sailors-killed-in-Indian-submarine-explosion.html.

[11] S. Unnithan and S. Aroor, "INS Chakra formally joins Indian Navy today," 4 April 2012. [Online]. Available: http://indiatoday.intoday.in/story/ins-chakra-formally-joins-indian-navy-nuclear-submarine/1/182889.html.

[12] Wikimedia Foundation, Inc., "List of submarine incidents since 2000," 9 March 2015. [Online]. Available: http://en.wikipedia.org/wiki/List_of_submarine_incidents_since_2000.

[13] Woodbank Communications Ltd, "Electropaedia," Woodbank Communications Ltd, 15 June 2005. [Online]. Available: http://www.mpoweruk.com/chemistries.htm. [Accessed 20 August 2015].

[14] BMT Design & Technology, *BMT Submarine Design & Engineering Course,* Melbourne: BMT Design & Technology, 2015.

[15] A. Janke and B. Nagorny, *Lithium-Ion Batteries for Submarine,* Kiel, Germany: ThyssenKrupp Marine Systems, 2015.

[16] N. Ponchaut, K. Marr, F. Colella, V. Somandepalli and Q. Horn, *Thermal Runaway and Safety of Large Lithium-Ion Battery Systems,* Natick, USA: Exponent Inc., 2015.

[17] A. W. Golubkov, S. Scheikl, R. Planteu, G. Voitic, W. Helmar, C. Stangl, G. Fauler, A. Thaler and V. Hacker, "Thermal runaway of commercial 18650 Li-ion batteries with LFP and NCA cathodes – impact of state of charge and overcharge," *RSC Advances,* vol. 5, no. 70, p. 57171 to 57186, 2015.

[18] P. Mullen, "Special Conditions: Honda Aircraft Company, Model HA-420 HondaJet, Lithium-Ion Batteries," Aircraft Certification Service, Kansas City, 2015.

[19] M. Jacoby, "Assessing The Safety Of Lithium-Ion Batteries," *Chemical & Engineering News,* vol. 91, no. 6, pp. 33 - 37, 2013.

[20] International Air Transport Association, *Lithium Batteries - Risk Mitigation Guidance for Operators 1st Edition,* Montreal, Canada: International Air Transport Association, 2014.

[21] Naval Sea Systems Command, *Lithium Battery Systems Navy Platform Integration Safety Manual,* Bethesda, USA: Naval Sea Systems Command, 2011.

[22] J. Battista, *High-speed imaging tests on batteries to help design safe enclosures on Navy ships,* Bethesda, USA: NAVSSES Public Affairs, Naval Sea Systems Command, 2015.

[23] J. Banner and C. Winchester, *Lithium Battery Safety in Support of Operation Fielding of Unmanned Underwater Vehicles,* Bethesda, USA: Naval Surface Warfare Center, Carderock Division, 2011.

[24] Saft, *Saft lithium battery systems - Cutting-edge technology for underwater vehicles,* Maryland, USA: Saft Specialty Battery Group, 2010.

[25] P. Kallender-Umezu, "Japan To Make Major Switch on Sub Propulsion," Gannett Government Media Corporation, 29 September 2014. [Online]. Available: http://archive.defensenews.com/article/20140929/DEFREG03/309290032/Japan-Make-Major-Switch-Sub-Propulsion. [Accessed 27 August 2015].

[26] Want China Times, "PLA developing submarines powered by lithium-ion batteries," WantChinaTimes.com, 31 May 2015. [Online]. Available: http://www.wantchinatimes.com/news-subclass-cnt.aspx?id=20150531000123&cid=1101. [Accessed 2015 August 27].

[27] S. Gregory, *The Use of Cone Calorimeter and Associated Techniques in the Determination of Fire Hazards,* London, UK: South Bank University, 2000.

CONFERENCE ABSTRACT

Design and Qualification of Submarine Engines

Arndt von Drathen, Principal Engineer Submarine and MCMV Applications
at MTU Friedrichshafen GmbH, mailto: Arndt.vonDrathen@mtu-online.com

Conventional submarines charge their batteries under snorkelling conditions. Short term snorkel closures and regionally varying wave heights and lengths significantly change the operating conditions of the charging unit's diesel engine. One of the primary design criteria for a submarine engine is to make sure that its thermal loading and the variation is kept to a minimum under such changing environmental conditions to ensure continuous and safe operation. Other primary design criteria are low acoustic emissions as well as a high reliability.

This paper focuses on the design and its verification of submarine specific engine characteristics. It describes design and qualification methods as well as the test bench setup to simulate intake air depression and exhaust gas backpressure variations applicable to engine operation on board submarines.

Author's Biography

A trained ship mechanic, Arndt von Drathen graduated in 2002 from Flensburg University of Applied Sciences as a marine engineer. After writing his thesis during his time with MTU's R&D department, he joined MTU's technical sales department focusing on propulsion system solutions.

From July 2006 until December 2014 Arndt headed the Submarine and Minehunter Group within the MTU Application Centre Marine, being responsible for the conceptual design, detailed engineering and processing of all naval propulsion system projects for these applications.

Due to his strong engineering background, Arndt has been appointed project leader for the Series 4000 submarine engine development project and acts as principal engineer for Submarine and Minehunter application projects.

Submarine Main Propulsion using Lithium-ion Batteries

Choong Yeon CHONG
cychong@kokam.com

I PREFACE

Lead-acid battery (LAB) technology, even with its drawbacks in power and energy density, has survived a century as a source of stored energy for main propulsion or as stand-by batteries for submarines. However increased endurance and speed demands have stimulated the development of a new generation of energy storage technology, based on mature Lithium-ion battery (LIB) technology. The strategic advantages from deployment of LIB compared to LAB include:

- Higher endurance ratio, 2 to 5 times improvement
- Lower indiscretion ratio, some 5 times improvement
- Maximum speed endurance can be extended over 7 times
- The ability to obtain these benefits, without the bulk and complexity of AIPS

Kokam, based in South Korea, is a world-renowned LIB solution provider especially for Defence applications. Over a decade Kokam has been deploying its highly specialized and qualified LIB in submersibles, marine vessels and combat torpedoes. Following years of research and development, involving the progression from specialized lithium-ion cell chemistry to complete onboard systems with step by step scale up; Kokam has now successfully installed large arrays of LIB for larger displacement submarines. This is now at the commissioning stage.

The developments include unique short-circuit resistive chemistry, ultra-high energy and power, faster charging (for the same capacity) and significantly longer cycle life, compared to not only lead-acid but also other lithium-ion technologies.

Furthermore, Kokam has developed modules and other LIB dedicated equipment such as battery management systems (BMS), which are fully optimized for its cell characteristics and facilitates replacement of existing lead-acid technology without major modification to the existing submarine design and legacy equipment. Kokam's LIB system design criteria are established in an order of priority with safety at the highest level. We ensure continuity of power supply to enable recovery from contingency events.

Remarkable advances in terms of endurance and indiscretion ratio with the use of LIB provide compelling strategic and tactical advantages, along with a high value proposition for both submarine operators and industry. The elimination of a number of auxiliary systems required for LAB, coupled with the use of maintenance free LIB cells, enables a large reduction in operating costs with less demands on both the boat and land based crews. The technologies gained from this submarine battery development are now also becoming available for full electric high tonnage marine vessels.

Figure 1 Artist's impression of a submarine with Kokam standard LIB

II LITHIUM-ION CELLS

A. Cell type

Kokam manufactures prismatic pouch cells as shown in Figure 2.

Figure 2 Generic types of Lithium-ion cell

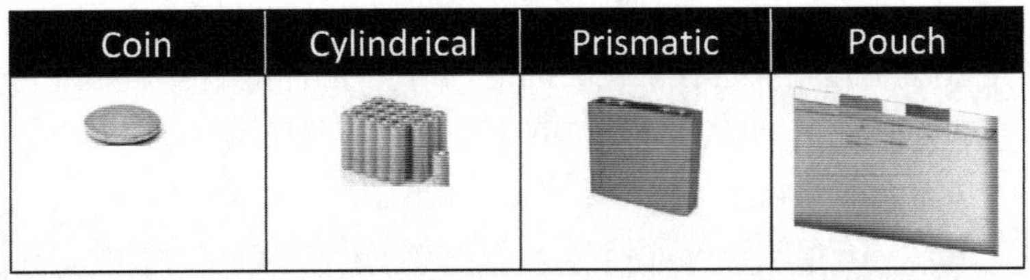

From its beginning in the early 2000's, Kokam has concentrated on chemistry for submarine applications, taking advantage of the remarkably lower internal impedance and extremely large capacity of its own cells, which stem from its patented Z-folding stacking and production know-how of large form factored cells. The "low impedance - low heat" initiative enabled a large arrayed, high capacity, LIB to work at very high efficiency with no cooling.

For ions/electrons, the distances to travel are shorter in Z-folded electrodes while longer in winded electrodes, as graphically shown by the red arrow in Figure 3.

Figure 3 Z-folding as stacking vs winding

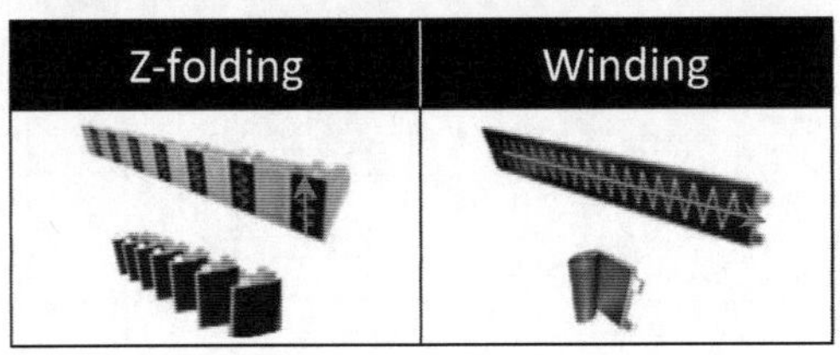

The longer the distance the larger the internal impedance. Because cylindrical and prismatic Lithium-ion cells have to use winding electrodes, they dissipate more heat in the same capacity and heat rapidly increases at higher charge/discharge rate.

Figure 4 and 5 shows the comparison of the heat dissipating from different type of Lithium-ion cells. It gets more heat if heat isn't evenly distributed and the heat dissipation area is smaller for the same capacity. The differences between Pouch and Prismatic or Cylindrical cells become more distinctive as the C-rate increases.

Figure 4 Heat dissipation comparison at 3C (20 minute discharge)

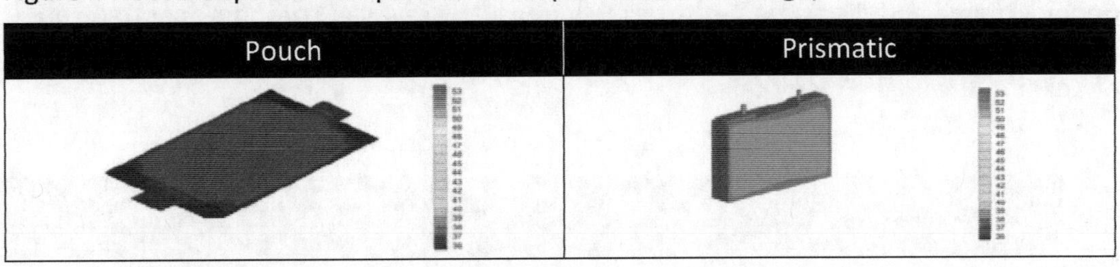

Figure 5 Heat dissipation comparison at 1C (1 hour discharge)

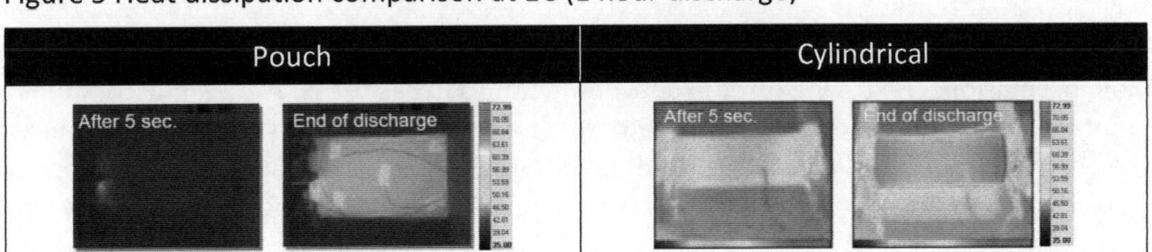

B. Cell chemistry

Lithium-ion cells are composed of cathode, anode, separator, electrolyte, case (packaging) and tabs. Amongst them, the first four components are categorized as key components.

Information about the many types of generic chemistry are all quoted from the public domain and the key characteristics of them are illustrated by the radar chart as shown Figure 6. It should be noted that the radar chart scale and the weightings by the radial length for each characteristic are just for relative comparison purpose, not actual scale or values.

Figure 6 Radar charts for characteristics of cathode and anode chemistry

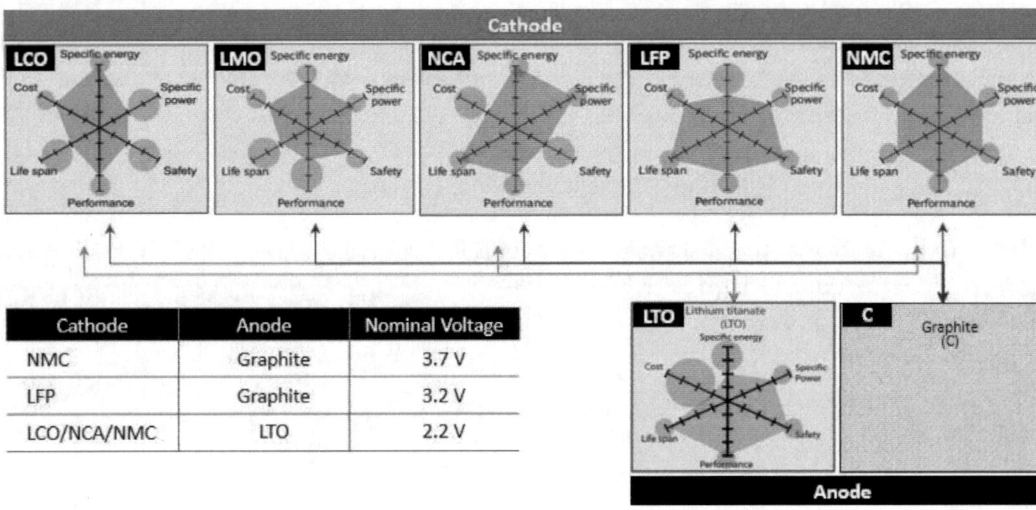

Cathode	Anode	Nominal Voltage
NMC	Graphite	3.7 V
LFP	Graphite	3.2 V
LCO/NCA/NMC	LTO	2.2 V

Kokam have used all of the chemistries above and their derivatives or mix of them depending on applications. NMC - graphite was selected for the reasons of a considers balance of the pros and cons with respect to submarine use. As Figure 6 shows, LFP (Lithium Iron Phosphate) and LTO (Lithium Titanic Oxide) have good safety characteristics but less energy density. LTO has long life but is more expensive.

Table 1 shows Kokam's definition of type of cells under sales in terms of maximum continuous charge and discharge C-rate. Energy density is reversely proportional to C-rate (I C is a discharge in one hour, 2C a discharge in 30 minutes). It reaches up to 270 Wh/kg for the case of Kokam Ultra-High Energy cells.

Table 1 Kokam's definition of type of cells (charge/discharge)

Ultra-high Energy	High Energy	Hybrid	High Power	Ultra-high Power	Super-high Power
0.5C / 1C	1C / 2C	2C / 5C	3C / 8C	5C / 10C	*/ 100C

Kokam selected High Energy Type cells and tuned the chemistry for use as submarine main propulsion. The tuning takes in account factors such as the operating environment temperature and safety. The customer's load profile and the required cycle life are also used for the chemistry optimization process. In the case of off-shore vessels with dynamic positioning systems, or container ships with thrusters, Kokam Ultra-High Power chemistry is offered.

III SYSTEM ARCHITECURE

LIB system offers flexible configuration schemes and such flexibility is supported by series-parallel connectivity of the cells via the layered system of components known as Modules, Strings, and Banks.

Each layer is supervised by the respective Battery Management System (BMS). Module BMS (M-BMS), the lowest elementary BMS monitors cell voltages and temperatures, and balances cell voltages, under supervision of the upper layered BMS systems.

Kokam LIB system is designed and constructed strictly with reference to US military standards and complies with US MIL-STDs and US NAVSEA S9310-AQ-SAF-010 (Technical manual for navy lithium battery safety program responsibilities and procedures).

Kokam's decade of deployment of LIB systems for multiple hundreds of submersibles, ranging from manned commercial and military submarines to unmanned underwater vehicles, featuring unique patented technologies enables it to offer LIB systems of unrivaled performance and inherent safety for ocean-going conventional and nuclear submarines.

Figure 7 Kokam's generic LIB system architecture

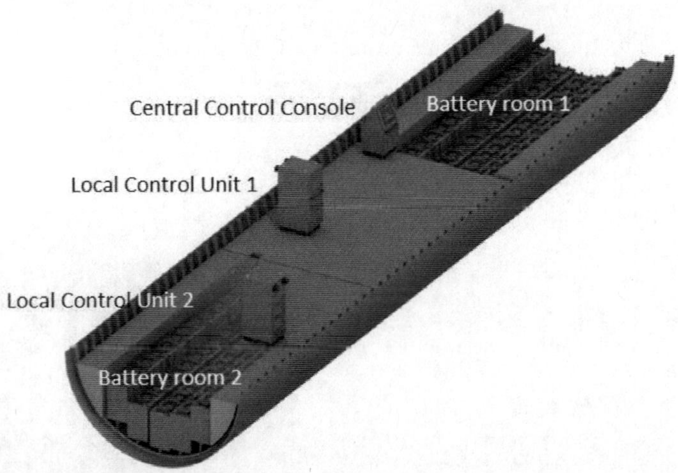

Kokam's generic LIB system architecture is simple. LIBs located in each battery room are monitored and controlled by the dedicated Local Control Unit (LCU).

Central Control Console (CCC) supervises complete LIB down to cell level via LCUs.

LCU comprises of a number of String Control Power Module (SCPM) that includes String BMS

The CCC comprises the System BMS (SYS-BMS) and a number of Bank BMS (B-BMS), and the Machine Interface (MMI) to monitor and control the complete LIB system and other necessary interfaces to submarine systems.

Industrially well-known and proven CAN bus is used for communication among types of BMSs. For external interfaces, CAN bus or Ethernet or any other systems are provided as per customer requirement. All switches are mechanical or electro-mechanical or power IGBT type (no electronic switches are used). Flexible bus-bars, instead of cables, are widely adopted for power transmission. All parts and components are Class certified.

Auxiliary supporting systems, as required for LAB, for such as cooling, hydrogen elimination, distilled water supply, acid circulation and battery room flushing are not required for LIB. By-pass bridging to decouple a faulty cell is also required for LIB. For these reasons LIB eliminates not only a substantial operating/maintenance burden for crew during the mission, but also design complexity.

A. Module

It has been Kokam's design strategy to maximize the use of non-conductive non-corrosive composite material for both trays and enclosures to minimize potential risk of leakage current and galvanic corrosion, offering a maintenance free battery module.

To enable existing submarines to replace used lead-acid cells with a LIB, the module (which is equivalent to a "cell" of LAB) is designed to be installed in the same way as lead-acid cells are installed.

Figure 8 shows a scaled view of conventional type 209 submarine Lead-acid cells, globally standardized and one of Kokam's standard LIB modules, of which capacity can be scaled with different capacity cells and/or number of cells in a submodule. The energy density of the LIB module for a typical coastal submarine is illustrated in Figure 8, being about 3X and 5X respectively for volume and weight, compared to the type 209 Lead-acid cell.

Figure 8 Scaled view of CL209 Lead-acid cell and Kokam's LIB equivalent

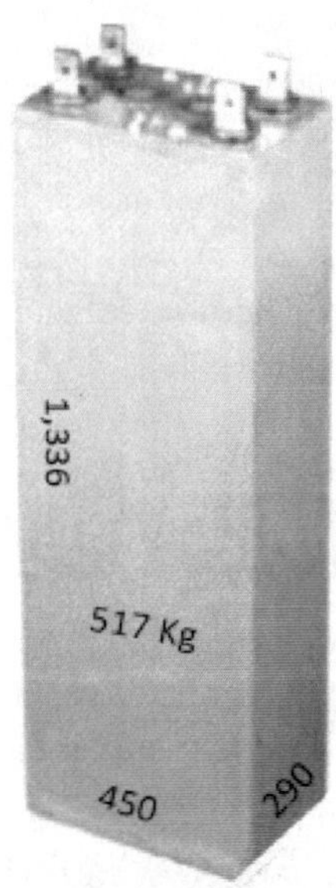

Energy	LAB	LIB
Volume density	1	≈ 2.9
Weight density	1	≈ 4.9
Capacity	1	≈ 2.7

Each module can be fitted with up to 4 submodules vertically arranged and connected in series or parallel. Each submodule consists of an array of trays which contain a single cell per tray.

Several sorts of chemistry, dependent on the customers tactical-technical requirements, with a standard cell capacity of 200Ah and form factor (foot print) are adopted.

Figure 9 shows an exploded view how a module is constructed.

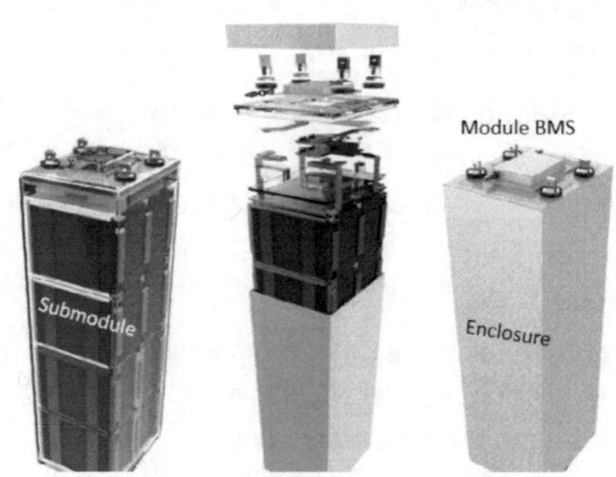

Figure 9 Explode view of module design

B. String

A LIB string is an electrical configuration consisting of a number of modules serially connected to reach the design propulsion network voltage, with one String Control Power Module (SCPM) which is remotely installed.

A number of SCPMs, each of which is composed fuses, power switches with driver, current limiter, and String BMS (S-BMS), are housed in LCU

SCPMs work autonomously in collaboration with the next higher-level Bank BMS (B-BMS) and lower BMS, M-BMS. It can be interrupted manually.

Any string that meets fault criteria is then automatically disconnected by power switches from the propulsion network. When an external short is projected the current limiter is autonomously activated to keep continuing power supply to the propulsion network until the external short circuit is removed.

Figure 10 LCU fitted with 7 SCPMs

C. Bank

A Bank is also an electrical configuration that consists of a number of Strings connected in parallel. It works with a B-BMS which is remotely located and installed in CCC. A Bank is comparable to the partial battery of LAB.

D. System

This is the highest level electrical configuration that consists of a number of Banks. It works with SYS-BMS which supervises overall system status and functions though B-BMS and provides interfaces to other submarine systems such as integrated machinery control system.

Figure 11 CCC

IV INTEGRITY OF SAFETY

Unlike surface vessels, submarines are a completely enclosed system in that crew have to breathe regenerated air when submerged. In this regard, not only fire but also smoke is fatal to safety.

As is well known to the public, some types of Lithium-ion cells are prone to ignite and explode and, an explosion of a cell can propagate rapidly to the adjacent cells (termed "thermal runaway"). The larger the capacity and higher the energy density, the more difficult thermal runaway becomes to control.

Hydrogen leaks from Lead-acid cells are monitored and eliminated by burning continuously, likewise, irregularities of Lithium-ion cell temperature and voltage, and string current are supervised by BMSs to detect, analyze and deactivate the respective components to secure safety.

Root causes of a LIB fire stem from two main streams; abuse and (progressive development of) latent defects. The former could be prevented or overcome by safety measures, while the latter may appear without warning. In general, it is hard to detect or forecast, and eliminate latent defects, once the cells are packaged and installed. For this reason, the most practical measures are "preventive" which includes diagnostic inspection, aging and grading of the cells and modules.

Table 2 lists Kokam's intrinsic key safety features to support the integrity of safety, in addition to general safety measures such as fuses, switches, SCPMs, BMS and the afore mentioned preventative measures

Table 2 Kokam LIB key features that support safety integrity

Element	Key features
String	• Smart current limiter
Module	• Non-conductive tray and enclosure
Cell	• Tab fuse • Ignition resistive chemistry

A. Smart current limiter

Submarine LIB must survive external short-circuits to keep the submarine operational. This is done by a current limiter when there are symptoms of an external short-circuit occurring.

This smart current limiter, as a key functionality of the SCPM, enables limited inrushing string current, when an external short-circuit is occurring and allowing continued supply of LIB power to the propulsion network to prevent the submarine from black-outs or damages to the legacy equipment, or the LIB itself, until the external short-circuit is removed or returned to normal state.

Figure 12 Generic configuration of Kokam SCPM

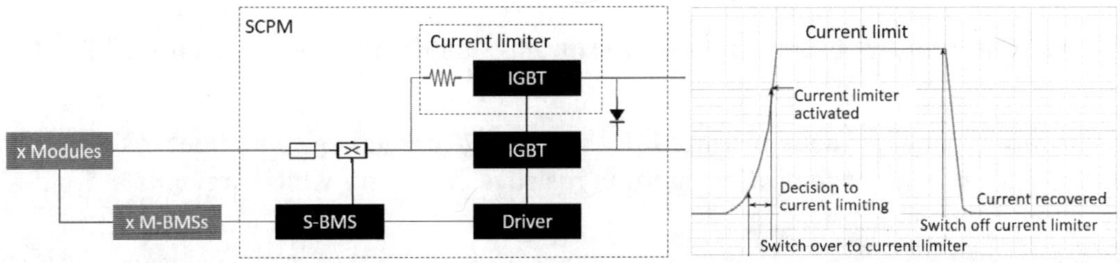

In Figure 12, The IGBT driver, supervising a differential voltage between input and output of the lower IGBT, determines in real-time whether to turn off the lower IGBT and turn on the upper IGBT to limit the inrushing current.

The limit current and time are determined in consideration of submarine specification and maximum allowable current per String and Bank. During this limiting time period, the external short-circuited current is removed then the current limit circuit is disconnected and the original power line, via the lower IGBT, is connected. This principle works in the reverse direction in the same manner to protect the battery.

Every String is equipped with a SCPM, autonomously working without report to or command from the String BMS. The fuse between the battery and SCPM will be blown if the SCPM does not function. Still another safety back-up feature is available – the cell tab fuse.

B. Non-conductive tray and enclosure

Accumulated flow of leakage currents may lead corrosion to cell packaging and the progressive development of corrosion may introduce electrolyte leaks when cells lose their mechanical integrity.

To minimize leakage currents between cells and cells to grounds, the tray for installation of a single cell and enclosure of group of trays are all made of non-conductive materials

C. Tab fuse

Tab fusing current and time are determined by the tab design which is also one of Kokam's in-house engineering capability.

Figure 13 Testing jig for tab fusing current and time

Test set-up	Test switch-on	Test switch-on	Test result
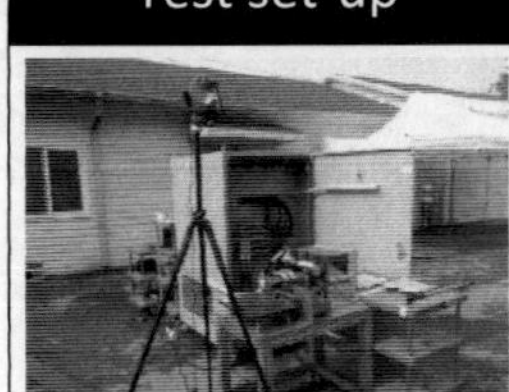	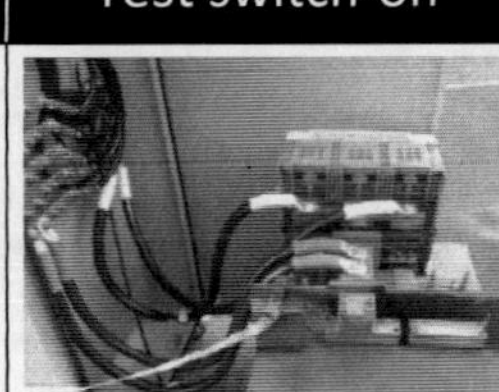		

D. Ignition resistive chemistry

The best safety measure is to provide LIB cells that do not ignite and burn, which to date have not been technically and commercially viable.

Kokam's globally unique patented "Nano coating" technology with heat resistive separator and gas observing agent offers Ignition resistive chemistry which prevents a fire if an internal short-circuit occurs.

This coating technology in the cell production stage transplants safety characteristics of Lithium Iron Phosphate (LFP) and Lithium Titanic Oxide (LTO) to the cathode and anode core material respectively, keeping the intrinsic characteristics of the cathode and anode active material unchanged.

Transplanting is done by coating nano size particle of LFP (nLFP) and LTO (nLTO) onto the surface of cathode and anode core material. Figure 14 shows this technology conceptually and the actual photograph of nano coating.

Figure 14 Nano coating concept and photo

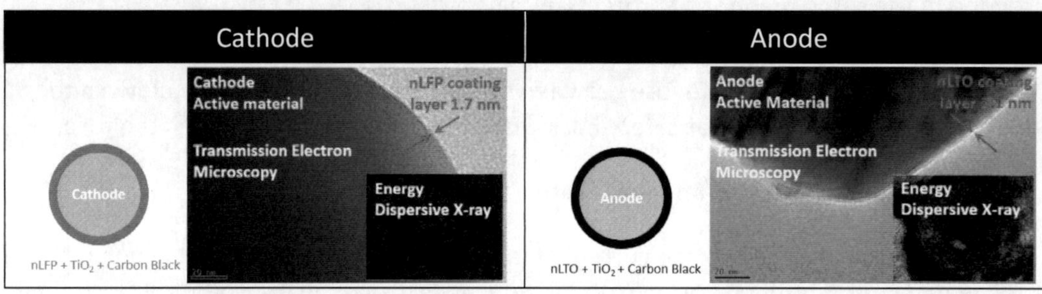

The results of the development tests were all far beyond expectations. Only a volume of non-toxic evaporated gas was emitted at the time and for a few seconds after the penetration was made and the short circuit established, as shown in Figure 15. Cells have been alive all the way but with some small reduction of voltage.

Figure 15 Nail penetration and abuse test

LIBs are more likely to be exposed to overcharge during charging, especially in mobility applications for which a high voltage is generally used, and if a collision occurs with an obstacle, in which case a foreign substance may penetrate into the battery to cause an electric short.

When charging, Lithium ions in the cathode are transferred to the anode and in the opposite direction when discharging a Lithium-ion cell.

In the case of excessive charging, the concentration of Lithium in the cathode is lowered below any critical limit, and the cathode material is then decomposed to generate oxygen, and the temperature is raised by the heat resulting from the decomposition process. When the anode has an unacceptable Lithium ion concentration, the Lithium is accumulated as a metal on the surface of the anode. If the cell voltage rises continuously, the electrolyte is then decomposed into various irreversible combustible gases. Now, oxygen, combustible gases and high-temperature composes an environment which is capable of causing fire or explosion.

The essence of the Kokam nano invention is in that; the surface of the cathode is coated with nano size particles of olivine-type Lithium Iron Phosphate Oxide which generates a very small amount of heat and hardly any oxygen in comparison to an uncoated cathode. In this way the decomposition of the cathode is delayed, thereby greatly improving the safety, even in an overcharging condition. The surface of the anode is coated with nano size particles of spinel-type Lithium Titanate Oxide to greatly lower electric resistance, which is advantageous for high output. If an anode is short-circuited to a cathode, Lithium Titanate Oxide is converted into a non-conductor to suppress the flow of current, thereby greatly improving the safety even in an internal short circuit condition.

Kokam proposes the nail penetration experiment as a representative test for replicating the internal short circuit. If a nail penetrates a cell so that the cathode and anode make direct contact in the cell, electric energy is rapidly focused at the shorted region, and thus high current flows. Heat caused by the flow of high current is hot enough to melt or burn substances of the cell and gives an influence on the electrolyte, which is an organic solvent with high flammability, thereby causing fire or explosion.

Kokam Nano coating technology can be summed up as ensuring safety of both LFP and LTO cell by coating nano-sized particles on the cathode and anode, while maintaining an inherent high energy density Nickel Manganese Cobalt (NMC) based cathode and carbon-based anode active materials.

After all, Kokam demonstrated the cell technology which it assured to support hybridization of energy and power as required for main propulsion for conventional and stand-by for nuclear submarines at a minimized connectivity among the cells given intrinsic safety with no cooling requirement.

Being a manned platform, with no exit for smoke being provided when submerged, submarines may be the worst platform at the risk of thermal runway. However, this unique Kokam Nano technology, including other key technologies for the enhancement of safety was the result of such effort to exclude potential risk of fire resulting from shorted electrodes.

Kokam Nano technology with different capacities and power ratings has been widely used for public transportation such as e-bus or e-tram and various types of military mobility, as well as for torpedoes (as the main source energy storage), thanks to the afore mentioned inherent safety features.

6T battery cells with Kokam Nano technology exceed the requirements of US MIL PRF 32565 (Performance specification battery, rechargeable, sealed, 6t lithium-ion) and will soon be introduced to market.

V QUANTUM LEAPS IN PRINCIPLE PERFORMANCE

A. Submerged Endurance

Submarine designers would have made it a rule to run a submarine about one hour at the maximum speed and up to four days at a lowest speed, until AIPS was introduced. A generic AIPS extended the submerged endurance at the lowest steerable speed by approximately a factor of about 3X to 4X. But this is once only during a mission period at sea. Most AIPS cannot extend the endurance when operating a little above one-third of maximum speeds.

Kokam LIB has been proven using LBTS to extend the submerged endurance by a minimum 2X at the lower speeds and 5X at the highest speed, compared to LAB; if they are installed in the same space. This is because of not only energy density but also the unmatchable efficiency of generic LIB at higher C-rates (proportional to submarine speeds).

Figure 16 shows the efficiency comparison from 100-hour rate (0.01C) to 1-hour rate (1C). It is clear that LIB offers increase of endurance about 2.5 times at maximum speed with the same capacity. Thanks to 2.7 times of energy density and 2.5 times of efficiency at the maximum, compared to LAB, the maximum speed endurance can be extended over 7 times.

Figure 16 Efficiency between Lead-acid cell and Kokam Lithium-ion module

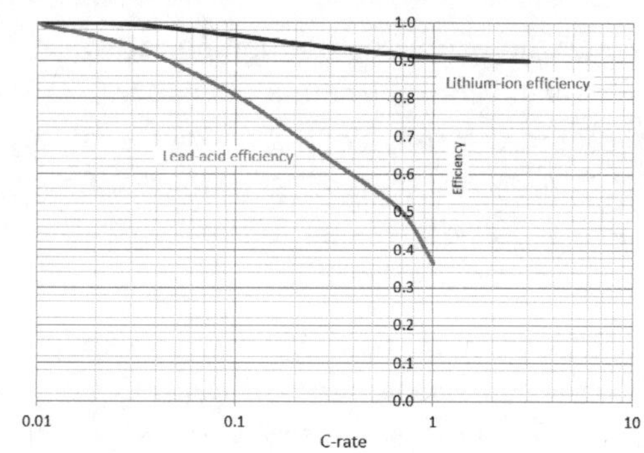

To be able to reach a better performance, minimization of contact and cable resistance is of concern, in addition to increases of cycle and calendar life. It is respected that only one-time replacement, on the basis of End of Life (EOL) at 80% remaining capacity would be necessary with existing Kokam technology This would be designed to be a midlife conversion and is based on Kokam's 10 years of field deployment experience.

B. Indiscretion Ratio

It has always been submariners wish to stay submerged longer, minimizing snorkeling time. As LIB, not like LAB, offers faster charging.

Unlike lead-acid battery, LIB can accept far faster charging compared to lead-acid, while it doesn't need third stage charging (constant power, CP) and even second stage charging (constant voltage, CV) at lower charging rates, and therefore it saves a lot of charging time.

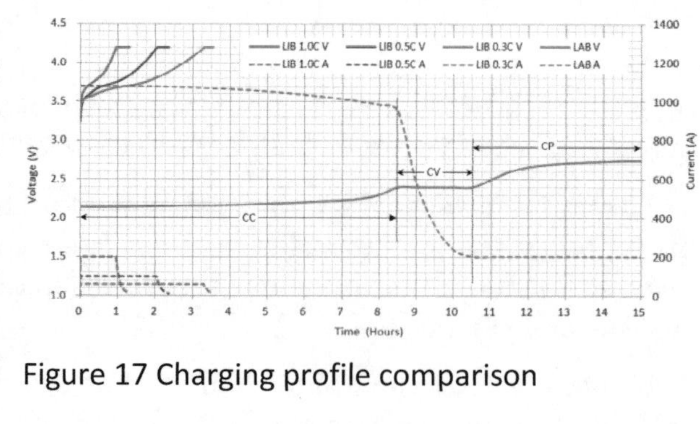

Figure 17 Charging profile comparison

To take advantage of LIB fully in the context of indiscretion ratio (stealth), it is advised to provide higher charging capacity.

VI TYPE TESTING AT LAND BASED TEST SITE

It is Kokam that has constructed a world first LIB land based test site (LBTS) for full scale Type and System Integration Test featuring actual legacy equipment such as charging generators. The propulsion motor was replaced with the equivalent load banks.

Safety function tests, in addition to the load following according to the load profiles given by the end-user and endurance tests, were repeated. Replication of battery rooms and LIB installations, with the same mechanical and electrical interfaces to that for actual submarine design, ensures the same performance as demonstrated at sea.

Figure 18 Kokam LBTS

VII SUMMARY

LIB technology for oceangoing submarines is now readily available and this technology promises quantum leaps in endurance and indiscretion ratio of conventional submarines. The intrinsic safety additionally supported by the unique "ignition resistive" Nano coating demonstrated state of the art of abuse capability by far more than expected and fully complies with the respective US-MIL-STDs (shock, vibration, noise), US NAVSEA S9310-AQ-SAF-010 (specifically overcharge) and other applicable industrial standards

The initial batch of Kokam LIB, without Nano coating, which was required to meet its national customer's requirement of exceptional DOD 90% operation, is already installed on Korean coastal submarines. These are in commissioning stage, now with 1000 hours of operation. Tests include endurance for full scale safety, load following and endurance at Kokam's LBTS. The next batch is planned with Nano coating for utmost safety enhancement.

The projected extra-ordinary performance characteristics of Kokam LIB were all realized and provide at least 5X longer spurt and 2X at a steerable minimum speed. The LBTS testing has been done in comparison with a virtual installation of LAB, in the same space considered for lead-acid technology.

PACIFIC 2015
International Maritime Conference
Sydney Exhibition Centre @ Glebe Island, Sydney Australia 6-8 October 2015

Paper for PACIFIC 2017
International Convention Centre Sydney from 3 to 5 October 2017

Type 26 Global Combat Ship: The Next Generation Warship

Chris Muskett, BAE Systems Maritime Naval Ships, chris.muskett@baesystems.com
Gavin Rudgley, BAE Systems Maritime Naval Ships, gavin.rudgley@baesystems.com

Abstract

The Type 26 Global Combat Ship program reached another significant milestone in 2017 – production. The program began a decade ago with the objective of replacing the well-regarded Type 23 Duke class. Whilst, over time, Royal Navy requirements evolved, the tenets of anti-submarine warfare capability, global reach and multi-mission adaptability have prevailed.

BAE Systems Naval Ships, as warship integrator working with the UK MoD, the Royal Navy and industry, has drawn together a number of technology advancements and learned from previous classes to develop a production-friendly, cost-effective solution, fit to serve a global navy for decades to come.

The design, now in production, uses a resilient CODLOG configuration to deliver high levels of survivability. Many years of research and development into low signature propellers and noise reduction features result in levels of performance never achieved in previous classes.

Central to multi-mission capability is the adaptable Mission Bay located immediately adjacent to the Hangar. This is available for the storage and operation of additional aviation assets, boats and other modular systems, and future proofs the platform by providing a basis for integration of new assets.

The highly capable integrated combat system is based around CMS-2 and utilises a Shared Infrastructure as its backbone. As well as connecting the existing sensors and effectors this provides a high degree of flexibility for future complex software and hardware installations.

Type 26 was developed using 3rd generation digital design and product development capability.

This paper summarises the past, present and the future of the Type 26 Global Combat Ship programme and, in doing so, offers an insight into this exciting new class of warship.

Contents

INTRODUCTION

The Type 26 Global Combat Ship program reached another significant milestone in 2017 – production. Steel was cut for the keel of GLASGOW, the first of the new City Class, on the 20 July 2017.

The programme began a decade ago with the objective of replacing the well-regarded Type 23 Duke class. Whilst, over time, Royal Navy requirements evolved, the tenets of anti-submarine warfare capability, global reach and multi-mission adaptability have prevailed.

BAE Systems Naval Ships, as warship integrator working with the UK MoD, the Royal Navy and industry, has drawn together a number of technology advancements and learned from previous classes to develop a production-friendly, cost-effective solution, fit to serve a global navy for decades to come.

The design, approved for production in 2017, uses a resilient CODLOG configuration to deliver high levels of survivability. Many years of research and development into low signature propellers and noise reduction features result in levels of performance hitherto never achieved in previous classes.

Central to multi-mission capability is the adaptable Mission Bay located immediately adjacent to the Hangar. This is available for the storage and operation of additional aviation assets, boats and other modular systems.

The highly capable integrated combat system is based around CMS-2 and utilises a Shared Infrastructure as its backbone. As well as connecting the existing sensors and effectors this provides a high degree of flexibility for future complex software and hardware installations.

This paper summarises the evolution of the Type 26 Global Combat Ship design, provides an overview of the design and sets out operational considerations. In doing so, the paper offers an insight into this exciting new class of warship.

DESIGN EVOLUTION

A class of 8 Type 26 GCS ships will replace the Royal Navy's current Type 23 fleet. They have been designed by a joint team comprising BAE Systems and the UK Ministry of Defence (UK MOD) and will enable maritime manoeuvre in support of task group and carrier strike, undertake maritime security and stabilisation operations, and be capable of independent operations worldwide.

The primary missions which are encapsulated in the specification of around 1,200 requirements are:

- Anti-Submarine Warfare (ASW) protecting task groups and in support of NATO;
- Warfighting by local area air defence, surface warfare and land attack;
- General Purpose roles including embarked forces operations, maritime security, humanitarian relief and defence diplomacy.

The program commenced in 2010 and a concept design developed to a level that enabled the UK MOD and BAE Systems to evaluate cost-capability trade-off options which led to a Capability Decision Point in 2011. The design has been progressively developed since then leading ulitmately to final design approval in early 2017 and cut steel on the 20 July 2017. The ships will be delivered through the 2020s and 2030s and will be in service until 2060.

Before summarising the design, it is worth considering the design drivers other than the requirements summarised above.

Pedigree – Evolution of Naval Ship Design

BAE Systems has a well-established track record of designing and building naval ships. Notable additions to the Royal Navy fleet that have had a strong influence over the naval ship design include the Type 23 frigates, the Type 45 destroyers, the Queen Elizabeth Class Aircraft Carriers (first of which has commenced trials), the Landing Platform Helicopter ship HMS OCEAN and the Bay Class Landing Ship Dock Auxiliary ships.

In designing, building and subsequently operating these ships, BAE Systems and the UK MOD have, with many other industry players, made a number of improvements in naval ship design. The following are examples.

- International engineering standards, namely Lloyd's Register Naval Ship Rules and ANEP 77 NATO Naval Ship Code. From the UK's perspective these have been developed in response to:
    - Operation Corporate. The lessons of the Falklands campaign were hard learned. These were captured by the UK MOD and many led to improvements in naval standards such as smoke containment and extraction.
    - HMS OCEAN. The acquisition of HMS OCEAN was the first time the UK MOD had sought to apply standards developed for merchant shipping to a warship. The results were published in a National Audit Office report in 1995 and the consequence was the development of Lloyd's Register Naval Ship Rules.
    - XV230. The loss of the Nimrod XV230 aircraft over Afghanistan resulted in an independent inquiry lead by Charles Haddon-Cave and subsequently to a marked shift in product safety management through life. In BAE Systems, this resulted in our four Product Safety principles which are now applied globally.
- Clarity of requirements. Naval ship acquisition involves balance, the goal being to get the right balance between performance and cost. In the past, the Royal Navy has seen in-service improvements for aspects such as davits for boat operations in high sea states and modifications for ship operation in extreme air and sea water temperatures. These lessons have been fully acknowledged and the consequence is that the Type 26 GCS is a true global warship.
- Modular multi-site manufacture. Within the UK, multi-site manufacture was developed during the Type 45 programme and expanded for the Queen Elizabeth Class. Although multi-site manufacture is not relevant for Type 26 GCS, these previous programmes have enabled BAE Systems to finesse manufacture of blocks, their subsequent integration and then float-off of the completed ship. These techniques have been very influential in the Type 26 GCS build strategy.
- Integration of complex combat systems. The integration of the Samson radar with the PAAMS missile system on Type 45 has paved the way for the integration of complex combat systems on Type 26 GCS which will now adopted a full Shared Infrastructure for ship and combat systems.
- Cutting-edge and flexible propulsion systems. The design of the Type 26 GCS propulsion system has benefitted from the all recent classes, notably with regard to electric propulsion. The initial foray in electric propulsion was the Type 23 frigate for which silent ASW operations are conducted on electric drive. Over the years, this technique evolved leading to full electrical propulsion (i.e. no gearbox) for the Queen Elizabeth Class Aircraft Carrier. Type 26 GCS has a blend; quiet and efficient electrical propulsion for ASW and fast cruise speeds, and then a gearbox and Rolls Royce gas turbine for high speed sprints.

Assured – A Mature Design

BAE Systems has taken the development of the design of Type 26 GCS to a greater level prior to cut steel than previous warships. From a design point of view, this is for two reasons. Firstly, to provide confidence that production will be right first time and secondly, to ensure that the ship achieves noise signatures required for a true ASW frigate. This has been achieved through a number of initiatives, notably including Visualisation.

Visualisation suites (three in Glasgow, one in Filton one in Portsmouth, and one in Canberra) are linked and enable a fully interactive multi-site reviews of the design, of production planning, of build safety, of in-service support and operation. This is now an active toolset made possible with a fully digital design bringing together live data in the CAD model and the Integrated Bill of Materials (iBOM). Combining this technology with a CAD strategy which sees pipes as small as 12 mm diameter modelled has allowed us to gain customer approval of compartments in detail before commencing build.

Other strategies have involved soft envelopes for ship-wide systems, Product Families, Zonal Integration reviews, Design Rules (capturing lessons from previous programmes), progressive Verification and Validation by BAE Systems, the MoD and Lloyd's Register.

Furthermore, Land Based Testing of the Combat Systems and certain elements of the propulsion system is ongoing to further de-risk the programme. This focuses on the integration of some equipment developed on the back of a long naval pedigree (e.g. gearbox, motors) and well-proven equipment (e.g. gas turbines).

Adaptable – Designed for a Modern Navy

The role of the Royal Navy continuously evolves in response to an ever changing global environment. Inevitably, this results in new or different equipment and people. In recognition of this, the Type 26 GCS design incorporates considerable margins for space (notably, the Mission Bay, described later in this paper), power, displacement, stability, chilled water. The Combat System will be able to accommodate equipment updates and modifications. The aviation facilities with the Flight Deck being able to accommodate a Chinook, and with the direct access through the Hangar into the Mission Bay where a second helicopter or UAVs can be stowed provides flexibility with regard to aviation assets. The Mission Bay may also accommodate a variety of other payloads. Finally, the accommodation arrangement is generous and the standard of accommodation is fit for a modern navy.

Producible - Designed for Efficient and Safe Build

Linked to the use of Visualisation as mentioned above, a number of other strategies have been employed to ensure the design is production friendly. These include:

- Functional grouping of, for example, accommodation;
- Technical galleries running port and starboard to facilitate system routing (and enhance survivability);
- Modularisation to enable off-ship assembly and testing with 285 large (such as Operations Complex) and 950 small modules (see image);

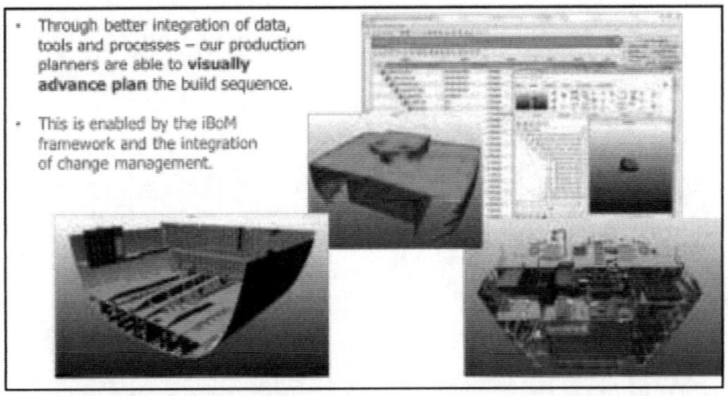

- Design for Build Safety assessments formally conducted to ensure access throughout the ship during production, and subsequent operation, is considerate to the worker and safe;
- Work Sequencing through Digital Planning, which again makes use of Visualisation and the integrated Bill of Materials; see image.

Supportable – Designed for Long Deployments

Although the nature of naval operations has evolved over the years, a central premise is the ability to undertake long deployments. With this in mind, the Type 26 GCS design has been developed to have the following characteristics:

- No special infrastructure required for deployed support;
- Low maintenance burden for the crew coupled with a policy for lean manning;
- All planned maintenance with a periodicity of less than 18 months to be conducted by the crew;
- On board support information available at point of use;
- 6 year docking cycle (aligned with the periodicity of Naval Classification Surveys) with a target to extend up to 9 years;
- Extensive storage for long unsupported missions.

Supportability has therefore been a key design driver. As with production engineers influencing the design for producibility, so have support engineers to ensure compliance with the above characteristics is achieved. This has been done by:

- Embedding support engineers in the design team;
- Conducting Human Factors assessments;
- Ensuring through-life cost is a true discriminator in system and equipment selection;
- Striving for minimal maintenance and husbandry burdens;
- Increasing automation through condition monitoring and enhanced diagnostics to a level well above that for previous ship classes;
- Selection of a Combined Diesel Electric or Gas (CODLOG) propulsion system (see below) then enables diesels to be run more efficiently at optimum power for efficiency and reduced wear;
- Selecting equipment with due consideration of mean time between maintenance;
- Integrating onboard electronic training and support information on the ship;
- Embedding removal routes in the design and using Visualisation with avatars and dynamic modelling to assess their effectiveness.

The result of all of these design drivers can be summed up in a simple statement; the Type 26 GCS will offer adaptable capability, reliability, cost-effective producibility, supportability and safety that is better than any previous warship for the Royal Navy.

PLATFORM DESIGN

Main Particulars and Characteristics

Principal Dimensions		Speed, Range and Endurance	
Length – Overall	149.90 m	Maximum Speed on Motors	about 18 knots
Length – Design Waterline	141.00 m	Maximum Speed on GT	about 29 knots
Breadth – Overall	20.75 m	Cruise Speed	15 knots
Breadth – Design Waterline	19.05 m	Range	7,000 nm
Depth Amidships – No.1 Deck	10.90 m	Total Accommodation	208
Displacement	6,900 tonnes	Core Complement	157

Classification and Standards

Type 26 GCS is being designed, built, equipped and tested in accordance with Lloyd's Register Rules and Regulations for the Classification of Naval Ships (LR), and will be entered and maintained in Naval Classification. The Lloyds Register Naval Classification notations are:

- Hull ✠100A1 NS2 Frigate SA1 AIR ESA2 RSA2 SDA FDA ES TA1 CM
- Machinery ✠LMC PSMR(L)* CCS RAS(ABV)
- Other ✠MD IB1 SH1 SP LA(NS) LAP POL*(I,IV,V,VI,AFS) CEPAC(2) ENV(A BW GW IHM NOx-2 NOx-03 OW RS SOx) *IWS FIRE**

NATO (ANEP 77 Naval Ship Code and other ANEPs and STANAGs associated with maritime interoperability), IMO Rules (SOLAS, MARPOL, including Tier 3 compliance) and other specific Regulations are applied.

Area of Operation and Environmental Conditions

The ship is designed for unrestricted global operation in the following environmental conditions:

- Sea water temperature: -2° C to 40° C;
- Air temperature: -30° C to 51° C;
- Relative humidity: 25% to 100%;

Structure

The hull structure is of all-welded, steel construction and generally longitudinally-stiffened. The superstructure is fully integrated with the main hull and will withstand the corresponding global hull loads. The structure is arranged to have good longitudinal continuity to ensure satisfactory strength and fatigue performance under the significant longitudinal bending loads in intact, extreme, and damaged conditions. The design will meet the required 25 year ship service life, with a longer 35 year life assumed for corrosion and fatigue sensitive areas to provide a margin and reduce through life support costs. The steel grade for the primary external structure (plate and secondary stiffening) is DH36.

Localised use of aluminium is made for upper deckhouses and bulwarks. The foremast above 06 Deck is of composite sandwich construction.

Hull Form, Stability, Seakeeping and Manoeuvring

The hull form has evolved from proven Royal Navy hullforms with clear linkage to the Type 23 class frigate and the Type 45 class destroyer. Maximising cavitation inception speed and minimising underwater radiated noise have been key aims for Type 26 GCS.

A comprehensive programme of physical model testing, including power and propulsion, manoeuvring and seakeeping tests, have been conducted to ensure required performance. This follows established RN practice for developing ships for worldwide service and has been supported with Computational Fluid Dynamics analysis.

The stability requirements of the UK MOD have been applied, and with watertight integrity to the requirements of Lloyd's Register.

The vessel is designed to perform the warfighting and general roles described below in Sea State 6 and below, and is designed to survive Sea State 10 with limited operational capability.

Ship Signatures

The ship is designed to minimise susceptibility to underwater and above water threats and to maximise the effectiveness of decoys and countermeasures.

The hull, superstructure, weapons/sensors and all external surfaces of the ship are designed to minimise the ship's Radar Cross Section (RCS). The hull sides above the waterline are sloped outwards and the superstructure sides are sloped inwards to achieve this. Micro geometry signature management includes avoiding trihedral corners where possible, screening of boat/ Mission Bay openings, recessed RAS points and bulwark screens. Topside equipment including doors, hatches, windows and ladders will also be designed to reduce the RCS.

The Infrared (IR) signature of the ship is managed by IR suppression on propulsion uptakes, careful funnel and uptake design, insulation, and HVAC design.

The acoustic signature of the ship is critical to its primary ASW role, and has had the greatest effort applied to minimise it. Type 23 is widely regarded to be a quiet ship, so a great deal of learning has been pulled through from that design, noting that because of the multi-role nature of Type 26 GCS she is a larger vessel and so a lot more effort in the design is required even to achieve the same acoustic signature as Type 23. The detail is classified, however some of the design features for a quiet ship embedded in the design from the outset, are:

- Design of hull appendages to minimise wake;
- Design of low signature propellers;
- Selection of propulsion architecture to enable main machinery isolation;
- Physical separation of main machinery with
- Fluid system design;
- Use of vibration attenuating mounting systems and acoustic enclosures;
- Flexible connectors and other techniques for avoidance of noise transfer;
- Integration of a hull vibration monitoring system (also enables machinery diagnostics).

Type 26 GCS is designed with a high level of shock protection to ensure survivability against underwater weapons. The platform will comply with RN shock standards to maintain hull integrity, damage control and functioning power and propulsion systems. All Combat System equipment is "shock hardened". All permanently attached equipment will remain captive during a shock event.

PROPULSION SYSTEM

A Combined Diesel Electric or Gas (CODLOG) propulsion system is fitted. The system consists of a single Rolls Royce MT30 marine gas turbine delivering power to twin propellers through a cross connect David Brown gearbox and port and starboard reduction gearboxes. Each shaft is then fitted with a GE electric propulsion motor aft of the gearbox and a thrust block fitted aft of each motor. The gearboxes incorporate clutches to allow independent operation of each shaft line and allow single shaft operation if one shaft line is out of action. The propellers are of a fixed pitch low signature design for improved cavitation performance. Shore-based integration testing is now in hand.

Four MTU Diesel Generators (DGs) are fitted, each rated for continuous operation and achieving IMO Tier 2 emission levels. The DG uptakes are fitted with Selective Catalytic Reducers (SCRs) to achieve IMO Tier 3 emission levels for the propulsion system. The gas turbine, gearboxes and electric motors are located in separate water tight compartments providing a high level of survivability.

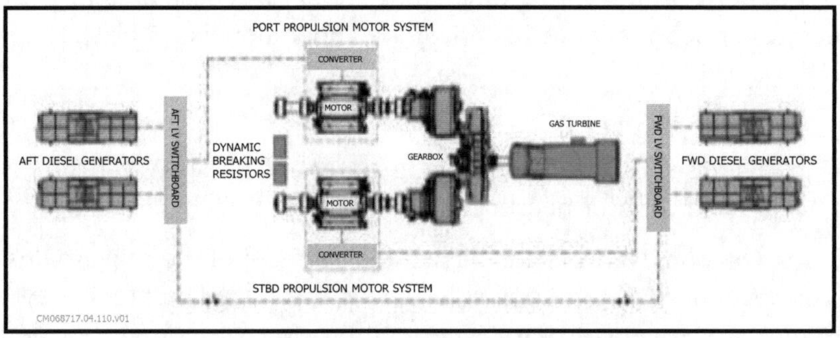

Power & Propulsion Functional Block Diagram

The power supply quality of the 690V and 440V generation and distribution systems are designed to meet Lloyd's Register requirements. Weapons and communications equipment are supplied from Zonal Power Supply Units (ZPSU) which are designed to meet the requirements of STANAG 1008 where applicable.

The benefit in the selection of a GT with high unitary power is that the choice of a simpler more cost effective CODLOG system is enabled. This allows for a cheaper and safer Low Voltage electrical distribution system as opposed to a more expensive High Voltage system. Crucially, the 'Or' system (CODLOG) can achieve ship top speed using only the GT, avoiding key dependencies on other system elements, and releasing 100% of the installed electrical power generating capacity to support ship service and weapons load demand, making the platform more adaptable in terms of its ability to accept incremental advanced weapons and sensors as they become available in future years.

AUXILIARY SYSTEMS

All auxiliary systems have been designed with separation, redundancy, and mounting strategies to deliver a level of survivability fit for a modern warship. Design of these systems for low acoustic signature is a key feature of Type 26 GCS and is delivered through use of several hundred design rules which articulate learning from Type 23, and several thousand positive design embodiment statements.

OUTFIT AND FURNISHINGS

Seamanship arrangements include:

- Two anchors and chain for anchoring in a depth of water up to 55m;
- Towing arrangement in accordance with Lloyds Register requirements;

- Replenishment at Sea (RAS) for solids and liquids to NATO standards (beam, over the bow and VERTREP);
- 4 Marine Evacuation Systems (MES) providing capacity for 200% of the ship's complement.

The Medical Complex is designed capable of providing Role 1 and Role 2 service depending on individual mission requirements, in accordance with the requirements of Def Stan 02-106 and is located on the same deck as the Flight Deck and Boat Bays for ease of access with casualties.

MISSION SYSTEMS

Mission Bay

The flexible Mission Bay gives Type 26 GCS the ability to readily adapt to a range of mission profiles.

Positioned on 1 Deck immediately forward of the Hangar and spanning the full width of the vessel, the Mission Bay can be used for the stowage, side launch and recovery of a variety of manned and unmanned vehicles as well as the storage of standard twenty-foot containers.

The space can accommodate up to 10 TEUs or 4 large boats. A deckhead mounted handling system developed by Rolls Royce from proven offshore technology enables the movement of a range of different loads and launching of craft from either side of the ship.

Launch and recovery of payloads will be through port and starboard side doors and will be affected by the Mission Bay Handling System (MBHS), which is fully adaptable to the loads to be handled. The MBHS can be used under normal circumstances in up to Sea State 6. The MBHS has an outreach of 16m from the centreline to port and starboard, sufficient to reach out to a quayside to self-load boats, containers, and mission modules.

Boat Bay

In addition to the Mission Bay, a single dedicated Boat Bay is provided to port, containing the ship's rescue boat, a 9.5m Rigid Hull Inflatable Boat (RHIB).

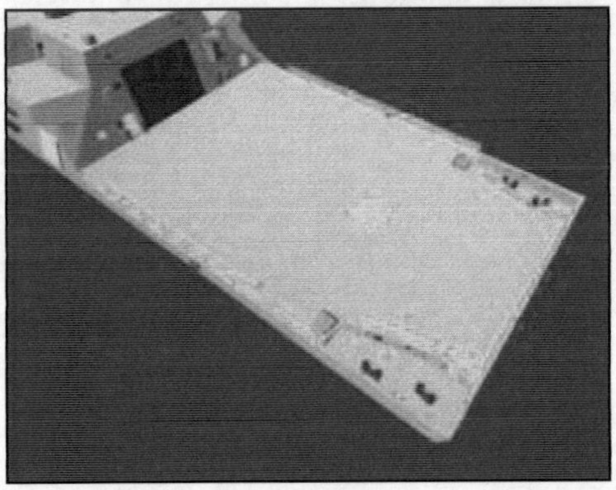

Helicopter Facilities

The large Flight Deck of Type 26 GCS can be used to operate a visiting Chinook heavy lift helicopter, or for organic operation of one Merlin HM2 or two Wildcat HMA2. The Hangar and Mission Bay are arranged for the operation, maintenance and stowage of medium sized helicopters and UAVs. A second helicopter up to Merlin HM2 size can be stowed in the Mission Bay with access to the Flight Deck through the Hangar/ Mission Bay connecting door.

COMMAND AND CONTROL SYSTEMS

Type 26 GCS is a flexible surface combatant with a comprehensive and flexible sensor and weapon suite to fulfil its multiple roles, supplemented by modern integrated command, control and communications infrastructure.

Type 26 GCS has been developed to accommodate a Combat System tailored to the specific requirements of the Royal Navy (RN). The underlying GCS design allows for alternative sensor and weapons fits to suit individual customer needs.

A platform-wide Shared Infrastructure (SI) is provided for data communications. The SI comprises four basic subsystems:

- Shared Computing Environment (SCE), which provides a distributed, high capacity processing capability on common hardware to all systems that require general-purpose computing facilities;
- Shared Network Infrastructure (SNI), which provides a resilient, high capacity and secure platform wide IT infrastructure for all the systems on the platform to use. It includes a structured fibre network, Internet Protocol (IP) networking, active network, application layer services and the data transfer service;
- Common Console (CC), which provides generic console hardware, human interface (screen, keyboard, pointing device(s)) and associated services;
- Shared Infrastructure Services (SIS), a set of services to help maintain the security of the IT infrastructure and its hosted systems and provide users with an integrated working environment.

The shared infrastructure provides a resilient, high capacity and secure platform-wide information and communications network available for all the systems on the platform to use.

The Combat Management System is BAE Systems' CMS-2, with HCI via. a number of Common Consoles located throughout the vessel. In addition, a number of multi-functional displays are installed to support command situational awareness, planning and briefing.

SENSORS

A comprehensive suite of sensors is fitted providing the vessel with the ability to determine situational awareness in the air, surface and underwater environment as follows:

- Radars:
  - BAE Systems Artisan (Type 997) 3D Multifunction Radar which provides both air and surface surveillance in addition to supporting missile mid course guidance;
  - I-band and EF-band navigation radars for navigational safety as well as supporting surface surveillance and helicopter operations;
- Identification Friend or Foe (IFF) interrogator and transponder outfit and Electronic Surveillance Measures (ESM);
- S2087 Active low frequency towed array sonar;
- S2150 Bow mounted medium frequency sonar;
- S2070 Surface Ship Torpedo Defence (SSTD) system which consists of a towed decoy and expendable countermeasures;
- Up to five Electro Optical Directors (EOD);
- Infra-Red Search and Track (IRST).

Key components of the sensor suite and the SI are being de-risked through the Type 23 Capability Sustainment Program (CSP) and will be proven before being cross-decked or installed on Type 26 GCS.

COMMUNICATIONS

GCS has an integrated set of communications systems that provide access to internal and external communications at three protective marking levels, both wired and wireless. A communications control subsystem allows access to all the tools required to configure, monitor and control the system. Other information systems include tactical data links (11, 16 and 22), defence information infrastructure, military tasking equipment, on-board warfare team trainer and UAV support.

ARMAMENTS

GCS weapons fit comprises a full gunnery solution, together with EW decoys, torpedo decoys and a local area defence missile system. Gunnery comprises a range of weapons from a BAE Systems 5" Mk.45 gun to 7.62mm GPMGs and includes the Phalanx CIWS.

A 24-cell strike length Mk.41 vertical launch system is also fitted although the Royal Navy has not yet determined the associated missile system(s).

A range of magazines and ready use lockers are provided for ammunition and soft-kill ordnance. The large magazines include handling systems. Primary and Secondary ammunition routes have been designed to support the movement and replenishment of ordnance.

REPLACING THE WORKHORSE

Type 26 GCS will replace the Type 23 Duke Class frigates as the "workhorse of the fleet" for the Royal Navy. As well as Type 23's globally acknowledged ASW capability which is retained as the primary role of Type 26, they will be deployed in a variety of roles across the world, including:

- Fleet Ready Escort - One frigate or destroyer is held at short notice in home waters and is ready to react when required in support of homeland defence.
- Counter-Terrorism - Working closely with other agencies, the Royal Navy provides access to the sea and, if required, a special forces interdiction capability.
- Counter-Piracy - The Royal Navy is supporting the growing coalition (EU, NATO and beyond) of countries acting to counter piracy in the Gulf of Aden and Indian Ocean.
- Counter Narcotics - The Royal Navy cooperates with a number of UK, EU and US agencies to stem the flow of narcotics worldwide. Operating off the coast of Africa and around the West Indies the Naval Service continues to intercept millions of pounds worth of drugs year on year.
- Territorial Integrity Patrols & Law Enforcement. The Royal Navy is relied upon to conduct operations worldwide which include;
    - Fishery Protection on behalf of Marine Management Organisation
    - Search and Rescue
    - General law enforcement

The reach and versatility of the Royal Navy enables the British Government to deliver on its foreign commitments, provided all year around and in concurrence with other current operational commitments:

- Dependent Territories – The UK has enduring responsibilities for 14 Overseas Dependent Territories for which the Royal Navy provides ongoing engagement and the ability to deliver emergency assistance during routine patrols.
- North Atlantic - The Royal Navy and Royal Fleet Auxiliary (RFA) provide seasonal support to British Overseas Territories and Commonwealth countries in the West Indies throughout the hurricane season.
- South Atlantic - The Royal Navy provides a permanent frigate or destroyer in the South Atlantic reinforcing the UK sovereignty rights of the South Atlantic Territories and providing reassurance to the Falkland Islands residents and Commonwealth nations in West Africa.
- Falkland Islands - The Royal Navy operates dedicated patrols based around the Islands which maintain the integrity of the surrounding waters and reassure the local government and its residents.
- British Antarctic - The Royal Navy continues to support the scientist and researchers of the British Antarctic Survey in this harsh, frontier region.

Fulfilment of these roles by Type 26 GCS is enabled by:

- Flexible Mission Bay with capability to re-role without dockside support in 24 hours;
- Design of systems and equipment for extreme hot and cold climates;
- Extended mission durations allowing increased time on task;
- Modern design standards minimising environmental impact in all parts of the world.

WARFIGHTING CAPABILITY

Anti Submarine Warfare (ASW)

Type 26 GCS is designed to have a very low acoustic signature as described above. This maximises the effectiveness of the sensors when operating in conjunction with a Task Group, and minimises the probability of detection when operating independently.

Preparation for ASW activities is considered by experts to be as important as the conduct of the operations. Oceanographic data is enabled by a highly capable METOC system, including integrated communications to allow data sharing between assets and national reach back.

The versatile ASW capabilities of Type 26 GCS in conjunction with the ASW design features described above enable the ship to undertake a variety of missions. The class is capable of independent ASW operations, searching, localising, identifying, tracking, and neutralising threats if required. A single Type 26 GCS is capable of acting as a deterrent around key strategic parts of the oceans, maintaining safe access to ports and conducting barrier operations. Type 26 GCS will also operate to protect carrier or amphibious Task Groups in conjunction with other vessels, providing defence at choke points and in theatre.

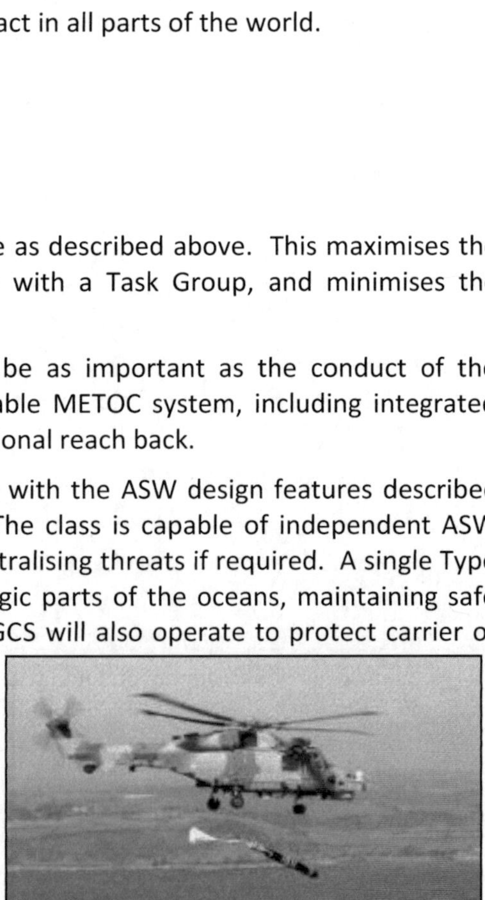

The S2087 low frequency variable depth towed sonar is the primary on-board ASW sensor and has active and passive capability, allowing long range detection/deterrent and long range identification respectively.

The Merlin HM2 features a highly capable S2089 dipping

sonar and sonobuoy capability, and is an important sensor for both detection and localisation of threats. Inbuilt interoperability ensures that in a Task Group environment where more Merlin HM2 aircraft may be available, they can be operated concurrently to build the underwater tactical picture.

The S2150 bow mounted sonar has active and passive capability and is capable of being used in the littoral environment where it may not be possible or effective to used towed sonar.

The design of the propulsion system is a key enabler of the effective use of these sensors, with flow noise minimised and cavitation-free electric motor drive available up to relatively high speeds.

Once a submarine or UUV threat is detected, Type 26 GCS has a number of response options. For self-defence purposes she has the S2170 Surface Ship Torpedo Defence system which features a combined towed array and decoy as well as acoustic countermeasures. The primary method of neutralising underwater threats is using helicopter launched Stingray torpedoes, which may be deployed be either the Merlin HM2 or the Wildcat HMA2 (pictured). Type 26 GCS also features provision for a pair of ship mounted torpedo launchers which may be used to neutralise or more likely out-manoeuvre a threat.

The Type 26 GCS Mission Bay provides a flexible way for the ASW capabilities to be upgraded as technology develops over the life of the class, with potential for off-board assets such as unmanned vehicles, unmanned sensors and mine clearance devices.

Anti Air Warfare (AAW)

Type 26 GCS has AAW capabilities designed to provide local area air defence, primarily own ship, with Task Group protection being provided by a specialised AAW vessel; for UK Task Groups this would be a Type 45 Destroyer. The Type 26 GCS design still has appropriate capability within her sensors and effectors to provide protection to merchant shipping or oil/gas installations, etc.

Type 26 GCS features an Electronic Surveillance Measures (ESM) suite along with the Artisan surveillance radar and a variety of other sensors including IRST and IFF, which along with the EOD can be used to detect, localise, and identify the threat.

Type 26 GCS features a range of responses against an air threat, and due to the very short timescales in the face of an air threat, the CMS includes the capability to pre-configure certain responses, termed "ploys". Where the hard kill of an airborne attack is appropriate, the launch of one or two Sea Ceptor missiles is likely to be appropriate, after which the CIWS and SCG may also be used. The design also features a range of decoys for radar or infra-red distraction of incoming missiles.

Anti Surface Warfare (ASuW)

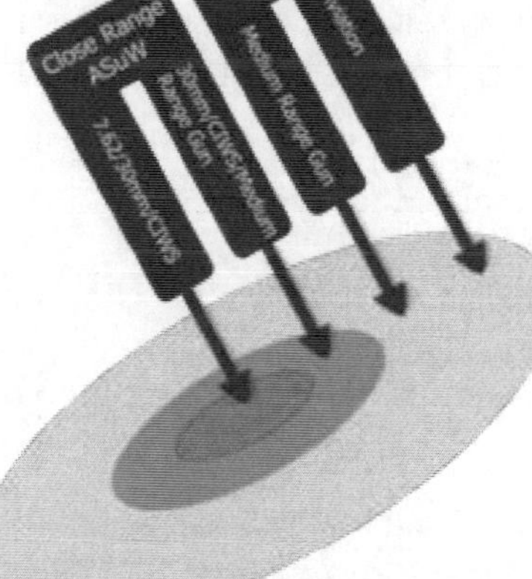

Type 26 GCS is capable of detecting, localising, and identifying surface threats of all shapes and sizes thanks to the range of sensors described against AAW. Threats on the surface will typically be slower moving, however particularly in the case of asymmetric threats in busy waters, such as conditions often experienced in the Persian Gulf, there may still be little time to respond.

A range of options are available to neutralise surface threats, perhaps easiest considered as a layered approach. The Royal Navy has not selected specific missiles for use with the Mk.41 VLS silo, but there is potential for a missile response against larger surface targets, as well as the 5" Mk.45 Medium Calibre Gun (MCG).

The Wildcat HMA2 helicopters, of which Type 26 GCS can carry two, are capable of carrying and deploying Sea Venom Surface Anti Surface Missile (ASM) and Martlet Lightweight Multi-role Missile (LMM). Combined with its specialised radar this helicopter will be highly effective against asymmetric surface threats in littoral environments.

Type 26 GCS ship-borne measures against asymmetric surface threats such as Fast Inshore Attack Craft (FIAC) include the CIWS, SCG, mini-guns, and excellent GPMG coverage for very close range.

Maritime and Naval Fires

Type 26 GCS will at times be expected to deliver Maritime Fires (attacks on hostile land forces threatening maritime craft, typically during amphibious operations) and Naval Fires (attacks on hostile land forces threatening friendly land forces). The friendly maritime craft or land forces may be coalition forces.

Essential to this capability is interoperability, which is enabled through provision of communications and data links with joint forces and coalition systems.

The traditional option for neutralising land forces is the MCG which is capable of sustained firing and carries a large capacity of ammunition. Type 26 GCS features up to 5 EODs allowing sustained use in support of land attack with the ship in littoral waters whilst maintaining adequate surveillance and control over the SCGs for self-defence.

Type 26 is capable of Missile attacks on land forces from the Mk.41 VLS subject to weapon selection.

The Wildcat HMA2 with its two missile types as described above is capable of neutralising land targets.

GENERAL PURPOSE ROLES

Embarked Forces

Type 26 GCS has an unparalleled capability to transport and deploy embarked forces for a ship of her size and type. Over 30% spare accommodation means that an embarked force of up to 35 can be carried along with the aircraft crew and support personnel, and in accommodation of the same standard as ship's staff rather than austere messes. If additional embarked forces are required, the Mission Bay has domestic FW, sewerage, and power to allow modular accommodation to be carried.

Embarked forces are provided with their own dedicated store and armoury, and the embarkation routes from accommodation, through messing, collecting weapons and ammunition, and boarding boats or aircraft, have all been designed for 95th percentile UK males wearing large packs, for example resulting in larger hatches and shallower ladder angles.

The Mission Bay is designed with adjustable boat chocks to allow carriage and deployment of embarked force special craft, and can also be reconfigured to carry vehicles.

Type 26 GCS features a medical complex described above which forms part of its capability to host embarked forces.

Maritime Security

Maritime trade is the lifeblood of the UK economy and industry. Part of the Type 26 GCS role is to act as a deterrent against maritime operations which threaten trade.

The Mission Bay supports deployment of up to four boats simultaneously whilst still retaining one dedicated rescue boat for the ship herself, meaning that multiple boarding operations can be conducted simultaneously to inspect cargoes or paperwork, to provide security to other vessels, or to investigate suspicious activity.

Type 26 GCS communications system has the flexibility to communicate with all types of maritime craft, military or civilian, without compromising her own security.

Humanitarian Relief

A key part of the Royal Navy's global activity is responding to natural or man-made disasters, during which the threat to human life is increased if a rapid response is not delivered against loss of infrastructure or shortage of food. The Type 26 GCS Mission Bay can accommodate up to 10 commercial TEU containers whilst simultaneously operating her organic aircraft, meaning that literally hundreds of tonnes of food, tools, vehicles, and even commercial mobile generators, can be delivered to disaster zones quickly and effectively. The spare accommodation capacity and ability to operate Chinook aircraft from the flight deck mean that a joint forces humanitarian response can be delivered from Type 26 GCS.

Defence Diplomacy

Building international relations and projecting a presence across the globe is a role performed by a wide range of Royal Navy ships and this will include Type 26 GCS. She has the facilities and interoperability required to cooperate with foreign armed forces and to host civilian and military personnel without compromising her security or UK national interests. Her long cruising range and global design standards mean that every corner of the world is accessible to Type 26 GCS.

SUMMARY

The Type 26 Global Combat Ship is a next generation Anti-Submarine Warfare and General Purpose Frigate. This new design inherits successful features of the world class Type 23 frigate and other ships of the Royal Navy, whilst capturing and incorporating the lessons learned.

The result is a ship fit for a global navy. Care and attention has been paid to making the ship production friendly, safe and cost-effective. Systems and equipment has been selected not just on performance and acquisition cost, but also availability, reliability and maintainability and through-life cost. The high level of automation and low maintenance and husbandry burden enables a lean manning policy to be realised.

The design has been developed prior to production to a level of detail previously unheard of, necessary to ensure that the ASW capability to detect ever prevalent conventional submarines is delivered.

The future proof Mission Bay, able to stow UAVs and secondary helicopters, with its direct access to the hangar and flight deck and with port and starboard doors for boat operation and embarkation of a wide range of mission equipment provides operational agility and flexibility fit for a truly "General Purpose" Frigate.

GLASGOW, the first of the new City Class, is now in production.

Formatting Issue - Best Available Paper/Graphic

Framing Considerations of Autonomous Naval Ships for the Royal Australian Navy

Gabrielle Anastasio[1], Joseph Cole[2], and Warren Smith[3]

[1] SBLT, UNSW Canberra at ADFA, Australia g.anastasio@student.unsw.edu.au
[2] NTB, Department of Defence, Australia joseph.cole@defence.gov.au
[3] UNSW Canberra at ADFA, Australia w.smith@adfa.edu.au

Autonomous ships can allow smaller, more effective, vessels to be utilised with the physical human presence removed, resulting in reduced costs, increased operational use and the prevention of the loss of life at sea. This work with a focus on considerations of autonomous ships for the Royal Australian Navy (RAN) is divided into two basic areas. The first is a theoretical reflection on a range of autonomous systems and an expression of some thoughts as to the feasibility of autonomous systems in the maritime environment, particularly in the RAN. The second is a summary of some initial steps taken in the framing of a case study to produce an autonomous ANZAC Class capable ship. This has included an examination of the existing ANZAC design, highlighting the human related compartments and systems that could be removed in the case of an autonomous vessel. This has led to a bounding exercise on the analysis of the effects on the vessel's size, steel weight and resistance.

Autonomous Surface Vessels (ASVs) present conceptually an opportunity to allow navies to operate where manning and cost restrictions prevent the use of traditional surface platforms. Ships without personnel onboard can operate without the consideration of human fatigue or misjudgment. They can also operate in conditions that may not be suitable for human life due to safety or environmental reasons. For the RAN, the use of an autonomous ship may be beneficial for increasing the efficiency, range, and length of operations. Research into the use and problems associated with existing autonomous systems are highlighted.

This paper reflects a work in progress as part of a 2017 undergraduate honours project by the lead author.

ACRONYMS AND NOMENCLATURE

Autonomous Surface Vessel (ASV)
Autonomy Level (AL)
Command Centre (CC)
Officer of the Watch (OOW)
Remotely Piloted Aircraft (RPA)
Technical Subject Code (TSC)

Length Overall (LOA) – m
Length Waterline (LWL) – m
Beam (B) – m
Draft (T) – m
Block Coefficient (C_b)

BACKGROUND

Naval autonomous surface vessels (ASVs), especially those based on traditional ship designs, are likely to be required to operate in a number of scenarios: in harbour and in open waters, alone and in company, and, in peacetime roles and combat situations. In order to operate in these ways a high level of control of the vessel and adaptability will be required, which would in principle be managed through a dynamic range of autonomy levels (ALs). The autonomy levels, as defined in Lloyd's Register's *Shipright Procedure – Autonomous Ships* [1] are reproduced in Table 1. The levels provide a scale of autonomy which can be applied to the consideration of a whole system or parts thereof [2].

Table 1. Lloyd's Register Autonomy Levels

AL	Name	Description
0	Manual-No autonomous function	All action and decision making is performed manually – i.e. a human controls all actions at the ship level. Note: systems on board may have a level of autonomy, with 'human in/on the loop'; for example, pms and engine control. Straight readouts, for example, gauge readings, wind direction and sea current, are not considered to be decision support.
1	On-Ship decision support	All actions at the ship level are taken by a human operator, but a decision support tool can present options or otherwise influence the actions chosen, for example DP Capability plots and route planning.
2	On and off-ship decision support	All actions at the ship level taken by human operator on board the vessel, but decision support tool can present options or otherwise influence the actions chosen. Data may be provided by systems on or off the ship, for example DP capability plots, OEM configuration recommendations, weather routing.
3	'Active' human in the loop	Decisions and actions at the ship level are performed autonomously with human supervision. High impact decisions are implemented in a way to give human operators the opportunity to intercede and over-ride them. Data may be provided by systems on or off the ship.
4	Human on the loop – operator/ supervisory	Decisions and actions are performed autonomously with human supervision. High impact decisions are implemented in a way to give human operators the opportunity to intercede and over-ride them.
5	Fully autonomous	Unsupervised or rarely supervised operation where decisions are made and actioned by the system, i.e. impact is at the total ship level.
6	Fully autonomous	Unsupervised operation where decisions are made and actioned by the system, i.e. impact is at the total ship level.

Concepts for ASVs involve an operator at a shore or ship based command centre (CC) determining what ALs the ASV will operate at based on inputs such as the type of mission, weather and other environmental and operational factors. The ASV will then operate at those ALs. Higher levels of autonomy, with low operator input, would be appropriate for simpler tasks such as navigation or contact tracking. Lower ALs, with high operator input, would be expected to be utilised for functions with potential safety implications such as close navigation, weapon engagement, or if the mission varies from what is predicted and the ship's control algorithms make it unsure how to proceed. It is expected that as systems are developed and the level of trust increases, the average ALs will move higher, with decreasing input from the operator. That is, the ASV would be able to perform more functions autonomously as technology improves over time with proven performance.

Land, Air and Sea Projects

The integration of systems with higher ALs is already evident in a number of areas, predominantly for Unmanned Aerial Vehicles (UAVs), in the automotive sector and increasingly at sea. Remotely Piloted Aircraft (RPAs) or 'drones' have features such as collision avoidance technology, automatic homing in the event of lost signal or low battery, and 'active track' that can be used to follow a predetermined object. These aspects are advantageous when applied in the maritime context, for instance the Anti-submarine Continuous Trail Unmanned Vessel (ACTUV) that aims to actively track diesel submarines [3]. UAVs have become commonplace in society and are used in a military context for imaging, surveillance, search and rescue, and delivery.

There are significant legislative and financial implications that need to be overcome in the acceptance and certification of autonomous systems. As an example, the drafting, endorsement and introduction of the Civil Aviation Safety Authority Part 101 Amendments allowed lighter and lower risk RPAs to be appropriately used [4]. However, cheaper manufacturing and better technology in sensors, navigation and imaging can be seen to be furthering their use and the challenge for the regulators is to keep up. Similar problems will likely occur with autonomous ships with the navigation and collision avoidance regulations needing to be addressed continually as new technological advances are made. The International Regulations for Preventing Collisions at Sea (COLREGs), as established by the International Maritime Organisation (IMO) in 1972, set the navigational rules or 'Rules of the Road' for vessels at sea [5]. In traditional ships these rules are interpreted by the Officer of the Watch (OOW), and decisions are made based on information from radars, sensors and by other personnel on-board acting as lookouts. Therefore, the traditional manned ship has a heavy dependency on the OOW's Situational Awareness, which is potentially a subjective interpretation of the COLREGs [6]. Though autonomous systems may reduce the subjectivity and prevent danger to personnel by removing them from the vessel, there are factors that the technology may not be able to accept, overcome or replicate, such as human intuition when removed from the physical cues of being in the real rather than a virtual environment. Navigation systems must be able to operate to at least the same level and fidelity of the human OOW and the lookout in terms of recognising a contact, categorising it and then making a plan of action based on this input data [7].

Autonomous Vessels will likely be held to the same standards as traditional ships, so collision avoidance systems aboard autonomous ships have to be equally as safe, if not safer than existing vessels, to be viable and considered acceptable. Alternatively, the path to the introduction of ASVs could be eased if the IMO were to approve regulations allowing ASVs to be given right of way for navigation in harbours and congested waters [8]. As a developing issue, at the 98th conference of the IMO Maritime Safety Committee an ongoing scoping exercise into Marine Autonomous Surface Systems has been tabled for further conferences to investigate how the IMO instruments can address issues of autonomous systems including the safety, security and environmental considerations for varying levels of autonomy [9]. There are, as can be imagined, many issues to be resolved not only for the ASV itself but for all stakeholders interacting with it.

In the automotive industry, vehicles operate at a range of autonomy levels from the fully autonomous self-driving car to smaller sub systems, which improve the safety and comfort of drivers. The sub systems such as cruise control, lane change and parking assist demonstrate the gradual integration of autonomous systems into our daily lives as drivers. Ships already feature sub systems that operate at higher autonomy levels such as Integrated Platform Management System, Electronic Chart Display and Information System, and dynamic positioning. These reduce human error and make life at sea safer and easier [10]. For ASV's, especially the non-naval variety, such as cargo ships or ferries the data transfer from ship to shore command is large during times it is in close proximity to other contacts. This occurs most when the vessel is in harbour or near shore for which land-based communications systems can be utilised to strengthen the communication bandwidth with the vessel. This communication aspect is discussed in the Technical University of Denmark's (DTU) *A Pre-analysis on Autonomous Ships* [11] where the duty lookout system can be achieved through data links to the shore CC whilst reduced manning can be achieved through the use of an electronic lookout. Autonomous navigation is achieved through the ships' interpretation of its own surroundings and a navigation officer being only required in more complex situations. Conversely, when the ship is in open water with fewer contacts a higher autonomy level can be utilised with a lowered requirement for data to be sent back to the CC. This is not a viable method for naval vessels which regularly conduct operations away from land where the communications are not as reliable, which could prove detrimental to the success of a mission [12]. As a side note, the protection of this data is important as noted by Lloyd's Register [1] and DTU [11] with the need to protect vessels from cyber attack and being used for purposes such as ransom, cargo theft or hostile actions. Producing a fully autonomous ship, whilst advantageous in theory, will not be practical in all circumstances especially for naval vessels in combat situations where control of weapons and target confirmation is required. In these circumstances the important factors are the balance and trust between human operators and the automated processes, as well as the compromise between the costs involved and benefits gained [13].

Organisations such as Rolls-Royce and DNV GL are investigating projects to further the large-scale use of ASVs. Projects include the *Maritime Unmanned Navigation through Intelligence in Networks* (MUNIN), and *Advanced Autonomous Waterborne Applications* (AAWA) using the ACTUV platform. These projects look into the feasibility of an autonomous vessel in various settings such as in coastal waters, merchant vessels in open-sea voyage, and battery powered freight transport [14]. The purpose behind these projects is based on exploring the potential benefits gained from having an autonomous ship. These include increased safety, preventing the loss of life at sea due to collisions and war like scenarios, reducing the risk of piracy, and allowing operations to continue longer. The advantages of having no crew on board and removing features such as the bridge, sleeping and recreation areas, sewerage and elements of the ventilation, firefighting and heating systems allows for a lighter, sleeker ship with reduced operating, building and fuel costs and increased space for payload [14].

ANZAC CLASS CASE STUDY

Ship Background

The RAN has eight ANZAC Class ships, the first of which was commissioned in 1996. They are 'long-range escorts with roles including air defence, anti-submarine warfare, surveillance, reconnaissance and interdiction' [15]. The ships are combined diesel or gas (CODOG) running one gas turbine and two diesel engines with a top speed of 27 knots and a cruising speed of 18 knots, achieving a range of 6000 nautical miles. The Class features a range of sensors, processing systems, electronic warfare and armament suites. The ships also have the capacity for an embarked helicopter and a number of RHIBs which allow for an increased range of operations such as anti-submarine operations, boardings and personnel transport [15]. In order to maintain ANZAC Class capability, or as close to as possible, the autonomous version of this vessel should be able to perform the roles mentioned above. The functions/methods for providing some capabilities in the ANZAC Class ships are summarised in **Error! Reference source not found.**.

Table 2. Examples of Ship Capability

Capability	Function / methods
Long Range Escorts	Seaworthy ship able to act in support of a carrier group, ability to RAS
Air and surface Defence	MK 41 VLS ESSM, Harpoon anti-ship missiles, 5 inch Mk 45 mod 2 automatic rapid fire gun, 50 calibre machine guns, SPS-49(V)8 ANZ Air search radar, CEAFAR phased array radar, Infrared Search and Track.
Anti-submarine	Helicopters, Mk 32 mod 5 triple mounted torpedo tubes, towed array and Spherion B sonar.
Interdiction (anti-smuggling)	Helicopters (S-70B-2 Seahawk or MH-60R), RHIBs

The design of an equivalent or near equivalent autonomous vessel would rely on careful planning of the mission as well as adaptability in the weapons systems as not all responses would be possible. This would require some adaptation of the on-board systems and ship configuration. For instance, weapons systems would have to be pre-loaded upon entering a combat situation, or have easily automated loading should the situation escalate. Additionally, as some other functions, such as the helicopter, RHIB and RAS capability are personnel intensive or dependent, they may have to be forfeited in an autonomous ship design configuration. This would lead to a loss of capability, particularly for anti-smuggling operations, an activity the current ANZAC Class vessels are heavily involved in when deployed to the Middle East [16]. A number of potential design concepts could be developed ranging from having a skeleton crew and facilities on board to operating as a fully autonomous vessel under remote control. As the technology to conduct autonomous operations advances, a phasing out of crew could be planned for and managed.

Compartment Analysis

This stage of the analysis involves a first pass examination of the General Arrangement (GA) drawings, Technical Subject Code (TSC) weight breakdown, and other documents to estimate which compartments and systems are human related and to what extent. A

Formatting Issue - Best Available Paper/Graphic

detailed analysis has not been conducted but it is obvious that removing crew will reduce the demand on space and weight. Baseline calculations suggest 30-50% of the ship supports human functions. For analysis and classification purposes the GA was simplified and the spaces such as galleys, recreation, offices etc, were identified as being predominantly human related and would not feature significantly in any of the unmanned/autonomous variants. In addition, there are a number of compartments and systems that require to be retained in principle but can be resized for various unmanned or autonomous scenarios. These areas are shown collectively in red in Figure 1. The areas in blue identify the aviation spaces that would potentially be removed in the case of a fully autonomous vessel due to manning restrictions. It can be seen from the GA and associated calculations in Table 3 that approximately 40% of the ship's deck area supports human activity with significant impact felt on 1 and 2 deck. The other decks were judged to have negligible areas as candidates for removal. Eliminating ANZAC's helicopter capability would likely lead to a vessel of approximately 80-85 m in length on the basis of maintaining comparable separation of masts and other effectors. These figures serve to bound the anticipated principal particulars for a comparable ASV to ANZAC. Perhaps a 80-85 m ASV could maintain a helicopter deck but not a hangar for long-term embarkation.

Figure 1. ANZAC GA Compartment Analysis

Table 3. Compartment Analysis Summary

Deck	% deck removed based on human functions
01	28
1	64
2	63
3	27
4	4
Average reduction in deck area	**37**

Weight Analysis

To gain a similar appreciation of 'sizing' from a weight perspective, a simple analysis of the ANZAC ship's TSC weight itemisations was conducted. The steel weight represented under the ANZAC TSC 100 Group was used to determine a cubic number weight coefficient and predictions of the hull weight for smaller vessels were subsequently made. The results are depicted in Table 4 for a range of vessels from 50-100% of ANZAC's displacement and they are plotted graphically, presenting a linear relationship, in Figure 2. The 50% displacement vessel has a 44% steel weight when compared to the full ANZAC. These assumptions reflect the ship in its original configuration in terms of the number of decks and superstructure etc. In reality an autonomous design will reduce the number of decks and potentially remove the superstructure completely as in concept vessels from Rolls-Royce [12]. For a naval vessel the upper deck weight provisions would focus on radars, weapons and other navigation equipment and a small emergency bridge in the early stages of concept development. To allow for autonomous operations, a substantial range of sensors, cameras and communication equipment will need to be added to ensure the safe and effective operation of the vessel. For this particular calculation the weight contribution of the additional equipment is deemed to be small when compared to the overall weight of the vessel.

Table 4. TSC100 Steel Weight Estimates

	50%	**60%**	**70%**	**80%**	**90%**	**ANZAC**
LWL	92.6	98.4	103.6	108.3	112.7	116.7
Beam	11.6	12.4	13	13.6	14.2	15.9
Depth	9	9.6	10.2	10.7	11.1	11.05
LBD/100	96.7	117.2	137.4	157.7	177.6	205.0
Total Steel weight (TSC 100)						**1097.0**
Cubic Coefficient $Steel\ Weight = Cubic\ coefficient \times \frac{LBD}{100}$						5.3501
New Steel weight	517.3	626.9	735.1	843.5	950.3	

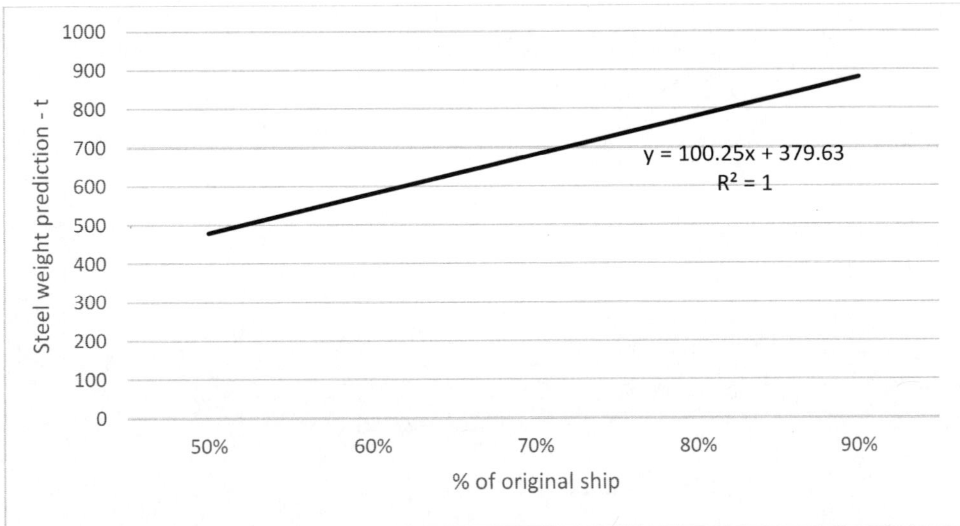

Figure 2. Steel weight estimates based on TSC 100 of ANZAC Class

Removal of the super structure will also lead to lower centre of gravity, reducing the need for ballasting – improving the reliability of the ship in an unmanned state due to reducing the problems with ballasting systems. This would also have the benefits of freeing up additional tanks for fuel for maintaining the same speed and range. The reconfiguration of machinery spaces and fuel tanks combined with the reduction in size of tanks in the lower decks such as potable water, greywater, blackwater and associated treatment facilities provides further weight savings.

The range of candidate ASVs hull form designs (40-100% of ANZAC's displacement) were then analysed utilising the Maxsurf suite of programs. Maxsurf is a naval architecture software program based on non-uniform rational basis spline surfaces to assist in the design and optimisation of hull forms [17]. The ship was modelled on the frigate design provided in the Maxsurf library and set to the parameters of the ANZAC Class. The trim was set to 0 with displacements ranging from 1440 to 3600 t. The C_b was fixed at 0.461, ensuring the fullness of form of the designs remained constant. The geometric similarity covers a range of lengths from 86-116.7 m with associated beams and drafts as shown in Figure 3.

The resistance of a select number of designs was evaluated using the Maxsurf Resistance program utilising the Fung method, which is a regression method utilising a large database applied to Transom Stern Hull vessels and has been applied to other naval vessels such as destroyers [18]. The results are presented in Figure 4 and it is evident that the smaller hull form represents a 27% reduction in resistance at 27 knots. This would clearly lead to a redesign of machinery with a significant reduction in installed power and corresponding scope for increasing range of an ASV, along with fuel tank configurations noted previously.

Formatting Issue - Best Available Paper/Graphic

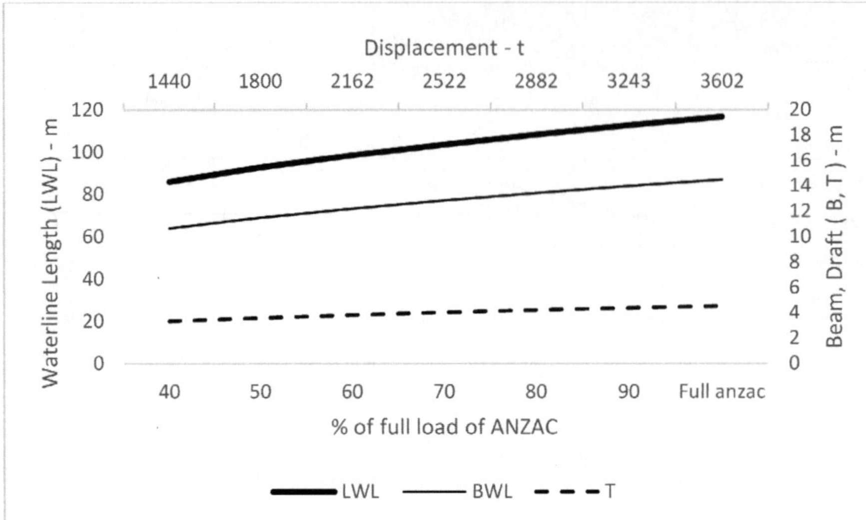

Figure 3. ANZAC Geosim Calculations using Maxsurf

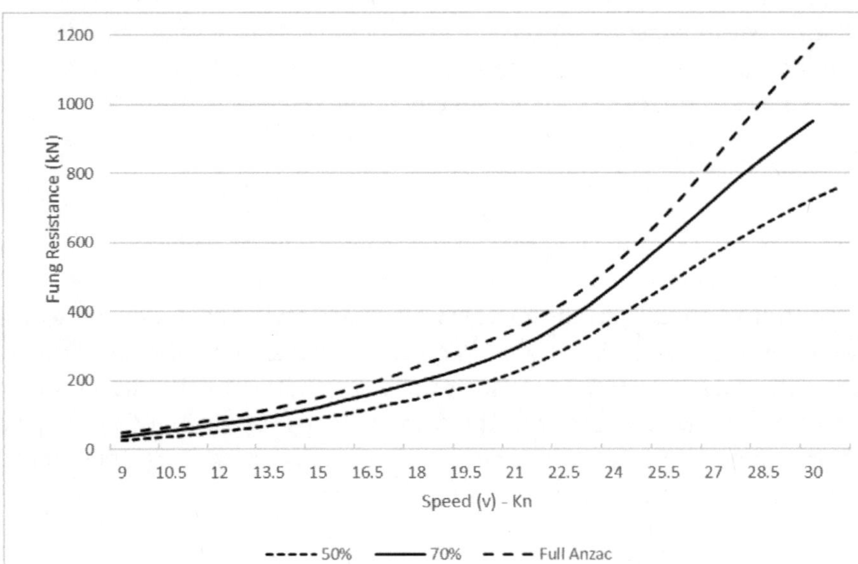

Figure 4. Resistance Comparative Results using Maxsurf

While the above analysis focuses on potential weight and volume reduction that could occur for an autonomous concept, there are likely to be aspects where autonomy drives an increase in the weight and volume of certain systems. In the first instance it is assumed that these increases will be small in comparison to the potential reductions identified previously. Greater capability, reliability and redundancy may be required for the control systems in order to achieve the objective of autonomous control. The requirement for greater reliability and redundancy may also spread to other critical systems, such as propulsion systems, as personnel are not available to support organic maintenance and repair. Some solutions posed in existing projects are the use of podded engine systems that are installed similar to a cargo shipping container, non-ballasting ships, and further machinery redundancies as discussed in [8] and [12]. For a naval vessel, the need for redundancy and/or repair is driven not only by equipment failures, but also by damage. Naval ships traditionally embody high levels of equipment hardening, zoning, separation and redundancy as controls for the risk of military damage. These levels may need to be increased further under an autonomous concept.

As the protection of human life aspect is no longer a core driver for overall design under an autonomous concept, this provides the opportunity for redundancy to be considered at a fleet level rather than at that of an individual ship. As the size and cost of individual ships is reduced it may become more cost effective to invest in additional redundant ships than in additional redundant systems in each ships.

FURTHER WORK

This paper reflects a work in progress as part of a 2017 undergraduate honours project. In this context, framing characteristics of candidate ASVs will continue to be further investigated to define their perceived limits within this project. Further modelling and investigation into the systems to be added to allow for autonomous operation will also be considered.

Beyond the thesis work, decisions for autonomous naval ships for the RAN will require much investigation, reasoned argument and demonstrable evidence. There is much to be done and achieved by all stakeholders.

CONCLUSION

Autonomous vessels are becoming a reality, with the inherent opportunities to improving the safety and efficiency of vessels at sea. The design and implementation comes with a number of considerations from a technical, electronic and legislative perspective. Considered in the work reported in this paper are some foundational thoughts related to the current and potential future use of autonomous and unmanned systems that are used in the land, air and sea. It is a significant future field of which the surface is just being scratched.

While much work remains to be done, it would appear to the authors that the future will include substantial RAN platforms operated remotely and/or autonomously. However, given this ultimate end state, it is impossible to predict or suggest a time scale for any phases of implementation.

ACKNOWLEDGEMENT

This work was conducted principally as an undergraduate honours thesis project by the lead author. The support of the work by NTB is acknowledged and appreciated.

Formatting Issue - Best Available Paper/Graphic

References

1. Lloyds, R., *ShipRight procedure – autonomous ships* in *Cyber-enabled ships*. 2016, Lloyds Register.
2. Royce, A.a.R., *Remote and Autonomous Ships: The Next Steps* 2016.
3. Littlefield, S. *Anti-Submarine Warfare (ASW) Continuous Trail Unmanned Vessel (ACTUV)*. 2016 [cited 2017 10/05]; Available from: http://www.darpa.mil/program/anti-submarine-warfare-continuous-trail-unmanned-vessel.
4. Authority, C.A.S. *Part 101 Amendments - Cutting red tape for remotely piloted aircraft*. 2016 [cited 2017 14/06]; Available from: https://www.casa.gov.au/aircraft/standard-page/part-101-amendments-cutting-red-tape-remotely-piloted-aircraft.
5. IMO, *Convention on the International Regulations for Preventing Collisions at Sea, 1972 (COLREGs)*. 1972.
6. Campbell, S., W. Naeem, and G.W. Irwin, *A review on improving the autonomy of unmanned surface vehicles through intelligent collision avoidance manoeuvres.* Annual Reviews in Control, 2012. **36**(2).
7. Benjamin, M.R. and J.A. Curcio. *COLREGS-Based Navigation of autonomous Marine Vehicle*. in *IEEE/OES Autonomous Underwater Vehicle*. 2004.
8. MUNIN, *New ship designs for autonomous vessels*, Ø.J. Rødseth, Editor. 2015, Marintek.
9. Shipping, A.B.o. *MSC 98 Brief*. 2017; Available from: https://absinfo.eagle.org/acton/attachment/16130/f-02d3/1/-/-/-/-/ABS%20MSC%2098.pdf?sid=TV2:IzWPKdVRG.
10. Kongsberg, *Automated Ships Ltd and Kongsberg to build first unmanned and fully-automated vessel for offshore operations*. 2016.
11. Blanke, M., M. Henriques, and J. Bang, *A Pre-analysis on Autonomous ships*. 2017, Technical University of Denmark. p. 27.
12. AAWA and Rolls-Royce, *Remote and Autonomous Ships: The Next Steps*. 2016, Rolls-Royce. p. 88.
13. Proud, R.W., J.J. Hart, and R.B. Mrozinski, *Methods for Determining the Level of Autonomy to Design into a Human Spaceflight Vehicle: A Function Specific Approach*. 2003, National Aeronautics And Space Administration: Houston Tx Lyndon B Johnson Space Center.
14. Levander, O., *Autonomous Ships on the High Seas.* IEEE Spectrum, 2017. **54**(2): p. 16.
15. Navy, R.A. *Frigate helicopter (FFH)*. 2016 [cited 2017 08/05/2017]; Available from: http://www.navy.gov.au/fleet/ships-boats-craft/ffh.
16. Griggs, C.C., *Arunta conquering the drug trade* 2017, Navy Daily.
17. Incorporated, B.S. *Maxsurf*. 2017; Available from: https://www.bentley.com/en/products/product-line/offshore-structural-analysis-software/maxsurf.
18. Fung, S.C., *Resistance and Powering Prediction for Transom Stern Hull Forms During Early Stage Ship Design* SNAME, 1991. **99**: p. 59.

Integrating Unmanned Surface Vessels into the Australian Defence and Border Protection Forces

Professor Derek Rogers
Head, Centre of Excellence in Autonomous Vessels, Saab Australia Pty Ltd.
derek.rogers@au.saabgroup.com

ABSTRACT

An offset in a military competition is a method of asymmetrically compensating for some form of disadvantage. Rather than match an opponent in an unfavourable competition, the approach then is to change the game to a more favourable situation. Unmanned and Autonomous Systems offer such game changing technology for Australia. The Australian Defence and Border Protection Forces need to be capable across a spectrum of operations (military, policing, humanitarian aid and disaster recovery), and complex maritime environments (including ocean, littoral, and riverine) under the constraints of cost and minimal personnel.

The United States of America (US) and other countries increasingly see the value in unmanned and autonomous systems applied for the dull, dirty and dangerous, providing force multiplication. The US has defined unmanned systems as a key component of its Third Offset Strategy. The First Offset Strategy from the 1950's emphasised nuclear deterrence, and the Second Offset Strategy post the Vietnam War emphasised technological superiority in the areas of Intelligence, Surveillance, and Reconnaissance (ISR), precision-guided weapons, stealth technologies, and space-based navigation and communication systems. However the US Third Offset Strategy also seeks to improve the US military's collaboration with innovative companies, akin to the inclusion of industry as a Fundamental Input to Capability in Australia.

This paper discusses multi-disciplinary work in the area of Unmanned and Autonomous Surface Vessels, how they could benefit the Australian Defence Force (Army and Navy) and Border Force, providing an equivalent offset strategy tailored to the specific Australian context and including Australian industry.

INTRODUCTION

Unmanned and Autonomous Surface Vessels

Neglecting the unmanned fire ships of early history that were boats set on fire and let to drift into enemy ships, the first development of a wireless remote controlled Unmanned Surface Vessel (USV) is generally attributed to Nicola Tesla, [1]. At New York's Madison Square Gardens in 1898, Tesla demonstrated his ability to manoeuvre and control the running lights of a 4 foot long toy ship on an indoor pool.

In more modern times Navies have used remote controlled USV primarily for target training. The U.S. Navy classifies these as a Seaborne Powered Target (SEPTAR) and as of 2013 these included, according to Tarantola in an article [2]: the Hammerhead usv-t; the Mobile Ship Target (MST); the QST-35; the Fast-Attack Craft Target (FACT); the High-Speed Manoeuvring Surface Target (HSMST); and the Ship-Deployable Surface Target (SDST). The Royal Australian Navy also has several target training vessels.

Tesla was ahead of his time in many ways though and Turi in reference [1] reports that: "*In a 1900 article from Century magazine, Tesla described a moment of self-realization, seeing his own mind and body as an automaton, reacting to external stimuli and situations. He stated that contemporary automatons were simply using a "borrowed mind," and responded to orders from a distant and intelligent operator. Tesla believed that one day we may be able to endow a machine with its "own mind," where it, too, can act on environmental stimuli of its own accord.*"

Tesla was in effect describing what we would now regard as autonomous behaviour and this is the focus of the majority of research and development for USV nowadays where commercial systems are starting to appear but are still fairly rudimentary in regard to their autonomous capabilities.

Autonomy

There are many different ways of describing autonomy as covered by this author in a paper for the International Maritime Conference of 2015, [3]. That information has been repeated below and will again be built upon.

The Society for Underwater Technology (United Kingdom) described six levels of autonomy in their consultation paper "Issues Concerning the Rules for the Operation of Autonomous Marine Vehicles (AMVs)", [4]. This categorisation as given in Table 1 is good for a coarse categorisation of the level of autonomy of a system.

Table 1: A System Level Categorisation of Autonomy

Level	Name	Description
0	Remote Control	All functions directly controlled by a control link i.e. remote control
1	Emergency Stop	Includes a simple emergency abort system that shuts down the vehicle in the event of the failure of a primary vehicle system
2	Waypoint Navigation	Can follow pre-programmed waypoints and automatically correct simple failures
3	Collision Detection and Avoidance	Includes obstacle avoidance algorithms
4	Mission Adaptation	Can adjust the current mission based on information retrieved by on-board sensors
5	Cooperative operation	Any behaviours that are beyond the other categories, such as cooperative behaviour between vessels

An alternative classification is from a working group of the European Defense Association on the Safety and Regulations for Unmanned Maritime Systems (SARUMS). Magnus Örnfelt, Chair of SARUMs, in a presentation for a workshop in Portugal in May 2014, [5], presented a five level control model as shown in Table 2 that built on earlier work of the US Department of Defence (US DoD), North Atlantic Treaty Organisation (NATO) and others. Shown by Table 2 this classification can be applied to specific functions of a USV, whereby some functions may be human directed, other functions human monitored, and other functions autonomous according to this model.

Table 2: An Operator Involvement (or Functional) Categorisation of Autonomy

Level	Control	Description
1	Human operated	Remote control – all control by human operator e.g. operator instructs USV to make a turn.
2	Human directed	Permissive – USV asks for permission e.g. USV requests to make a turn and operator confirms, declines or does not respond.
3	Human delegated	Declarational – USV declares intention e.g. USV indicates it will make a turn and operator can overrule or remain silent.
4	Human monitored	Reportive – USV reports action e.g. USV informs operator that it has made a turn.
5	Autonomous	Autonomous – USV does not report action e.g. USV makes the turn.

Another way to look at autonomy for a USV commences with the most fundamental building block of the use of autonomy to improve the operational envelope[1], or sea-keeping, of the vessel.

Much of the research associated with USV to date has focussed on autonomous collision detection and avoidance however for this to be successful relies on the sea-keeping ability of the vessel. Indeed, in some cases, the solution that a collision detection and avoidance algorithm may provide could well endanger a vessel leading to a situation of pitch-poling or the broaching of the vessel, [6]. For many researchers, and developments, this can be neglected because of the use of an enclosed hull. However, when the USV in question does not have an enclosed hull, sea-keeping decision making should in general take precedence over collision detection and avoidance. Even with an enclosed hull the use of autonomy to improve sea-keeping is important in order to avoid or minimise events that can jeopardise the mechanical integrity of the vessel, for example, slams, shudders, and with certain hulls, torsional events.

Taking this model further, collision detection and avoidance should in general take decision precedence over mission adaptation, which should take decision precedence over cooperative operation.

Finally in the special case of autonomous weaponised platforms, a whole discussion by themselves, this author proposes that even more fundamental (a higher priority) than sea-keeping in terms of decision making, should be the International Humanitarian Law (IHL) principles most notably distinction and proportionality. An interested reader is referred to reference [7] for further information on the IHL principles.

Applications

Perhaps more widely understood now than at the time of the paper presented by the author at the International Maritime Conference of 2015, [3], are the applications for USV including:

- Mine countermeasures;
- Anti-Submarine Warfare;

[1] The term operational envelope refers to the maximum sea state and wind conditions in which the vessel can operate.

- Intelligence, Surveillance and Reconnaissance;
- Target Training;
- Rapid Environmental Assessment;
- Maritime Electronic Warfare and Deception; and
- Logistics.

A report by the RAND Corporation (U.S.), [8] for example, entitled "U.S. Navy Employment Options for Unmanned Surface Vessels" goes into far greater depth on these and other defence applications.

This author in reference [3] also listed the civil security applications such as anti-piracy, search and rescue, offshore and coastal asset protect including port and harbor protection, oil-spill clean-up, and fire-fighting. Likewise scientific uses such as environmental monitoring including oceanography and hydrograph were listed and to this can be added commercial applications especially in the fishing / aquaculture industry.

The Australian Need

The Australian Defence Force (ADF) has contributed to major conflicts since World War I and with a small population has relied on the integration with allies and force multiplication to maintain an advantage.

Australia has the largest maritime area of any nation covering more than 10% of the surface of the Earth, according to the publication Australian Maritime Doctrine by the SeaPower Centre – Royal Australian Navy (RAN) [9]. The SeaPower Centre publication also highlighted that Australia has two major maritime vulnerabilities. Firstly that shipping intended only for Australia or New Zealand can be identified a considerable distance away. Secondly and more relevant to this paper that 40 percent of exports and 32 percent of imports by value pass through key choke points such as the Indonesian archipelago and the Torres Strait at the top of Australia.

Expanding upon this Lee Cordner in an article "Managing Maritime Security Risks in the Indian Ocean", [10] identified 19 maritime security risks for Australia including:

- Offshore oil and gas safety and security incidents;
- Disruption of energy cargoes at sea;
- Illegal exploitation of marine living resources, in areas of national jurisdiction;
- Non-state actors impinging upon freedom of navigation (piracy, maritime terrorism);
- Transgressions of sovereignty in the territorial sea or Exclusive Economic Zone (EEZ); and
- Law and order at sea transgressions: crime, piracy, robbery, smuggling, trafficking, illegal immigration, and Illegal, Unregulated and Unreported (IUU) fishing.

Collectively these risks are addressed by the RAN and the Australian Border Force (ABF), as well as other organisations such as the Australian Federal Police (AFP), and various government and independent authorities often working in collaboration. A discussion of the specific allocation of responsibilities was covered by this author in [11] which also highlighted unmanned and autonomous surface vessels as complementary to other currently used technologies such as maritime patrol aircraft, over the horizon radar, and the emerging use of unmanned aerial vehicles.

However Cordner's threats are all associated with the Australian mainland, whilst the ADF is involved with coalition activities that can be anywhere in the world and Australia has a wider interest with our shipping passing through littoral regions such as the Gulf of Aden, the Indonesian archipelago, and the South China Sea.

Related to this Kilcullen in a report for the Australian Army, [12], highlighted the rise of the networked, urban, littoral environment distinguishing this from previous experiences in urban, littoral environments. Kilcullen wrote that since 2001 Australia has operated in primarily rural landlocked environments, but much of the conflict in the world over this time has been in the urban littorals: *"In short, urban littoral combat has been more common this century than Australia's (primarily Afghan) experience suggests*."

With most people living in coastal cities, many of these now becoming megacities, and conflict typically occurring where people live, Kilcullen then highlights that future conflict is most likely to occur in these environments and the classic concepts of manoeuvre operations developed for the urban littoral prior to the 1990's are no longer likely to apply.

"*Australia's primary operational environment (POE) is overwhelmingly littoral and increasingly urbanised. But we no longer face the littoral of which Ralph Peters or Charles Krulak wrote in the 1990s – in the pre-mobile phone era, before significant penetration of the internet into the developing world. Today we face an urban, networked littoral.*"

Kilcullen also describes how as cities grow there is a disproportionate increase in slum populations and the consequences of this. "*Huge numbers of disillusioned, excluded, under-served slum-dwellers create openings for non-state groups – militias, religious or political movements, or criminals – who step into the economic and governance vacuum to offer what populations need and governments cannot provide. Thus growing cities find their cores surrounded by peri-urban no-go areas, afflicted by violent crime and controlled by non-state armed groups.*"

Cities also become city clusters or mega regions. Kilcullen describes a report from the Economist, 23rd of September 2012 about how India's western seaboard could turn into a mega region and remarks "*... this has significant military implications because it prevents a force from isolating urban objectives in a traditional way.*"

Tellingly though is Kilcullen's description of a number of recent conflicts and the effects of connectivity, for example: "*During the 2008 Mumbai terrorist attacks, controllers in Karachi monitored Twitter feeds from civilians trapped in the Taj Mahal hotel. They identified civilians' locations in real time then passed these by SMS and voice to assault teams in the hotel, leading them to hiding places.*"

Kilcullen reports that "*Australians are rapidly losing the technological edge on which they traditionally relied to compensate for lack of numbers.*"

This is the same conclusion that the US came to and led to the development of a Defence Innovation Initiative also known as the Third Offset Strategy for which unmanned and autonomous systems are critical.

OFFSET STRATEGY

An offset in a military competition is a method of asymmetrically compensating for some form of disadvantage. Rather than match an opponent in an unfavourable competition, the approach then is to change the game to a more favourable situation.

In November 2014 the then US Secretary of Defence, Chuck Hagel delivered the Reagan National Defence Forum Keynote, [13] where he announced the Defence Innovation Initiative, more generally known as the US Third Offset Strategy and in reality a series of offset strategies as elaborated upon by Bob Work, the then Deputy Secretary for Defence in a later speech, [14].

In his speech Chuck Hagel explained that "*Technologies and weapons that were once the exclusive province of advanced nations have become available to a broad range of militaries and non-state actors ... all clear threats to the United States and its allies. ... In the 1950's, President Eisenhower successfully offset the Soviet Union's conventional superiority through his New look build-up of America's nuclear deterrent. In the 1970's, Secretary of Defence Harold Brown, working closely with Undersecretary – and future Defence Secretary – Bill Perry, shepherded their own offset strategy, establishing the Long-Range Research and Development Planning Program that helped develop and field revolutionary new systems, such as extended-range precision-guided munitions, stealth aircraft, and new intelligence, surveillance, and reconnaissance platforms. ... Today I'm announcing a new Defence Innovation Initiative – an initiative that we expect to develop into a game-changing third 'offset' strategy.*"

Hagel's focus for this third offset strategy included robotics, autonomous systems, miniaturisation, big data and advanced manufacturing, which encompasses 3D printing. Equally importantly though Hagel's approach was to actively engage with the private sector, recognising that, as Bob Work subsequently said in [14] "*... in the 1950's and the 1970's, generally these advances were military capabilities that were brought along by military labs. But now ... all of those are being driven by the commercial sector.*"

There are many parallels here for Australia. In defence terms Australia has always sought to maintain technological superiority in the region and our region is one that faces a growing number of challenges with the networked, urban, and littoral. However even in terms of border protection, other risks such as illegal fishing, the protection of maritime oil and gas platforms, Australia has a large maritime area but only limited personnel available to guard against the security threats. Autonomous systems then can play a valuable role in cost effectively meeting Australia's many maritime security concerns. Similar to the US though and especially in regard to USV, Australian industry can also play a key role in developing and sustaining the technology.

AUSTRALIAN INDUSTRY AS A FUNDAMENTAL INPUT TO CAPABILITY

The Australian Federal Government's Defence Industry Policy Statement, 2016, [15], for the first time formally recognised Industry as a Fundamental Input to Capability (FIC) where Defence defines capability as "*... the capacity or ability to achieve an operational effect*", [16]. This was a significant change and aligned with a new policy in defence innovation also announced in the Industry Policy Statement, [15]:

"*Investment in innovation helps to ensure Defence remains resilient to emerging threats, including the possible use of disruptive technologies by adversaries. It also enables us to take advantage of new or developing areas of technology that have the potential to provide a capability edge for Australia's relatively small force.*

Australia benefits from an industrial base consisting of a large number of agile small to medium enterprises strengthened by a vital collection of global companies—with access to global knowledge networks, supply chains networks and export markets.

Together our small, medium and large enterprises (along with Australian research institutions and academics) have a track record of producing some of the world's most innovative defence technologies. However, these enterprises have often found it difficult to engage with Defence due to the fragmented nature of innovation programs and complex entry processes.

The Government is committed to implementing a new approach to Defence innovation that will address these barriers and more effectively access the potential of Australian defence industry to innovate."

This policy change also included the Next Generation Technologies Fund, where the government focussed on a number of emerging technology areas including trusted autonomous systems. Whilst not specifically referring to a Third Offset Strategy, Australia has thus set a direction consistent with the US Third Offset Strategy. A comment here though is that not only can this benefit the Australian Defence Force, but also the Australian Border Force, and provide spill over benefits to the wider maritime industry in Australia.

AN EXAMPLE OF AUSTRALIAN INDUSTRY WORK

The work of Saab Australia in USV is a good example of the potential of Australian industry. Internally funded Research and Development (R&D) since October 2014, in October 2015 Saab Australia demonstrated a proof of concept mission system named Bonefish™ suitable for a USV at the Pacific 2015 event.

Significant in regard to this R&D has been the use of a large number of Australian organisations in manufacturing, supplying equipment, and supplying technical expertise. Other papers have covered the basic features and technical details of Bonefish™, [3], [11] and [17], for example, which continues to be enhanced in regard to autonomy, machine cognition, and human-machine teaming, in collaboration with the Defence Science and Technology Group (DST Group) and several Australian universities.

However Saab Australia's work, whilst leading in many technical areas, is more an exemplar because of its multidisciplinary approach including a focus on System of Systems (SoS) integration, that is, integration with naval Command and Control (C2) systems, Launch and Recovery (L&R), and Sustainment.

The following figure illustrates the connection of the Bonefish™ USV Control Station to the latest Saab 9LV Combat Management System as installed on a number of vessels around the world including the RAN's ANZAC frigates, amphibious assault ships, and the soon to be constructed two new replenishment ships under the SEA 1654 program. Bonefish™ can be integrated with any C2 system though and for the exercise Autonomous Warrior 2018, integration has been undertaken with the Open Architecture Combat System (OACS) expected to be used with that event.

Figure 1: Saab 9LV Combat Management System with the Bonefish™ USV Control Station and network management laptop

C2 integration involves the provision of situational awareness information from the USV to the C2 system, and command information from the C2 system to the USV.

Situational awareness information includes: USV position; and other USV sensor data such as radar tracks, radar plots or the radar picture, sonar and depth information, video and still imagery, and meteorological information. Open standards exist for some of this information but not for all. Status (alarm) information also needs to be conveyed where some of the status information may be filtered if there is an operator at the USV Control Station. Noting that there may or may not be an operator at a USV Control Station depending on the mode of autonomy and whether or not multiple assets are controlled (or more correctly supervised) from the C2 system.

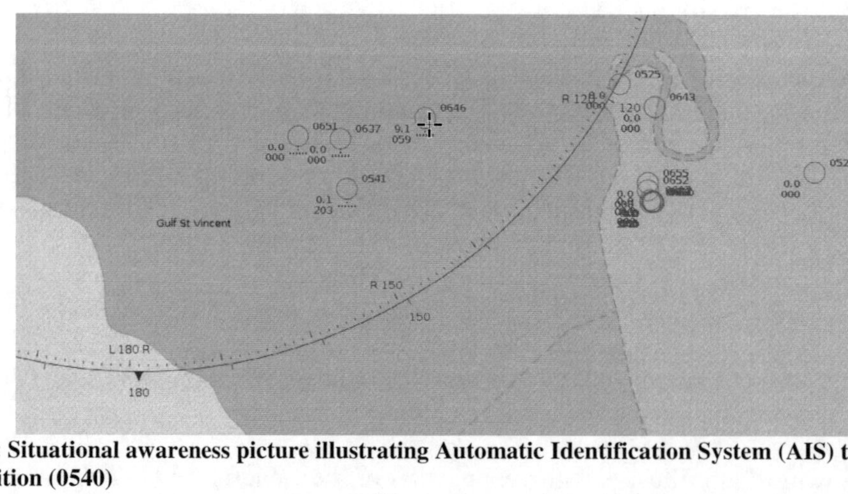

Figure 2: Situational awareness picture illustrating Automatic Identification System (AIS) tracks with USV position (0540)

Command information from the C2 system to the USV will also vary based on the applicable mode of Table 2 (human-controlled, human-directed, human-delegated, human-monitored, and autonomous). In some of these modes the USV may be 'programmed' with a mission or its underlying rules of operation. The establishment of these rules of operation in regard to autonomy for sea-keeping and collision avoidance but also mission adaptation and cooperative operation are on-going areas of research.

With respect to L&R, this is currently a fairly complex, dangerous, and labour intensive task. The DST Group has been actively working on a set of solutions for USV, collectively named LAURA. LAURA is a joint industry program between DST Group, CMN, Babcock, BAE Systems, Marin, MacTaggart Scott, and the Ministry of Defence in the UK, and aims to establish under STANAG an interface specification and proposed standard for L&R. Other vendors around the world also propose automated L&R solutions but some of these require significant ship modifications, which limits their usefulness.

Initially Bonefish™ is intended to be applied to a Rigid Hulled Inflatable Boat (RHIB) because of the greater ease of technology adoption. As outlined in earlier papers, Bonefish™, can allow manned or unmanned operations and so implementing Bonefish™ with an existing RHIB will add capability but not result in any loss of capability. The use of a familiar platform such as a RHIB also brings advantages in terms of support as does the extensive use of Commercial off The Shelf (COTS) equipment for Bonefish™, leveraging existing C2 systems for integration, and the use of technology such as LAURA. The result is that any additional support costs for a USV RHIB will be negligible when compared with existing RHIBs.

Understanding the broader context of SoS integration, and the factors involved in technology adoption is crucial then to Australian innovation succeeding with the Government's new policy. The Government's policy has asked Defence to recognise Industry as an FIC but equally Industry has to demonstrate its value by proposing innovative solutions that don't impose a greater cost burden.

Saab Australia's work in USV, for example, has also been very cognisant of the regulatory environment that often lags technology development, working within the existing frameworks and engaging with regulatory and other government bodies in areas of uncertainty, and in regard to tests and trials.

The design philosophy of Bonefish™ that created a very flexible, low cost solution leveraging COTS equipment has also recognised that Australia does not have the budgets of countries such as the United Kingdom or the United States, but through this approach can still produce very capable and cost effective solutions – understanding the economics and cost constraints of the ADF and ABF.

Finally, taking into account the social and ethical discussions in regard to autonomy, Saab has spent time exploring potential concerns that people may have with respect to safety, the potential loss (or more likely change in the nature of) jobs, and weaponisation, something that is not always necessary but is an area of concern for society as illustrated by the Open Letter on Autonomous Weapons that has been endorsed by a number of influential individuals, [18].

CONCLUDING REMARKS

Increasingly the ADF and ABF recognise the value of unmanned if not autonomous systems for dull, dirty and dangerous operations. Much of the focus to date has been on Unmanned Aerial Vehicles (UAV) and to a lesser extent Unmanned Underwater Vehicles

(UUV) but there is growing recognition of the roles that USV can play; some complementary and some of their own accord.

Australia has always been constrained in terms of personnel and sought military advantage by technological superiority. However the same philosophy applies with respect to other security threats and even contributions to Humanitarian Aid and Disaster Relief (HADR).

The unmanned and autonomous systems industry is in its infancy in Australia and the USV industry even more embryonic. Nonetheless all of the pieces for a USV industry to be successful exist in Australia – through the use of COTS equipment the skills required are integration more so than expensive risky development. There are good opportunities to establish and sustain an Australian ecosystem, the supply chain. There are many also research problems to solve to attain higher levels of autonomy, which provide increased economic and social benefits. However collaboration between Australian industry and Australian research organisations such as DST Group and universities can collectively work to solve these problems. The new Defence Cooperative Research Centre in Trusted Autonomous Systems should serve a focus for this.

Saab Australia has led in adopting a multidisciplinary approach towards the development of USV in Australia, worked with a large number of local suppliers and manufacturers, and collaborated closely with DST Group and Australian universities in realising a solution that is not only leading edge, but integrated in a wider SoS context.

Australia has thus set in place the pieces for its own Third Offset Strategy with respect to autonomous systems and Industry as an FIC. The next challenge though will be realising this potential.

REFERENCES

[1]	Turi J., "Tesla's toy boat: A drone before its time", Engadget, https://www.engadget.com/2014/01/19/nikola-teslas-remote-control-boat/ January 2014
[2]	Tarantola A., "Monster Machines What the Navy Shoots for Targets", Gizmodo, https://www.gizmodo.com.au/2013/07/monster-machines-what-the-us-navy-shoots-for-target-practice/ July 2013
[3]	Rogers D., "On the Integration of Unmanned Surface Vessels (USV) with Naval Systems and Operations", International Maritime Conference, Sydney Australia, October 6-8, 2015.
[4]	"Issues Concerning the Rules for the Operation of Autonomous Marine Vehicles (AMVs)" – Consultation Paper, Society for Underwater Technology, Underwater Robotics Group.
[5]	M. Örnfelt, " Safety and Regulations for Unmanned Maritime Systems - SARUMS", prepared for UMS Workshop, Port Portugal, May 2014, at http://slideplayer.com/slide/3555936/
[6]	Rogers D., "Unmanned Surface vessels (USV)", Autonomous Warrior 2018 workshop – DSTG, Navy and Industry, held at Saab Australia in Adelaide, February 3rd 2017
[7]	"International Humanitarian Law", Wikipedia, https://en.wikipedia.org/wiki/International_humanitarian_law#Principle_of_distinction
[8]	Savitz et. al., "U.S. Navy Employment Options for Unmanned Surface Vessels (USV)", 2013, ISBN 978-0-08330-8143-8 at http://www.rand.org/pubs/research_reports/RR384.html
[9]	"Australian Maritime Doctrine RAN Doctrine 1", Sea Power Centre – Australia, Royal Australian Navy, 2010.
[10]	Cordner L., "Managing Maritime Security Risks in the Indian Ocean" in "Protecting the

	Ability to Trade in the Indian Ocean Maritime Economy", Forbes A., (Editor), Proceedings of the Indian Ocean Naval Symposium Seminar 2014, Sea Power Centre – Australia, Royal Australian Navy, 2014.
[11]	Rogers D., "The Use of Unmanned and Autonomous Surface Vessels to Improve Maritime Security", CIVSEC 2016, Melbourne Australia, May 31st – June 1st, 2016.
[12]	Kilcullen D., "The Australian Army, in the Urban, Networked Littoral", 2014, ISBN 9780992547400; https://www.army.gov.au/sites/g/files/net1846/f/kilcullen_11august.pdf
[13]	Hagel C., Secretary of Defence Speech, Reagan National Defence Forum Keynote, California, November 15, 2014, https://www.defense.gov/News/Speeches/Speech-View/Article/606635
[14]	Work B., "The Third U.S. Offset Strategy and its Implications for Partners and Allies", Washington D.C., January 28th, 2015, https://www.defense.gov/News/Speeches/Speech-View/Article/606641/the-third-us-offset-strategy-and-its-implications-for-partners-and-allies
[15]	"Defence Industry Policy Statement", Commonwealth of Australia, 2016, http://www.defence.gov.au/whitepaper/Docs/2016-Defence-Industry-Policy-Statement.pdf
[16]	"Defence Capability Handbook", Commonwealth of Australia, 2012, http://www.defence.gov.au/publications/DefenceCapabilityDevelopmentHandbook2012.pdf
[17]	Rogers D., "Unmanned Surface Vessels", Maritime Unmanned Systems Conference, October 6 -7, Sydney Australia 2015
[18]	"Autonomous Weapons: An open letter from AI & Robotics Researchers", Future of Life Institute, July 28, 2016, https://futureoflife.org/open-letter-autonomous-weapons/

Application of the Lloyd's Register Unmanned Marine System Code to a Small Remotely Controlled Survey Vessel

Anderson CHAPLOW, Matthew PALMER, Jim GORTON,
Lloyd's Register EMEA, mod-office@lr.org

ABSTRACT

In response to the growing demand for regulation of unmanned and autonomous marine vessels, Lloyd's Register has recently developed a goal-based standard for the certification of surface and sub-surface unmanned marine systems (UMS) – the LR Code for Unmanned Marine Systems. The use of a goal-based format allows the Code to capture the rapid advances of technology in unmanned vessels in an environment where prescriptive requirements do not exist or are unsuitable.

The Code provides a framework for the certification of these vessels and this paper will describe the process by which the goal-based requirements can be applied to vessels of all sizes and autonomy levels based on establishment of system-level solution and integrity requirements and application of appropriate verification processes based on these requirements.

This paper will discuss the experience gained during the initial certification of a small survey vessel and the development of supporting verification guidance which will particularly useful to small UMS designers, manufacturers, operators and regulators in understanding how safe, dependable, capable and resilient vessels can be certified via a pragmatic and effective framework.

NOMENCLATURE

ANEP	Allied Naval Engineering Publications
COLREGS	International Regulations for Preventing Collisions at Sea
ConOps	Concept of Operations
IMO	International Maritime Organisation
INSA	International Naval Safety Association
LR	Lloyd's Register
MASRWG	Marine Autonomous Systems Regulatory Working Group
MARPOL	IMO Marine Pollution Convention
MCA	Maritime and Coastguard Agency
SOLAS	IMO Safety of Life at Sea Convention
TRL	Technology Readiness Level
UMS	Unmanned Marine System

1. INTRODUCTION

The unmanned vessel sector has been growing rapidly over the past few years, and it is expected that Unmanned Marine Systems (UMS) will soon be in widespread use in many aspects of the maritime industry. Lloyd's Register (LR) has been active in understanding how to best serve industry in this new and expanding technology in the marine area through the certification of these UMS.

The current maritime regulatory framework is developed on the basis of mitigating the risks for vessels that are manned. The risks for the operation of UMS differ materially from those for a manned vessel and so any equivalent regulatory framework developed for UMS needs to take these into account.

It was concluded that a simple adaption of Classification Society Rules within the existing regulatory framework would not meet the requirements of the Regulator, nor Owners and Operators, for the range of vessels envisaged, and that a different approach would be needed for the assurance of UMS. Due to the rapid advances in technology prescriptive rules would not be able to keep up with industry and their limited scope would be too narrow to cover most UMS.

As a result LR has developed a goal-based code - the LR Code for Unmanned Marine Systems (the 'Code') [1] - to provide a framework for the assurance of safety, and mitigation of risk for UMS. This LR Code for Unmanned Marine Systems is written to complement a rapidly developing regulatory framework for UMS that provides equivalence to that for manned vessels.

The Code is written to be applicable to all types of unmanned vessels operating above and below the surface in a range of autonomy levels. Application is dependent upon the Owner defining their operational requirements and risk tolerance, which will then determine the verification requirements through the required 'Level of Integrity' process which LR has developed as one of the features of this Code.

As a part of the development process LR has worked closely with industry in several working groups to develop the Code and this paper discusses the first application of the Code to a small unmanned survey vessel, the challenges involved in certifying the UMS, in particular determining how to effectively apply the Code and the process developed as a result.

2. REGULATION OF MANNED AND UNMANNED VESSELS

2.1 How do we regulate manned vessels?

The main source of regulation for manned vessels is the International Maritime Organisation (IMO). The IMO does this through the application of predominantly prescriptive requirements defined within International Conventions, Classification Society Rules and Regulations, International and National Codes and Standards. These requirements can be loosely broken down into three elements: safe design, environmental protection and safe operation; with each of the elements supported by a hierarchy of regulation working together to provide a 'safe system' – this is shown in Figure 1.

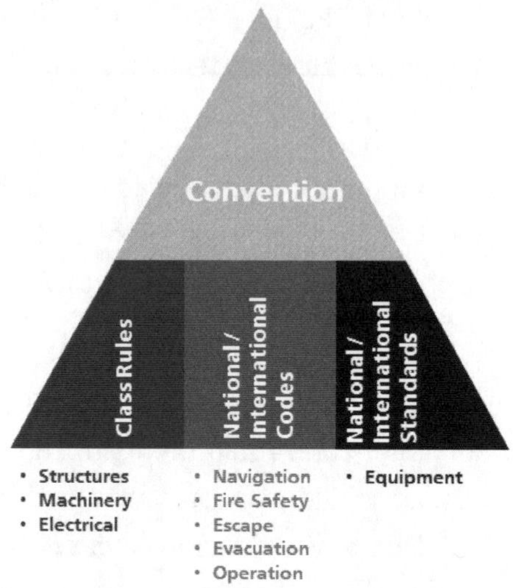

Figure 1: Current Regulatory Framework

2.2 How can we regulate unmanned vessels?

When reviewing the current regulatory regime for manned vessels it was agreed, that for UMS, there was a need to show equivalence to the existing statutory regime – taking as the basis that a UMS should be 'at least as safe as' a manned vessel. However application of existing regulation would prove difficult as the prescriptive requirements are predicated on the presence of persons onboard (both as a hazard and as a mitigative measure) and, significantly, would fail to recognise the unique risks associated with UMS.

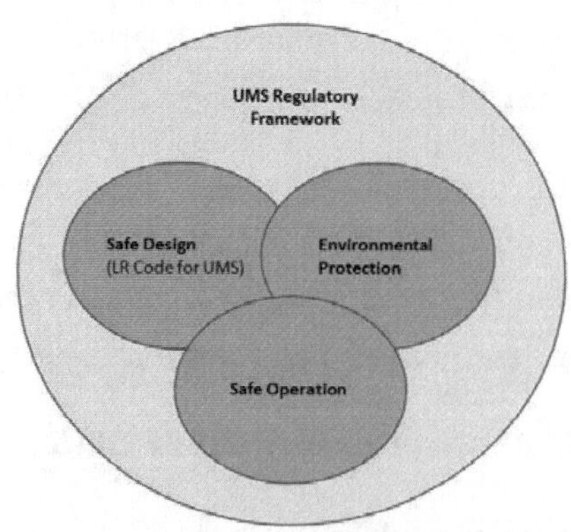

Figure 2: UMS Regulatory Framework

Extending the three elements of safe design; environmental protection and safe operation an equivalent regulatory framework for UMS can be established and appropriate regulation developed to support each element – this is shown in Figure 2.

In addition to the development of the LR UMS Code, LR has also worked closely with the UK Marine Autonomous Systems Regulatory Working Group (MASRWG). This group is made up of stakeholders from industry working together to create a regulatory framework for UMS. It is expected that this framework will establish appropriate requirements in each of the three element areas, but particularly Safe Operation', and will provide a complimentary and equivalent regime to that for manned vessels.

2.3 The LR UMS Code – a Regulation

The LR Code for Unmanned Marine Systems is intended to cover the Safe Design element in its entirety and it therefore seeks to assure the safe design, build and maintenance of UMS against an established framework that minimises the effort required by an Owner or Operator to achieve certification. It is intended to be complimentary to any regulation developed by the IMO or Flag States however in the absence of such a regulatory framework has extended its scope to cover aspects traditional out with the scope of Classification, e.g. stability, fire and navigation, in order to provide a complete solution.

By studying the current regulatory regime and looking at industry needs, it was decided that the LR Code for Unmanned Marine Systems would have to be equivalent to Statutory Codes and sit above traditional Classification Rules as shown in Figure 3.

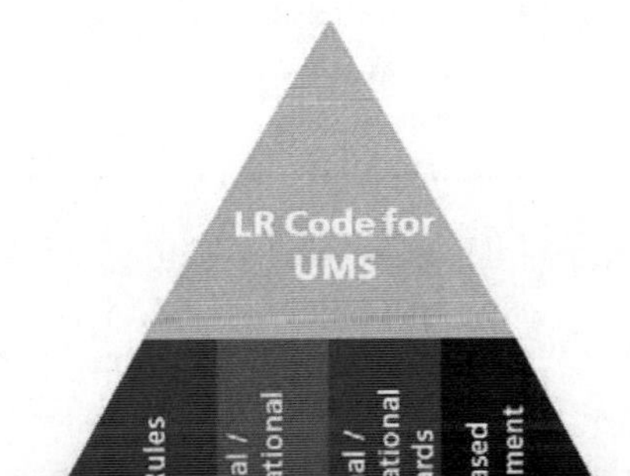

Figure 3: LR UMS Code structure

Safe Operation and Environmental Protection are out with the scope of the Code. The requirements for safe operation of UMS are expected to be initially prescribed locally by Flag States or Port Authorities, such as the requirements in the MASWRG framework. Whilst the requirements for Environmental Protection are considered to be broadly applicable as currently written, although some interpretation will be required. The Code also does not cover the risks resulting from embarked cargo or mission specific equipment; however these can be considered within an overall certification process using existing requirements as necessary.

3. THE STRUCTURE OF THE LR CODE FOR UNMANNED MARINE SYSTEMS

The LR Code for Unmanned Marine Systems is a goal-based code. The benefits of using the goal-based structure is that it defines an ultimate safety objective whilst allowing for the consideration of alternative designs and solutions that meet the safety objective; thereby supporting innovation in an area that is developing rapidly. Whilst initially targeted at small sized UMS, including naval systems, it is scalable and is capable of application to larger, more complex vessels as technology and regulation develops

LR has experience in both developing and applying goal-based codes through its involvement in the International Naval Safety Association (INSA), authors of the Naval Ship Code [2] and the Naval Submarine Code; and also as a member of the IMO working groups developing goal-based requirements for commercial vessels.

Figure 4: LR Code for Unmanned Marine Systems Structure

The structure of the LR UMS Code follows the goal-based structure also utilised in the INSA Naval Ship and Submarine Codes. With reference to Figure 4; the tiers of the Code are as follows, with the increasing width of lower tiers representing increased level of detail:

- Tier 0 is the overall Aim of the Code which is that:

  "The Unmanned Marine System (UMS) shall be safe, dependable, capable and resilient in all Reasonably Foreseeable Operating Conditions."

- Tier 1 defines high-level Goals for each hazard areas - which form the chapters of the Code. These Goals align to support the overall Aim.
- Tier 2 Functional Objectives identify the functional requirements to be satisfied in order to meet the Tier 1 Goals for specific hazard areas.
- Tier 3 Performance Requirements define the detailed requirements to meet Tier 2 Functional Objectives and are qualitative to allow them to be met by a range of solutions.

- Tier 4 Solutions describe the specified design solution and how it is to be verified. These will need to be justified against the Tier 3 Performance Requirements to show that they are appropriate.

The Goals, Functional Objectives and Performance requirements have been divided into the following key hazard areas forming the main technical chapters of the Code:

- Structure
- Stability
- Control System
- Electrical System
- Navigation System
- Propulsion and Manoeuvring
- Fire
- Auxiliary Systems

The Functional Objectives and Performance Requirements are written to cover identified hazards relating to the operation of UMS. To ensure completeness, these have been validated against existing Conventions, Codes and Standards including; Classification Rules and Regulations, SOLAS, and ANEP77 and also by engagement with existing UMS projects.

Tier 4 solutions are not defined within the Code. This is because of the potential diversity in design and operation of UMS. The Code expects an appropriate solution be selected from a tailored combination of: Classification Rules, National or International Codes and Standards or where standards do not exist or are not appropriate, through the application of risk-based assessment methodologies that demonstrate the performance requirements are achieved. This tailoring of solutions is shown in Figure 3 and also discussed further below in relation to a small remotely controlled UMS.

4. CERTIFICATION OF A SMALL REMOTELY CONTROLLED UMS

4.1 Introduction

So far we have discussed a regulatory framework for UMS and how the LR Code for UMS can support this, we have also looked at the structure of the Code. What we know, however, is that the application of goal-based standards in the absence of exemplar 'solutions' can be extremely difficult and that the development of a 'code' does not necessarily guarantee the ability to certify a UMS. The next part of the paper will discuss the development of a process developed to certify a UMS against the Code and the lessons learnt through the process.

In the current absence of 'enabling' regulation, it is recognised that the initial uptake of UMS technology on a commercial basis will be in small-scale operations conducted in local or limited waters under the control of one Regulator permitting use on a case-by-case basis. To this end the Code needs to be delivered through a robust but practical and cost-effective process commensurate with the size, complexity and associated risk presented by the vessel.

It is therefore appropriate then that the first vessel presented for Code certification was a seven metre survey vessel operated by remote control within line of sight from a mothership and deploying interchangeable sensor packages in support of pipeline installations. Certification was required to fulfil the requirements of the Charterer in order to demonstrate the structured management of risk against a transparent independently assured framework.

4.2 Certification Process

To achieve certification for the project a process needed to be rapidly developed which met the requirements of the Client but importantly the requirements of the Code. Using our experience of application of goal-based standards, together with a series of existing tools, it was determined to develop the process concurrently with the project and to amend it as required along the way.

The final process developed is shown in Figure 5, this will be referred to throughout this section and demonstrates a structured approach to certification which defines five key stages and the necessary inputs required to progress through each stage. Key elements of this process will be discussed further below.

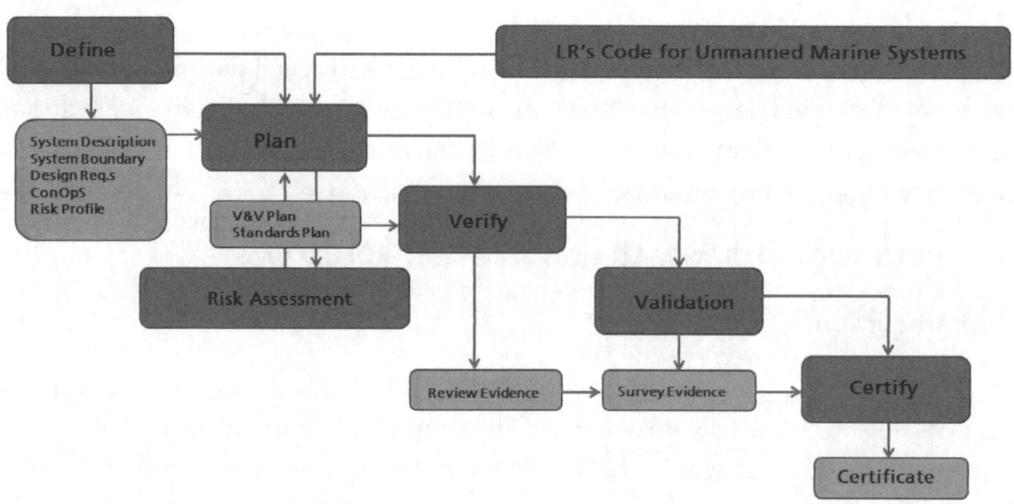

Figure 5 UMS Certification process

4.3 Stage One - Define

The first stage in the process is to establish the system requirements and define the boundaries of the certification assessment. This is important as it establishes the context of the subsequent certification processes and enables the selection of appropriate verification methods.

In the first instance the Owner describes the system, defines the certification boundaries and key design requirements and uses this information to compile a Concept of Operations (ConOps) for the UMS. The ConOps is a key requirement for appropriate application of the Code and a template is provided in Annex A (of the Code); ultimately the scope of certification is bounded to operation of the UMS in accordance with the defined ConOps and therefore this needs to consistent with the Owners requirements.

The ConOps is used to establish the risk profile of the UMS by defining where and how it will be operated, maintained and ultimately disposed of; this information is then used in the next stage to establish the design solution and verification requirements. For example a small UMS operating inshore under remote control will have a different risk profile to a large UMS operating autonomously offshore and each of these UMS will have a different set of design solutions and verification requirements but ultimately both achieving compliance with the LR UMS Code.

Establishment of the Stage One inputs for the small UMS was relatively straightforward. It is intended to be remotely operated in a controlled water space, within line of sight from the mothership with a nominated failsafe mode in the event of signal loss and ultimately with an emergency override. Its size presented limited hazards in relation to its environment and it was readily recoverable in the event of a system failure. The design requirements were well established and completion of the ConOps was straightforward.

4.4 Stage Two - Plan

The second stage is delivered in the form of a workshop; initially, for the small UMS, this followed an existing Technology Qualification (TQ) process established by LR to enable the certification of novel technologies however parts of this were found to be inappropriate and a modified process has been developed which better achieves the requirements of the Code.

This was partly because the workshop was held towards the end of the construction phase and therefore issues associated with technology readiness had already been resolved through the Owners development phase. In addition, whilst aspects of the TQ process were useful and have been retained, it was found that the extent of 'novelness' of a small remotely controlled boat is effectively limited to the control system and so adequate definition of the 'problem' was difficult.

The modified workshop process is as follows:

4.4.1 Review of the CONOPS, Design Requirements, System Boundary
The first activity of the workshop is to review the inputs from Stage One – the system information, design requirements and Concept of Operations to ensure a complete understanding of the UMS under assessment.

4.4.2 System Decomposition

The next step is to conduct a system decomposition of the UMS, which is the process of breaking it down into recognisable systems and sub-systems. These will then be used to inform the establishment of design requirements, solutions, levels of integrity and verification processes.

4.4.3 Technology Assessment

Once the system decomposition is complete a Technology Assessment is conducted, using a screening matrix (Figure 6) to assess the 'uncertainty' rating for each system on the basis of Technology Readiness against experience of the application of the technology in the designed system. This will look at the maturity and reliability of the system or component, and experience of its use in the defined context, for example in the marine environment.

The results of this screening will put the system or component into a band (red, amber, green) to determine whether and to what extent further analysis, or risk based assessment, is required. For items presenting a high level of uncertainty (red) a risk assessment will be required to identify the associated hazards and controls required to mitigate the risk appropriately.

It is expected that most systems and components found on UMS will be well-proven and supported by existing marine standards and will therefore fall into the 'green' bands. Equipment with expected reliability issues (requiring periodic human intervention) or remote actuation systems will typically be 'amber'; whilst autonomous control systems will be 'red'.

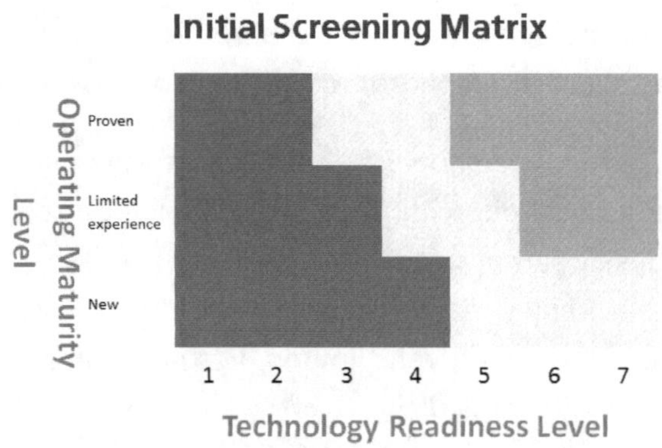

Figure 6 Initial Screening Matrix

4.4.4 Required Levels of Integrity

Following the system decomposition and technology assessment processes the required Level of Integrity is determined for each UMS system. The level of integrity is to be determined by assessing the effect on the UMS, considered as a system of systems, of all reasonably foreseeable system failures and by considering their consequence on the ability of the UMS to achieve the Aim of the Code in the following areas:

- System Safety Consequences
  a. To people onboard
  b. To people/objects in the vicinity
  c. To the environment
- System Operational Consequences
  a. To capability
  b. To resilience

Against these consequence criteria the following basic levels will be assigned

- High; failure is not acceptable
- Medium; failure may be acceptable subject to the presence of controls measures – either preventative or mitigative.
- Low; failure is acceptable

For more complex systems this approach will be too simplistic and higher level risk assessments will need to be conducted.

4.4.5 Performance Requirement Mapping

Each chapter of the LR UMS Code contains Performance Requirements supporting different Functional Objectives. These are then mapped against the system decomposition to establish the applicable Performance Requirements for each system – this is the key step in demonstrating that the Code requirements have been included in the system design and tracked in the verification process.

4.4.6 Design Solutions/Risk Assessments

All of the above information is then used to determine the appropriate design solutions required in order to achieve compliance with the Performance Requirements of the LR UMS Code. These will be a mix of existing standards, bespoke solutions or the results of the specific risk assessment requirements identified as part of the Technology Assessment – these will need to be conducted external to the workshop.

For the small UMS this was found to be a mix of bespoke solutions, existing small boat design standards such as ISO 12215 and codes such as the UK MCA Workboat Code. These were found to fulfil the Code requirements commensurate to the risk and also equivalent to the standards required for manned vessels.

4.4.7 Verification Requirements

The final activity is to determine the required verification requirements in order to ensure that the design solutions are achieved and the Performance Requirements met. These are applied with reference to the Required Levels of Integrity to ensure the proportionate and efficient engagement of LR commensurate to the size and complexity of the UMS. A table of verification requirements is given in Annex B of the Code.

The workshop process was found to be an effective way of establishing the Verification Plan for the UMS through the collaborative generation of inputs and as a result of the work done for the small UMS a templated Verification Plan has been developed as an output which can then be finalised by the Owner and submitted for acceptance as the controlling document for completion of the certification process. This enables a stepped approach to be taken to UMS projects which enables the Owner to phase the scope of engagement and to establish a clear path to certification.

In accordance with the defined ConOps and Risk Profile many of the vessel systems were screened as 'green' and the assigned Levels of Integrity found as 'Low' therefore the scope of the subsequent verification and validation activities was reduced – this was deemed to be appropriate and confirmation that the proportionate approach had been taken.

4.5 Stage Three - Verify

At the end of Stage Two a Verification Plan is created and approved to show how the Tier 3 Performance Requirements are met, including the details and justification of the Tier 4 Solutions, the method of assessment of these and the means of demonstrating conformance. This is achieved through design review - 'verification'; and physical inspection - 'validation'.

The verification method is determined by required Level of Integrity for each system and a schedule for verification based on this has been proposed in Annex B of the Code. This breaks the process down into the following seven lifecycle phases and defines appropriate verification processes based on each of the three Levels of Integrity.

- Design Review
- Materials, Equipment and Components
- Hull Construction
- System Installation
- Trials and Commissioning
- In-service
- Damage, Modifications and Repairs

For the small UMS under review the Verification Plan included a mix of third party review, manufacturers certification and design statements – this was driven by the predominantly low Level of Integrity requirements resulting from the minimal risk profile. In particular it was considered that the required Level of Integrity for the control system was low – necessitating minimal verification activities – this was due to the fact it is remotely controlled in a restricted area for which the consequence of failure is limited and an independent failsafe emergency stop function provided. For a different operating context, say operation in confined waters or over-the-horizon this would not be the case.

4.6 Stage Four - Validate

One of the key areas of the Verification Plan is the survey aspects; this is where the material state of the UMS is validated in accordance with the Verification Plan and as above the validation activities are applied according to Annex B of the Code commensurate with the required Level of Integrity.

The small UMS under review was surveyed by LR Surveyors during and following construction according to processes defined in the UK MCA Workboat Code – the nominated solution for much of the system design and assurance. Following which a sea trial was conducted in order to demonstrate compliance both with the requirements of the ConOps and the LR UMS Code Performance Requirements – due to the limited complexity of the control system this was a relatively straightforward matter. By identifying the Performance Requirements able to be achieved through trials, an existing small boat checklist was modified to capture the necessary scope and this was able to be provided to the Owner to enable the development and execution of trials procedures to demonstrate these.

It is clear that validation processes for autonomous control systems is still an area that requires a significant amount of work and that a repeatable, dependable and reliable approach needs to be developed in order to facilitate the certification of UMS on a wider scale. LR is engaged in the MAchine eXecutable Collision regulations for Marine Autonomous Systems (MAXCMAS) project which has as its goal the development of a method to interpret COLREGs through a compliant path planning system for autonomous vessels. This involved simulator testing to prove that the software worked for a variety of scenarios which was repeated many times to provide assurance that the software worked.

Simulators trials are a very good way of proving that a system does what it says it will do in a stable environment but real-life operations add extra variables that cannot be accounted for in a simulator – most specifically environmental conditions. It is expected that validation methods for autonomous controls systems will ultimately involve simulator testing, hardware certification and physical trials to prove effective integration and performance and LR is actively engaged in developing these.

4.7 Stage Five - Certify

On completion of the required verification and validation activities, the evidence package is submitted for review against the approved Verification Plan and the completion of all activities confirmed. The collation of evidence is shown in Figure 7

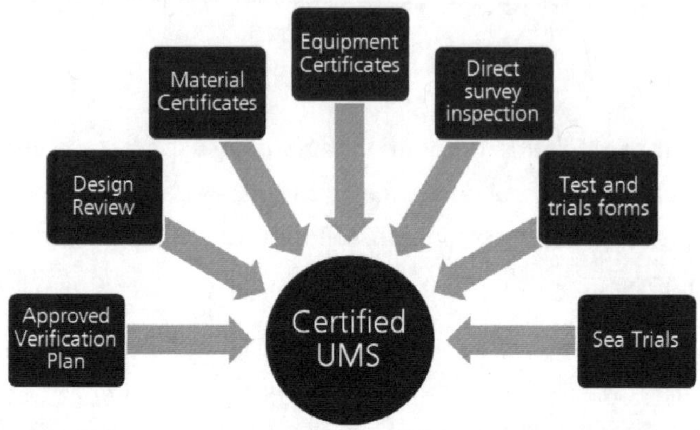

Figure 7: Evidence for the certification of a UMS

Completion of the processes required in order to demonstrate compliance with the LR UMS Code will result in certification of the UMS to the Code, this is considered to be a whole-of-system certification which can be used to demonstrate to external parties that the UMS design has been found safe, resilient, capable and dependable against a transparent set of requirements by an internationally recognised independent third party – Lloyd's Register.

Due to the bespoke application of the Code as a combination of goal-based requirements, system complexity and risk profiles, the Certificate will be intrinsically linked to the defined Concept of Operations (ConOps) and the activities undertaken to achieve the required Levels of Integrity. This will ensure that the vessel is appropriately certified for the activities it is required to undertake and for the environments it will undertake them in.

The certificate can be delivered either as a one-off statement on completion of all activities or as an ongoing assurance process in accordance with an agreed schedule of ongoing surveys and inspections.

At the time of writing the small remotely controlled UMS is not yet certified; certification is pending collation of the required evidence package and it is expected that this will be completed by early October.

5. THE HUMAN ELEMENT

During the assessment the role of the human in the safe design and in particular safe operation of the UMS became more prominent. What we need to recognise is that humans appear frequently in interactions with UMS even though by definition they are unmanned or autonomous.

Even if the UMS is not designed to periodically carry people onboard, people will still be required to interact with the UMS on an occasional basis. For example: to carry out maintenance; repairs; data recovery or launch and recovery of the UMS; or, in the event of failure, to retrieve, repair or restart the UMS at sea. As such UMS design still needs to consider interactions with humans at certain points.

It is not sufficient to assume that 'manned' codes will cover all human element activities expected to be undertaken onboard UMS, for example we may find that where a UMS is only 'occasionally' manned, such as for emergency recovery, trials or shore-side maintenance, existing 'manned' codes may not be relevant or legally applicable. These scenarios will be covered by domestic health and safety legislation but again these may not be appropriate.

To address this LR intends to develop a Chapter within the Code dealing with the Human Interactions, this will ensure that a set of high level requirements for human occupancy and interaction are defined which ensure that where appropriate 'manned' codes are complied with or where not appropriate or applicable alternative solutions are sought to ensure the safety and comfort of human/UMS interaction.

6. CONCLUSIONS

In conclusion Lloyd's Register has developed a Code for the certification of 'safe design' aspects of unmanned marine systems – the LR UMS Code. This is intended to be complimentary to any emerging UMS regulatory regimes but also to provide a transparent and equivalent level of safety to that required for manned vessels.

The Code has been successfully applied to a small remotely controlled UMS and this initial application of the Code has been used to develop a certification process which is proportionate to the size, complexity and risk profile of the UMS. This has found to be an effective demonstration of the ability of the Code to be appropriately tailored to different types of vessel.

The learning gained as a result of this application has been used to inform the process described here and to establish delivery methods, templates and tools to support further applications of the Code, particularly to small and relatively straightforward UMS.

Further work is required on verification methods for complex autonomous control systems and this is a challenge that Lloyd's Register (and many others) are currently working to understand. In addition further development is required on the LR UMS Code to include requirements for human/UMS interactions and to capture the feedback that has been generated since the Code was initially published.

7. REFERENCES

1. Lloyd's Register's Code for Unmanned Marine Systems, February 2017

2. NATO, Naval Ship Code, ANEP 77, Edition F Version 1, dated Aug 2014.

3. Jesus Mediavilla Varas, Spyros Hirdaris, Renny Smith, Paolo Scialla, Walter Caharija

   MAXCMAS project. Autonomous COLREGs compliant ship navigation

8. AUTHORS BIOGRAPHY

Anderson Chaplow BEng AMRINA is a Lead Specialist and Projects Team Leader in the Lloyd's Register's Naval Liaison Office. His team are responsible for coordinating the feedback and development of Class Rules, Naval Codes and Standards and managing consultancy tasks covering design review, technical support and advice on standards management. Anderson is now managing the delivery of the LR Code for Unmanned Maritime Systems to industry and participates in the UK's Marine Industry Alliance, Marine Autonomous Systems Regulatory Working Group. Anderson has been with Lloyd's Register for nearly eleven years with previous roles in the survey and certification of equipment & components, complex new construction projects and existing vessels.

Matthew Palmer BEng CEng MIMechE holds the current position of Senior Specialist in Lloyd's Register's Naval Liaison Office. Matthew works on developing Goal-based Standards and Rules and is contributing to the development of the LR Code for Unmanned Marine Systems for UMS. Matthew is also the part of the Secretariat for the INSA Naval Ship and Submarine Code. Matthew has been leading the industry engagement, particularly in the coordination and delivery of the two trial projects run as a part of the LR Code for Unmanned Marine Systems development.

Jim Gorton AMRINA is Lloyd's Register's Naval Strategy Manager responsible for coordinating and supporting the naval business across LR's six geographical areas. Prior to joining Lloyd's Register in 2007, Jim was a Chief Officer with the Royal Fleet Auxiliary Service serving for 27 years. His last appointment was as the Head of Maintenance Strategy and Policy for the Flotilla based in MOD Abbey Wood. In his 27 years Jim served on most classes of RFA ship initially as a Communications Officer and later as a Systems Engineer Officer. He also served in a number of shore appointments in London, Bath and Bristol.

Knowledge-based Naval Mission & Combat Systems

Alain H. Carof
NAVAL Group / deputy CTO for Mission & Combat Systems
alain.carof@naval-group.com

ABSTRACT

This paper addresses NAVAL Group (previously DCNS) approach to their new generation combat system for Surface combatant (SETIS) and Submarines (SUBTICS), as part of a broader NAVAL Group innovation initiative, called "Smart Digitized Ship", to tackle this global challenge. These Systems will incrementally integrate High-efficiency techniques coming from various domains dealing with modern Knowledge-based Decision support Systems (Multi-criteria Optimization, Big Data, Artificial Intelligence, knowledge-based decision making,).
Some results and lessons learnt will be provided on a few critical domains, and perspectives will be given to the way NAVAL Group is open to potential collaboration in these areas

EVOLUTION OF COMBAT SYSTEMS

Naval Group (previously DCNS) has developed, for several decades, several families of Surface ships and Submarines combat systems, for both French Navy and export Navies. These combat systems have permanently been upgraded to cope with continuous evolution in missions, threats, sensors and effectors.
Recent combat systems installed on board Anti-Air Horizon Frigates and the Charles de Gaulle Aircraft carrier, as well as the BPC (multi-mission helicopter launching deck Platform) are good examples of powerful systems, bringing substantial military contributions to allied forces operations. And, last but definitively not the least, the latest generation of surface ship combat system – called SETIS - developed for the FREMM Frigates, is now offering an impressive multi-Mission capability to Client Navies, encompassing :

- Anti-Submarine Warfare (ASW),
- Anti-Air Warfare (AAW),
- Anti-Surface Warfare (ASuW)
- Asymmetric Warfare (AsyW)
- Coastal Land attack Capability
- Deep Strike Attack capability

A compact version of SETIS combat system has just been operationally installed and qualified on board the GOWIND Corvette class

The SUBTICS system, installed on-board SCORPENE class submarines, is also a good representative of modern Submarine combat systems, offering several capabilities in terms of Sonar information processing, and a fast & accurate Situation assessment with

Automatic Target motion Analysis, Multi-sensor fusion, as well as a capability to launch multiple Torpedoes and Missiles.

Combat system generic architecture

A generic SETIS Combat system architecture is indicated below, with various sensors and effectors interconnected via a powerful high speed data network, fully redundant and reconfigurable. It has to be noted that the Data Network is not limited to the combat system, but it has been extended to some critical ship functions, in order to optimize the overall ship operational capabilities

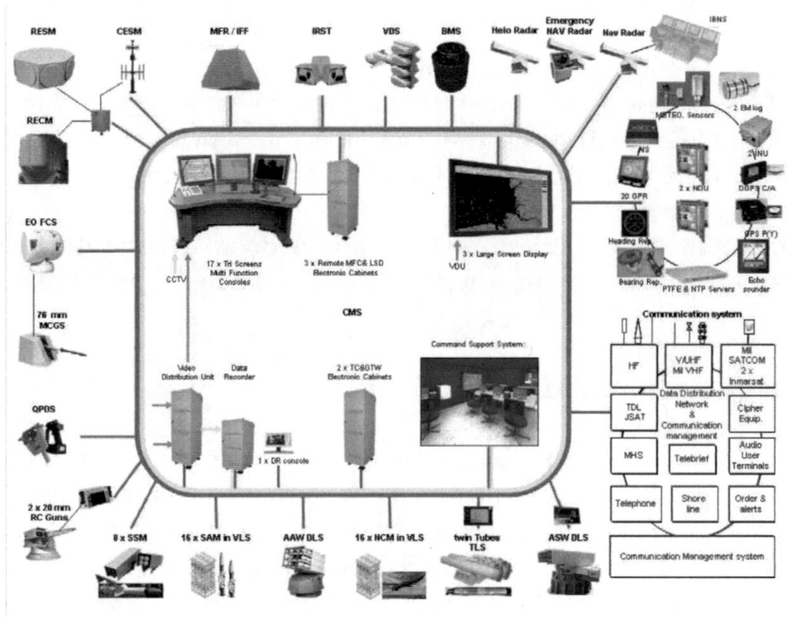

In terms of functional architecture, the generic scheme can be represented as a classical "OODA-like" loop (Observe, Orient, Decide and Act) for each warfare domain, with a global tactical situation Management

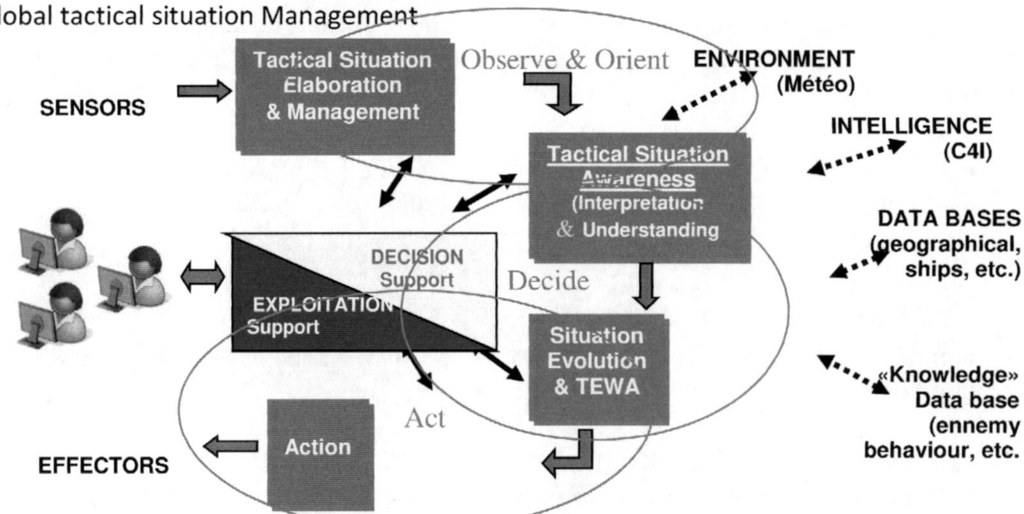

A more detailed architecture would indicate additional functions and interfaces, such as tactical data links, C4I, etc. ... However, this is not necessary for this paper's purpose.

Need for faster evolution

Although existing combat systems are still operationally efficient, an acceleration of their capability to evolve is nevertheless required by Navies, in order to maintain tactical advantage against enemies who can acquire modern technologies more and more quickly, benefitting from civil world development (Internet, Telecommunications, etc.).

Rapidly changing technologies in Software, computers, communication networks and (semi) autonomous vehicles have led to increasingly capable enemies equipped with powerful sensors and weapons. Emerging challenges, such as growing threats in asymmetric warfare or cyber-warfare, do require drastic changes in the way Naval Forces will operate to maintain information dominance and decision superiority, in particular in difficult maritime environmental conditions

Robust-to-change, resilient & cyber-secured computing infrastructure, based on service-oriented software, data servers and high speed data network, is mandatory to allow fast incremental functional improvements. However, such technologies can be obtained in the civil world and the only real Operational differentiator remains at the functional level, in particular at the Information providing & Decision making levels. Modern weapons are useless without pertinent information, delivered on time, and faster and better decision-making support systems.

Naval Group has been developing and evaluating several techniques combining powerful real-time computational algorithms (Tracking, Fusion, Optimisation), Kowledge-based decisions and Artificial Intelligence techniques (learning, etc.) to tackle various challenges:

- High-Level Perception of the environment & Automatic Classification
- Knowledge-based Tactical Situation Interpretation & battle-space management
    - Multi-Platform Multi-sensors heterogeneous data & information Fusion
    - Suspicious behaviour detection and Threat Evaluation
- Decision support system
    - Optimal resource management and allocation
    - Optimal planning
- Exploitation Support system
    - Virtual assistance for (reduced) manning under stress and fatigue conditions,
    - Enhanced collaborative system Exploitation through Virtual and Augmented Reality

KNOWLEDGE-BASED SYSTEMS

High Level Perception and Automatic Classification

Most sensors on-board naval platforms, such as Radars, EO/IR or Sonars, have been continuously enhanced, due to the increased performances of electronic and computing technologies. However, although some improvement have been done in robotic sensing, automatic target classification is still far from achieved, due to the challenging issues of learning systems, such as their training (reference data base) and the absence of explanation of wrong decisions.

Optronics sensor example

HD Cameras are nowadays widely used in maritime surveillance applications to detect small non-cooperative boats. Classical approaches for automatic target "extraction" have shown limitations and have been made more and more complex to deal with global non-static backgrounds. These approaches taking into account inter-pixel correlations are accurate only when the background is regularly textured in space and time, which is clearly not the case in maritime surveillance application (sea, landscape, clouds, harbor structures, etc.).
A simple, efficient and auto adaptive video extractor & tracker has been developed by NAVAL Group and implemented on a standard computer. The results obtained so far are good, and have been further improved with a wake suppression module.
An example is given below, on a full HD video (1920 x 1080). The tracking process, here, is done in the Cartesian plane. (see [1] [2] & [3] for more details on NAVAL Group process)

Raw detections on the video extractor

Tracks on the video extractor

The next step towards scene interpretation is the image classification of the moving object. Several techniques have been studied and evaluated. However, two main difficulties result from the learning phase:

- the creation of the reference data base needs a sufficient number of appropriate images of small ships in maritime environment, which is not the case in existing (and available) web image bases; furthermore, the influence of the environment is highly critical and has to be part of the training: "Knowledge" has to be injected in the classification system.
- Any error in the decision making can provide a fatal issue in operational situation where a fast reaction can be using a weapon... Again, additional knowledge must be integrated in the decision support system, such as trained information fusion on multiple (heterogeneous) sources: successive object image attributes and/or classification, kinematics and selected environment attributes (sea state, waves, sun light, ..)

Some results are given for a simple case with 5 classes in two different maritime environments: Mediterranean sea (near Toulon) and Atlantic sea (near Lorient). The technique is based on a “Bag of words” classification scheme with supervised learning & SVM techniques, which has proven to be more efficient than convolutional neuronal networks for a small number of classes (typically 5 to 10), when ship average images correspond to about 100-200 pixels and when the number of examples is limited (less than 100 per class).

Typical examples of the (5) classes reference data base used for the classifier Training

The two following tables correspond to the ratio of correct allocation between the predicted class and the real one, in the two different environmental conditions.

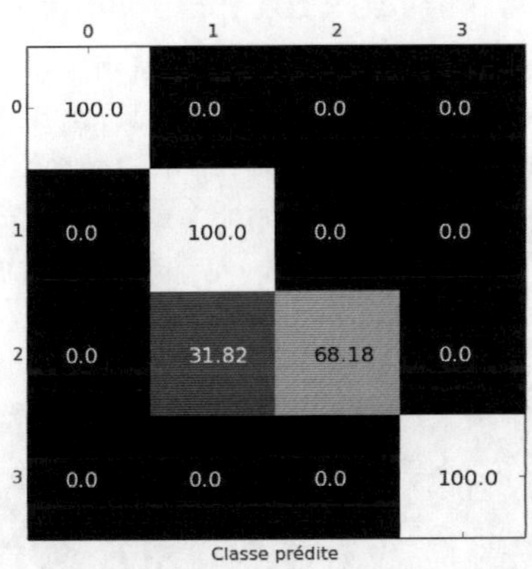

Mediterranean Sea conditions (Toulon)

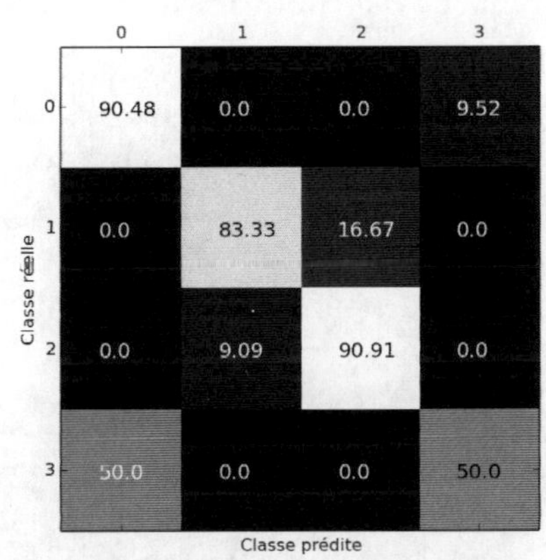

Atlantic sea conditions (Lorient)

Now, an extension of this technique to a much larger number of ship classes is not simple, due to the effort to establish a representative reference set. Classification methods based on learning techniques have to be carefully analysed when the number of (tagged) examples is low, or when a wrong decision results in a potentially dangerous action.

GLOBAL BATTLESPACE MANAGEMENT

The French Multiplatform Tactical Picture capability (MTP)

The MTP capability, also called PLDF (Plot Level Data Exchange and Fusion), aims at elaborating a real-time Air and surface tactical picture based on sensor plots (raw decisions) from all sensors of the entire naval force. The MTP units exchanges sensors data by Internet Protocol network, and each MTP unit achieves a real time data fusion between local and remote sensors to elaborate a "local" tactical situation.

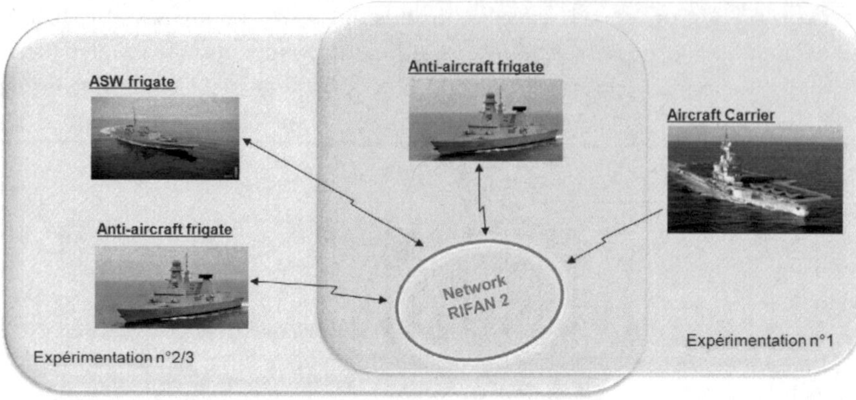

The overall process is complex, and has required ten years of study, development and evaluation, as part of the TSMPF French program. Strong algorithm expertise and operational knowledge have been used to achieve:

- Composite multiplatform tracking, robust to sensors and platform location errors.
- Accurate estimation of sensors and platform location errors, based on plot picture correlation with automatic correction capability.
- Composite multiplatform tracking, able to process delayed plots. In particular, it is compatible with the latency of satellite link exchanges.
- Reduced bandwidth, with only the best sensor plots being sent to others platforms

The MTP operational capability has been evaluated for several years (see [4] for more information) and it clearly demonstrates a true advantage due to sensors complementarity. Improvements for the tactical picture, particularly in terms of continuity and tracking accuracy are significant and essential for increasing performances of several Combat systems capabilities such as automatic identification, Tactical Data Link correlation and weapons engagement.

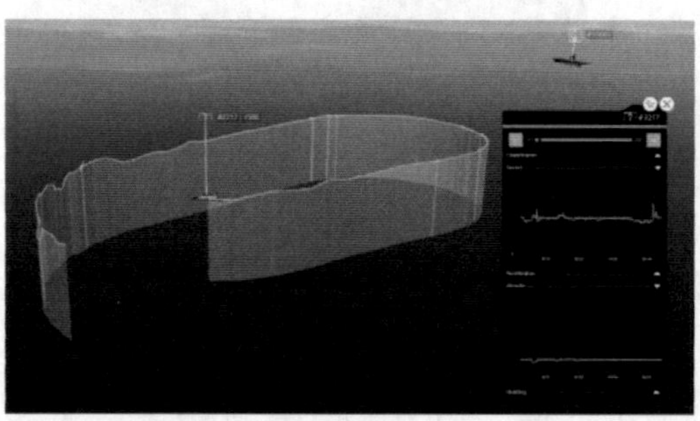

Multi Platform (MTP) Tracking result

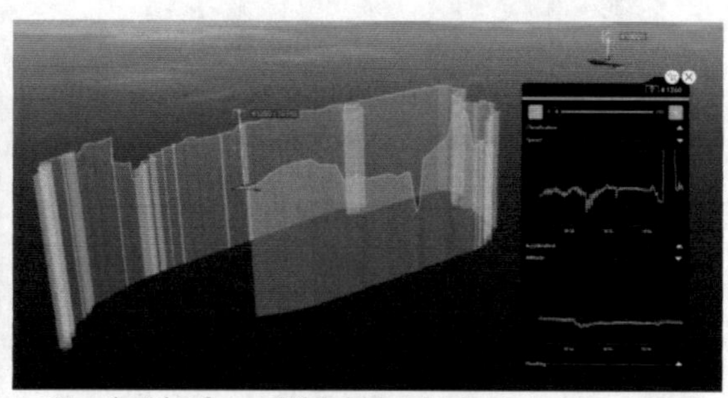

Single Platform (Ship) Tracking

With these highly positive results, NAVAL Group is now envisaging a fully operational deployment of the MTP capability on the main French Navy ships.

In addition, cooperation work at European level is carried out under the Multi Platform Engagement Capability Working group, an inter-allied evaluation is planned for the second half of 2017 between France and Netherlands. A new study should also begin in 2017 to assess the contribution of MPF capability in joint context, with a experiment between Navy and Air force in 2018.

MTP Extended capability

Following the promising results of the MTP capability, work has already started by analyzing the integration of new platforms and sensors, such as Shore-based sensors (Maritime Surveillance), Maritime Patrol Aircrafts sensors (Radars, etc.) and Drones Mission payload
.Additional knowledge has to be taken into account to deal with their specificities.

Maritime Surveillance : Global tactical situation management

Between 2010 and 2014, Naval Group (then DCNS) has been coordinating an important European Union Project, called I2C (*). This project was aimed to be a proof of concept for a future European external border surveillance system (EUROSUR).
The overall aim is to provide for reliable, continuous and all weather surveillance of any vessel tracks and activities over wide maritime territories, in order to successfully fight against illegal activities and serious crimes at sea. Data were obtained from a mix of shore-based radars (HF, FMCW, conventional), Radars and Optronic sensors on-board Airship (Zeppelin), ship sensors, AIS communication means, and networks (WAN satellite 3G broadband, secured VPN).
See [5] for a more detailed description of I2C Project

I2C sensors used in the Final demonstration in Toulon (summer 2011)

- (*) I2C: Interoperable sensors & Information sources for Common detection of Abnormal vessel behaviours and Collaborative suspect events analysis

The experimental system developed in I2C enabled the team to test methods of data fusion from:

- a set of new sensors (i.e. long range detection & detection of small boats) and conventional sensors deployed ashore and on offshore assets, and various other available data such as meteorological forecasts, vessel and harbour files, regulated zones and geo data, vessel black listings and
- Intelligence resources, and to set up and share a multi layered intelligent situational picture including documented alerts on detected suspicious vessel behaviours.

Global Situation awareness with automatic alerts (abnormal behaviours)

KNOWLEDGE-BASED DECISION SUPPORT SYSTEM

Suspicious behaviour detection and Threat Evaluation

The I2C project has also allowed Naval Group to develop and evaluate two-steps algorithms for automatic detection of suspicious behaviour and corresponding threat identification. As shown on the overall architecture, here below:

- the first step corresponds to an automatic anomaly detection, such as ship at null speed, strange kinematics profiles, AIS anomalies, etc. This is based on various (heterogeneous) attributes such as kinematics, obtained both by sensors (when available) and AIS ship information, and information from different data bases (meteorological and geographical bases, European ship information system, Lloyds registration base, etc.)

- the second step corresponds to the mission driven interpretation of the abnormal behaviour. Operational Knowledge-based rules are required for threat interpretation, which depends on the mission (Maritime safety and security, fisheries control, customs, border control, law enforcement, etc.)

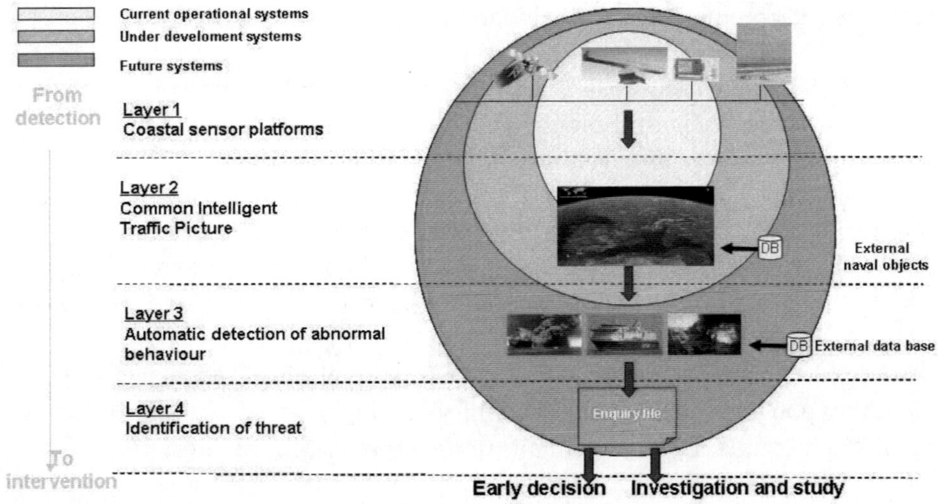

Multi-layered I2C Architecture for situation elaboration and automatic alert (abnormal behaviors)

Target Identification

This function has to decide whether a moving objet that has just been classified (e.g. type of aircraft or ship) is a potential friend of foe. Although some sensors are dedicated to this function (IFF or AIS), errors can occur either intentionally (e.g. AIS spoofing) or not... and the consequences can be dramatic if they are not detected and corrected.

Knowledge can be introduced at this critical level through a Bayesian Network, assuming that "rules" have been checked with operational experts [.]

Fusion of expertise is also a good way to detect potential conflict between sensors elementary decisions (e.g. identification conflict between radar and IR camera)

Threat Evaluation and Weapon Assignment (TEWA)

This critical function of a combat system has a lot of similarities with the previously described abnormal anomaly detection. However, the reaction usually deals with weapons and the choice of decision algorithms is therefore much more constrained:

- Any decision must be explained, which induces severe limitation to learning systems, such as convolutional neuronal networks or deep learning systems
- Some decisions may also have to be made without examples (new incoming missile)
- Final Decision is usually depending on multiple criteria and multiple objectives,
- If successive recommendations to human operators are made (e.g. dynamic programming for optimal weapons planning) frequent changes of the optimal priority are not recommended
- In most cases, a Real-time (optimal) decision is requested, which then requires very efficient multi-criteria optimization techniques

Optimal resource management & Optimal Planning

Better Decision-Making does obviously require better information providing. This in turn usually requires some capability to reconfigure various system resources, such as sensors and/or computing, to acquire more accurate data about the tracks.
However, such capability has to be done in real time, to cope with high speed and/or highly maneuvering threats, such as modern missiles or in asymmetric warfare
Specific Algorithms (Multi-criteria/Multi-Objectives algorithms,) have been developed and evaluated by the NAVAL Group Research center for various operational applications such as:

- Optimal sensors management (e.g. Radars in Multi-Platform Tactical Picture)
- Optimal Weapon Assignment (with dynamic planning)
- Optimal Radio Frequency management (under strict RADAZ constraints)

These Algorithms provide real-time solutions that should allow operational deployment in the short-term:

- Sensor allocation < 1 sec. for 10^{30} possible solutions and 50 tracks to manage
- Weapon assignment (recommendation to the operator): Updated every 1 sec.
- 30 sec. for 30 transmitters (each one with 100 parameters)

Operator expertise has often to be taken into account in various applications, and dedicated algorithms based on partial preference models have been developed for integrating Expert rules. An example of such partial preference information model applied to Multi Objective Optimization" is given in [6]
Such approach, developed with Paris-Dauphine University (LAMSADE), has shown very competitive performances both on computation time and quality of the generated preferred set. Comparison has been made on two classical problems: the Multi-Objective Knapsack Problem (MOKP) and the Multi-Objective Assignment Problem (MOAP) and substantial computation gains have been achieved.

Additional applications have been made for Optimal Route and Trajectory planning (Drone & SM), which promisingly allows for efficient re-planning in Obstacle avoidance functions. However, computation time is not the only issue in some difficult problems such as recommendation of an optimal trajectory for a submarine to avoid being detected by the enemy. As passive sonars can only provide uncertain target ranges, and both the own ship and the enemy are moving, an approach based on stochastic control has been evaluated through a joint study the French Research center INRIA (see [7])

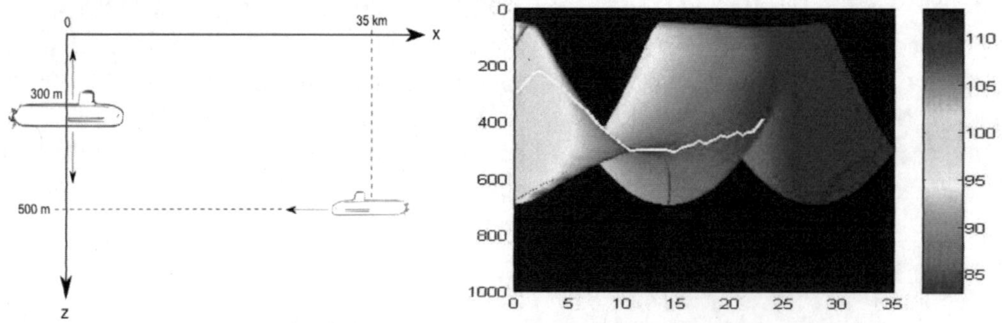

Example of SM trajectory recommendation in acoustic duel

HUMAN FACTOR and ENHANCED CREW EXPLOITATION

Decision-Making is, in most cases, under Human operator's responsibility. If automatic detection and tracking systems are now well accepted, Decision support systems are only providing "recommendations" which can –or not – be accepted by the operators. Even in the case of anti-missile defense, where extremely fast decision must be taken to cope with the very high speed of the threat, it is up to operators to put the system into a fully reflexive mode.
A promising field of R&D is also associated to Natural Language Processing, both for aural (spoken) discussions, and for textual interpretation ("chat", open database queries, etc.) although the error rate may still be a limiting factor, in application where a wrong interpretation can lead to a dangerous (re)action

The domain of Human-Machine Interaction (HMI) is therefore of key importance to provide the most efficient combat system, and operators' knowledge must be taken into account, generally through different learning phases.

Modern HMI Tools : RA, RV, Helmet , ...

Various technologies have been developed in the field of Human-Machine interactions, coming from a very broad spectrum of applications (Video game, car industry, internet, etc.) and some of these technologies, such as Augmented Reality, Virtual Reality, 3D representations, head-up displays, etc. have already been evaluated for military purpose, often with pro's and con's, and no final conclusion can be made : each technology have to be carefully evaluated before any real implementation within a combat system. Technologies for Aircraft may not be suitable for a surface ship ...
Furthermore, the operator's behavior during shore-based training sessions may be very different from at sea real operations. Additional "learning" approaches have to be made, for instance by recording the activity during real operations and applying post-mission data analysis, assuming an appropriate understanding of the operational context.

Stress and Fatigue Management

Increased Combat System complexity and information overload are becoming an issue for operators, especially in real operations where fatigue and stress can perturb the decision making process. A promising approach to cope with this critical problem is based on combining bio-monitoring sensors, operator activity and human mental representation (cognitive space) to check whether the operator can efficiently perform its task, and –if not- to activate (or recommend) some HMI changes (load balancing, adaptive command and control, automatic process of low level tasks, decision-making process, etc.)

Collaborative techniques

Decision-making in Combat systems is actually based on a team work, either located on a single ship or distributed on multiple platforms and specific techniques/technologies for improving information exchange and decision-making

CONCLUSION and PERSPECTIVES

This paper has addressed NAVAL Group (previously DCNS) approach to their new generation combat system for Surface ship combatant and Submarines, with specific focus on the development and incremental integration of knowledge-based algorithms and artificial intelligence to cope with new threats and emerging information warfare.
Similar work is also underway to deal with other critical domains, such as ship control, Navigation decision support, System Management & reconfiguration, safety and Cyber-security, e-maintenance, simulation & training, etc.
However, most underlying technologies and techniques of potential interest are coming from different civil domains (Internet, Video Games, Telecommunication, Transport industry, etc.). R&D activity is thus recommended to be done with various academic and industrial partners both in France and in countries where Naval Group has established efficient links, in particular through their (ex DCNS) Research center
International partnership is thus a driving factor for NAVAL Group in the coming years, through joint Research & innovation centers (e.g. in Chili) or project-based collaborations.

ACKNOWLEDGMENTS

The author would like to give special thanks to Jean-Michel VanPeperstraete and Dann Laneuville, Information Processing experts, from the NAVAL Group Research Center

REFERENCES

[1] D.Laneuville, A.Negre, *Video Extraction and Tracking of moving Targets with a camera Camera,* Proc. of IEEE Aerospace Conference, Big Sky, MT, March 2017.

[2] D. Laneuville, A. Nègre and P. Dufour, *4D Cartesian State Estimation of Sea Surface Targets with a Single Camera*, Proc. of IEEE Aero. Conference, Big Sky, MT, March 2016.

[3] A. Nègre, J. Houssineau, D. Laneuville, *Passive Surveillance with Geographically Distributed Cameras and GM-PHD filter,* Proc. of IEEE Passive Conference, Brest, 2010.

[4] Serge FRANCOIS and Antoine BOUQUET , *New evaluation of the French Multiplatform Tactical Picture capability,* MAST Asia (Tokyo) June 2017

[5] Michel MOREL & Stéphane CLAISSE, *Integrated System for Interoperable sensors & Information sources for Common abnormal vessel behavior detection & Collaborative identification of threat (I2C)* IEEE Conference Publishing, Piscataway, NJ, USA

[6] S. Kaddani, S. Tamby, D. Vanderpooten, J-M. Vanpeperstraete, *Partial preference models using translated cones in discrete multi-objective optimization,* IEEE Symposium on Computational Intelligence, 2016

[7] Huilong Zhang, Benoıte de Saporta, François Dufour, Dann Laneuville, Adrien Nègre *Optimal Trajectories for Underwater Vehicles by Quantization and Stochastic Control,* FUSION 2017 , Singapore

The Aegis Weapons System – Providing Proven and Extendable Maritime Capabilities in Meeting Australian Defence White Paper Objectives

Alan B. (Brad) Hicks, Lockheed Martin, RADM USN (Retired)
Gary Feldman, Lockheed Martin, gary.feldman@lmco.com
Submitted for the PACIFIC 2017 International Maritime Conference, 3-5 October 2017

OVERVIEW

This paper examines the Royal Australian Navy's selection of the Aegis Weapons System for the Hobart class destroyer in terms of delivered capability, future capability and growth opportunities, training, and support of Australia's continuous shipbuilding program.

A short history of the Aegis Weapons System program is provided highlighting innovative actions, investments and enhancements that have taken place to maintain currency, develop new capabilities and expanded mission areas, in the United States, Australia and internationally. The paper also identifies potential Commonwealth of Australia opportunities to leverage the selection of the Aegis Weapons System from a defence and industry standpoint.

Finally, the paper highlights how the Australian Defence Force can harness Aegis technology to achieve stated 2016 Defence White Paper objectives of:

- Broadening and deepening alliance with the United States.
- Maintaining technology and capability superiority.
- Emphasising the joint force in bringing together different capabilities to make the Australian Defence Force more rapid and effective.

Figure 1. Hobart class destroyer at Sea. Photo Credit: Department of Defence.

AUSTRALIAN SURFACE NAVY COMBAT SYSTEM INTRODUCTION

Australia is undergoing a once in a generation maritime recapitalisation effort. The past five years has seen the Commonwealth of Australia (henceforth 'Commonwealth') procure MH-60R Seahawk; the Royal Australian Navy's (RAN's) next generation submarine hunter and anti-surface warfare helicopter, P-8 Poseidon maritime patrol aircraft, three Aegis-based Hobart class destroyers, a new class of submarines (with Lockheed Martin as the Combat System Integrator), as well as being one of the first nations to receive the F-35 Joint Strike Fighter with its inherent fifth generation fighter capabilities. This section of the paper provides an overview of the Aegis Weapon System (AWS) and its ability to support the Commonwealth's continuous shipbuilding program.

With the Aegis Weapon System now at sea with the Royal Australian Navy, the Hobart class destroyer joins the global Aegis fleet of over 120 ships including all U.S. Navy (USN) Ticonderoga class cruisers, Arleigh Burke destroyers, and twenty-seven international ships from the navies of Japan, the Republic of Korea, Norway and Spain. The Aegis fleet will continue to grow through the selection of Aegis as the combat system for the USN's Future Frigate. In Australia, Aegis will be a critical command and control node in the future of the Australian Defence Force's (ADF's) networked and integrated battlespace through capabilities such as Cooperative Engagement Capability (CEC) and potentially, Naval Integrated Fire Control – Counter Air (NIFC-CA) and Ballistic Missile Defence under an Integrated Air and Missile Defence (IAMD) construct. Aegis is the maritime command and control system that integrates other ADF platforms such as the MH-60R Seahawk, F-35 Joint Strike Fighter with its inherent fifth generation fighter capabilities, F/A-18E/F Super Hornet, P8 Poseidon maritime patrol aircraft and E-7A Wedgetail airborne early warning & control aircraft.

The Hobart class destroyers combat system architecture is a unique blend of weapons, sensors and equipment to meet the specific operational and warfighting needs of the RAN. Aegis is integrated with the Australian Tactical Interface and other Australian domestic sensors and weapons into the Hobart platform by the AWD Alliance. Lockheed Martin is the Aegis Combat System Engineering Agent (CSEA) through the USN's FMS program. As the Aegis CSEA Lockheed Martin develops, integrates, tests and delivers Aegis (and the MK-41 Vertical Launch System (VLS)). Lockheed Martin also provides integrated logistics for Aegis. From a ship integration and test standpoint, Lockheed Martin assists the AWD Alliance in grooming, installation and combat system test. An experienced Lockheed Martin Integration and Test Team has been on-site in South Australia since January 2016 helping to ensure the successful integration and test of Hobart. Lockheed Martin also provides the Synthetic Combat Operator Trainer and provides engineering support for the Aegis Command Team Trainer. Overall, Lockheed Martin supports the successful delivery and support of the Hobart class destroyers. Through a Technology Assistance Agreement, Lockheed Martin has trained Australian engineers and technicians on the Aegis system to enable integration, test and sustainment support of Aegis to be performed utilising Australian industry (Lockheed Martin Australia).

The Aegis Weapons System is an advanced, multi-mission combat system at sea conducting force and area anti-submarine, anti-surface, anti-air and strike warfare. Aegis has been developed, upgraded and modernised over the past 40+ years. It has had the benefit in terms of funding and time for extensive research and development, integration, test and certification by the USN and five international navies. Aegis ships have over 1,200 years of cumulative at sea service and over 3,800 live missile firings across the Aegis fleet.

The Aegis Weapons System, currently on its ninth baseline (BL 9) has continuously improved and added new capabilities to meet new threats, execute new missions and integrate technology advancement. Accordingly, the scalable nature of Aegis already presents a growth path for the RAN in terms of more advanced capabilities. But perhaps most importantly, the Aegis Weapon System is the USN's maritime command and control hub for U.S. joint integration, network centric warfare and international interoperability. Aegis is based these operational cornerstones:

- Coverage superiority – 360 degrees coverage to the maximum height and range.
- Fastest reaction time – time from initial detection to 'missiles away' is minimised.
- Superior firepower – controlling numerous missiles in flight simultaneously.
- Environmental resistance – overcome man-made and environmental interference.
- Continuous availability – must be capable of operating 24/7 in all warfare areas.

Figure 2. Aegis International Fleet.

Cooperative Engagement Capability (CEC)

Cooperative Engagement Capability combines sensor data from multiple sources into a single, shared picture of tracked aerial threats. This allows CEC equipped vessels to fire missiles at aircraft or at cruise missiles that don't appear on their own sensors. It's also utilised in situations where multiple sensors can triangulate a target, reducing the effectiveness of adversaries' jamming or other electronic countermeasures. CEC is a component of the USN's NIFC-CA capability, a system which integrates sensors and weapons across air and sea platforms. Only Aegis provides this capability.

Aegis Baseline 9 and Common Source Library

To maximise Aegis upgradeability and responsiveness to new and emerging threats and to reduce through-life support costs, Aegis BL 9:

- Incorporates an open architecture computing environment framework.
- Is entirely hosted on Commercial-off-the Shelf (COTS) hardware. The use of COTS reduces development costs and allows advances in commercial computing to be translated into additional capability. COTS obsolescence challenges are overcome through pre-planned refreshes on a four-year cycles known as Technology Insertions (TI): TI 2012, TI 2016, TI 2020, etc.
- Implements the Common Source Library (CSL) to manage software components across the range of Aegis configurations. Through the CSL, Aegis capabilities are componentised and leveraged across the fleet. Existing and new capabilities are contained in the CSL, including associated specifications, test plans/procedures and software, eliminating or significantly reducing development and test costs through reuse.

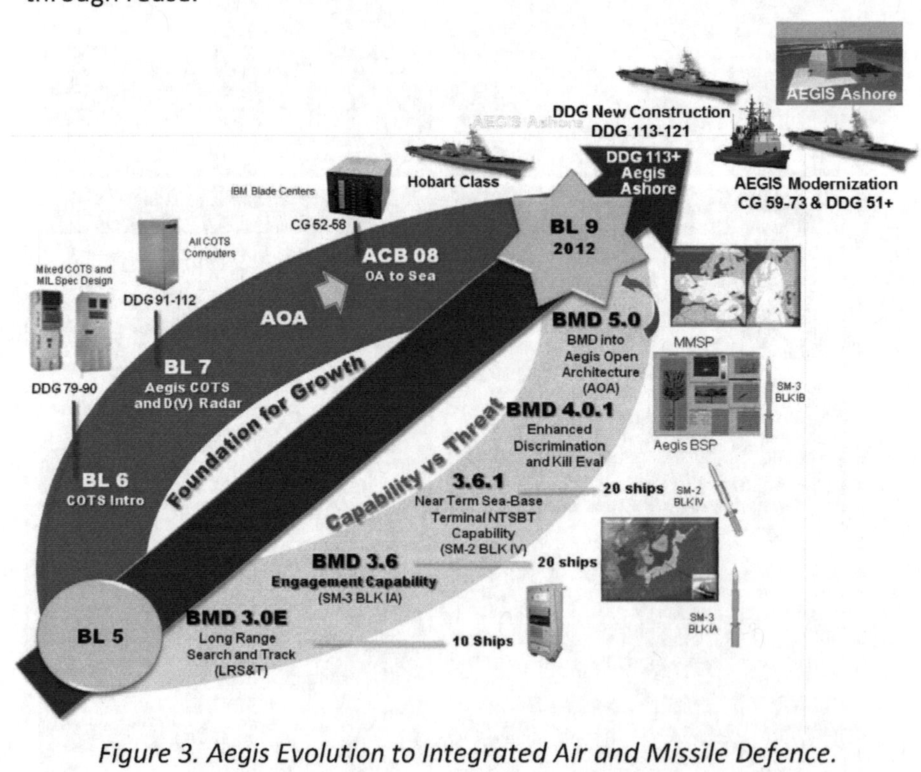

Figure 3. Aegis Evolution to Integrated Air and Missile Defence.

Aegis started utilising a CSL approach for code and specifications in 2008. The CSL approach allows resources to be consolidated and employed more effectively across multiple programs. The CSL has been a proven success, having supported Naval, Coast Guard, and land test site (e.g. White Sands) programs. The benefits of the CSL include:

- Maximise reuse of code and requirements
- Aligns well with Continuous Shipbuilding Plan
- Leverage testing across multiple users
- Share the cost of development of new functions across many users
- Share corrections and their verification
- Stay current with new threats and countermeasures
- Fast turnaround of fleet input
- Significantly reduces thru-life support costs

The CSL is a common repository for source code, specifications and test plans and procedures. It utilises variation technology to maintain a superset of requirements from all users. Each user maintains a unique Profile which identifies the specific functionality and configuration of that particular product. An actuation process then creates specifications and source code that contains only the capabilities identified by that Profile.

The CSL development approach is to deliver a build approximately every 4 months. New functionality and corrections are scheduled for each build across the user base. At each build point, a load is delivered and each user actuates a load based on the unique Profile. Users select an entry and departure point in the CSL where they will begin adding or modifying functionality and a departure point when they will depart the CSL to groom the load for final delivery. Throughout a continuous shipbuilding program, there will be multiple opportunities to create deliveries out of the CSL, providing the program a continuous stream of updates, increases in capability and maintenance corrections. The CSL process is depicted in figure 4 below.

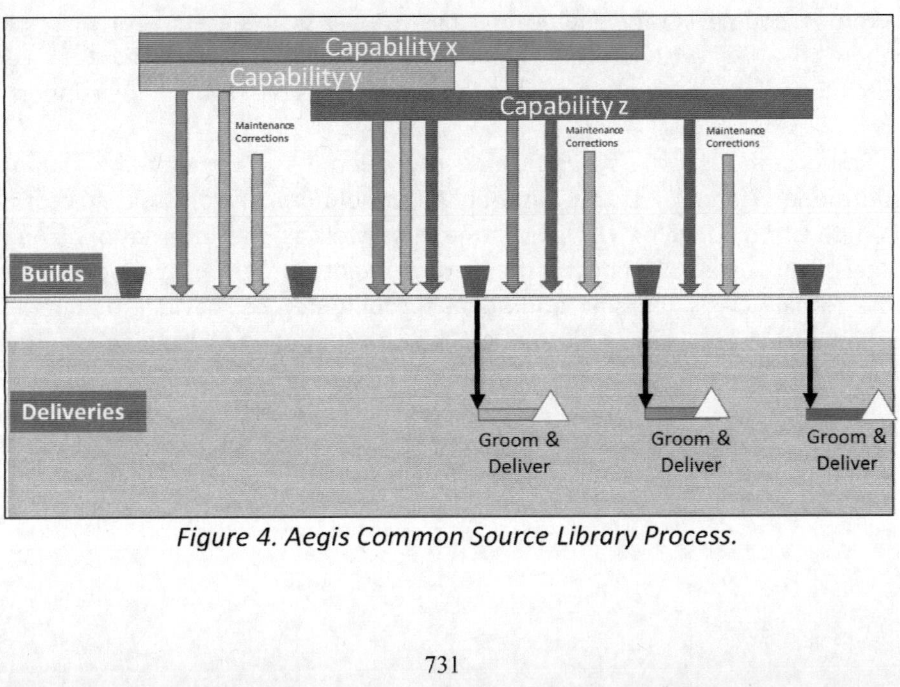

Figure 4. Aegis Common Source Library Process.

Interoperability
Aegis enables maximum interoperability with coalition forces by fully integrating with advanced capabilities such as the Cooperative Engagement Capability network and MH-60R Seahawk submarine hunter through Hawklink datalink. Aegis provides a growth path to NIFC-CA and integration of the F-35 Joint Strike Fighter.

Leveraging Aegis for ADF assets captures the tremendous depth of testing and fleet experience with data links and networks in complex, large scale networks. Years of deployment and testing with CEC and coalition data links has yielded a wealth of lessons learned. The USN has studied the results and incorporated design changes and corrections across multiple systems to improve the quality of the data provide to the Warfighter. These changes have been incorporated into the Common Source Library and are available to current and future programs. Aegis provides an opportunity to extend maritime network centric warfare to the AIR 6500 Integrated and Networked Battlespace program. Future design refinements continue to be folded into the CSL code base and are available to current and future Aegis ships.

While ships of other nations can communicate with U.S. maritime forces (e.g. through tactical data links like Link 16), what is far more important is the extent and impact of true interoperability in terms of data sharing and the ability to contribute and serve as a vital node in a networked environment. It is only through genuine interoperability that the synergies anticipated through a networked and integrated battlespace can be leveraged across joint capabilities both now and into the future. Aegis integrates into a U.S. Cooperative Engagement Capability environment and fully integrates allied MH-60R naval combat helicopters through Hawklink and Link 16 (e.g. when performing theatre/force level anti-submarine warfare). By selecting Aegis, the ADF ensures that its assets are interoperable with U.S. and coalition forces.

An Affordable Through Life Cost Solution
Utilisation of the Aegis BL 9 offers the most cost efficient, affordable approach for the Royal Australian Navy. Aegis BL 9 has already been developed and certified with future upgrades now funded and being further developed by the USN. The open architecture and common source library enable low cost, low risk integration of new capabilities throughout the life cycle of the ship class. Through BL 9 and the CSL, continuous capability upgrades can be integrated at the lowest possible through life support cost.

Most important is the significant reduction in through-life support and upgrade/modernisation costs through an Aegis solution. Beyond direct cost, Australia also benefits from ongoing USN investment as well as the international Aegis family investment. An Aegis solution for the Future Frigate program leverages commonality with the Hobart class, realising additional through-life cost savings by simplifying the support infrastructure, sparing, integrated logistics support, and training requirements.

ADVANCING AUSTRALIAN INDUSTRY

The Aegis international model has a strong history of cooperation to include the addition of domestic content, technology transfer and strategic industrial partnerships. Aegis integrated with an Australian interface not only helps to secure Australia's maritime approaches, but also secures Australian jobs and Australia's industrial base. For Australia, this means the incorporation of innovative Australian technologies such as the CEAFAR radar. Integration of Australian industry products with Aegis unites world class maritime capabilities and introduces substantial global export opportunities for Australian innovators.

As is being demonstrated on the Hobart destroyer program, the Hobart combat system which integrates Aegis with the Australian Tactical Interface (ATI) has created and sustained hundreds of Australian high-tech positions. This safeguards Australia's current expertise, while at the same time advancing Australian industrial capability through up-skilling of Australian engineers and technicians working on the advanced technology that comes with Aegis. Integration of new systems (sensors, weapons, and communications) is being performed by Australian industry through the AWD Alliance. Further, as new Australian capabilities come online, the development, integration, test and sustainment of these will be performed by Australians.

Lockheed Martin Australia, under a U.S. Government Technology Assistance Agreement, has trained local Australian personnel on the Aegis system. The agreement and training enable's the Lockheed Martin Australian workforce to support Aegis integration, test and sustainment. This ensures sovereign control while maximising capability and reducing thru-life support costs.

AEGIS TRAINING SYSTEM

The Aegis Training System is a significant capability to assist the Warfighter in preparing to fight and win at sea. It is one of the five "Aegis Cornerstone" measurements (Continuous Availability) and has prepared the USN through the Aegis Program for the past 40+ years. Rear Admiral Wayne E. Meyer, the father of Aegis, understood the importance of collaboration and integration of Government, Academia, and Industry in bringing Aegis program to fruition. Not only did the USN, Academia, and Industry engineer and produce the most advanced multi-mission combat system, they also had the foresight to ensure sustainability and life cycle of the Aegis platform. That Aegis platform "life cycle" included a robust training program aimed at creating and maintaining crew readiness to fight the ship and win. For over four decades now, the Aegis Training System (Program) has prepared Aegis crews for success at sea.

When partners choose Aegis and the Aegis Training Program, as Australia has joined the Aegis global fleet through the Hobart class destroyer, they benefit from a myriad of corporate knowledge and skills. The Aegis Training Program leverages the ability across baselines to provide the Aegis global fleet the benefit of USN training method advancements, curriculum products, and proven system employment lessons learned.

Three key tenets of the USN Aegis Training Program are highlighted in the following paragraphs.

Initial Crew and Sustainment Training

Leveraging both USN investment and experience, Australia has access to the USN training Centre of Excellence, the Centre for Surface Combat Systems (CSCS) in Dahlgren Virginia. Australia has been provided with a country specific training and readiness sustainment plan. The plan for maintenance and operational training requirements and execution is based on the investment and experience of the USN advanced baselines. USN curriculum, including training materials, tools, and devices, are leveraged into the RAN's training requirements and philosophy. The RAN is already benefiting from this capability with its CSCS partnership for initial crew training for the Hobart class destroyer. The RAN will continue to benefit from this partnership with the USN over the Hobart class destroyer 30+ years of life cycle with sustainment training, baseline upgrade training, and replacement crew training addressing crew attrition, rotation and sustainment. This life cycle training partnership will consist of continued USN investment and application of the proven training sequence pictured below: Initial Crew Training (Schoolhouse, Onboard) ----CSSQT----Waterfront Sustainment Training & Readiness (figure 5).

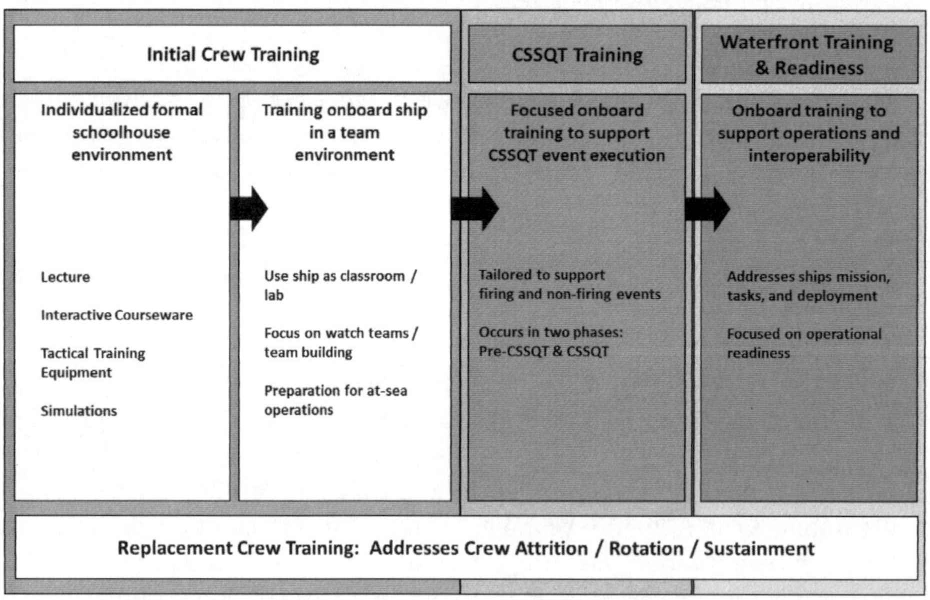

Figure 5. Aegis Training Approach to Support Continuous Availability.

On Board Training Systems

Introduced in the late 1970s in the initial design of Aegis, the USN fielded an organic and embedded capability that provided its Sailors with the ability to conduct individual and team training within the lifelines onboard the ship. What made this capability unique at the time was that the Aegis on board trainer device was embedded in the tactical combat system. Tactical Sensor, Command and Control (C2), and Weapon Systems contained training code embedded within their subsystem. Therefore, every Aegis cruiser, and subsequently Aegis destroyer, has a self-contained training capability to conduct combat system watch-stander and watch-team training using the computer programs and equipment used tactically to fight the ship.

Fast forward to 2017, advanced Aegis on board training systems are fielded on all USN Ticonderoga class cruisers, Arleigh Burke destroyers, and twenty-seven international ships in five partner nations, including the Hobart class destroyer. The advanced Aegis on board training system is the enabler for sustained combat readiness within, and outside the lifelines of the ship. Aegis Sailors train as they would fight, so they will fight as they are trained using their actual equipment to proof and prove the latest fighting doctrine and Tactics, Techniques, and Procedures (TTPs).

The USN continues to evolve the Aegis on board training system to enable all levels of Joint and Coalition training across various mission areas, e.g., NIFC-CA, IAMD, ASW, EW, employing advanced secure networking and media technologies that are adaptive, agile and affordable.

Designing in Lower Training Life Cycle Costs

The USN continues looking forward in current and future ship combat systems to design in lower training life cycle cost by employing COTS technologies and system commonality. COTS remove and replace maintenance concepts combined with having common systems / equipment across platforms eliminates the need for lengthy and complex maintenance training. Commonality of system operations and employment allows sailors to move from one platform to another without the need for re-training.

The Commonwealth is already benefiting from these life cycle training concepts in many ways. RAN exchange Officers are routinely aboard USN Aegis ships or in the Aegis "Schoolhouse" where they learn the operation and maintenance of Aegis along with the employment of TTPs and fighting doctrine. Some of these same exchange Officers are now able to walk aboard the Hobart class destroyer and function in a 'plug and play' mode.

The Hobart class destroyer Program was able to leverage proven USN curriculum which includes technical and operational training courses, documentation, and training tools in support of the Training Technology Spectrum such as Interactive Courseware (ICW) and training simulators like the Synthetic Combat Operational Trainer (SCOT) and the Command Team Trainer (CTT) deployed at HMAS Watson. As the Hobart class destroyer is updated with future Aegis baseline capability and the USN Training Technology Spectrum continues to be improved and updated, the Commonwealth will benefit from leveraging USN curriculum and operational knowledge and skill transfer.

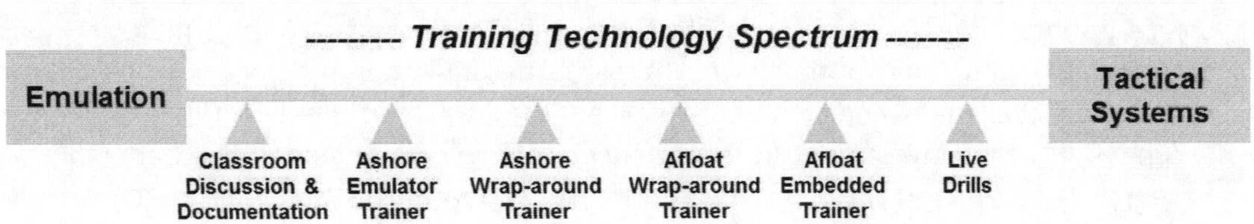

Figure 6. Aegis Full Spectrum Training Reduces Training Life Cycle Costs.

AEGIS BENEFITS TO AUSTRALIA'S CONTINUOUS SHIPBUILDING APPROACH

Aegis capability and scalability integrated with innovative Australian systems provides Australia with the most versatile naval combat system with broad utility across the spectrum of maritime and joint operations. The Aegis program has supported USN continuous shipbuilding programs for over 35 years on Ticonderoga class cruisers, Arleigh Burke class destroyers, and LCS programs. The USN continues to make substantial investments in capabilities and infrastructure on Aegis which can be leveraged for Australia's continuous shipbuilding program. There are over 120 active Aegis ships at sea that will continue to be supported for the foreseeable future, demonstrating the U.S. and international Navy's long term commitment.

Hobart successfully demonstrated the ability to integrate Aegis with Australian domestic capabilities. Open-architecture and CSL enhances combat system scalability up for multi-mission platforms (IAMD) and scalability down for more focused missions (such as LCS), affording Australia the opportunity to define the capabilities necessary to meet its platform requirements. Through the Aegis CSL, open architecture, Australian interface and on-going U.S. and international Aegis investment, the Commonwealth has the ability to increase capability and support new missions as additional ships are built, deployed and upgraded throughout the continuous shipbuilding program. By leveraging deployed and proven Aegis warfighting capabilities, new capabilities/missions are integrated into the RAN's fleet systematically and with very low risk. Unique Australian combat system capabilities can be developed and integrated with Aegis by Australian industry integrators and Australian combat system product providers ensuring that the RAN's ships are continuously ready to meet new threats with innovative Australian technologies.

A successful continuous shipbuilding capability provides Australia with a national asset, which includes a strong and growing industrial base and the ability to recruit and retain a high tech workforce. Through technology transfer and the integration of Australian capabilities on Hobart, Aegis supports a strong, sustainable and innovative shipbuilding industry throughout Australia. The technology transfer began during the Hobart class destroyer program design and continues to provide Australia with the skills and capabilities needed to maintain and enhance a strong domestic combat system industrial base. Australian industry is trained and is building the expertise to fully sustain Australia's surface combatants. For more than 50 years, Lockheed Martin Australia has shared in the nation's goals to develop a highly-skilled, sustainable domestic workforce. Our investments in Science, Technology, Engineering and Mathematics (STEM) initiatives create opportunities for thousands of Australians. To support a long-term strategy and continuous shipbuilding approach, Lockheed Martin understands the need to understand tomorrow's challenges and pursue leading edge solutions in meeting those challenges. Lockheed Martin's investment in the Science, Technology, Engineering, Leadership and Research Laboratory (STELaRLab) in Melbourne supports this objective by constituting Lockheed Martin's national Research & Development operations centre for collaboration with Australian universities, Australian industry and the Commonwealth.

ACHIEVING DEFENCE WHITE PAPER OBJECTIVES – AEGIS CONTRIBUTION

Broadening and Deepening Alliance with the United States

With the Aegis Weapon System now at sea with the Royal Australian Navy, the Hobart class destroyer joins the global Aegis fleet of over 120 ships including all USN Ticonderoga class cruisers, Arleigh Burke class destroyers, and twenty-seven international ships from the navies of Japan, the Republic of Korea, Norway and Spain. Aegis forms the mainstay of the U.S Navy's Surface Combatant Fleet. Ticonderoga class cruisers, Arleigh Burke class destroyers, and Freedom class Littoral Combat Ships all benefit from the warfighting capabilities that Aegis provides.

A common Aegis solution across the Hobart class and the Future Frigates provides the Commonwealth with 12 Aegis capable ships as a strategic defence partner in Indo-Asia-Pacific and enhances regional security to the defence against global threats. The interoperability inherent in Aegis and the continuous upgrades ensure a deep alliance with the U.S. Commonality and interoperability strengthens bi-lateral government to government and Navy to Navy relationships. An Aegis Combat System allows Australia's surface combatants to operate seamlessly with USN and coalition forces.

Maintaining Technology and Capability Superiority

The Aegis CSL allows Australia to implement continuous capability enhancements, maintaining pace with evolving threats. Leveraging the USN's investment in Aegis, capability upgrades and support of new missions are developed, integrated, tested and available across the global Aegis enterprise. Aegis BL 9 is leading the USN's initiatives of network centric warfare and distributed lethality through the implementation of new capabilities such as CEC, NIFC-CA and F-35 integration. The Australian interface in the Hobart Combat System architecture enables Australian industry to develop and sustain unique warfighting capabilities required by the RAN.

Each additional Aegis platform becomes a force multiplier for Australia and allied nations. Aegis meets the advanced warfighting, interoperability, distributed lethality, long-term growth path, and commonality needs of the Royal Australian Navy; and those capabilities are ready, battle proven and at sea today.

Emphasising the Joint Force in Bringing Together Different Capabilities to Make the Australian Defence Force More Rapid and Effective

Combining strengths of the Australia acquisitions together to provide increased situational awareness, support better time-critical decision making, reduced reaction time, greater engagement envelopes and the ability to complement capabilities from the air, land and sea domains. This requires a core command and control functionality that is proven, interoperable for both joint and combined operations, and provides a means for maritime capabilities to contribute and pace the growth (P-8, F-35, and E-7A) of the joint force.

Aegis is the USN's proven and fielded command and control hub for all maritime aspects of U.S. joint integration and ballistic missile defence, and as a result, Aegis features as a central component to Australia's aspirational view of an integrated and networked battlespace through projects such as CEC, NIFC-CA, F-35 and AIR 6500.

CONCLUSION

Aegis is the world's most advanced, lethal and technologically evolved maritime combat system. Aegis is proven at sea, and its performance has been successfully demonstrated during Hobart's Sea Trials. Aegis commonality across the RAN's fleet provides Australia with an extremely capable, low risk solution that optimises acquisition and lifecycle costs.

An Aegis solution aligns Australia and the U.S. operationally, broadens the relationship and enables seamless joint and combined operations while providing opportunities to grow the Australian Industrial Capability by expanding the maintenance, sustainment, and training workforce.

Lockheed Martin Australia
Lockheed Martin is an industry leader in defence and technology, with a strong presence in Australia of more than 760 direct employees across every mainland state and territory and more than an additional 2,900 Australians employed by other companies as a direct result of Lockheed Martin programs and products. Lockheed Martin is devoted to fostering innovation in Australia and partners with Australia's research and industry communities as well as SMEs to take advantage of our global supply chains, provide opportunities for technology transfer, drive growth through export, develop new high-skilled jobs and encourage business growth at all levels.

Specific to naval combat systems, Lockheed Martin has a strong pedigree having designed and integrated combat systems in ships and submarines around the globe for over 40 years. In the Australian context, Lockheed Martin is strategically, operationally and tactically involved in building Australia's future maritime capability through our role as the Combat System Integrator for Australia's Future Submarine program and is the provider of the Aegis Weapon System for the Hobart class destroyer. Lockheed Martin is committed to helping to build Australia's maritime capability.

The Genesis of Australia's Future Naval Enterprise Combat System Capability

Dave Hanley, Raytheon Australia, David.Hanley@Raytheon.com.au

The Air Warfare Destroyer program is the most complex naval ship Program undertaken in Australia.

The core of the AWD capability is its Combat System – representing over 90% of the ship's strategic value to Australia's security.

Recent successful completion of Sea Acceptance Trials marked the first time the AWD's full suite of combat systems were tested together as a complete mission system – aptly demonstrating the next-generation of Naval capability for the RAN.

The success of complex Combat System delivery begins with its architectural definition. What follows is the systems engineering process to finalise physical design, procure equipment / systems – leading to integration and final acceptance testing and trials. This process is reliant on a systems of systems approach by a Combat System Integration capability capable of deploying necessary experience and skills including Technical, Program Management, and Supply Chain resources.

In the case of AWD, development of the robust architecture for the Combat System was the cornerstone for the programs success giving the navy a stable capability platform for the life of the Hobart Class. This stability will allow for graceful evolution of the capability to meet emerging threats over the lifetime of the Class – including the ability to deliver capability updates / upgrades without destabilising the overall design and suffering the inherent costs and risks.

Looking beyond AWD, the approach taken on the AWD Program has enabled characterisation of an Australian Naval Enterprise Common Combat System Architecture that can support the realisation of future Naval capability for the Australian fleet.

The Royal Institution of Naval Architects
8-9 Northumberland Street
London, WC2N 5DA
United Kingdom

ISBN 978-1-5108-8305-5

CURRAN ASSOCIATES INC.
proceedings.com

Workshop on Open Infrastructures and Analysis Frameworks for HLT 2014

Held at COLING 2014

Dublin, Ireland
23 August 2014

ISBN: 978-1-63439-220-4